CRITICAL
STABILITY
CONSTANTS

Volume 5: First Supplement

CRITICAL STABILITY CONSTANTS

CRITICAL STABILITY CONSTANTS

Volume 5: First Supplement

by Arthur E. Martell
and Robert M. Smith

Department of Chemistry
College of Science
Texas A & M University
College Station, Texas

PLENUM PRESS · NEW YORK AND LONDON

ISBN 0-306-41005-2

PREFACE

Over the past twenty years the Commission on Equilibrium Data of the Analytical Division of the International Union of Pure and Applied Chemistry has been sponsoring a noncritical compilation of metal complex formation constants and related equilibrium constants. This work was extensive in scope and resulted in publication of two large volumes of *Stability Constants* by the Chemical Society (London). The first volume, edited by L. G. Sillen (for inorganic ligands) and by A. E. Martell (for organic ligands), was published in 1964 and covered the literature through 1962. The second volume, subtitled Supplement No. 1, edited by L. G. Sillen and E. Hogfeldt (for inorganic ligands), and A. E. Martell and R. M. Smith (for organic ligands), was published in 1971 and covered the literature up to 1969. These two large compilations attempted to cover all papers in the field related to metal complex equilibria (heats, entropies, and free energies). Most recently a noncritical compilation of organic ligands by D. D. Perrin (Pergamon Press) extended coverage of the literature through 1973. A similar volume for inorganic ligands by E. Hogfeldt should be in print soon. Since it was the policy of the Commission during that period to avoid decisions concerning the quality and reliability of the published work, the compilation would frequently contain from ten to twenty values for a single equilibrium constant. In many cases the values would differ by one or even two orders of magnitude, thus frustrating readers who wanted to use the data without doing the extensive literature study necessary to determine the correct value of the constant in question.

Because of difficulties of this nature, and because of the general lack of usefulness of a noncritical compilation for teaching purposes and for scientists who are not sufficiently expert in the field of equilibrium to carry out their own evaluation, we have decided to concentrate our efforts in this area toward the development of a critical and unique compilation of metal complex equilibrium constants. Although it would seem that decisions between available sets of data must sometimes be arbitrary and therefore possibly unfair, we have found the application of reasonable guidelines leads directly to the elimination of a considerable fraction of the published data of doubtful value. Additional criteria and procedures that were worked out to handle the remaining literature are described in the *Introduction* of this book. Many of these methods are quite similar to those used in other compilations of critical data.

In cases where a considerable amout of material has accumulated, it is felt that most of our critical constants will stand the test of time. Many of the data listed, however, are based on only one or a very few literature references and are subject to change when better data come along. It should be fully understood that this compilation is a continually changing and growing body of data, and will be revised from time to time as new results of these systems appear in the literature. The present volume represents the first supplement to the previous four volumes, and covers the literature that has appeared through 1979.

The scope of these tables includes the heats. entropies, and free energies of all metal ion complexation reactions involving organic and inorganic ligands. Volume 1 (1974), 2 (1975), 3 (1977), and 4 (1976) covered the range of different types of ligands of binary complexation reactions in aqueous solutions through 1973, mid-1974, 1975 and 1974 respectively. The present volume updates the coverage of the four volumes through 1979. The critical surveys of EDTA by G. Anderegg (77Aa), of 1,10-phenanthroline, 2,2'-bipyridyl, and related compounds by W. A. E. McBryde (78M), of fluoride complexes by A. M. Bond and G. T. Hefter (80BH), and the noncritical survey of organic ligands from 1969 through 1973 by D. D. Perrin (79P) were of significant aid in making the coverage more complete.

Texas A&M University
College Station, Texas

Arthur E. Martell
Robert M. Smith

CONTENTS

CONTENTS

INTRODUCTION

Purpose

This compilation of metal complex equilibrium (formation) constants and the corresponding enthalpy and entropy changes represent the authors' selection of the most reliable values among those available in the literature. In many cases wide variations in published constants for the same metal complex equilibrium indicate the presence of one or more errors in ligand purity, in the experimental measurements, or in calculations. Usually, the nature of these errors is not readily apparent in the publication, and the reader is frequently faced with uncertainties concerning the correct values. In the course of developing noncritical compilations of stability constants, the authors have long felt that these wide variations in published work constitute a serious impediment to the use of equilibrium data. Thus these critical tables were developed in order to satisfy what is believed to be an important need in the field of coordination chemistry.

Scope

These tables include all organic and inorganic ligands for which reliable values have been reported in the literature. The present volume supplements the first four volumes to make the coverage more current.

New ligands and revisions of previous critical values are included. When new values require revision or additional values at other ionic strengths become available, the entire new set of values is repeated for that metal ion with that ligand, and supercedes the previous set. These new sets now become the recommended critical values. When a new set of metal constants is included, the proton ligand constants are also included, even if they have not been revised. Reference to the ligand page in the previous volume is given just below the ligand name.

Selection Criteria

When several workers are in close agreement on a particular value, the average of their results has been selected for that value. Values showing considerable scatter have been eliminated. In cases where the agreement is poor and few results are available for comparison, more subtle methods were needed to select the best value. This selection was often guided by a comparison with values obtained for other metal ions with the same ligand and with values obtained for the same metal ion with similar ligands.

While established trends among similar metal ions and among similar ligands were valuable in deciding between widely varying values, such guidelines were used cautiously, so as not to overlook occasionally unexpected real examples of specificity of anomalous behavior.

When there was poor agreement between published values and comparison with other metal ions and ligands did not suggest the best value, the results of more experienced research groups who had supplied reliable values for other ligands were selected. When such assurances were lacking, it was sometimes possible to give preference to values reported by an investigator who had published other demonstrably reliable values obtained by the same experimental method.

In some cases the constants reported by several workers for a given group of metal ions would have similar relative values, but would differ considerably in the absolute magnitudes of the constants. Then a set of values from one worker near the median of all values reported were selected as the best constants.

By this method it is believed that internal consistency was preserved to a greater extent than would be obtained by averaging reported values for each individual metal ion. When an important constant was missing from the selected set of values, but was available in another set of values not selected for this compilation, the missing constant was obtained by adjusting the nonselected values by a common factor, which was set so as to give the best agreement between the two groups of data.

Values reported by only one investigator are included in these tables unless there was some reason to doubt their validity. It is recognized that some of these values may be in error, and that such errors will probably not be detected until the work is repeated by other investigators, or until more data become available for analogous ligands or other closely related metal ions. Some values involving unusual metal ions have been omitted because of serious questions about the form of their complexes.

Papers deficient in specifying essential reaction conditions (e.g., temperature, ionic strength, nature of supporting electrolyte) were not employed in this compilation. Also used as a basis for disqualification of published data is lack of information on the purity of the ligand. Frequent deficiencies are lack of calibration of potentiometric apparatus, and failure to define the equilibrium quotients reported in the paper. Papers in which both temperature and ionic strength are not controlled have been omitted from the bibliography.

A bibliography for each ligand is included so that the reader may determine the completeness of the literature search employed in the determination of critical values. The reader may also employ these references to make his own evaluation if he has any questions or reservations concerning this compilation.

Arrangement

The arrangement of the tables is based on the placement of similar ligands together. Ligands containing carboxylic acid functional groups are placed together except for aminocarboxylic acids and for phenolic carboxylic acids, which are listed with the phenols. Within each group of tables, ligands with a smaller number of coordinating groups are placed before those with a larger number of coordinating groups. Next there is a table of protonation constants for ligands for which no stability constants or only questionable metal stability constants are reported. Finally, there is a list of other ligands considered but not included in the tables for various reasons. Macrocyclic polyamines have been grouped together in a seperate heading because of a surge of interest in these ligands.

Metal Ions

The metal ions within each table are arranged in the following order: hydrogen, alkali metals, alkaline earth metals, lanthanides (including Sc and Y), actinides, transition metals, and posttransition metals. Within each group the arrangement is by increasing oxidation state of the metal, and within each oxidation state the arrangement follows the periodic table from top to bottom and from left to right. An exception is that Cu^+, Ag^+, Pd^{2+}, and Pt^{2+} are included with the posttransition metals.

Equilibrium

An abbreviation equilibrium quotient expression in the order products/reactants is included for each constant, and periods are used to separate distinct entities. Charges have been omitted as these can be determined from the charge of the metal ion and the abbreviated ligands formulas (such as HL) given after the name. Water has not been included in the equilibrium expressions since all of the values cited are for aqueous solutions. For example, $ML_2/M \cdot L^2$ for Cu^{2+} and acetic acid would represent the equilibrium: $Cu^{2+} + 2CH_3CO_2^- = (CH_3CO_2)_2Cu$. The symbol M represents the metal ion given in the first column and may include more than one atom as in the case of Hg_2^{2+}. The symbol H_{-1} (H_{-2}, etc.) is used for the ionization from the ligand of a proton that would not be ionized in the absence of the metal ion.

Equilibria involving protons are written as stability constants (protonation constants) rather than as ionization constants to be consistent with the metal complex formation constants. Consequently the ΔH and ΔS values have signs opposite to those describing ionization constants.

Solids and gases are identified by (s) and (g) respectively and are included for identification purposes even though they are not involved in the equilibrium quotient.

Log K Values

The log K values are the logarithms of the equilibrium quotients given in the second column at the specified conditions of temperature and ionic strength. The selected values are those considered to be the most reliable of the ones available. In some cases the value is the median of several values and in other cases it is the average of two or more values. The range of other values considered reliable is indicated by + or − quantities describing the algebraic difference between the other values and the selected values. The symbol ±0.00 indicates that there are one or more values which agree exactly with the stated value to the number of significant figures given. Values considered to be of questionable validity are enclosed in parentheses. Such values are included when the evidence available is not strong enough to exclude them on the basis of the above criteria. Values concerning which there is considerable doubt have been omitted.

The log K values are given for the more commonly reported ionic strengths. The ionic strengths most used are 0.1, 0.5, 1.0, 2.0, 3.0, and 0. Zero ionic strength is perhaps more important from a theoretical point of view, but several assumptions are involved in extrapolating or calculating from the measured values. The Davies equation is often used to calculate constants to zero from low-ionic-strength measurements. It was established from results obtained with monovalent and divalent ions and its extension to trivalent ions is extremely questionable. Values listed at 0.1 ionic strength may also include ionic strengths from 0.05 to 0.2, especially when results of several workers are averaged. Footnotes give conditions for values measured under conditions differing from those listed at the top of the table. Letters for footnotes, in the majority of cases, are uniform throughout the volume and refer to the same conditions.

The temperature of 25 °C was given preference in the tables because of its widespread use in equilibrium measurements and reporting other physical properties. When available, enthalphy changes (ΔH) were used to calculate log K at 25 °C when only measurements at other temperatures were available.

Other temperatures frequently employed are 20 °C, 30 °C, and 37 °C. These are not included in the tables when there is a lack of column space and ΔH is available, since they may be calculated using the ΔH value. Values at other temperatures, especially those at 20 °C and 30 °C, were converted to 25 °C to facilitate quantitative comparisons with the 25 °C values listed.

Equilibria involving protons have been expressed as concentration constants in order to be more consistent with the metal ion stability constants which involve only concentration terms. Concentration constants may be determined by calibrating the electrodes with solutions of known hydrogen ion concentrations or by conversion of pH values using the appropriate hydrogen ion activity coefficient. When standard buffers are used, mixed constants (also known as Bronsted or practical constants) are obtained which include both activity and concentration terms. Literature values expressed as mixed constants have been converted to concentration constants by using the hydrogen ion activity coeffcents determined in KCl solution before inclusion in the tables. In some cases, papers were omitted because no indication was given as to the use of concentration or mixed constants. Some papers were retained despite this lack of information when it could be ascertained which constant was used by comparing to known values or by personal communication with the authors. For those desiring to convert the listed protonation constants to mixed constants, the following values should be added to the listed values at the appropriate ionic strength (the tabulation applies only to single proton association constants):

Ionic strength	Increase in log K
0.05	0.09
0.10	0.11
0.15	0.12
0.2	0.13
0.5	0.15
1.0	0.14
2.0	0.11
3.0	0.07

The values in the tables have not been corrected for complexation with medium ions for the most part. There are insufficient data to make corrections for most of the ligands, and in order to make values between ligands more comparable, the correction has not been made in the few cases where it could be made. In general the listed formation constants at constant ionic strength include competition by ions from KNO_3 and $NaC10_4$ and are somewhat smaller than they would be if measured in solutions of tetraalkylammonium salts.

Protonation constants with a log value higher than possible for the reported ionic strength in basic solutions, or lower than possible in acidic solutions, have been placed in parentheses or discarded because of their possibly incorrect ionic strength, or considerably reduced accuracy if based on extrapolation of measurements at other acidities or alkalinities. Variation of activity coefficients for hydroxide ion and hydrogen ion at high concentrations reduces the accuracy of the calculated ionic strength. Values measured at relatively high ionic strength and then corrected or extrapolated to zero ionic strength have been retained.

Equilibria involving B(III), As(III), Ge(IV), and Te(VI) complexes with polyhydroxy ligands have been written showing the loss of a proton on complex formation. Thus the equilibrium quotient $M(OH)_2(H_{-2}L) \cdot H/M(OH)_3 \cdot L$ is employed for the reaction of B(III) with glycerol as a representation for the reaction. These equilibria are often reported in the literature with the ionization constant of the metal species $(H_2MO_3 \cdot H/M(OH)_3)$ divided into this quotient, thus eliminating the proton from the complex formation reaction $(M(OH)_2(H_{-2}L)/H_2MO_3 \cdot L)$.

Enthalpy Values

The enthalpy of complexation values (ΔH) listed in the tables have the units kcal/mole because of the widespread use of these units by workers in the field. These may be converted to SI units of kj/mole by multiplying the listed values by 4.184.

Calorimetrically determined values and temperature-variation determined values from cells without liquid junction were considered of equal validity for the tables. Other temperature-variation determined values were rounded off to the nearest kcal/mole and were enclosed in parentheses because of their reduced accuracy. Other values considered to be reliable but differing from the listed value were indicated by + or − quantities describing the algebraic difference between the other values and the selected values.

The magnitude of ΔH may vary with temperature and ionic strength, but usually this is less than the variation between different workers and little attempt has been made to show ΔH variation with changing conditions except for certain carefully measured equilibria such as the protonation of hydroxide ion and of ammonia. These ΔH values may be used for estimating log K values at temperatures other than those listed, using the relationship

$$\frac{\Delta H}{2.303RT^2} = \frac{d \log K}{dT}$$

or, at 25 °C

$$\log K_2 = \log K_1 + \Delta H(T_2 - T_1)(0.00246).$$

This assumes that $\Delta C_p = 0$, which is not necessarily the case. The greater the temperature range employed, the greater the uncertainty of the calculated values.

Entropy Values

The entropy of complexation values (ΔS) listed in the tables have the units cal/mole/degree and have been calculated from the listed log K and ΔH values, using the expression

$$G = \Delta H - T \Delta S$$

or, at 25 °C

$$\Delta S = 3.30(1.303 \log K + \Delta H).$$

These entropy values have been rounded off to the nearest cal/mole/degree, except in cases where ΔH values were quite accurate.

Bibliography

The references considered in preparing each table are given at the end of the table. The more reliable references are listed after the ions for which values are reported. In some tables groups of similar metal ions have been grouped together for the bibliography. The term "Other references" is used for those reporting questionable values, or values at conditions considerably different from those used in the tables, or values for metal ions not included in the tables because of questionable knowledge about the forms of their complexes. These additional references are cited to inform the reader of the extent of the literature search made in arriving at the selected values. Some values in this volume are repeated from a previous volume but the references are not repeated unless there was an error in the previous volume. Therefore those desiring a complete set of references must also consult the Bibliography of the previous volume.

The bibliographical symbols used represent the year of the reference and the first letter of the surnames of the first two listed authors. In cases of duplication, letters a, b, c, etc., or the first letter of the third author's name are employed. The complete reference is given in the bibliography at the end of each volume.

Miscellaneous Comments

The formulation of polynuclear complexes is often made on the basis of improving the fit to the experimental data and not on experimental evidence for their existence. They should therefore be used with caution since other sets of complexes might equally fit the data.

A knowledge of the optical activity of potentially optically active ligands in binuclear complexes is necessary for the characterization of the complexes. When the optical activity is not stated or a DL-mixture is used, there is considerable doubt as to the precise nature of the complex and the values are placed in parentheses.

Hydrolysis constants are usually expressed as proton ionizations in the tables except when the author gives them as hydroxide stability constants and fails to give the value of the constant for the ionization of water employed in the calculations.

In a work of this magnitude, there will certainly be errors and a few pertinent publications will have been overlooked by the compilers. We should like to request those who believe they have detected errors in the selection process, know of publications that were omitted, or have any suggestions for improvement of the tables, write to:

A. E. Martell
Department of Chemistry
Texas A&M University
College Station, Texas 77843, U.S.A.

It is the intention of the authors to publish more complete and accurate revisions of these tables as demanded by the continually growing body of equilibrium data in the literature.

$$\begin{array}{c} NH_2 \\ | \\ CH_2CO_2H \end{array}$$

$C_2H_5O_2N$ Aminoacetic acid (glycine)   HL
                    (Other values on Vol.1, p.1)

Metal ion	Equilibrium	Log K 25°, 0.1	Log K 25°, 1.0	Log K 25°, 0	ΔH 25°, 0	ΔS 25°, 0
H^+	HL/H.L	$9.56_b \pm 0.02$	9.63 ± 0.06	9.778 ± 0.00	-10.6 ± 0.0	9
		$9.55^b \pm 0.02$	$9.22^n \pm 0.05$	$10.14^e \pm 0.07$	-12.2^e	
	$H_2L/HL.H$	$2.36_b \pm 0.04$	2.37 ± 0.06	2.350 ± 0.00	-1.0 ± 0.1	7
		$2.36^b \pm 0.04$	$2.30^n \pm 0.04$	$2.74^e \pm 0.06$	-2.2^e	
Mg^{2+}	ML/M.L	1.34^b			$(1)^r$	$(10)^{b*}$
Sr^{2+}	ML/M.L	0.6		0.91		
Mn^{2+}	ML/M.L	$2.80_b \pm 0.05$	2.71^n	3.19 ± 0.02	-0.3 ± 0.0	14
		$2.60^b \pm 0.05$				
	$ML_2/M.L^2$	$4.5_b \pm 0.2$	4.76^n			
	$ML_3/M.L^3$	$5.3^b \pm 0.4$	5.5^n			
	MHL/ML.H		7.3^n			
Co^{2+}	ML/M.L	$4.67_b \pm 0.04$		$5.07 \; -0.05$	$-2.8^a \pm 0.7$	12^a
		$4.55^b \pm 0.04$				
	$ML_2/M.L^2$	$8.46_b \pm 0.04$		9.04 ± 0.05	$-6.4^a +1$	17^a
		$8.22^b \pm 0.06$				
	$ML_3/M.L^3$	$10.8_b \pm 0.1$		11.6	-9.8	20
		$10.7^b \pm 0.1$				
Ni^{2+}	ML/M.L	$5.78_b \pm 0.05$	$5.70 \; -0.01$	$6.18 \; -0.05$	$-4.5^a \pm 0.5$	11^a
		$5.68^b \pm 0.05$				
	$ML_2/M.L^2$	$10.58_b \pm 0.07$	10.52 ± 0.03	11.13 ± 0.01	$-9.1^a \pm 0.4$	18^a
		$10.48^b \pm 0.03$				
	$ML_3/M.L^3$	$14.0_b \pm 0.2$	14.0 ± 0.1	(14.2)	$-14.7^a \pm 0.2$	15^a
		$14.0^b \pm 0.0$				
Cu^{2+}	ML/M.L	$8.13_b \pm 0.07$	$8.00^n \pm 0.02$	8.56 ± 0.06	$-6.3^a \pm 0.5$	16^a
		$8.14^b \pm 0.02$				
	$ML_2/M.L^2$	15.0 ± 0.1	$14.69^n \pm 0.04$	15.64 ± 0.05	$-13.1^a \pm 0.5$	25^a
		$15.0^b \pm 0.1$				
	MHL/ML.H	2.92	$2.9^n \pm 0.3$			
CH_3Hg^+	ML/M.L	7.88^u	7.52			
	$ML_2/M.L^2$		9.5			
Zn^{2+}	ML/M.L	4.96 ± 0.03	$4.90^n \pm 0.02$	$5.38 \; +0.1$	$-2.7^a \pm 0.7$	14^a
	$ML_2/M.L^2$	9.19 ± 0.08	$8.98^n \pm 0.03$	$9.81 \; +0.2$	$-5.8^a \pm 0.6$	23^a
	$ML_3/M.L^3$	11.6 ± 0.1	$11.29^n \pm 0.02$	12.3	-9.4	24
	MHL/ML.H		$4.6^n \pm 0.0$			
	ML/MOHL.H		$8.8^n \pm 0.1$			
Cd^{2+}	ML/M.L	4.24 ± 0.03	4.14	$4.69 \; +0.1$	-2.1	14
		4.18^b				
	$ML_2/M.L^2$	7.71 ± 0.02	7.60	$8.40 \; +0.4$	-5.4	20
		7.50^b				
	$ML_3/M.L^3$	9.76^b	9.74	10.7	-8.6	20

---

^{a} 25°, 0.1; ^{b} 25°, 0.5; ^{e} 25°, 3.0; ^{n} 37°, 0.15; ^{r} 0-30°, 0.09; ^{u} 25°, 0.25;   *assuming
ΔH for 0.09=ΔH for 0.5

Glycine (continued)

Metal ion	Equilibrium	Log K 25°, 0.1	Log K 25°, 1.0	Log K 25°, 0	ΔH 25°, 0	ΔS 25°, 0
Pb^{2+}	$ML/M.L$	4.36^b	4.78	5.28^e	$(-3)^s$	$(14)^e$
	$ML_2/M.L^2$	7.7 7.62^b	7.66	8.32^e		
	$MHL/ML.H$		5.97	6.13^e	$(-3)^s$	$(18)^e$
	$ML/MOHL.H$			$(7.64)^e$	$(-7)^s$	$(11)^e$
	$M(HL)_2/ML(HL).H$		6.5			
	$ML(HL)/ML_2.H$		7.0			
Ga^{3+}	$ML/M.L$	9.33^t		9.60^e		
	$MHL/ML.H$			2.63^e		

^{b}25°, 0.5; ^{e}25°, 3.0; ^{s}10-40°, 3.0; ^{t}22°, 0.1

Bibliography:

H^+ 69CP,73GS,74CPS,74GNF,75BHP,75CM,75DOd,
          75FL,75HV,75IP,76GMa,78JIa,78L,78RM,
          78VV,79EB,79EM,79HJ,79MTN,79SP,79VK

Mg^{2+} 69HL

Sr^{2+} 52SL

Mn^{2+} 69CP,69HL,74MMN

Co^{2+} 69HL,75IP

Ni^{2+} 74MMN,75IP,75SGP,76DOC,79EB,79SG

Cu^{2+} 61DR,69CP,69PP,69YH,70GS,73GS,73HR,75FL,
          75IP,75NW,75SS,76PSa,77DOa,78FM,78RM,
          79MB,79SP

CH_3Hg^+ 74RO,78JIa

Pd^{2+} 76AM

Zn^{2+} 69CP,74MMN,75CM,75DOd,79SP

Cd^{2+} 69HL,75IP,77SFb

Pb^{2+} 69HL,76CWa,78BS,79MTN

Ga^{3+} 75BHP

Other references: 67K,68CWI,68GS,68KR,68RK,
          70CB,70CBa,70FM,70FMb,70GS,71KS,72UT,
          73BF,73DR,73FA,73H,73RD,73RK,73SKa,
          73TG,74DB,74FA,74FAa,74FAb,74FLa,74KH,
          74KNP,74KU,74SK,74SS,74Wa,75CB,75JB,
          76BBC,76HS,76KFA,76KVP,76NF,76TG,77KDK,
          77KKc,77MSc,77PU,77RS,78AE,78SKG,79BBG,
          79BK,79FS,79JK,79KC,79KKK,79KM,79NL,
          79RRS

$$\begin{array}{c} NH_2 \\ | \\ CH_3CHCO_2H \end{array}$$

$C_3H_7O_2N$	L-2-Aminopropanoic acid (alanine)	HL

(Other values on Vol.1, p.4)

Metal ion	Equilibrium	Log K 25°, 0.1	Log K 25°, 1.0	Log K 25°, 0	ΔH 25°, 0	ΔS 25°, 0
H^+	HL/H.L	9.70 ±0.02 9.65[b] ±0.05	9.72 ±0.05 9.38[n]	9.867 10.25[e] ±0.05	-10.8 ±0.2 -11.8[c]	9
	H_2L/HL.H	2.31 ±0.04 2.31[b] ±0.02	2.30 +0.01	2.348 2.75[e] ±0.04	-0.7 ±0.1	8
Mn^{2+}	ML/M.L	2.50 ±0.05	2.39[n]	3.02		
	$ML_2/M.L^2$		4.29[n]			
	$ML_3/M.L^3$		5.7[n]			
Ni^{2+}	ML/M.L	5.40 ±0.08 5.31[b] ±0.1	5.40 -0.01	5.83 ±0.1	-3.6[a] ±0.4 -4.0[c]	13[a]
	$ML_2/M.L^2$	9.9[y] ±0.1 9.73[b] ±0.2	9.91 +0.01	10.5 ±0.1	-7.8[a] ±0.9 -8.9[c]	19[a]
	$ML_3/M.L^3$	12.9 -0.1	13.0 +0.1		-13.5[c]	14[c]
Cu^{2+}	ML/M.L	8.15 ±0.07 8.14[b] ±0.07	8.02[n]	8.55 ±0.04	-5.2[a] ±0.4	20[a]
	$ML_2/M.L^2$	14.9[y] ±0.1 14.9[b] ±0.1	14.6[n]	15.5 ±0.1	-11.9[a] ±0.4	28[a]
	MHL/ML.H		2.6[n]			
Ag^+	ML/M.L			3.64		
	$ML_2/M.L^2$	(7.02)[w]		(7.18)[w]		
CH_3Hg^+	ML/M.L		7.52			
	$ML_2/M.L^2$		(9.5)[w]			
Zn^{2+}	ML/M.L	4.56 ±0.06 4.56[b] ±0.01	4.55 4.57[n]	4.95 +0.2	-1.5[a]	16[a]
	$ML_2/M.L^2$	8.55[y] ±0.05 8.54[b] ±0.02	8.54 8.56[n]	9.23 +0.2	-4.3[a]	25[a]
	$ML_3/M.L^3$	10.6[b] ±0.0	10.6[n]			
	ML/MOHL.H	8.2	8.5[n]			
Pb^{2+}	ML/M.L	4.15[i]		5.17[e]		
	$ML_2/M.L^2$			(8.13)[e,x]		
	MHL/ML.H			6.41[e]		

_____

[a] 25°, 0.1; [b] 25°, 0.5; [c] 25°, 1.0; [e] 25°, 3.0; [i] 20°, 0.4; [n] 37°, 0.15; [w] optical isomerism not stated; [x] DL-mixture; [y] L-, D-, and DL-isomers had same value.

Bibliography:

H^+	69CP,71KS,73GS,75HV,78RM,78JIa,78L,79EB, 79EM,79MTN	Zn^{2+}	69CP,79GKD
Mn^{2+}	69CP,70GP	Pb^{2+}	79MTN
Ni^{2+}	75SGP,77GK,79EB,79GKD		
Cu^{2+}	69CP,70GM,70GP,73GS,74GNK,78RM		
Ag^+	79MST		
CH_3Hg^+	78JIa		

Other references: 68GS,70CB.70CBa,70FM, 70FMb,73BS,73FA,73RK,73SKa,73VB,74FA, 74FAa,74FAb,74FLa,74KH,74KU,74SK,74SS, 75FN,75JB,76BBC,76KFA,76KVP,76NF,77KDK, 77KKc,77RRa,78AE,78CK,78KZa,78MST, 79RRS

$$\overset{\displaystyle NH_2}{\underset{}{CH_3CH_2\overset{|}{C}HCO_2H}}$$

| $C_4H_9O_2N$ | DL-2-Aminobutanoic acid (Other values on Vol.1, p.6) | | | | | HL |

DL-2-Aminobutanoic acid
(Other values on Vol.1, p.6)

Metal ion	Equilibrium	Log K 25°, 0.1	Log K 25°, 1.0	Log K 25°, 0	ΔH 25°, 0	ΔS 25°, 0
H⁺	HL/H.L	9.63 ±0.03	(9.52)	9.831	-10.8[a]±0.1	8[a]
	H₂L/HL.H	2.32 ±0.03	2.30	2.284	-0.4 -0.6	9
Ni²⁺	ML/M.L	5.30 ±0.08			-4.1[a]	11[a]
	ML₂/M.L²	(9.7)[w]±0.2			-8.1[a]	17[a]
Cu²⁺	ML/M.L	8.07 ±0.06			-5.5[a]+1	19[a]
	ML₂/M.L²	(14.8)[w]±0.1			-11.5[a]+2	29[a]

[a]25°, 0.1;   [w]optical isomerism not stated.

Bibliography:
H⁺,Cu²⁺ 73GS
Ni²⁺ 75SGP

Other references: 75SS,76SSe,77SJS,77SSd, 78SS,79NS,79RRS

$$\overset{\displaystyle NH_2}{\underset{}{CH_3CH_2CH_2\overset{|}{C}HCO_2H}}$$

| $C_5H_{11}O_2N$ | DL-2-Aminopentanoic acid (norvaline) (Other references on Vol.1, p.7) | | | | HL |

DL-2-Aminopentanoic acid (norvaline)
(Other references on Vol.1, p.7)

Metal ion	Equilibrium	Log K 25°, 0.1	Log K 25°, 0	ΔH 25°, 0	ΔS 25°, 0
H⁺	HL/H.L	9.64 ±0.01	9.806	-10.9 ±0.0	8
	H₂L/HL.H	2.32 ±0.02	2.318	-0.5 ±0.1	9
Co²⁺	ML/M.L	4.22 ±0.07			
	ML₂/M.L²	(7.7)[x]±0.1			
Ni²⁺	ML/M.L	5.35 ±0.07		-4.3[a]	10[a]
	ML₂/M.L²	(9.7)[x]±0.1		-8.5[a]	16[a]
Cu²⁺	ML/M.L	8.12 ±0.05		-5.1[a]	20[a]
	ML₂/M.L²	(14.9)[x]±0.1		-12.0[a]	28[a]
Ag⁺	ML/M.L	3.08			
	ML₂/M.L²	(6.27)[x]			
Zn²⁺	ML/M.L	4.42			
	ML₂/M.L²	(8.52)[x]			
Cd²⁺	ML/M.L	3.73			
	ML₂/M.L²	(7.03)[x]			

[x]DL-mixture;   [a]25°, 0.2

Bibliography:
H⁺,Cu²⁺ 73GS,75IP
Co²⁺,Ag⁺-Cd²⁺75IP

Ni²⁺ 75IP,75SGP

Other references: 73BS,79PG

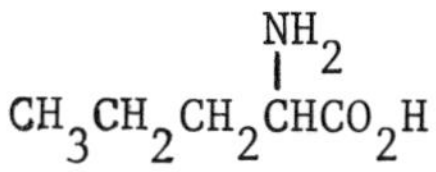

$$\overset{\displaystyle NH_2}{\underset{\displaystyle |}{CH_3CH_2CH_2CHCO_2H}}$$

C₆H₁₃O₂N		DL-2-Aminohexanoic acid (norleucine) (Other references on Vol.1, p.8)				HL

Metal ion	Equilibrium	Log K 25°, 0.1	Log K 25°, 0	ΔH 25°, 0	ΔS 25°, 0
H^+	HL/H.L	9.66 ±0.01	9.833	−10.9	8
	$H_2L/HL.H$	2.30	2.334	−0.5 ±0.1	9
Co^{2+}	$ML/M.L$	4.30 ±0.04			
	$ML_2/M.L^2$	(7.84)^x ±0.05		−5.9[a]	16[a]
Ni^{2+}	$ML/M.L$	5.35 ±0.08		−4.0[a]	11[a]
	$ML_2/M.L^2$	(9.7)^x ±0.2		−7.8[a]	18[a]
Cu^{2+}	$ML/M.L$	8.18			
	$ML_2/M.L^2$	(14.9)^x			
Ag^+	$ML/M.L$	3.21			
	$ML_2/M.L^2$	(6.71)^x			
Zn^{2+}	$ML/M.L$	4.59			
	$ML_2/M.L^2$	(8.93)^x			
Cd^{2+}	$ML/M.L$	3.86			
	$ML_2/M.L^2$	(7.33)^x			

[a]25°, 0.05; [x]DL-mixture

Bibliography:
$H^+,Co^{2+},Cu^{2+}-Cd^{2+}$ 75IP Ni^{2+} 75IP,75SGP

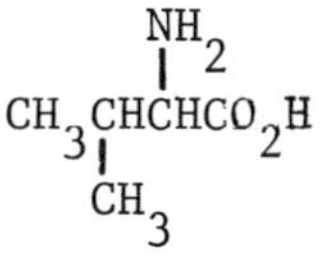

$$\overset{\displaystyle NH_2}{\underset{\displaystyle |}{\underset{\displaystyle \underset{|}{CH_3}}{CH_3CHCHCO_2H}}}$$

C₅H₁₁O₂N		L-2-Amino-3-methylbutanoic acid (valine) (Other values on Vol.1, p.9)					HL

Metal ion	Equilibrium	Log K 25°, 0.1	Log K 25°, 1.0	Log K 25°, 0	ΔH 25°, 0	ΔS 25°, 0
H^+	HL/H.L	9.49 ±0.03 9.44[b] ±0.05	9.50 ±0.06 9.20[n]	9.718	−10.8 ±0.2	8
	$H_2L/HL.H$	2.26 ±0.02 2.26[b] ±0.06	2.34 ±0.04	2.286	−0.1	10
Mn^{2+}	$ML/M.L$		2.34[n]			
	$ML_2/M.L^2$		4.0[n]			
	$ML_3/M.L^3$		5.2[n]			
	MHL/ML.H		8.2[n]			
Cu^{2+}	$ML/M.L$	8.09 ±0.04	7.95[n]		−5.5[a]	19[a]
	$ML_2/M.L^2$	14.9[y] ±0.1	14.6[n]		−11.5[a] ±0.2	30[a]
	MHL/ML.H		2.7[n]			

[a]25°, 0.1; [b]25°, 0.5; [n]37°, 0.15; [y]L- and DL-isomers had same value.

Valine (continued)

Metal ion	Equilibrium	Log K 25°, 0.1	Log K 25°, 1.0	Log K 25°, 0	ΔH 25°, 0	ΔS 25°, 0
CH_3Hg^+	$ML/M.L$		7.27			
	$ML_2/M.L^2$		$(9.2)^x$			
Zn^{2+}	$ML/M.L$		4.44^n			
	$ML_2/M.L^2$		8.24^n			
	$ML_3/M.L^3$		10.62^n			
	$ML/MOHL.H$		8.6^n			

^{n}37°, 0.15; ^{x}DL-mixture

Bibliography:

H^+ 69CP,71KS,77BP,78JIa,79EM

Mn^{2+},Zn^{2+} 69CP

Cu^{2+} 69CP,69PP,77BP

CH_3Hg^+ 78JIa

Other references: 70FM,70FMb,70GM,73FA,78MS, 73SC,74BF,74FA,74FAa,74FAb,74KH,74RO, 74SK,76KFA,76NF,77KDK,77RRa,78AE,78KZa

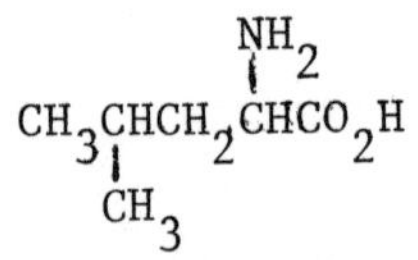

$$CH_3CHCH_2CHCO_2H$$

with NH_2 and CH_3 substituents

| | $C_6H_{13}O_2N$ | L-2-Amino-4-methylpentanoic acid   (leucine) (Other values on Vol.1, p.11) | | | | HL |

Metal ion	Equilibrium	Log K 25°, 0.1	Log K 25°, 1.0	Log K 25°, 0	ΔH 25°, 0	ΔS 25°, 0
H^+	$HL/H.L$	$9.28_b \pm 0.02$ $9.59^b \pm 0.01$	(9.48) 9.24^n	9.747	$-10.8^a \pm 0.1$	8^a
	$H_2L/HL.H$	$2.27_b \pm 0.09$ $2.34^b \pm 0.08$	2.36^n 2.24^n	2.329	-0.4	9
Cu^{2+}	$ML/M.L$	8.2 ± 0.1	8.04^n	8.51	-5.6^a	19^a
	$ML_2/M.L^2$	$15.0^y \pm 0.2$	14.7^n		-11.5^a	30^a
	$MHL/ML.H$		2.5^n			
Zn^{2+}	$ML/M.L$		4.51^n			
	$ML_2/M.L^2$		8.56^n			
	$ML/MOHL.H$		8.64^n			

^{a}25°, 0.1; ^{b}25°, 0.5; ^{n}37°, 0.15; ^{y}L-and DL-isomers had about the same value.

Bibliography:

H^+ 71HP,71KS,77BP,78L

Cu^{2+} 71HP,75BPb,77BP

Zn^{2+} 71HP

Other references: 70FM,70FMb,73FA,73SC, 74FA,74AAa,74FAb,74KH,75JB,76KFA, 76NF,79PG

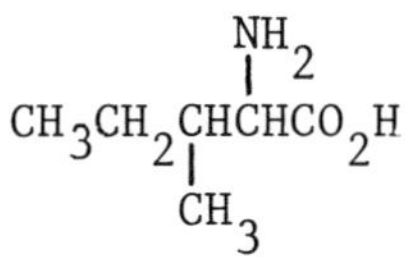

$$
\begin{array}{c}
NH_2 \\
| \\
CH_3CH_2CHCHCO_2H \\
| \\
CH_3
\end{array}
$$

$C_6H_{13}O_2N$ L-2-Amino-3-methylpentanoic acid   (isoleucine)   HL

(Other values on Vol.1, p.12)

Metal ion	Equilibrium	Log K 25°, 0.1	Log K 25°, 1.0	Log K 25°, 0	ΔH 25°, 0	ΔS 25°, 0
H^+	HL/H.L	9.56 ±0.03	9.55	9.754	-10.8	8
		9.55^b	9.24^n	10.15^e		
	$H_2L/HL.H$	2.21 ±0.04		2.319	-0.3	10
		$(2.51)^b$	2.24^n	2.84^e		
Co^{2+}	$ML/M.L_2$	(4.59)			$(-4.4)^a$	$(6)^a$
	$ML_2/M.L^2$	(8.94)			-5.3^a	$(23)^a$
Cu^{2+}	ML/M.L	(8.45)±0.06	(8.5)		$(-6.3)^a$	$(17)^a$
			7.95^n			
	$ML_2/M.L^2$	(15.4) +0.4	(15.4)		$(-10.6)^a$	$(35)^a$
			14.7^n			
	ML/MOHL.H		7.51^n			
Zn^{2+}	$ML/M.L_2$	4.49^b	4.40^n			
	$ML_2/M.L^2_3$	8.49^b	8.08^n			
	$ML_3/M.L^3$	10.9^b				
	ML/MOHL.H		7.90^n			
Ga^{3+}	ML/M.L			9.60^e		
	MHL/ML.H			2.59^e		

^{a}25°, 0.1; ^{b}25°, 0.5; ^{e}25°, 3.0; ^{n}37°, 0.15

Bibliography:

H^+ 71HP,75BHP,81IS Zn^{2+} 71HP

Co^{2+} 78IS Ga^{3+} 75BHP

Cu^{2+} 71HP,78IS   Other references: 74FAa,74FAb,74KH,75JB

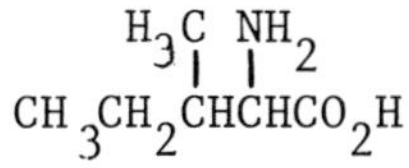

$$
\begin{array}{c}
H_3C \;\; NH_2 \\
| \;\;\;\; | \\
CH_3CH_2CHCHCO_2H
\end{array}
$$

$C_6H_{13}O_2N$ D-allo-Isoleucine   HL

Metal ion	Equilibrium	Log K 25°, 0.1	ΔH 25°, 0.1	ΔS 25°, 0.1
H^+	HL/H.L	9.51	(-10.4)	(9)
	$H_2L/HL.H$	2.09		
Co^{2+}	$ML/M.L_2$	4.10	(-3.0)	(9)
	$ML_2/M.L^2$	7.46	-4.0	21
Cu^{2+}	$ML/M.L_2$	8.09	-5.7	18
	$ML_2/M.L^2$	15.0	(-10.0)	(35)

Bibliography: 78IS,81IS

$$\overset{\overset{\displaystyle NH_2}{\displaystyle |}}{CH_2=CHCH_2CHCO_2H}$$

| C₅H₉O₂N | | DL-2-Amino-4-pentenoic acid | HL |

$C_5H_9O_2N$ — DL-2-Amino-4-pentenoic acid — HL

Metal ion	Equilibrium	Log K 25°, 0.1
H^+	HL/H.L	9.28
Co^{2+}	ML/M.L $ML_2/M.L^2$	4.21 (7.65)[x]
Ni^{2+}	ML/M.L $ML_2/M.L^2$	5.31 (9.89)[x]
Cu^{2+}	ML/M.L $ML_2/M.L^2$	8.00 (14.63)[x]
Ag^+	ML/M.L $ML_2/M.L^2$	4.22 (7.38)[x]
	MHL/M.HL	1.20
Zn^{2+}	ML/M.L $ML_2/M.L^2$	4.50 (8.51)[x]
Cd^{2+}	ML/M.L $ML_2/M.L^2$	3.77 (7.13)[x]

[x]DL-mixture.

Bibliography: 75IP

$$\overset{\overset{\displaystyle NH_2}{\displaystyle |}}{CH_2=CHCH_2CH_2CHCO_2H}$$

$C_6H_{11}O_2N$ — DL-2-Amino-5-hexenoic acid — HL

Metal ion	Equilibrium	Log K 25°, 0.1
H^+	HL/H.L	9.43
Co^{2+}	ML/M.L $ML_2/M.L^2$	4.24 (7.75)[x]
Ni^{2+}	ML/M.L $ML_2/M.L^2$	5.38 (9.89)[x]
Cu^{2+}	ML/M.L $ML_2/M.L^2$	8.09 (14.90)[x]
Ag^+	ML/M.L $ML_2/M.L^2$	3.81 (6.74)[x]
	MHL/M.HL	1.42
Zn^{2+}	ML/M.L $ML_2/M.L^2$	4.49 (8.60)[x]
Cd^{2+}	ML/M.L $ML_2/M.L^2$	3.76 (7.16)[x]

[x]DL-mixture.

Bibliography: 75IP

A. PRIMARY AMINES

$$\overset{\displaystyle NH}{\underset{\displaystyle |}{CH_2=CHCH_2CH_2CH_2CHCO_2H}}$$

$C_7H_{13}O_2N$ DL-2-Amino-6-heptenoic acid   HL

Metal ion	Equilibrium	Log K 25°, 0.1
H[+]	HL/H.L	9.52
Co[2+]	ML/M.L	4.22
	ML₂/M.L²	(7.68)[x]
Ni[2+]	ML/M.L	5.32
	ML₂/M.L²	(9.72)[x]
Cu[2+]	ML/M.L	8.09
	ML₂/M.L²	(14.91)[x]
Ag[+]	ML/M.L	3.34
	ML₂/M.L²	(6.41)[x]
	MHL/M.HL	1.73
Zn[2+]	ML/M.L	4.45
	ML₂/M.L²	(8.63)[x]
Cd[2+]	ML/M.L	3.75
	ML₂/M.L²	(7.13)[x]

[x]DL-mixture

Bibliography: 75IP

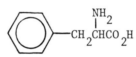

$C_9H_{11}O_2N$ L-2-Amino-3-phenylpropanoic acid   (phenylalanine)   HL
(Other values on Vol.1, p.18)

Metal ion	Equilibrium	Log K 25°, 0.1	Log K 25°, 1.0	Log K 25°, 0	ΔH 25°, 0.1	ΔS 25°, 0.1
H[+]	HL/H.L	9.09 ±0.04 9.06[b] ±0.02	9.05	9.31 9.61[e]	-10.5 ±0.2	6
	H₂L/HL.H	2.17 ±0.04 2.22[b] ±0.05	2.20	2.20 2.75[e]	(-1)[r]	(7)
Cu[2+]	ML/M.L	7.90 ±0.04		8.25[e]	-5.0 ±0.2	19
	ML₂/M.L²	14.8[y] ±0.1		15.6[e]	-11.4 ±0.4	30
Cd[2+]	ML/M.L	3.7 ±0.1		4.36[e]		
	ML₂/M.L²	6.9[y] ±0.2		7.94[e,x]		
	ML₃/M.L³			11.1[e,x]		
Pb[2+]	ML/M.L	4.01[i]		4.63[e]		
	ML₂/M.L²	(8.84)[i,y]		(8.35)[e,x]		

[b]25°, 0.5; [e]25°, 3.0; [i]20°, 0.37; [r]0-40°, 0; [x]DL-mixture;

[y]L-,D- and DL-isomers had same value.

Phenylalanine   (continued)

Bibliography:

H^+ 71KS,74GNK,77BP,77IH,78L,78RMa,79EM   Other references: 73BS,73RD,73SKa,74FA,

Cu^{2+} 74GNK,77BP,78RMa   74FAa,74FAb,74KH,74RO,75PN,76KFA,

Cd^{2+} 74WW,78MST   76NF,77RRa,78AE

Pb^{2+} 73CT

$$H_2NCH_2CH_2CO_2H$$

$C_3H_7O_2N$		3-Aminopropanoic acid (β-alanine) (Other values on Vol.1, p.20)				HL

Metal ion	Equilibrium	Log K 25°, 0.1	Log K 25°, 1.0	Log K 25°, 0	ΔH 25°, 0	ΔS 25°, 0
H^+	HL/H.L	10.10_b ±0.02 10.16^b ±0.04	10.12 ±0.03	10.295	-11.3 ±0.0	9
	$H_2L/HL.H$	3.53_b ±0.03 3.55^b ±0.09		3.551	-1.1 ±0.1	13
Ni^{2+}	$ML/M.L$	4.54 ±0.04	7.46	4.99	-3.8_t	10
	$ML_2/M.L^2$	7.87 ±0.09	7.84		-6.1^t	16^a
	$ML_3/M.L^3$	9.7	9.55			
CH_3Hg^+	$ML/M.L$	7.25^u				
	$MHL/ML.H$	5.21^u				

^{b}25°, 0.5; ^{t}22°, 0.1; ^{u}25°, 0.25; ^{a}25°, 0.1

Bibliography:

H^+ 69YH,71KS,76DOC,78JIa   Other references: 68GS,70CB,70CBa,70FM,

Ni^{2+} 76DOC   70FMb,73BS,73RD,73SKa,74DB,74FA,

Cu^{2+} 69YH   74FAa,74FAb,74KH,74SK,75JB,76KFA,

CH_3Hg^+74RO   76KVP,77YKU,79RRS

$$\overset{\displaystyle CH_3}{\underset{\displaystyle H_2NCHCH_2CO_2H}{|}}$$

$C_4H_9O_2N$		DL-3-Aminobutanoic acid (Other values on Vol.1, p.21)	HL

Metal ion	Equilibrium	Log K 25°, 0.1	ΔH 25°, 0.1	ΔS 25°, 0.1
H^+	HL/H.L	10.04 ±0.02	-11.0_s	9
	$H_2L/HL.H$	3.43 -0.01	(0) ^{s}	(16)
Co^{2+}	$ML/M.L$	3.53 -0.01	(-2) ^{s}	(9)
	$ML_2/M.L^2$	$(5.88)^w$	(-5) ^{s}	(10)
Cu^{2+}	$ML/M.L$	7.13 ±0.05	-4.9	16
	$ML_2/M.L^2$	$(12.84)^w$±0.07	-10.5	24

^{s}15-40°, 0.2; ^{w}optical isomerism not stated.

Bibliography: 75SS

$$H_2NCH_2CH_2CH_2CO_2H$$

$C_4H_9O_2N$ 4-Aminobutanoic acid   HL
(Other values on Vol.1, p.21)

Metal ion	Equilibrium	Log K 25°, 0.1	Log K 25°, 0.5	Log K 25°, 0	ΔH 25°, 0	ΔS 25°, 0
H^+	HL/H.L	10.28	10.31 +0.04	10.556	-12.2 ±0.2	7
	H_2L/HL.H	4.06	3.99[u]	4.031	-0.4 ±0.0	17
Cu^{2+}	ML/M.L	5.47				
CH_3Hg^+	ML/M.L		7.54[u]			
	MHL/ML.H		5.7[u]			

[u]25°, 0.25

Bibliography:
H^+ 74RO,75SS; Cu^{2+} 75SS; CH_3Hg^+ 74RO; Other references: 69MG,76KVP,79RRS

$$H_2NCH_2CH_2CH_2CH_2CO_2H$$

$C_5H_{11}O_2N$ 5-Aminopentanoic acid   HL
(Other values on Vol.1, p.22)

Metal ion	Equilibrium	Log K 25°, 0.5	Log K 25°, 0	ΔH 25°, 0	ΔS 25°, 0
H^+	HL/H.L	10.51 +0.08	10.77	-13.0	6
	H_2L/HL.H	4.20[u]	4.26±0.01	0.2	20
CH_3Hg^+	ML/M.L	7.75[u]			
	MHL/ML.H	6.04[u]			

[u]25°, 0.25;

Bibliography: 74RO

$$H_2NCH_2CH_2CH_2CH_2CH_2CO_2H$$

$C_6H_{13}O_2N$ 6-Aminohexanoic acid   HL
(Other values on Vol.1, p.22)

Metal ion	Equilibrium	Log K 25°, 0.1	Log K 25°, 0.5	Log K 25°, 0	ΔH 25°, 0	ΔS 25°, 0
H^+	HL/H.L	10.52 ±0.02	10.62 ±0.00	10.804 / 11.45[e]	-13.6	4
	H_2L/HL.H		4.33[u]	4.373	0.2 ±0.2	21
Cu^{2+}	MHL/M.HL			5.06[e]		
				1.76[e]		
	M(HL)$_2$/M.(HL)$_2$			3.04[e]		
Cu^+	M_4L_4/M^4.L^4			42.7[e]		
CH_3Hg^+	ML/M.L		7.83[u]			
	MHL/ML.H		6.11[u]			

[e]25°, 3.0;   [u]25°, 0.25;

Bibliography: H^+,Cu^{2+} 72OT; Cu^+ 72OS; CH_3Hg^+ 74RO

$$\overset{\displaystyle NH_2}{\underset{\displaystyle HO_2CCH_2CHCO_2H}{|}}$$

$C_4H_7O_4N$	L-Aminobutanedioic acid (aspartic acid) $\hspace{4cm}$ H_2L

(Other values on Vol.1, p.24)

Metal ion	Equilibrium	Log K 25°, 0.1	Log K 25°, 1.0	Log K 25°, 0	ΔH 25°, 0.2	ΔS 25°, 0.1
H^+	$HL/H.L$	$9.62^b\pm0.02$ $9.50^b\pm0.04$	9.25 ± 0.05 9.27^n	10.002^e 10.01	-9.4	13
	$H_2L/HL.H$	$3.70^b\pm0.02$ $3.64^b\pm0.08$	3.66 ± 0.1 3.60^n	3.900^e 4.07	-1.1	13
	$H_3L/H_2L.H$	$1.94^b\pm0.02$ $1.86^b\pm0.01$	1.9 ± 0.1 1.95^n	1.990^e 2.35	-1.4	4
Ni^{2+}	$ML/M.L$	7.15 ± 0.02 6.90^b				
	$ML_2/M.L^2$	$12.39^y\pm0.09$ 12.26^b				
	$MHL/ML.H$	4.05				
Cu^{2+}	$ML/M.L$ $_2$	8.89 ± 0.09			-6.1	20
	$ML_2/M.L^2$	$15.93^y\pm0.07$			-12.4	31
	$MHL/ML.H$	3.70 ± 0.05			-2.4	9
Zn^{2+}	$ML/M.L$ $_2$	$5.58^b\pm0.08$	$(6.01)^n$			
	$ML_2/M.L^2$	$10.1^b\pm0.1$	$(10.10)^n$			
	$MHL/ML.H$		$(5.9)^n$			
Cd^{2+}	$ML/M.L$ $_2$	4.35 ± 0.05		5.01^e		
	$ML_2/M.L^2$	7.55		9.12^e		
Pb^{2+}	$ML/M.L$ $_2$			6.67^e		
	$ML_2/M.L^2$			9.4^e		
	$MHL/ML.H$			$(5.61)^e$		

---

^{b}25°, 0.5; ^{e}25°, 3.0; ^{n}37°, 0.15; ^{y}L- and DL-isomers had same value.

Bibliography:

H^+ 68F,73SK,74NGF,75HV,76MT,79GR,79KC,79MB Other references: 70MS,72SS,72SSa,72SSb,
$\hspace{11.5cm}$ 72SSc,73H,73IY,73RD,73RK,73SKa,74Ab,
Ni^{2+} 68F,74GNF, $\hspace{6cm}$ 74ST,75JB,75KU,76AMa,76KN,76KVP,77BS,
$\hspace{11cm}$ 77RG,77ZG,77ZV,78JS,78SJ,79BS,79FW,79S,
Cu^{2+} 74NGF,75Ra,77BP,78SYN,79MB $\hspace{4cm}$ 79SJb

Zn^{2+} 68F,76MT,79GR

Cd^{2+} 74WW

Pb^{2+} 73CT

$$\overset{\overset{\textstyle NH_2}{\textstyle |}}{HO_2CCH_2CH_2CHCO_2H}$$

| $C_5H_9O_4N$ | L-2-Aminopentanedioic acid (glutamic acid) (Other values on Vol.1, p.27) | | | | | H_2L |

Metal ion	Equilibrium	Log K 25°, 0.1	Log K 25°, 1.0	Log K 25°, 0	ΔH 25°, 0.2	ΔS 25°, 0.1
H^+	HL/H.L	$9.59^b \pm 0.09$ $9.41^b -0.01$	9.42 ± 0.01 9.27^n	$9.95 +0.01$ 9.89^e	-9.8 ± 0.3	11
	$H_2L/HL.H$	$4.20^b \pm 0.08$ $4.1^b \pm 0.1$	4.20 $(4.03)^n$	4.42 ± 0.1 4.54^e	-0.8 ± 0.2	17
	$H_3L/H_2L.H$	$2.18^b \pm 0.1$ $2.3^b \pm 0.2$	(2.39)	2.23 ± 0.07 2.57^e	-0.3 ± 0.5	9
Cu^{2+}	$ML/M.L$	8.33 ± 0.06	$(8.74)^n$		-5.0	21
	$ML_2/M.L^2$	$14.84^y \pm 0.10$	$(14.91)^n$		$-11.7 +0.3$	29
	MHL/ML.H	4.15 ± 0.05	4.05^n		-1.6	14
Ag^+	$ML/M.L$	3.79				
	$M_2L/ML.M$	2.76				
Zn^{2+}	$ML/M.L$	4.49^b	$(4.76)^n$			
	$ML_2/M.L^2$	8.25^b	$(8.54)^n$			
	$ML_3/M.L^3$	9.8^b				
	MHL/ML.H		$(8.03)^n$			
Pb^{2+}	$ML/M.L$	4.70^u				
	$ML_2/M.L^2$	7.55^u				
Ga^{3+}	$ML/M.L$	11.15^t		11.30^e		
	MHL/ML.H			2.89^e		
	$MH_2L/MHL.H$			2.54^e		

---

^{b}25°, 0.5; ^{e}25°, 3.0; ^{n}37°, 0.15; ^{t}22°, 0.1; ^{u}25°, 0.3; ^{y}L- and DL-isomers had the same value.

Bibliography:

H^+ 71HP,73SK,74NGF,74SC,76BH,77BP,78SYN, 79MB,

Cu^{2+} 71HP,74NGF,75Ra,77BP,77BS,78SYN,79MB

Ag^+ 79MST

Zn^{2+} 71HP

Pb^{2+} 74K

Ga^{3+} 76BH

Other references: 69KA,72SS,72SSa,72SSb, 73H,73RD,73RK,73SKa,74FLa,75KU,76AK, 76GPP,77RG,78Aa,78JS,78SJ,78ZG,79BS, 79FW,79S,79SJb

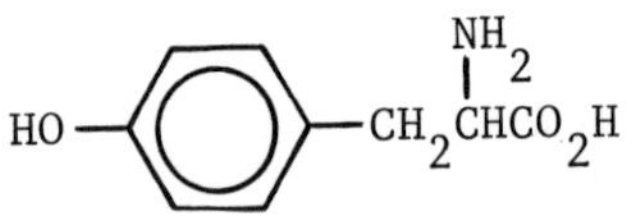

$$HO-\langle C_6H_4\rangle-CH_2\overset{\overset{\displaystyle NH_2}{|}}{C}HCO_2H$$

$C_9H_{11}O_3N$ __L-2-Amino-3-(4-hydroxyphenyl)propanoic acid__   (tyrosine) H_2L

(Other values on Vol.1, p.31)

Metal ion	Equilibrium	Log K 25°, 0.1	Log K 25°, 3.0	Log K 25°, 0	ΔH 25°, 0.1	ΔS 25°, 0.1
H^+	$HL/H.L$	10.11 ±0.1	10.39	10.47	-5.8	27
	$H_2L/HL.H$	9.04 ±0.05	9.43	9.19	-10.1	7
	$H_3L/H_2L.H$	2.17 ±0.05	2.81			
Ni^{2+}	$MHL/M.HL$	5.0 ±0.1			-2.0	16
	$M(HL)_2/M.(HL)_2$	9.4 ±0.2			-4.8	27
	$M(HL)_3/M.(HL)_3$	12.1				
	$M(HL)_2/ML(HL).H$	9.32				
	$ML(HL)/ML_2.H$	9.92				
	$M(HL)_3/ML(HL)_2.H$	9.04				
	$ML(HL)_2/ML_2(HL).H$	10.41				
Cu^{2+}	$MHL/M.HL$	7.8 ±0.1			-5.4 -0.5	18
	$M(HL)_2/M.(HL)_2$	14.7 ±0.2			-11.1 -1	30
	$M(HL)_2/MHL_2.H$	9.17			-6.1	22
	$MHL_2/ML_2.H$	10.06			-6.4	25
Ga^{3+}	$MHL/M.HL$		8.7			
	$MH_2L/MHL.H$		3.3			

Bibliography:

H^+, Cu^{2+} 74GNK

Ni^{2+} 79GKD

Ga^{3+} 76BHa

Other references: 75PN,76S,76Sa,76Sb,76SSf, 77Sh,**78**CK,79PG

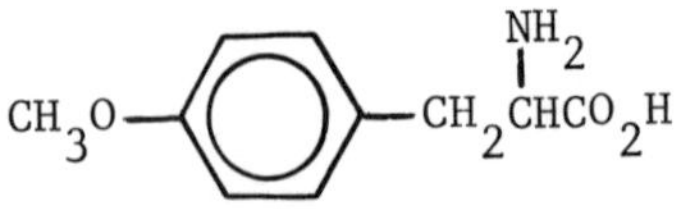

$$CH_3O-\langle C_6H_4\rangle-CH_2\overset{\overset{\displaystyle NH_2}{|}}{C}HCO_2H$$

$C_{10}H_{13}O_3N$ __L-0-Methyltyrosine__   HL

(Other reference on Vol.1, p.395)

Metal ion	Equilibrium	Log K 25°, 0.16	Log K 20°, 0.37
H^+	$HL/H.L$	(9.15)	(9.16)
	$H_2L/HL.H$		2.35
Cu^{2+}	$ML/M.L$		7.89
	$ML_2/M.L_2$		(11.63)

Bibliography: 74W

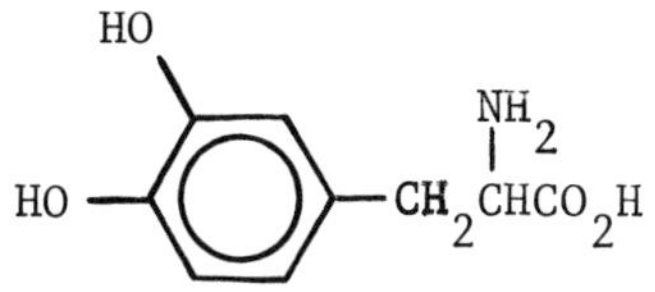

$C_9H_{11}O_4N$ L-2-Amino-3-(3,4-dihydroxyphenyl)propanoic acid   (L-DOPA) H_3L

(Other values on Vol.1, p.33)

Metal ion	Equilibrium	Log K 25°, 0.1	Log K 20°, 0.37	Log K 25°, 1.0	ΔH 25°, 0.1	ΔS 25°, 0.1
H^+	HL/H.L	(13.4) ±0.0		13.40	(-10)[s]	(28)
	$H_2L/HL.H$	9.78 ±0.07	9.87	9.74	(-7)[s]	(21)
	$H_3L/H_2L.H$	8.72 ±0.08	8.81	8.71	(-9)[s]	(10)
	$H_4L/H_3L.H$	2.20 ±0.02	(2.04)	2.31		
Ni^{2+}	$MH_2L/M.H_2L$	4.88 ±0.03		4.96		
	$M(H_2L)_2/M.(H_2L)^2$	8.9 ±0.3	8.73	9.16		
	$MH_2L/MHL.H$	7.9 ±0.3				
	$M(H_2L)_2/M(HL)(H_2L).H$	7.87				
	$M(HL)(H_2L)/M(HL)_2.H$	9.00				
	$M(HL)_2/ML(HL).H$	9.62				
	$ML(HL)/ML_2.H$	11.47				
Cu^{2+}	$MH_2L/M.H_2L$	7.55 ±0.03	7.60	7.6		
	$M(H_2L)_2/M.(H_2L)^2$	(14.15)[x]	14.51			
	$M(H_2L)_2/M(HL)(H_2L).H$	(6.80)[x]				
	$M(HL)(H_2L)/M(HL)_2.H$	(8.47)[x]				
	$M(HL)_2/ML(HL).H$	(9.51)[x]				
	$ML(HL)/ML_2.H$	(10.35)[x]				
	$M_2L_2/(ML)^2$	(26.85)[x]		27.04		
Fe^{3+}	$ML.H^2/M.H_2L$			-1.29		
Zn^{2+}	$MH_2L/M.H_2L$	3.77	(4.4)			
	$MH_2L/MHL.H$	6.77				
	$M(HL)(H_2L)/M(HL)_2.H$	8.59				
	$M(HL)_2/ML(HL).H$	9.67				
	$ML(HL)/ML_2.H$	10.42				
Cd^{2+}	$MH_2L/M.H_2L$		3.61			
Pb^{2+}	$MH_2L/M.H_2L$		5.56			

[s]1-45°, 0.1;   [x]DL-mixture

Bibliography:

H^+ 71M,74GS,76GK,77IH,78RM Fe^{3+} 76MPS
Ni^{2+} 79GKD Zn^{2+} 74GS,78RM,79GKD
Cu^{2+} 74GS,76GK,76RM,78RM Cd^{2+},Pb^{2+} 74GS

                                   Other references: 73KL,78RMa

$$\begin{array}{c} NH_2 \\ | \\ HOCH_2CHCO_2H \end{array}$$

$C_3H_7O_3N$ **L-2-Amino-3-hydroxypropanoic acid**   (serine)   HL
(Other values on Vol.1, p.35)

Metal ion	Equilibrium	Log K 25°, 0.1	Log K 25°, 1.0	Log K 25°, 0	ΔH 25°, 0.1	ΔS 25°, 0.1
H^+	HL/H.L	9.05 ±0.03	9.04	9.209	-10.3 ±0.2	7
		8.99[b]	8.72[n]	9.45[e] +0.2		
	$H_2L/HL.H$	2.13 ±0.04	2.13	2.187	-1.2 -0.1	6
		2.24[b]	2.06[n]	2.41[e] +0.2		
Mn^{2+}	ML/M.L	2.50	2.32[n]	2.89[e]		
	$ML_2/M.L^2$	(3.98)[w]		4.8[e]	(-1)[s]	(15)
Co^{2+}	ML/M.L	4.33 ±0.05	4.20[n]	4.58[e]	-2.7	11
	$ML_2/M.L^2$	(7.8)[w] ±0.2	7.56[n]	8.57[e]	-4.9	20
	$ML_3/M.L^3$		9.81[n]	11.55[e]		
Ni^{2+}	ML/M.L	5.40 ±0.08	5.21[n]	5.63[e]	-3.8 ±0.0	12
	$ML_2/M.L^2$	9.9[y] ±0.1	9.59[n]	10.62[e]	-8.3 ±0.3	18
	$ML_3/M.L^3$	13.1[y] ±0.4	12.5[n]	14.2[e]	-13.3 ±0.0	16
Cu^{2+}	ML/M.L	7.89 ±0.04	7.95	8.95[e]	-5.5 ±0.0	18
			7.57[n]			
	$ML_2/M.L^2$	14.5[y] ±0.1	14.7	16.2[e]	-11.7 ±0.1	27
			14.0[n]			
	MHL/ML.H	3.55				
	$ML_2/MOHL_2.H$	9.81	10.25			
Zn^{2+}	ML/M.L	4.60 ±0.06	4.47[n]	4.90[e]	-2.3	14
	$ML_2/M.L^2$	(8.5)[w] ±0.2	8.31[n]	9.28[e]	-4.9 -1	23
	$ML_3/M.L^3$		10.6[n]	11.9[e]		
Cd^{2+}	ML/M.L			4.15[e]		
	$ML_2/M.L^2$			7.86[e]		
	$ML_3/M.L^3$			10.22[e]		
Pb^{2+}	ML/M.L		4.41[n]	5.05[e]		
	$ML_2/M.L^2$		7.51[n]	8.27[e]		
	$ML_3/M.L^3$			10.0[e]		
Ga^{3+}	ML/M.L			9.0[e]		
	MHL/ML.H			2.4[e]		

---

[b] 25°, 0.5;   [e] 25°, 3.0;   [n] 37°, 0.15;   [s] 15-40°, 0.2;   [w] optical isomerism not stated;
[y] L- and DL-isomers had the same value.

Bibliography:
H^+ 67S,73GM,73SK,76PSa,78VV,59FO
Mn^{2+} 67S
Co^{2+},Zn^{2+} 67S,69PS
Ni^{2+} 67S,76PSa
Cu^{2+} 67S,71GN,73GM,73KSa,76PSa
Cd^{2+} 74WW

Pb^{2+} 67S,73CT
Ga^{3+} 76BHa

Other references: 70FM,70FMb,70GM,73BS,73FA,
73KSa,73RK,73SKa,74FA,74FAa,74FAb,74FLa,
74KH,74KU,74PN,74SK,75HV,76KFA,79KC,
79PG,79SGa,79SSa

$$\begin{array}{cc} \text{HO} & \text{NH}_2 \\ | & | \\ \text{CH}_3\text{CHCHCO}_2\text{H} & \end{array}$$

$C_4H_9O_3N$ <u>L-2-Amino-3-hydroxybutanoic acid</u>   (threonine)   HL

(Other values on Vol.1, p.37)

Metal ion	Equilibrium	Log K 25°, 0.1	Log K 25°, 1.0	Log K 25°, 0	ΔH 25°, 0.1	ΔS 25°, 0.1
H^+	HL/H.L	8.96 ±0.03	8.97	9.100	-10.0 ±0.2	7
		8.92[b]	8.59[n]	9.35[e]		
	H_2L/HL.H	2.20 ±0.04	2.16	2.088	-1.3 ±0.1	6
		2.21[b]	2.08[n]	2.41[e]		
Mn^{2+}	ML/M.L	2.58	(2.07)[n]			
	$ML_2/M.L^2$	(3.96)[w]			(-1)[s]	(15)
Co^{2+}	ML/M.L	4.34[y] ±0.09	(4.16)[n]		-1.9 ±0.8	14
	$ML_2/M.L^2$	8.0[y] ±0.2	(7.45)[n]		-5.0 ±0.5	20
	$ML_3/M.L^3$		(8.82)[n]			
Ni^{2+}	ML/M.L	5.46[y] ±0.04	(5.15)[n]		-3.5 ±0.3	13
	$ML_2/M.L^2$	10.01[y] ±0.07	(9.37)[n]		-8.3 ±0.2	18
	$ML_3/M.L^3$	13.3[y] ±0.1	(11.8)[n]		-13.4	16
Cu^{2+}	ML/M.L	8.00 ±0.06	7.95 ±0.1	8.43	-5.5 ±0.2	18
			(7.55)[n]	8.61[e]	(-4.3)[e]	
	$ML_2/M.L^2$	14.69[y] ±0.08	14.8 ±0.1	15.4	-11.6 ±0.2	28
			(14.0)[n]	16.1[e]	(-11.2)[e]	
	MHL/ML.H	3.59				
	$ML_2/MOHL_2$.H	9.84	9.89			
	$MOHL_2/M(OH)_2L_2$.H	10.78	11.19			
Zn^{2+}	ML/M.L	4.70 ±0.04	(4.43)[n]		-2.5	13
	$ML_2/M.L^2$	(8.6)[w] ±0.1	(8.14)[n]		-5.3	22
	$ML_3/M.L^3$		(10.1)[n]			
Pb^{2+}	ML/M.L		(4.43)[n]			
	$ML_2/M.L^2$		(7.20)[n]			

[b] 25°, 0.5;   [e] 25°, 3.0;   [n] 37°, 0.15;   [s] 15-40°, 0.2;   [w] optical isomerism not stated;
[y] L- and DL-isomers had same values.

Bibliography:

H^+ 67S,73GM,73SK,76PSa,77DOa,78VV

Mn^{2+},Zn^{2+},Pb^{2+} 67S

Co^{2+} 67S,78IS

Ni^{2+} 67S,76PSa

Cu^{2+} 67S,70JP,71GN,73GM,75BW,76PSa,77DOa, 78IS

Other references: 73RGa,73SCa,73SKa,74FA, 74FAa,74FAb,74KH,74PN,74SK,75HV,75SC,

$$\begin{array}{c} NH_2 \\ | \\ CH_3CHCHCO_2H \\ | \\ OH \end{array}$$

$C_4H_9O_3N$

L-allo-Threonine
(Other reference on Vol.1, p.395)

HL

Metal ion	Equilibrium	Log K 25°, 0.1	Log K 25°, 1.0	Log K 25°, 0	ΔH 25°, 0	ΔS 25°, 0
H^+	HL/H.L	8.89	8.92	9.096	-10.2 ±0.2	7
	$H_2L/HL.H$	(1.85)	2.09	2.108	-0.7	7
Co^{2+}	$ML/M.L$	4.00			-2.0[a]	12[a]
	$ML_2/M.L^2$	7.21			-3.1[a]	23[a]
Cu^{2+}	$ML/M.L$	7.46	7.59		-5.5[a]	16[a]
	$ML_2/M.L^2$	13.94	14.17		(-10.0)[a]	(30)[a]

---

[a] 25°, 0.1

Bibliography:

H^+ 73GM,81IS

Co^{2+} 78IS

Cu^{2+} 73GM,78IS

$$\begin{array}{ccc} O & & NH_2 \\ \| & & | \\ H_2NCCH_2 & & CHCO_2H \end{array}$$

$C_4H_8O_3N_2$

L-2-Aminobutanedioic acid 4-amide (asparagine)
(Other values on Vol.1, p.40)

HL

Metal ion	Equilibrium	Log K 25°, 0.1	Log K 25°, 1.0	Log K 25°, 3.0	ΔH 25°, 0.1	ΔS 25°, 0.1
H^+	HL/H.L	8.72 ±0.02	8.67	9.30	-9.8 ±0.1	7
		8.64[b]			-12.1[e]	2[e]
	$H_2L/HL.H$	2.15 ±0.01	2.09	2.59	-1.1	6
		2.11[b]			-1.2[e]	8[e]
Mn^{2+}	$ML/M.L$			3.10	-1.7[e]	8[e]
	$ML_2/M.L^2$			5.22	-3.3[e]	13[e]
Fe^{2+}	$ML/M.L$			4.37		
	$ML_2/M.L^2$			7.57		
	$ML_3/M.L^3$			10.26		
Co^{2+}	$ML/M.L$	4.51 ±0.04		4.90	-2.9[e]	13[e]
	$ML_2/M.L^2$	8.01 ±0.1		9.03	-6.4[e]	20[e]
	$ML_3/M.L^3$	9.96		11.86	-8.6[e]	25[e]
Ni^{2+}	$ML/M.L$	5.68		6.15	-4.1[e]	14[e]
	$ML_2/M.L^2$	10.23		11.16	-10.4[e]	16[e]
	$ML_3/M.L^3$			14.55	-15.2[e]	16[e]

---

[b] 25°, 0.5; [e] 25°, 3.0

Asparagine (continued)

Metal ion	Equilibrium	Log K 25°, 0.1	Log K 25°, 1.0	Log K 25°, 3.0	ΔH 25°, 0.1	ΔS 25°, 0.1
Cu^{2+}	ML/M.L	7.83 ±0.04		8.68	-6.3 [e]-6.6	15 [e]18
	$ML_2/M.L^2$	14.36 ±0.07		16.05	-12.1 ±0.8 [e]-14.8	25 [e]24
	$ML_2/M(H_{-1}L)L.H$	10.45				
	$M(H_{-1}L)L/M(H_{-1}L)_2.H$	12.0				
Zn^{2+}	ML/M.L			5.07	-2.5 [e]	15 [e]
	$ML_2/M.L^2$			9.43	-5.5 [e]	25 [e]
	$ML_3/M.L^3$			12.30	-6.7 [e]	34 [e]
Pd^{2+}	MHL/M.HL			2.81		
	$M(H_{-1}L).H/M.L$			9.1		
Cd^{2+}	ML/M.L			4.07		
	$ML_2/M.L^2$			7.58		
	$ML_3/M.L^3$			9.61		
Pb^{2+}	ML/M.L			4.91		
	$ML_2/M.L^2$			7.82		
	$ML_3/M.L^3$			8.8		

[e] 25°, 3.0

Bibliography:

H^+ 73TS,75GNF,77L Pb^{2+} 73CT

Mn^{2+}-Ni^{2+},Zn^{2+} 74BW   Other references: 70FM,70FMb,73FA,73TSa,
Cu^{2+} 74BW,75GNF   73TSb,74FA,74FAa,74FAb,74TS,76KFA,
Pd^{2+} 74GW   77MSc,77RRa
Cd^{2+} 74WW

$$\overset{O}{\overset{\|}{H_2NCCH_2}}CH_2\overset{NH_2}{\overset{|}{CH}}CO_2H$$

| $C_5H_{10}O_3N_2$ | | L-2-Aminopentanedioic acid 5-amide (glutamine)(Other values on Vol.1, p.41) | | | | HL |

Metal ion	Equilibrium	Log K 25°, 0.1	Log K 37°, 0.15	Log K 25°, 3.0	ΔH 25°, 0.2	ΔS 25°, 0.1
H^+	HL/H.L	9.96 ±0.058.94 [b]	8.71	9.64	-9.9 [e]-12.2	8 [e]3
	$H_2L/HL.H$	2.16 ±0.012.14 [b]	2.03	2.72	-0.8 [e]-1.1	7 [e]9
Ni^{2+}	ML/M.L	5.16		5.56	-3.2 [e]	15 [e]
	$ML_2/M.L^2$	9.42		10.28	-8.6 [e]	18 [e]
	$ML_3/M.L^3$			13.82	-13.1 [e]	19 [e]
Cu^{2+}	ML/M.L	7.76 ±0.01	7.24	9.05	-5.6 [e]-3.9	17 [e]28
	$ML_2/M.L^2$	14.23 +0.4	13.40	16.54	-11.7 [e]-10.3	26 [e]41

[b] 25°, 0.5;   [e] 25°, 3.0

Glutamine (continued)

Metal ion	Equilibrium	Log K 25°, 0.1	Log K 37°, 0.15	Log K 25°, 3.0	ΔH 25°, 0.2	ΔS 25°, 0.1
Zn^{2+}	$ML/M.L$		4.27	4.83		
	$ML_2/M.L^2$		7.94	9.17		
	$ML_3/M.L^3$			11.84		
Cd^{2+}	$ML/M.L$			4.10		
	$ML_2/M.L^2$			7.66		
	$ML_3/M.L^3$			10.00		
Pb^{2+}	$ML/M.L$			4.70		
	$ML_2/M.L^2$			8.4		
	$ML_3/M.L^3$			10.1		

Bibliography:

H^+ 71HP,73KSa,75GNF,77L

Co^{2+} 73W

Ni^{2+} 73W,74BW

Cu^{2+} 71HP,73KSa,73W,74BW,74GNF

Zn^{2+} 71HP

Cd^{2+} 74WW

Pb^{2+} 73CT

Other references: 69LC,73TS,73TSa,73TSb,
                   74SCa,74TS,77MSc

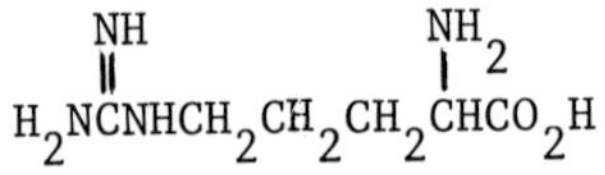

$$\overset{NH}{\overset{\|}{H_2NCNHCH_2CH_2CH_2}}\overset{NH_2}{\overset{|}{CHCO_2H}}$$

$C_6H_{14}O_2N_4$	L-2-Amino-5-guanidopentanoic acid (arginine) (Other values on Vol.1, p.43)					HL

Metal ion	Equilibrium	Log K 25°, 0.1	Log K 25°, 1.0	Log K 25°, 0	ΔH 25°, 0	ΔS 25°, 0
H^+	$HL/H.L$	(12.1) ±0.0				
	$H_2L/HL.H$	9.02 ±0.03	9.21 8.70^n	8.991	-10.7	5
	$H_3L/H_2L.H$	2.03 ±0.08	2.19 1.70^n	1.823	-1.0	5
Co^{2+}	$MHL/M.HL$	3.87 +0.1				
	$M(HL)_2/M.(HL)^2$	7.05 ±0.2				
	$M(HL)_3/M.(HL)^3$	9.25 ±0.1				
Ni^{2+}	$MHL/M.HL$	5.05 ±0.1				
	$M(HL)_2/M.(HL)^2$	9.10 ±0.2				
	$M(HL)(H_2L)/M(HL)_2.H$	15.23				
	$ML(HL)_2/M(HL)_2.L$	4.47				
Cu^{2+}	$MHL/M.HL$	7.45 ±0.1	7.38^n			
	$M(HL)_2/M.(HL)^2$	13.85 ±0.2	13.66^n			
	$M_2L_2/M^2.L^2$		3.17^n			
Zn^{2+}	$MHL/M.HL$	4.15 ±0.04	4.03^n			
	$M(HL)_2/M.(HL)^2$	8.10 ±0.03	7.56^n			
	$MHL/ML.H$		4.2^n			

^{n}37°, 0.15

Arginine (continued)

Bibliography:

H^+, Cu^{2+} 71HP,76BP,78SYN	Zn^{2+} 71HP	
Co^{2+}, Ni^{2+}76BP	Other reference: 79PG	

$$\underset{HSCH_2CHCO_2H}{\overset{NH_2}{|}}$$

$C_3H_7O_2NS$ <u>L-2-Amino-3-mercaptopropanoic acid</u> (cysteine) H_2L
(Other values on Vol.1, p.47)

Metal ion	Equilibrium	Log K 25°, 0.1	Log K 25°, 1.0	Log K 25°, 0	ΔH 25°, 0.1	ΔS 25°, 0.1
H^+	HL/H.L	10.29 ± 0.08	9.95 $10.11^n\ 0.00$	$10.77^{\pm 0.01}$ $10.70^e \pm 0.02$	-8.4 -9.7^e	19 16^e
	$H_2L/HL.H$	8.16 ± 0.05	8.07 $7.96^n \pm 0.01$	8.36 ± 0.03 $8.71^e \pm 0.07$	-8.5 -9.3^e	9 9^e
	$H_3L/H_2L.H$	1.91 ± 0.06	$1.9^n \pm 0.1$	(1.7) $2.40^e \pm 0.04$	-0.1^e	11^e
Ni^{2+}	ML/M.L	9.8 $9.4^h \pm 0.04$				
	$ML_2/M.L^2$	20.07^y $20.18^h \pm 0.03$			$(-14)^s$	(45)
	MHL/M.HL	4.96^h				
	$M_2L_3/M^2.L^3$	33.0^h				
	$M_3L_4/M^3.L^4$	45.7^h				
Zn^{2+}	$ML_2/M.L^2$	$18.12^y \pm 0.07$ 18.21^h	$17.95^n \pm 0.04$	19.39^e		
	MHL/M.HL	4.5	4.51^n			
	$ML(HL)/ML_2.H$	$6.34^y \pm 0.04$ $(6.58)^h$	$6.3^n\ -0.1$	6.47^e		
	$M(HL)_2/ML(HL).H$	5.50 5.8^h	5.6^n	6.02^e		
	$M_3HL_4/M^2.H.L^4$	49.0 49.5^h	48.5 ± 0.2	52.5^e		
	$M_3HL_4/M_3L_4.H$	6.90	6.03^n	6.3^e		
	$M_3H_2L_4/M_3HL_4.H$		5.77^n			
Cd^{2+}	ML/M.L			$(12.88)^e$		
	$ML_2/M.L^2$			19.6^e		
Pb^{2+}	ML/M.L	11.6		12.21^e	$(-10)^t$	$(22)^e$
	$ML_2/M.L^2$			18.57^e		
	MHL/M.HL			6.64^e	$(-4)^t$	$(17)^e$
	$MHL_2/M.H.L^2$			27.3^e	$(-27)^t$	$(34)^e$
	$ML_2/MOHL_2.H$			2.9^e		

^{e}25°, 3.0; ^{h}20°, 0.1; ^{n}37°, 0.15; ^{s}15-40°, 0.2; ^{t}10-40°, 3.0; ^{y}L- and DL-isomers
had the same value.

Cysteine (continued)

Metal ion	Equilibrium	Log K 25°, 0.1	Log K 25°, 1.0	Log K 25°, 0	ΔH 25°, 0.1	ΔS 25°, 0.1
Ga^{3+}	ML/M.L			16.1^e		
	MHL/ML.H			2.9^e		
	$MH_2L/MHL.H$			2.7^e		
In^{3+}	$ML/M.L$	14.12^h				
	$ML_2/M.L^2$	27.26^h				
	$ML_3/M.L^3$	32.2^h				
	MHL/ML.H	4.34^h				
	$ML(HL)/ML_2.H$	4.52^h				
	$M(HL)_2/ML(HL).H$	3.96^h				

^{e}25°, 3.0; ^{h}21°, 0.1

Bibliography:

H^+ 71HP,76BHa,76KSa,78BMW,79PBG Ga^{3+} 76BHa

Ni^{2+} 68PS,72RJ,79SGH In^{3+} 76KSa

Zn^{2+} 68PS,71HP,72RJ,76CWW,78BMW,79SGH

Cd^{2+} 74WW   Other references: 70FMa,73CT,73H,74PN,
                                                          74RM,75MM,75RM,75ZK,76HS,76ZN,77La,
Pb^{2+} 76CWa,76CWW   78BKS,79ZN

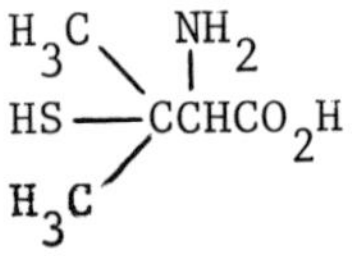

$C_5H_{11}O_2NS$ D-2-Amino-3-mercapto-3-methylbutanoic acid (penicillamine) H_2L
                     (Other values on Vol,1, p.48)

Metal ion	Equilibrium	Log K 25°, 0.1	Log K 25°, 0.5	Log K 25°, 3.0
H^+	HL/H.L	10.6^h ±0.1	10.50	11.01
		10.7^h ±0.1		
	$H_2L/HL.H$	7.92^h ±0.04	7.92	8.60
		8.01^h ±0.03		
	$H_3L/H_2L.H$	1.90^h ±0.1	1.95	2.43
		1.90^h ±0.07		
Ni^{2+}	ML/M.L	10.63^h		
		10.75^h		
	$ML_2/M.L^2$	22.97 ±0.05		
		$(22.92)^x$		
		$(22.89)^{h,x}$		
	$ML(HL)/ML_2.H$	4.14		
Cu^+	$M(HL)_2/M.(HL)^2$		18.18	
	$M_5L_4/M^5.L^4$		(101.50)	

^{h}20°, 0.1; ^{x}DL-mixture

Penicillamine (continued)

Metal ion	Equilibrium	Log K 25°, 0.1	Log K 25°, 0.5	Log K 25°, 3.0
Zn^{2+}	ML/M.L	9.5 ±0.1		
		9.59^h		
	$ML_2/M.L^2$	19.40 ±0.02		20.52
		19.56^h		
	MHL/ML.H	5.14		
	$ML(HL)/ML_2.H$	6.0^y ±0.1		6.27
		5.99^h		
	$M(HL)_2/ML(HL).H$	5.38^y±0.05		5.93
		5.8^h		
	$ML_2/MOHL_2.H$			(11.96)
	$M_3HL_4/M^3.H.L^4$			53.8
	$M_3HL_4/M_3L_4.H$			6.25
Cd^{2+}	ML/M.L	10.6 ±0.3		12.68
	$ML_2/M.L^2$			20.68
	MHL/ML.H			4.47
	$ML(HL)/ML_2.H$			7.63
	$M(HL)_2/ML(HL).H$			6.22
	$ML_2/MOHL_2.H$			11.54
Hg^{2+}	ML/M.L	16.3 ±0.5		14.32
Pb^{2+}	ML/M.L	12.3 ±0.0		19.05
	$ML_2/M.L^2$			
	MHL/ML.H			3.40
	$ML(HL)/ML_2.H$			8.93
	$M(HL)_2/ML(HL).H$			6
	$ML_2/MOHL_2.H$			(11.50)
In^{3+}	ML/M.L	15.33^h		
	$ML_2/M.L^2$	$(29.79)^{h,x}$		
	MHL/ML.H	3.53^h		
	ML/MOHL.H	9.8^h		
	$ML(HL)/ML_2.H$	$(3.49)^{h,x}$		

---

^{h}20°, 0.1; ^{x}DL-mixture; ^{y}D- and DL-isomers had same values.

Bibliography:

H^+	70SY,76CWW,76KS,79LP,79OL,79PBG	Cd^{2+},Pb^{2+}	70SY,76CWW
Ni^{2+}	68PS,79SGH	Hg^{2+}	70SY
Cu^+	79OL	In^{3+}	76KS
Zn^{2+}	68PS,76CWW,79SGH		

Other references: 76HS,77HSa,78BKS,78RM

$$\overset{\displaystyle NH_2}{\underset{\displaystyle CH_3SCH_2CH_2CHCO_2H}{|}}$$

| $C_5H_{11}O_2NS$ | | <u>DL-2-Amino-4-(methylthio)butanoic acid</u> (methionine) | | | | HL |

(Other values on Vol.1, p.50)

Metal ion	Equilibrium	Log K 25°, 0.1	Log K 25°, 1.0	Log K 25°, 3.0	ΔH 25°, 0.1	ΔS 25°, 0.1
H^+	HL/H.L	9.06 ±0.02	9.00	9.69	$(-10)^s$	(8)
		9.07^b	8.79^n			
	$M_2L/ML.H$	2.10 ±0.08	2.26	2.70	$(0)^s$	(10)
		2.10^b	2.14^n			
Co^{2+}	$ML/M.L_2$	4.14 ±0.02				
	$ML_2/M.L^2$	7.58^y±0.02				
Ni^{2+}	$ML/M.L_2$	5.33 ±0.01			-3.2	14
	$ML_2/M.L^2$	9.89 ±0.00			-8.4	17
		$(9.99)^x$			$(-8.6)^x$	$(17)^x$
	$ML_3/M.L^3$	11.6				
		$(11.9)^x$				
Cu^{2+}	$ML/M.L_2$	7.86 ±0.01	7.67^n			
	$ML_2/M.L^2$	14.6^y±0.1	14.08^n			
Zn^{2+}	$ML/M.L_2$	4.38 ±0.01	4.22^n			
	$ML_2/M.L^2$	8.33^y±0.03	6.93^n			
Cd^{2+}	$ML/M.L_2$	3.69 ±0.02				
	$ML_2/M.L^2$	7.00^y±0.03				
Ga^{3+}	ML/M.L			8.9		
	MHL/ML.H			2.6		

---

^{b}25°, 0.5; ^{n}37°, 0.15; ^{s}10-40°, 0.09; ^{x}DL-mixture; ^{y}L-,D-, and DL-isomers had about the same value.

Bibliography:

H^+ 71HP,75IP,76BHa,78L Ga^{3+} 76BHa

Co^{2+},Cd^{2+} 75IP   Other references: 70FMa,73BS,73FA,73RS,
Ni^{2+} 75IP,76SP   74FA,74FAa,74FAb,74PN,76KFA,77PU,78CK
Cu^{2+},Zn^{2+} 71HP,75IP

$$NH_2$$
$$HO_2CCH_2SCH_2CHCO_2H$$

| $C_5H_9O_4NS$ | L-2-Amino-3-(carboxymethylthio)propanoic acid (S-carboxymethylcysteine) (Other reference on Vol.1, p.396) | | | H_2L |

Metal ion	Equilibrium	Log K 25°, 0.1	Log K 25°, 1.0	Log K 25°, 2.0
H^+	HL/H.L	8.89	9.01	9.13
	H_2L/HL.H	3.36	3.33	3.55
	H_3L/H_2L.H	1.99	1.91	
Mn^{2+}	ML/M.L		(2.58)	
Co^{2+}	ML/M.L	4.90		
	ML_2/M.L^2	8.52		
Ni^{2+}	ML/M.L	6.22	(6.44)	6.40
	ML_2/M.L^2	11.16	(11.70)	11.60
	MHL/ML.H		(4.3)	
Cu^{2+}	ML/M.L	8.15	(8.60)	
	ML_2/M.L^2		(15.24)	
	MHL/ML.H	2.97	(3.16)	
Zn^{2+}	ML/M.L	5.12		
	ML_2/M.L^2	9.26		
Pb^{2+}	ML/M.L		(5.78)	
	MHL/ML.H		(4.70)	
	ML/MOHL.H		(9.60)	

Bibliography:
H^+,Ni^{2+} 74NBA,76AH,79CS Co^{2+},Zn^{2+} 74NBA
Mn^{2+},Pb^{2+} 79CS Cu^{2+} 74NBA,79CS

$$NH_2$$
$$HO_2CCH_2CH_2SCH_2CHCO_2H$$

| $C_6H_{11}O_4NS$ | L-2-Amino-3-(2-carboxyethylthio)propanoic acid [S-(2-carboxyethyl)cysteine] | H_2L |

Metal ion	Equilibrium	Log K 25°, 2.0
H^+	HL/H.L	9.05
	H_2L/HL.H	4.36
Ni^{2+}	ML/M.L	5.65
	ML_2/M.L^2	10.10
	ML_3/M.L^3	12.77

Bibliography: 76AH

$$\overset{\displaystyle NH_2}{\underset{\displaystyle |}{H_2NCH_2CH_2SCH_2CHCO_2H}}$$

$C_5H_{12}O_2N_2S$	L-2-Amino-3-(2-aminoethylthio)propanoic acid [S-(2-aminoethyl)cysteine] (Other values on Vol.1, p.53)	HL

Metal ion	Equilibrium	Log K 25°, 0.1
H^+	HL/H.L	9.67
	$H_2L/HL.H$	8.32
	$H_3L/H_2L.H$	1.70
Co^{2+}	MHL/M.HL	3.46
	MHL/ML.H	6.94
	$MHL_2/ML.HL$	2.61
	$MHL_2/ML_2.H$	9.09
Ni^{2+}	MHL/M.HL	4.38
	MHL/ML.H	5.57
	$MHL_2/ML.HL$	3.71
	$MHL_2/ML_2.H$	9.04
Cu^{2+}	MHL/M.HL	7.22
	MHL/ML.H	6.30
	$MHL_2/ML.HL$	6.14
	$MH_2L_2/MHL_2.H$	7.82
	$MHL_2/ML_2.H$	9.27
Zn^{2+}	MHL/M.HL	3.71
	MHL/ML.H	6.82

Bibliography: 74NBA

$$\overset{\displaystyle H_2N \qquad\qquad NH_2}{\underset{\displaystyle |\qquad\qquad\quad |}{HO_2CCHCH_2SSCH_2CHCO_2H}}$$

$C_6H_{12}O_4N_2S_2$	L-Dithiobis(2-amino-3-propanoic acid) (cystine) (Other references on Vol.1, p.55)	H_2L

Metal ion	Equilibrium	Log K 20°, 0.15	Log K 37°, 0.15
H^+	HL/H.L	8.80	8.57
	$H_2L/HL.H$	8.03	7.83
Cu^{2+}	MHL/M.HL		7.51
	$M_2L_2/M^2.L^2$		28.07

Bibliography: 71HP

$$\begin{array}{c} NH_2 \\ | \\ H_2NCH_2CHCO_2H \end{array}$$

$C_3H_8O_2N_2$		DL-2,3-Diaminopropanoic acid (Other values on Vol.1, p.56)				HL

Metal ion	Equilibrium	Log K 25°, 0.1	Log K 25°, 0.5		ΔH 25°, 0.2	ΔS 25°, 0.1
H$^+$	HL/H.L	9.40 −0.01	9.35		−10.0	9
	H$_2$L/HL.H	6.68 ±0.02	6.55		−9.5	−1
	H$_3$L/H$_2$L.H	1.31 +0.01			−0.5	4
Co^{2+}	ML/M.L	6.28	6.2			
	ML$_2$/M.L^2	(11.36)^x	11.2			
	MHL/ML.H	6.02				
	ML(HL)/ML$_2$.H	(6.60)^x				
Ni^{2+}	ML/M.L	8.16 +0.3				
	ML$_2$/M.L^2	(15.22)^x±0.06				
	MHL/ML.H	5.27 −0.2				
	M(HL)$_2$/ML(HL).H	(5.24)^x				
	ML(HL)/ML$_2$.H	(5.90)^x				
Cu^{2+}	ML/M.L	10.56 ±0.06			−10.0	15
	ML$_2$/M.L^2	(19.82)^x±0.01			(−21.5)^x	(19)^x
	MHL/ML.H	5.01 ±0.03			−5.9	3
	M(HL)$_2$/ML(HL).H	(4.75)^x±0.03			(−5.4)^x	(4)^x
	ML(HL)/ML$_2$.H	(5.51)^x−0.01			(−6.3)^x	(4)^x

---

^{x}DL-mixture

Bibliography:

H$^+$,Cu^{2+} 76BP,78GFN   Co^{2+},Ni^{2+} 76BP

$$\begin{array}{c} NH_2 \\ | \\ H_2NCH_2CH_2CHCO_2H \end{array}$$

$C_4H_{10}O_2N_2$		L-2,4-Diaminobutanoic acid		HL

Metal ion	Equilibrium	Log K 25°, 0.1	ΔH 25°, 0.2	ΔS 25°, 0.1
H$^+$	HL/H.L	10.21 ±0.02	−11.4	8
	H$_2$L/HL.H	8.19 ±0.04	−10.2	3
	H$_3$L/H$_2$L.H	1.7 ±0.2	−1.3	3
Co^{2+}	ML/M.L	6.75		
	ML$_2$/M.L^2	12.00		
	MHL/ML.H	6.85		
	ML(HL)/ML$_2$.H	7.92		
Ni^{2+}	ML/M.L	8.91		
	ML$_2$/M.L^2	15.97		
	MHL/ML.H	5.83		
	ML(HL)/ML$_2$.H	6.87		

L-2,4-Diaminobutanoic acid (continued)

Metal ion	Equilibrium	Log K 25°, 0.1	ΔH 25°, 0.2	ΔS 25°, 0.1
Cu^{2+}	$ML/M.L$	10.56 ±0.06	-9.6	16
	$ML_2/M.L^2$	19.02		
		$(18.61)^x$	$(-20.6)^x$	$(16)^x$
	$MHL/ML.H$	6.65 -0.01	-8.1	3
	$M(HL)_2/ML(HL).H$	6.27	$(-6.6)^x$	$(7)^x$
		$(6.35)^x$		
	$ML(HL)/ML_2.H$	7.91	$(-8.6)^x$	$(9)^x$
		$(8.28)^x$		

---

^{x}DL-mixture

Bibliography:

H^+ 72HM,76BP,78GFN Cu^{2+} 76BP,78GFN

Co^{2+},Ni^{2+} 76BP   Other reference: 52A

$$\overset{\overset{\displaystyle NH_2}{\displaystyle |}}{H_2NCH_2CH_2CH_2CHCO_2H}$$

$C_5H_{12}O_2N_2$	L-2,5-Diaminopentanoic acid (ornithine) (Other values on Vol.1, p.57)				HL

Metal ion	Equilibrium	Log K 25°, 0.1	Log K 25°, 1.0	Log K 25°, 0	ΔH 25°, 0.2	ΔS 25°, 0.1
H^+	$HL/H.L$	10.54 ±0.02	10.46_n 10.10^n	10.755	-12.4	7
	$H_2L/HL.H$	8.78 ±0.05	8.58_n 8.43^n	8.690	-11.3	2
	$H_3L/H_2L.H$	1.9 ±0.2	1.80^n	1.705	-0.9	6
Co^{2+}	$MHL/M.HL$	3.65	3.52			
	$M(HL)_2/M.(HL)^2$	6.76	6.32			
	$MHL/ML.H$	9.16				
	$M(HL)_2/ML(HL).H$	9.13				
	$ML(HL)/ML_2.H$	10.06				
Ni^{2+}	$MHL/M.HL$	4.73	4.44			
	$M(HL)_2/M.(HL)^2$	8.74	8.12			
	$M(HL)_3/M.(HL)^3$	11.73				
	$MHL/ML.H$	8.14				
	$M(HL)_2/ML(HL).H$	8.32				
	$ML(HL)/ML_2.H$	9.43				
	$M(HL)_3/ML(HL)_2.H$	9.20				
	$ML(HL)_2/ML_2(HL).H$	9.56				
	$ML_2(HL)/ML_3.H$	10.17				

---

^{n}37°, 0.15

Ornithine (continued)

Metal ion	Equilibrium	Log K 25°, 0.1	Log K 25°, 1.0	Log K 25°, 0	ΔH 25°, 0.2	ΔS 25°, 0.1
Cu^{2+}	MHL/M.HL	7.35 ±0.07	7.17[n] 7.20[n]		-6.1	13
	$M(HL)_2/M.(HL)^2$	13.51 ±0.10 (13.56)[x]	13.31 13.20[n]		(-12.9)[x]	(19)[x]
	MHL/ML.H		7.5[n]			
	ML/MOHL.H		8.9[n]			
	$M(HL)_2/ML(HL).H$	8.92 (9.20)[x]	8.79[n]		(-11.2)[x]	(5)[x]
	$ML(HL)/ML_2.H$	10.02 (10.12)[x]	9.84[n]		(-12.6)[x]	(4)[x]
Zn^{2+}	MHL/M.HL		3.60 3.86[n]			
	$M(HL)_2/M.(HL)^2$		7.18[n]			
	MHL/ML.H		8.00[n]			
	ML/MOHL.H		8.55[n]			
	$M(HL)_2/ML(HL).H$		8.19[n]			

[n]37°, 0.15; [x]DL-mixture

Bibliography:

H^+ 71HP,76BP,78GFN,79GP Cu^{2+} 71HP,76BP,78GFN,78SYN
Co^{2+},Ni^{2+} 76BP Zn^{2+} 71HP

$$\overset{\displaystyle NH_2}{\underset{}{|}}$$
$$H_2NCH_2CH_2CH_2CH_2CHCO_2H$$

$C_6H_{14}O_2N_2$		L-2,6-Diaminohexanoic acid (lysine) (Other references on Vol.1, p.58)				HL

Metal ion	Equilibrium	Log K 25°, 0.1	Log K 25°, 1.0	Log K 37°, 0.15	ΔH 25°, 0.2	ΔS 25°, 0.1
H^+	HL/H.L	10.68 ±0.02	10.56	10.25 ±0.03	-12.8	6
	$H_2L/HL.H$	9.12 ±0.03	9.05	8.86 ±0.04	-11.3	4
	$H_3L/H_2L.H$	2.19 ±0.04	2.23	2.20	-0.5	8
Co^{2+}	MHL/M.HL	3.84	3.62			
	$M(HL)_2/M.(HL)^2$	7.07	6.68			
	$M(HL)_3/M.(HL)^3$	9.44				
	$M(HL)_2/ML(HL).H$	9.91				
	$ML(HL)/ML_2.H$	10.04				
	$M(HL)_3/ML(HL)_2.H$	9.9				
Ni^{2+}	MHL/M.HL	4.93				
	$M(HL)_2/M.(HL)^2$	9.15	9.00[y]			
	$M(HL)_3/M.(HL)^3$	12.04				
	MHL/ML.H	9.85				
	$M(HL)_2/ML(HL).H$	10.06				
	$ML(HL)/ML_2.H$	10.09				
	$M(HL)_3/ML(HL)_2.H$	9.79				

[y]L- and DL-isomers had the same value.

Lysine (continued)

Metal ion	Equilibrium	Log K 25°, 0.1	Log K 25°, 1.0	Log K 37°, 0.15	ΔH 25°, 0.2	ΔS 25°, 0.1
Cu^{2+}	MHL/M.HL	7.64 ±0.03			-6.2	14
	M(HL)$_2$/M.(HL)^2	14.0 ±0.1			-12.3	23
	M(HL)$_2$/ML(HL).H	10.00 ±0.08			-12.7	3
	ML(HL)/ML$_2$.H	10.52 ±0.05			-12.9	5

Bibliography:

H$^+$ 42KH,76BP,76RS,77OK,78GFN,78SYN,79GP   Cu^{2+} 76BP,78GFN,78SYN

Co^{2+},Ni^{2+} 76BP   Other references: 75NM,76KVP

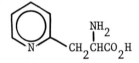

C$_8$H$_{10}$O$_2$N$_2$		L-2-Amino-3-(2-pyridyl)propanoic acid		HL
		(3-pyridylalanine)		

Metal ion	Equilibrium	Log K 25°, 0.1
H$^+$	HL/H.L	8.95
	H$_2$L/HL.H	3.89
Co^{2+}	ML/M.L	5.30
	ML$_2$/M.L^2	9.80
		(10.61)^x
Ni^{2+}	ML/M.L	7.06
	ML$_2$/M.L^2	12.96
		(13.90)^x
Cu^{2+}	ML/M.L	8.26
	ML$_2$/M.L^2	15.09
		(15.34)^x
Zn^{2+}	ML/M.L	4.93
	ML$_2$/M.L^2	9.05
		(9.74)^x

^{x}DL-mixture

Bibliography: 76RN

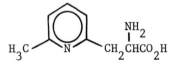

C$_9$H$_{12}$O$_2$N$_2$		D-2-Amino-3(6-methyl-2-pyridyl)propanoic acid		HL
		(3-(6-methy-2-pyridyl)alanine)		

Metal ion	Equilibrium	Log K 25°, 0.1
H$^+$	HL/H.L	8.95
	H$_2$L/HL.H	4.78

3-(6-Methyl-2-pyridyl)alanine (continued)

Metal ion	Equilibrium	Log K 25°, 0.1
Co^{2+}	$ML/M.L$	3.98
	$ML_2/M.L^2$	7.10
		$(7.69)^x$
Ni^{2+}	$ML/M.L$	5.05
	$ML_2/M.L^2$	9.07
		$(9.68)^x$
Cu^{2+}	$ML/M.L$	6.82
	$ML_2/M.L^2$	12.38
		$(12.96)^x$
Zn^{2+}	$ML/M.L$	4.22
	$ML_2/M.L^2$	8.21
		$(8.67)^x$

^{x}DL-mixture

Bibliography: 76RN

$C_6H_9O_2N_3$	L-2-Amino-3-(4-imidazolyl)propanoic acid (histidine) (Other values on Vol.1, p.61)				HL

Metal ion	Equilibrium	Log K 25°, 0.1	Log K 25°, 1.0	Log K 25°, 0	ΔH 25°, 0.1	ΔS 25°, 0.1
H^+	$HL/H.L$	$9.09_b \pm 0.04$	$9.08_n \pm 0.03$	9.28_e	$-10.5 +0.1$	6
		$9.03^b \pm 0.02$	$8.79^n \pm 0.03$	9.63^e		
	$H_2L/HL.H$	$6.02_b \pm 0.04$	$6.1_n \pm 0.1$	5.97_e	$-7.0 +0.1$	4
		$6.03^b \pm 0.08$	$5.80^n \pm 0.01$	6.97^e		
	$H_3L/H_2L.H$	1.7 ± 0.1	1.8		-0.7	5
		1.8^b	$1.9^n \pm 0.3$	2.28^e		
Mn^{2+}	$ML/M.L$	$3.34 +0.2$	3.24^n	3.91^e	-2.7^e	9^e
	$ML_2/M.L^2$		6.16^n	$(6.61)^e$	-5.2^e	13^e
Co^{2+}	$ML/M.L$	6.87 ± 0.06	$6.70^n \pm 0.02$	7.44^e	-5.6^e	15^e
	$ML_2/M.L^2$	12.38 ± 0.04	$12.06^n \pm 0.00$	$(13.48)^e$	-11.8 ± 0.2	17
		$(12.53)^x \pm 0.03$			$(-12.0)^x$	$(17)^x$
	$MHL/ML.H$	3.97				
	$ML(HL)/ML_2.H$	5.99				
		$(5.96)^x$				
Ni^{2+}	$ML/M.L$	8.66 ± 0.03	8.43^n	9.20^e	-8.1	12
	$ML_2/M.L^2$	15.52 ± 0.06	15.14^n	$(16.65)^e$	-16.5 ± 0.3	16
		$(15.77)^x \pm 0.07$				
	$MHL/ML.H$	$3.62 -0.2$				
	$ML(HL)/ML_2.H$	$5.03 -0.3$				
		$(5.00)^x$				

^{b}25°, 0.5; ^{e}25°, 3.0; ^{n}37°, 0.15; ^{x}DL-mixture

Histidine (continued)

Metal ion	Equilibrium	Log K 25°, 0.1	Log K 25°, 1.0	Log K 25°, 0	ΔH 25°, 0.1	ΔS 25°, 0.1
Cu^{2+}	$ML/M.L$	10.16 ± 0.06	$9.79^{n} \pm 0.01$	$(10.09)^{e}$	-11.0 ± 0.6	10
	$ML_2/M.L^2$	$18.10^{y} \pm 0.1$	$17.43^{n} \pm 0.02$	19.03^{e}	-20.6 ± 0.7	
	$MHL/ML.H$	3.97 ± 0.05	$4.09^{n} \pm 0.03$	$(5.53)^{e}$	-5.5^{e}	$(7)^{e}$
	$ML/M(H_{-1}L).H$	8.1 ± 0.1	$7.6^{n} \pm 0.2$	(6.6)		
	$M(HL)_2/ML(HL).H$	3.4 ± 0.2	3.39^{n}	$(4.87)^{e}$	-5.3^{e}	$(4)^{e}$
	$ML(HL)/ML_2.H$	5.7 ± 0.1	$5.6^{n} \pm 0.1$	$(6.85)^{e}$	-5.7^{e}	$(12)^{e}$
	$ML_2/MOHL_2.H$	11.0 ± 0.3				
	$M_2(OH)_2L_2.H^2/M^2.L^2$	8.0 ± 0.0				
VO^{2+}	$ML/M.L$	9.05				
	$ML_2/M.L^2$	15.49 $(15.58)^{x}$				
	$ML/MOHL.H$	5.59				
	$M(HL)_2/ML(HL).H$	4.6				
	$ML(HL)/ML_2.H$	5.91 $(5.74)^{x}$				
Zn^{2+}	$ML/M.L$	6.51 ± 0.06	$6.34 + 0.01$	7.07^{e}	-5.5^{e}	14^{e}
	$ML_2/M.L^2$	12.04 ± 0.03 $(12.18)^{x} \pm 0.02$	$11.70^{n} \pm 0.03$	12.74^{e}	$-11.4 -0.2$ $(-11.8)^{x}$	17 $(16)^{x}$
	$MHL/ML.H$	4.90 ± 0.05	$4.62^{n} \pm 0.02$			
	$M(HL)_2/ML(HL).H$	5.67				
	$ML(HL)/ML_2.H$	5.80 ± 0.01 $(5.96)^{x}$	5.5^{n}			
Cd^{2+}	$ML/M.L$			6.48^{e}		
	$ML_2/M.L^2$			11.11^{e}		
Pb^{2+}	$ML/M.L$	(5.94)	5.96^{n}	6.90^{e}		
	$ML_2/M.L^2$	$(10.11)^{y}$	9.0^{n}	9.81^{e}		
	$M(HL)_2/ML(HL).H$	6.22^{y}				
	$ML(HL)/ML_2.H$	7.04 $(7.40)^{x}$				

---

^{e}25°, 3.0; ^{n}37°, 0.15; ^{x}DL-mixture; ^{y}L- and DL-isomers had the same value.

Bibliography:

H^+ 67PS,73KS,75Apa,76BPa,76DO,76PS,78VZ, 79SP

Mn^{2+} 67PS

Co^{2+} 67PS,75APa,76PS,76R

Ni^{2+} 67PS,76BPa,76DOb,76PS,76R

Cu^{2+} 67PS,73KS,75Ra,76DO,76BPa,76PS,77BS, 79SP

VO^{2+} 76PS

Zn^{2+} 67PS,75APa,76BPa,76PS,76R,77DO,79SP

Cd^{2+} 74WW

Pb^{2+} 67PS,73CT,76PS

Other references: 74L,74PS,74YA,75PN,78BT, 78SKG

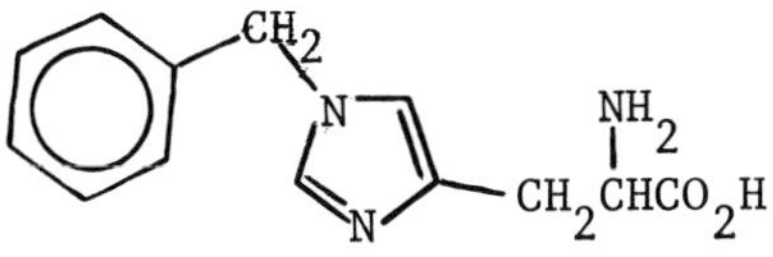

| $C_{13}H_{15}O_2N_3$ | N(3)-Benzyl-L-histidine | | HL |

$C_{13}H_{15}O_2N_3$

N(3)-Benzyl-L-histidine   HL

Metal ion	Equilibrium	Log K 25°, 0.1
H^+	HL/H.L	9.21
	H_2L HL.H	5.53
	$H_3L/H_2L.H$	1.94
Ni^{2+}	$ML/M.L$	8.82
	$ML_2/M.L^2$	16.11
	MHL/M.HL	2.32
	$MHL_2/ML_2.H$	4.07
Cu^{2+}	$ML/M.L$	10.19
	$ML_2/M.L^2$	18.52
	MHL/M.HL	4.53
	$M(HL)_2/M.(HL)^2$	8.63
	$MHL_2/ML_2.H$	5.10
Zn^{2+}	$ML/M.L$	6.58
	$ML_2/M.L^2$	12.17

Bibliography: 76BPa   Other reference: 74YA

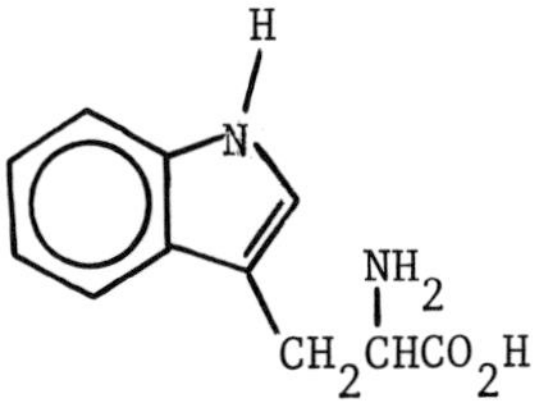

$C_{11}H_{12}O_2N_2$ L-2-Amino-3-(3-indolyl)propanoic acid   (tryptophan)   HL
(Other values on Vol.1, p.63)

Metal ion	Equilibrium	Log K 25°, 0.1	Log K 25°, 1.0	Log K 25°, 3.0	ΔH 25°, 0.1	ΔS 25°, 0.1
H^+	HL/H.L	9.32 ±0.05	9.30	9.92	-10.7	7
		9.32^b	8.97^n			
	$H_2L/HL.H$	2.35	2.39	2.75	$(0)^s$	(11)
		2.23^b	2.34^n			
Mn^{2+}	$ML/M.L$	2.50		2.84		
	$ML_2/M.L^2$			(5.15)		
	$ML_3/M.L^3$			(8.0)		
Cu^{2+}	$ML/M.L$	8.25 ±0.03	8.05^n	8.71	-5.5	19
	$ML_2/M.L^2$	$(15.4)^x$ ±0.1	$(15.32)^n$	(16.66)	-12.7	28
Zn^{2+}	$ML/M.L$	4.69	4.50^n	5.01		
	$ML_2/M.L^2$		8.76^n	(9.78)		
	$ML_3/M.L^3$			(13.50)		

^{b}25°, 0.5; ^{n}37°, 0.15; ^{s}2-44°, 0.1; ^{x}DL-mixture

Tryptophan (continued)

Metal ion	Equilibrium	Log K 25°, 0.1	Log K 25°, 1.0	Log K 25°, 3.0	ΔH 25°, 0.1	ΔS 25°, 0.1
Cd^{2+}	$ML/M.L$			4.48		
	$ML_2/M.L^2$			(8.58)		
	$ML_3/M.L^3$			(12.03)		
Pb^{2+}	$ML/M.L$			(4.9)		
	$ML_2/M.L^2$			(10.27)		

Bibliography:

H^+	71HP,73BS,76SN,78L	Cd^{2+}	74WW
Mn^{2+}	76SN	Pb^{2+}	73CT
Cu^{2+}	71HP, 76SN,77BP	Other references:	72La,75PN,76BF,79PG
Zn^{2+}	71HP,76SN		

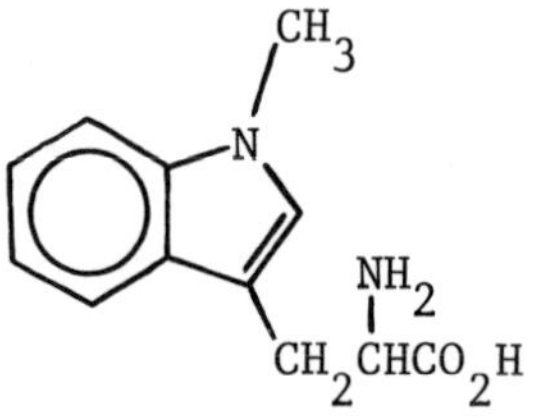

$C_{12}H_{14}O_2N_2$ L-1-Methyltryptophan   HL

Metal ion	Equilibrium	Log K 20°, 0.37
H^+	$HL/H.L$	9.50
	$H_2L/HL.H$	2.32
Cu^{2+}	$ML/M.L$	8.32
	$ML_2/M.L^2$	12.14

Bibliography: 74W

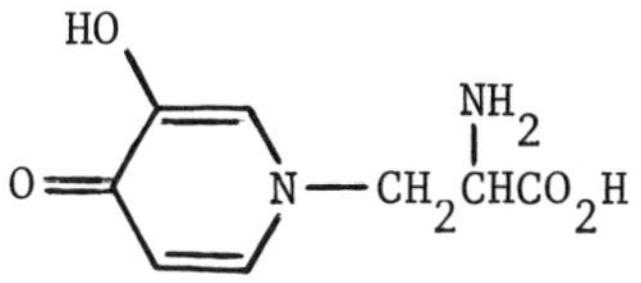

| $C_8H_{10}O_4N_2$ | L-2-Amino-3-(3-hydroxy-4-oxo-1,4-dihydro-1-pyridinyl)propanoic acid | H_2L |
| | (L-mimosine) | |

Metal ion	Equilibrium	Log K 37°, 0.15
H^+	HL/H.L	8.74
	$H_2L/HL.H$	6.88
	$H_3L/H_2L.H$	2.50
	$H_4L/H_3L.H$	1.0
Cu^{2+}	MHL/M.HL	7.62
	$M(HL)_2/M.(HL)^2$	13.78
	$MH_2L/MHL.H$	1.7
	MHL/ML.H	(6.88)
	$M(HL)_2/ML(HL).H$	6.86
	$ML(HL)/ML_2.H$	7.59
	$M_2L/M^2.L$	15.70
	$M_2L_2/M^2.L^2$	29.52
	$M_2L_3/M^2.L^3$	32.2
Zn^{2+}	MHL/M.HL	5.10
	$M(HL)_2/M.(HL)^2$	9.29
	$M(HL)_3/M.(HL)^3$	11.8
	MHL/ML.H	7.3
	$M(HL)_2/ML(HL).H$	7.01
	$ML(HL)/ML_2.H$	7.49
	$M(HL)_3/ML(HL)_2.H$	6.8
	$ML(HL)_2/ML_2(HL).H$	7.8
	$ML_2(HL)/ML_3.H$	8.6
	$M_2L_2/M^2.L^2$	19.01
	$M_2L(HL)/M_2L_2.H$	4.87
	$M_2L_3/M^2.L^3$	22.8
	$M_2L_2(HL)/M_2L_3.H$	7.4
	$M_2L(HL)_2/M_2L_2(HL).H$	6.7
Cd^{2+}	MHL/M.HL	4.17
	$M(HL)_2/M.(HL)^2$	7.52
	MHL/ML.H	7.0
	$M(HL)_2/ML(HL).H$	7.1
	$ML(HL)/ML_2.H$	7.7
	$M_2L/M^2.L$	8.6
	$M_2L_2/M^2.L^2$	14.8

L-Mimosine (continued)

Metal ion	Equilibrium	Log K 37°, 0.15
Pb^{2+}	$MHL/M.HL$	6.69
	$M(HL)_2/M.(HL)^2$	10.44
	$MH_2L/MHL.H$	1.9
	$MHL/ML.H$	6.93
	$M(HL)_2/ML(HL).H$	6.85
	$ML(HL)/ML_2.H$	7.61
	$M_2L/M^2.L$	11.4
	$M_2L_2/M^2.L^2$	19.40
	$M_2HL_2/M^2L^2.H$	6.27

Bibliography: 79SPT

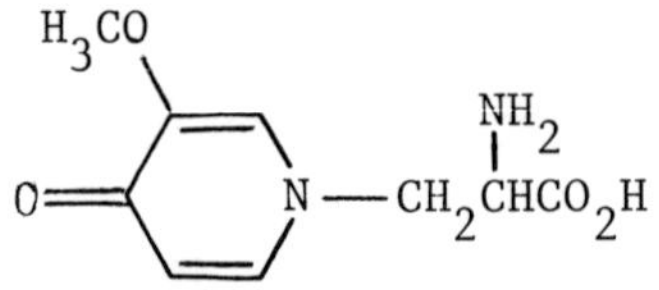

$C_9H_{12}O_4N_2$ L-2-Amino-3-(3-methoxy-4-oxo-1,4-dihydro-1-pyridinyl)propanoic acid   HL
                                    (mimosine methyl ether)

Metal ion	Equilibrium	Log K 37°, 0.15
H^+	$HL/H.L$	6.86
	$H_2L/HL.H$	2.19
Cu^{2+}	$ML/M.L$	6.26
	$ML_2/M.L^2$	11.51
Zn^{2+}	$ML/M.L$	3.39
	$ML_2/M.L^2$	6.07
	$ML_3/M.L^3$	8.0
	$ML/MOHL.H$	8.4
Cd^{2+}	$ML/M.L$	2.73
	$ML_2/M.L^2$	4.98
Pb^{2+}	$ML/M.L$	3.0
	$ML_2/M.L^2$	5.5

Bibliography: 79SPT

$$CH_3NHCH_2CO_2H$$

| $C_3H_7O_2N$ | | N-Methylglycine (sarcosine) | | | | HL |
| | | (Other values on Vol.1, p.66) | | | | |

Metal ion	Equilibrium	Log K 25°, 0.1	Log K 25°, 1.0	Log K 25°, 0	ΔH 25°, 0	ΔS 25°, 0
H^+	HL/H.L	9.97 ±0.04[b] 10.00[b] ±0.01	9.96 ±0.06	10.200	-9.8 ±0.1	14
	$H_2L/H.HL$	2.15[b] ±0.02 2.14[b] ±0.01	2.16	2.11	(-1.7)	(4)
Ni^{2+}	ML/M.L	5.39[b] 5.24[b]				
	$ML_2/M.L^2$	9.75[b] 9.54[b]				
	$ML_3/M.L^3$	12.6[b] 12.4[b]				
Cu^{2+}	ML/M.L	7.68	7.78	8.14	-4.6	22
	$ML_2/M.L^2$	14.16	14.23	14.97	-10.0	35
Zn^{2+}	ML/M.L	4.53[b] 4.31[b]				
	$ML_2/M.L^2$	8.33[b] 8.3[b]				

[b] 25°, 0.5

Bibliography:

H^+	75DOd,78L		Zn^{2+}	75DOd
Ni^{2+}	76DOC		Other reference:	77KDK
Cu^{2+}	77DOa			

| $C_5H_9O_2N$ | | L-Pyrrolidine-2-carboxylic acid (proline) | | | | HL |
| | | (Other values on Vol.1, p.69) | | | | |

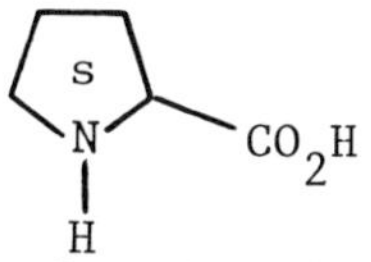

Metal ion	Equilibrium	Log K 25°, 0.1	Log K 25°, 1.0	Log K 25°, 0	ΔH 25°, 0	ΔS 25°, 0
H^+	HL/H.L	10.41[b] ±0.1 10.41[b] ±0.1	10.39 10.20[n] ±0.03	10.640[e] 11.09[e]	-10.35 ±0.04 -10.5[b]	14
	$H_2L/HL.H$	1.9[b] ±0.1 1.9[b] ±0.1	2.02 1.93[n] ±0.04	1.952[e] 2.33[e]	-0.31 ±0.03 -0.3[b]	8
Mn^{2+}	ML/M.L	3.34	2.84[n]			
	$ML_2/M.L^2$		5.53[n]			
	$ML_3/M.L^3$		6.7[n]			
	MHL/ML.H		9.0[n]			
	$ML(HL)/ML_2.H$		9.2[n]			

[b] 25°, 0.5;   [e] 25°, 3.0;   [n] 37°, 0.15

Proline   (continued)

Metal ion	Equilibrium	Log K 25°, 0.1	Log K 25°, 1.0	Log K 25°, 0	ΔH 25°, 0	ΔS 25°, 0
Cu^{2+}	ML/M.L	$8.84_b \pm 0.02$ $8.74_b \pm 0.02$	8.67^n		-7.4^a	16^a
	$ML_2/M.L^2$	$16.36_b \pm 0.08$ $16.33_b \pm 0.02$	16.00^n		$-14.7^a \pm 0.5$	26^a
	MHL/ML.H		2.6^n			
	ML(HL)/ML_2.H		4.1^n			
Zn^{2+}	ML/M.L		$5.13^n + 0.6$			
	$ML_2/M.L^2$		$9.69^n + 0.5$			
	$ML_3/M.L^3$		11.3^n			
	ML/MOHL.H		$8.35^n - 0.5$			
	$ML_2/MOHL_2.H$		9.73^n			

$^a 25°, 0.1;$ $^b 25°, 0.5;$ $^n 37°, 0.15$

Bibliography:

H^+ 69CP,71KS,73KP,75FST,76MT,77BP,78L,
        78VKS,78VV

Mn^{2+} 69CP

Cu^{2+} 69CP,69PP,73KP,75FST,77BP

Zn^{2+} 69CP,76MT

Other references: 70FM,70FMb,72La,73FA,
    73SC,73SKa,74FA,74FAa,74FAb,74KH,
    74SK,75PN,76KFA,77KDK,78KKd,78KZa

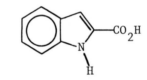

$C_9H_7O_2N$ 2-Indolecarboxylic acid   HL

Metal ion	Equilibrium	Log K 25°, 0.1	Log K 25°, 1.0	Log K 25°, 0
H^+	HL/H.L	3.68	3.68	3.88 ±0.02
Ca^{2+}	ML/M.L		0.57	
Co^{2+}	ML/M.L		0.44	
Ni^{2+}	ML/M.L		-0.10	
Cu^{2+}	ML/M.L		1.08	

Bibliography:

H^+ 62LS,72LPa

$Ca^{2+}-Cu^{2+}$ 72LPa

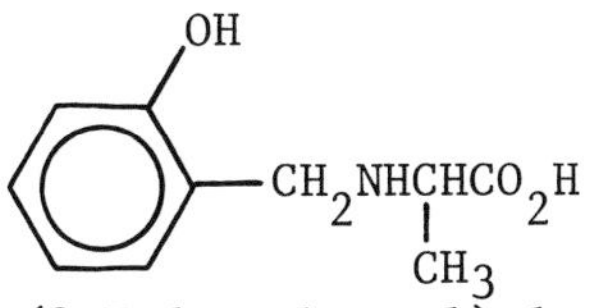

$$\text{L-N-(2-Hydroxybenzyl)alanine}$$

(structure: phenol ring with OH, attached $CH_2NHCHCO_2H$ with CH_3 branch)

$C_{10}H_{13}O_2N$ L-N-(2-Hydroxybenzyl)alanine H_2L

Metal ion	Equilibrium	Log K 25°, 0.1
H^+	HL/H.L	10.94
	$H_2L/HL.H$	8.60
	$H_3L/H_2L.H$	2.11
Co^{2+}	$ML/M.L$	8.15
	$ML_2/M.L^2$	13.54 (13.35)[x]
	$MHL_2/ML_2.H$	8.01 (8.14)[x]
Ni^{2+}	$ML/M.L$	9.00
	$ML_2/M.L^2$	15.02 (14.81)[x]
	MHL/ML.H	6.19
	$MHL_2/ML_2.H$	8.22 (8.29)[x]
Cu^{2+}	$ML/M.L$	14.12
	$ML_2/M.L^2$	18.45[y]
	MHL/ML.H	3.90
	$MHL_2/ML_2.H$	9.45 (9.36)[x]
Zn^{2+}	$ML/M.L$	8.75
	$ML_2/M.L^2$	13.24 (13.20)[x]
	ML/MOHL.H	9.50 9.10 (9.23)[x]
	$MHL_2/ML_2.H$	

[x]DL-mixture; [y]L- and DL-isomers had same value.

Bibliography: 75R

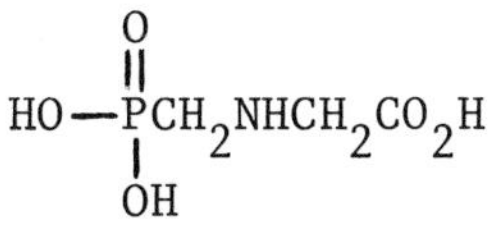

(structure: $HO-\overset{O}{\underset{OH}{P}}CH_2NHCH_2CO_2H$)

$C_3H_8O_5NP$ N-(Phosphonomethyl)glycine (glyphosate) H_3L

Metal ion	Equilibrium	Log K 25°, 0.1
H^+	HL/H.L	10.14
	$H_2L/HL.H$	5.46
	$H_3L/H_2L.H$	2.16
Mg^{2+}	ML/M.L	3.25
Ca^{2+}	ML/M.L	3.25
Mn^{2+}	ML/M.L	5.53
Cu^{2+}	ML/M.L	11.92
Zn^{2+}	ML/M.L	8.4

Bibliography: 78MCG

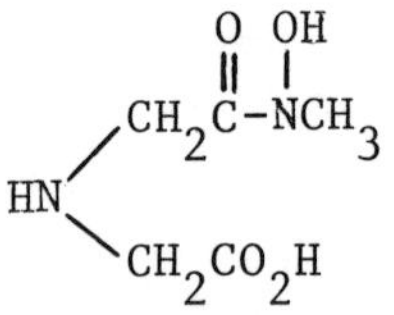

$$\begin{array}{c} \quad\quad\quad\quad \overset{O}{\overset{\|}{}} \;\; \overset{OH}{\overset{|}{}} \\ CH_2\text{-}C\text{-}NCH_3 \\ HN \\ \quad\quad CH_2CO_2H \end{array}$$

$C_5H_{10}O_4N_2$ (Carboxymethylamino)aceto-N-methylhydroxamic acid H_2L

Metal ion	Equilibrium	Log K 20°, 0.1
H^+	HL/H.L	9.22
	H_2L/HL.H	7.58
	H_3L/H_2L.H	1.90
Cu^{2+}	ML/M.L	13.8
	MHL/ML.H	3.88
	ML/MOHL.H	6.65

Bibliography: 78KPJ

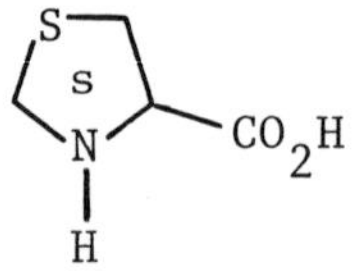

$C_4H_7O_2N$ DL-1,3-Thiazoline-4-carboxylic acid   (thiaproline)   HL

Metal ion	Equilibrium	Log K 25°, 0.15
H^+	HL/H.L	6.11
	H_2L/HL.H	1.51
Co^{2+}	ML/M.L	3.03
	$ML_2/M.L^2$	$(5.35)^w$
Ni^{2+}	ML/M.L	3.93
	$ML_2/M.L^2$	$(7.21)^w$
	$ML_3/M.L^3$	$(8.83)^w$
Cu^{2+}	ML/M.L	6.02
	$ML_2/M.L^2$	$(11.22)^w$
	MHL/M.HL	1.83
Zn^{2+}	ML/M.L	3.10
	$ML_2/M.L^2$	$(5.63)^w$

^{w}optical isomerism not stated.

Bibliography: 76FJ

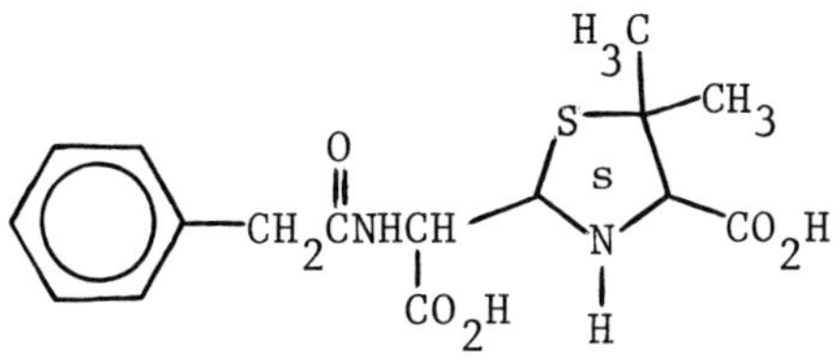

$C_{16}H_{20}O_5N_2S$ DL-Benzylpenicilloic acid H_2L

Metal ion	Equilibrium	Log K 25°, 0.15
H^+	HL/H.L	5.19
	$H_2L/HL.H$	2.30
	$H_3L/H_2L.H$	1.76
Ni^{2+}	ML/M.L	3.07
Cu^{2+}	ML/M.L	6.70

Bibliography: 76FJ

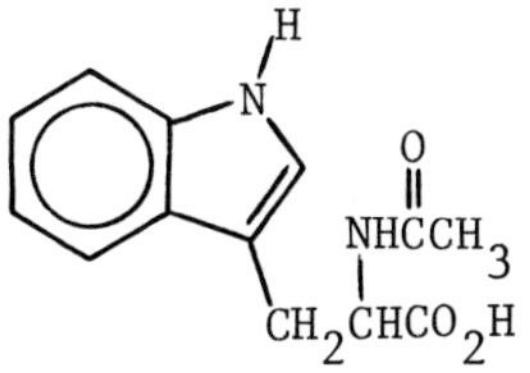

$C_{13}H_{13}O_3N_2$ L-N-Acetyltryptophan   HL

Metal ion	Equilibrium	Log K 20°, 0.37
H^+	HL/H.L	3.23
Cu^{2+}	$ML/M.L$	7.59
	$ML_2/M.L^2$	12.20
	$ML/M(H_{-1}L).H$	4.14
	$M(H_{-1}L)L/M(H_{-1}L).L$	2.93
	$M(H_{-1}L)L/M(H_{-1}L)_2.H$	7.42
	$ML_2/M(H_{-1}L)L.H$	4.79

Bibliography: 74W

$$H_2NCH_2CH_2NHCH_2CO_2H$$

$C_3H_{10}O_2N_2$ N-2-Aminoethylglycine (EDMA)   HL
                                  (Other values on Vol.1, p.76)

Metal ion	Equilibrium	Log K 25°, 0.1	Log K 25°, 1.0
H^+	HL/H.L	10.0 ±0.1	9.97
	$H_2L/HL.H$	6.67 ±0.02	6.74
	$H_3L/H_2L.H$	1.9 ±0.02	2.05
Cu^{2+}	$ML/M.L$	13.40	
	$ML_2/M.L^2$	21.44	
	MHL/ML.H	3.15	
	ML/MOHL.H	9.07	
Cd^{2+}	$ML/M.L$	8.48	6.86
	$ML_2/M.L^2$	12.23	12.32

Bibliography:

H^+ 76RS,77OH Cd^{2+} 77OH

Cu^{2+} 73YB

$$H_2NCH_2CH_2NHCH_2CH_2NHCH_2CO_2H$$

$C_6H_{15}O_2N_3$ Diethylenetriamine-N-acetic acid   HL

Metal ion	Equilibrium	Log K 25°, 0.1
H^+	HL/H.L	9.92
	$H_2L/HL.H$	8.70
	H_3L/H_2L H	4.32
	$H_4L/H_3L.H$	1.3
Co^{2+}	$ML/M.L$	12.04
	MHL/ML.H	5.90
	ML/MOHL.H	10.40
Ni^{2+}	$ML/M.L$	14.40
	MHL/ML.H	4.00
Cu^{2+}	$ML/M.L$	17.29
	MHL/ML.H	3.82
	ML/MOHL.H	9.32

Bibliography: 75MMH

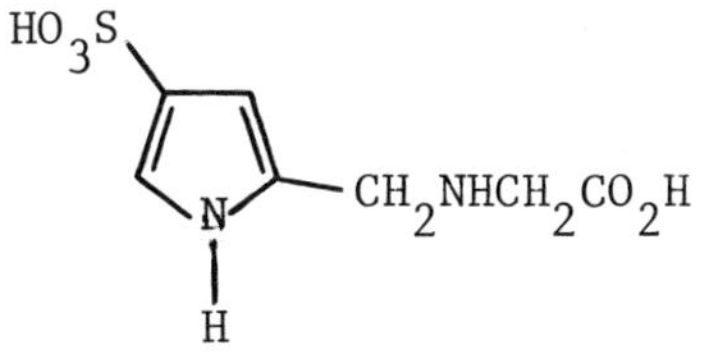

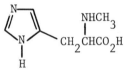

$C_7H_{10}O_5N_2S$ 2-(Carboxymethylaminomethyl)azole-4-sulfonic acid H_2L
(N-(4-sulfonyl-2-pyrrylmethyl)glycine)

Metal ion	Equilibrium	Log K 25°, 1.0
H^+	HL/H.L	8.71
	H_2L/HL.H	2.25
Cu^{2+}	$ML/M.L_2$	6.90
	$ML_2/M.L$	13.03
	$ML/ML_{-1}L.H$	6.75
	$MH_{-1}L/MOH(H_{-1}L).H$	9.71
	$ML_2/MOHL_2.H$	9.03
	$MOHL_2/M(OH)_2L_2.H$	12.2
	$M_2OH(H_{-1}L)_2/MH_{-1}L.MOH(H_{-1}L)$	2.2

Bibliography: 76SA

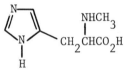

$C_7H_{11}O_2N_3$ L-2-(Methylamino)-3-(4-imidazolyl)propanoic acid   (N(α)-methylhistidine)   HL

Metal ion	Equilibrium	Log K 25°, 0.1
H^+	HL/H.L	9.32
	H_2L/HL.H	5.93
	$H_3L/H_2L.H$	1.42
Co^{2+}	$ML/M.L_2$	6.82
	$ML_2/M.L^2$	12.10 (12.42)[x]
Ni^{2+}	$ML/M.L_2$	8.54
	$ML_2/M.L^2$	15.05 (15.56)[x]
Cu^{2+}	$ML/M.L_2$	9.62
	$ML_2/M.L^2$	16.89[y]
	MHL/ML.H	3.83
	$MHL_2/ML_2.H$	5.37 (5.31)[x]
Zn^{2+}	$ML/M.L_2$	6.37
	$ML_2/M.L^2$	11.34 (11.54)[x]

[x]DL-mixture; [y]L- and DL-isomers had same value.

Bibliography: H^+-Ni^{2+},Zn^{2+} 76R; Cu^{2+} 75Ra

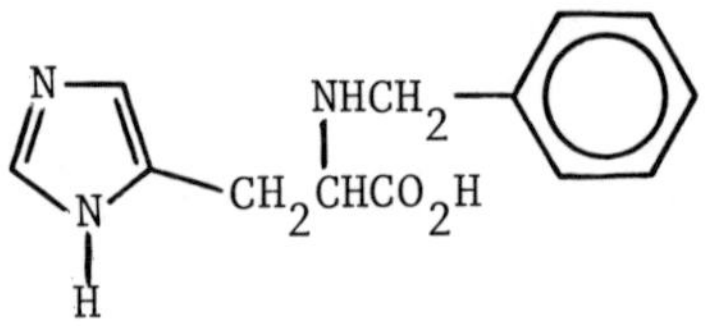

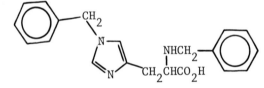

$C_{13}H_{15}O_2N_3$ L-2-(Benzylamino)-3-(4-imidazolyl)propanoic acid   HL
                                  (N(α)-benzylhistidine)

Metal ion	Equilibrium	Log K 25°, 0.1
H$^+$	HL/H.L	8.32
	H$_2$L/HL.H	5.94
	H$_3$L/H$_2$L.H	1.30
Co^{2+}	ML/M.L	6.24
	ML$_2$/M.L^2	11.89
		(11.99)^x
Ni^{2+}	ML/M.L	7.87
	ML$_2$/M.L^2	14.47
		(14.81)^x
Cu^{2+}	ML/M.L	8.86
	ML$_2$/M.L^2	15.87
		(15.79)^x
	MHL/ML.H	3.93
	MHL$_2$/ML$_2$.H	5.33
		(5.28)^x
Zn^{2+}	ML/M.L	5.79
	ML$_2$/M.L^2	11.06
		(11.2)^x

_____

^{x}DL-mixture

Bibliography: H$^+$-Ni^{2+},Zn^{2+} 76R;   Cu^{2+} 75Ra

_____

$C_{20}H_{21}O_2N_3$ L-2-(Benzylamino)-3-(3-benzyl-5-histidyl)propanoic acid   HL
                                  (N(α),N(3)-dibenzyl-L-histidine)

Metal ion	Equilibrium	Log K 25°, 0.1
H$^+$	HL/H.L	8.47
	H$_2$L/HL.H	5.50
	H$_3$L/H$_2$L.H	1.97
Ni^{2+}	ML/M.L	8.01
	ML$_2$/M.L^2	15.14
	MHL$_2$/ML$_2$.H	4.18
Cu^{2+}	ML/M.L	8.96
	ML$_2$/M.L^2	16.76
	MHL/ML.H	3.50
	M(HL)$_2$/ML(HL).H	4.32
	MHL$_2$/ML$_2$.H	4.26

N(α),N(1)-Dibenzyl-L-histidine (continued)

Metal ion	Equilibrium	Log K 25°, 0.1
Zn^{2+}	$ML/M.L$	5.74
	$ML_2/M.L^2$	11.49
	$MHL_2/ML_2.H$	5.43

Bibliography: 76BPa

$$HO_2CCH_2NHCH_2CH_2NHCH_2CO_2H$$

$C_6H_{12}O_4N_2$	Ethylenediiminodiacetic acid (EDDA) (Other values on Vol.1, p.86)				H_2L

Metal ion	Equilibrium	Log K 25°, 0.1	Log K 25°, 1.0	Log K 25°, 3.0	ΔH 25°, 0.1	ΔS 25°, 0.1
H^+	$HL/H.L$	9.60 ±0.03	9.64 ±0.05	10.06	-7.7 ±0.2 / -8.9[c]	18 / 14[c]
	$H_2L/HL.H$	6.53 ±0.06	6.71 ±0.02	7.18	-7.5 ±0.2 / -8.9[c]	5 / 8[c]
	$H_3L/H_2L.H$	2.36 -0.2	2.37	2.98	$(0)^s$	$(11)^c$
	$H_4L/H_3L.H$	1.3	1.7 ±0.1	1.5	$(0)^s$	$(8)^c$
Ce^{3+}	$ML/M.L$	7.48			-1.7^c	29^*
	$ML_2/M.L^2$	12.40			-3.1^c	46^*
Pr^{3+}	$ML/M.L$	7.84			-2.1^c	29^*
	$ML_2/M.L^2$	13.07			-3.8^c	47^*
Ho^{3+}	$ML/M.L$	8.42			-0.5^c	37^*
	$ML_2/M.L^2$	15.42			-3.3^c	59^*
Tm^{3+}	$ML/M.L$	8.75			-0.4^c	39^*
	$ML_2/M.L^2$	16.39			-3.8^c	62^*
Mn^{2+}	$ML/M.L$	7.0 ±0.1			-0.8 ±0.1	29
Co^{2+}	$ML/M.L$	11.23 ±0.03			-5.8	32
	$MHL/ML.H$	4.20				
	$ML/MOHL.H$	10.60				
VO_2^+	$ML/M.L$		14.5	16.0	$(-14)^s$	$(19)^c$
	$ML/HVO_4.L.H^3$		29.7			
Zn^{2+}	$ML/M.L$	1.11 ±0.1			-6.0 ±0.2	31
	$ML/MOHL.H$	10.56			-14.5	0
Cd^{2+}	$ML/M.L$	9.1 ±0.3			-4.7 ±0.7	26
Pb^{2+}	$ML/M.L$	10.7 -0.3			-6.7	27
	$ML/MOHL.H$	11.02			-9.2	20
Ga^{3+}	$ML/M.L$	19.12				
	$ML/MOHL.H$	4.70				
	$MOHL/M(OH)_2L.H$	7.6				

---

^{c}25°, 1.0; ^{s}15-35°, 1.0; *assuming ΔH for 1.0=ΔH for 0.1

Bibliography:

H^+ 74GG,75MMb,76HM,76YN,79GM,79ZL

Ce^{3+}-Ho^{3+} 74GG

Mn^{2+},Zn^{2+},Cd^{2+} 79GM

Co^{2+} 75MMb

VO_2^+ 76YN,79ZL

Pb^{2+} 74K,79GM

Ga^{3+} 76HM

Other references: 73NH,74SJ,76BG,79KKT

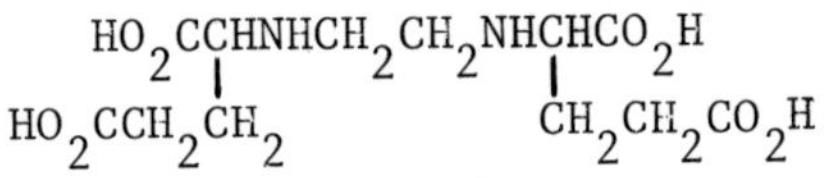

$$HO_2CCHNHCH_2CH_2NHCHCO_2H$$
$$HO_2CCH_2CH_2 \qquad CH_2CH_2CO_2H$$

$C_{12}H_{20}O_8N_2$ Ethylenediiminodi-2-pentanedioic acid   (EDDG) H_4L
                          (Other values on Vol.1, p.94)

Metal ion	Equilibrium	Log K 25°, 0.1
H^+	HL/H.L	9.46
	$H_2L/HL.H$	6.81
	$H_3L/H_2L.H$	4.25
	$H_4L/H_3L.H$	3.28
Mn^{2+}	ML/M.L	6.74
Co^{2+}	ML/M.L	10.64
Eu^{2+}	ML/M.L	2.50
	MHL/ML.H	8.41
	$MH_2L/MHL.H$	7.3
Tl^+	ML/M.L	2.4
	MHL/ML.H	8.9
Pb^{2+}	ML/M.L	8.45
	MHL/ML.H	5.93
	$MH_2L/MHL.H$	4.6
In^{3+}	ML/M.L	20.55
	MHL/ML.H	5.1

Bibliography:

Mn^{2+} 74SG Pb^{2+} 73GSK

Co^{2+} 73SG In^{3+} 73GKS

Eu^{2+} 74Ga   Other reference: 74ST

Tl^+ 73GK

$$CH_3$$
$$HO_2CCHNHCHCHNHCHCO_2H$$
$$HO_2C \qquad CH_3 \quad CO_2H$$

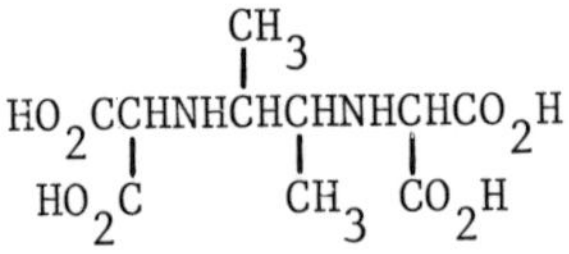

$C_{10}H_{16}O_8N_2$ meso-(1,2-Dimethylethylene)diiminodipropanedioic acid H_4L
                          (meso-2,3-butylenediamine-N,N'-dimalonic acid)

Metal ion	Equilibrium	Log K 25°, 0.1
H^+	HL/H.L	9.94
	$H_2L/HL.H$	6.06
	$H_3L/H_2L.H$	2.75
	$H_4L/H_3L.H$	2.34
Mg^{2+}	ML/M.L	5.09
	MHL/ML.H	5.42
	$M_2L/ML.M$	2.10
Ca^{2+}	ML/M.L	5.70
	MHL/ML.H	5.96
	$M_2L/ML.M$	2.05

meso-(1,2-Dimethylethylene)diiminodipropanedioic acid   (continued)

Metal ion	Equilibrium	Log K 25°, 0.1
Sr^{2+}	ML/M.L	4.50
	MHL/ML.H	7.00
	$M_2L/ML.M$	1.91
Ba^{2+}	ML/M.L	4.04
	MHL/ML.H	7.46
	$M_2L/ML.M$	1.48
Cu^{2+}	ML/M.L	(18.60)
	MHL/ML.H	3.04
	MOHL/ML.OH	7.64
Hg^{2+}	ML/M.L	18.60
	MHL/ML.H	4.61
	MOHL/ML.OH	6.77

Bibliography: H^+,Cu^{2+} 77GS; $Mg^{2+}-Ba^{2+}$ 78SG; Hg^{2+} 78SGa

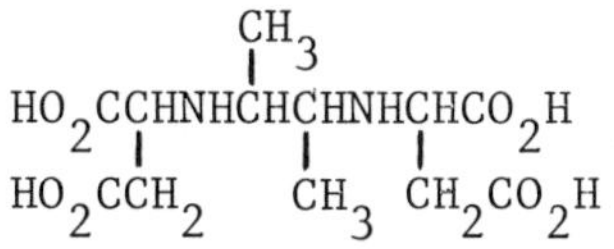

$$\begin{matrix} & & CH_3 & & \\ & & | & & \\ HO_2CCHNHCHCHNHCHCO_2H & & & \\ | & | & | & \\ HO_2CCH_2 & CH_3 & CH_2CO_2H & \end{matrix}$$

$C_{12}H_{20}O_8N_2$ meso-(1,2-Dimethylethylene)diiminodibutanedioic acid H_4L
                         (meso-2,3-butylenediamine-N,N'-disuccinic acid)

Metal ion	Equilibrium	Log K 25°, 0.1
H^+	HL/H.L	10.59
	$H_2L/HL.H$	6.55
	$H_3L/H_2L.H$	3.64
	$H_4L/H_3L.H$	2.80
Mg^{2+}	ML/M.L	5.75
	MHL/ML.H	6.68
	$M_2L/ML.M$	2.23
Ca^{2+}	ML/M.L	5.13
	MHL/ML.H	6.96
	$M_2L/ML.M$	2.34
Sr^{2+}	ML/M.L	3.75
	MHL/ML.H	8.45
	$M_2L/ML.M$	1.24
Ba^{2+}	ML/M.L	3.31
	MHL/ML.H	8.51
	$M_2L/ML.M$	1.04
Cu^{2+}	ML/M.L	16.76
	MHL/ML.H	3.65
	MOHL/ML.OH	8.72
Hg^{2+}	ML/M.L	17.10
	MHL/ML.H	4.59
	$MH_2L/MHL.H$	2.90
	MOHL/ML.OH	6.52

Bibliography: H^+ 77GS,79GSN; $Mg^{2+}-Ba^{2+}$ 78SG; Cu^{2+} 77GS; Hg^{2+} 78SGa

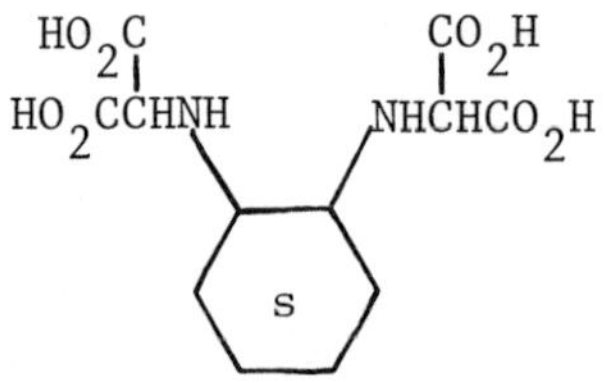

$$HO_2C\text{-}...\quad CO_2H$$

HO₂C ― ┃ ― CO₂H
HO₂CCHNH ― NHCHCO₂H
(ring labeled S)

$C_{12}H_{18}O_8N_2$ trans-1,2-Cyclohexylenediiminodipropanedioic acid H_4L
                (trans-1.2-diaminocyclohexane-N,N'-dimalonic acid)

Metal ion	Equilibrium	Log K 25°, 0.1
H^+	HL/H.L	10.49
	$H_2L/HL.H$	6.27
	$H_3L/H_2L.H$	2.83
	$H_4L/H_3L.H$	1.72
Hg^{2+}	ML/M.L	19.62
	MHL/ML.H	3.18
	$MH_2L/MHL.H$	3.7
	MOHL/ML.OH	5.76

Bibliography: 78SGb

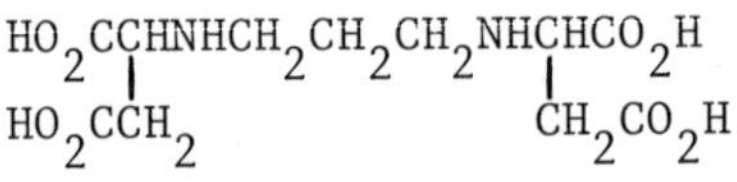

HO₂CCHNHCH₂CH₂CH₂NHCHCO₂H
┃   ┃
HO₂CCH₂   CH₂CO₂H

$C_{11}H_{18}O_8N_2$ DL-1,3-Diaminopropane-N,N'-dibutanedioic acid H_4L
                (1,3-propylenediamine-N,N'-disuccinic acid)

Metal ion	Equilibrium	Log K 25°, 0.1
H^+	HL/H.L	10.21
	$H_2L/HL.H$	8.79
	$H_3L/H_2L.H$	4.26
	$H_4L/H_3L.H$	3.20
Y^{3+}	ML/M.L	10.18
La^{3+}	ML/M.L	8.52
Ce^{3+}	ML/M.L	8.77
Pr^{3+}	ML/M.L	8.26
Nd^{3+}	ML/M.L	9.37
Sm^{3+}	ML/M.L	9.82
Eu^{3+}	ML/M.L	10.08
Gd^{3+}	ML/M.L	9.96
Tb^{3+}	ML/M.L	10.46
Dy^{3+}	ML/M.L	10.55
Ho^{3+}	ML/M.L	10.76
Er^{3+}	ML/M.L	10.91
Tm^{3+}	ML/M.L	11.12
Yb^{3+}	ML/M.L	11.19

1,3-Propylenediamine-N,N'-disuccinic acid (continued)

Metal ion	Equilibrium	Log K 25°, 0.1
Lu³⁺	ML/M.L	11.17
Cu²⁺	ML/M.L	12.14
	MHL/ML.H	6.97
	MH₂L/MHL.H	4.79
Hg²⁺	ML/M.L	17.09
	MHL/ML.H	5.9
	MH₂L/MHL.H	5.6
	MOHL/ML.OH	7.8
Pb²⁺	ML/M.L	9.02
In³⁺	ML/M.L	22.02
	MHL/ML.H	4.3
Tl³⁺	ML/M.L	26.32
	MHL/ML.H	4.34

Bibliography:

H⁺	76KG	Pb²⁺	76KGa
Y³⁺-Lu³⁺	76GKb	Tl³⁺	76DGa
Cu²⁺	76KG,79GKN	In³⁺	76GD
Hg²⁺	76GKa		

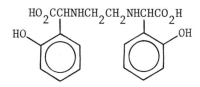

$C_{18}H_{20}O_6N_2$	Ethylenediiminobis[(2-hydroxyphenyl)acetic acid] (EHPG) H_4L

(Other values on Vol.1, p.96)

Metal ion	Equilibrium	Log K 25°, 0.1	Log K 20°, 0.1
H⁺	HL/H.L	11.68	11.85
	H₂L/HL.H	10.24	10.56
	H₃L/H₂L.H	8.64	8.78
	H₄L/H₃L.H	6.32	6.39
Cu²⁺	ML/M.L		23.94
	MHL/ML.H	8.04	8.06
	MH₂L/MHL.H	4.98	4.98
Pb²⁺	ML/M.L		15.09
	MHL/ML.H		9.67
	MH₂L/MHL.H		7.29
Ga³⁺	ML/M.L	33.6	

Bibliography:
Cu²⁺,Pb²⁺ 73NO
Ga³⁺ 76HM

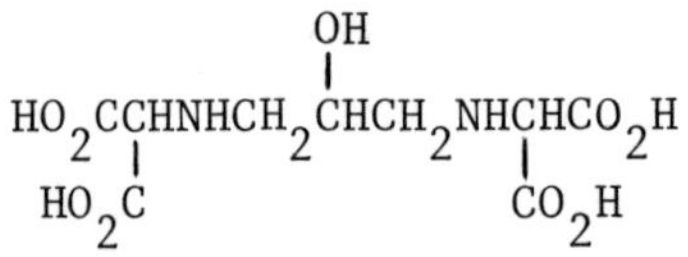

$$\begin{array}{c} \text{OH} \\ | \\ \text{HO}_2\text{CCHNHCH}_2\text{CHCH}_2\text{NHCHCO}_2\text{H} \\ | \qquad\qquad\qquad | \\ \text{HO}_2\text{C} \qquad\qquad\qquad \text{CO}_2\text{H} \end{array}$$

$C_{19}H_{14}O_9N_2$ <u>1,3-Diamino-2-hydroxypropane-N,N'-dipropanedioic acid</u> H_4L
<u>(2-hydroxy-1,3-propylenediamine-N,N'-dimalonic acid)</u>

Metal ion	Equilibrium	Log K 25°, 0.1
H^+	HL/H.L	10.40
	$H_2L/HL.H$	8.90
	$H_3L/H_2L.H$	3.40
	$H_4L/H_3L.H$	2.41
Mg^{2+}	ML/M.L	3.96
	MHL/ML.H	9.64
	$M_2L/ML.M$	1.95
Ca^{2+}	ML/M.L	3.08
	MHL/ML.H	9.37
Sr^{2+}	ML/M.L	3.00
	MHL/ML.H	9.35
Ba^{2+}	ML/M.L	2.00
	MHL/ML.H	10.05
Cu^{2+}	ML/M.L	12.67
	MHL/ML.H	8.8
	$MH_2L/MHL.H$	5.4
Tl^+	ML/M.L	3.29
	MHL/ML.H	9.13
	$M_2L/ML.M$	1.5
Hg^{2+}	ML/M.L	19.3
	MOHL/ML.OH	7.2
In^{3+}	ML/M.L	24.24
	MHL/ML.H	3.4
Tl^{3+}	ML/M.L	34.95
	MHL/ML.H	2.57

Bibliography:

H^+-Ba^{2+} 75KG Hg^{2+} 76GKa

Cu^{2+} 79GKN In^{3+} 76GD

Tl^+ 76DG Tl^{3+} 76DGa

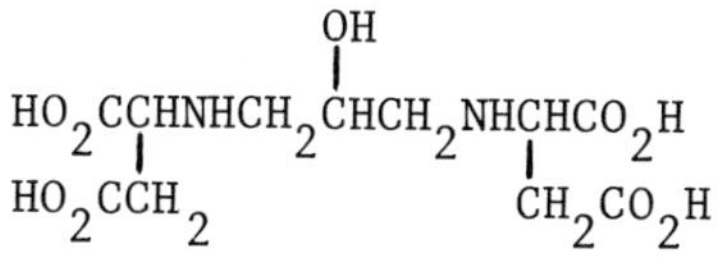

$$HO_2CCHNHCH_2\overset{\overset{\displaystyle OH}{|}}{C}HCH_2NHCHCO_2H$$
$$HO_2CCH_2 \qquad\qquad CH_2CO_2H$$

$C_{11}H_{18}O_9N_2$ **DL-1,3-Diamino-2-hydroxypropane-N,N'-dibutanedioic acid**
(2-hydroxy-1,3-propylenediamine-N,N'-disuccinic acid) H_4L

Metal ion	Equilibrium	Log K 25°, 0.1
H^+	HL/H.L	10.02
	$H_2L/HL.H$	8.78
	$H_3L/H_2L.H$	4.08
	$H_4L/H_3L.H$	3.16
Mg^{2+}	ML/M.L	3.67
	MHL/ML.H	8.79
Ca^{2+}	ML/M.L	2.94
	MHL/ML.H	9.19
Sr^{2+}	ML/M.L	2.80
	MHL/ML.H	9.10
Ba^{2+}	ML/M.L	1.95
	MHL/ML.H	9.21
Y^{3+}	ML/M.L	11.05
La^{3+}	ML/M.L	9.29
Ce^{3+}	ML/M.L	9.54
Pr^{3+}	ML/M.L	9.9
Nd^{3+}	ML/M.L	10.21
Sm^{3+}	ML/M.L	11.1
Eu^{3+}	ML/M.L	11.3
Gd^{3+}	ML/M.L	11.0
Tb^{3+}	ML/M.L	11.58
Dy^{3+}	ML/M.L	12.0
Ho^{3+}	ML/M.L	12.0
Er^{3+}	ML/M.L	12.1
Tm^{3+}	ML/M.L	12.34
Yb^{3+}	ML/M.L	12.39
Lu^{3+}	ML/M.L	12.15
Cu^{2+}	ML/M.L	12.84
	MHL/ML.H	7.3
	$MH_2L/MHL.H$	4.1
	ML/MOHL.H	9.47
Tl^+	ML/M.L	3.12
	MHL/ML.H	8.92
	$M_2L/ML.M$	1.6
Hg^{2+}	ML/M.L	17.73
	MHL/ML.H	5.2
	MOHL/ML.OH	7.4
In^{3+}	ML/M.L	23.75
	MHL/ML.H	3.3

DL-1,3-Diamino-2-hydroxypropane-N,N'-dibutanedioic acid (continued)

Metal ion	Equilibrium	Log K 25°, 0.1
Tl^{3+}	ML/M.L	29.90
	MHL/ML.H	2.16

Bibliography:

H^+-Ba^{2+}	74KG	Hg^{2+}	76GKa
Y^{3+}-Lu^{3+}	76GKb	In^{3+}	76GD
Cu^{2+}	76KG	Tl^{3+}	76DGa
Tl^+	76DG		

$$HO_2CCHNHCH_2CH_2OCH_2CH_2NHCHCO_2H$$
$$HO_2C \qquad\qquad\qquad CO_2H$$

$C_{10}H_{16}O_9N_2$ Oxybis(ethyleneiminomalonic acid) H_4L

Metal ion	Equilibrium	Log K 25°, 0.1
H^+	HL/H.L	9.52
	$H_2L/HL.H$	8.46
	$H_3L/H_2L.H$	3.20
	$H_4L/H_3L.H$	1.75
Mg^{2+}	ML/M.L	3.24
	MHL/ML.H	8.24
Ca^{2+}	ML/M.L	5.13
	MHL/ML.H	7.18
Sr^{2+}	ML/M.L	4.30
	MHL/ML.H	7.55
Ba^{2+}	ML/M.L	4.28
	MHL/ML.H	7.66
Co^{2+}	ML/M.L	10.18
Ni^{2+}	ML/M.L	11.99
Cu^{2+}	ML/M.L	14.84
	MHL/ML.H	5.34
	MOHL/ML.OH	6.68
Zn^{2+}	ML/M.L	10.53
Cd^{2+}	ML/M.L	11.17
Hg^{2+}	ML/M.L	19.90
	MHL/ML.H	4.86
	MOHL/ML.OH	5.84

Bibliography:

H^+-Ba^{2+}	79KB
Co^{2+}-Hg^{2+}	79KBa

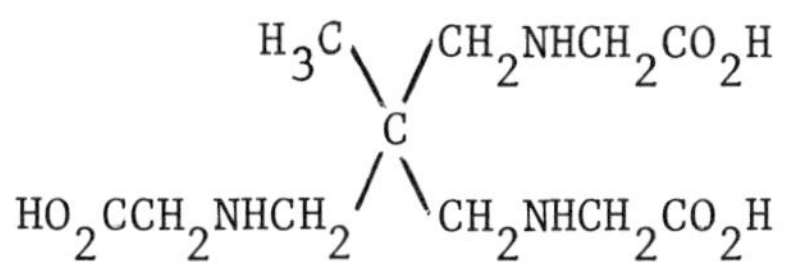

$$H_3C \diagdown \diagup CH_2NHCH_2CO_2H$$
$$C$$
$$HO_2CCH_2NHCH_2 \diagup \diagdown CH_2NHCH_2CO_2H$$

| $C_{11}H_{21}O_6N_3$ | 1,1,1-Tris(aminomethyl)ethane-N,N',N"-triacetic acid | H_3L |

Metal ion	Equilibrium	Log K 25°, 0.1
H^+	HL/H.L	10.73
	$H_2L/HL.H$	7.73
	$H_3L/H_2L.H$	4.69
Co^{2+}	ML/M.L	12.56
	MHL/ML.H	5.81
Ni^{2+}	ML/M.L	15.67
	MHL/ML.H	4.68
Cu^{2+}	ML/M.L	16.32
	MHL/ML.H	7.15
Zn^{2+}	ML/M.L	12.78
	MHL/ML.H	5.85

Bibliography: 77HZ

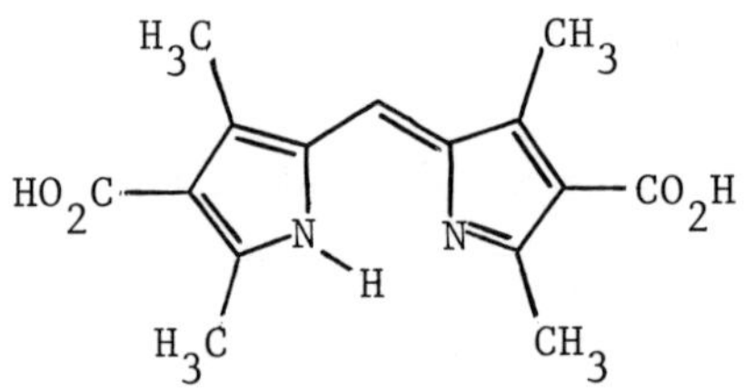

$C_{15}H_{14}O_4N_2$ 3,3',5,5'-Tetramethyldipyrromethene-4,4'-dicarboxylic acid H_2L

Metal ion	Equilibrium	Log K 25°, 1.0
H^+	HL/H.L	(16.1)
	$H_2L/HL.H$	8.03
	$H_3L/H_2L.H$	4.2
	$H_4L/H_3L.H$	3.6
Ni^{2+}	$ML_2/M.L^2$	24.4
Cu^{2+}	$ML_2/M.L^2$	31.0

Bibliography: 79SA

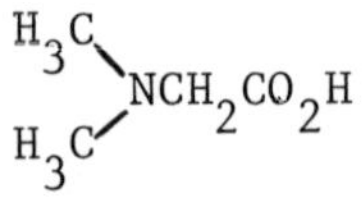

$C_4H_9O_2N$ N,N-Dimethylglycine   HL
(Other values on Vol.1, p.100)

Metal ion	Equilibrium	Log K 25°, 0.1	Log K 25°, 0.5	Log K 25°, 0	ΔH 25°, 0	ΔS 25°, 0
H^+	HL/H.L	9.80 +0.01	9.73	9.940	-7.7	20
	$H_2L/HL.H$	1.99 ±0.09	1.99			
Cu^{2+}	ML/M.L	7.25 ±0.05				
	$ML_2/M.L^2$	13.71 ±0.06				

Bibliography:
H^+ 75FST,78L
Cu^{2+} 75FST

Other reference: 77KDK

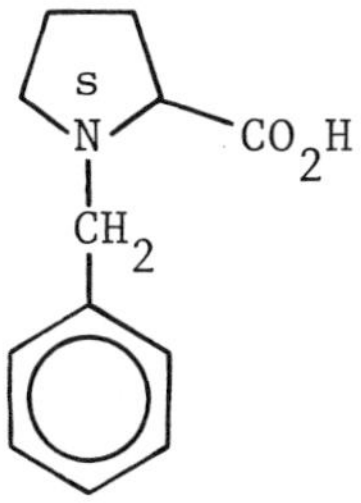

$C_{12}H_{13}O_3N$

L-N-Benzylproline
(Other reference on Vol.1, p.101)

HL

Metal ion	Equilibrium	Log K 25°, 0.1	ΔH 25°, 0.1	ΔS 25°, 0.1
H^+	HL/H.L	9.90	$(-10)^r$	(12)
	H_2L/HL.H	2.00	$(-1)^r$	(6)
Cu^{2+}	$ML/M.L_2$	7.07	$(-9)^r$	(2)
	$ML_2/M.L^2$	13.36	$(-12)^{r,x}$	(21)
		$(14.36)^x$	$(-11)^{r,x}$	$(29)^x$

^{r}15-35°, 0.1; ^{x}DL-mixture

Bibliography: 78KZ   Other references: 77KDK,78KZ,78KZa

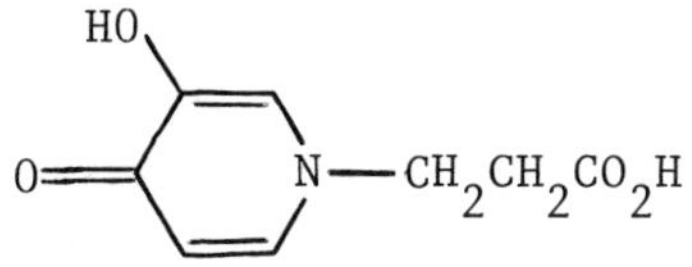

$C_8H_9O_4N$ 3-(3-Hydroxy-4-oxo-1,4-dihydro-1-pyridinyl)propanoic acid   (mimosinic acid) H_2L

Metal ion	Equilibrium	Log K 37°, 0.15
H^+	HL/H.L	8.79
	H_2L/HL.H	3.84
	H_3L/H_2L.H	2.80
Cu^{2+}	$ML/M.L_2$	9.48
	$ML_2/M.L^2$	16.96
	MHL/ML.H	3.76
	MHL_2/ML_2.H	3.43
	MH_2L_2/MHL_2.H	4.27
Zn^{2+}	$ML/M.L_2$	6.86
	$ML_2/M.L^2$	12.31
	MHL/ML.H	3.9
Cd^{2+}	$ML/M.L_2$	5.89
	$ML_2/M.L_3$	10.19
	$ML_3/M.L^3$	12.0
Pb^{2+}	$ML/M.L_2$	8.57
	$ML_2/M.L^2$	13.50
	MHL/ML.H	3.77

Bibliography: 79SPT

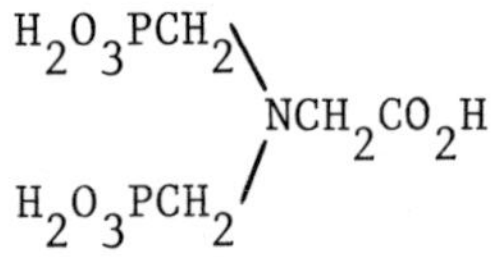

$$H_2O_3PCH_2$$
$$NCH_2CO_2H$$
$$H_2O_3PCH_2$$

$C_4H_{11}O_8NP_2$	N,N-Bis(phosphonomethyl)glycine (Other values on Vol.1, p.103)				H_5L

Metal ion	Equilibrium	Log K 25°, 0.1	Log K 25°, 0	ΔH 25°, 0	ΔS 25°, 0
H^+	HL/H.L	10.81 ±0.01	11.9	−6.6	32
	$H_2L/HL.H$	6.39 ±0.02			
	$H_3L/H_2L.H$	5.03 ±0.03			
	$H_4L/H_3L.H$	1.98 ±0.03			
	$H_5L/H_4L.H$	1.73			
Fe^{3+}	ML/M.L	14.65 ±0.0			
	MHL/ML.H	4.80 ±0.0			
	ML/MOHL.H	7.20 −0.2			

Bibliography:

H^+ 74NKD,79VKR   Other references: 73KZ,77NF,78KPS,79EFa

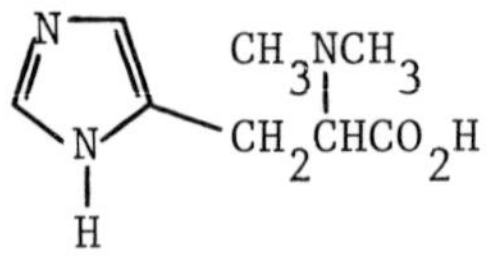

Fe^{3+} 73KSD

$$CH_3NCH_3$$
$$CH_2CHCO_2H$$

$C_8H_{13}O_2N_3$	L-2-(Dimethylamino)-3-(4-imidazolyl)propanoic acid (α-N,N-dimethylhistidine)		HL

Metal ion	Equilibrium	Log K 25°, 0.1
H^+	HL/H.L	8.88
	$H_2L/HL.H$	6.01
	$H_3L/H_2L.H$	1.14
Co^{2+}	$ML/M.L_2$	6.88
	$ML_2/M.L^2$	10.30 (10.83)[x]
Ni^{2+}	$ML/M.L_2$	8.48
	$ML_2/M.L^2$	12.14 (13.04)[x]
Cu^{2+}	$ML/M.L_2$	9.05
	$ML_2/M.L^2$	13.75 (14.01)[x]
	MHL/ML.H	3.78
	$MHL_2/ML_2.H$	7.12 (6.89)[x]
Zn^{2+}	$ML/M.L_2$	6.34
	$ML_2/M.L^2$	9.04 (9.68)[x]
	ML/MOHL.H	9.58 (9.54)[x]

[x]DL-Mixture

Bibliography: H^+-Ni^{2+},Zn^{2+} 76R; Cu^{2+} 75Ra

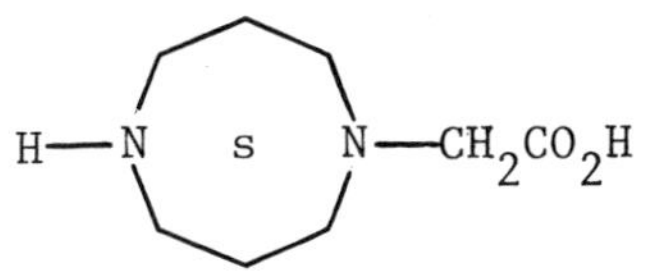

H—N s N—CH$_2$CO$_2$H

C$_8$H$_{16}$O$_2$N$_2$ 1,5-Diazacyclooctane-N-acetic acid   HL

Metal ion	Equilibrium	Log K 25°, 0.13
H$^+$	HL/H.L	11.8
	H$_2$L/HL.H	5.18
	H$_3$L/H$_2$L.H	1.7
Cu^{2+}	ML/M.L	15.3

Bibliography: 76WL

HO$_2$CCH$_2$—N s N—CH$_2$CO$_2$H

C$_{10}$H$_{18}$O$_4$N$_2$ 1,5-Diazacyclooctane-N,N'-diacetic acid   H$_2$L

Metal ion	Equilibrium	Log K 25°, 0.1
H$^+$	HL/H.L	11.5
	H$_2$L/HL.H	4.80
	H$_3$L/H$_2$L.H	2.0
Co^{2+}	ML/M.L	8.6
	ML/MOHL.H	9.27
Ni^{2+}	ML/M.L	10.3
	ML/MOHL.H	10.1
Cu^{2+}	ML/M.L	18.6
Zn^{2+}	ML/M.L	11.3
	ML/MOHL.H	8.59

Bibliography: 75Bb

Other reference: 75CK

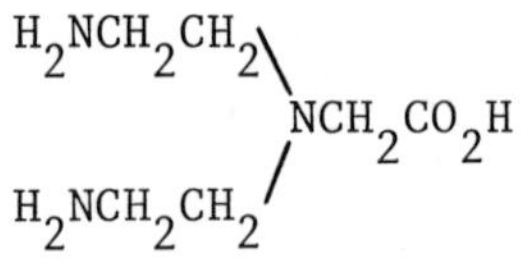

$$\begin{array}{c} H_2NCH_2CH_2 \\ \diagdown \\ \diagup \\ H_2NCH_2CH_2 \end{array} NCH_2CO_2H$$

$C_6H_{15}O_2N_3$	N,N-Bis(2-aminoethyl)glycine	HL
	(Other values on Vol.1, p.108)	

Metal ion	Equilibrium	Log K 25°, 0.1
H^+	HL/H.L	10.81
	$H_2L/HL.H$	9.59
	$H_3L/H_2L.H$	3.24
Ga^{3+}	$MH_2L/M.H_2L$	4.51
	ML/MOHL.H	2.6

Bibliography: 76HM   Other reference: 75MMH

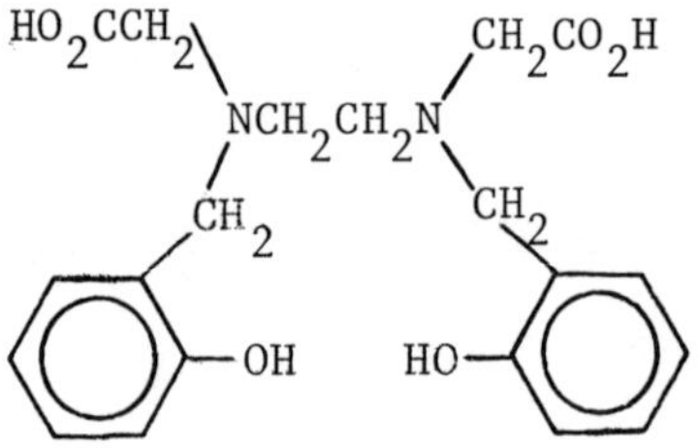

$$\begin{array}{c} HO_2CCH_2 \qquad\qquad CH_2CO_2H \\ \diagdown \qquad\qquad\qquad \diagup \\ NCH_2CH_2N \\ \diagup \qquad\qquad\qquad\quad \diagdown \\ CH_2 \qquad\qquad\qquad\qquad CH_2 \end{array}$$

$C_{20}H_{24}O_6N_2$	N,N'-Bis(2-hydroxybenzyl)ethylenedinitrilo-N,N'-diacetic acid	(HBED)	H_4L
	(Other values on Vol.1, p.112)		

Metal ion	Equilibrium	Log K 25°, 0.1
H^+	HL/H.L	(12.46)
	$H_2L/HL.H$	11.00
	$H_3L/H_2L.H$	8.32
	$H_4L/H_3L.H$	4.64
Ga^{3+}	ML/M.L	39.57

Bibliography: 76HM

$$\text{HN} \diagup^{\text{CH}_2\text{CO}_2\text{H}}_{\diagdown \text{CH}_2\text{CO}_2\text{H}}$$

| $C_4H_7O_4N$ | Iminodiacetic acid (IDA) | | | | H_2L |
| | (Other values on Vol.1, p.116) | | | | |

Metal ion	Equilibrium	Log K 25°, 0.1	Log K 25°, 1.0	Log K 25°, 0	ΔH 25°, 0.1	ΔS 25°, 0.1
H^+	HL/H.L	9.32 ±0.04	9.30 ±0.09	9.79 ±0.00	-7.9 ±0.2	16
		9.16[b]±0.02		9.68[e]	-8.5[c]	
	H_2L/HL.H	2.61 ±0.03	2.57 ±0.06	2.84 +0.1	-0.76	9
		2.62[b]±0.06		2.77[e]	-0.81[c]	
	H_3L/H_2L.H	1.82[b]	1.87 ±0.03		-0.86	5
		1.76[b]		1.92[e]	-1.01[c]	
Ca^{2+}	ML/M.L	2.59 +0.1	2.09	3.40[o]	0.3[h]	13
Th^{4+}	ML/M.L	9.32[b]				
UO_2^{2+}	ML/M.L	8.96	8.71			
			8.66[j]			
NpO_2^{2+}	ML/M.L		8.72[j]			
PuO_2^{2+}	ML/M.L		8.50[j]			
Cu^{2+}	$ML/M.L$	10.5 ±0.1			-4.2 ±0.3	34
	$ML_2/M.L^2$	16.3 ±0.3			-10.6 ±0.3	39
	MHL/ML.H	2.3 ±0.4			-0.5	9
	ML/MOHL.H	9.4				
VO^{2+}	ML/M.L	9.30[b]				
	MHL/ML.H	1.95[b]				
	ML/MOHL.H	5.50[b]				
	$M_2L_2/M_2(OH)_2L_2.H^2$	8.21[b]				
VO_2^+	$ML/M.L$			11.7[e]		
	$ML_2/M.L^2$			22.2[e]		
Pd^{2+}	$ML/M.L$		17.5[j]			
	$ML_2/M.L^2$		26.8[j]			
	MHL/ML.H		0.75[j]			
Ga^{3+}	ML/M.L	12.76				
	ML/MOHL.H	3.5				

[b] 25°, 0.5;　[c] 25°, 1.0;　[e] 25°, 3.0;　[h] 20°, 0.1;　[j] 20°, 1.0;　[o] 20°, 0

Bibliography:

H^+　　73CBP,73SK,76AM,76GMa,79BC,79ZL

Ca^{2+}　　68KS

Th^{4+}　　73SK

UO_2^{2+}-PuO_2^{2+}　　73CBP

Fe^{2+}　　72Nb

Ni^{2+}　　70CM

Cu^{2+}　　73YB,75NW,79BC

VO^{2+}　　66KF,73NP

VO_2^+　　79ZL

Pd^{2+}　　76AM

Ga^{3+}　　76HM

Other references:　66KT,67TKR,69AS,70KMa,
73DR,73H,73T,73YB,74KMS,75KPS,75CGa,
75LB,75NF,76BBC,76BG,76KI,76TB,76ZK,
77JKS,77NF,77SFb,78RS,79FS,79KCa,
79SSc,79TKK

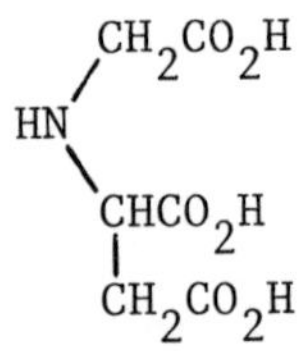

$$\begin{array}{c} CH_2CO_2H \\ HN \diagdown \\ \quad CHCO_2H \\ \quad CH_2CO_2H \end{array}$$

$C_6H_9O_6N$ 2-Carboxymethyliminodiacetic acid (N-carboxymethylaspartic acid) H_3L

Metal ion	Equilibrium	Log K 25°, 0.1
H^+	HL/H.L	9.65
	$H_2L/HL.H$	3.85
	$H_3L/H_2L.H$	2.58
Mg^{2+}	ML/M.L	4.57
Ca^{2+}	ML/M.L	3.71
Sr^{2+}	ML/M.L	3.32
Ba^{2+}	ML/M.L	3.21
Y^{3+}	ML/M.L	9.38
La^{3+}	ML/M.L	8.36
Ce^{3+}	ML/M.L	8.51
Pr^{3+}	ML/M.L	8.60
Nd^{3+}	ML/M.L	8.90
Sm^{3+}	ML/M.L	9.24
Eu^{3+}	ML/M.L	9.30
Gd^{3+}	ML/M.L	9.32
Tb^{3+}	ML/M.L	9.38
Dy^{3+}	ML/M.L	9.40
Ho^{3+}	ML/M.L	9.54
Er^{3+}	ML/M.L	9.60
Tm^{3+}	ML/M.L	9.71
Yb^{3+}	ML/M.L	9.85
Lu^{3+}	ML/M.L	9.90
Cu^{2+}	ML/M.L	12.80
	MHL/ML.H	7.65
Hg^{2+}	ML/M.L	13.12

Bibliography: H^+-Ba^{2+} 75GN; Y^{3+}-Lu^{3+},Hg^{2+} 76NGb; Cu^{2+} 78GN

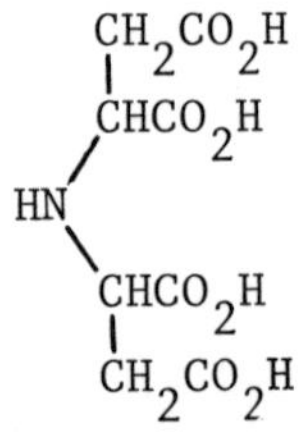

$$\begin{array}{c} CH_2CO_2H \\ CHCO_2H \\ HN \diagdown \\ \quad CHCO_2H \\ \quad CH_2CO_2H \end{array}$$

$C_8H_{11}O_8N$ 2,2'-Bis(carboxymethyl)iminodiacetic acid (iminodisuccinic acid) H_4L

Metal ion	Equilibrium	Log K 25°, 0.1
H^+	HL/H.L	10.12
	$H_2L/HL.H$	4.83
	$H_3L/H_2L.H$	3.84
	$H_4L/H_3L.H$	2.96

Iminodisuccinic acid (continued)

Metal ion	Equilibrium	Log K 25°, 0.1
Mg^{2+}	ML/M.L	5.50
Ca^{2+}	ML/M.L	4.42
Sr^{2+}	ML/M.L	3.37
Ba^{2+}	ML/M.L	2.18

Bibliography: 78MNG   Other references: 79BEM,79MM

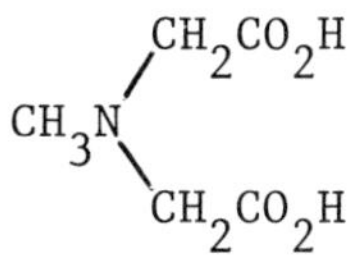

$$CH_3N \begin{array}{c} CH_2CO_2H \\ \\ CH_2CO_2H \end{array}$$

$C_5H_9O_4N$		**N-Methyliminodiacetic acid (MIDA)**				H_2L
		(Other values on Vol.1, p.124)				

Metal ion	Equilibrium	Log K 25°, 0.1	Log K 25°, 1.0	Log K 20°, 0	ΔH 25°, 0	ΔS 25°, 0.1
H^+	HL/H.L	9.60 ±0.04	9.48	10.088	-6.9^h+0.1	21
		9.42^b	9.50^u		$(-8)^r$	
	$H_2L/HL.H$	2.2 ±0.1	2.36	2.146	0.2	11
		2.36^b	2.52^u		$(0)^r$	
	$H_3L/H_2L.H$		1.57		$(0)^r$	$(7)^c$
Ni^{2+}	ML/M.L	8.67 −0.3			-4.7^h	24
	$ML_2/M.L^2$	16.0 ±0.1			-7.7^h	47
VO^{2+}	ML/M.L	9.44^b				
	ML/MOHL.H	5.76^b				
	$(MOHL)_2.H^2/(ML)^2$	9.05^b				
VO_2^+	ML/M.L		10.2		$(-7)^r$	$(23)^c$
	ML/MOHL.H		6.13			
	$ML/HVO_4.L.H^3$		25.9			
Zn^{2+}	ML/M.L	7.69 ±0.06	7.44^u		-2.2^h	28
	$ML_2/M.L^2$	14.00 ±0.01	13.61^u		-5.8^h	45
Hg^{2+}	ML/M.L	$(5.47)^{h,v}$				
	$ML_2/M.L^2$	$(9.15)^{h,v}$				
	ML/MOHL.H	9.18^h				

^{b}25°, 0.5; ^{c}25°, 1.0; ^{h}20°, 0.1; ^{r}15-35°, 1.0; ^{u}35°, 2.0; ^{v}not corrected for Cl^-

Bibliography:

H^+	76YN,77MG,77N		VO_2^+	76YN
Ni^{2+}	70CM		Zn^{2+}	77MG
Cu^{2+}	69LA			
VO^{2+}	77N		Other references:	72KNT,73H,79MMK

$$\text{HO}_2\text{CCH}_2\text{N} \overset{\displaystyle \text{CH}_2\text{CO}_2\text{H}}{\underset{\displaystyle \text{CH}_2\text{CO}_2\text{H}}{\Big\langle}}$$

$C_6H_9O_6N$ **Nitrilotriacetic acid (NTA)** H_3L
(Other values on Vol.1, p.139)

Metal ion	Equilibrium	Log K 25°, 0.1	Log K 25°, 1.0	Log K 20°, 0	ΔH 20°, 0.1	ΔS 25°, 0.1
H^+	HL/H.L	9.65 ± 0.02 $9.88^z \pm 0.07$	$8.91_b \pm 0.01$ $8.96^b \pm 0.02$	10.334 9.17^e	-4.6 ± 0.1	29
	H_2L/HL.H	2.48 ± 0.02 $2.5^z \pm 0.1$	$2.3_b \pm 0.1$ $2.30^b \pm 0.03$	2.940 2.63^e	$0.2^o + 0.4$	12^*
	H_3L/H_2L.H	1.9 ± 0.1 $1.9^z \pm 0.2$	$1.9_b \pm 0.1$ $1.7^b \pm 0.1$	1.650 2.0^e	$0.2^o + 0.4$	9^*
	H_4L/H_3L.H	(0.8)	1.2 ± 0.2 1.1^j	1.3^e		
K^+	ML/M.L	0.4^z				
Be^{2+}	ML/M.L	7.86			$(6)^r$	(56)
Cu^{2+}	ML/M.L	13.1 ± 0.1			$-1.9 +0.1$	54
	$ML_2/M.L^2$	17.5 ± 0.1			$-8.9 +0.6$	50
	MHL/ML.H	1.6 ± 0.3				
	ML/MOHL.H	9.2 ± 0.2				
VO^{2+}	ML/M.L		12.30^b			
	ML/MOHL.H	7.38	7.15^b			
VO_2^+	ML/M.L		13.8	13.78^e	$(0)^t$	$(63)^c$
	$ML/HVO_4.L.H^3$		28.3			
Zr^{4+}	ML/M.L	20.8	$19.8^j \pm 0.3$	18.6^k		
Ag^+	ML/M.L	5.36				
Pd^{2+}	ML/M.L		17.0^j			
	$ML_2/M.L^2$		23.7^j			
	MHL/ML.H		7.82^j			
	MH_2L/MHL.H		0.5^j			
	$M_2L_2/(ML)^2$		2^j			
	M_2OHL_2/ML.MOHL		3.1^j			
Hg^{2+}	ML/M.L	$14.3 +0.3$	13.5^b			
Ga^{3+}	ML/M.L	$13.8^h \pm 0.2$				
	ML/MOHL.H	4.27				
	$MOHL/M(OH)_2L$.H	7.64				
Bi^{3+}	ML/M.L	18.2	17.5^j			
	$ML_2/M.L^2$		26.0^j			

$^b25°, 0.5;$ $^c25°, 1.0;$ $^e25°, 3.0;$ $^h20°, 0.1;$ $^j20°, 1.0;$ $^k20°, 2.0;$ $^o20°, 0;$
$^r20-40°, 0.1;$ $^s17-70°, 0.1;$ $^t15-35°, 1.0;$ $^z(CH_4)_4N^+$ salt used as background electrolyte.
*assuming ΔH for 0.0=ΔH for 0.1

NTA (continued)

Bibliography:

H^+ 73Ha,73MMC,75LL,76CL,76GMa,76YN,77N

K^+ 73Ha

Be^{2+} 77DA

Y^{3+}-Lu^{3+} 77CGG,77GGC

Cu^{2+} 73Ha,73MMC,74HS,77GNa,79Sb

VO^{2+} 77N

VO_2^+ 75LL,76YN

Zr^{4+},Pd^{2+} 76AM

Ag^+ 79MST

Hg^{2+} 77GGC,77GNa

Ga^{3+} 76HM

Bi^{3+} 76EN

Other references: 66EA,66KFa,66KR,66LP,67H,
    67MP,68MT,69AS,69CA,69Ma,69RK,70KB,
    71IK,71Mb,71S,73DR,73RB,73YPa,74PI,
    75LB,75NF,75RM75TP,76VPa,76ZK,77KMK,
    77KS,77NF,77SFb,78KV,78KVb,78KVc,
    78MGD,79BK,79EF,79FS,79ZL

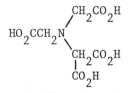

$C_7H_{11}O_6N$		N-(2-Carboxyethyl)iminodiacetic acid				H_3L
		(Other values on Vol.1, p.146)				

Metal ion	Equilibrium	Log K 25°, 0.1	Log K 20°, 0.1	ΔH 25°, 0.1	ΔS 25°, 0.1
H^+	HL/H.L	9.60 +0.01	9.66	−5.2	26
	H_2L/HL.H	3.71	3.69		
	H_3L/H_2L.H	2.1 ±0.2	2.1		
Mn^{2+}	ML/M.L	7.33		1.1	37
	MHL/ML.H	3.79			
Co^{2+}	ML/M.L	10.1 +0.1		−1.7	40
Cu^{2+}	ML/M.L	12.6 +0.6		−4.1	44
Zn^{2+}	ML/M.L	10.0 +0.1	10.1	−1.1	42
Cd^{2+}	ML/M.L	8.3 −0.6			
	MHL/ML.H	3.45			

Bibliography: Mn^{2+}-Zn^{2+} 70MU; Cd^{2+} 73KU

$$HO_2CCH_2N \begin{matrix} CH_2CO_2H \\ CH_2CO_2H \\ CO_2H \end{matrix}$$

$C_7H_9O_8N$		2-Carboxynitrilotriacetic acid	H_4L
		(N,N-bis(carboxymethyl)aminomalonic acid)	

Metal ion	Equilibrium	Log K 25°, 0.1
H^+	HL/H.L	8.70
	H_2L/HL.H	3.94
	H_3L/H_2L.H	3.74
	H_4L/H_3L.H	2.93

N,N-Bis(carboxymethyl)aminomalonic acid   (continued)

Metal ion	Equilibrium	Log K 25°, 0.1
Mg^{2+}	ML/M.L	5.15
Ca^{2+}	ML/M.L	7.10
Sr^{2+}	ML/M.L	6.02
Ba^{2+}	ML/M.L	5.50

Bibliography: 76NGa

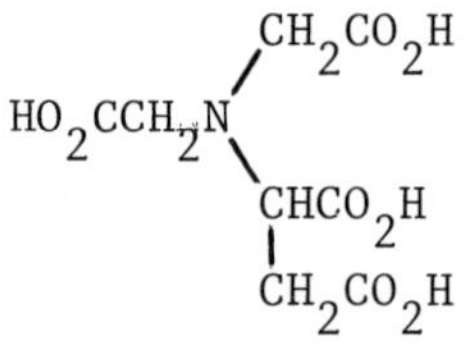

$$HO_2CCH_2N \begin{array}{c} CH_2CO_2H \\ CHCO_2H \\ CH_2CO_2H \end{array}$$

$C_8H_{11}O_8N$	DL-2-(Carboxymethyl)nitrilotriacetic acid (N,N-bis(carboxymethyl)aspartic acid)	H_4L

Metal ion	Equilibrium	Log K 25°, 0.1
H^+	HL/H.L	9.18
	$H_2L/HL.H$	4.26
	$H_3L/H_2L.H$	2.79
	$H_4L/H_3L.H$	2.46
Mg^{2+}	ML/M.L	5.92
Ca^{2+}	ML/M.L	5.81
Sr^{2+}	ML/M.L	4.50
Ba^{2+}	ML/M.L	4.02
Cu^{2+}	ML/M.L	12.69
	MHL/ML.H	**7.39**

Bibliography: H^+-Ba^{2+} 75NG; Cu^{2+} 78GN

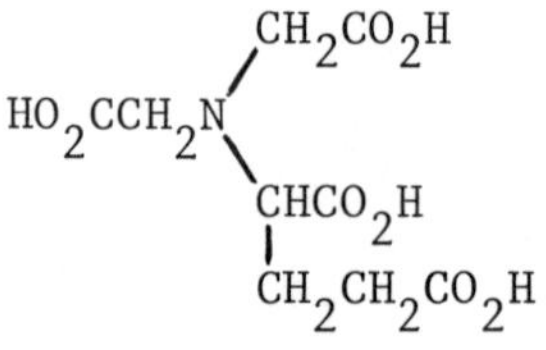

$$HO_2CCH_2N \begin{array}{c} CH_2CO_2H \\ CHCO_2H \\ CH_2CH_2CO_2H \end{array}$$

$C_9H_{13}O_8N$	DL-2-(Carboxyethyl)nitrilotriacetic acid (N,N-bis(carboxymethyl)glutamic acid)	H_4L

Metal ion	Equilibrium	Log K 25°, 0.1
H^+	HL/H.L	9.36
	$H_2L/HL.H$	5.03
	$H_3L/H_2L.H$	3.49
	$H_4L/H_3L.H$	2.56

N,N-Bis(carboxymethyl)glutamic acid   (continued)

Metal ion	Equilibrium	Log K 25°, 0.1
Mg^{2+}	ML/M.L	5.18
Ca^{2+}	ML/M.L	5.93
Sr^{2+}	ML/M.L	4.06
Ba^{2+}	ML/M.L	3.54
Cu^{2+}	ML/M.L	13.09
	MHL/ML.H	7.91
Hg^{2+}	ML/M.L	14.33

Bibliography: H^+-Ba^{2+} 76NGa; Cu^{2+},Hg^{2+} 77GNa

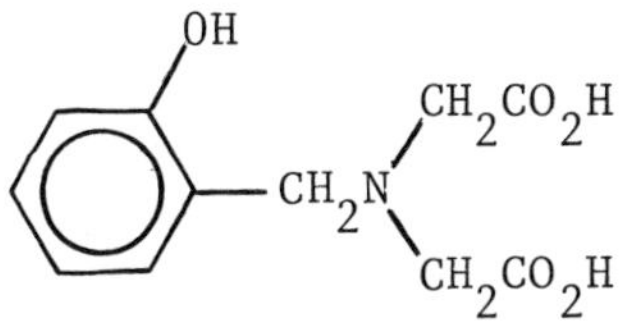

OH / CH_2CO_2H / CH_2N / CH_2CO_2H

$C_{11}H_{13}O_5N$ N-(2-Hydroxybenzyl)iminodiacetic acid H_3L
(Other values on Vol.1, p.161)

Metal ion	Equilibrium	Log K 25°, 0.1	Log K 20°, 0.1	Log K 25°, 1.0
H^+	HL/H.L	11.71	11.79	
	$H_2L/HL.H$	8.07	8.17	8.05
	$H_3L/H_2L.H$	2.34	2.2	2.36
Co^{2+}	ML/M.L	12.87		
	$ML_2/M.L^2$	24.41		
	MHL/M.HL	6.97		
Ni^{2+}	ML/M.L	13.83		
	MHL/M.HL	8.24		
Cu^{2+}	ML/M.L	16.11		
	MHL/M.HL	10.91		
Fe^{3+}	ML/M.L	22.4		
	ML/MOHL.H			5.71
	$MOHL/M(OH)_2L.H$			9.1
Zn^{2+}	ML/M.L	12.99		
	$ML_2/M.L^2$	24.52		
	MHL/M.HL	7.07		
Ga^{3+}	ML/M.L	22.5		
	ML/MOHL.H	5.75		
	$MOHL/M(OH)_2L.H$	8.53		

Bibliography:

H^+-Zn^{2+} 75HM

Ga^{3+} 76HM

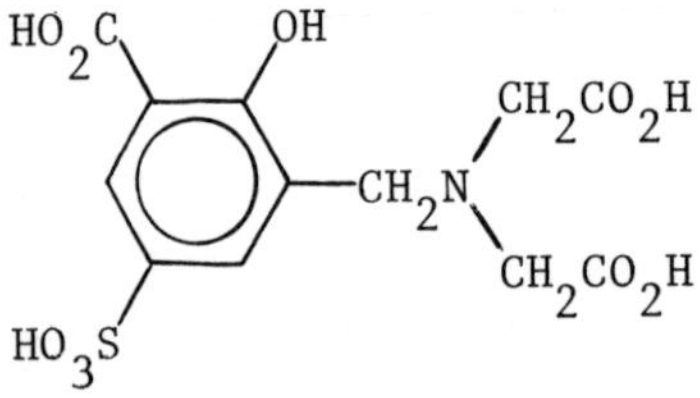

$C_{12}H_{13}O_{10}NS$ N-(3-Carboxy-2-hydroxy-5-sulfophenylmethyl)iminodiacetic acid H_5L
(3-[bis(carboxymethyl)aminomethyl]-5-sulfosalicylic acid)

Metal ion	Equilibrium	Log K 25°, 0.1
H^+	HL/H.L	(12.6)
	$H_2L/HL.H$	9.46
	$H_3L/H_2L.H$	3.47
	$H_4L/H_3L.H$	2.71
	$H_5L/H_4L.H$	1.73
Mg^{2+}	ML/M.L	8.2
Co^{2+}	ML/M.L	13.4
	MHL/M.HL	7.8
Ni^{2+}	ML/M.L	14.0
	MHL/M.HL	8.3
Cu^{2+}	ML/M.L	15.8
	MHL/M.HL	10.6
Fe^{3+}	ML/M.L	19.5

Bibliography: 78TZ

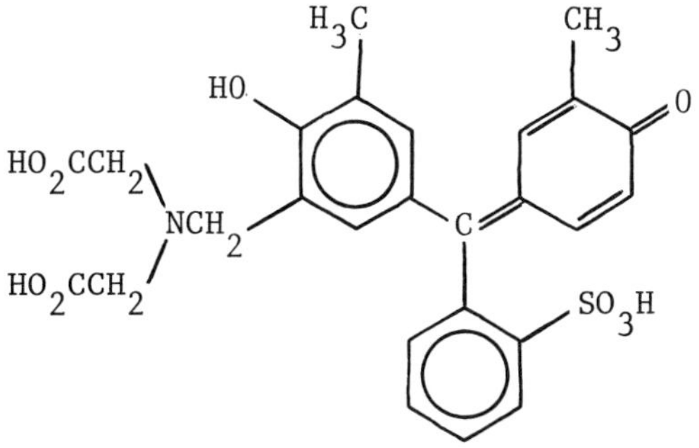

$C_{26}H_{24}O_9NS$

5'-Bis(carboxymethyl)aminomethyl-3,3'-dimethyl-
4'hydroxyfuchson-2"-sulfonic acid   (Semi-Xylenol Orange)
(Other reference on Vol.1, p.399)

H_4L

Metal ion	Equilibrium	Log K 25°, 0.1
H^+	HL/H.L	10.9
	$H_2L/HL.H$	7.44
	$H_3L/H_2L.H$	2.60
	$H_4L/H_3L.H$	1.5
Mg^{2+}	ML/M.L	6.89
	MHL/ML.H	6.92
	MOHL/ML.OH	2.43
Ca^{2+}	ML/M.L	6.53
	MHL/ML.H	7.16
Sr^{2+}	ML/M.L	5.30
	MHL/ML.H	7.73
Ba^{2+}	ML/M.L	4.75
	MHL/ML.H	7.97
Zn^{2+}	ML/M.L	11.84
	$ML_2/M.L^2$	18.44
	MHL/ML.H	7.13
	$MH_2L/MHL.H$	2.70
Al^{3+}	ML/M.L	16.7
	MHL/ML.H	3.3
	ML/MOHL.H	7.2

Bibliography:

Mg^{2+}-Ba^{2+} 74YO

Zn^{2+} 74YM

Al^{3+} 79M

Other reference: 78SYM

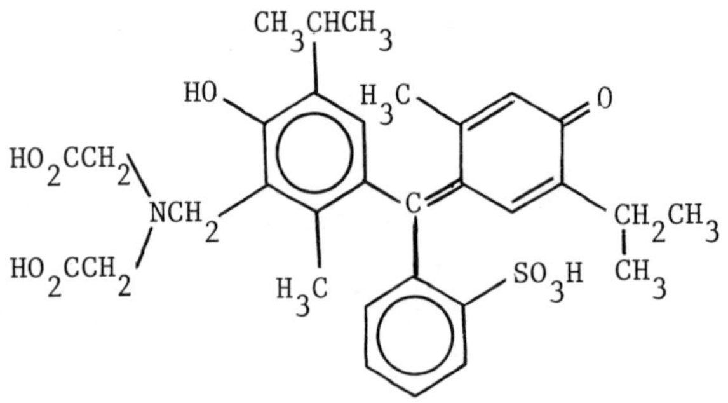

CH$_3$CHCH$_3$

HO

H$_3$C

O

HO$_2$CCH$_2$

NCH$_2$

HO$_2$CCH$_2$

C

CH$_2$CH$_3$

CH$_3$

H$_3$C

SO$_3$H

C$_{32}$H$_{37}$O$_9$NS   5'-Bis(carboxymethyl)aminomethyl-3,3'-bis(methylethyl)-   H$_4$L
                         6,6'-dimethyl-4'-hydroxyfuchson-2''-sulfonic acid
                                  (Semi-Methylthymol Blue)

Metal ion	Equilibrium	Log K 25°, 0.1
H$^+$	HL/H.L	(12.12)
	H$_2$L/HL.H	7.61
	H$_3$L/H$_2$L.H	2.81
	H$_4$L/H$_3$L.H	2.0
Mg^{2+}	ML/M.L	7.05
	MHL/ML.H	7.55
	MOHL/ML.OH	2.35
Ca^{2+}	ML/M.L	6.52
	MHL/ML.H	7.82
Sr^{2+}	ML/M.L	5.34
	MHL/ML.H	8.38
Ba^{2+}	ML/M.L	4.54
	MHL/ML.H	8.87
Co^{2+}	ML/M.L	12.75
	MHL/ML.H	6.53
	MH$_2$L/MHL.H	2.90
Ni^{2+}	ML/M.L	12.37
	MHL/ML.H	7.00
	MH$_2$L/MHL.H	2.48
Cu^{2+}	ML/M.L	13.5
	MHL/ML.H	6.3
	MOHL/ML.OH	3.6
Zn^{2+}	ML/M.L	12.92
	MHL/ML.H	6.20
	MH$_2$L/MHL.H	3.24
Al^{3+}	ML/M.L	17.9
	MHL/ML.H	3.5
	ML/MOHL.H	7.6

Bibliography:

H$^+$,Co^{2+}-Zn^{2+} 74YMa   Al^{3+} 79M

Mg^{2+}-Ba^{2+} 74YO   Other references: 76KMY,76MFT

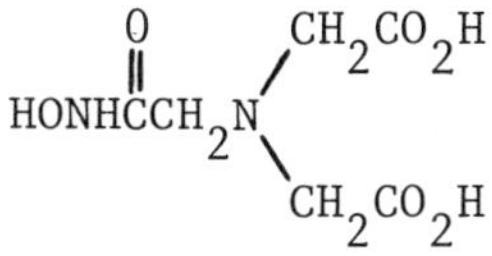

$$\text{HONHCCH}_2\text{N} \underset{\displaystyle \text{CH}_2\text{CO}_2\text{H}}{\overset{\displaystyle \text{CH}_2\text{CO}_2\text{H}}{\diagdown}}$$

$C_6H_{10}O_6N_2$ N,N-Bis(carboxymethyl)aminoacetohydroxamic acid H_3L

Metal ion	Equilibrium	Log K 20°, 0.1
H^+	HL/H.L	9.42
	H_2L/HL.H	6.16
	H_3L/H_2L.H	2.44
Cu^{2+}	ML/M.L	14.72
	MHL/ML.H	4.15
	ML/MOHL.H	7.90
	MOHL/M(OH)$_2$L.H	9.35
Fe^{3+}	ML/M.L	16.22
	MHL/ML.H	3.50
	ML/MOHL.H	5.70
	MOHL/M(OH)$_2$L.H	9.18

Bibliography: 77KJ

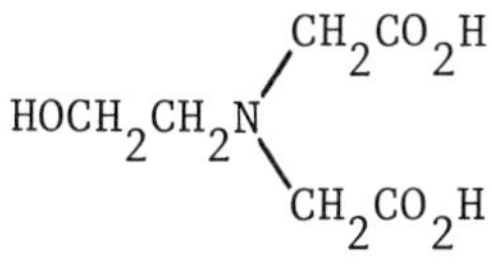

$$\text{HOCH}_2\text{CH}_2\text{N} \underset{\displaystyle \text{CH}_2\text{CO}_2\text{H}}{\overset{\displaystyle \text{CH}_2\text{CO}_2\text{H}}{\diagdown}}$$

$C_6H_{11}O_5N$ N-(2-Hydroxyethyl)iminodiacetic acid   (HIDA) H_2L
(Other values on Vol.1, p.163)

Metal ion	Equilibrium	Log K 25°, 0.1	Log K 25°, 0.5	Log K 25°, 1.0	ΔH 25°, 0.1	ΔS 25°, 0.1
H^+	HL/H.L	8.68 ±0.04	8.52	8.69 ±0.03	−5.3	22
	H_2L/HL.H	2.20 ±0.05	2.28	2.22	−0.6	8
VO^{2+}	ML/M.L		9.26			
	ML/MOHL.H		5.11			
Hg^{2+}	ML/M.L$_2$	(5.48)[h,v]				
	ML$_2$/M.L^2	(8.83)[h,v]				
	ML/MOHL.H	9.56[h]				
	MOHL/M(OH)$_2$L.H	10.51[h]				
Ga^{3+}	ML/M.L	11.33				
	ML/MOHL.H	3.25				
	MOHL/M(OH)$_2$L.H	5.48				

[h]20°, 0.1; [v]not corrected for Cl$^-$

Bibliography:

H^+,VO^{2+} 77N Ga^{3+} 76HM

Co^{2+} 79KV

Ni^{2+} 76JP   Other references: 71EV,71RN,74CM,75KKI, 76ZK,77MF,78K,78KVa,78RS,78SPS

$C_8H_9O_7N$ **Uramil-N,N-diacetic acid** H_3L
(Other values on Vol.1, p.183)

Metal ion	Equilibrium	Log K 25°, 0.1	Log K 20°, 0.1	Log K 20°, 0	ΔH 20°, 0.1	ΔS 25°, 0.1
H^+	HL/H.L	9.52	9.63	10.33		
		$9.65^z \pm 0.07$			-3.7^z	32^z
	$H_2L/HL.H$	$2.72^z \pm 0.05$	2.67	3.1		
	$H_3L/H_2L.H$	$1.9^z \pm 0.2$	1.7	1.9		
Li^+	ML/M.L	4.88^z	4.90^z	5.61	-1.8^z	16^z
Na^+	ML/M.L	2.71^z	2.72^z	3.33	-1.1^z	9^z
K^+	ML/M.L	1.22^z	1.23^z	1.94	-0.4^z	4^z
Be^{2+}	ML/M.L	10.13	10.36		$(-3)^r$	(36)
	MHL/ML.H		2.71			
Mg^{2+}	ML/M.L	8.13 ±0.05	8.19		0.6^z	39
	$ML_2/M.L^2$	8.35^z	11.81			
Ca^{2+}	ML/M.L	8.21 ±0.06	8.31		-3.2^z	27
		8.40^z				
	$ML_2/M.L^2$	13.90^z	13.58			
Sr^{2+}	ML/M.L	6.86 ±0.04	6.93		-2.9^z	22
		7.02^z				
	$ML_2/M.L^2$		10.99			
Ba^{2+}	ML/M.L	6.06 ±0.04	6.13		-2.7^z	19
		6.16^z				
	$ML_2/M.L^2$		9.81			
La^{3+}	ML/M.L	12.82^z				
	$ML_2/M.L^2$	22.64^z				
Mn^{2+}	ML/M.L	9.95				
		10.28^z				
	$ML_2/M.L^2$	14.04^z				
	MHL/ML.H	3.05				
Fe^{2+}	ML/M.L	10.56				
Co^{2+}	ML/M.L	11.84				
Ni^{2+}	ML/M.L	13.12				
		14.19^z				
	$ML_2/M.L^2$	17.3^z				
Cu^{2+}	ML/M.L	14.10				
		15.54^z				
	$ML_2/M.L^2$	19.4				
	MHL/ML.H	2.46				

$^z (CH_3)_4 N^+$ salt used as background electrolyte; ^{r}20-40°, 0.1

Uramil-N,N-diacetic acid   (continued)

Metal ion	Equilibrium	Log K 25°, 0.1	Log K 20°, 0.1	Log K 20°, 0	ΔH 20°, 0.1	ΔS 25°, 0.1
Tl^+	ML/M.L	5.92^z	5.99^z		-5.8^z	8^z
Zn^{2+}	ML/M.L	12.21 13.39^z				
	$ML_2/M.L^2$	16.7^z				
Cd^{2+}	ML/M.L	10.81 11.64^z				
	$ML_2/M.L^2$	18.28^z				
	MHL/ML.H	2.90				
	MOHL/ML.OH	4.58				
Pb^{2+}	ML/M.L	12.73				
	$ML_2/ML.L$	3.74^z				

$^z (CH_3)_4 N^+$ salt used as background electrolyte.

Bibliography:

H^+, La^{3+} 75JT

Li^+-K^+, Tl^+ 76A

Be^{2+} 77DA

$Mg^{2+}-Ba^{2+}$ 63IDa,75JT,76A,77DA

$Mn^{2+}-Cu^{2+}, Zn^{2+}-Pb^{2+}$ 72FA,75JT

Other reference: 74PI

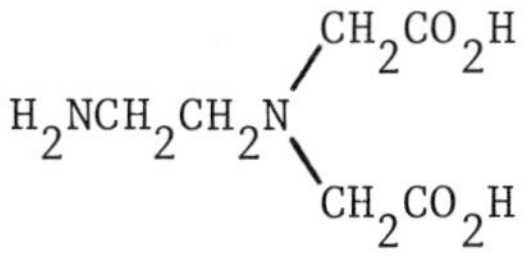

$$H_2NCH_2CH_2N \begin{array}{c} CH_2CO_2H \\ \\ CH_2CO_2H \end{array}$$

$C_6H_{12}O_4N_2$ <u>N-(2-Aminoethyl)iminodiacetic acid</u> (ethylenediamine-N,N-diacetic acid) H_2L
(Other values on Vol.1, p.194)

Metal ion	Equilibrium	Log K 25°, 0.1	Log K 20°, 0.1
H^+	HL/H.L	10.87	11.05
	$H_2L/HL.H$	5.53	5.58
Co^{2+}	ML/M.L	11.59	11.78
	$ML_2/M.L^2$		15.91
	MHL/ML.H	4.95	4.22
	ML/MOHL.H	10.75	
Ag^+	ML/M.L	6.85^u	
Hg^{2+}	ML/M.L	$(9.75)^{h,v}$	
	$ML_2/M.L^2$	$(15.80)^{h,v}$	
Ga^{3+}	ML/M.L	16.75	
	ML/MOHL.H	3.96	
	$MOHL/M(OH)_2L.H$	7.6	

^{h}20°, 0.1; ^{u}25°, 1.5; ^{v}not corrected for Cl^-.

Bibliography: Co^{2+} 75MMb; Ag^+ 77NF; Ga^{3+} 76HM

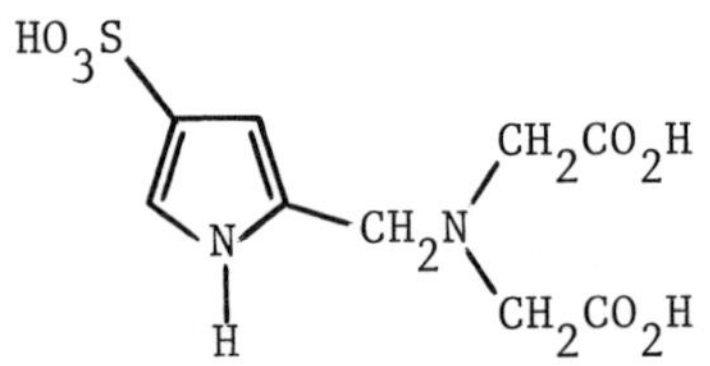

$C_9H_{12}O_7N_2S$ 2-[Bis(carboxymethyl)aminomethyl]azole-4-sulfonic acid H_3L
                      (N-(4-sulfonyl-2-pyrrylmethyl)iminodiacetic acid)

Metal ion	Equilibrium	Log K 25°, 1.0
H^+	HL/H.L	8.30
	$H_2L/HL.H$	2.30
	$H_3L/H_2L.H$	1.46
Cu^{2+}	$ML/M.L$	9.6
	$ML_2/M.L^2$	15.4
	$ML/MH_{-1}L.H$	8.01
	$MH_{-1}L/MOH(H_{-1}L).H$	9.91
	$M_2OH(H_{-1}L)_2/MH_{-1}L.MOH(H_{-1}L)$	1.6

Bibliography: 76SA

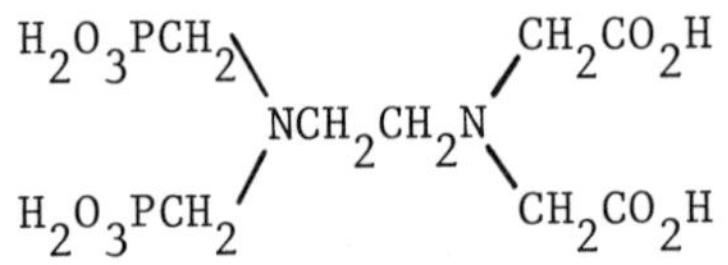

$C_8H_{18}O_{10}N_2P_2$ Ethylenedinitrilo-N,N-diacetic-N',N'-bis(methylenephosphonic) acid H_6L

Metal ion	Equilibrium	Log K 25°, 0.1
H^+	HL/H.L	11.63
	$H_2L/HL.H$	10.91
	$H_3L/H_2L.H$	6.52
	$H_4L/H_3L.H$	5.30
	$H_5L/H_4L.H$	2.61
	$H_6L/H_5L.H$	2.25
Mg^{2+}	$MH_2L/M.H_2L$	3.7
Nd^{3+}	$MH_2L/M.H_2L$	6.34
Tb^{3+}	$MH_2L/M.H_2L$	7.50
Co^{2+}	$ML/M.L$	(16.03)
	$MH_2L/M.H_2L$	3.70
Ni^{2+}	$ML/M.L$	15.23
	$MH_2L/M.H_2L$	3.70
Cu^{2+}	$ML/M.L$	17.20
	$MH_2L/M.H_2L$	8.45
Fe^{3+}	$MH_2L/M.H_2L$	10.41

Bibliography: $H^+, Mg^{2+}-Tb^{3+}, Fe^{3+}$ 76TI; $Co^{2+}-Cu^{2+}$ 75IT, 76TI

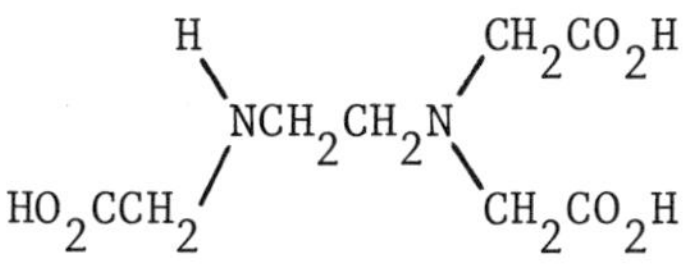

$$H_2CO_2H$$

$C_8H_{14}O_6N_2$ Ethyleneiminonitrilotriacetic acid H_3L
N-(2-carboxymethyliminoethyl)iminodiacetic acid)

Metal ion	Equilibrium	Log K 25°, 0.1
Cr(III)	ML/MOHL.H	6.25

Bibliography: 750W

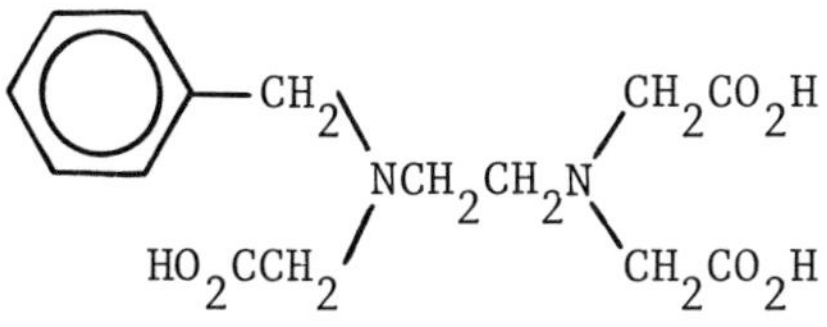

$C_{15}H_{20}O_6N_2$ N-Benzylethylenedinitrilo-N,N',N'-triacetic acid H_3L
(Other values on Vol.1,p. 198)

Metal ion	Equilibrium	Log K 25°, 0.1
H^+	HL/H.L	10.08
	$H_2L/HL.H$	5.24
	$H_3L/H_2L.H$	2.54
	$H_4L/H_3L.H$	1.67
Y^{3+}	ML/M.L	12.69
La^{3+}	ML/M.L	10.81
Ce^{3+}	ML/M.L	11.28
Pr^{3+}	ML/M.L	11.69
Nd^{3+}	ML/M.L	11.82
Sm^{3+}	ML/M.L	12.19
Eu^{3+}	ML/M.L	12.35
Gd^{3+}	ML/M.L	12.40
Tb^{3+}	ML/M.L	12.79
Dy^{3+}	ML/M.L	13.02
Ho^{3+}	ML/M.L	13.26
Er^{3+}	ML/M.L	13.47
Tm^{3+}	ML/M.L	13.65
Yb^{3+}	ML/M.L	13.85
Lu^{3+}	ML/M.L	13.93

Bibliography: 78MP

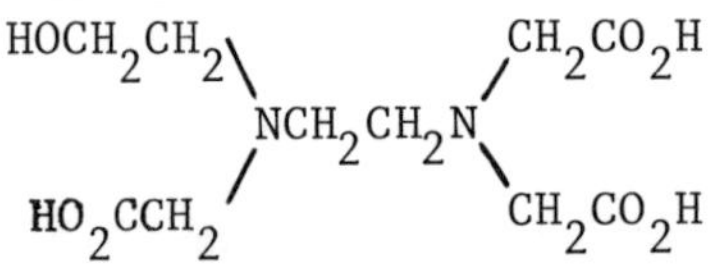

$$\begin{array}{c}
HOCH_2CH_2\backslash \qquad \qquad /CH_2CO_2H \\
NCH_2CH_2N \\
HO_2CCH_2/ \qquad \qquad \backslash CH_2CO_2H
\end{array}$$

$C_{10}H_{18}O_7N_2$ <u>N-(2-Hydroxyethyl)ethylenedinitrilo-N,N',N'-triacetic acid</u>   (HEDTA) H_3L

(Other values on Vol.1,p.199)

Metal ion	Equilibrium	Log K 25°, 0.1	Log K 25°, 1.0	ΔH 25°, 0.1	ΔS 25°, 0.1
H^+	HL/H.L	9.81 ±0.1	9.20 +0.01	-6.7 ±0.0	22
	H_2L/HL.H	5.39 ±0.02	5.49 ±0.04	-3.1	14
	H_3L/H_2L.H	2.6 ±0.1	2.33 ±0.03	1.1	16
	H_4L/H_3L.H		1.5		
Cr^{3+}	ML/MOHL.H	6.13			
Ag^+	ML/M.L	6.67 ±0.04			
Cd^{2+}	ML/M.L	13.1	13.2	-10.3	25
	MHL/ML.H		2.30		
Hg^{2+}	ML/M.L	20.1 ±0.0	19.4[b]	-20.0	25
	ML/MOHL.H	8.4			
Pb^{2+}	ML/M.L	15.5	14.8	-12.6	29
	MHL/ML.H		2.14		
Ga^{3+}	ML/M.L	18.1			
	ML/MOHL.H	4.38			

[b] 25°, 0.5

Bibliography:

H^+ 76MGD,76OSM

Cr^{3+} 75OW

Y^{3+}-Lu^{3+} 73BBB,77CGG,77GGC

Ag^+ 79MST

Cd^{2+},Pb^{2+} 76OSM

Hg^{2+} 77GGC

Ga^{3+} 76HM

Other references: 72TB,73KK,74NKK,74PI,
74YKP,75AZ,75TP,76KNI,76NGT,76PTS,
77GMD,77TK,78MGD,78RS

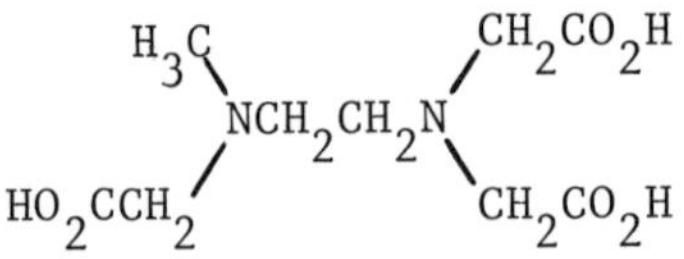

$$\begin{array}{c}
H_3C\backslash \qquad \qquad /CH_2CO_2H \\
NCH_2CH_2N \\
HO_2CCH_2/ \qquad \qquad \backslash CH_2CO_2H
\end{array}$$

$C_9H_{16}O_6N_2$ <u>N-Methylethylenedinitrilo-N,N',N'-triacetic acid</u> H_3L

Metal ion	Equilibrium	Log K 25°, 0.1
H^+	HL/H.L	10.31
	H_2L/HL.H	5.42
	H_3L/H_2L.H	2.45
	H_4L/H_3L.H	1.93
Y^{3+}	ML/M.L	13.35
La^{3+}	ML/M.L	11.50
Ce^{3+}	ML/M.L	11.87
Pr^{3+}	ML/M.L	12.33

N-Methylethylenedinitrilo-N,N',N'-triacetic acid   (continued)

Metal ion	Equilibrium	Log K 25°, 0.1
Nd^{3+}	ML/M.L	12.51
Sm^{3+}	ML/M.L	12.86
Eu^{3+}	ML/M.L	12.96
Gd^{3+}	ML/M.L	12.98
Tb^{3+}	ML/M.L	13.35
Dy^{3+}	ML/M.L	13.61
Ho^{3+}	ML/M.L	13.81
Er^{3+}	ML/M.L	14.04
Tm^{3+}	ML/M.L	14.31
Yb^{3+}	ML/M.L	14.43
Lu^{3+}	ML/M.L	14.51
Cr(III)	ML/MOHL.H	6.25

Bibliography: H^+-Lu^{3+} 73PJ; Cr(III)   75OW

$$\begin{array}{c} HO_2CCH_2 \\ \diagdown \\ NCH_2CH_2N \\ \diagup \\ HO_2CCH_2 \end{array} \begin{array}{c} CH_2CO_2H \\ \diagup \\ \\ \diagdown \\ CH_2CO_2H \end{array}$$

$C_{10}H_{16}O_8N_2$ Ethylenedinitrilotetraacetic acid   (EDTA) H_4L
                             (Other values on Vol.1, p.204)

Metal ion	Equilibrium	Log K 25°, 0.1	Log K 25°, 1.0	Log K 25°, 0	ΔH 25°, 0	ΔS 25°, 0.1
H^+	HL/H.L	$10.17\ ^{}\pm0.02$ $10.38^z\pm0.01$	$8.78^u\pm0.00$ $9.88^v\pm0.1$ $10.15^z\pm0.1$	11.014 $9.05^{u,e}\pm0.01$	$-5.59^a\pm0.2$ $-5.68^h-0.2$ $-6.3^c\ -0.1$	28
	$H_2L/HL.H$	$6.11\ \pm0.02$ $6.13^z\pm0.03$	$6.23^u\pm0.03$ $6.24^v\pm0.06$ $6.10^z\pm0.08$	6.320 $7.20^{u,e}\pm0.02$	$-4.3^a\ -0.1$ $-4.5^h\ \pm0.1$ $-5.6^c\ \pm0.1$	14
	$H_3L/H_2L.H$	$2.68\ \pm0.02$	$2.38\ \pm0.08$ $2.73^z\pm0.03$	$2.54^e\pm0.04$	1.3^a 1.4^h	17
	$H_4L/H_3L.H$	$1.95\ \pm0.05$	$2.00\ \pm0.05$ 2.2^z	$2.20^e\pm0.07$	0.3^a 0.2^h	10
	$H_5L/H_4L.H$	$1.5\ \pm0.0$	$1.5\ \pm0.1$	1.7^e	0.5^h	9
	$H_6L/H_5L.H$		$0.0\ \pm0.1$	0.6^e 0.4^e	-0.5^h	-2^{c*}
Li^+	ML/M.L	$2.79\ \pm0.06$			$0.2^h\ -0.1$	13
Na^+	ML/M.L	$1.64\ \pm0.2$			$-1.7^a\ \pm0.5$	2
K^+	ML/M.L	$0.8\ \pm0.2$			0^a	4
Be^{2+}	ML/M.L	9.63			$(10)^r$	(78)

^{a}25°, 0.1; ^{c}25°, 1.0; ^{e}25°, 3.0; ^{h}20°, 0.1; ^{r}20-40°, 0.1; $^u Na^+$ salt used as background electrolyte; $^v K^+$ salt used as background electrolyte; $^z(CH_3)_4N^+$ salt used as background electrolyte; *assuming ΔH for 0.1=ΔH for 1.0

EDTA (continued)

Metal ion	Equilibrium	Log K 25°, 0.1	Log K 25°, 1.0	Log K 25°, 0	ΔH 25°, 0	ΔS 25°, 0.1
Mg^{2+}	ML/M.L	8.83 ±0.1		9.12[o]	3.5[h] ±0.6	52
	MHL/ML.H	3.85[h]				
Ca^{2+}	ML/M.L	10.61 ±0.1	9.68[v]	11.00[o]	-6.6[h] ±0.1	26
Th^{4+}	ML/M.L	23.2			-3.1[a]	
	MHL/ML.H	1.98			0.1[a]	
	ML/MOHL.H	7.04				
	$(ML)^2/M_2(OH)_2L_2.H^2$	9.82[h] 10.03				
	$M_2(OH)_2L_2/(MOHL)^2$	4.3				
Ti^{3+}	ML/M.L	(21.3)				
	MHL/ML.H	2.02				
	ML/MOHL.H	8.64				
	$MOHL/M(OH)_2L.H$	11.61				
Fe^{3+}	ML/M.L	25.0[h] ±0.0	(25.1)[j]		-2.7[a]	105
	MHL/ML.H	1.3[h] ±0.1			-0.1[a]	6
	ML/MOHL.H	7.49[h]	7.53 ±0.05		(-10)[s]	(1)[c]
	$MOHL/M(OH)_2L.H$	9.41[h]				
	$(ML)^2/M_2(OH)_2L_2.H^2$		12.21		(-5)[s]	(40)[c]
	$M_2(OH)_2L_2/(MOHL)^2$		2.8 ±0.2		(-15)[s]	(-38)[c]
Zr^{4+}	ML/M.L	29.4 +0.1	27.7 +1	32.8	-0.6	133*
	ML/MOHL.H	6.2	6.1[j]			
	$M_2(OH)_2L_2/(MOHL)^2$	3.5				
Hf^{4+}	ML/M.L	29.5		33.7	-0.2	134*
VO_2^+	ML/M.L	15.5[h]		15.5[e]		
	MHL/ML.H	4.31[h]		4.3[e]		
	$MH_2L/MHL.H$	3.49[h]		3.7[e]		
	$MH_3L/MH_2L.H$	1.4[h]				
Ag^+	ML/M.L	7.22[h] ±0.05			-7.5[h]	8
	MHL/ML.H	6.01[h] +0.6				
Tl^+	ML/M.L	6.41 ±0.01			-10.4[h]	-6
	MHL/ML.H	5.77[h]				
Pd^{2+}	ML/M.L		24.5[j]			
	MHL/ML.H		3.01[j]			
	$MH_2L/MHL.H$		2.31[j]			
	$MH_3L/MH_2L.H$		0.9[j]			
Zn^{2+}	ML/M.L	16.44 ±0.1		14.87[e]	-4.9[h] ±0.8	59
	MHL/ML.H	3.0		3.10[e]	-2.2[a]	6
	MOHL/ML.OH	2.1			(0)[h]	(10)
Cd^{2+}	ML/M.L	16.36 ±0.1		14.68[e]	-9.1[h] -1.0	44
	MHL/ML.H	2.9		2.75[e]	-0.4[a]	12

---

[a] 25°, 0.1; [c] 25°, 1.0; [e] 25°, 3.0; [h] 20°, 0.1; [o] 20°, 0; [j] 20°, 1.0; [s] 0-42°, 1.0;

[v] K^+ salt used as background electrolyte; *assuming ΔH for 0.0=ΔH for 0.1

EDTA (continued)

Metal ion	Equilibrium	Log K 25°, 0.1	Log K 25°, 1.0	Log K 25°, 0	ΔH 25°, 0	ΔS 25°, 0.1
Hg^{2+}	ML/M.L	21.5 ±0.1	20.8[b]		-18.9[h]-0.3	35
	MHL/ML.H	3.1[h]				
	ML/MOHL.H	9.11[h]				
Pb^{2+}	ML/M.L	17.88 ±0.01		(15.19)[e]	-13.2[h]±0.9	38
	MHL/ML.H	2.8[h]	2.49	2.82[e]		
Ga^{3+}	ML/M.L	21.0 ±0.7				
	MHL/ML.H	1.8 ±0.1				
	ML/MOHL.H	5.52 5.65[h]				

[b]25°, 0.5; [e]25°, 3.0; [h]20°, 0.1

Bibliography:

H^+ 68KS,73CS,74VK,75BK,76A,76AM,76CWW,
        76GMa,76VKO,77CB,77VL,77VLa,78VKO,
        79LM,79VKO

Li^+-K^+ 76A,76VBa,77VL

Be^{2+} 77DA

Mg^{2+} 75VB

Ca^{2+} 68KS,73HR,76VB

$Y^{3+}-Lu^{3+}$ 73O,74BK,75BK,75Sa,76GMa,77CGG,77GGC

Th^{4+} 78D

Co^{2+},Ni^{2+} 76VBb

Cu^{2+} 76HM,77VB

Ti^{3+} 66PP

Fe^{3+} 77CB,78Da

Cr^{3+} 75OW,77ABJ

Zr^{4+} 78VL

Hf^{4+} 78VLa

VO_2^+ 78LL

Ag^+ 76A

Tl^+ 63ID,76A

Pd^{2+} 76AM

Zn^{2+},Cd^{2+},Pb^{2+} 73HR,76CWW

Hg^{2+} 77GGC

Al^{3+} 79M

Ga^{3+} 76HM

Other references: 52BK,52MP,53RL,55EH,57FSa,
        58CL.59GM,59GMA,59MA,59MK,59NT,60MS,
        60SS,62IN,62KP,62YO,66KR,66LP,67LP,
        68LP,69BH,69KK,69TK,70MA,71CL,71EZ,
        71GBG,71KP,71KTa,71MA,71MMW,71SSa,
        72KT,72LH,72RK,72TK,72V,73AV,73BW,
        73CG,73DR,73KI,73YP,74B,74G,74KM,
        74KNT,74LA,74NP,74NS,74TN,75APB,
        75CGb,75IYa,75Kd,75Ke,75LB,75LN,
        75PP,75TK,75TP,75VKF,76CJ,76OT,
        76VP,77Aa,77GD,77HA,77HS,77KL,
        77KST,77MP,77OM,77P,78J,78K,78Kb,
        78KVb,78KVc,78MGD,78RS,78SSa,78TSK,
        79JPC,79JPC,79JPP,79KKT,79MP,79SFD,
        79ZL

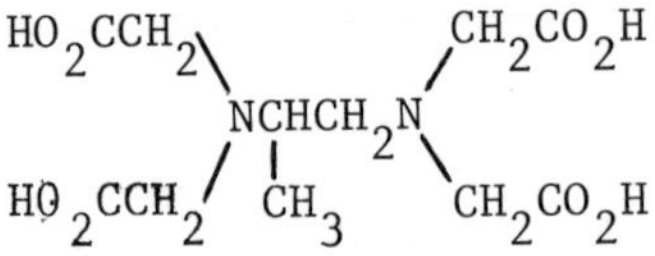

$$\text{HO}_2\text{CCH}_2 \diagdown \qquad \text{CH}_2\text{CO}_2\text{H}$$
$$\text{NCHCH}_2\text{N}$$
$$\text{HO}_2\text{CCH}_2 \diagup \; \text{CH}_3 \qquad \text{CH}_2\text{CO}_2\text{H}$$

| $C_{11}H_{18}O_8N_2$ | DL-(Methylethylene)dinitrilotetraacetic acid   (PDTA) | | | | H_4L |

(Other values on Vol.1, p.212)

Metal ion	Equilibrium	Log K 25°, 0.1	Log K 20°, 0	ΔH 25°, 0.2	ΔS 25°, 0.1
H^+	HL/H.L	10.84 ±0.05	11.17	-5.6[a]±0.5	31
	$H_2L/HL.H$	6.20 ±0.01	6.40	-2.6[h]	20
	$H_3L/H_2L.H$	2.78[h]±0.01			
	$H_4L/H_3L.H$	1.87[h]±0.02			
Mn^{2+}	ML/M.L	14.9		-5.3	50
Co^{2+}	ML/M.L	17.3		-4.9	63
Ni^{2+}	ML/M.L	19.6		-8.9	60
Cu^{2+}	ML/M.L	19.8 ±0.0		-9.3	59
Zn^{2+}	ML/M.L	17.3		-6.2	58

[a]Average of values at 20°, 0.1 and 20°, 1.0;   [h]20°, 0.1

Bibliography: 75SGa   Other reference: 73BW

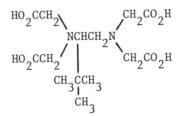

| $C_{14}H_{24}O_8N_2$ | DL-(1,1-Dimethylethylethylene)dinitrilotetraacetic acid | H_4L |

(t-butylethylenediaminetetraacetic acid)

Metal ion	Equilibrium	Log K 25°, 0.1
H^+	HL/H.L	11.6
	$H_2L/HL.H$	5.29
	$H_3L/H_2L.H$	3.62
	$H_4L/H_3L.H$	2.1
Mg^{2+}	ML/M.L	11.0
Ca^{2+}	ML/M.L	11.8
Sr^{2+}	ML/M.L	9.4

Bibliography:

H^+ 72YY

Mg^{2+}-Sr^{2+} 74YY

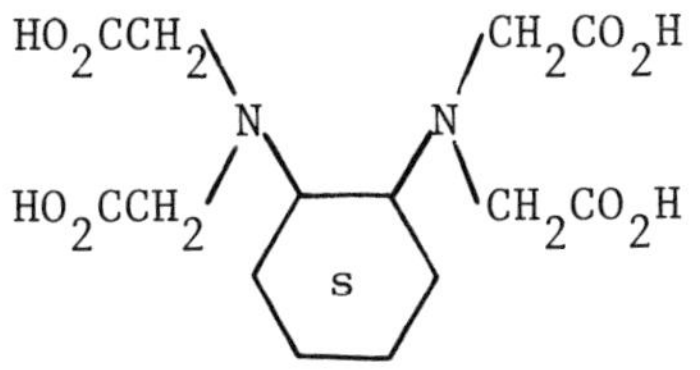

$C_{14}H_{22}O_8N$ <u>trans-1,2-Cyclohexylenedinitrilotetraacetic acid</u>   <u>(CDTA)</u> H_4L
(Other values on Vol.1, p.236)

Metal ion	Equilibrium	Log K 25°, 0.1	Log K 25°, 1.0	Log K 25°, 3.0	ΔH 20°, 0.1	ΔS 25°, 0.1
H^+	HL/H.L	(12.3)	9.21^u	9.90^u	-6.7 +0.1	34
	$H_2L/HL.H$	6.12 ±0.03	5.84^u	6.72^u	-2.1	21
	$H_3L/H_2L.H$	3.53^h ±0.02	3.52^j	3.65		
	$H_4L/H_3L.H$	2.42^h ±0.02	2.41^j	3.21		
	$H_5L/H_4L.H$		1.7^j			
Ag^+	ML/M.L	8.40 ±0.01				
Ga^{3+}	ML/M.L	22.8^h ±0.3				
	MHL/ML.H	2.7 2.42^h				
	ML/MOHL.H	7.48 6.46^h				
Hg^{2+}	ML/M.L	24.8 +0.3	23.3^b		-16.6 +0.6	58
	MHL/ML.H	3.1^h +0.4				
	ML/MOHL.H	10.46^h				

^{b}25°, 0.5; ^{h}20°, 0.1; ^{j}20°, 1.0; ^{u}Na$^+$ salt used as background electrolyte.

Bibliography:

Y^{3+}-Lu^{3+} 77CGG,77GGC

Ag^+ 79MST

Hg^{2+} 77GGC

Ga^{3+} 76HM

Other references: 66GJ,69Ma,69NK,70KC,
70PL,71EZ,71KTa,71S,72MSS,73BW,73KI,
73KK,73TKa,75IY,75IYa,73Ka 75LB,76NM,
77HA,78MGD,78VP,79HW,79VN

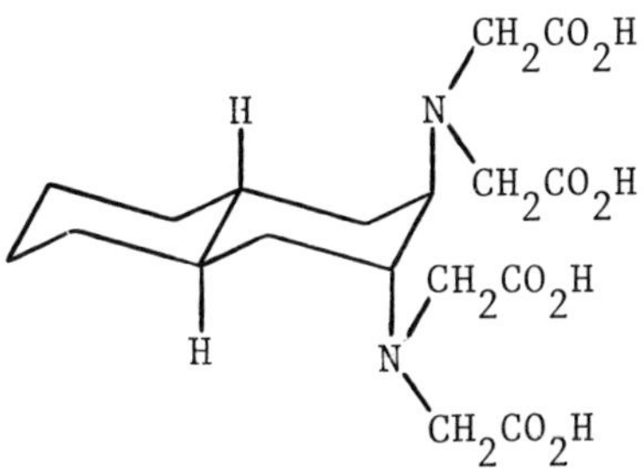

$C_{18}H_{28}O_8N_2$ <u>trans-Decahydronaphthylene-trans-2,3-bis(iminodiacetic acid)</u> H_4L

Metal ion	Equilibrium	Log K 25°, 0.1
H^+	HL/H.L	10.61
	$H_2L/HL.H$	6.47
	$H_3L/H_2L.H$	3.04
	$H_4L/H_3L.H$	2.26

trans-Decahydronaphthylene-trans-2,3-bis(iminodiacetic acid) (continued)

Metal ion	Equilibrium	Log K 25°, 0.1
Mg^{2+}	ML/M.L	10.36
Ca^{2+}	ML/M.L	11.54
Sr^{2+}	ML/M.L	9.34

Bibliography: 74YKU

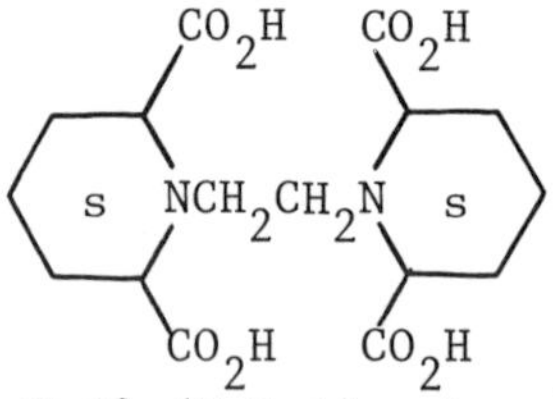

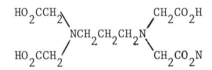

$C_{16}H_{24}O_8N_2$ Ethylenebis-N,N'-(2,6-dicarboxy)piperidine* H_4L

Metal ion	Equilibrium	Log K 25°, 0.1
H^+	HL/H.L	10.30
	$H_2L/HL.H$	6.98
	$H_3L/H_2L.H$	2.64
	$H_4L/H_3L.H$	2.0
Mg^{2+}	ML/M.L	6.36
Ca^{2+}	ML/M.L	7.28
Sr^{2+}	ML/M.L	4.94
Ba^{2+}	ML/M.L	4.14
Mn^{2+}	ML/M.L	11.20
Ni^{2+}	ML/M.L	17.43
Cu^{2+}	ML/M.L	18.24
Zn^{2+}	ML/M.L	15.08
Cd^{2+}	ML/M.L	13.56
Pb^{2+}	ML/M.L	16.05

*Isomer not stated.

Bibliography: 79PB

$C_{11}H_{18}O_8N_2$ Trimethylenedinitrilotetraacetic acid   (TMDTA) H_4L
                        (Other values on Vol.1, p.244)

Metal ion	Equilibrium	Log K 25°, 0.1	Log K 20°, 1.0	ΔH 20°, 0.1	ΔS 25°, 0.1
H^+	HL/H.L	10.39 -0.2	10.23 9.80^u	-5.2	30
	$H_2L/HL.H$	7.96 -0.1	7.81 ±0.01	-4.4	22
	$H_3L/H_2L.H$	2.57^h +0.1	2.53 ±0.03		
	$H_4L/H_3L.H$	1.88^h +0.1	2.1 ±0.2		

^{h}20°, 0.1; ^{u}NaBr used as background electrolyte.

TMDTA (continued)

Metal ion	Equilibrium	Log K 25°, 0.1	Log K 20°, 1.0	ΔH 20°, 0.1	ΔS 25°, 0.1
Cr^{2+}	ML/M.L	11.12[h]			
Pd^{2+}	ML/M.L		28.8[u]		

[h]20°, 0.1; [u]NaBr used as background electrolyte.

Bibliography:
H^+, Pd^{2+} 76AM   Other reference: 68EM
Cr^{2+} 74R

$$HO_2CCH_2 \diagdown \atop HO_2CCH_2 \diagup NCH_2CH_2CH_2CH_2N \diagup CH_2CO_2H \atop \diagdown CH_2CO_2H$$

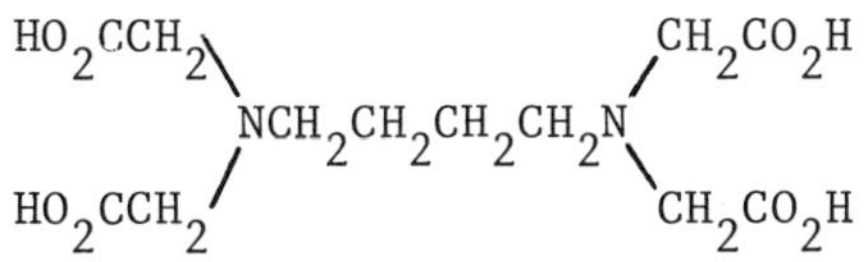

$C_{12}H_{20}O_8N_2$ Tetramethylenedinitrilotetraacetic acid H_4L
(Other values on Vol.1, p.247)

Metal ion	Equilibrium	Log K 25°, 0.1	Log K 20°, 1.0	ΔH 20°, 0.1	ΔS 25°, 0.1
H^+	HL/H.L	10.58 -0.2	10.35 10.09[u]	-6.7	26
	$H_2L/HL.H$	8.98 +0.02	8.92 ±0.04	-5.8	22
	$H_3L/H_2L.H$	2.45[h]+0.2	2.5 ±0.1		
	$H_4L/H_3L.H$	1.90[h]+0.02	2.3 ±0.1		
Cr^{2+}	ML/M.L	10.55[h]			
Pd^{2+}	ML/M.L		25.8[u]		

[h]20°, 0.1; [u]NaBr used as background electrolyte.
Bibliography: H^+, Pd^{2+} 76AM; Cr^{2+} 74R

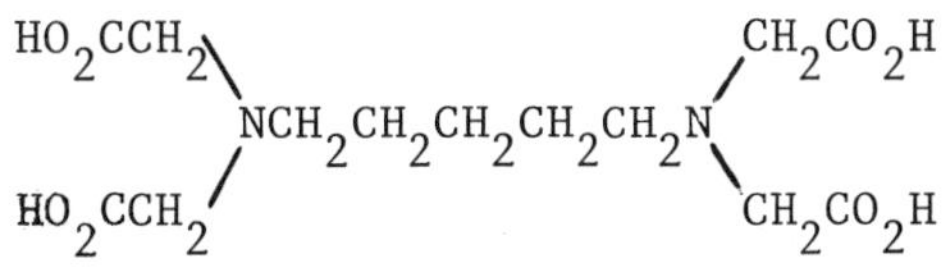

$$HO_2CCH_2 \diagdown \atop HO_2CCH_2 \diagup NCH_2CH_2CH_2CH_2CH_2N \diagup CH_2CO_2H \atop \diagdown CH_2CO_2H$$

$C_{13}H_{22}O_8N_2$ Pentamethylenedinitrilotetraacetic acid H_4L
(Other values on Vol.1, p.249)

Metal ion	Equilibrium	Log K 25°, 0.1	Log K 20°, 1.0	ΔH 20°, 0.1	ΔS 25°, 0.1
H^+	HL/H.L	10.60 -0.1	10.50 10.27[u]	-7.5	23
	$H_2L/HL.H$	9.44 -0.02	9.51 ±0.05	-6.3	22
	$H_3L/H_2L.H$	2.7[h] ±0.0	2.6 ±0.0		
	$H_4L/H_3L.H$	2.3[h] -0.1	2.2 ±0.2		
Pd^{2+}	ML/M.L		26.4[u]		

[h]20°, 0.1; [u]NaBr used as background electrolyte.

Bibliography: 76AM

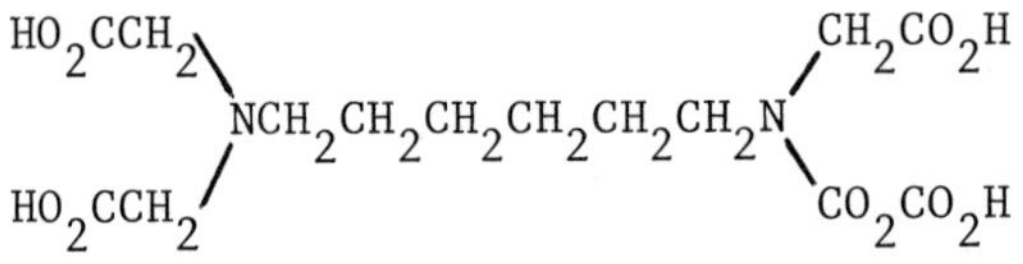

$$HO_2CCH_2 \qquad\qquad\qquad\qquad\qquad CH_2CO_2H$$
$$NCH_2CH_2CH_2CH_2CH_2CH_2N$$
$$HO_2CCH_2 \qquad\qquad\qquad\qquad\qquad CO_2CO_2H$$

$C_{14}H_{24}O_8N_2$	Hexamethylenedinitrilotetraacetic acid	H_4L
	(Other values on Vol.1, p.251)	

Metal ion	Equilibrium	Log K 25°, 0.1	Log K 20°, 1.0	ΔH 20°, 0.1	ΔS 25°, 0.1
H^+	HL/H.L	10.71 -0.2	10.56 10.39^u	-7.9	23
	$H_2L/HL.H$	9.71 ±0.04	9.65 ±0.04	-6.2	24
	$H_3L/H_2L.H$	$2.70^h\pm0.0$	2.5 ±0.1		
	$H_4L/H_3L.H$	$2.20^h\pm0.0$	2.3 ±0.1		
Sc^{3+}	MHL/M.HL	10.20			
Y^{3+}	MHL/M.HL	6.87			
La^{3+}	MHL/M.HL	6.28			
Pr^{3+}	MHL/M.HL	6.40			
Nd^{3+}	MHL/M.HL	6.43			
Sm^{3+}	MHL/M.HL	6.64			
Eu^{3+}	MHL/M.HL	6.71			
Gd^{3+}	MHL/M.HL	6.66			
Dy^{3+}	MHL/M.HL	6.93			
Ho^{3+}	MHL/M.HL	6.96			
Er^{3+}	MHL/M.HL	7.07			
Tm^{3+}	MHL/M.HL	7.18			
Yb^{3+}	MHL/M.HL	7.30			
Lu^{3+}	MHL/M.HL	7.46			
Th^{4+}	MHL/M.HL	10.92			
Cu^{2+}	ML/M.L	15.4^h			
Fe^{3+}	ML/M.L	16.5^h			
Pd^{2+}	ML/M.L		26.3^u		
Cr^{2+}	ML/M.L	9.27^h			

^{h}20°, 0.1; ^{u}NaBr used as background electrolyte.

Bibliography:

H^+,Pd^{2+} 76AM Cr^{2+} 74R

Sc^{3+}-Th^{4+} 74KPS Fe^{3+} 77KKM

Cu^{2+} 75KAM   Other references: 71KA,71S,78SP

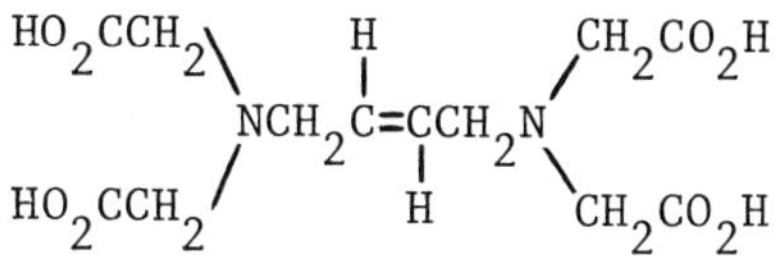

$$HO_2CCH_2 \diagdown \quad\quad\quad \overset{H}{\underset{|}{}} \quad\quad\quad CH_2CO_2H$$

$$\diagup NCH_2C = CCH_2N \diagdown$$

$$HO_2CCH_2 \diagup \quad\quad\quad \overset{|}{H} \quad\quad\quad CH_2CO_2H$$

$C_{12}H_{18}O_8N_2$ trans-1,4-Diaminobut-2-enetetraacetic acid H_4L

Metal ion	Equilibrium	Log K 20°, 0.1
H^+	HL/H.L	9.80
	$H_2L/HL.H$	8.29
	$H_3L/H_2L.H$	2.92
	$H_4L/H_3L.H$	1.96
Mg^{2+}	ML/M.L	3.83
	MHL/ML.H	9.12
	$M_2L/ML.M$	2.9
Ca^{2+}	ML/M.L	3.98
	MHL/ML.H	9.07
	$M_2L/ML.M$	2.9
Sr^{2+}	ML/M.L	3.18
	MHL/ML.H	9.37
	$M_2L/ML.M$	2.5
Ba^{2+}	ML/M.L	2.86
	MHL/ML.H	9.48
	$M_2L/ML.M$	2.3
Y^{3+}	ML/M.L	9.10
	MHL/ML.H	7.04
	$M_2L/ML.M$	4.9
Ce^{3+}	ML/M.L	8.57
	MHL/ML.H	7.31
	$M_2L/ML.M$	4.6
Mn^{2+}	ML/M.L	6.94
	MHL/ML.H	7.51
	$M_2L/ML.M$	3.9
Cu^{2+}	ML/M.L	14.13
	MHL/ML.H	6.25
	$M_2L/ML.M$	6.5
Zn^{2+}	ML/M.L	10.80
	MHL/ML.H	6.52
	$M_2L/ML.M$	5.4

Bibliography:

H^+ 74PPS

$Mg^{2+}-Zn^{2+}$ 76TT

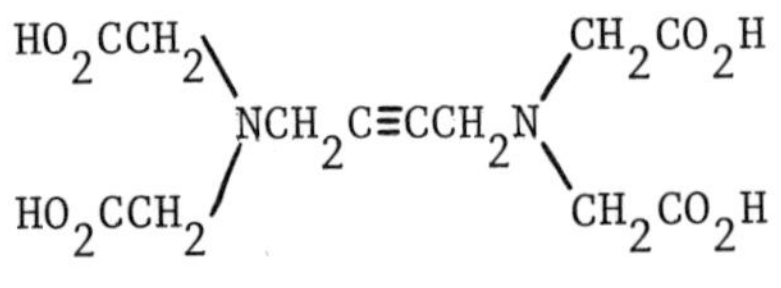

$$\text{HO}_2\text{CCH}_2 \diagdown \atop \text{HO}_2\text{CCH}_2 \diagup \text{NCH}_2\text{C} \equiv \text{CCH}_2\text{N} {\diagup \text{CH}_2\text{CO}_2\text{H} \atop \diagdown \text{CH}_2\text{CO}_2\text{H}}$$

| $\text{C}_{12}\text{H}_{16}\text{O}_8\text{N}_2$ | 1,4-Diaminobut-2-ynetetraacetic acid | | H_4L |

Metal ion	Equilibrium	Log K 20°, 0.1
H^+	HL/H.L	8.27
	$\text{H}_2\text{L/HL.H}$	6.80
	$\text{H}_3\text{L/H}_2\text{L.H}$	3.07
	$\text{H}_4\text{L/H}_3\text{L.H}$	1.89
Mg^{2+}	ML/M.L	3.31
	MHL/ML.H	7.77
	$\text{M}_2\text{L/ML.M}$	2.6
Ca^{2+}	ML/M.L	3.93
	MHL/ML.H	7.70
	$\text{M}_2\text{L/ML.M}$	3.1
Sr^{2+}	ML/M.L	2.81
	MHL/ML.H	7.92
	$\text{M}_2\text{L/ML.M}$	2.2
Ba^{2+}	ML/M.L	2.59
	MHL/ML.H	7.96
	$\text{M}_2\text{L/ML.M}$	1.9
Y^{3+}	ML/M.L	8.12
	MHL/ML.H	6.20
	$\text{M}_2\text{L/ML.M}$	5.5
Ce^{3+}	MHL/M.HL	5.80
Mn^{2+}	ML/M.L	5.65
	MHL/ML.H	7.11
	$\text{M}_2\text{L/ML.M}$	4.2
Cu^{2+}	ML/M.L	11.28
	MHL/ML.H	5.18
	$\text{M}_2\text{L/ML.M}$	7.5
Zn^{2+}	ML/M.L	9.30
	MHL/ML.H	5.45
	$\text{M}_2\text{L/ML.M}$	5.8

Bibliography: 79TS

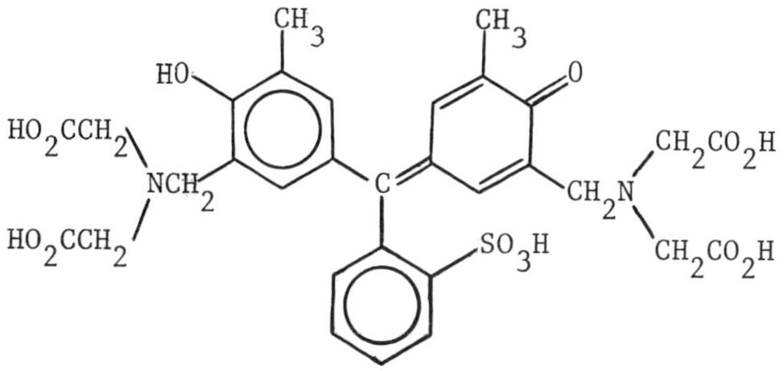

$C_{31}H_{32}O_{13}N_2S$ 5,5'-Bis[bis(carboxymethyl)aminomethyl]-3,3'-dimethyl- H_6L
4'-hydroxyfuchson-2"-sulfonic acid   (Xylenol Orange)
(Other values on Vol.1, p.259)

Metal ion	Equilibrium	Log K 25°, 0.1
H^+	HL/H.L	(12.23)
	$H_2L/HL.H$	10.39
	$H_3L/H_2L.H$	6.67
	$H_4L/H_3L.H$	2.85
	$H_5L/H_4L.H$	2.3
Mg^{2+}	ML/M.L	9.02
	MHL/ML.H	10.3
	$MH_2L/MHL.H$	6.56
	MOHL/ML.OH	2.43
	$M_2L/ML.M$	6.14
	$M_2HL/M_2L.H$	6.8
	$M_2OHL/M_2L.OH$	3.21
Ca^{2+}	ML/M.L	8.65
	MHL/ML.H	10.4
	$MH_2L/MHL.H$	6.71
	$M_2L/ML.M$	6.02
	$M_2HL/M_2L.H$	6.9
Sr^{2+}	ML/M.L	7.71
	MHL/ML.H	10.4
	$MH_2L/MHL.H$	7.36
	$M_2L/ML.M$	4.89
	$M_2HL/M_2L.H$	7.6
Ba^{2+}	ML/M.L	6.67
	MHL/ML.H	10.5
	$MH_2L/MHL.H$	7.54
	$M_2L/ML.M$	4.57
	$M_2HL/M_2L.H$	8.0
UO_2^{2+}	$MH_2L/M.H_2L$	7.18
	$M(H_2L)_2/M.(H_2L)^2$	11.36
Fe^{3+}	MHL/M.HL	18.8
	$M_2L/ML.M$	12.5
	MHL/ML.H	6.4
	ML/MOHL.H	8.7
	$(M_2L)^2/(MOH)_2L.H^2$	11.8

Xylenol Orange   (continued)

Bibliography:

Mg^{2+}-Ba^{2+} 74YOa

UO_2^{2+} 74Bb

Fe^{3+} 79YM

Other references: 66DM,69BK,71KN,72BB,72KNK,
76BS,76Y,77KBa,77KKh,77KM,78KKK,78KM,
78SYM,78TGK,79SSb

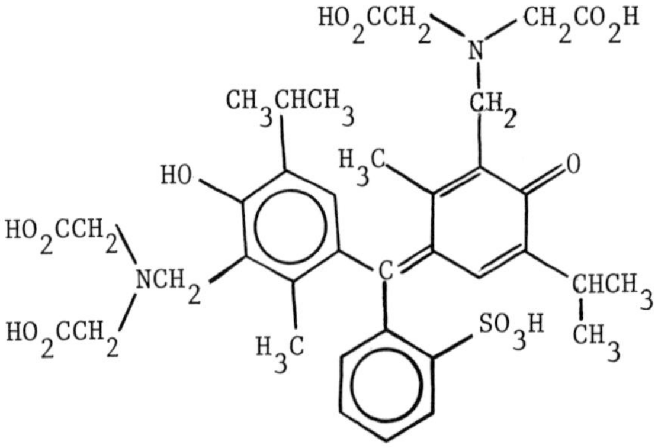

$C_{37}H_{44}O_{13}N_2S$ 5,5'-Bis[bis(carboxymethyl)aminomethyl]-3,3'-bis(methylethyl)- H_6L
6,6'-dimethyl-4'-hydroxyfuchson-2"-sulfonic acid
(Methylthymol Blue)

Metal ion	Equilibrium	Log K 25°, 0.1
H^+	HL/H.L	(12.94)
	$H_2L/HL.H$	11.14
	$H_3L/H_2L.H$	6.85
	$H_4L/H_3L.H$	3.04
	$H_5L/H_4L.H$	2.0
	$H_6L/H_5L.H$	1.8
Mg^{2+}	ML/M.L	8.87
	MHL/ML.H	10.8
	$MH_2L/MHL.H$	7.04
	MOHL/ML.OH	2.69
	$MOHL_2/ML_2.OH$	3.44
	$M_2L/ML.M$	5.80
	$M_2HL/M_2L.H$	7.3
Ca^{2+}	ML/M.L	8.25
	MHL/ML.H	10.9
	$MH_2L/MHL.H$	7.35
	$M_2L/ML.M$	5.38
	$M_2HL/M_2L.H$	7.7
Sr^{2+}	ML/M.L	7.05
	MHL/ML.H	11.1
	$MH_2L/MHL.H$	7.93
	$M_2L/ML.M$	4.58
	$M_2HL/M_2L.H$	8.3

Methylthymol Blue   (continued)

Metal ion	Equilibrium	Log K 25°, 0.1
Ba^{2+}	ML/M.L	6.93
	MHL/ML.H	11.2
	$MH_2L/MHL.H$	8.06
	$M_2L/ML.M$	(4.65)
	$M_2HL/M_2L.H$	8.5
Co^{2+}	ML/M.L	12.69
	MHL/ML.H	10.84
	$MH_2L/MHL.H$	6.33
	$MH_3L/MH_2L.H$	2.10
	$M_2L/ML.M$	11.0
	$M_2HL/M_2L.H$	3.8
Ni^{2+}	ML/M.L	12.66
	MHL/ML.H	10.56
	$MH_2L/MHL.H$	6.59
	$MH_3L/MH_2L.H$	1.78
	$M_2L/ML.M$	11.26
	$M_2HL/M_2L.H$	3.51
Cu^{2+}	ML/M.L	14.05
	MHL/ML.H	10.18
	$MH_2L/MHL.H$	6.08
	$M_2L/ML.M$	12.4
	$M_2HL/M_2L.H$	2.2
	$M_2(OH)_2L/M_2L.(OH)^2$	2.9
Fe^{3+}	ML/M.L	17.7
	$M_2L/ML.M$	12.1
	MHL/ML.H	6.7
	ML/MOHL.H	9.3
	$(M_2L)^2/(MOH)_2L.H^2$	**12.6**
Zn^{2+}	ML/M.L	13.31
	MHL/ML.H	10.56
	$MH_2L/MHL.H$	6.11
	$MH_2L/MH_2L.H$	2.30
	$M_2L/ML.M$	11.1
	$M_2HL/M_2L.H$	3.3

Bibliography:
$H^+, Co^{2+}-Zn^{2+}$ 74YI
$Mg^{2+}-Ba^{2+}$ 74YOa
Fe^{3+} 79YM

Other references: 68AN,68MF,69BK,69NN,69PKK,
        69SB,70KN,70SB,70SBa,72CP,73CPM,76ZL,
        77ZL,77ZLa,78SSS,79KGT

$$\begin{array}{ccc}
\text{HO}_2\text{CCH}_2\backslash & \overset{\text{OH}}{} & /\text{CH}_2\text{CO}_2\text{H} \\
& \text{NCH}_2\text{CHCH}_2\text{N} & \\
\text{HO}_2\text{CCH}_2/ & & \backslash\text{CH}_2\text{CO}_2\text{H}
\end{array}$$

$C_{11}H_{18}O_9N_2$ (2-Hydroxytrimethylene)dinitrilotetraacetic acid H_4L
            (2-hydroxy-1,3-propylenedinitrilotetraacetic acid)
                    (Other values on Vol.1, p.260)

Metal ion	Equilibrium	Log K 25°, 0.1	Log K 20°, 0.1
H^+	HL/H.L	9.49 -0.01	9.56 ±0.1
	$H_2L/HL.H$	6.96 ±0.07	7.00 ±0.1
	$H_3L/H_2L.H$	2.56 ±0.04	2.52 ±0.06
	$H_4L/H_3L.H$	1.6 ±0.0	1.7 ±0.1
Ag^+	ML/M.L	5.28	
	MHL/ML.H	7.60	
	$M_2L/ML.M$	3.0	
Hg^{2+}	ML/M.L	18.35	
	MHL/ML.H	3.85	
	$M_2L/ML.M$	8.0	

Bibliography: 75H   Other reference: 68EM

$$\begin{array}{ccc}
\text{HO}_2\text{CCH}_2\backslash & & /\text{CH}_2\text{CO}_2\text{H} \\
& \text{NCH}_2\text{CH}_2\text{OCH}_2\text{CH}_2\text{OCH}_2\text{CH}_2\text{N} & \\
\text{HO}_2\text{CCH}_2/ & & \backslash\text{CH}_2\text{CO}_2\text{H}
\end{array}$$

$C_{14}H_{24}O_{10}N_2$ Ethylenebis(oxyethylenenitrilo)tetraacetic acid   (EGTA) H_4L
                    (Other values on Vol.1, p.269)

Metal ion	Equilibrium	Log K 25°, 0.1	Log K 25°, 1.0	Log K 25°, 3.0	ΔH 20°, 0.1	ΔS 25°, 0.1
H^+	HL/H.L	9.40 ±0.02	9.15	9.36	-5.8	24
		8.89[b]				
	$H_2L/HL.H$	8.78 -0.01	8.60	8.61	-5.8	21
		8.40[b]				
	$H_3L/H_2L.H$	2.66 ±0.02	2.5	3.00		
		2.50[b]				
	$H_4L/H_3L.H$	2.0 ±0.0	2.10[b]	2.73		
VO^{2+}	ML/M.L	14.02[b]				
	MHL/ML.H	5.20[b]				
Zn^{2+}	ML/M.L	12.6 ±0.2		11.49	-4.3 ±0.7	43
	MHL/ML.H	4.96[h]±0.02		5.86		
	$M_2L/ML.M$	3.3[h]				
Cd^{2+}	ML/M.L	16.5 ±0.6		15.02	-14.8 ±0.7	26
	MHL/ML.H	3.47[h]±0.03		3.65		
Ag^+	ML/M.L	(7.05 ±0.01)				
		(6.88)[h]				

[b] 25°, 0.5;   [h] 20°, 0.1

EGTA (continued)

Bibliography:

H^+ 75N,76CWW Zn^{2+},Cd^{2+}76CWW

VO^{2+}75N   Other references: 67BR,69NK,73BW,73HR,73Ka,
                                                                       75KAL,75LB,79Ma
Ag^+ 79MST

$$HO_2CCH_2 \diagdown \qquad\qquad O \qquad\qquad O \qquad\qquad CH_2CO_2H$$
$$\overset{\displaystyle}{\underset{HO_2CCH_2 \diagup}{}} NCH_2\overset{\|}{C}NHCH_2CH_2NH\overset{\|}{C}CH_2N \overset{\displaystyle CH_2CO_2H}{\diagup}$$

$C_{14}H_{22}O_{10}N_4$	Diglycylethylenediaminetetraacetic acid (DGENTA)	H_4L
	(Other values on Vol.1, p.273)	

Metal ion	Equilibrium	Log K 25°, 0.1
H^+	HL/H.L	7.30 +0.01
	$H_2L/HL.H$	6.17 ±0.02
	$H_3L/H_2L.H$	2.68 ±0.02
	$H_4L/H_3L.H$	1.95 ±0.03
	$H_5L/H_4L.H$	1.2
Ga^{3+}	ML/M.L	13.26
	MHL/ML.H	3.75

Bibliography: 76HM

$$HO_2CCH_2 \diagdown \qquad\qquad\qquad CH_2CO_2H \quad CH_2CO_2H$$
$$\underset{HO_2CCH_2 \diagup}{} NCH_2CH_2\overset{\displaystyle |}{N}CH_2CH_2N \overset{\displaystyle CH_2CO_2H}{\diagup}$$

$C_{14}H_{23}O_{10}N_3$	Diethylenetrinitrilopentaacetic acid (DTPA)	H_5L
	(Other values on Vol.1, p.281)	

Metal ion	Equilibrium	Log K 25°, 0.1	Log K 25°, 0.5	Log K 25°, 1.0	ΔH 20°, 0.1	ΔS 25°, 0.1
H^+	HL/H.L	10.49 ±0.07 10.62^z	9.9	9.38 ±0.00 10.36^z	-8.0 -0.4	21
	$H_2L/HL.H$	8.60 ±0.05 8.64^z	8.32	8.20 ±0.04 8.35^z	-4.3 +2	25
	$H_3L/H_2L.H$	4.28 ±0.04	4.10 ±0.02	4.15 ±0.03	-1.7 +0.8	14
	$H_4L/H_3L.H$	2.64 ±0.10	2.6 ±0.2	2.6 ±0.1	-0.5^a	10
	$H_5L/H_4L.H$	2.0 ±0.2	2.1 ±0.1	2.2 ±0.2	0.3^a	9
	$H_6L/H_5L.H$	1.6 ±0.1		1.7		
	$H_7L/H_6L.H$	0.7		0.9		
Cu^{2+}	ML/M.L	21.4 ±0.3			-13.6 +0.2	52
	MHL/ML.H	4.80 ±0.01	4.77^h±0.03			
	$MH_2L/MHL.H$	2.96 ±0.08				
	$MH_3L/ML_2.H$	2.56				
	$M_2L/ML.M$	6.79 ±0.00	5.54^h			

^{a}27°, 0.1; ^{h}20°, 0.1; ^{z}$(CH_3)_4$NCl used as background electrolyte.

DTPA (continued)

Metal ion	Equilibrium	Log K 25°, 0.1	Log K 25°, 0.5	Log K 25°, 1.0	ΔH 20°, 0.1	ΔS 25°, 0.1
Ag^+	ML/M.L	8.06 ±0.06				
Pd^{2+}	ML/M.L			29.7[j]		
	MHL/ML.H			3.49[j]		
	$MH_2L/MHL.H$			2.93[j]		
	$MH_3L/MH_2L.H$			2.56[j]		
	$MH_4L/MH_3L.H$			1.93[j]		
VO^{2+}	$ML/M.L$		16.31			
	$ML_2/M.L^2$		23.3			
	MHL/ML.H		6.00			
Hg^{2+}	ML/M.L	26.4 ±0.6	26.1		-23.7 +0.1	41
	MHL/ML.H	4.24[h] -0.1				
Ga^{3+}	ML/M.L	24.3	24.5[h]			
	MHL/ML.H	4.16	4.35[h]			
	ML/MOHL.H	7.51	7.43[h]			

[h]20°, 0.1; [j]20°, 1.0

Bibliography:

H^+	75N,76AM,77GG,78MGD,79LM	Hg^{2+}	77GGC
$Y^{3+}-Lu^{3+}$	77CGG,77GGC	Pb^{2+}	71LW
Cu^{2+}	74B	Ga^{3+}	76HM,77KVa
VO^{2+}	75N		
Ag^+	79MST		
Pd^{2+}	76AM		
Cd^{2+}	71LW,75LW		

Other references: 66LP,69Ma,69NK,70PR,70VM, 71Mc,71Mb,71PR,71S,72KI,72LW,73BW, 73CC,73KB,73KBa,73NK,73TK,73YPa, 74LK,74MB,74NS,75LB,75TP,76GA,76NGT, 77BK,77HA

$C_{18}H_{30}O_{12}N_4$ Triethylenetetranitrilohexaacetic acid   (TTHA) H_6L
                              (Other values on Vol.1, p.286)

Metal ion	Equilibrium	Log K 25°, 0.1	Log K 20°, 0.1	Log K 30°, 0.1
H^+	HL/H.L	10.5 ±0.3	10.65	10.43
	$H_2L/HL.H$	9.5 ±0.1	9.54	9.35
	$H_3L/H_2L.H$	6.17 ±0.01	6.10	5.98
	$H_4L/H_3L.H$	4.08 ±0.02	4.03	4.00
	$H_5L/H_4L.H$	2.7 ±0.1	2.7	2.5
	$H_6L/H_5L.H$	2.2 -0.1	2.3	2.5
	$H_7L/H_6L.H$	1.8		
	$H_8L/H_7L.H$	1.5		

TTHA (continued)

Metal ion	Equilibrium	Log K 25°, 0.1	Log K 20°, 0.1	Log K 30°, 0.1
Ag^+	ML/M.L	8.7		
	MHL/ML.H	8.9		
	$MH_2L/MHL.H$	6.2		
	$M_2L/ML.M$	5.3		
	$M_2HL/M_2L.H$	6.5		
	$M_2H_2L/M_2HL.H$	5.1		
	$M_3L/M_2L.M$	3.0		
Ga^{3+}	MHL/ML.H	5.6		
	$MH_2L/MHL.H$	3.62		
	$MH_3L/MH_2L.H$	2.58		

Bibliography:

H^+ 76HM,79LM Pb^{2+} 71LWa

Cu^{2+} 73HK Ga^{3+} 76HM

Ag^+ 72RH

                                          Other references: 73HK,76GA,78MN,78MNa,
                                                              79NKK

$$\underset{\text{H}_2\text{NCH}_2\overset{\text{O}}{\overset{\|}{\text{C}}}\text{NHCH}_2\text{CO}_2\text{H}}{}$$

$C_4H_8O_3N_2$	Glycylglycine (diglycine) (Other values on Vol.1, p.294)					HL

Metal ion	Equilibrium	Log K 25°, 0.1	Log K 25°, 1.0	Log K 25°, 0	ΔH 25°, 0	ΔS 25°, 0.1
H^+	HL/H.L	8.08 ±0.03	8.10 ±0.02 7.74[n]±0.00	8.259 ±0.007 8.56[e]	-10.5 ±0.1 -10.6[a]±0.1 -11.7[e]	2
	$H_2L/HL.H$	3.13 ±0.05	3.16 ±0.05 3.11 -0.01	2.144 ±0.004 3.51[e]	-0.2 ±0.2 -0.3[a] -1.3[e]	14
Cu^{2+}	ML/M.L	5.50 ±0.07	5.51 ±0.09 5.34[n]	6.04	-6.5[a]±0.4	3
	$ML/M(H_{-1}L).H$	4.07 ±0.02	4.27 ±0.1 3.98[n]		-7.2[a]±0.4	-6
	$M(H_{-1}L)/MOH(H_{-1}L).H$	9.28 ±0.04	9.45 ±0.1 8.85[n]		-10.3[a]	8
	$MOH(H_{-1}L)/M(OH)_2(H_{-1}L).H$	12.8				
	$M(H_{-1}L)L/M(H_{-1}L).L$	3.14 ±0.07	3.05 ±0.1 3.11[n]		-7.0[a]±0.4	-9
	$M_2OH(H_{-1}L)_2/M(H_{-1}L).MOH(H_{-1}L)$	2.16 ±0.04	2.14 ±0.07 1.87[n]		-2.6[a]	1

[a]25°, 0.1; [e]25°, 3.0; [n]37°, 0.15

Glycylglycine   (continued)

Metal ion	Equilibrium	Log K 25°, 0.1	Log K 25°, 1.0	Log K 25°, 0	ΔH 25°, 0	ΔS 25°, 0.1
Zn^{2+}	ML/M.L	3.44 ±0.01	$(3.1)^q$ 3.24^n	3.80	-3.1^a	5
	$ML_2/M.L^2$	6.31	5.88^n	6.57	-8.0^a	2
	MHL/ML.H		5.6^q			
	ML/MOHL.H		8.24^n			
Pb^{2+}	ML/M.L		3.0^q	3.23 3.82^e	$(-3)^s$	$(7)^e$
	$ML_2/M.L^2$			5.93		
	MHL/ML.H		6.4^q	6.19^e	$(-12)^s$	$(-12)^e$

---

^{a}25°, 0.1; ^{e}25°, 3.0°, ^{n}37°, 0.15; ^{q}25°, 0.8; ^{s}10-40°, 3.0

Bibliography:

H^+ 69YH,72AP,75CM,75K,75KM,75S,76CWa,77GN, Cu^{2+} 72AP,75BP,75KM,75S,77GN
77HM Zn^{2+} 72AP
Co^{2+} 77HM Pb^{2+} 76CWa
Ni^{2+} 75KM
                                            Other references: 69MM,71MM,74NB,76PN,77RR,
                                                                79SBa,79SBd

---

$$H_2NCH_2\overset{\overset{\text{O}}{\|}}{C}NHCH_2CH_2CO_2H$$

$C_5H_{10}O_3N_2$ Glycyl-β-alanine   HL
                        (Other values on Vol.1, p.296)

Metal ion	Equilibrium	Log K 25°, 0.1
H^+	HL/H.L	8.09
	$H_2L/HL.H$	3.91
Ni^{2+}	ML/M.L	4.19 +0.05
	$ML_2/M.L^2$	7.53 +0.3
	$ML_3/M.L^3$	9.7
	$ML/M(H_{-1}L).H$	9.24
	$ML_2/M(H_{-1}L)_2.H^2$	18.9
Cu^{2+}	ML/M.L	5.70 −0.01
	$ML/M(H_{-1}L).H$	4.60 ±0.04
	$M(H_{-1}L)/MOH(H_{-1}L).H$	10.13
	$M(H_{-1}L)L/M(H_{-1}L).L$	2.87
	$M_2OH(H_{-1}L)_2/M(H_{-1}L).MOH(H_{-1}L)$	2.46

Bibliography:
Ni^{2+},Cu^{2+} 75BP   Other reference: 76PN

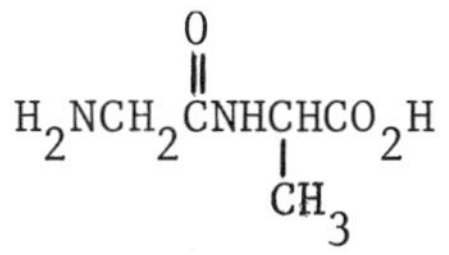

A. DIPEPTIDES

$$\begin{array}{c} O \\ \| \\ H_2NCH_2CNHCHCO_2H \\ | \\ CH_3 \end{array}$$

$C_5H_{10}O_3N_2$ Glycyl-L-alanine   HL
(Other values on Vol.1, p.297)

Metal ion	Equilibrium	Log K 25°, 0.1	Log K 25°, 0	ΔH 25°, 0.1	ΔS 25°, 0.1
H^+	HL/H.L	8.11 ±0.02	8.331	-10.5 ±0.5	2
	H_2L/HL.H	3.07 ±0.04	3.153	-0.2 ±0.2	13
Ni^{2+}	ML/M.L	4.23	-0.01		
	$ML_2/M.L^2$	7.60[y] ±0.1			
	$ML_3/M.L^3$	9.7			
	$ML/M(H_{-1}L).H$	8.79			
	$ML_2/M(H_{-1}L)_2.H^2$	19.84			
Cu^{2+}	ML/M.L	5.77 ±0.03		-6.6 +0.1	4
	$ML/M(H_{-1}L).H$	4.09 ±0.05		-6.7 ±0.4	-4
	$M(H_{-1}L)/MOH(H_{-1}L).H$	9.42 ±0.02		-10.5	8
	$M(H_{-1}L)L/M(H_{-1}L).L$	3.15 ±0.07		-6.5	-7
	$M_2OH(H_{-1}L)_2/M(H_{-1}L).MOH(H_{-1}L)$	2.25 ±0.04		-3.3	-8

---

[y]L- and DL-isomers had about the same value.

Bibliography:

H^+ 75BP,75K,75S,77GN Cu^{2+}75BP,75S,77GN

Ni^{2+}75BP   Other reference: 76PN

$$\begin{array}{c} O \\ \| \\ H_2NCH_2CNHCHCO_2H \\ | \\ CH_3CHCH_3 \end{array}$$

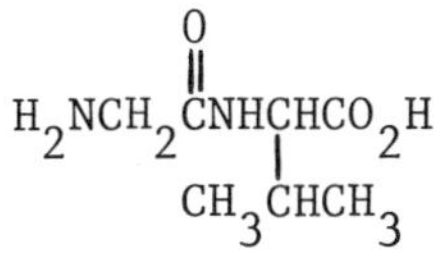

$C_7H_{14}O_3N_2$ Glycyl-L-valine   HL
(Other values on Vol.1, p.298)

Metal ion	Equilibrium	Log K 25°, 0.1
H^+	HL/H.L	8.09 ±0.03
	H_2L/HL.H	3.08 ±0.06
Ni^{2+}	ML/M.L	4.25 ±0.05
	$ML_2/M.L^2$	7.79 ±0.0
	$ML_3/M.L^3$	10.4
	$ML/M(H_{-1}L).H$	9.44

Glycyl-L-valine   (continued)

Metal ion	Equilibrium	Log K 25°, 0.1
Cu^{2+}	$ML/M.L$	5.74 ± 0.06
	$ML_2/M.L^2$	11.26
	$ML/M(H_{-1}L).H$	4.68 ± 0.05
	$M(H_{-1}L)/MOH(H_{-1}L).H$	9.24 ± 0.06
	$MOH(H_{-1}L)/M(OH)_2(H_{-1}L).H$	11.9
	$M(H_{-1}L)L/M(H_{-1}L).L$	3.36
		3.28^w
	$M_2OH(H_{-1}L)_2/M(H_{-1}L).MOH(H_{-1}L)$	2.70
		2.92^w

————————

^{w}D-isomer

Bibliography: 75BP,75BPb   Other references: 76PN,79SK

——

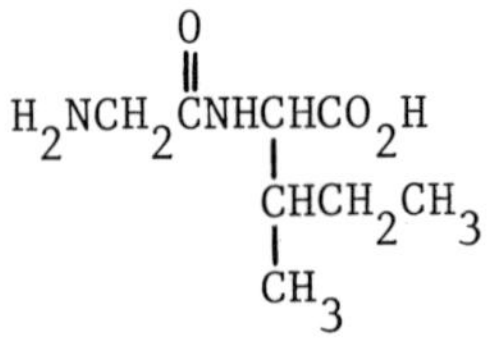

$$H_2NCH_2\overset{\overset{\text{O}}{\|}}{C}NHCHCO_2H$$
$$\underset{\underset{CH_3}{|}}{\underset{|}{CHCH_2CH_3}}$$

$C_8H_{16}O_3N_2$	Glycyl-L-isoleucine (Other reference on Vol.1, p.410)		HL

Metal ion	Equilibrium	Log K 25°, 0.1
H^+	$HL/H.L$	8.15
	$H_2L/HL.H$	3.00
Cu^{2+}	$ML/M.L$	5.83
	$ML/M(H_{-1}L).H$	4.71

————————

Bibliography: 75S

——

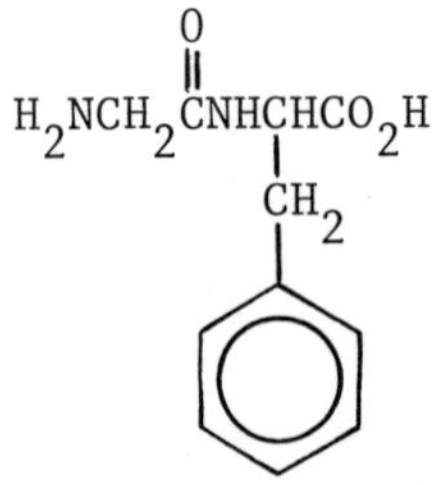

$$H_2NCH_2\overset{\overset{\text{O}}{\|}}{C}NHCHCO_2H$$
$$\underset{CH_2}{|}$$

(benzene ring)

$C_{11}H_{14}O_3N_2$	Glycyl-L-3-phenylalanine (Other values on Vol.1, p.300)		HL

Metal ion	Equilibrium	Log K 25°, 0.1
H^+	$HL/H.L$	8.11 ± 0.05
	$H_2L/HL.H$	3.09 ± 0.1

Glycyl-L-3-phenylalanine   (continued)

Metal ion	Equilibrium	Log K 25°, 0.1
Ni^{2+}	$ML/M.L$	4.03
	$ML_2/M.L^2$	7.49
	$ML_3/M.L^3$	9.8
	$ML/M(H_{-1}L).H$	8.59
	$ML_2/M(H_{-1}L)_2.H^2$	18.34
Cu^{2+}	$ML/M.L$	5.86 ±0.04
	$ML/M(H_{-1}L).H$	3.92 ±0.04
	$M(H_{-1}L)/MOH(H_{-1}L).H$	9.38 ±0.01
	$M(H_{-1}L)L/M(H_{-1}L).L$	3.27[y] ±0.02
	$M_2OH(H_{-1}L)_2/M(H_{-1}L).MOH(H_{-1}L)$	2.50[y] ±0.04

[y]D- and L-isomers had about the same value

Bibliography: 75BP,75BPb

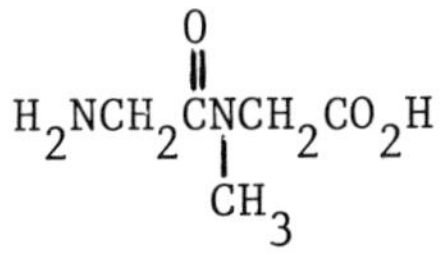

$$H_2NCH_2\overset{\overset{O}{\|}}{\underset{\underset{CH_3}{|}}{C}}NCH_2CO_2H$$

$C_5H_{10}O_3N_2$	Glycylsarcosine (Other values on Vol.1, p.301)				HL

Metal ion	Equilibrium	Log K 25°, 0.1	ΔH 25°, 0.1	ΔS 25°, 0.1
H^+	$HL/H.L$	8.46 ±0.02	-10.2	4
	$H_2L/HL.H$	2.76 ±0.09		
Cu^{2+}	$ML/M.L$	6.35 ±0.07	-6.0	9
	$ML_2/M.L^2$	11.4 ±0.1	-11.4	14

Bibliography: 75S

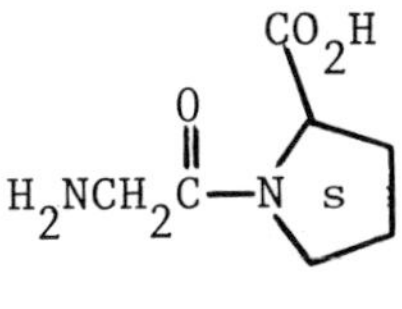

$$H_2NCH_2\overset{\overset{O}{\|}}{C}-N\underset{s}{\diagdown}\quad^{CO_2H}$$

$C_7H_{12}O_3N_2$	Glycyl-L-proline (Other references on Vol.1, p.302)				HL

Metal ion	Equilibrium	Log K 25°, 0.1	ΔH 25°, 0.1	ΔS 25°, 0.1
H^+	$HL/H.L$	8.41 ±0.05	-10.6	3
	$H_2L/HL.H$	2.77 ±0.09		

Glycyl-L-proline (continued)

Metal ion	Equilibrium	Log K 25°, 0.1	ΔH 25°, 0.1	ΔS 25°, 0.1
Ni^{2+}	$ML/M.L$	4.69 ± 0.07	-10.6	3
	$ML_2/M.L^2$	8.5 ± 0.1		
	$ML_3/M.L^3$	11.3 ± 0.3		
	$ML/MOHL.H$	10.6		
Cu^{2+}	$ML/M.L$	6.44 ± 0.06	-6.2	8
	$ML_2/M.L$	11.5 ± 0.1	-13.2	8

Bibliography:

H^+, Cu^{2+} 75BP,75S   Other references: 79SKa

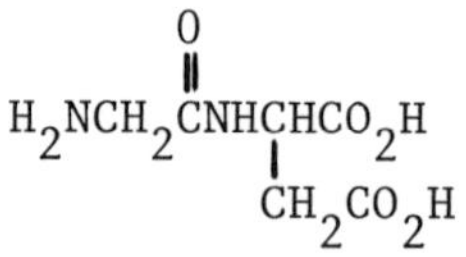

Ni^{2+} 75BP

$$\underset{\underset{CH_2CO_2H}{|}}{H_2NCH_2\overset{\overset{O}{\|}}{C}NHCHCO_2H}$$

$C_6H_{10}O_5N_2$ Glycyl-L-aspartic acid H_2L
                       (Other reference on Vol.1, p.410)

Metal ion	Equilibrium	Log K 25°, 0.1
H^+	$HL/H.L$	8.35
	$H_2L/HL.H$	4.29
	$H_3L/H_2L.H$	2.81
Co^{2+}	$ML/M.L$	2.57
	$ML/M(H_{-1}L).H$	9.26

Bibliography: 77HM   Other reference: 77BS

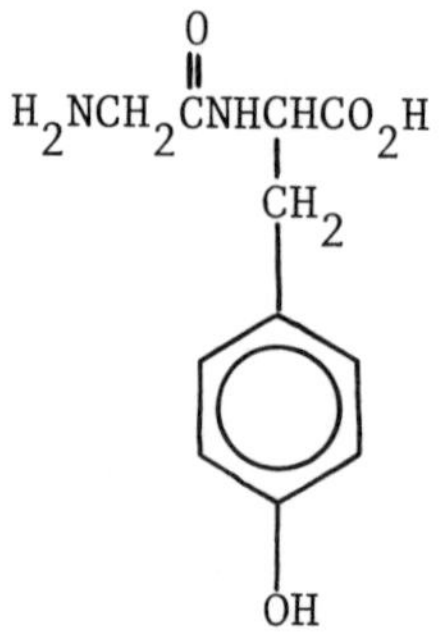

$$\underset{\underset{\underset{\underset{OH}{}}{C_6H_4}}{\underset{|}{CH_2}}}{H_2NCH_2\overset{\overset{O}{\|}}{C}NHCHCO_2H}$$

$C_{11}H_{14}O_4N_2$ Glycyl-L-tyrosine H_2L
                       (Other references on Vol.1, p.401)

Metal ion	Equilibrium	Log K 25°, 0.16
H^+	$HL/H.L$	9.92 ± 0.01
	$H_2L/HL.H$	8.12 ± 0.06
	$H_3L/H_2L.H$	2.92 ± 0.02

Glycyl-L-tyrosine (continued)

Metal ion	Equilibrium	Log K 25°, 0.16
Co^{2+}	MHL/M.HL	3.02
	$M(HL)_2/M.(HL)^2$	5.38
	$M(HL)_2/ML(HL).H$	9.04
	$ML(HL)/ML_2.H$	9.50
	$ML_2/MOHL_2.H$	9.98
	$MOHL_2/M(OH)_2L_2.H$	10.24
Zn^{2+}	MHL/M.HL	3.43
	$M(HL)_2/M.(HL)^2$	6.20
	$M(HL)_2/ML(HL).H$	8.52
	$ML(HL)/ML_2.H$	9.55
	$ML_2/MOHL_2.H$	9.66
	$MOHL_2/M(OH)_2L_2.H$	10.43

Bibliography: 79AK   Other reference: 77BS

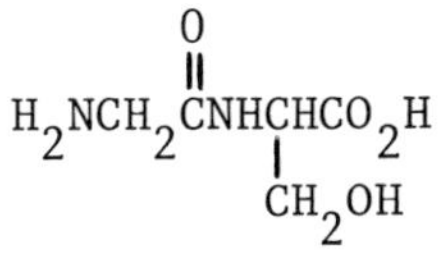

$$H_2NCH_2\overset{\overset{\displaystyle O}{\|}}{C}NHCHCO_2H$$
$$\underset{\displaystyle CH_2OH}{|}$$

$C_5H_{10}O_4N_2$ Glycyl-L-serine   HL
(Other references on Vol.1, p.401)

Metal ion	Equilibrium	Log K 25°, 0.1	Log K 25°, 0	ΔH 25°, 0	ΔS 25°, 0
H^+	HL/H.L	8.14 ±0.04			
	$H_2L/HL.H$	2.97 ±0.05	2.981	−0.2	13
Co^{2+}	ML/M.L	3.08			
	$ML/M(H_{-1}L).H$	8.77			
Cu^{2+}	ML/M.L	5.66			
	$ML/M(H_{-1}L).H$	3.8			

Bibliography:

H^+ 77HM,77SN

Co^{2+} 77HM

Cu^{2+} 77SN

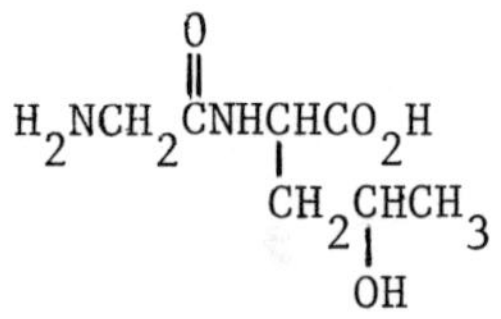

$$\begin{array}{c} O \\ \| \\ H_2NCH_2CNHCHCO_2H \\ | \\ CH_2CHCH_3 \\ | \\ OH \end{array}$$

| $C_7H_{14}O_4N_2$ | | Glycyl-L-threonine | HL |

Metal ion	Equilibrium	Log K 25°, 0.1
H^+	HL/H.L	8.14
	$H_2L/HL.H$	3.00
Cu^{2+}	ML/M.L	5.57
	$ML(M(H_{-1}L).H$	4.14

Bibliography: 77SN

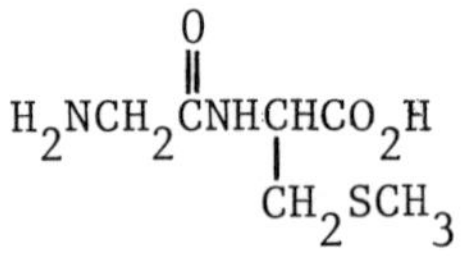

$$\begin{array}{c} O \\ \| \\ H_2NCH_2CNHCHCO_2H \\ | \\ CH_2SCH_3 \end{array}$$

| $C_6H_{12}O_3N_2S$ | | Glycyl-S-methyl-L-cysteine | HL |

Metal ion	Equilibrium	Log K 25°, 0.1
H^+	HL/H.L	8.12
	$H_2L/HL.H$	2.90
Cu^{2+}	ML/M.L	5.7
	$ML/M(H_{-1}L).H$	3.8

Bibliography: 77SN

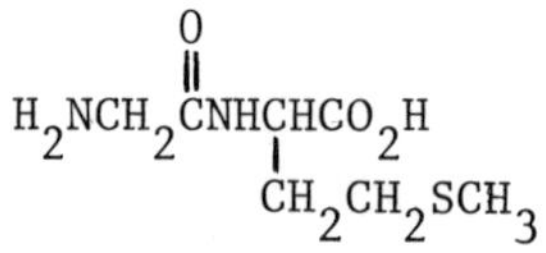

$$\begin{array}{c} O \\ \| \\ H_2NCH_2CNHCHCO_2H \\ | \\ CH_2CH_2SCH_3 \end{array}$$

| $C_7H_{14}O_3N_2S$ | | Glycyl-L-methionine (Other references on Vol.1, p.304) | HL |

Metal ion	Equilibrium	Log K 25°, 0.1
H^+	HL/H.L	8.15 ±0.04
	$H_2L/HL.H$	2.96 ±0.05
Co^{2+}	ML/M.L	3.13
	$ML_2/M.L^2$	$(5.83)^x$
Ni^{2+}	ML/M.L	4.15
	$ML_2/M.L^2$	$(7.67)^x$

^{x}DL-mixture

Glycyl-L-methionine   (continued)

Metal ion	Equilibrium	Log K 25°, 0.1	Log K 25°, 0.1
Cu^{2+}	$ML/M.L$		5.69 ± 0.08
	$ML_2/M.L^2$		10.7
	$ML/M(H_{-1}L).H$		$3.96 +0.01$
	$M(OH)(H_{-1}L).H/M.OH.L$		6.55
	$M(H_{-1}L)L/M(H_{-1}L).L$		3.34

Bibliography:

H^+	76PN,77SN,79AE
Co^{2+},Ni^{2+}	76PN
Cu^{2+}	77SN,79AE

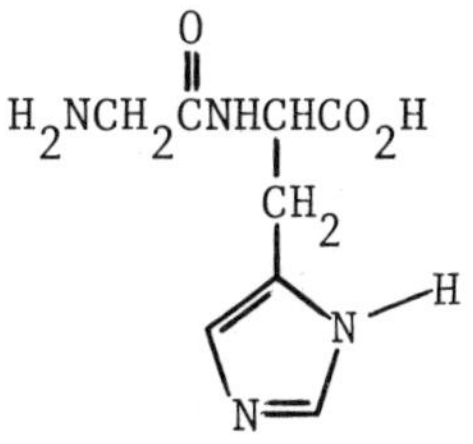

$C_8H_{12}O_3N_4$

Glycyl-L-histidine
(Other values on Vol.1, p.305)

HL

Metal ion	Equilibrium	Log K 25°. 0.1	Log K 37°, 0.15
H^+	$HL/H.L$	8.11 ± 0.05	7.85
	$H_2L/HL.H$	6.68 ± 0.05	6.49
	$H_3L/H_2L.H$	2.45 ± 0.10	2.54
Co^{2+}	$ML/M.L$	3.32	3.37
	$ML_2/M.L^2$		6.26
	$MHL/ML.H$		6.71
	$ML/M(H_{-1}L).H$	7.24	7.07
Ni^{2+}	$ML/M.L$	(3.9)	
	$ML_2/M.L^2$	8.82	
	$ML_3/M.L^3$	11.57	
	$ML/M(H_{-1}L).H$	(5.5)	
	$M(H_{-1}L)/MOH(H_{-1}L).H$	11.0	
	$MOH(H_{-1}L)/M(OH)_2(H_{-1}L).H$	11.2	

Glycyl-L-histidine   (continued)

Metal ion	Equilibrium	Log K 25°, 0.1	Log K 37°, 0.15
Cu^{2+}	$ML/M.L$	9.14	8.68
	$ML_2/M.L^2$	16.53	15.41
	$MHL/ML.H$		3.45
	$ML/M(H_{-1}L).H$	4.26	4.14
	$M(H_{-1}L)/MOH(H_{-1}L).H$	9.72	
			9.48
	$MHL_2/ML_2.H$		4.92
	$ML_2/M(H_{-1}L)L.H$		7.73
	$M(H_{-1}L)L/M(H_{-1}L)_2.H$	(3.4)	
	$M(H_{-1}L)L/M(H_{-1}L).L$	3.56	
Zn^{2+}	$ML/M.L$		3.65
	$ML_2/M.L^2$		6.89
	$M_2L_2/(ML)^2$		3.30
	$M_2L_2/M_2(H_{-1}L)L.H$		7.39
	$M_2(H_{-1}L)L/M_2(H_{-1}L)_2.H$		6.19

Bibliography:

H^+ 75AP,75BPa,77BS,77HM Zn^{2+} 75APa

Co^{2+} 75APa,77HM   Other reference: 74AYa

Ni^{2+} 75BPa

Cu^{2+} 75AP,75BPa

$$H_2NCHCNHCH_2CO_2H$$
with O (double bond) above the C, and CH_3 below the first CH

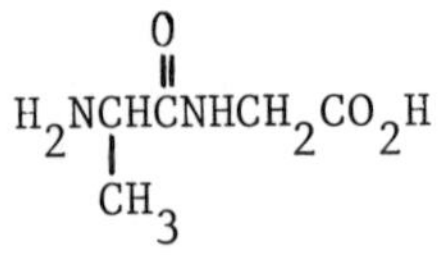

$C_5H_{10}O_3N_2$

L-Alanylglycine   HL
(Other values on Vol.1, p.306)

Metal ion	Equilibrium	Log K 25°, 0.1	ΔH 25°, 0.1	ΔS 25°, 0.1
H^+	$HL/H.L$	8.09 ±0.03	−10.6 ±0.4	1
	$H_2L/HL.H$	3.11 ±0.03	−0.7 ±0.3	12
Cu^{2+}	$ML/M.L$	5.29 ±0.05	−6.4 ±0.4	3
	$ML/M(H_{-1}L).H$	3.8 ±0.1	−6.6 ±0.1	−5
	$M(H_{-1}L)/MOH(H_{-1}L).H$	9.51	−10.5	8
	$M(H_{-1}L)L/M(H_{-1}L).L$	(2.60)[x]	(−5.7)[x]	(−7)[x]
	$M_2OH(H_{-1}L)_2/M(H_{-1}L).MOH(H_{-1}L)$	(2.15)[x]	(−2.6)[x]	(1)[x]

---

[x] DL-mixture

Bibliography: H^+ 74MS,75S,77GN; Cu^{2+} 75S,77GN

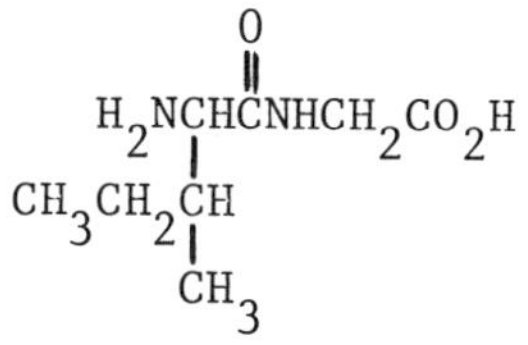

$$\overset{\displaystyle O}{\overset{\displaystyle \|}{H_2NCHCNHCH_2CO_2H}}$$
$$CH_3CH_2CH$$
$$|$$
$$CH_3$$

$C_8H_{16}O_3N_2$	L-Isoleucylglycine		HL

Metal ion	Equilibrium	Log K 25°, 0.1
H^+	HL/H.L	7.96
	$H_2L/HL.H$	3.04
Cu^{2+}	ML/M.L	4.75
	$ML/M(H_{-1}L).H$	3.26

Bibliography: 75S

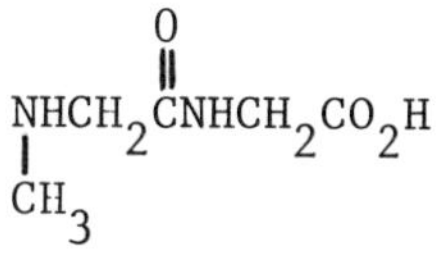

$$\overset{\displaystyle O}{\overset{\displaystyle \|}{NHCH_2CNHCH_2CO_2H}}$$
$$|$$
$$CH_3$$

$C_5H_{10}O_3N_2$	Sarcosylglycine (Other references on Vol.1, p.310)		HL

Metal ion	Equilibrium	Log K 25°, 0.1
H^+	HL/H.L	8.44 −0.01
	$H_2L/HL.H$	3.00 ±0.03
Cu^{2+}	ML/M.L	5.32
	$ML/M(H_{-1}L).H$	3.96

Bibliography: 75S

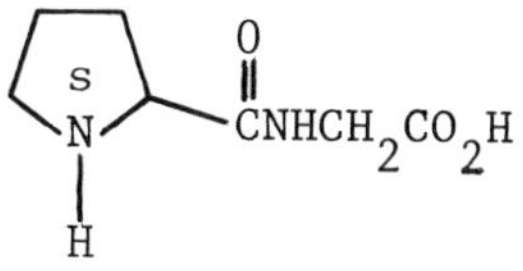

$$\overset{\displaystyle O}{\overset{\displaystyle \|}{CNHCH_2CO_2H}}$$
(pyrrolidine ring with N–H)

$C_7H_{12}O_3N_2$	L-Prolylglycine (Other references on Vol.1, p.311)		HL

Metal ion	Equilibrium	Log K 25°, 0.1
H^+	HL/H.L	8.86 ±0.01
	$H_2L/HL.H$	3.06 ±0.02
Cu^{2+}	ML/M.L	6.50 ±0.01
	$ML/M(H_{-1}L).H$	3.80 ±0.04
	$M(H_{-1}L)/MOH(H_{-1}L).H$	9.22
	$M(H_{-1}L)L/M(H_{-1}L).L$	2.66
	$M_2OH(H_{-1}L)_2/M(H_{-1}L).MOH(H_{-1}L)$	1.88

Bibliography: 75S

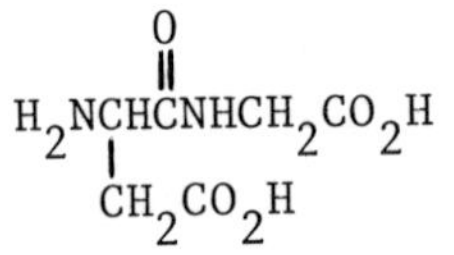

$$\underset{\underset{\displaystyle CH_2CO_2H}{|}}{H_2NCHCNHCH_2CO_2H}$$

C$_6$H$_{10}$O$_5$N$_2$ L-Aspartylglycine   H$_2$L

Metal ion	Equilibrium	Log K 25°, 0.1
H$^+$	HL/H.L	8.03
	H$_2$L/HL.H	3.67
	H$_3$L/H$_2$L.H	2.82
Co^{2+}	ML/M.L	4.10

Bilbiography: 77HM

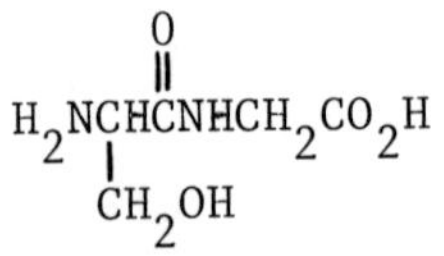

$$\underset{\underset{\displaystyle CH_2OH}{|}}{H_2NCHCNHCH_2CO_2H}$$

C$_5$H$_{10}$O$_4$N$_2$ L-Serylglycine   HL
                                    (Other reference on Vol.1, p.401)

Metal ion	Equilibrium	Log K 25°, 0.1
H$^+$	HL/H.L	7.33 ±0.00
	H$_2$L/HL.H	3.15 ±0.00
Cu^{2+}	ML/M.L	4.19
	ML/M(H$_{-1}$L).H	3.60

Bibliography: 77SN

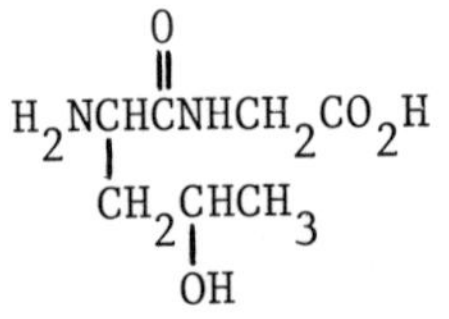

$$\underset{\underset{\displaystyle CH_2CHCH_3}{\underset{\displaystyle \underset{\displaystyle OH}{|}}{|}}}{H_2NCHCNHCH_2CO_2H}$$

C$_7$H$_{14}$O$_4$N$_2$ L-Threonylglycine   HL

Metal ion	Equilibrium	Log K 25°, 0.1
H$^+$	HL/H.L	7.34
	H$_2$L/HL.H	3.14
Cu^{2+}	ML/M.L	5.1
	ML/M(H$_{-1}$L).H	3.6

Bibliography: 77SN

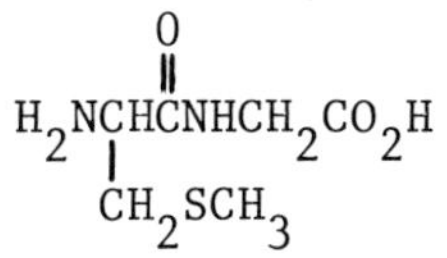

$$
\begin{array}{c}
\overset{\displaystyle O}{\overset{\displaystyle \|}{}} \\
H_2NCHCNHCH_2CO_2H \\
| \\
CH_2SCH_3
\end{array}
$$

$C_6H_{12}O_3N_2S$		S-Methyl-L-cysteinylglycine	HL

Metal ion	Equilibrium	Log K 25°, 0.1
H^+	HL/H.L	7.11
	$H_2L/HL.H$	3.11
Cu^{2+}	ML/M.L	5.00
	$ML/M(H_{-1}L).H$	3.44

Bibliography: 77SN

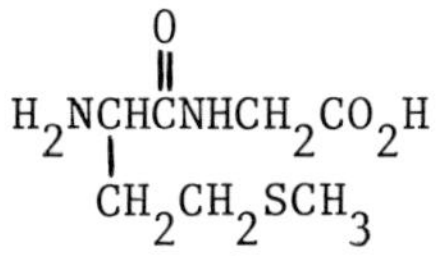

$$
\begin{array}{c}
\overset{\displaystyle O}{\overset{\displaystyle \|}{}} \\
H_2NCHCNHCH_2CO_2H \\
| \\
CH_2CH_2SCH_3
\end{array}
$$

$C_7H_{14}O_3N_2S$		L-Methionylglycine	HL

Metal ion	Equilibrium	Log K 25°, 0.1
H^+	HL/H.L	7.56
	$H_2L/HL.H$	3.13
Cu^{2+}	ML/M.L	4.67
	$ML/M(H_{-1}L).H$	3.2

Bibliography: 77SN

$C_8H_{12}O_3N_4$ L-Histidylglycine   HL
(Other reference on Vol.1, p.313)

Metal ion	Equilibrium	Log K 25°, 0.1	Log K 37°, 0.15
H^+	HL/H.L	7.60 −0.01	(7.15)
	$H_2L/HL.H$	5.81 ±0.02	(5.39)
	$H_3L/H_2L.H$	2.71	(2.32)
Co^{2+}	$ML/M.L$	5.19	(4.54)
	$ML_2/M.L^2$		(8.16)
	MHL/M.HL		(2.17)
	$ML/M(H_{-1}L).H$	7.15	
Ni^{2+}	$ML/M.L$	6.84	
	$ML_2/M.L^2$	12.39	
Cu^{2+}	$ML/M.L$	8.83	(8.02)
	$ML_2/M.L^2$		(14.15)
	MHL/M.HL	4.30	
	$ML/M(H_{-1}L).H$	8.07	(6.30)
	$M(H_{-1}L)/M(H_{-2}L).H$	9.49	
	$M(H_{-1}L)L.H/M.L^2$	5.88	(5.66)
	$M(HL)L/ML_2.H$		(4.41)
	$ML_2/M(H_{-1}L)L.H$		(8.49)
	$M(H_{-1}L)L/M(H_{-1}L)_2.H$		(9.76)
	$(ML)_2/M_2(H_{-1}L)_2.H^2$		(9.15)
Zn^{2+}	$ML/M.L$		(4.25)
	$ML_2/M.L^2$		(8.46)
	MHL/M.HL		(2.37)

Bibliography:

H^+ 75AP,75BPa,77HM Cu^{2+} 75AP,75BPa

Co^{2+} 75APa,77HM Zn^{2+} 75APa

Ni^{2+} 75BPa   Other references: 74AY,74YA

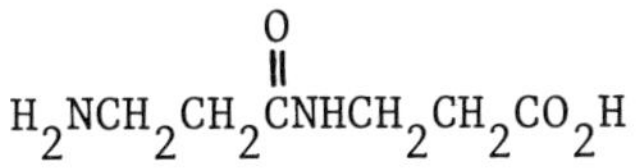

$$\overset{\displaystyle O}{\overset{\displaystyle \|}{H_2NCH_2CH_2CNHCH_2CH_2CO_2H}}$$

β-Alanyl-β-alanine   HL
(Other values on Vol.1, p.313)

$C_6H_{12}O_3N_2$

Metal ion	Equilibrium	Log K 25°, 0.1
H^+	HL/H.L	9.29 ±0.01
	$H_2L/HL.H$	3.91 ±0.04
Co^{2+}	ML/M.L	3.00
Ni^{2+}	ML/M.L	3.94
	$ML_2/M.L^2$	6.49

Bibliography: 76PN

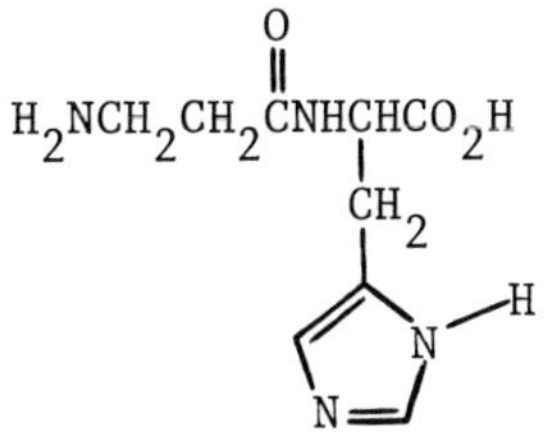

$$\overset{\displaystyle O}{\overset{\displaystyle \|}{H_2NCH_2CH_2CNHCHCO_2H}}$$
$$\underset{\displaystyle CH_2}{}$$

β-Alanyl-L-histidine (L-carnosine)   HL
(Other values on Vol.1, p.314)

$C_9H_{14}O_3N_4$

Metal ion	Equilibrium	Log K 25°, 0.1	Log K 37°, 0.15
H^+	HL/H.L	9.36 ±0.04	8.92
	$H_2L/HL.H$	6.76 ±0.03	6.46
	$H_3L/H_2L.H$	2.49 −0.01	2.52
Co^{2+}	ML/M.L	3.69	3.22
	MHL/ML.H		7.68
Ni^{2+}	ML/M.L	4.30	
	MHL/ML.H	7.99	
	$ML/M(H_{-1}L).H$	7.45	
	$M(H_{-1}L)L/M(H_{-1}L).L$	2.11	
Cu^{2+}	$ML/M.L$	8.52	8.14
	$ML_2/M.L^2$		14.39
	$MH_2L/MIL.H$	4.80	
	MHL/ML.H	5.46	4.76
	$ML/M(H_{-1}L).H$	5.60	6.24
	$M(H_{-1}L)/MOH(H_{-1}L).H$	11.20	
	$M(H_{-1}L)L/M(H_{-1}L).L$	2.45	
	$M(HL)_2/M.(HL)^2$	7.9	
	$MHL_2/ML_2.H$		3.74
	$ML_2/M(H_{-1}L)L.H$		8.69
	$M_2L_2/(ML)^2$		2.28
	$(ML)^2/M_2(H_{-1}L)_2.H^2$		8.32

L-Carnosine   (continued)

Metal ion	Equilibrium	Log K 25°, 0.1	Log K 37°, 0.15
Zn^{2+}	ML/M.L		3.86
	MHL/ML.H		7.24

Bibliography:

H^+ 75AP,75BPa,79VH Ni^{2+} 75BPa

Co^{2+},Zn^{2+} 75APa Cu^{2+} 75AP,75BPa

$C_{10}H_{16}O_3N_4$ β-Alanyl-L-1-methylhistidine   (L-anserine)   HL

Metal ion	Equilibrium	Log K 25°, 0.1
H^+	HL/H.L	9.33
	H_2L/HL.H	7.03
Ni^{2+}	MHL/ML.H	7.5
Cu^{2+}	MHL/ML.H	4.9

Bibliography: 60M

$C_6H_{12}O_3N_2$ L-Alanyl-L-alanine   HL
                        (Other references on Vol.1, p.316)

Metal ion	Equilibrium	Log K 25°, 0.1	ΔH 25°, 0.1	ΔS 25°, 0.1
H^+	HL/H.L	8.06 ±0.03	-10.6	1
	H_2L/HL.H	3.19 ±0.01	-1.0	11
Co^{2+}	ML/M.L	2.58 ±0.05		
	ML_2/M.L^2	4.42		
Ni^{2+}	ML/M.L	3.51		
	ML_2/M.L^2	6.56		
	MHL/ML.H	6.90		
	ML/M(H_{-1}L).H	8.67		
Cu^{2+}	ML/M.L	5.35 ±0.04		
	ML/M(H_{-1}L).H	3.56 ±0.05		
	M(H_{-1}L)/MOH(H_{-1}L).H	9.48		
	M(H_{-1}L)L/M(H_{-1}L).L	2.96		
	M_2OH(H_{-1}L)$_2$/M(H_{-1}L).MOH(H_{-1}L)	2.36		

Bibliography: H^+ 74NA,75KMa,75S,76PN; Co^{2+},Ni^{2+} 76PN; Cu^{2+} 75KMa,75S

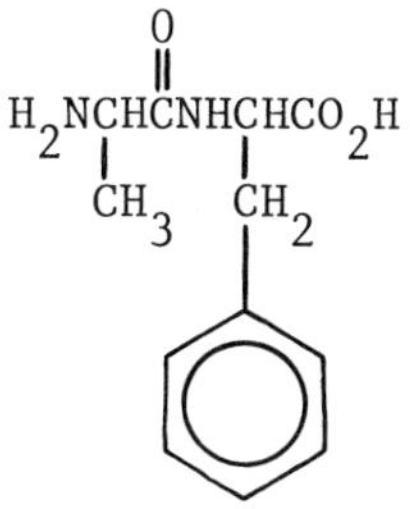

$$H_2NCHCNHCHCO_2H$$

with structure showing O (double bond), CH_3 and CH_2 groups, and phenyl ring

| $C_{12}H_{16}O_3N_2$ | | L-Alanyl-L-phenylalanine | HL |

Metal ion	Equilibrium	Log K 25°, 0.1
H^+	HL/H.L	7.87
	H_2L/HL.H	3.25
Cu^{2+}	ML/M.L	5.20
	ML/M(H_{-1}L).H	3.44
Zn^{2+}	ML/M.L$_2$	3.38
	ML$_2$/M.L^2	6.20

Bibliography: 74NA   Other reference: 74KH

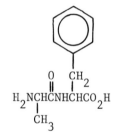

$$H_2NCHCNHCHCO_2H$$

with structure showing phenyl ring, CH_2, O (double bond), and CH_3 group

| $C_{12}H_{16}O_3N_2$ | | L-Alanyl-D-phenylalanine | HL |

Metal ion	Equilibrium	Log K 25°, 0.1
H^+	HL/H.L	8.08
	H_2L/HL.H	3.02
Cu^{2+}	ML/M.L	5.42
	ML/M(H_{-1}L).L	3.93
Zn^{2+}	ML/M.L$_2$	3.61
	ML$_2$/M.L^2	6.55

Bibliography: 74NA

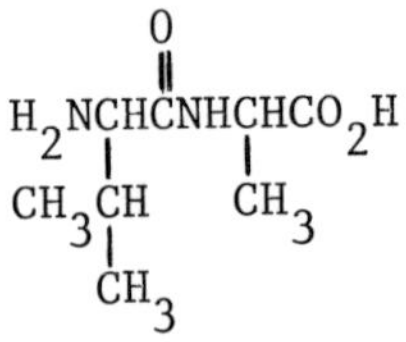

$$
\begin{array}{c}
O \\
\parallel \\
H_2NCHCNHCHCO_2H \\
| \qquad\qquad | \\
CH_3CH \quad\; CH_3 \\
| \\
CH_3
\end{array}
$$

L-Valyl-L-alanine
(Other references on Vol.1, p.401)

$C_8H_{16}O_3N_2$ HL

Metal ion	Equilibrium	Log K 25°, 0.1
H^+	$HL/H.L$	7.92 ± 0.02
	$H_2L/H.L^2$	3.32
Ni^{2+}	$ML/M.L$	3.47
	$ML_2/M.L^2$	6.28
	$ML/MOHL.H$	8.83
Cu^{2+}	$ML/M.L$	5.61
	$ML/M(H_{-1}L).H$	3.79
	$M(H_{-1}L)/MOH(H_{-1}L).H$	9.57
	$M(H_{-1}L)L/M(H_{-1}L).L$	2.6

Bibliography: 75BP

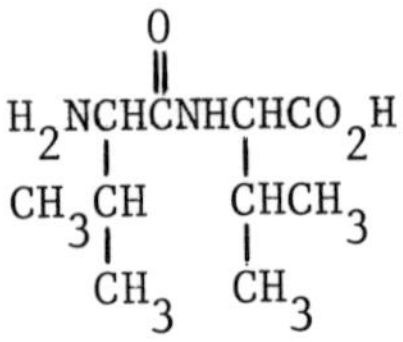

$$
\begin{array}{c}
O \\
\parallel \\
H_2NCHCNHCHCO_2H \\
| \qquad\qquad | \\
CH_3CH \quad\; CHCH_3 \\
| \qquad\qquad | \\
CH_3 \quad\;\; CH_3
\end{array}
$$

L-Valyl-L-valine   HL

$C_{10}H_{20}O_3N_2$

Metal ion	Equilibrium	Log K 25°, 0.1
H^+	$HL/H.L$	$7.83 +0.01$
	$H_2L/HL.H$	3.29 ± 0.02
Co^{2+}	$ML/M.L$	2.21
Ni^{2+}	$ML/M.L$	3.06 ± 0.07
	$ML_2/M.L^2$	5.8 ± 0.1
	$ML/M(H_{-1}L).H$	9.16
	$ML_2/M(H_{-1}L)_2.H$	18.3
Cu^{2+}	$ML/M.L$	5.20
	$ML/M(H_{-1}L).H$	3.85
	$M(H_{-1}L)/MOH(H_{-1}L).H$	9.35
	$M(H_{-1}L)L/M(H_{-1}L).L$	2.8

Bibliography: H^+,Ni^{2+} 75BP,76PN; Co^{2+} 76PN; Cu^{2+} 75BP

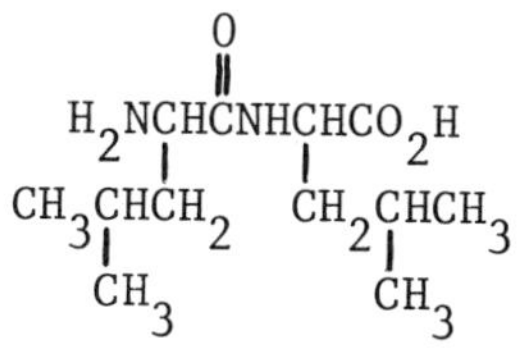

$$\begin{array}{c} O \\ \parallel \\ H_2NCHCNHCHCO_2H \\ | \qquad\qquad | \\ CH_3CHCH_2 \quad CH_2CHCH_3 \\ | \qquad\qquad\qquad | \\ CH_3 \qquad\qquad CH_3 \end{array}$$

$C_{12}H_{24}O_3N_2$ L-Leucyl-L-leucine   HL

Metal ion	Equilibrium	Log K 25°, 0.1
H^+	HL/H.L	7.80 ±0.01
	$H_2L/HL.H$	3.38 ±0.04
Cu^{2+}	ML/M.L	5.22 ±0.02
	$ML/M(H_{-1}L).H$	3.88 ±0.02
	$M(H_{-1}L)/MOH(H_{-1}L).H$	9.46 ±0.01
	$M_2OH(H_{-1}L)_2/M(H_{-1}L).MOH(H_{-1}L)$	2.43 ±0.01

The same values were obtained for D-leucyl-D-leucine.

Bibliography: 74NA,75BP,75BPb

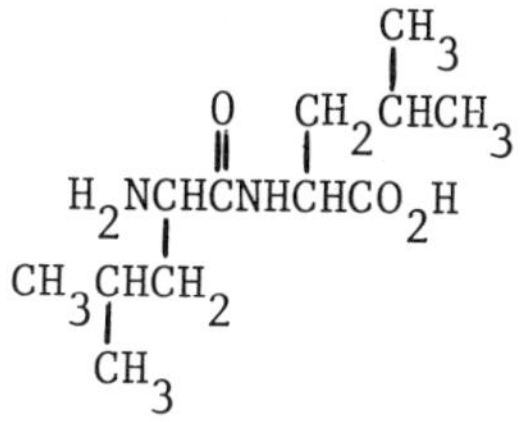

$$\begin{array}{c} \qquad\qquad CH_3 \\ \qquad\qquad | \\ O \quad CH_2CHCH_3 \\ \parallel \quad | \\ H_2NCHCNHCHCO_2H \\ | \\ CH_3CHCH_2 \\ | \\ CH_3 \end{array}$$

$C_{12}H_{24}O_3N_2$ D-Leucyl-L-leucine   HL

Metal ion	Equilibrium	Log K 25°, 0.1
H^+	HL/H.L	8.17 ±0.03
	$H_2L/HL.H$	3.04 ±0.02
Cu^{2+}	ML/M.L	5.47 ±0.02
	$ML/M(H_{-1}L).H$	4.89 ±0.01
	$M(H_{-1}L)/MOH(H_{-1}L).H$	9.46 ±0.01
	$M_2OH(H_{-1}L)_2/M(H_{-1}L).MOH(H_{-1}L)$	2.44 ±0.01

The same values were obtained for L-leucyl-D-Leucine.

Bibliography: 74NA,75BPb

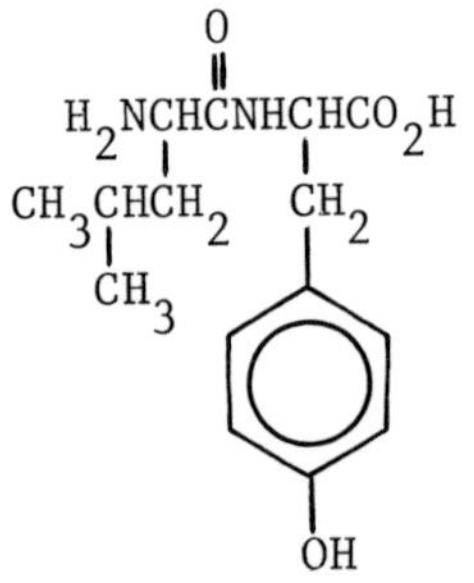

$$H_2NCHCNHCHCO_2H$$

L-Leucyl-L-tyrosine
(Other values on Vol.1, p.318)

$C_{15}H_{22}O_4N_2$ H_2L

Metal ion	Equilibrium	Log K 25°, 0.1
H^+	HL/H.L	10.03 ±0.06
	$H_2L/HL.H$	7.72 ±0.01
	$H_3L/H_2L.H$	3.15 ±0.05
Cu^{2+}	MHL/M.HL	5.17 ±0.02
	$MHL/MH(H_{-1}L).H$	3.32 ±0.06
	$MH(H_{-1}L)/M(H_{-1}L).H$	9.04
	$M(H_{-1}L)/MOH(H_{-1}L).H$	10.3
	$M_2OH(H_{-1}L)_2/M(H_{-1}L).MOH(H_{-1}L)$	2.4
Zn^{2+}	MHL/M.HL	3.36
	$M(HL)_2/M.(HL)^2$	6.32

Bibliography:

H^+, Zn^{2+} 74NA

Ni^{2+} 75KMa

Cu^{2+} 74NA,75KMa

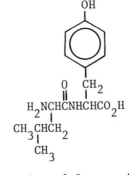

D-Leucyl-L-tyrosine
(Other values on Vol.1, p.319)

$C_{15}H_{22}O_4N_2$ H_2L

Metal ion	Equilibrium	Log K 25°, 0.1
H^+	HL/H.L	10.30 ±0.07
	$H_2L/HL.H$	8.26 ±0.07
	$H_3L/H_2L.H$	2.92 ±0.08

D-Leucyl-L-tyrosine   (continued)

Metal ion	Equilibrium	Log K 25°, 0.1
Cu^{2+}	MHL/M.HL	5.37 ±0.03
	MHL/MH(H_{-1}L).H	4.08 +0.01
	MH(H_{-1}L)/M(H_{-1}L).H	9.09
	M(H_{-1}L)/MOH(H_{-1}L).H	10.3
	M_2OH(H_{-1}L)$_2$/M(H_{-1}L).MOH(H_{-1}L)	2.4
Zn^{2+}	MHL/M.HL	3.89
	M(HL)$_2$/M.(HL)^2	7.09

Bibliography:

H^+, Zn^{2+} 74NA

Ni^{2+} 75KMa

Cu^{2+} 74NA,75KMa

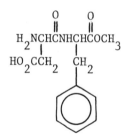

$$H_2NCHCNHCHCOCH_3$$

L-Aspartyl-L-phenylalanine methylester   HL

$C_{14}H_{18}O_5N_2$

Metal ion	Equilibrium	Log K 37°, 0.15
H^+	HL/H.L	7.37
	H_2L/HL.H	2.96
Zn^{2+}	ML/M.L	3.80
	ML$_2$/M.L^2	6.87
	MHL/ML.H	5.53
	MHL$_2$/ML$_2$.H	6.3
	ML/MOHL.H	8.6

Bibliography: 76MT

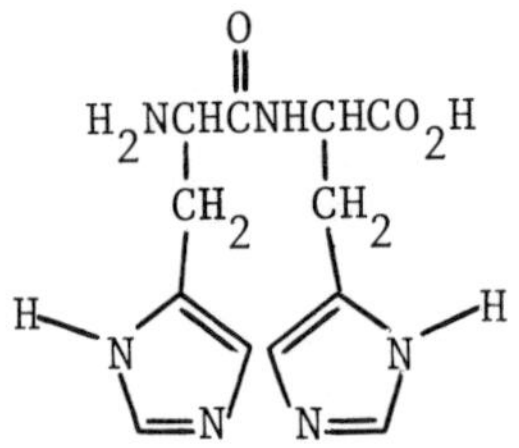

$$C_{12}H_{16}O_3N_6$$ L-Histidyl-L-histidine   HL

Metal ion	Equilibrium	Log K 25°, 0.1
H$^+$	HL/H.L	7.52
	H$_2$L/HL.H	6.70
	H$_3$L/H$_2$L.H	5.64
	H$_4$L/H$_3$L.H	2.58
Co^{2+}	ML/M.L	5.49
	MHL/ML.H	6.20
	ML/M(H$_{-1}$L).H	7.8

Bibliography: 77HM

(see volume 1, p.410 and this volume, p. 459 for references to DL-histidyl-DL-histidine.)

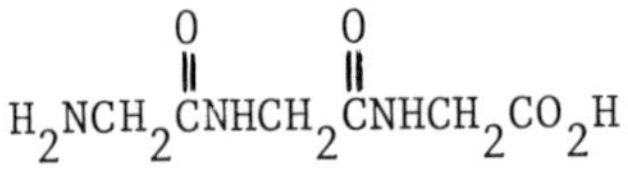

$$C_6H_{11}O_4N_3$$ Glycylglycylglycine (triglycine)   HL
                                    (Other values on Vol.1, p.322)

Metal ion	Equilibrium	Log K 25°, 0.1	Log K 25°, 1.0	Log K 25°, 0	ΔH 25°, 0.1	ΔS 25°, 0.1
H$^+$	HL/H.L	7.89 ±0.02	(8.01)±0.05	8.09	10.1	2
			7.58^n+0.01	8.57^e±0.03	-12.0^e	
	H$_2$L/HL.H	3.21 ±0.04	3.26 ±0.1	3.22	-0.2	14
			3.19^n-0.08	3.6^e ±0.2	-1.0^e	
Cu^{2+}	ML/M.L	5.08 ±0.05	5.1 ±0.3	5.41$_e$	-6.1 ±0.2	2
			4.88^n	5.66^e		
	ML$_2$/M.L^2	9.6	9.66	10.17^e		
	MHL/ML.H	4.3	4.35	4.47^e		
	ML/M(H$_{-1}$L).H	5.13 ±0.07	5.30 ±0.04	5.79^e	-7.5	-2
			5.07^n			
	M(H$_{-1}$L)/M(H$_{-2}$L).H	6.72 ±0.06	6.67 ±0.02	6.73^e	-7.4	6
			6.38^n			
	M(H$_{-2}$L)/MOH(H$_{-2}$L).H	11.7 ±0.3	11.5			
	M(H$_{-1}$L)L/M(H$_{-1}$L).L	3.50	3.38			
			2.93^n			

^{e}25°, 3.0; ^{n}37°, 0.15

Triglycine (continued)

Metal ion	Equilibrium	Log K 25°, 0.1	Log K 25°, 1.0	Log K 25°, 0	ΔH 25°, 0.1	ΔS 25°, 0.1
Cu^{2+} (cont.)	$M(HL)_2/ML_2 \cdot H^2$			8.8^e		
	$ML_2/M(H_{-1}L)L \cdot H$	6.7		6.26^e		
	$M(H_{-1}L)L/M(H_{-1}L)_2 \cdot H$			8.72^e		
	$(ML)_2/(ML)^2$			1.80^e		
	$M_2OH(H_{-1}L)_2/M(H_{-1}L) \cdot MOH(H_{-1}L)$		2.53			
	$M_2HL_2/(ML)_2 \cdot H$			4.2^e		
	$(MHL)_2/M_2HL_2 \cdot H$			3.7^e		
	$(ML)_2/M_2(H_{-1}L)_2 \cdot H^2$			11.7^e		
	$M_2(H_{-1}L)_2/M_2(H_{-2}L)_2 \cdot H^2$			14.8^e		
Pd^{2+}	$ML/M(H_{-1}L) \cdot H$		1.5			
	$M(H_{-1}L)/M(H_{-2}L) \cdot H$		2.2			
Zn^{2+}	$ML/M \cdot L$	3.18	$(3.1)^q$ 3.00^n	3.33		
	$ML_2/M \cdot L^2$		5.34^n	6.32		
	$MHL/ML \cdot H$		5.8^q			
	$ML/MOHL \cdot H$		8.35^n			

$^e 25°, 3.0;$ $^n 37°, 0.15;$ $^q 25°, 0.8$

Bibliography:

H^+ 72AP,75BP,75CM,75KM,76CWa Pd^{2+} 78CW

Ni^{2+} 75BP,75KM Zn^{2+} 72AP

Cu^{2+} 72AP,75BP,75KM   Other references: 78GG,79GG

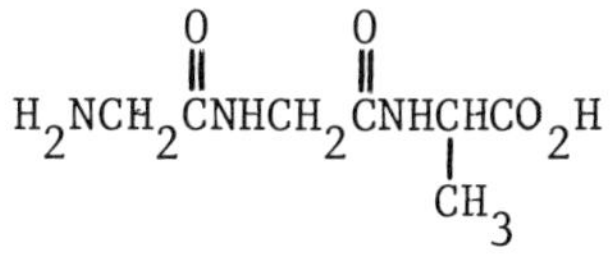

$$\text{H}_2\text{NCH}_2\overset{\overset{\displaystyle O}{\|}}{\text{C}}\text{NHCH}_2\overset{\overset{\displaystyle O}{\|}}{\text{C}}\text{NHCHCO}_2\text{H}$$
$$\underset{\text{CH}_3}{|}$$

$C_7H_{13}O_4N_3$ Glycylglycyl-L-alanine   HL
                    (Other values on Vol.1, p.325)

Metal ion	Equilibrium	Log K 25°, 1.0
Pd^{2+}	$ML/M(H_{-1}L) \cdot H$	2
	$M(H_{-1}L)/M(H_{-2}L) \cdot H$	3.2

Bibliography: 78CW

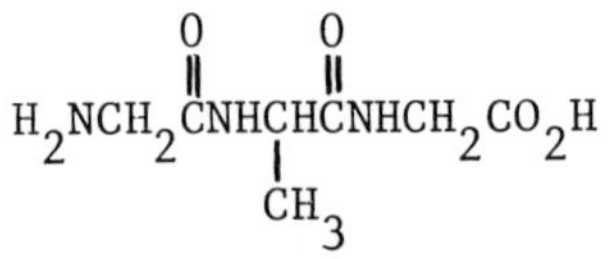

$$H_2NCH_2\overset{\overset{\displaystyle O}{\|}}{C}NHCHCNHCH_2CO_2H$$

$$\underset{CH_3}{|}$$

| $C_7H_{13}O_4N_3$ | | Glycyl-L-alanylglycine | HL |
| | | (Other values on Vol.1, p.326) | |

Metal ion	Equilibrium	Log K 25°, 1.0
Pd^{2+}	$ML/M(H_{-2}L).H^2$	4.0

Bibliography: 78CW

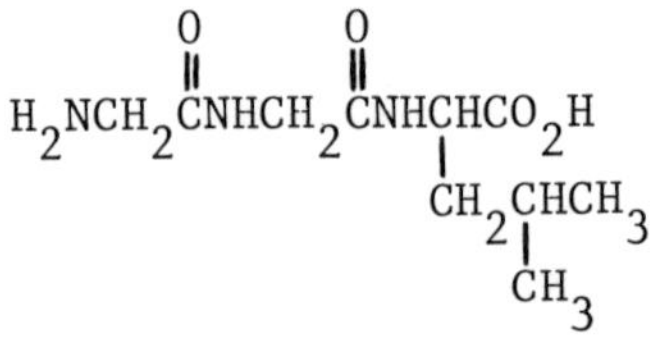

$$H_2NCH_2\overset{\overset{\displaystyle O}{\|}}{C}NHCH_2\overset{\overset{\displaystyle O}{\|}}{C}NHCHCO_2H$$

$$\underset{\underset{CH_3}{|}}{CH_2CHCH_3}$$

| $C_{10}H_{19}O_4N_3$ | | Glycylglycyl-L-leucine | HL |
| | | (Other reference on Vol.1, p.327) | |

Metal ion	Equilibrium	Log K 25°, 0.1
H^+	HL/H.L	7.90 ±0.00
Ni^{2+}	$ML/M.L$	3.7
	$ML_2/M.L^2$	7.0
	$ML/M(H_{-2}L).H^2$	16.03
Cu^{2+}	$ML/M.L$	5.1
	$ML/M(H_{-1}L).H$	5.03
	$M(H_{-1}L)/M(H_{-2}L).H$	7.20 +0.01

Bibliography: 78B

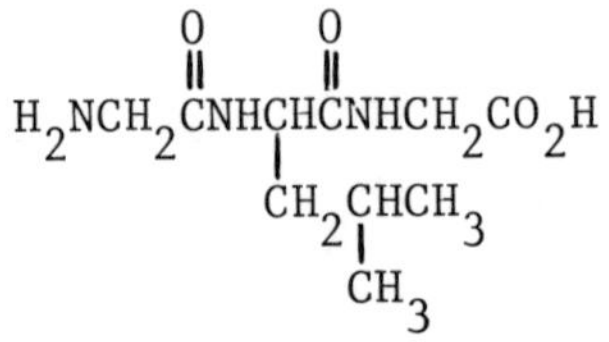

$$H_2NCH_2\overset{\overset{\displaystyle O}{\|}}{C}NHCHCNHCH_2CO_2H$$

$$\underset{\underset{CH_3}{|}}{CH_2CHCH_3}$$

| $C_{10}H_{19}O_4N_3$ | | Glycyl-L-leucylglycine | HL |
| | | (Other reference on Vol.1, p.327) | |

Metal ion	Equilibrium	Log K 25°, 0.1
H^+	HL/H.L	7.95 +0.1
Ni^{2+}	$ML/M.L$	3.7
	$ML/M(H_{-2}L).H^2$	16.31
Cu^{2+}	$ML/M.L$	5.37
	$ML/M(H_{-1}L).H$	5.87
	$M(H_{-1}L)/M(H_{-2}L).H$	6.43 ±0.00

Bibliography: 78B

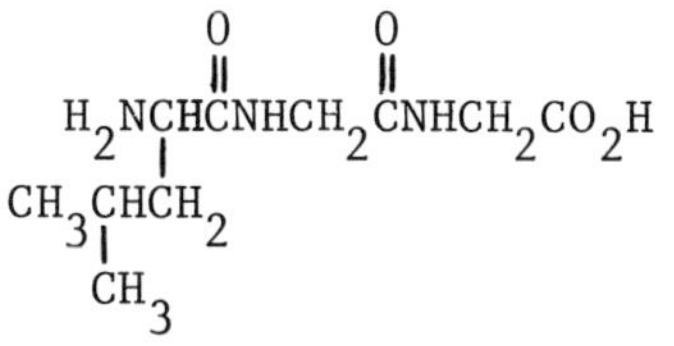

$$H_2NCHCNHCH_2CNHCH_2CO_2H$$

(structural formula)

$$\overset{O}{\overset{\|}{}} \quad \overset{O}{\overset{\|}{}}$$

CH$_3$CHCH$_2$

CH$_3$

$C_{10}H_{19}O_4N_3$	L-Leucylglycylglycine	HL

(Other references on Vol.1, p.403)

Metal ion	Equilibrium	Log K 25°, 0.1
H$^+$	HL/H.L	7.75 ±0.01
	H$_2$L/HL.H	3.20
Ni^{2+}	ML/M.L	3.0
	ML$_2$/M.L^2	5.4
	ML/M(H$_{-2}$L).H^2	15.48
	M(H$_{-2}$L)/MOH(H$_{-2}$L).H	(13.0)
Cu^{2+}	ML/M.L	4.1
	ML/M(H$_{-1}$L).H	4.17
	M(H$_{-1}$L)/M(H$_{-2}$L).H	6.72

Bibliography: 78B

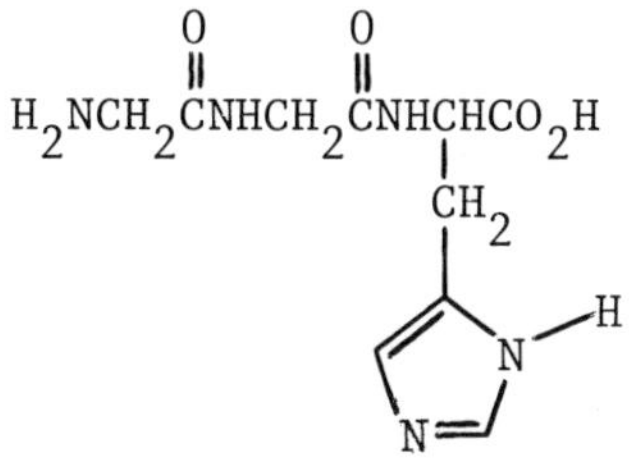

$$H_2NCH_2CNHCH_2CNHCHCO_2H$$

(structural formula)

$$\overset{O}{\overset{\|}{}} \quad \overset{O}{\overset{\|}{}}$$

CH$_2$

(imidazole ring) N—H

N

$C_{10}H_{15}O_4N_5$	Glycylglycyl-L-histidine	HL

(Other values on Vol.1, p.330)

Metal ion	Equilibrium	Log K 25°, 0.1	Log K 20°, 0.1	Log K 37°, 0.15
H$^+$	HL/H.L	8.02 ±0.08	8.22	7.39
	H$_2$L/HL.H	6.77 ±0.03	6.87	6.40
	H$_3$L/H$_2$L.H	2.72 ±0.00	2.84	2.31
Cu^{2+}	ML/M.L			7.04
	M(H$_{-2}$L).H^2/M.L	−1.98 −0.1		−2.11
	MHL/M.HL			4.19
	ML/M(H$_{-1}$L).H	4.63		
	ML/M(H$_{-2}$L).H^2	9.41		9.15
	M(H$_{-1}$L)L.H/M.L^2			8.70
	M(H$_{-1}$L)L/M(H$_{-1}$L)$_2$.H			7.40

Glycylglycyl-L-histidine   (continued)

Metal ion	Equilibrium	Log K 25°, 0.1	Log K 20°, 0.1	Log K 37°, 0.15
Zn^{2+}	ML/M.L			3.31
	MHL/M.HL			2.57
	$M_2L_2/(ML)^2$			3.15
	$M_2L_2/M_2(H_{-1}L)L.H$			6.46
	$M_2(H_{-1}L)L/M_2(H_{-1}L)_2.H$			7.80

Bibliography:

H^+ 74LKS,77AP,78S,79SN Zn^{2+} 77AP

Cu^{2+} 74LKS,77AP,79SN   Other references: 74AYa,74YA,76KLb,76WC

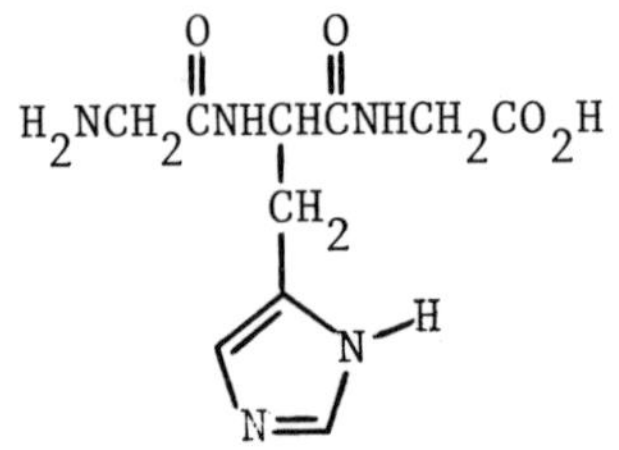

$C_{10}H_{15}O_4N_5$

Glycyl-L-histidylglycine
(Other references on Vol.1, p.331)

HL

Metal ion	Equilibrium	Log K 37°, 0.15	Log K 25°, 1.0	Log K 25°, 3.0
H^+	HL/H.L	7.56	7.96	8.62
	$H_2L/HL.H$	6.24	6.92	7.54
	$H_3L/H_2L.H$	2.91	3.26	3.57
Cu^{2+}	ML/M.L	8.52		
	$ML_2/M.L^2$	15.78		(19.95)
	$MH_2L/M.H_2L$			1.50
	$MH_2L/M(H_{-1}L).H^3$			11.40
	$ML/M(H_{-1}L).H$	3.20		
	$M(H_{-1}L)/M(H_{-2}L).H$	9.01		
	$MHL_2/MH_2L.M(H_{-1}L)$			4.50
	$ML_2/M(H_{-1}L)L.H$	7.37		
	$M_3L_4/M^3.L^4$			(53.7)

Bibliography: 75AP

Other references: 74AYa,74YA,75OS

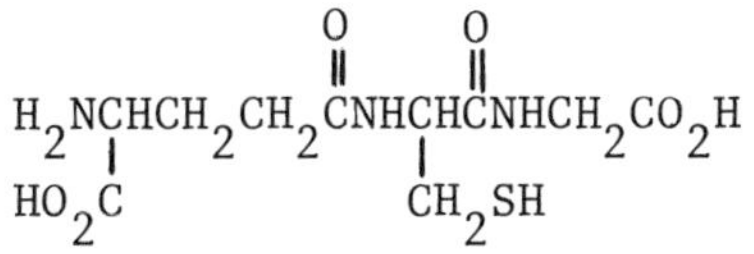

$$\underset{HO_2C}{H_2NCHCH_2}CH_2\overset{O}{\overset{\|}{C}}NH\underset{CH_2SH}{CH}C\overset{O}{\overset{\|}{C}}NHCH_2CO_2H$$

$C_{10}H_{17}O_6N_3S$ L-5-Glutamyl-L-cysteinylglycine (glutathione) H_3L

(Other references on Vol.1, p.403)

Metal ion	Equilibrium	Log K 25°, 0.15	Log K 37°, 0.15	Log K 25°, 3.0	ΔH 25°, 3.0	ΔS 25°, 3.0
H^+	HL/H.L	9.53 ±0.03	9.27 ±0.03	9.88	(-9)[s]	(15)
		9.43[b]				
	H_2L/HL.H	8.64 ±0.03	8.39 ±0.02	9.16	(-8)[s]	(15)
		8.62[b]				
	H_3L/H_2L.H	3.48 ±0.02	3.47 ±0.01	3.82	(-1)[s]	(14)
		3.48[b]				
	H_4L/H_3L.H	2.08 ±0.03	2.07 ±0.03	2.59	(-1)[s]	(9)
		(2.34)[b]				
Ca^{2+}	ML/M.L		3.8			
	MHL/ML.H		9.1			
	MH_2L/MHL.H		7.8			
	ML/MOHL.H		10.3			
La^{3+}	MHL/M.HL		3.55			
	MH_2L/MHL.H		7.2			
	MHL/MOHL.H^2		16.38			
Cu^+	MHL/M.HL	15.5[b]				
	$M(HL)_2$/M.$(HL)^2$	19.9[b]				
Zn^{2+}	ML/M.L		7.98	8.57		
	ML_2/M.L^2		12.5	13.59		
	MHL/ML.H		6.1	6.19		
	ML/MOHL.H		8.7	8.64		
	$M(HL)_2$/ML(HL).H			7.35		
	ML(HL)/ML_2.H		8.86	9.68		
	ML_2/$MOHL_2$.H		9.38	10.0		
Cd^{2+}	ML/M.L			10.18		
	ML_2/M.L^2			15.35		
	MHL/ML.H			6.84		
	ML/MOHL.H			9.9		
	$M(HL)_2$/ML(HL).H			7.94		
	ML(HL)/ML_2.H			9.74		
	ML_2/$MOHL_2$.H			12		
Pb^{2+}	ML/M.L			10.6	(-16)[s]	(-5)
	ML_2/M.L^2			15.0		
	MHL/ML.H			6.91	(-3)[s]	(22)
	$M(HL)_2$/ML(HL).H			8.91	(-10)[s]	(7)
	MHL_2/MHL.L			6.58	(-4)[s]	(17)
	ML_2/$MOHL_2$.H			10		

[b]25°, 0.5; [s]10-40°, 3.0

Glutathione (continued)

Bibliography:

H^+ 76AE,76CWa,76CWW,76TW,79OL,79PBG

Ca^{2+},La^{3+} 76TW

Cu^+ 79OL

Zn^{2+} 76CWW,76TW

Cd^{2+} 76CWW

Pb^{2+} 76CWa,76CWW

Other references: 65KD,76HS,76KL

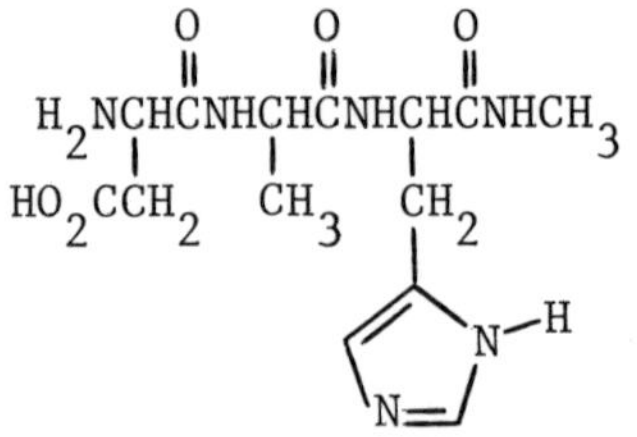

$$\underset{\substack{HO_2CCH_2}}{H_2NCHCNHCH}\underset{\substack{CH_3}}{CNHCH}\underset{\substack{CH_2}}{CNHCH_3}$$

$C_{14}H_{22}O_5N_6$ L-Aspartyl-L-alanyl-L-histidine methylamide   HL

Metal ion	Equilibrium	Log K 25°, 0.15	ΔH 25°, 0.15	ΔS 25°, 0.15
H^+	HL/H.L	7.73 ±0.00	-8.8	6
	$H_2L/HL.H$	6.47 ±0.09	-6.6	8
	$H_3L/H_2L.H$	2.95 ±0.03	-0.4	12
Co^{2+}	$M(HL)_2/M.(HL)^2$	5.41		
	$MH_{-1}L.H/M.L$	-5.27		
	$M(H_{-1}L)L/M(H_{-1}L).L$	3.75		
	$M(H_{-1}L)L/M(H_{-1}L)_2.H$	9.03		
Cu^{2+}	ML/M.L	8.39	-8.9	9
	$ML/M(H_{-2}L).H^2$	7.64	-7.7	9
Zn^{2+}	ML/M.L	4.75		
	$ML_2/M.L^2$	8.15		
	MHL/ML.H	5.20		
	ML/MOHL.H	9.09		
	$ML_2/M(OH)_2L_2.H^2$	16.93		

Bibliography:

H^+ 79AR,79LS

Co^{2+},Zn^{2+} 79LS

Cu^{2+} 79AR

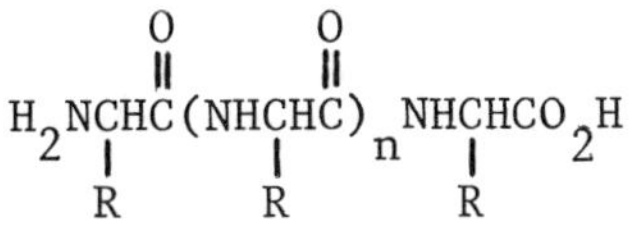

$$H_2NCHC(NHCHC)_nNHCHCO_2H$$

with structure showing:
$$\overset{O}{\underset{R}{\|}} \quad \overset{O}{\underset{R}{\|}} \quad \underset{R}{}$$

$C_xH_yO_zN_w$		Polypeptides		HL
R,n	Metal ion	Equilibrium	Log K 25°, 1.0	
R=H,n=2 Tetraglycine ($C_8H_{14}O_5N_4$) (Other values on Vol.1, p.332)	Cu(III)	$M(H_{-3}L)/M(H_{-4}L).H$	12.1	
R=CH$_3$,n=2 Tetraalanine ($C_{12}H_{22}O_5N_4$) (H^+ values on Vol.1, p.404)	Cu(III)	$M(H_{-3}L)/M(H_{-4}L).H$	12.0	
R=H,n=3 Pentaglycine ($C_{10}H_{17}O_6N_5$) (Other values on Vol.1, p.335)	Cu(III)	$M(H_{-3}L)/M(H_{-4}L).H$	11.6	
R=H,n=4 Hexaglycine ($C_{12}H_{20}O_7N_6$)	Cu(III)	$M(H_{-3}L)/M(H_{-4}L).H$	11.4	

Bibliography: 79NKC

Other reference: 75KM

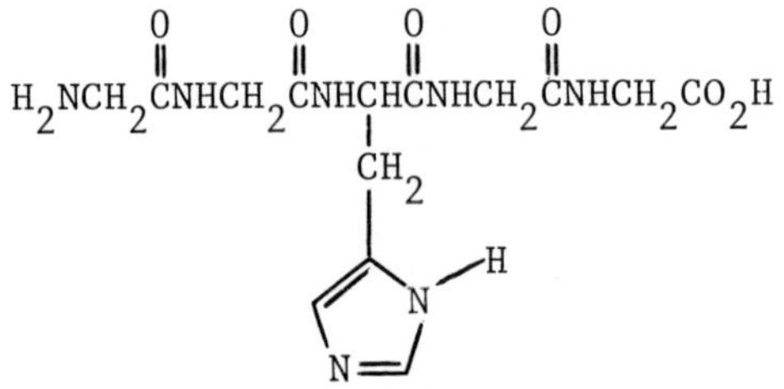

$$H_2NCH_2\overset{O}{\overset{\|}{C}}NHCH_2\overset{O}{\overset{\|}{C}}NHCHCNHCH_2\overset{O}{\overset{\|}{C}}NHCH_2CO_2H$$

Glycylglycyl-L-histidylglycylglycine   HL

$C_{14}H_{21}O_6N_7$

Metal ion	Equilibrium	Log K 37°, 0.15
H^+	HL/H.L	7.42
	H_2L/HL.H	6.03
	H_3L/H_2L.H	3.14
Cu^{2+}	ML/M.L	6.83
	MHL/ML.H	4.25
	ML/M(H_{-2}L).H^2	7.72
	M(H_{-1}L)L.H/M.L^2	9.38
	M(H_{-1}L)L/M(H_{-1}L)$_2$.H	7.22
Zn^{2+}	ML/M.L	2.91
	MHL/ML.H	6.43
	ML/MOHL.H	7.12
	M_2L_2/(ML)^2	3.24
	M_2L_2/M_2OHL_2.H	6.82

Bibliography: 77AP

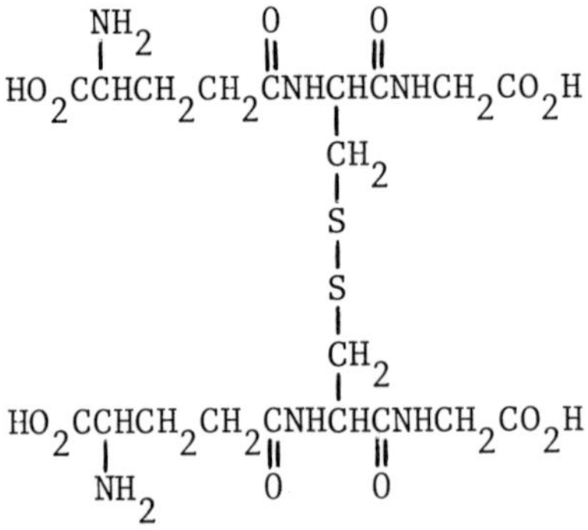

$$HO_2CCHCH_2CH_2\overset{O}{\overset{\|}{C}}NHCHCNHCH_2CO_2H$$

N,N-[N'',N''''-Di(L-5-glutamyl)-L-cystinyl]di(glycine) H_4L
(oxidized glutathione)
(Other reference on Vol.1, p.404)

$C_{20}H_{32}O_{12}N_6S_2$

Metal ion	Equilibrium	Log K 37°, 0.15
H^+	HL/H.L	9.17
	H_2L/HL.H	8.61
	H_3L/H_2L.H	3.92
	H_4L/H_3L.H	3.14
	H_5L/H_4L.H	2.60
	H_6L/H_5L.H	1.5

Oxidized glutathione (continued)

Metal ion	Equilibrium	Log K 37°, 0.15
Cu^{2+}	ML/M.L	13.72
	$ML_2/M.L^2$	17.48
	MHL/ML.H	3.97
	$MH_2L/MHL.H$	3.80
	$M_2L/ML.M$	2.3

Bibliography: 78MMW

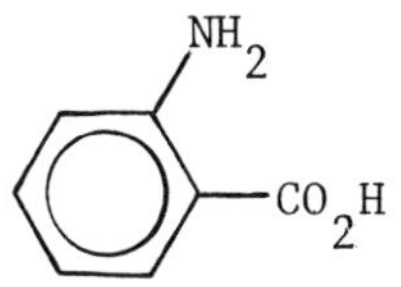

$C_7H_7O_2N$	2-Aminobenzoic acid (anthranilic acid) (Other values on Vol.1, p.338)				HL

Metal ion	Equilibrium	Log K 25°, 0.1	Log K 25°, 0.5	Log K 25°, 0	ΔH 25°, 0	ΔS 25°, 0.1
H^+	HL/H.L	4.79 ±0.03	4.78 ±0.02	4.96 ±0.01	-2.8	13
	$H_2L/HL.H$	2.00 ±0.03	(2.22)	2.08 ±0.03	-3.8	-4
Be^{2+}	ML/M.L		1.96			
	$M_2OHL_2.H/M^2.L^2$		1.51			
	$M_3(OH)_3L.H^3/M^3.L$		-7.33			
	$M_3(OH)_3L_2.H^3/M^3.L^2$		-5.62			
La^{3+}	ML/M.L	3.00[u]				
Pr^{3+}	ML/M.L	3.08[u]				
Sm^{3+}	ML/M.L	3.38[u]				
Eu^{3+}	ML/M.L	3.51[u]				
Gd^{3+}	ML/M.L	3.11[u]				
Tb^{3+}	ML/M.L	3.04[u]				
Dy^{3+}	ML/M.L	3.08[u]				
Er^{3+}	ML/M.L	3.08[u]				
Yb^{3+}	ML/M.L	3.20[u]				
Lu^{3+}	ML/M.L	3.36[u]				

[u] 12.5°, 0.2

Bibliography:

H^+ 73SK,75DBT

Be^{2+} 75DBT

La^{3+}-Lu^{3+} 69SF

Other references: 73SK,75STa,77ANY

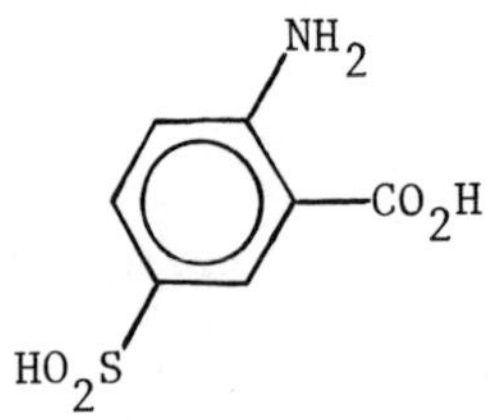

$C_7H_7O_5NS$ 2-Amino-5-sulfobenzoic acid H_2L
                          (Other value on Vol.1, p.339)

Metal ion	Equilibrium	Log K 25°, 0.1
H^+	HL/H.L	4.29 ±0.01
	H_2L/HL.H	1.56
Fe^{3+}	ML/M.L	3.65

Bibliography: 69BL

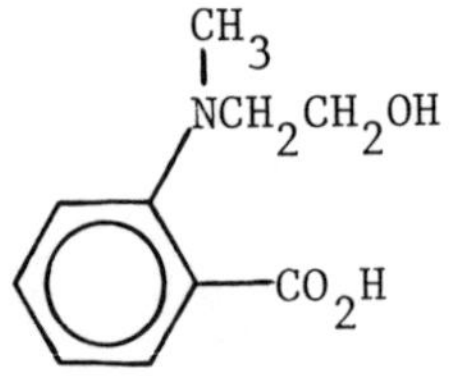

$C_{10}H_{13}O_3N$ N-2-Hydroxyethyl-N-methyl-2-aminobenzoic acid   HL

Metal ion	Equilibrium	Log K 22°, 0.1
H^+	HL/H.L	8.24
Co^{2+}	ML/M.L	3.05
Ni^{2+}	ML/M.L	3.80
	ML_2/M.L^2	6.90
Cu^{2+}	ML/M.L	5.90
	ML_2/M.L^2	10.15
	MHL/ML.H	(6.65)
Zn^{2+}	ML/M.L	2.90

Bibliography: 65UG

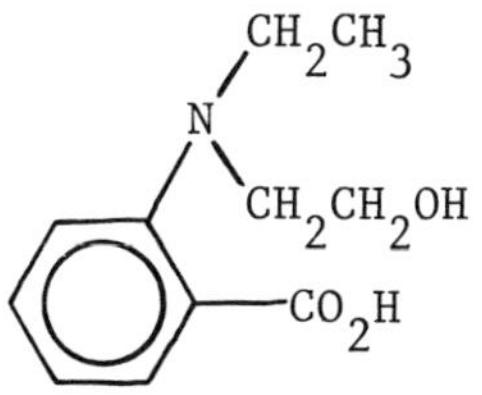

$C_{11}H_{15}O_3N$ 2-(N-2-Hydroxyethyl-N-ethylamino)benzoic acid   HL

Metal ion	Equilibrium	Log K 22°, 0.1
H^+	HL/H.L	8.99
Ni^{2+}	ML/M.L	2.45
	ML_2/M.L^2	6.25
Cu^{2+}	ML/M.L	5.85
	MHL/ML.H	(5.85)

Bibliography: 65UG

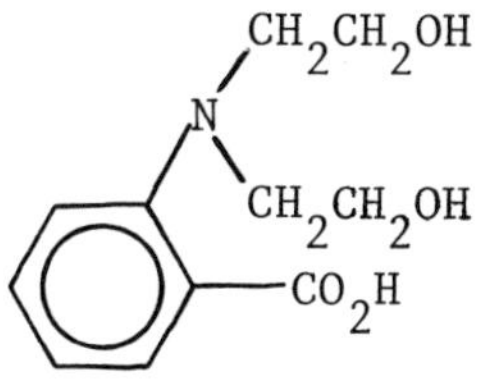

$C_{11}H_{15}O_4N$ 2-[Bis(2-hydroxyethyl)amino]benzoic acid   HL

Metal ion	Equilibrium	Log K 22°, 0.1
H^+	HL/H.L	7.49
Co^{2+}	ML/M.L	2.40
Ni^{2+}	ML/M.L	3.55
	ML_2/M.L^2	5.8
Cu^{2+}	ML/M.L	5.35
	ML_2/M.L^2	8.65
	MHL/ML.H	(6.30)
Zn^{2+}	ML/M.L	2.50

Bibliography: 65UG

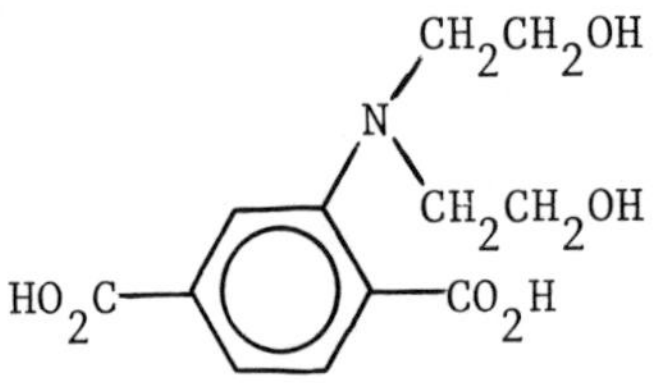

$C_{12}H_{14}O_6N$ 2-[Bis(2-hydroxyethyl)amino]benzene-1,4-dicarboxylic acid H_2L

Metal ion	Equilibrium	Log K 25°, 0.1
H^+	HL/H.L	7.00
	H_2L/HL.H	3.10
Co^{2+}	ML/M.L	2.35
Ni^{2+}	ML/M.L	3.51
Cu^{2+}	ML/M.L	5.30
	MHL/ML.H	3.38
	ML/MOHL.H	6.25
Zn^{2+}	ML/M.L	2.40

Bibliography: 73WU

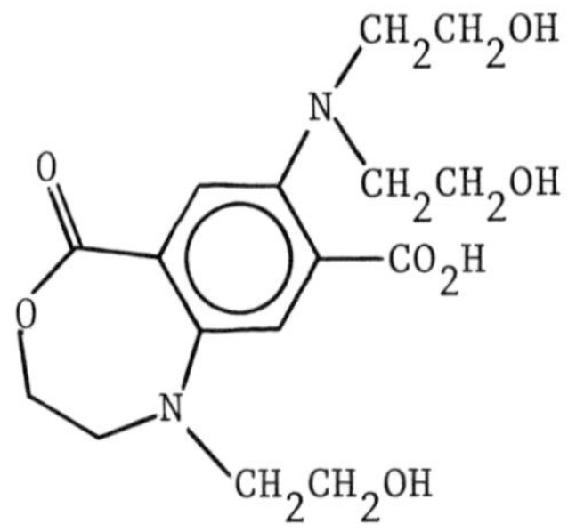

$C_{16}H_{22}O_7N_2$ 2,5-Bis[bis(2-hydroxyethyl)amino]benzene-1,4-dicarboxylic acid monolactone   HL

Metal ion	Equilibrium	Log K 25°, 0.1
H^+	HL/H.L	6.84
Co^{2+}	ML/M.L	2.05
Ni^{2+}	ML/M.L	3.22
Cu^{2+}	ML/M.L	4.89
	ML/MOHL.H	6.3

Bibliography: 73WU

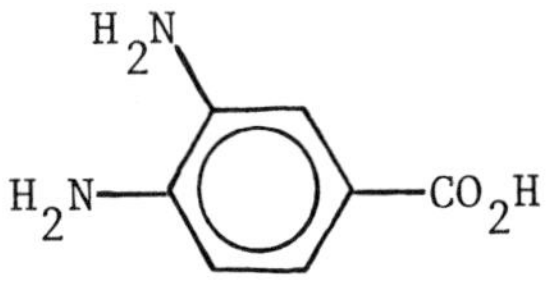

| $C_7H_8O_2N_2$ | | 3,4-Diaminobenzoic acid | HL |

Metal ion	Equilibrium	Log K 25°, 0.1
H^+	HL/H.L	4.75
	H_2L/HL.H	3.69
Cu^{2+}	MHL/ML.H	3.84

Bibliography: 71KT

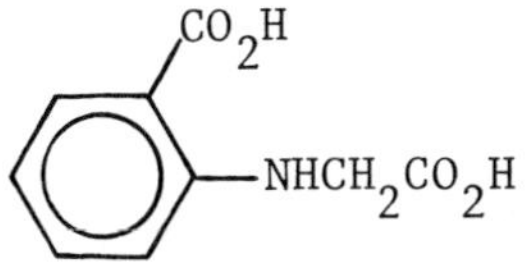

| $C_9H_9O_4N$ | N-(2-Carboxyphenyl)glycine (2-(carboxymethylamino)benzoic acid) | H_2L |

(Other values on Vol.1, p.348)

Metal ion	Equilibrium	Log K 25°, 0.1
H^+	HL/H.L	4.90 −0.01
	H_2L/HL.H	3.33 ±0.03
Co^{2+}	ML/M.L	3.20
Ni^{2+}	ML/M.L	4.45 ±0.0
	ML_2/M.L^2	7.6
Cu^{2+}	ML/M.L	6.65 −0.1
	ML_2/M.L^2	9.5
Zn^{2+}	ML/M.L	3.05 ±0.0

Bibliography: 73UW

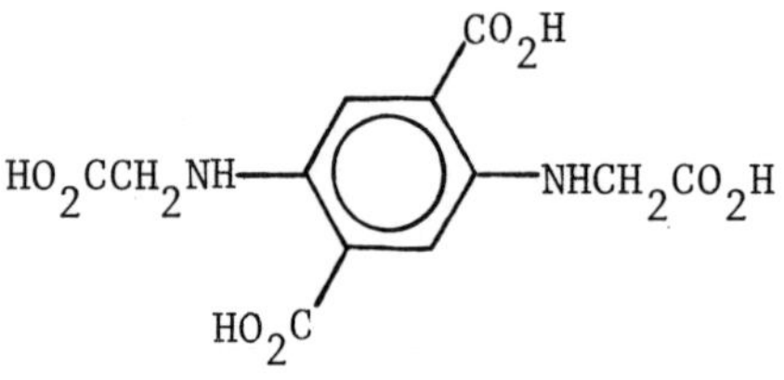

$$HO_2CCH_2NH-\underset{HO_2C}{\overset{CO_2H}{\bigcirc}}-NHCH_2CO_2H$$

$C_{12}H_{12}O_8N_2$ <u>2,5-Bis(carboxymethylamino)benzene-1,4-dicarboxylic acid</u> H_4L

Metal ion	Equilibrium	Log K 25°, 0.1
H^+	HL/H.L	6.94
	$H_2L/HL.H$	4.26
	$H_3L/H_2L.H$	3.32
	$H_4L/H_3L.H$	2.32
Co^{2+}	ML/M.L	5.80
	MHL/ML.H	4.15
	$MH_2L/MHL.H$	3.25
	$M_2L/ML.M$	2.50
Ni^{2+}	ML/M.L	7.35
	MHL/ML.H	3.95
	$MH_2L/MHL.H$	2.70
	$M_2L/ML.M$	3.40

Bibliography: 73UW

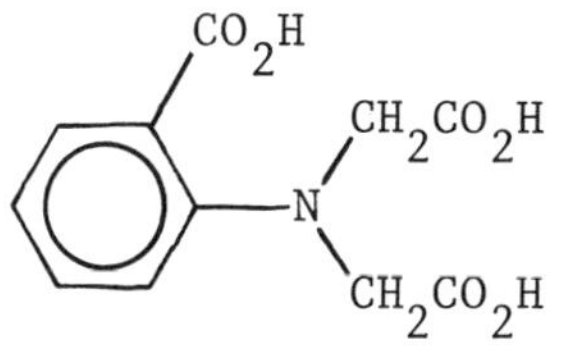

$C_{11}H_{11}O_6N$ N-(2-Carboxyphenyl)iminodiacetic acid H_3L
                          (Other values on Vol.1, p.353)

Metal ion	Equilibrium	Log K 25°, 0.1	Log K 20°, 0.1	ΔH 25°, 0.1	ΔS 25°, 0.1
H^+	HL/H.L	7.73	7.75 ±0.04	-1.7	30
	H_2L/HL.H	2.98	2.98 ±0.00		
	H_3L/H_2L.H	2.33	2.34 -0.01		
Mn^{2+}	ML/M.L	5.85			
Co^{2+}	ML/M.L	8.42		4.7	42
Cu^{2+}	ML/M.L	10.93		3.3	50
Zn^{2+}	ML/M.L	8.42		1.8	56
				3.1	49

Bibliography: 70MU   Other references: 70DP, 70DPa

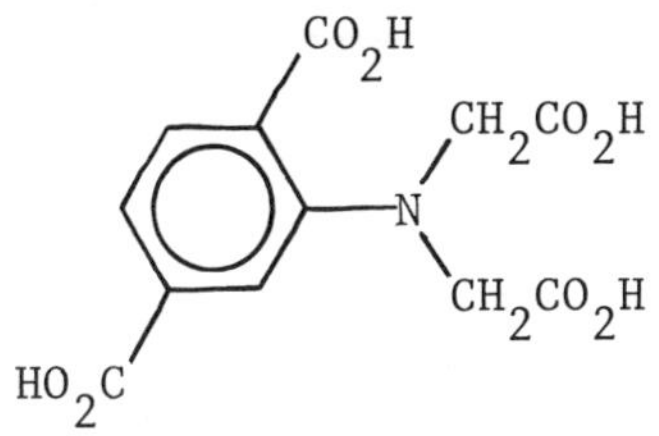

$C_{12}H_{11}O_8N$ N-(2,5-Dicarboxyphenyl)iminodiacetic acid H_4L

Metal ion	Equilibrium	Log K 25°, 0.1	ΔH 25°, 0.1	ΔS 25°, 0.1
H^+	HL/H.L	7.54	-0.4	33
Mg^{2+}	ML/M.L	4.14	8.3	47
Ca^{2+}	ML/M.L	5.05	1.6	29
Sr^{2+}	ML/M.L	4.12	1.4	24
Co^{2+}	ML/M.L	8.83	3.8	53
Ni^{2+}	ML/M.L	9.95	1.8	50
Cu^{2+}	ML/M.L	10.85	2.2	57
Zn^{2+}	ML/M.L	8.86	3.4	52
Cd^{2+}	ML/M.L	7.54		
	MHL/ML.H	3.86		

Bibliography:
H^+-Zn^{2+}70MUa

Cd^{2+} 73KU

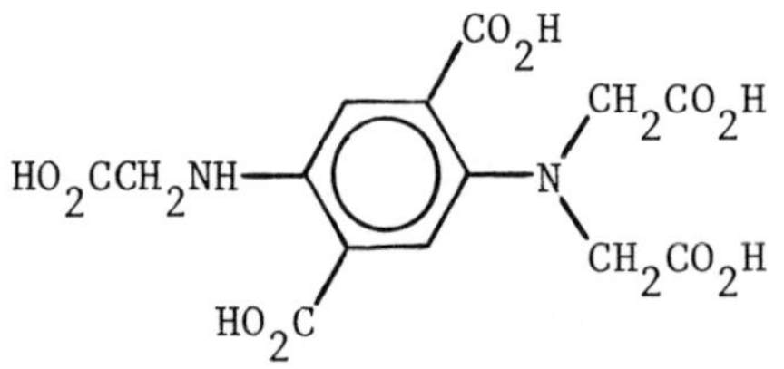

$C_{14}H_{14}O_{10}N_2$ <u>2,5-Dicarboxy-4-(carboxymethylamino)phenyliminodiacetic acid</u> H_5L
<u>(2-carboxymethylamino-5-bis(carboxymethyl)aminobenzene-1,4-dicarboxylic acid)</u>

Metal ion	Equilibrium	Log K 25°, 0.1
H^+	HL/H.L	9.40
	$H_2L/HL.H$	4.48
	$H_3L/H_2L.H$	3.57
	$H_4L/H_3L.H$	2.67
	$H_5L/H_4L.H$	1.8
Mg^{2+}	ML/M.L	5.25
Ca^{2+}	ML/M.L	5.75
Sr^{2+}	ML/M.L	4.60
Ba^{2+}	ML/M.L	4.20
Co^{2+}	ML/M.L	9.80
	MHL/ML.H	4.45
	$MH_2L/MHL.H$	3.1
	$M_2L/ML.M$	2.60
Ni^{2+}	MHL/ML.H	4.20
	$MH_2L/MHL.H$	3.35
	$M_2L/ML.M$	4.10
Cu^{2+}	ML/M.L	12.40
	MHL/ML.H	4.35
	$MH_2L/MHL.H$	3.20
	$M_2L/ML.M$	5.50
Cd^{2+}	ML/M.L	8.55
	MHL/ML.H	4.65
	$MH_2L/MHL.H$	3.38

Bibliography: 73UW

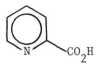

$C_6H_5O_2N$	Pyridine-2-carboxylic acid (picolinic acid)				HL
	(Other values on Vol.1, p.367)				

Metal ion	Equilibrium	Log K 25°, 0.1	Log K 25°, 1.0	Log K 25°, 0	ΔH 25°, 0.1	ΔS 25°, 0.1
H^+	HL/H.L	5.21 ±0.01		5.39 ±0.01	-2.5 ±0.0	15
		5.17[b] ±0.01	5.34[d]			
	H_2L/HL.H	1.03		1.01 ±0.02	-0.5	3
		0.86[b] ±0.02				
Ni^{2+}	ML/M.L	6.72	6.6	7.63	-6.1[c]	10[c]
	$ML_2/M.L_2$	12.44	12.2	(12.45)	-11.2[c]	18[c]
	$ML_3/M.L_3$	17.07	16.7		-12.1[c]	36[c]
Cu^{2+}	ML/M.L	7.87[b]	7.7		-6.4[c]	14[c]
	$ML_2/M.L_2$	14.78	14.5		-13.3[c]	22[c]
		14.70[b] +0.2				
VO^{2+}	ML/MOHL.H	5.03				
	ML_2/MOHL.H	6.95				
Zn^{2+}	ML/M.L	5.25	5.1	5.75	-4.0[c]	10[c]
	$ML_2/M.L_2$	9.52	9.3	10.01	-8.0[c]	16[c]
	$ML_3/M.L_3$	12.75	12.5		-12.3[c]	15[c]

[b]25°, 0.5; [c]25°, 1.0; [d]25°, 2.0

Bibliography:
Y^{3+}-Lu^{3+} 77CGG, 77GGC
Ni^{2+}, Cu^{2+}, Zn^{2+} 79A
VO^{2+} 66KF

Other references: 68AM,74RK,75AM,75STb, 76BC,76JPB,77SM,78ZP,78KCS,78MMa, 79ASJ,79PGa,79SJc

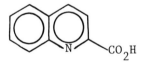

$C_{10}H_7O_2N$	Quinoline-2-carboxylic acid (quinaldic acid)			HL
	(Other values on Vol.1, p.372)			

Metal ion	Equilibrium	Log K 25°, 0.1	Log K 25°, 1.0	Log K 25°, 0
H^+	HL/H.L	4.77 ±0.02	4.68	4.97
	H_2L/HL.H	1.9		
CH_3Hg^+	ML/M.L	5.75		
$C_6H_5Hg^+$	ML/M.L	6.35		

Bibliography: 74Aa

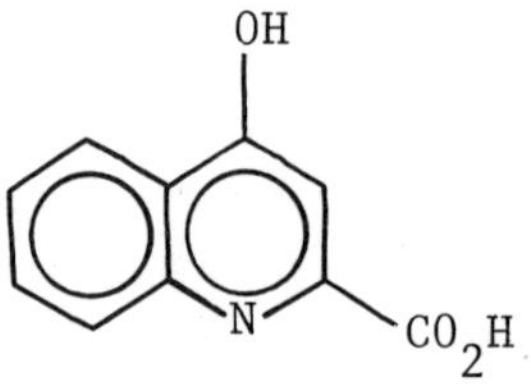

$C_{10}H_7O_3N$		4-Hydroxyquinoline-2-carboxylic acid	H_2L

Metal ion	Equilibrium	Log K 25°, 0.1
H^+	HL/H.L	11.0
	$H_2L/HL.H$	2.65
CH_3Hg^+	ML/M.L	8.4
Hg^{2+}	$ML_2/M.L^2$	22.8

Bibliography: 74Aa

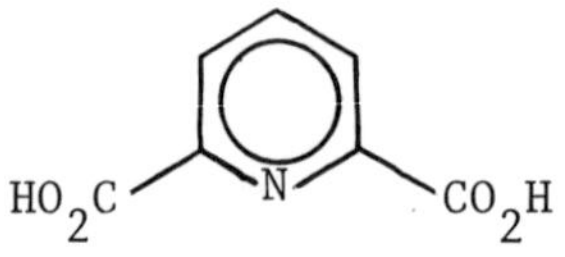

$C_7H_5O_4N$		Pyridine-2,6-dicarboxylic acid (dipicolinic acid)				H_2L
		(Other values on Vol.1, p.377)				

Metal ion	Equilibrium	Log K 25°, 0.1	Log K 25°, 1.0	Log K 25°, 0	ΔH 25°, 0.1	ΔS 25°, 0.1
H^+	HL/H.L	4.69 ±0.06		5.07	1.0[b]	25[b]
		4.51[b] ±0.05			0.2[b]	21[b]
	$H_2L/HL.H$	2.09 ±0.1				
		2.13[b] ±0.1		2.23	-1.1[b]	6[b]
Ni^{2+}	ML/M.L	6.91	6.6		-2.9[c]	20[c]
	$ML_2/M.L^2$	13.42	13.1		-6.3[c]	39[c]
Cu^{2+}	ML/M.L	9.10 +0.1	8.8		-2.8	32
		8.88[b]			-3.8[c]	28[c]
	$ML_2/M.L^2$	16.39 ±0.0	16.1		-9.8	42
		16.17[b]			-11.3[c]	36[c]
	MHL/ML.H	3.0			-0.5	7
	ML/MOHL.H	7.5				
Zn^{2+}	ML/M.L	6.32	6.0		-2.1[c]	20[c]
	$ML_2/M.L^2$	11.81	11.5		-5.4[c]	35[c]

[b] 25°, 0.5;   [c] 25°, 1.0

Bibliography: H^+,Cu^{2+} 79A,79BC; Ni^{2+},Zn^{2+} 79A

Other references: 73YB,77SM,78SJ,79S,79SJb,79ASJ

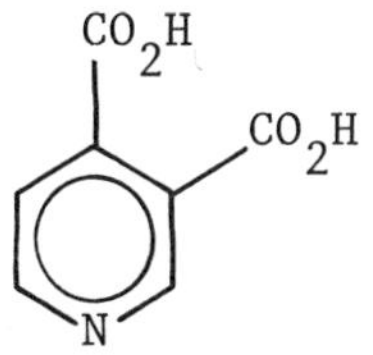

$$CO_2H$$
$$CO_2H$$

| $C_7H_5O_4N$ | Pyridine-3,4-dicarboxylic acid (cinchomeronic acid) | | H_2L |

Metal ion	Equilibrium	Log K 25°, 0.5	Log K 25°, 1.0
H^+	HL/H.L	4.90	
	H_2L/HL.H	2.70	
	H_3L/H_2L.H		0.6
Cr^{3+}	MHL/M.HL	(4.30)	
	$M(HL)_2/M.(HL)^2$	(5.90)	

Bibliography: 76CD

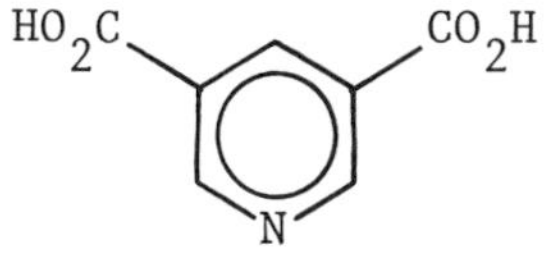

$$HO_2C \qquad CO_2H$$

| $C_7H_5O_4N$ | Pyridine-3,5-dicarboxylic acid (dinicotinic acid) | H_2L |

Metal ion	Equilibrium	Log K 25°, 0.5
H^+	HL/H.L	4.30
	H_2L/HL.H	2.10
Cr^{3+}	MHL/M.HL	(2.2)

Bibliography: 74DC

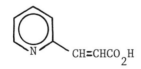

| $C_8H_7O_2N$ | 3-Pyridylpropenoic acid | HL |

Metal ion	Equilibrium	Log K 25°, 0.1
H^+	HL/H.L	4.56
	H_2L/HL.H	2.83
Cu^{2+}	ML/M.L	1.0

Bibliography: 74IL

$$CH_3NH_2$$

CH$_5$N   Methylamine   L
(Other values on Vol.2, p.1)

Metal ion	Equilibrium	Log K 25°, 0.2	Log K 25°, 0.5	Log K 25°, 0	ΔH 25°, 0	ΔS 25°, 0
H$^+$	HL/H.L	10.67	10.69 ±0.03	10.64 ±0.02	-13.2 ±0.1	4
				10.80^d		
Cu^{2+}	ML/M.L			4.11^d		
	ML$_2$/M.L^2			7.51^d		
	ML$_3$/M.L^3			10.21^d		
	ML$_4$/M.L^4			12.08^d		
CH$_3$Hg$^+$	ML/M.L		7.57			

^{d}25°, 2.0 CH$_3$NH$_3^+$NO$_3^-$ used as background electrolyte.

Bibliography:

H$^+$ 74RO,76IB,77BO   CH$_3$Hg$^+$ 74RO

Cu^{2+} 76IB   Other references: 78SL,79FS

$$CH_3CH_2NH_2$$

C$_2$H$_7$N   Ethylamine   L
(Other values on Vol.2, p.2)

Metal ion	Equilibrium	Log K 25°, 0.5	Log K 25°, 0	ΔH 25°, 0	ΔS 25°, 0
H$^+$	HL/H.L	10.66 ±0.01	10.636 ±0.000	-13.65±0.07	2.9
CH$_3$Hg$^+$	ML/M.L	7.64			

Bibliography: 74RO   Other reference: 76PB

$$\begin{array}{c} CH_3 \\ | \\ CH_3CHNH_2 \end{array}$$

C$_3$H$_9$N   2-Propylamine (isopropylamine)   L
(Other references on Vol.2, p.331)

Metal ion	Equilibrium	Log K 25°, 0.5	Log K 25°, 0	ΔH 25°, 0	ΔS 25°, 0
H$^+$	HL/H.L	(10.61)	10.67	-13.95 ±0.03	2.0
CH$_3$Hg$^+$	ML/M.L	7.56			

Bibliography: 74RO   Other references: 68PMa,70PMa

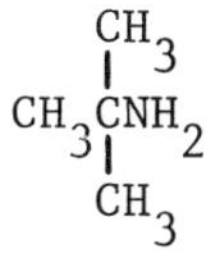

$$\begin{array}{c} CH_3 \\ | \\ CH_3CNH_2 \\ | \\ CH_3 \end{array}$$

$C_4H_{11}N$	2-Methyl-2-propylamine (t-butylamine)				L
	(Other references on Vol.2, p.331)				

Metal ion	Equilibrium	Log K 25°, 0.5	Log K 25°, 0	ΔH 25°, 0	ΔS 25°, 0
H^+	HL/H.L	(10.66)	10.685	-14.39 ±0.04	0.6
CH_3Hg^+	ML/M.L	7.52			

Bibliography: 74RO

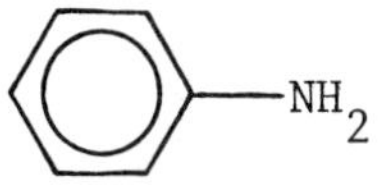

C_6H_7N	Aminobenzene (aniline)				L
	(Other values on Vol.2, p.8)				

Metal ion	Equilibrium	Log K 25°, 0.1	Log K 25°, 1.0	Log K 25°, 0	ΔH 25°, 0	ΔS 25°, 0
H^+	HL/H.L	4.65 ±0.03	4.82	4.601 ±0.005	-7.31 ±0.07	-3.5
		4.71[b]			-7.64[b]	-4.1[b]
Ni^{2+}	ML/M.L	1.6			0.1[a]	8[a]
Ag^+	ML/M.L	1.40[b]				
	$ML_2/M.L^2$	3.28[b]				

[a] 25°, 0.1; [b] 25°, 0.5

Bibliography:

H^+	73MB,79HM	Ag^+	73MB
Ni^{2+}	79HM		Other references: 59ZP,72VG

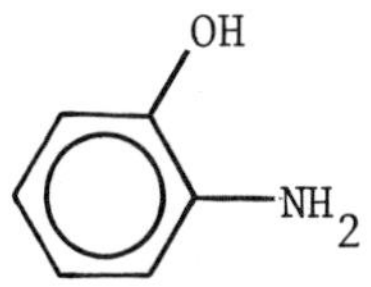

C_6H_7ON	2-Aminophenol	HL

Metal ion	Equilibrium	Log K 25°, 0.1	Log K 25°, 1.0
H^+	HL/H.L	9.87	9.67
	$H_2L/HL.H$	4.74	4.89
Cu^{2+}	ML/M.L	8.49	8.08
	$ML_2/M.L^2$	15.52	14.60

Bibliography: 75SP,75BG,75SP

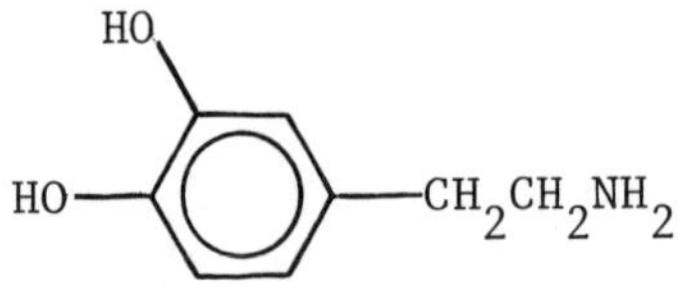

$C_8H_{11}O_2N$	2-(3,4-Dihydroxyphenyl)ethylamine (dopamine)			H_2L
	(Other references on Vol.2, p.332)			

Metal ion	Equilibrium	Log K 25°, 0.1	Log K 20°, 0.37	Log K 25°, 2.0
H^+	HL/H.L	(13.1)		13.4
	$H_2L/HL.H$	10.36 ±0.05	10.45	
	$H_3L/H_2L.H$	8.88 ±0.02	8.91	
Ni^{2+}	MHL/M.HL	6.27	6.32	
	$M(HL)_2/M.(HL)^2$	9.46	9.95	
	MHL/ML.H	9.96		
	$M(HL)_2/MHL_2.H$	10.05		
	$M(HL)_2/ML_2.H^2$	20.86	19.89	
Cu^{2+}	MHL/M.HL	11.12	11.24	
	$M(HL)_2/M.(HL)^2$	19.63	20.16	
	MHL/ML.H	7.6	7.13	
	$M(HL)_2/MHL_2.H$	10.17		
	$M(HL)_2/ML_2.H^2$	20.79	20.49	
Zn^{2+}	MHL/M.HL	7.11	7.28	
	$M(HL)_2/M.(HL)^2$	12.73		
	$M(HL)_2/MHL_2.H$	10.26		
	$M(HL)_2/ML_2.H^2$	21.14		
Cd^{2+}	MHL/M.HL		5.94	
Pb^{2+}	MHL/M.HL		10.18	

Bibliography:

H^+-Zn^{2+} 74GS,79KG; Cd^{2+},Pb^{2+} 74GS   Other references: 71RD,78IH

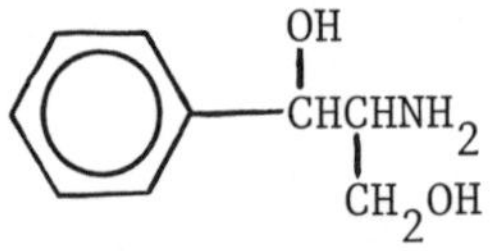

$C_9H_{13}O_2N$	L(+)-threo-2-Amino-1-phenylpropane-1,3-diol	HL

Metal ion	Equilibrium	Log K 25°, 0.1
H^+	HL/H.L	12.0
	$H_2L/HL.H$	8.45
Cu^{2+}	ML/M.L	10.3
	$ML_2/M.L^2$	18.0
	MHL/ML.H	6.19
	ML/MOHL.H	7.42
	$ML_2/MOHL_2.H$	11.9

Bibliography: 79DM

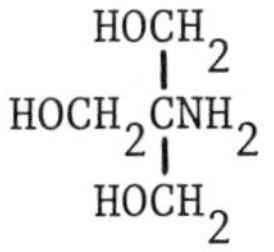

$$\begin{array}{c} HOCH_2 \\ | \\ HOCH_2CNH_2 \\ | \\ HOCH_2 \end{array}$$

$C_4H_{11}O_3N$ <u>2-Amino-2-(hydroxymethyl)-1,3-propanediol</u>   L
<u>(tris(hydroxymethyl)aminomethane, THAM, TRIS)</u>
(Other values on Vol.2, p.20)

Metal ion	Equilibrium	Log K 25°, 0.1	Log K 25°, 0.5	Log K 25°, 0	ΔH 25°, 0	ΔS 25°, 0
H^+	HL/H.L	8.09	8.15 8.198[q]	8.075 8.65[e]	-11.36 ±0.03 -11.46[q]	-1.2 -0.9[q]
Ni^{2+}	$ML/M.L$	2.63		3.20[e]		
	$ML_2/M.L^2$	4.5		5.73[e]		
	$M^2.L^3/M_2(OH)_3L_3.H^3$	27.0				
	$M^3.L^3/M_3(OH)_5L_3.H^5$	13.4				
	$M_3(OH)_5L_2.H^5/M^3.L^2$			30.06[e]		
	$M_4(OH)_8L_4.H^8/M^4.L^4$			46.65[e]		
	$M_4(OH)_8L_4/M_4(OH)_9L_4.H$			8.20[e]		
	$M_4(OH)_9L_4/M_4(OH)_{10}L_4.H$			8.75[e]		

[e] 25°, 3.0;   [q] 25°, 0.67

Bibliography:

H^+ 77RC,78Fa

Ni^{2+} 78Fa

Other references: 71O,75MBa,76RG,78PSb

$$\overset{\displaystyle O}{\underset{\displaystyle \|}{}}$$

$$H_2NCH_2\overset{O}{\overset{\|}{C}}NH_2$$

| $C_2H_6ON_2$ | Glycinamide (Other values on Vol.2, p.26) | | | | L |

Metal ion	Equilibrium	Log K 25°, 0.1	Log K 25°, 1.0	ΔH 25°, 0.1	ΔS 25°, 0.1
H^+	HL/H.L	7.94 ±0.02	8.19	-10.3 ±0.5	2
Ni^{2+}	$ML/M.L$	3.80			
	$ML_2/M.L^2$	6.88			
	$ML_3/M.L^3$	9.3			
	$ML_2/M(H_{-1}L)_2.H^2$	19.01			
Cu^{2+}	$ML/M.L$	5.35 ±0.06	5.42	-6.0 ±0.5	4
	$ML_2/M.L^2$	9.52 ±0.1	9.83	-11.9 ±0.7	4
	$ML/M(H_{-1}L).H$	6.83 ±0.08			
	$ML_2/M(H_{-1}L)L.H$	6.91 ±0.00			
	$ML_2/M(H_{-1}L)_2.H^2$	15.03 +0.01		-16.6	13

Bibliography: 75DB

$$H_2NCH_2\overset{O}{\overset{\|}{C}}NHCH_2\overset{O}{\overset{\|}{C}}NH_2$$

| $C_4H_9O_2N_3$ | Glycylglycinamide (Other references on Vol.2, p.28) | | L |

Metal ion	Equilibrium	Log K 25°, 0.1
H^+	HL/H.L	7.78 ±0.02
Ni^{2+}	$ML/M.L$	3.42
	$ML_2/M.L^2$	6.21
	$ML_3/M.L^3$	8.6
	$ML/M(H_{-1}L).H$	8.52
	$M(H_{-1}L)/M(H_{-2}L).H$	9.34
	$M(H_{-2}L)/MOH(H_{-2}L).H$	10.53
Cu^{2+}	$ML/M.L$	4.88 ±0.1
	$ML/M(H_{-1}L).H$	5.03 ±0.04
	$M(H_{-1}L)/M(H_{-2}L).H$	7.97 ±0.05
	$M(H_{-2}L)/MOH(H_{-2}L).H$	9.78 ±0.05

Bibliography: 75DB

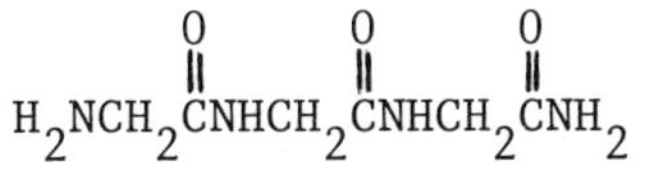

$$
\begin{matrix}
& \overset{O}{\underset{\|}{}} & & \overset{O}{\underset{\|}{}} & & \overset{O}{\underset{\|}{}} \\
\end{matrix}
$$

H$_2$NCH$_2$CNHCH$_2$CNHCH$_2$CNH$_2$

C$_6$H$_{12}$O$_3$N$_4$	Glycylglycylglycinamide	L

Metal ion	Equilibrium	Log K 25°, 0.1	Log K **25°, 1.0**
H$^+$	HL/H.L	7.75	
Ni^{2+}	ML/M.L	3.42	
	ML$_2$/M.L^2	6.29	
	ML/M(H$_{-3}$L).H^3	23.79	
Cu^{2+}	ML/M.L	4.77	
	ML/M(H$_{-1}$L).H	5.28	
	M(H$_{-1}$L)/M(H$_{-2}$L).H	6.99	
	M(H$_{-2}$L)/M(H$_{-3}$L).H	8.69	
Cu(III)	M(H$_{-3}$L)/M(H$_{-4}$L).H		12.3

Bibliography:
H$^+$-Cu^{2+} 75DB   Cu(III) 79NKC

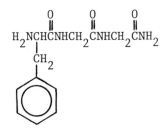

C$_{13}$H$_{18}$O$_3$N$_4$	L-3-Phenylalanylglycylglycinamide	L

Metal ion	Equilibrium	Log K 25°, 1.0
Cu(III)	M(H$_{-3}$L)/M(H$_{-4}$L).H	11.9

Bibliography: 79NKC

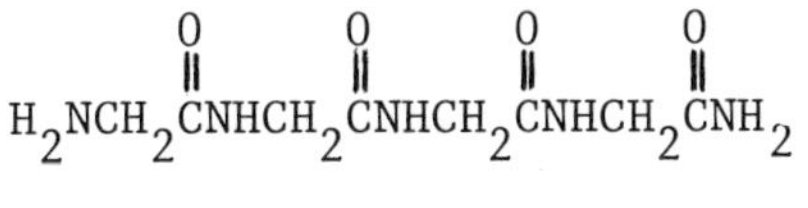

$$
\begin{matrix}
& \overset{O}{\underset{\|}{}} & & \overset{O}{\underset{\|}{}} & & \overset{O}{\underset{\|}{}} & & \overset{O}{\underset{\|}{}} \\
\end{matrix}
$$

H$_2$NCH$_2$CNHCH$_2$CNHCH$_2$CNHCH$_2$CNH$_2$

C$_8$H$_{15}$O$_4$N$_5$	Glycylglycylglycylglycinamide	L

Metal ion	Equilibrium	Log K 25°, 1.0
Cu(III)	M(H$_{-3}$L)/M(H$_{-4}$L).H	11.4

Bibliography: 79NKC

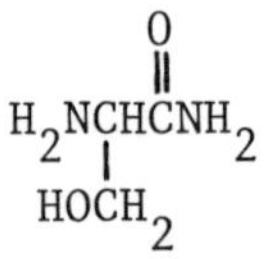

$$H_2NCHCNH_2$$ with O double bond above C, and $HOCH_2$ below

$C_3H_8O_2N_2$ Serine amide   L

Metal ion	Equilibrium	Log K 25°, 0.1
H^+	HL/H.L	7.18
Cu^{2+}	ML/M.L	4.61
	$ML_2/M.L^2$	8.21
	$ML/M(H_{-1}L).H$	6.56
	$M(H_{-1}L)/MOH(H_{-1}L).H$	7.57
	$M(H_{-1}L)L/M(H_{-1}L).L$	3.76
	$M_2OH(H_{-1}L)_2/M(H_{-1}L).MOH(H_{-1}L)$	2.82

Bibliography: 75BP

$$HSCH_2CH_2NH_2$$

C_2H_7NS 2-Aminoethanethiol   (2-mercaptoethylamine)   HL
                  (Other values on Vol.2, p.32)

Metal ion	Equilibrium	Log K 25°, 0.1	Log K 30°, 1.0	Log K 25°, 0	ΔH 25°, 0.01	ΔS 25°, 0.1
H^+	HL/H.L	10.71 ±0.02 10.69^i	10.69			
	$H_2L/HL.H$	8.21 ±0.03 8.27^i	8.28	8.23	-7.43	12.7*
In^{3+}	ML/M.L	13.94^i				

^{i}20°, 0.5;   *assuming ΔH for 0.01=ΔH for 0.1
Bibliography: In^{3+} 78KSa
H^+ 76CJ,78KSa   Other reference: 73GS,78BKS

$$CH_3SCH_2CH_2NH_2$$

C_3H_9NS 2-Methylthioethylamine   L
                  (Other values on Vol.2, p.34)

Metal ion	Equilibrium	Log K 25°, 0.5	Log K 25°, 0	ΔH 25°, 0.5	ΔS 25°, 0.5
H^+	HL/H.L	9.47	9.34	-12.88	0.1
Ni^{2+}	ML/M.L	3.30		-5.6	-4
	$ML_2/M.L^2$	6.10		-11.6	-11
	$ML_3/M.L^3$	7.73		-17.7	-24
Cu^{2+}	ML/M.L	5.57	5.51	-8.0	-1
	$ML_2/M.L^2$	10.65	10.68	-17.1	-9

2-Methylthioethylamine   (continued)

Metal ion	Equilibrium	Log K 25°, 0.5	Log K 25°, 0	ΔH 25°, 0.5	ΔS 25°, 0.5
Ag^+	$ML/M.L$	4.88			
	$ML_2/M.L^2$	9.29		-19.6	-23
	$MHL/M.HL$	2.64		-6.5	-10
	$M(HL)_2/M.(HL)^2$	4.06		-13.3	-26
	$MHL_2/M.HL.L$	7.56		-17.6	-24
	$M_2L/M^2.L$	6.86		-14.9	-19
	$M_2L_2/M^2.L^2$	13.01		-29.3	-39

Bibliography:

H^+ 77HGE,77TGE,77TGS Ag^+ 77TGE,77TGS

Ni^{2+},Cu^{2+} 77HGE

$$CH_3CH_2SCH_2CH_2NH_2$$

$C_4H_{11}NS$ 2-Ethylthioethylamine   L

Metal ion	Equilibrium	Log K 25°, 0.5	ΔH 25°, 0.5	ΔS 25°, 0.5
H^+	$HL/H.L$	9.44	-12.88	0.0
Ag^+	$ML/M.L$	5.1		
	$ML_2/M.L^2$	9.66	-20.3	-24
	$MHL/M.HL$	2.99	-7.4	-11
	$M(HL)_2/M.(HL)^2$	4.66	-13.9	-24
	$ML(HL)/M.HL.L$	8.09	-18.4	-25
	$M_2L/M^2.L$	7.42	-17.2	-24
	$M_2L_2/M^2.L^2$	13.66	-30.4	-40

Bibliography: 77TGE,77TGS

$$CH_3CH_2CH_2SCH_2CH_2NH_2$$

$C_5H_{13}NS$ 2-(Propylthio)ethylamine   L

Metal ion	Equilibrium	Log K 25°, 0.5	ΔH 25°, 0.5	ΔS 25°, 0.5
H^+	$HL/H.L$	9.45	-12.80	0.3
Ag^+	$ML/M.L$	5.29		
	$ML_2/M.L^2$	9.70	-20.2	-23
	$MHL/M.HL$	2.95	-7.5	-12
	$M(HL)_2/M.(HL)^2$	4.61	-14.1	-26
	$MHL_2/M.HL.L$	8.08	-18.6	-25
	$M_2L/M^2.L$	7.40	-17.4	-25
	$M_2L_2/M^2.L^2$	13.59	-31.13	-42

Bibliography: 77TGE,77TGS

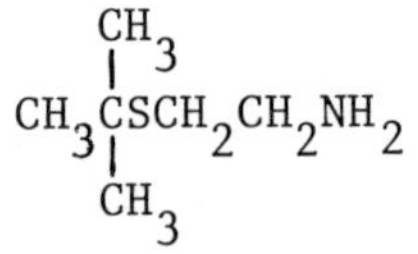

$$\begin{array}{c} CH_3 \\ | \\ CH_3CSCH_2CH_2NH_2 \\ | \\ CH_3 \end{array}$$

$C_6H_{15}NS$ 2-(1,1-Dimethylethylthio)ethylamine   L

Metal ion	Equilibrium	Log K 25°, 0.5	ΔH 25°, 0.5	ΔS 25°, 0.5
H^+	HL/H.L	9.34	-12.39	1.1
Ag^+	$ML/M.L$	5.27		
	$ML_2/M.L^2$	9.99	-19.7	-20
	MHL/M.HL	3.35	-8.4	-13
	$M(HL)_2/M.(HL)^2$	5.60	-14.7	-24
	$MHL_2/M.HL.L$	8.41	-17.9	-22
	$M_2L/M^2.L$	7.88	-16.6	-20
	$M_2L_2/M^2.L^2$	14.17	-31.3	-40

Bibliography: 77TGE,77TGS

$$CH_3SCH_2CH_2CH_2NH_2$$

$C_4H_{11}NS$ 3-Methylthiopropylamine   L

Metal ion	Equilibrium	Log K 25°, 0.5	ΔH 25°, 0.5	ΔS 25°, 0.5
H^+	HL/H.L	10.10	-13.38	1.3
Ag^+	$ML/M.L$	4.8		
	$ML_2/M.L^2$	7.82	-17.4	-23
	MHL/M.HL	3.33	-7.6	-10
	$M(HL)_2 M.(HL)^2$	5.60	-14.1	-22
	$MHL_2/M.HL.L$	7.13	-16.6	-23
	$M_2L/M^2.L$	7.20	-14.5	-16
	$M_2L_2/M^2.L^2$	13.69	-29.6	-37

Bibliography: 77TGE,77TGS

$$HOCH_2CH_2SCH_2CH_2NH_2$$

$C_4H_{11}ONS$ 2-(2-Aminoethylthio)ethanol   L
(Other references on Vol.2, p.35)

Metal ion	Equilibrium	Log K 25°, 0.5	Log K 25°, 1.0	Log K 25°, 0	ΔH 25°, 0.5	ΔS 25°, 0.5
H^+	HL/H.L	9.37	(9.18)	9.26	-12.66	0.4
Ni^{2+}	$ML/M.L$	3.21	3.34		4.8	-1
	$ML_2/M.L^2$	5.76	6.14		-10.2	-8
	$ML_3/M.L^3$		7.84		$(-10)^s$	$(2)^c$
Cu^{2+}	$ML/M.L$	5.24	5.54	5.32	-7.6	-2
	$ML_2/M.L^2$	10.01	10.62	10.23	-16.2	-9
	$MOHL_2/ML_2.OH$	4.39			-6	0
Ag^+	$ML/M.L$	4.81	4.64	4.88	$(-9)^r$	$(-8)^*$
	$ML_2/M.L^2$	9.21	7.70	8.90	-19.1	-22
	MHL/M.HL	2.55			-6.7	-11
	$M(HL)_2/M.(HL)^2$	4.11			-11.1	-18
	$MHL_2/M.HL.L$	7.54			-16.8	-22
	$M_2L/M^2.L$	7.05			-16.0	-21
	$M_2L_2/M^2.L^2$	12.82			-28.2	-36

^{c}25°, 1.0; ^{r}10-40°, 0; ^{s}30-50°, 1.0;   *assuming ΔH for 0.0=ΔH for 0.5

Bibliography:

H^+ 77HGE,77TGE,77TGS

Ni^{2+},Cu^{2+} 77HGE

Ag^+ 77TGE,77TGS

Other reference: 78UV

$$H_2NCH_2CH_2NH_2$$

$C_2H_8N_2$

Ethylenediamine (en)   L
(Other values on Vol.2, p.36)

Metal ion	Equilibrium	Log K 25°, 0.1	Log K 25°, 1.0	Log K 25°, 0	ΔH 25°, 0.1	ΔS 25°, 0.1
H^+	HL/H.L	9.89[b] ±0.07	10.20 ±0.04	9.928	-11.9[c] ±0.1	5
		10.03[b] ±0.06	9.70[n]	10.74[e]	-12.2[c] ±0.0	6[c]
	$H_2L/HL.H$	7.08[b] ±0.04	7.45[d] ±0.04	6.848	-10.9 ±0.2	-4
		7.31[b] +0.01	7.71[d] 0.03	7.93[e] ±0.00	-10.9[c] ±0.4	-2
			6.93[n]			
Mn^{2+}	ML/M.L	2.74	2.77		-2.8[c]	3[c]
	$ML_2/M.L^2$	4.8	4.87		-6.0[c]	2[c]
	$ML_3/M.L^3$		5.81		-11.1[c]	-11[c]
Co^{2+}	ML/M.L	5.5 ±0.1	(5.96)±0.03		-6.9[c]	4[c]
			5.30[n]			
	$ML_2/M.L^2$	10.1 ±0.1	10.7 +0.1		-14.0[c]	2[c]
			9.6[n]			
	$ML_3/M.L^3$	13.2 ±0.6	14.0 ±0.1		-22.2[c]	-10[c]
			12.0[n]			
Ni^{2+}	ML/M.L	7.31[b] ±0.08	7.56 ±0.05	7.32 +0.08	-9.0[c] ±0.1	4[c]
		7.41[b] ±0.06				
	$ML_2/M.L^2$	13.4[b] ±0.1	14.0 ±0.1	13.5 +0.1	-18.3[c] ±0.1	3[c]
		13.7[b] ±0.1				
	$ML_3/M.L^3$	17.5[b] ±0.2	18.4 ±0.2	17.6 +0.3	-28.0[c] ±0.1	-10[c]
		18.0[b] ±0.1				
Cu^{2+}	ML/M.L	10.50 ±0.06	10.74 ±0.02	10.48 +0.04	-12.6[b]	6[b]
		10.62[b] ±0.1	11.02[d]		-13.1[c] ±0.2	5[c]
	$ML_2/M.L^2$	19.6 ±0.2	20.2 ±0.2	19.5 +0.5	-25.2[b]	7[b]
		19.9[b] ±0.1	20.6[d]		-25.5[c] ±0.2	7[c]
	MOHL/ML.OH	0.73[b]				
Cr^{3+}	$ML_3/ML_2.L$		6.43			
	$ML_3/M.L^3$		19.5			
	$ML_2/HOHL_2.H$[y]	4.86[p]	4.75			
	$MOHL_2/M(OH)_2L_2.H$[y]	7.34[p]	7.35			
	$ML_2/MOHL_2.H$[z]		4.12			
	$MOHL_2/M(OH)_2L_2.H$[z]		7.71			
Ag^+	ML/M.L	4.70[h]	5.06	6.13[e]	-11.7[c]	-16[c]
	$ML_2/M.L^2$	7.70[h]	7.7		-12.5[c]	-7[c]
		7.64[h]			-13.1[b]	-9[b]
	MHL/M.HL	2.35[b]	2.4	2.91[e]	-6.1[b]	-10[b]
		2.34[b]				
	$M(HL)_2/M.(HL)^2$	4.90[b]	5.1	6.07[e]	-12.1[b]	-18[b]
	ML/MOHL.H			10.08[e]		
	$MHL_2/ML_2.H$	1.17[b]			1.1[b]	9[b]
	$M_2L/ML.M$	1.8[h]	1.2	1.5[e]		
	$M_2L_2/M^2.L^2$	13.2[h]	13.2	14.53[e]	-23.2[b]	-18[b]
		13.15[b]			-25.8[c]	-26[c]

---

[b] 25°, 0.5; [c] 25°, 1.0; [d] 25°, 2.0; [e] 25°, 3.0; [h] 20°, 0.1; [n] 37°, 0.15; [p] 4°, 0.1;
[y] cis isomer; [z] trans isomer

Ethylenediamine (continued)

Metal ion	Equilibrium	Log K 25°, 0.1	Log K 25°, 1.0	Log K 25°, 0	ΔH 25°, 0.1	ΔS 25°, 0.1
Ag^+ (cont.)	$M_2L_2/(ML)^2$	3.8^h	3.1 -0.1	2.3^e	-2.4^c	6^c
Zn^{2+}	ML/M.L	5.7 ±0.1 5.75^b	5.86 ±0.06 6.15^d 5.53^n	5.66	-7.0^c	3^c
	$ML_2/M.L^2$	10.6 ±0.1 10.8^b	11.1 11.5^d 10.3^n	10.6	-11.9^c	11^c
	$ML_3/M.L^3$	12.6 ±0.5 12.8^b ±0.3	12.9 12.7^n	13.9	-17.1^c	2^c
Cd^{2+}	ML/M.L	5.4 ±0.1	5.62 ±0.06 8.84^d	5.41	$(-6)^r$	(5)
	$ML_2/M.L^2$	9.6 ±0.2	10.2 ±0.0 10.6^d	9.9	$(-13.3)^s$	(-1)
	$ML_3/M.L^3$	11.6 ±0.1	12.3 ±0.00 12.7^d		$(-19.7)^s$	(-13)

---

^{b}25°, 0.5; ^{c}25°, 1.0; ^{d}25°, 2.0; ^{h}20°, 0.1; ^{n}37°. 0.15; ^{r}10-40°, 0; ^{s}not corrected for Cl⁻.

Bibliography:

H^+	69PS,72CV,75BDT,76M,76PB,76SG,77BP, 78BSa	Zn^{2+}	69PS,70DN,78SK
Mn^{2+}	70DN,77SF	Cd^{2+}	70DN,79KZ
Co^{2+}	69PS,70DN		
Ni^{2+}	70DN,75LM,78SKG,79SG		

Other references: 61CM,68SP,68VB,72NB, 75BWa,74MK,74MKa,74PB,74SKT,75M,76BMc, 76ES,76H,77GB,77HS,77KVF,78SGd,78SL, 78Y, 79GBS,79GSK

Cu^{2+} 62SY,70DN,70GS,75NW,76SG,77BP,78BSa, 78FM,79MBb

Cr^{3+} 57H,75AB,76MM

Ag^+ 73OI,76V,78MS,79BBa

$$\underset{\displaystyle H_2NCHCH_2NH_2}{\overset{\displaystyle \overset{\textstyle CH_3}{|}}{}}$$

$C_3H_{10}N_2$ <u>DL-Methylethylenediamine</u> <u>(1,2-propylenediamine, pn)</u>   L
(Other values on Vol.2, p.39)

Metal ion	Equiliubrium	Log K 25°, 0.1	Log K 25°, 0.5	Log K 25°, 0	ΔH 25°, 0	ΔS 25°, 0
H^+	HL/H.L	9.78 +0.09	9.90 ±0.09	9.72 +0.10 $(9.99)^e$	-11.9_b -12.3^b	5_b 4^b
	$H_2L/HL.H$	6.85 -0.01	7.06 -0.01	6.61 ±0.00 7.64^e	-9.7_b -10.9^b	-2_b -4^b

---

^{b}25°, 0.5; ^{e}25°, 3.0

1,2-Propylenediamine   (continued)

Metal ion	Equilibrium	Log K 25°, 0.1	Log K 25°, 0.5	Log K 25°, 0	ΔH 25°, 0	ΔS 25°, 0
Ag^+	ML/M.L			5.52^e		
	MHL/M.HL			2.73^e		
	$M(HL)_2/M.(HL)^2$			$(5.86)^{e,w}$		
	ML/MOHL.H			9.7^e		
	$M_2L/ML.M$			2.1^e		
	$M_2L_2/M^2.L^2$			$(13.47)^{e,w}$		

---

^{e}25°, 3.0; ^{w}optical isomerism not stated.

Bibliography:

H^+,Ag^+ 730I   Other references: 68L,74PB,75MJ,77WB,78MG,
Cu^{2+} 61NM   78SGe,79GBG

$$\begin{array}{c} CH_3 \\ | \\ H_2NCHCHNH_2 \\ | \\ CH_3 \end{array}$$

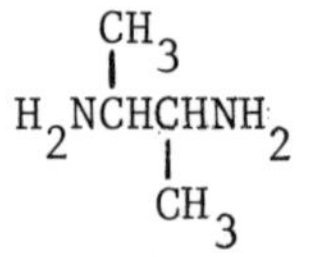

$C_4H_{12}N_2$ <u>DL-1,2-Dimethylethylenediamine</u>   <u>(DL-2,3-butylenediamine)</u>   L
                      (Other references on Vol.2, p.41)

Metal ion	Equilibrium	Log K 25°, 0.1	Log K 25°, 0.65	ΔH 25°, 0.1	ΔS 25°, 0.1
H^+	HL/H.L	9.85	9.85	-11.42	6.8
	$H_2L/HL.H$	6.67	6.76	-11.11	-6.8
Mn^{2+}	ML/M.L	2.94			
Co^{2+}	ML/M.L	5.58			
	$ML_2/M.L^2$	$(10.08)^x$			
Ni^{2+}	ML/M.L	7.64	7.71	-7.8	9
	$ML_2/M.L^2$	14.01^y	14.19	-14.8	14
	$ML_3/M.L^3$	18.0	18.5	-19.5	17
	MHL/ML.H	4.70			
Cu^{2+}	ML/M.L	11.27	11.39	-12.6	9
	$ML_2/M.L^2$	20.92^y	21.21	-23.9	16
	MHL/ML.H	2.90			
Zn^{2+}	ML/M.L	5.94		-4.3	13
	$ML_2/M.L^2$	$(11.26)^x$		-7.1	28
	$ML_3/M.L^3$	$(14.76)^x$			
	MHL/ML.H	5.15			
Pb^{2+}	ML/M.L	5.35			
	$ML_2/M.L^2$	$(10.4)^x$			

---

^{x}DL-mixture; ^{y}D- and DL-isomers had the same value.

Bibliography: 77PS

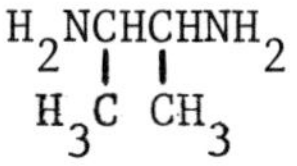

$$\underset{\underset{H_3C\ \ CH_3}{|\ \ \ |}}{H_2NCHCHNH_2}$$

C$_4$H$_{12}$N$_2$ meso-1,2-Dimethylethylenediamine (meso-2,3-butylenediamine)   L
                        (Other references on Vol. 2, p.42)

Metal ion	Equilibrium	Log K 25°, 0.1	Log K 25°, 0.46	Log K 25°, 0.65	ΔH 25°, 0.1	ΔS 25°, 0.1
H$^+$	HL/H.L	9.82	9.80	9.82	-11.83	5.2
	H$_2$L/HL.H	6.67	6.80	6.77	-10.80	-5.7
Mn^{2+}	ML/M.L	2.64	2.72			
	ML$_2$/M.L^2		4.32			
Co^{2+}	ML/M.L	4.84				
	ML$_2$/M.L^2	8.88				
Ni^{2+}	ML/M.L	6.73		7.04	-6.7	8
	ML$_2$/M.L^2	12.31		12.74	-12.4	15
	ML$_3$/M.L^3	14.86		15.63	-15.1	17
	MHL/ML.H	6.0				
Cu^{2+}	ML/M.L	10.54		10.72	-11.2	11
	ML$_2$/M.L^2	19.79		20.06	-23.1	13
	MHL/ML.H	3.37				
Zn^{2+}	ML/M.L	5.45	6.06		-3.3	14
	ML$_2$/M.L^2	10.77	11.15		-7.2	25
	ML$_3$/M.L^3	14.63				
	MHL/ML.H	6.2				
Pb^{2+}	ML/M.L	5.45				
	ML$_2$/M.L^2	10.2				

Bibliography: 77PS

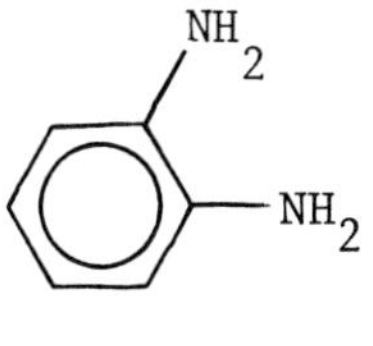

C$_6$H$_8$N$_2$ 1,2-Diaminobenzene   L
                        (Other references on Vol.2, p.49)

Metal ion	Equilibrium	Log K 25°, 0.1	ΔH 25°, 0.3	ΔS 25°, 0.1
H$^+$	HL/H.L	4.63 ±0.00	(6)^r	(1)*
Cu^{2+}	ML/M.L	4.50 ±0.05		
	ML$_2$/M.L^2	7.79 ±0.07		

^{r}15-40°, 0.3; *assuming ΔH for 0.3=ΔH for 0.1

Bibliography: 75SP   Other references: 79AM

$$H_2NCH_2CH_2CH_2NH_2$$

$C_3H_{10}N_2$ <u>Trimethylenediamine (1,3-propylenediamine)</u>   L
(Other values on Vol.2, p.51)

Metal ion	Equilibrium	Log K 25°, 0.1	Log K 25°, 1.0	Log K 25°, 0	ΔH 25°, 0.1	ΔS 25°, 0.1
H^+	HL/H.L	10.52 ±0.04	10.77 ±0.04	10.49 ±0.02	-13.0 ±0.2	5
		10.65^b+0.01	10.96^d	11.11^e-0.2	-13.1^b±0.1	5^b
	$H_2L/HL.H$	8.74 ±0.02	9.12 ±0.04	8.48 +0.01	-12.3 ±0.2	-1
		8.95^b±0.02	9.30^d	9.47^e+0.2	-12.7^b±0.1	-2^b
Ni^{2+}	ML/M.L	6.31 ±0.1	6.47 ±0.08	6.29 ±0.01	-7.6	3
		6.42^b			-7.8^c	3^c
	$ML_2/M.L^2$	10.5 ±0.1	11.0 ±0.2	10.54 ±0.06	-15.0	-2
		10.8^b			-15.0^c	0^c
	$ML_3/M.L^3$	12.9 ±0.5	(12.0)		-21.3^c	$(-17)^c$
Cu^{2+}	ML/M.L	9.66 ±0.01	10.01 ±0.03	9.61 ±0.03	-11.4 ±0.4	6
		9.84^b				
	$ML_2/M.L^2$	16.8 ±0.1	17.4 ±0.2	16.7	-22.4 ±0.4	2
		17.1^b			-22.8^c	3^c
	ML/MOHL.H	7.66		7.42		
	$MOHL/M(OH)_2L.H$	11.70		11.68		
	$M_2(OH)_2L_2/(MOHL)^2$	2.41		2.17		
Ag^+	ML/M.L	5.85^h		5.71	$(-14)^r$	(-21)
				6.59^e		
	MHL/M.HL	2.51^h		3.39^e		
	$M(HL)_2/M.(HL)^2$			7.04^e		
	ML/MOHL.H			10.16^e		
	$M(HL)_2/M(HL)L.H$			9.65^e		
	$M(HL)L/ML_2.H$			9.2^e		
	$M_2L/ML.M$	0.6^h				
	$M_2L_2/M^2.L^2$			15.9^e		

---

^{b}25°, 0.5; ^{c}25°, 1.0; ^{d}25°, 2.0; ^{e}25°, 3.0; ^{h}20°, 0.1; ^{r}10-40°, 0

Bibliography:

H^+ 76GS,76M Ag^+ 77OC

Ni^{2+},Cu^{2+} 76GS   Other references: 68L,72NB,75PB,77GB,77HS,
                                         77WB

$$H_2NCH_2CH_2CH_2CH_2NH_2$$

$C_4H_{12}N_2$ <u>Tetramethylenediamine</u> <u>(1,4-butylenediamine)</u>   L
                              (Other values on Vol.2, p.54)

Metal ion	Equilibrium	Log K 25°, 0.1	Log K 25°, 1.0	Log K 25°, 0	ΔH 25°, 0	ΔS 25°, 0
H^+	HL/H.L	10.72 ±0.07	11.11 ±0.04	10.69 ±0.04	-13.6[b]	3[b]
		10.87[b] ±0.07		(11.05)[e]	-13.6[b]	5[b]
	H_2L/HL.H	9.44 ±0.05	9.91 ±0.03	9.22 ±0.02	-13.2[b]	-2[b]
		9.69[b] ±0.06		(10.39)[e]	-13.2[b]	1[b]
Ag^+	ML/M.L	5.9[h]		5.5	(-14)[r]	(-22)
				6.4[e]		
	MHL/M.HL	3.0[h]		3.63[e]		
	$M(HL)_2/M.(HL)^2$			7.73[e]		
	ML/MOHL.H			10.8[e]		
	$M(HL)_2/M(HL)L.H$			10.7[e]		
	$M(HL)L/ML_2.H$			10.5[e]		
	$M_2L_2/M^2.L^2$			15.3[e]		

[b] 25°, 0.5; [e] 25°, 3.0; [h] 20°, 0.1; [r] 10-40°, 0.

Bibliography: 77OC   Other reference: 77HS

$$\overset{\displaystyle CH_2}{\underset{\displaystyle H_2NCH_2CCH_2NH_2}{\|}}$$

$C_4H_{10}N_2$ <u>2-Methylenetrimethylenediamine</u> <u>(1,3-diamino-2-methylenepropane)</u>   L

Metal ion	Equilibrium	Log K 25°, 0.5
H^+	HL/H.L	10.09
	H_2L/HL.H	8.33
Co^{2+}	ML/M.L	4.45
	$ML_2/M.L^2$	8.0
Ni^{2+}	ML/M.L	7.19
	$ML_2/M.L^2$	12.58
	$ML_3/M.L^3$	16.9
Cu^{2+}	ML/M.L	9.27
	$ML_2/M.L^2$	16.2
Pd^{2+}	ML/M.L	13.64
	$ML_2/M.L^2$	25.3
Zn^{2+}	ML/M.L	4.63
	$ML_2/M.L^2$	9.35
Cd^{2+}	ML/M.L	4.32
	$ML_2/M.L^2$	7.6

Bibliography: 75HS

$$H_2NCH_2CH_2OCH_2CH_2NH_2$$

$C_4H_{12}ON_2$ 4-Oxa-1,7-diazaheptane (oxybis(ethyleneamine))   L
(Other values on Vol.2, p.58)

Metal ion	Equilibrium	Log K 25°, 0.5	Log K 25°, 0	ΔH 25°, 0.5	ΔS 25°, 0.5
H^+	$HL/H.L$	9.89	9.75	-12.11	4.6
	$H_2L/H.HL$	9.15	8.90	-12.96	-1.6
Ni^{2+}	$ML/M.L$	5.90	5.62	-6.7	5
	$ML_2/M.L^2$	9.52	9.01	-13.2	-1
Cu^{2+}	$ML/M.L$	8.97	8.70	-9.5	9
	$ML_2/M.L^2$	12.75	13.1	-14.0	11
	$MOHL/ML.OH$	5.48		(-9.3)	(-6)
Zn^{2+}	$ML/M.L$	5.74		-4.8	10
	$ML_2/M.L^2$	9.86		-12.1	4
	$MOHL/ML.OH$	5.11		0.4	25
	$M(OH)_2L/ML.(OH)^2$	8.37			

Bibliography: 74BV

$$H_2NCH_2CH_2OCH_2CH_2OCH_2CH_2NH_2$$

$C_6H_{16}O_2N_2$ 1,10-Diaza-4,7-dioxadecane (ethylenebis(oxyethyleneamine))   L
(Other values on Vol.2, p.58)

Metal ion	Equilibrium	Log K 25°, 0.1	Log K 25°, 0	ΔH 25°, 0.1	ΔS 25°, 0.1
H^+	$HL/H.L$	9.71[z]	9.73	-11.9[z]	5[z]
	$H_2L/HL.H$	8.91[z]	8.75	-12.1[z]	-1[z]
Ag^+	$ML/M.L$	7.7[z]	7.9	-13.8[z]	-11[z]
Cd^{2+}	$ML/M.L$	5.68[z]			
Hg^{2+}	$ML/M.L$	18.55[z]		-24.5[z]	3[z]

[z] $(CH_3)_4NCl$ used as background electrolyte.

Bibliography: 75A

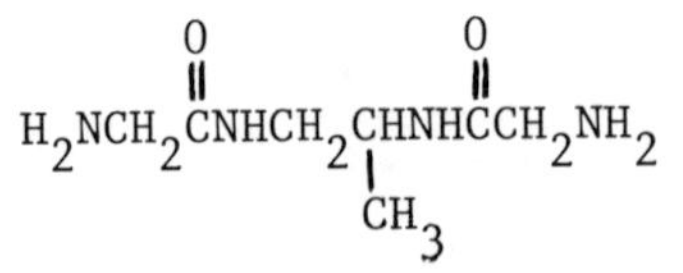

$$H_2NCH_2\overset{O}{\overset{\|}{C}}NHCH_2\underset{\underset{CH_3}{|}}{C}HNH\overset{O}{\overset{\|}{C}}CH_2NH_2$$

$C_7H_{16}O_2N_4$ N,N'-Diglycyl-1,2-propylenediamine   L

Isomer	Metal ion	Equilibrium	Log K 25°, 0.1
D-	H^+	$HL/H.L$	8.22
		$H_2L/HL.H$	7.38
	Cu^{2+}	$ML/M.L$	7.8
		$(MH_{-1}L)_2.H^2/(ML)^2$	-10.1
		$(MH_{-2}L)^2.H^2/(MH_{-1}L)_2$	-16.71

A. PRIMARY AMINES

N,N'-Diglycyl-1,2-propylenediamine (continued)

Isomer	Metal ion	Equilibrium	Log K 25°, 0.1
L-	H^+	HL/H.L	8.37
		$H_2L/HL.H$	7.33
	Cu^{2+}	ML/M.L	8.1
		$(MH_{-1}L)_2.H^2/(ML)^2$	-9.9
		$(MH_{-2}L)^2.H^2/(MH_{-1}L)_2$	-16.64

Bibliography: 74MR   Note: D-and L-isomers should have the same constants.

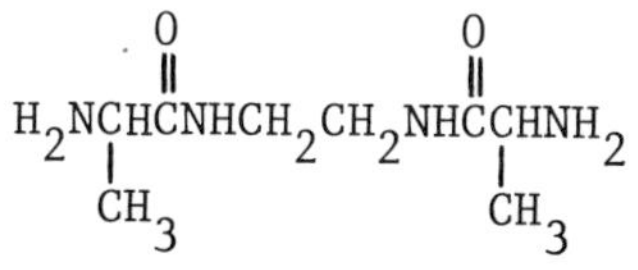

$$H_2NCHCNHCH_2CH_2NHCCHNH_2$$

with $\overset{O}{\overset{\|}{C}}$ groups and CH_3 substituents

$C_8H_{18}O_2N_4$ N,N'-Di-L-alanylethylenediamine   L

Metal ion	Equilibrium	Log K 25°, 0.1
H^+	HL/H.L	8.37
	$H_2L/HL.H$	7.26
Cu^{2+}	ML/M.L	7.4
	$(MH_{-1}L)_2.H^2/(ML)^2$	-9.0
	$(MH_{-2}L)^2.H^2/(MH_{-1}L)_2$	-17.2

Bibliography: 74MR

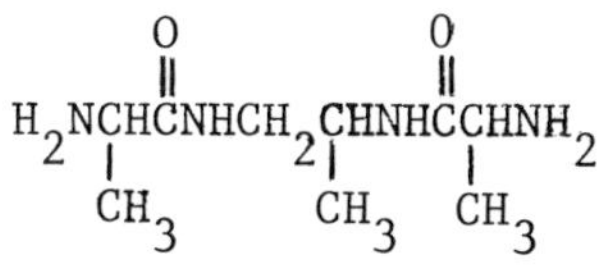

$$H_2NCHCNHCH_2CHNHCCHNH_2$$

with $\overset{O}{\overset{\|}{C}}$ groups and CH_3 substituents

$C_9H_{20}O_2N_4$ N,N'-Di-L-alanyl-1,2-propylenediamine   L

Propylene Isomer	Metal ion	Equilibrium	Log K 25°, 0.1
D-	H^+	ML/H.L	8.32
		$H_2L/HL.H$	7.38
	Cu^{2+}	ML/M.L	7.3
		$(MH_{-1}L)_2.H^2/(ML)^2$	-10.1
		$(MH_{-2}L)^2.H^2/(MH_{-1}L)_2$	-16.23

N,N'-Di-L-alanyl-1,2-propylenediamine   (continued)

Propylene Isomer	Metal ion	Equilibrium	Log K 25°, 0.1
L-	H$^+$	HL/H.L	8.35
		H$_2$L/HL.H	7.24
	Cu^{2+}	ML/M.L	7.4
		(MH$_{-1}$L)$_2$.H^2/(ML)^2	-9.5
		(MH$_{-2}$L)^2.H^2/(MH$_{-1}$L)$_2$	-16.73

Bibliography: 74MR

$$H_2NCH_2CH_2SCH_2CH_2NH_2$$

C$_4$H$_{12}$N$_2$S   4-Thia-1,7-diazaheptane (thiobis(ethyleneamine))   L
                         (Other values on Vol.2, p.65)

Metal ion	Equilibrium	Log K 25°, 0.5	Log K 25°, 1.0	ΔH 25°, 0.5	ΔS 25°, 0.5
H$^+$	HL/H.L	9.68	9.67	-12.93	0.9
	H$_2$L/HL.H	8.82	8.86	-12.74	-2.4
Ni^{2+}	ML/M.L	7.38	7.39	-10.0	0
	ML$_2$/M.L^2	13.52	13.63	-21.6	-11
Cu^{2+}	ML/M.L	9.02	9.21	-12.3	0
	ML$_2$/M.L^2	14.26	14.46	-20.3	-3
	MOHL/ML.OH	5.90		-4.3	13

Bibliography: H$^+$ 77HG; Ni^{2+},Cu^{2+} 79HGD,79HGE; Other reference: 78UV

$$H_2NCH_2CH_2CH_2SCH_2CH_2NH_2$$

C$_5$H$_{14}$N$_2$S   4-Thia-1,8-diazaoctane   L

Metal ion	Equilibrium	Log K 25°, 0.5	ΔH 25°, 0.5	ΔS 25°, 0.5
H$^+$	HL/H.L	10.14	-13.38	1.5
	H$_2$L/HL.H	9.16	-13.10	-2.1
Ni^{2+}	ML/M.L	5.99	-9.3	-4
	ML$_2$/M.L^2	9.85	-18.6	-17
	MHL/M.HL	2.75	-5.1	-5
	MHL$_2$/ML.HL	1.88	-7.1	-15
Cu^{2+}	ML/M.L	10.04	-14.1	-1
	ML$_2$/M.L^2	12.90	-21.9	-15
	MHL/M.HL	4.66	-7.0	-2
	MHL$_2$/ML.HL	2.44	-7.6	-14
	MOHL/ML.OH	4.60	-3.2	10

Bibliography:
H$^+$ 77HG   Ni^{2+},Cu^{2+} 79HGD,79HGE

$$H_2NCH_2CH_2CH_2CH_2SCH_2CH_2NH_2$$

$C_6H_{16}N_2S$		4-Thia-1,9-diazanonane		L

Metal ion	Equilibrium	Log K 25°, 0.5	ΔH 25°, 0.5	ΔS 25°, 0.5
H[+]	HL/H.L	10.44	-13.77	1.5
	H$_2$L/HL.H	9.25	-13.17	-1.9
Cu^{2+}	ML/M.L	9.49		
	MHL/M.HL	5.99		

Bibliography: H[+] 77HG; Cu^{2+} 79HGE

$$H_2NCH_2CH_2CH_2SCH_2CH_2CH_2NH_2$$

$C_6H_{16}N_2S$	4-Thia-1,9-diazanonane	(thiobis(trimethyleneamine))		L

Metal ion	Equilibrium	Log K 25°, 0.5	ΔH 25°, 0.5	ΔS 25°, 0.5
H[+]	HL/H.L	10.37	-13.50	2.1
	H$_2$L/HL.H	9.63	-13.36	-0.8
Cu^{2+}	ML/M.L	9.79		

Bibliography: H[+] 77HG Cu^{2+} 79HGE

$$H_2NCH_2CH_2SCH_2CH_2SCH_2CH_2NH_2$$

$C_6H_{16}N_2S_2$	4,7-Dithia-1,10-diazadecane	(ethylenebis(thioethyleneamine))			L

(Other references on Vol.2, p.67)

Metal ion	Equilibrium	Log K 25°, 0.1	Log K 30°, 1.0	Log K 25°, 0	ΔH 25°, 0	ΔS 25°, 0
H[+]	HL/H.L	9.53	9.47	9.46	(-12)[r]	(3)
	H$_2$L/HL.H	8.70	8.86	8.54	(-12)[r]	(-1)
Co^{2+}	ML/M.L	4.50	4.89			
	ML$_2$/M.L^2	7.61				
Ni^{2+}	ML/M.L	7.41	7.90		(-11)[s]	(-1)[c]
Cu^{2+}	ML/M.L	10.70	11.32	10.63	(-15)[r]	(-2)
	MHL/M.HL	5.6				
Ag[+]	ML/M.L	8.34				
	MHL/M.HL	5.79				
Zn^{2+}	ML/M.L	4.96				
	ML$_2$/M.L^2	9.06				
Cd^{2+}	ML/M.L	5.31	5.61			
	ML$_2$/M.L^2		8.05			
Pb^{2+}	ML/M.L	5.78				
	MHL/M.HL	4.1				

[r] 10-40°, 0; [s] 0-50°, 1.0; [c] 25°, 1.0

Bibliography: 77AS

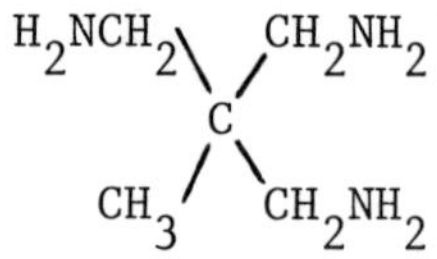

$$\begin{array}{c} H_2NCH_2 \diagdown \quad \diagup CH_2NH_2 \\ C \\ CH_3 \diagup \quad \diagdown CH_2NH_2 \end{array}$$

$C_5H_{15}N_3$ <u>1,3-Diamino-2-aminomethyl-2-methylpropane</u>   L

Metal ion	Equilibrium	Log K 20°, 0.1
H^+	HL/H.L	10.48
	$H_2L/HL.H$	8.48
	$H_3L/H_2L.H$	5.80
Ni^{2+}	ML/M.L	10.76
	MHL/ML.H	5.37
	$MH_2L/MHL.H$	4.62
Cu^{2+}	ML/M.L	11.55
	MHL/ML.H	7.26
	$MH_2L/MHL.H$	2.16
Zn^{2+}	ML/M.L	7.47
	MHL/ML.H	6.83
	$MH_2L/MHL.H$	6.52
Cd^{2+}	ML/M.L	5.81
	MHL/ML.H	7.89
	$MH_2L/MHL.H$	6.79

Bibliography: 70KAT

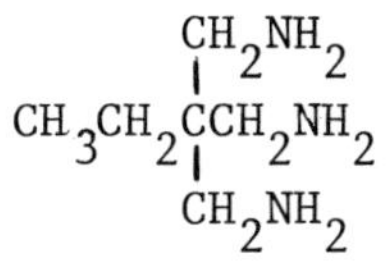

$$CH_2NH_2$$
$$|$$
$$CH_3CH_2CCH_2NH_2$$
$$|$$
$$CH_2NH_2$$

$C_6H_{17}N_3$	2,2-Bis(aminomethyl)butylamine	(1,1,1-tris(aminomethyl)propane)	L

Metal ion	Equilibrium	Log K 25°, 0.5
H^+	HL/H.L	10.23
	$H_2L/HL.H$	8.17
	$H_3L/H_2L.H$	5.53
Ni^{2+}	$ML/M.L$	10.47
	$ML_2/M.L^2$	17.97
	MHL/ML.H	4.91
	$MHL_2/ML_2.H$	6.34
Cu^{2+}	$ML/M.L$	11.17
	$ML_2/M.L^2$	19.32
	MHL/ML.H	7.26
	$MHL_2/ML_2.H$	8.22
	$MH_2L_2/MHL_2.H$	7.03
	MOHL/ML.OH	5.77
Zn^{2+}	$ML/M.L$	6.90
	$ML_2/M.L^2$	11.25
	MHL/ML.H	6.67
	MOHL/ML.OH	5.02
	$M(OH)_2L/MOHL.OH$	3.24

Bibliography: 77MSV

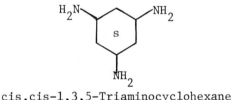

$C_6H_{15}N_3$	cis,cis-1,3,5-Triaminocyclohexane (Other reference on Vol.2, p.70)	L

Metal ion	Equilibrium	Log K 25°, 0.1	ΔH 25°, 0.1	ΔS 25°, 0.1
H^+	HL/H.L	10.16	-12.4	5
	$H_2L/HL.H$	8.66	-11.7	0
	$H_3L/H_2L.H$	7.17	-11.0	4
Ni^{2+}	$ML/M.L$	9.88		
Cu^{2+}	$ML/M.L$	10.55		
	MOHL/ML.OH	6.08		
Zn^{2+}	$ML/M.L$	6.90	0.0	31
	MOHL/ML.OH	5.85	-4.2	13

Bibliography:

H^+,Zn^{2+} 71CW,76FM   Other reference: 79MSI

Ni^{2+},Cu^{2+} 71CW

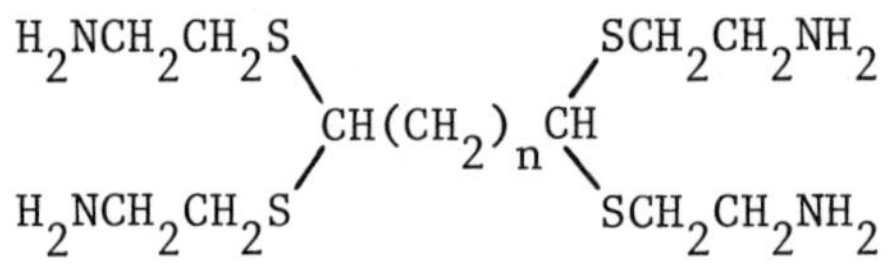

$$\begin{array}{c} H_2NCH_2CH_2S \qquad\qquad SCH_2CH_2NH_2 \\ \diagdown \qquad\qquad\qquad \diagup \\ CH(CH_2)_nCH \\ \diagup \qquad\qquad\qquad \diagdown \\ H_2NCH_2CH_2S \qquad\qquad SCH_2CH_2NH_2 \end{array}$$

$C_xH_yN_4S_4$ <u>Alkane-diylidenetetrakis(thioethyleneamine)</u>   L

n=	Metal ion	Equilibrium	Log K 25°, 0.1
0	H^+	$HL/H.L$	9.43
(ethane-)		$H_2L/HL.H$	9.12
$(C_{10}H_{26}N_4S_4)$		$H_3L/H_2L.H$	8.47
		$H_4L/H_3L.H$	8.01
	Mg^{2+}	$MH_2L/M.H_2L$	3.93
	Ca^{2+}	$MH_2L/M.H_2L$	3.86
	Co^{2+}	$MH_2L/M.H_2L$	3.90
	Ni^{2+}	$MH_2L/M.H_2L$	5.18
	Cu^{2+}	$MH_2L/M.H_2L$	7.67
	Zn^{2+}	$MH_2L/M.H_2L$	4.67
	Cd^{2+}	$MH_2L/M.H_2L$	4.05
	Hg^{2+}	$MH_2L/M.H_2L$	10.35
		$ML/M.L$	18.57
1	H^+	$HL/H.L$	9.43
(propane-1,3-)		$H_2L/HL.H$	9.17
$(C_{11}H_{28}N_4S_4)$		$H_3L/H_2L.H$	8.58
		$H_4L/H_3L.H$	8.08
	Hg^{2+}	$ML/M.L$	18.79
2	H^+	$HL/H.L$	9.51
(butane-1,4-)		$H_2L/HL.H$	9.38
$(C_{12}H_{30}N_4S_4)$		$H_3L/H_2L.H$	8.71
		$H_4L/H_3L.H$	8.26
	Hg^{2+}	$ML/M.L$	19.36
3	H^+	$HL/H.L$	9.90
(pentane-1,5-)		$H_2L/HL.H$	9.38
$(C_{13}H_{32}N_4S_4)$		$H_3L/H_2L.H$	8.86
		$H_4L/H_3L.H$	8.31
	Hg^{2+}	$ML/M.L$	19.55

Bibliography: 76CJ

$$CH_3NHCH_3$$

C_2H_7N			Dimethylamine			L
			(Other values on Vol.2, p.72)			

Metal ion	Equilibrium	Log K 25°, 0.2	Log K 25°, 0.5	Log K 25°, 0	ΔH 25°, 0	ΔS 25°, 0
H^+	HL/H.L	10.80	10.87	10.774-0.04	-12.0 ±0.1	9
CH_3Hg^+	ML/M.L		6.76			

Bibliography:

H^+ 74RO,77BO

CH_3Hg^+76RO,

Other references: 68PM,79FS

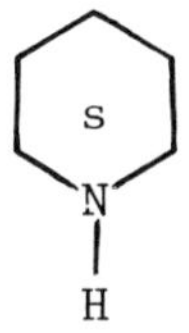

$C_5H_{11}N$			Piperidine (pentamethyleneimine)			L
			(Other references on Vol.2, p.73)			

Metal ion	Equilibrium	Log K 25°, 0.2	Log K 25°, 0.5	Log K 25°, 0	ΔH 25°, 0	ΔS 25°, 0
H^+	HL/H.L	11.1	11.12 ±0.01	11.125±0.003	-12.70 ±0.06 / -13.17[b]±0.02	8.3[b] / 6.7[b]
Ag^+	ML/M.L		3.10 ±0.07		-5.3[b]	-4[b]
	$ML_2/M.L^2$		6.52 ±0.1		-12.0[b]	-10[b]
Hg^{2+}	ML/M.L		8.74		-12[b]	0[b]
	$ML_2/M.L^2$		17.4		-18[b]	19[b]

[b]25°, 0.5

Bibliography:

H^+ 76LMR,78MH

Ag^+ 75EB

Hg^{2+} 72B,78MH

Other references: 68PM,78PSb,78SR

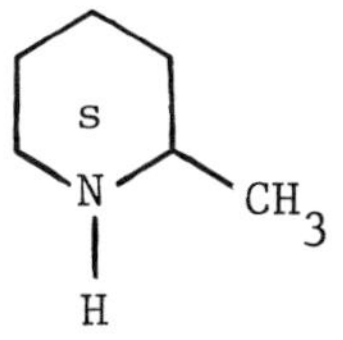

$C_6H_{13}N$		DL-2-Methylpiperidine			L
		(Other references on Vol.2, p.74)			

Metal ion	Equilibrium	Log K 25°, 0.5	ΔH 25°, 0.5	ΔS 25°, 0.5
H^+	HL/H.L	11.06	-14.03	3.5
Ag^+	ML/M.L	3.45	-5.5	-3
	$ML_2/M.L^2$	(6.94)[x]	(-12.4)[x]	(-10)[x]

[x]DL-mixture.

Bibliography: 75EB

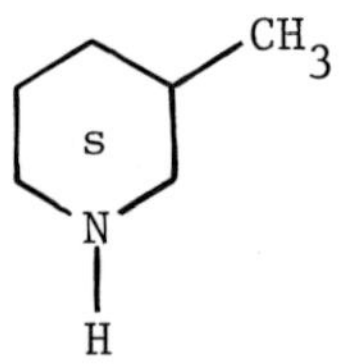

$C_6H_{13}N$ DL-3-Methylpiperidine   L
                                  (Other references on Vol.2, p.75)

Metal ion	Equilibrium	Log K 25°, 0.5	ΔH 25°, 0.5	ΔS 25°, 0.5
H^+	HL/H.L	11.05	-13.75	4.4
Ag^+	ML/M.L	3.04	-5.5	-5
	$ML_2/M.L^2$	$(6.43)^x$	$(-12.2)^x$	$(-12)^x$

―――――――――

[x] DL-mixture.

Bibliography: 75EB

―――

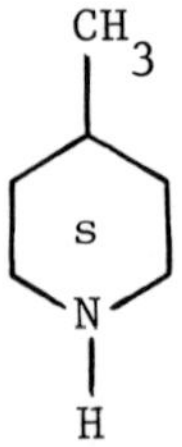

$C_6H_{13}N$ 4-Methylpiperidine   L
                                  (Other references on Vol.2, p.76)

Metal ion	Equilibrium	Log K 25°, 0.5	ΔH 25°, 0.5	ΔS 25°, 0.5
H^+	HL/H.L	11.08	-13.48	5.5
Ag^+	ML/M.L	3.20	-5.4	-3
	$ML_2/M.L^2$	6.50	-12.0	-11

―――――――――

Bibliography: 75EB

―――

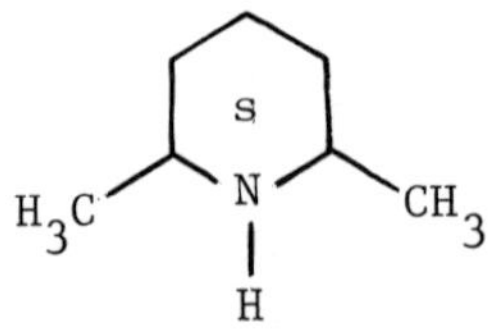

$C_7H_{15}N$ DL-2,6-Dimethylpiperidine   L
                                  (Other references on Vol.2, p.76)

Metal ion	Equilibrium	Log K 25°, 0.5	ΔH 25°, 0.5	ΔS 25°, 0.5
H^+	HL/H.L	11.11	-14.53	2.1
Ag^+	ML/M.L	3.96	-5.8	-1
	$ML_2/M.L^2$	$(7.71)^x$	$(-11.6)^x$	$(-4)^x$

―――――――――

[x] DL-mixture.

Bibliography: 75EB

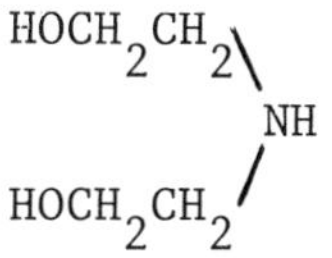

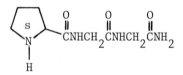

HOCH$_2$CH$_2$\
　　　　　　NH
HOCH$_2$CH$_2$/

| C$_4$H$_{11}$O$_2$N | Iminodi-2-ethanol (diethanolamine) (Other values on Vol.2, p.80) | | | | | L |

Metal ion	Equilibrium	Log K 25°, 0.1	Log K 25°, 0.5	Log K 25°, 0	ΔH 25°, 0	ΔS 25°, 0
H$^+$	HL/H.L	8.90	9.00 ±0.00	8.883	-10.07 ±0.06	6.8
Ag$^+$	ML/M.L		2.69		-5[b]	-4[b]
	ML$_2$/M.L^2		5.48	5.80[o]	-12[b]	-15[b]
Cd^{2+}	ML/M.L	2.40			-2[b]	4[a]
	ML$_2$/M.L^2	4.52			-4[b]	7[a]
Hg^{2+}	ML/M.L		7.84		-19[b]	-28[b]
	ML$_2$/M.L^2		15.66		-29[b]	-25[b]

[a] 25°, 0.1;　[b] 25°, 0.5;　[o] 20°, 0

Bibliography: 78MH　　　　　　　　　　　Other references:　68DP,73BSa,74UP

| C$_9$H$_{16}$O$_3$N$_4$ | L-Prolylglycylglycinamide | L |

Metal ion	Equilibrium	Log K 25°, 1.0
Cu(III)	M(H$_{-3}$L)/M(H$_{-4}$L).H	11.7

Bibliography:　79NKC

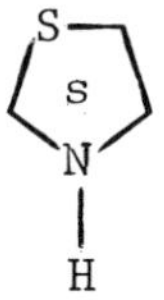

| C$_3$H$_7$NS | 1,3-Thiazolidine (1-thia-3-azacyclopentane) | | | L |

Metal ion	Equilibrium	Log K 25°, 0.5	ΔH 25°, 0.5	ΔS 25°, 0.5
H$^+$	HL/H.L	6.54	-7.41	5.1
Ag$^+$	ML/M.L	8.84		
	ML$_2$/M.L^2	12.1		

Bibliography:
H$^+$　77BE;　Ag$^+$　79BB

$$CH_3NHCH_2CH_2NHCH_3$$

| $C_4H_2N_2$ | | N,N'-Dimethylethylenediamine (Other values on Vol.2, p.88) | | | | L |

Metal ion	Equilibrium	Log K 25°, 0.1	Log K 25°, 1.0	Log K 25°, 0	ΔH 25°, 0.5	ΔS 25°, 0.5
H^+	HL/H.L	10.05 ±0.03	10.34 ±0.08	10.03		
		10.17[b]±0.08			-10.5 ±0.3	11
	$H_2L/HL.H$	7.05 ±0.04	7.55 ±0.02	6.80		
		7.31[b]±0.05			-9.7 ±0.6	1
Ni^{2+}	ML/M.L	6.83 ±0.07		6.84	-8.1[a]	5*
		7.04[b]±0.02				
	$ML_2/M.L^2$	10.80 ±0.04		10.69	-13.6[a]	6*
		11.19[b]				
	$ML_3/M.L^3$	13.1	13.3[t]			
Cu^{2+}	ML/M.L	9.96 ±0.06	10.41	9.96		
		10.15[b]±0.05			-11.1 ±0.0	9
	$ML_2/M.L^2$	16.9 ±0.1	17.9	16.9		
		17.4[b] ±0.1			-20.6 ±0.2	10
	MOHL/ML.OH	5.67 ±0.02				
	$M(OH)_2L/ML.(OH)^2$	8.31[b]			-7.4	13
	$M_2(OH)_2L_2/(ML)^2.(OH)^2$	15.02[b]			-11.2	31

[a]25°, 0.2; [b]25°, 0.5; [t]25°, 0.65; *assuming ΔH for 0.2=ΔH for 0.5

Bibliography:

H^+ 70K,72CV,76SG Cu^{2+} 70K,76SG

Ni^{2+} 79SG

$$CH_3CH_2NHCH_2CH_2NHCH_2CH_3$$

| $C_6H_{16}N_2$ | | N,N'-Diethylethylenediamine (Other values on Vol.2, p.89) | | | | L |

Metal ion	Equilibrium	Log K 25°, 0.1	Log K 25°, 1.0	Log K 25°, 0	ΔH 25°, 0.5	ΔS 25°, 0.5
H^+	HL/H.L	10.16 ±0.06	10.56	10.16	-10.9	11
		10.33[b]±0.06				
	$H_2L/HL.H$	7.28 ±0.04	7.73	6.96	-10.5	-1
		7.50[b]±0.03				
Cu^{2+}	ML/M.L	8.85	9.28	8.79	-8.2	14
		9.05[b]				
	$ML_2/M.L^2$	14.6[b]		14.4	-15.5	15
	MOHL/ML.OH	(0.8)[b]			(-5)	(-13)
	$M_2(OH)_2L_2/(ML)^2.(OH)^2$	(0.6)[b]			(-11)	(-34)

[b]25°, 0.5

Bibliography: 75BF

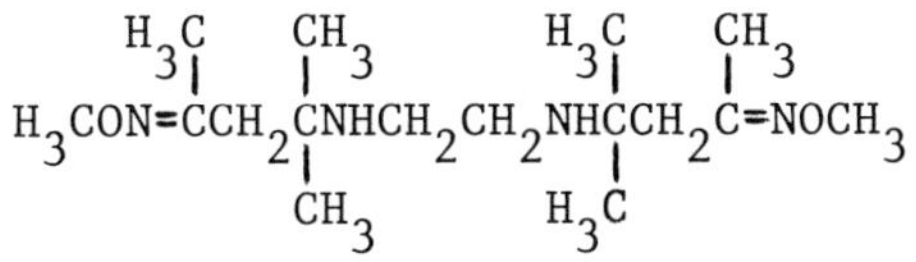

$$H_3CON=CCH_2CNHCH_2CH_2NHCCH_2C=NOCH_3$$

with methyl groups H_3C, CH_3, H_3C, CH_3 above and CH_3, H_3C below

| $C_{16}H_{24}O_2N_4$ | 4,4,9,9-Tetramethyl-5,8-diazadodecane-2,11-dione dioxime dimethylether | | | | L |

Metal ion	Equilibrium	Log K 25°, 0.1	ΔH 25°, 0.1	ΔS 25°, 0.1
H^+	HL/H.L	9.34	-8.58	13.9
	H_2L/HL.H	6.19	-8.73	-1.2
Cu^{2+}	ML/M.L	9.10	(-4.3)	(27)

Bibliography:

H^+ 77PR Cu^{2+} 78DMP

$$H_3CON=CCH_2CNHCH_2CH_2NHCCH_2C=NOH$$

with methyl groups H_3C, CH_3, H_3C, CH_3 above and CH_3, H_3C below

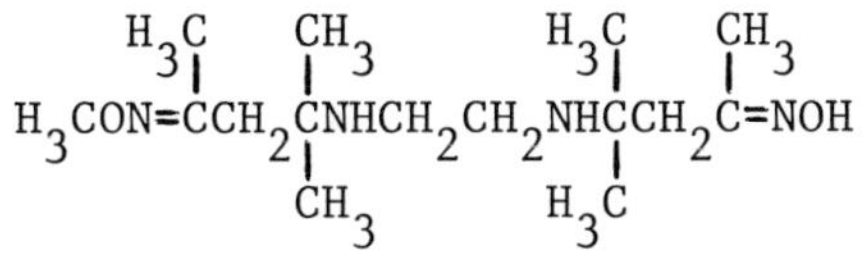

| $C_{15}H_{32}O_2N_4$ | 4,4,9,9-Tetramethyl-5,8-diazadodecane-2,11-dione dioxime methylether | | | | L |

Metal ion	Equilibrium	Log K 25°, 0.1	ΔH 25°, 0.1	ΔS 25°, 0.1
H^+	HL/H.L	9.38	-9.06	12.5
	H_2L/HL.H	6.28	-9.28	-2.4
Co^{2+}	ML/M.L	5.7		
	MOHL/ML.OH	4.4		
Ni^{2+}	ML/M.L	6.6		
	ML/M(H_{-1}L).H	8.2		
Cu^{2+}	ML/M.L	12.11	-10	22
	ML/M(H_{-1}L).H	6.76	-5.2	13
Zn^{2+}	ML/M.L	5.34		
	ML_2/M.L^2	11.23		
	ML/M(H_{-1}L).H	9.19		

Bibliography:

H^+ 77PR Cu^{2+} 78DMP

Co^{2+}, Ni^{2+}, Zn^{2+} 78PR

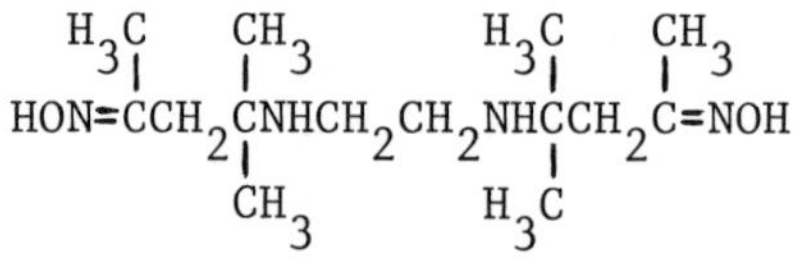

```
           H3C    CH3        H3C   CH3
            |      |          |     |
     HON=CCH2CNHCH2CH2NHCCH2C=NOH
            |                 |
           CH3               H3C
```

$C_{14}H_{30}O_2N_4$ 4,4,9,9-Tetramethyl-5,8-diazadodecane-2,11-dione dioxime   L

(Other values on Vol.2, p.94)

Metal ion	Equilibrium	Log K 25°, 0.1	Log K 25°, 0.2	Log K 25°, 0	ΔH 25°, 0	ΔS 25°, 0
H^+	HL/H.L	9.41	9.47	9.35	-9.5	11
	$H_2L/HL.H$	6.35	6.46	5.89	-9.4	-5
Fe^{2+}	$M(H_{-1}L)H/M.L$	-3.51				
Co^{2+}	$M(H_{-1}L).H/M.L$	-0.58				
	$MOH(H_{-1}L)/M(H_{-1}L).OH$	3.32				
Ni^{2+}	$M(H_{-1}L).H/M.L$	2.9				
Zn^{2+}	$M(H_{-1}L).H/M.L$	-0.31				
	$MOH(H_{-1}L)/M(H_{-1}L).OH$	3.87				

Bibliography:

Fe^{2+},Ni^{2+},Zn^{2+} 78PR Co^{2+} 78KP

$$HOCH_2CH_2NHCH_2CH_2NH_2$$

$C_4H_{12}ON_2$ 2-(2-Aminoethylamino)ethanol   (N-(2-hydroxyethyl)ethylenediamine)   L

(Other values on Vol.2, p.95)

Metal ion	Equilibrium	Log K 25°, 0.1	Log K 25°, 0.5	Log K 25°, 0	ΔH 25°, 0.5	ΔS 25°, 0.5
H^+	HL/H.L	9.59 ±0.03	9.74 -0.01	9.56	-11.1	7
	$H_2L/HL.H$	6.60 -0.01	6.85 ±0.1	6.34	-10.1	-3
Ni^{2+}	$ML/M.L$	6.82	6.97 -0.3	6.76	-7.8	6
	$ML_2/M.L^2$	12.44	12.80 -0.3	12.28	-16.1	5
Cu^{2+}	$ML/M.L$	10.09 ±0.02	10.33	10.02	-11.1	10
	$ML_2/M.L^2$	17.62 ±0.04	18.09 -0.1	17.45	-12	43
	$MOHL/ML.OH$	6.46 ±0.02		6.91	$(-7)^r$	$(6)^a$
	$M(OH)_2L/ML.(OH)^2$	10.29	17.64	11.07	-12.3	39
	$(MOHL)_2/(MOHL)^2$	2.15 ±0.04	2.29	1.87	$(-4)^r$	$(-4)^a$

---

[a] 25°, 0.1;   [r] 0-43°, 0.1

Bibliography: 75B   Other references: 70K,76H,76RR

$$CH_3SCH_2CH_2NHCH_2CH_2NH_2$$

$C_5H_{14}N_2S$ 7-Thia-1,4-diazaoctane   (N-(2-methylthioethyl)ethylenediamine)   L

Metal ion	Equilibrium	Log K 25°, 0.5	ΔH 25°, 0.5	ΔS 25°, 0.5
H+	HL/H.L	9.60	-11.40	5.7
	H₂L/HL.H	6.63	-10.31	-4.3
Ni²⁺	ML/M.L₂	7.06	-8.7	3
	ML₂/M.L²	12.92	-19.3	-6
Cu²⁺	ML/M.L₂	11.38	-14.0	5
	ML₂/M.L²	17.68	-24.8	-2

Bibliography: 77HGE

$$CH_3CH_2SCH_2CH_2NHCH_2CH_2NH_2$$

$C_6H_{16}N_2S$ 7-Thia-1,4-diazanonane   (N-(2-ethylthioethyl)ethylenediamine)   L

Metal ion	Equilibrium	Log K 25°, 0.5	ΔH 25°, 0.5	ΔS 25°, 0.5
H+	HL/H.L	9.61	-11.14	6.6
	H₂L/HL.H	6.61	-10.16	-3.9
Ni²⁺	ML/M.L₂	7.23	-9.0	3
	ML₂/M.L²	12.86	-19.9	-8
Cu²⁺	ML/M.L₂	11.60	-14.3	5
	ML₂/M.L²	17.90	-24.6	-7

Bibliography: 79HG

$$CH_3 \overset{\overset{\displaystyle CH_3}{|}}{\underset{\underset{\displaystyle CH_3}{|}}{C}} SCH_2CH_2NHCH_2CH_2NH_2$$

$C_8H_{20}N_2S$ 8,8-Dimethyl-7-thia-1,4-diazanonane   L
          (N-[2-(1,1-dimethylethylthio)ethyl]ethylenediamine)

Metal ion	Equilibrium	Log K 25°, 0.5	ΔH 25°, 0.5	ΔS 25°, 0.5
H+	HL/H.L	9.61	-11.57	5.1
	H₂L/HL.H	6.53	-9.68	-2.6
Ni²⁺	ML/ML₂	6.52	-8.7	1
	ML₂/M.L²	11.18	-17.4	-7
Cu²⁺	ML/M.L₂	10.84	-12.8	7
	ML₂/M.L²	17.23	-23.6	-4

Bibliography: 79HG

$$CH_3SCH_2CH_2NHCH_2CH_2CH_2NH_2$$

$C_6H_{16}N_2S$ <u>8-Thia-1,5-diazanonane</u>   (N-(2-methylthioethyl)trimethylenediamine)   L

Metal ion	Equilibrium	Log K 25°, 0.5	ΔH 25°, 0.5	ΔS 25°, 0.5
H^+	HL/H.L	10.26	-12.83	1.2
	H_2L/HL.H	8.19	-11.40	-8.0
Ni^{2+}	ML/M.L	6.88	-8.1	4
Cu^{2+}	ML/M.L	11.01	-13.2	6
	MOHL/ML.OH	5.15	-2.0	14

Bibliography: 79HG

$$CH_3NHCH_2CH_2SCH_2CH_2OCH_2CH_2SCH_2CH_2NHCH_3$$

$C_{10}H_{24}ON_2S_2$ <u>8-Oxa-5,11-dithia-2,14-diazapentadecane</u>   L

Metal ion	Equilibrium	Log K 25°, 0.1	ΔH 25°, 0.1	ΔS 25°, 0.1
H^+	HL/H.L	9.84	-10.3	10
	H_2L/HL.H	9.06	-12.3	0
Co^{2+}	$ML_2/M.L^2$	8.22		
Ni^{2+}	ML/M.L	4.78		
	$ML_2/M.L^2$	8.0		
Cu^{2+}	ML/M.L	9.15	-10	8
	$ML_2/M.L^2$	14.6		
Ag^+	ML/M.L	7.32		
	MHL/ML.H	7.81		
	M_2L/ML.M	2.9		
Zn^{2+}	ML/M.L	5.6		
	$ML_2/M.L^2$	10.8		
Cd^{2+}	ML/M.L	4.40		
Pb^{2+}	ML/M.L	7.49	-10	1

Bibliography: 79ASJ,79AS   Other reference: 75ASa

$$H_2NCH_2CH_2NHCH_2CH_2NH_2$$

$C_4H_{13}N_3$ <u>1,4,7-Triazaheptane</u>   (<u>diethylenetriamine, dien, 2,2-tri</u>)   L
(Other values on Vol.2, p.101)

Metal ion	Equilibrium	Log K 25°, 0.1	Log K 25°, 1.0	Log K 25°, 0	ΔH 25°, 0.1	ΔS 25°, 0.1
H[+]	HL/H.L	9.84 ±0.05 9.88[b] -0.04	9.92 ±0.03	9.80	-11.2	7
	$H_2L/HL.H$	9.02 ±0.06 9.09[b] -0.03	9.24 ±0.00	8.74	-12.0	1
	$H_3L/H_2L.H$	4.23 ±0.03 4.47[b] ±0.03	4.73 ±0.00	3.64	-7.2	-5
Ag[+]	ML/M.L	6.10	6.21[t]			
		6.1[h]				
	MHL/ML.H	7.0[h]				
	$M_2L/ML.M$	1.4[h]				
Cd[2+]	ML/M.L	8.3 -0.1			(-10)[s]	(4)*
Pb[2+]	ML/M.L	7.4			(-9)[s]	(4)*

[b] 25°, 0.5;   [h] 20°, 0.1;   [s] 15-35°, 0.2;   [t] 25°, 1.5;   *assuming ΔH for 0.2=ΔH for 0.1

Bibliography:

Ag[+]   77NF,79MST   Ohter references: 75WT,75WO,76PB

Cd[2+],Pb[2+]   78KKa

$$H_2NCH_2CH_2NHCH_2CH_2CH_2NH_2$$

$C_5H_{15}N_3$ <u>1,4,8-Triazaoctane</u>   (<u>2,3-tri</u>)   L
(Other references on Vol.2, p.103)

Metal ion	Equilibrium	Log K 25°, 0.1	Log K 25°, 0.5	ΔH 25°, 0.5	ΔS 25°, 0.5
H[+]	HL/H.L	10.21 ±0.01	10.44	-12.2	7
	$H_2L/HL.H$	9.17 +0.01	9.36	-12.1	2
	$H_3L/H_2L.H$	6.10 ±0.03	6.37	-10.0	-4

1,4,8-Triazaoctane   (2,3-tri) (continued)

Metal ion	Equilibrium	Log K 25°, 0.1	Log K 25°, 0.5	ΔH 25°, 0.5	ΔS 25°, 0.5
Co^{2+}	$ML/M.L$	$8.27 +0.2$	6.37	-10.0	-4
	$ML_2/M.L^2$	13.2		$(-10)^s$	$(4)^a$
Ni^{2+}	$ML/M.L$	11.01 ± 0.0	11.23	-13.5	6
	$ML_2/M.L^2$	18.0	18.29	-27.2	-8
	$MHL/M.HL$		5.86	-7.8	1
Cu^{2+}	$ML/M.L$	$16.30 +0.1$	16.60	-19.2	12
	$ML_2/M.L^2$	19.8	19.88	-25.3	6
	$MHL/M.HL$		9.02	-12	1
	$MHL_2/ML.HL$		2.51	-5.9	-8
	$MOHL/ML.OH$		4.72	-2.3	14
Zn^{2+}	$ML/M.L$	8.62 ± 0.0	8.77	-8.4	12
	$ML_2/M.L^2$	12.4	12.57	-16.9	1
	$MOHL/ML.OH$		4.99	-0.1	23
Cd^{2+}	$ML/M.L$	7.8			
	$ML_2/M.L^2$	11.5			
	$MHL/M.HL$	4.1			

---

^{a}25°, 0.1; ^{s}15-40°, 0.1

Bibliography: Cd^{2+} 73AH

H^+ 73AH,73BFP,73SM,74BFP Ni^{2+}-Zn^{2+} 73AH,73BFP,73SM,74BFP

Co^{2+} 73AH,73SM   Other reference: 75WO

$$H_2NCH_2CH_2NHCH_2CH_2CH_2CH_2NH_2$$

$C_6H_{17}N_3$	1,4,9-Triazanonane (2,4-tri)				L

Metal ion	Equilibrium	Log K 25°, 0.1	Log K 25°, 0.5	ΔH 25°, 0.1	ΔS 25°, 0.1
H^+	$HL/H.L$	10.65		-12.2^b	8^b
			10.67	-12.9^b	6^b
	$H_2L/HL.H$	9.42		-11.6^b	4^b
			9.59	-12.1^b	3^b
	$H_3L/H_2L.H$	6.71		-9.8^b	-2^b
			6.98	-10.7	-4^b
Ni^{2+}	$ML/M.L$		7.8		
	$ML_2/M.L^2$		11.8		
	$MHL/M.HL$		6.03		
	$M(HL)_2/M.(HL)^2$		10.60		
	$MHL_2/M.HL.L$		11.62		

---

^{b}25°, 0.5

2,4-**Tri**   (continued)

Metal ion	Equilibrium	Log K 25°, 0.1	Log K 25°, 0.5	ΔH 25°, 0.1	ΔS 25°, 0.1
Cu^{2+}	ML/M.L	13.44		-15.3^{b}	10^{b}
			13.05	-15.3^{b}	8^{b}
	$ML_2/M.L^2$		17.66	-22.2^{b}	6^{b}
	MHL/M.HL	8.94		-10.7^{b}	5^{b}
			9.19	-11.7^{b}	3^{b}
	$M(HL)_2/M.(HL)^2$		16.59	-23.9^{b}	-4^{b}
	$MHL_2/M.HL.L$		17.43	-23.4^{b}	1^{b}
	MOHL/ML.OH	4.42		-1.3^{b}	16^{b}
			4.36	-2.6^{b}	11^{b}
Zn^{2+}	ML/M.L		6.3		
	$ML_2/M.L^2$		10.4		
	MHL/M.HL		4.36		
	$M(HL)_2/M.(HL)^2$		8.31		
	$MHL_2/M.HL.L$		9.72		
	MOHL/ML.OH		5.48		
	$M(OH)_2L/ML.(OH)^2$		8.29		

^{b}25°, 0.5

Bibliography:

H^+,Cu^{2+} 74PPa,75BPV,76BBb,76GP Ni^{2+},Zn^{2+} 75BPV

$$H_2NCH_2CH_2CH_2NHCH_2CH_2CH_2NH_2$$

$C_6H_{17}N_3$ 1,5,9-Triazanonane   (3,3-tri)   L
                    (Other values on Vol.2, p.104)

Metal ion	Equilibrium	Log K 25°, 0.1	Log K 25°, 1.0	ΔH 25°, 0.1	ΔS 25°, 0.1
H^+	HL/H.L	10.65 ±0.10	10.74 ±0.01	-12.3	7
	$H_2L/HL.H$	9.57 ±0.04	9.79 ±0.02	-13.0	0
	$H_3L/H_2L.H$	7.69 ±0.04	8.08 ±0.02	-10.5	0
Ni^{2+}	ML/M.L	9.2			
Cu^{2+}	ML/M.L	14.2 ±0.1	14.6 ±0.1	-16.1	11
	MOHL/ML.OH	4.15 ±0.05	4.21	-2.3	11
Ag^+	$M_2L_2/M^2.L^2$		8.60	$(-16)^{s}$	$(-14)^{c}$
Zn^{2+}	ML/M.L	7.92	8.36	-5.4	18
	MOHL/ML.OH	5.23	5.57	-3.8	11
Cd^{2+}	ML/M.L	6.6	6.85	$(-6)^{s}$	$(11)^{c}$

^{s}25-55°, 1.0; ^{c}25°, 1.0

Bibliography:

H^+ 73AH,74DF,75KH Ag^+,Zn^{2+} 74DF

Ni^{2+} 73AH Cd^{2+} 73AH,74DF

Cu^{2+} 73AH,74DF,75KH   Other reference: 75WO

$$H_2NCH_2CH_2CH_2NHCH_2CH_2CH_2CH_2NH_2$$

$C_7H_{19}N_3$ 1,5,10-Triazadecane (3,4-tri)   L

Metal ion	Equilibrium	Log K 25°, 0.1		ΔH 25°, 0.1	ΔS 25°, 0.1
H⁺	HL/H.L	10.89		−13.6	4
	$H_2L/HL.H$	9.81		−12.8	2
	$H_3L/H_2L.H$	8.34		−11.7	−1
Cu²⁺	ML/M.L	11.61			
	MHL/M.HL	6.8			
	MOHL/ML.OH	4.48			

Bibliography:

H⁺   74PPa,76GP   Cu²⁺   74PPa

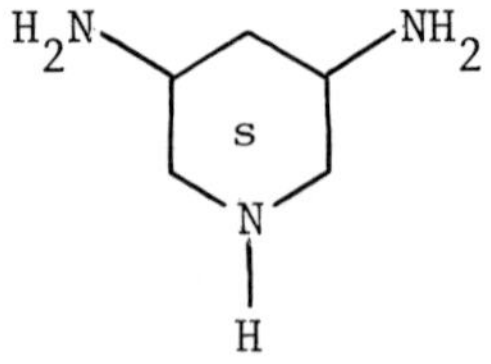

$C_5H_{13}N_3$ cis-3,5-Diaminopiperidine   L

Metal ion	Equilibrium	Log K 20°, 0.1
Pd²⁺	$M(HL)L/ML_2.H$	6.35
	$M(HL)_2/M(HL)L.H$	4.16

Bibliography: 79MSI

$$H_2NCH_2CH_2CH_2NHCH_2CH_2CH_2NHCH_3$$

$C_7H_{19}N_3$ 1,5,9-Triazadecane   L

Metal ion	Equilibrium	Log K 25°, 0.1
H⁺	HL/H.L	10.68
	$H_2L/HL.H$	9.69
	$H_3L/H_2L.H$	7.03
Cu²⁺	ML/M.L	12.94
	MOHL/ML.OH	3.95

Bibliography: 75KH

$$H_2NCH_2CH_2NHCH_2CH_2NHCH_2CH_2NH_2$$

$C_6H_{18}N_4$ <u>1,4,7,10-Tetraazadecane</u> (triethylenetetramine, <u>trien</u>, <u>2,2,2-tet</u>)   L

(Other values on Vol.2, p.105)

Metal ion	Equilibrium	Log K 25°, 0.1	Log K 25°, 0.5	Log K 25°, 1.0	ΔH 25°, 0.1	ΔS 25°, 0.1
H[+]	HL/H.L	9.74 ±0.06	9.87 ±0.09	10.02	-11.0	8
	H_2L/HL.H	9.07 ±0.03	9.21 ±0.10	9.39	-11.3	4
	H_3L/H_2L.H	6.56 ±0.02	6.87 ±0.01	7.00	-9.5	-2
	H_4L/H_3L.H	3.25 ±0.03	3.71 ±0.05	3.87	-6.8	-8
Ni[2+]	ML/M.L	14.0 ±0.3	14.4	14.5	-14.0 +0.1	17
	ML_2/M.L^2	(20.6)	18.6			
	M_2L_3/M^2.L^3	(40.0)	36.9			
	MHL/ML.H	4.8[h]				
	ML(yellow)/ML(blue)	-1.9			(3)[r]	(3)
Cu[2+]	ML/M.L	20.1 -0.1		20.9	-21.5 ±0.1	20
	MHL/ML.H	3.6 ±0.1				
	ML/MOHL.H	10.7 ±0.1				
Cr[3+]	ML/M.L	4.58[p]		4.47		
	MOHL/M(OH)$_2$L.H	7.21[p]		7.14		
Ag[+]	ML/M.L	7.50 [h] 7.65 [h]				
	MHL/ML.H	8.0 [h]				
	MH_2L/MHL.H	6.2 [h]				
	M_2L/ML.M	2.4 [h]				
Hg[2+]	ML/M.L	24.8 ±0.3	25.3 [i]		(-30)[s]	(13)
	MHL/ML.H		5.5 [i]			
Pb[2+]	ML/M.L	10.4 -0.1			(-8)[s]	(21)

<hr>

[h]20°, 0.1; [i]20°, 0.5; [p]4°, 0.1; [r]10-40°, 0.1; [s]15-35°, 0.2

Bibliography:

H[+]   76PB   Ag[+]   79MST

Ni[2+] 74HM,79LP   Hg[2+] 75LB,76KKd

Cu[2+] 76HM,77LS   Pb[2+] 77KKa

Cr[3+] 57H,76MM   Other references: 71P,78WNc,79GBG

<hr>

$$H_2NCH_2CH_2NHCH_2CH_2CH_2NHCH_2CH_2NH_2$$

$C_7H_{20}N_4$ <u>1,4,8,11-Tetraazaundecane</u> (2,3,2-tet)   L

(Other reference on Vol.2, p.107)

Metal ion	Equilibrium	Log K 25°, 0.1	Log K 25°, 0.5	ΔH 25°, 0.5	ΔS 25°, 0.5
H[+]	HL/H.L	9.99	10.25	-11.0	10
	H_2L/HL.H	9.29	9.50	-11.3	5
	H_3L/H_2L.H	7.46	7.28	-10.0	0
	H_4L/H_3L.H	5.49	6.02	-9.2	-3
Co[2+]	ML/M.L	12.2 +2			
Ni[2+]	ML/M.L	16.0 +0.1	16.4	-17.9 -1	15
	ML_2/M.L^2		20.1	-22.0	18
	ML(yellow)/ML(blue)	-0.54-0.01		(3)[r] ±0	(9)[a]

<hr>

[r]19-56°, 0.1; [a]25°, 0.1

2,3,2-Tet (continued)

Metal ion	Equilibrium	Log K 25°, 0.1	Log K 25°, 0.5	ΔH 25°, 0.5	ΔS 25°, 0.5
Cu^{2+}	ML/M.L	23.2 ±0.0	23.9	-27.7	16
Zn^{2+}	ML/M.L	12.6 +0.1	12.8	-11.9	19
Cd^{2+}	ML/M.L	11.1			
Hg^{2+}	ML/M.L	22.1[a]		(-27)[s]	(9)[a]
Pb^{2+}	ML/M.L	7.8[a]		(-7)[s]	(11)[a]

[a]25°, 0.2; [s]15-35°, 0.2

Bibliography:

H^+	73SMa	Hg^{2+}	76KKd
$Co^{2+}, Cu^{2+}-Zn^{2+}$	73SMa,75APB	Pb^{2+}	77KKa
Ni^{2+}	73SMa,74HM,75APB,77AF,78FP	Other references:	76NG,79KK
Cd^{2+}	75APB		

$H_2NCH_2CH_2CH_2NHCH_2CH_2CH_2CH_2NHCH_2CH_2CH_2NH_2$

$C_{10}H_{26}N_4$ 1,5,10,14-Tetraazatetradecane   (3,4,3-tet)   L

Metal ion	Equilibrium	Log K 25°, 0.1	ΔH 25°, 0.1	ΔS 25°, 0.1
H^+	HL/H.L	10.80	-13.1	5
	$H_2L/HL.H$	10.02	-12.4	4
	$H_3L/H_2L.H$	8.85	-12.4	-1
	$H_4L/H_3L.H$	7.96	-11.5	-2
Cu^{2+}	ML/M.L	14.70	(-21.0)	(-3)
	MHL/M.HL	9.99		

Bibliography:

H^+ 74PP,74PPa Cu^{2+} 74PP,76GP

$H_2NCH_2CH_2NHCH_2CH_2NHCH_2CH_2NHCH_2CH_2NH_2$

$C_8H_{23}N_5$ 1,4,7,10,13-Pentaazatridecane   (tetren)   L
                        (Other values on Vol.2, p.111)

Metal ion	Equilibrium	Log K 25°, 0.1	ΔH 25°, 0.1	ΔS 25°, 0.1
H^+	HL/H.L	9.70 ±0.04	-10.8	8
	$H_2L/HL.H$	9.14 ±0.04	-11.3	4
	$H_3L/H_2L.H$	8.05 ±0.03	-10.7	1
	$H_4L/H_3L.H$	4.70 ±0.02	-7.9	-5
	$H_5L/H_4L.H$	2.97 ±0.07	-6.8	-9
Co^{2+}	ML/M.L	13.3 ±0.2	-13.9	14
	MHL/ML.H	5.43 ±0.03		
Hg^{2+}	ML/M.L	26 ±1	(-33)[s]	(8)[*]
Pb^{2+}	ML/M.L	10.2 ±0.3	(-9)[s]	(16)[*]

[s]15-35°, 0.2; [*]assuming ΔH for 0.2=ΔH for 0.1;

Bibliography: Co^{2+} 78HM Hg^{2+}, Pb^{2+} 78KKa

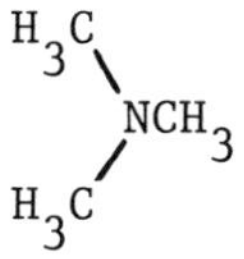

$$H_3C \diagdown$$
$$N CH_3$$
$$H_3C \diagup$$

<table>
<tr><td>C_3H_9N</td><td colspan="5">Trimethylamine
(Other references on Vol.2, p.337)</td><td>L</td></tr>
</table>

Metal ion	Equilibrium	Log K 25°, 0.2	Log K 25°, 0.5	Log K 25°, 0	ΔH 25°, 0	ΔS 25°, 0
H^+	HL/H.L	9.85	9.90	9.800 −0.05	−8.80 ±0.06	15.2
CH_3Hg^+	ML/M.L		5.05			

Bibliography: 74RO

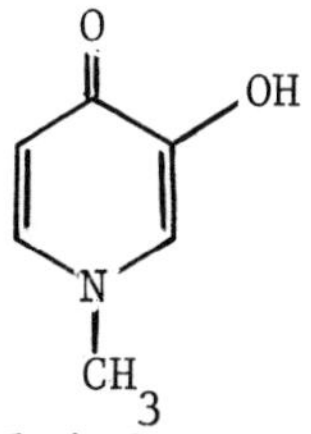

<table>
<tr><td>$C_6H_7O_2N$</td><td>3-Hydroxy-1-methyl-1,4-dihydropyridin-4-one</td><td>HL</td></tr>
</table>

Metal ion	Equilibrium	Log K 37°, 0.15
H^+	HL/H.L	8.68
	$H_2L/HL.H$	3.23
Cu^{2+}	ML/M.L	9.35
	$ML_2/M.L^2$	16.93
	MHL/ML.H	1.6
Zn^{2+}	ML/M.L	6.68
	$ML_2/M.L^2$	12.31
	$ML_3/M.L^3$	14.4
Cd^{2+}	ML/M.L	5.77
	$ML_2/M.L^2$	10.25
	$ML_3/M.L^3$	12.6
Pb^{2+}	ML/M.L	8.44
	$ML_2/M.L^2$	13.57
	MHL/ML.H	2.3

Bibliography: 79SPT

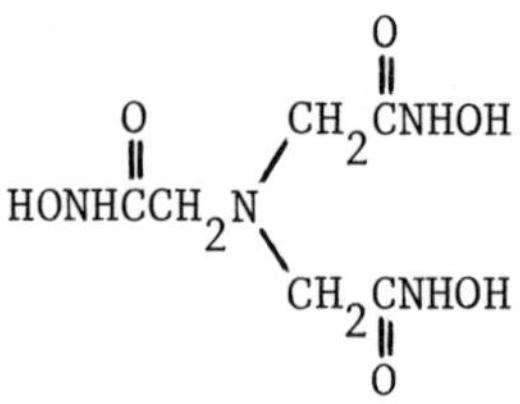

$$\text{HONHCCH}_2\text{N}\begin{array}{c}\text{O}\\ \text{CH}_2\text{CNHOH}\\ \\ \text{CH}_2\text{CNHOH}\end{array}$$

$C_6H_{12}O_6N_4$	Nitrilotriacethydroxamic acid			H_3L

Metal ion	Equilibrium	Log K 25°, 0.1	Log K 25°, 1.0	Log K 25°, 0
H^+	HL/H.L	11.72	11.82	12.39
		11.74[h]	11.63[b]	
	$H_2L/HL.H$	8.82[h]	8.83[b]	9.22
		8.88[h]	8.75[b]	
	$H_3L/H_2L.H$	6.83[h]	6.85[b]	7.02
		6.75[h]	6.74[b]	
	$H_4L/H_3L.H$	5.92[h]	6.04[b]	5.94
		5.90[h]	5.94[b]	
Cu^{2+}	ML/M.L	21.1[h]		
	ML/MOHL.H	9.25[h]		
	MHL/ML.H	4.80[h]		
	$MH_3L/MHL.H^2$	6.90[h]		
Fe^{3+}	ML/M.L	19.4[h]		
	$ML/MOHL.H_2$	9.10[h]		
	$MH_2L/ML.H^2$	10.58[h]		

[b] 25°, 0.5; [h] 20°, 0.1
Bibliography:
H^+ 75Ka,79LN
Cu^{2+},Fe^{3+} 75Ka

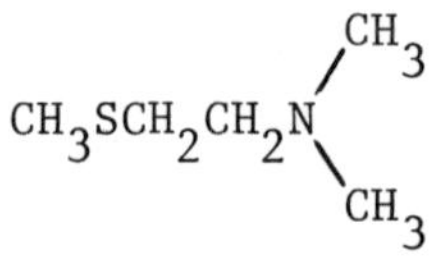

$$CH_3SCH_2CH_2N\begin{array}{c}CH_3\\ \\ CH_3\end{array}$$

$C_5H_{13}NS$	N,N-Dimethyl-2-(methylthio)ethylamine			L

Metal ion	Equilibrium	Log K 25°, 0.5	ΔH 25°, 0.5	ΔS 25°, 0.5
H^+	HL/H.L	9.01	-9.20	10.4
Ag^+	$ML/M.L_2$	4.30		
	$ML_2/M.L^2$	8.42	-16.6	-17
	MHL/M.HL	2.52	-6.3	-10
	$M(HL)_2/M.(HL)^2$	3.84	-12.5	-24
	$MHL_2/ML_2.H$	7.27	-8.4	5
	$M_2L/M^2.L$	5.96	-15	-23
	$M_2L_2/M^2.L^2$	11.37	-21.6	-21

Bibliography: 77TGE,77TGS

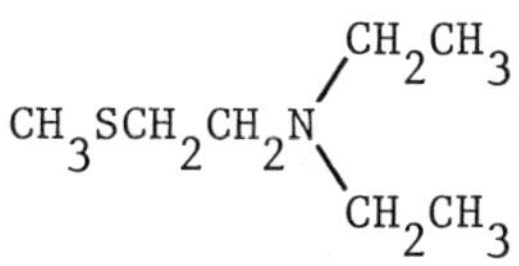

$$CH_3SCH_2CH_2N\begin{array}{c}CH_2CH_3\\ \\CH_2CH_3\end{array}$$

C$_7$H$_{17}$NS	N,N-Diethyl-2-(methylthio)ethylamine			L

Metal ion	Equilibrium	Log K 25°, 0.5	ΔH 25°, 0.5	ΔS 25°, 0.5
H$^+$	HL/H.L	9.56	-10.1	10
Ag$^+$	ML/M.L	4.88		
	ML$_2$/M.L^2	8.66	-15.2	-11
	MHL/M.HL	2.59	-6.5	-10
	M(HL)$_2$/M.(HL)^2	3.98	-12.8	-25
	MHL$_2$/ML$_2$.H	7.90	-8.7	7
	M$_2$L/M^2.L	6.16	-8.3	0
	M$_2$L$_2$/M^2.L^2	11.23	-15.4	0

Bibliography: 77TGE, 77TGS

$$CH_3SCH_2CH_2N\begin{array}{c}CH_3\\ |\\CHCH_3\\ \\CHCH_3\\ |\\CH_3\end{array}$$

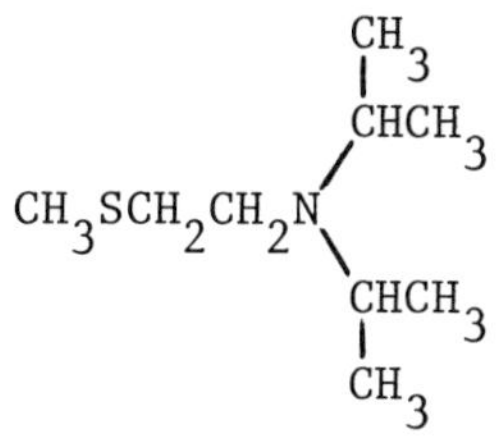

C$_9$H$_{21}$NS	N,N-Bis(1-methylethyl)-2-(methylthio)ethylamine			L

Metal ion	Equilibrium	Log K 25°, 0.5	ΔH 25°, 0.5	ΔS 25°, 0.5
H$^+$	HL/H.L	9.87	-10.49	10.0
Ag$^+$	ML/M.L	4.33		
	ML$_2$/M.L^2	7.62		
	MHL/M.HL	2.61	-7.1	-12
	M(HL)$_2$/M.(HL)^2	4.40	-10.6	-15
	MHL$_2$/ML$_2$.H	8.56		
	M$_2$L/M^2.L	6.25		

Bibliography: 77TGE,77TGS

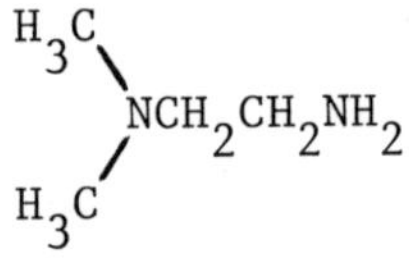

$$H_3C \diagdown$$
$$\qquad NCH_2CH_2NH_2$$
$$H_3C \diagup$$

$C_4H_{12}N_2$		N,N-Dimethylethylenediamine (Other Values on Vol.2, p.120)				L
Metal ion	Equilibrium	Log K 25°, 0.1	Log K 25°, 1.0	Log K 25°, 0	ΔH 25°, 0.5	ΔS 25°, 0.5
H^+	HL/H.L	9.60 ±0.02 [b] 9.72[b] ±0.03	9.90 ±0.07	9.54	-13.3 ±0.1	10
	$H_2L/HL.H$	6.49 ±0.03 [b] 6.72[b] ±0.04	6.96 ±0.03	6.18	-8.5 ±0.1	2
Ni^{2+}	ML/M.L	5.76 ±0.00 5.94[b]	6.09	5.68	-6.9[a]	3[a]
	$ML_2/M.L^2$	9.62 -0.01 9.89[b]	10.31	9.59	-12.4[a]	2[a]
	$ML_3/M.L^3$	11.2 ±0.1 11.5[b]	12.0	10.9		
Cu^{2+}	ML/M.L	9.19 ±0.04 [b] 9.33[b] ±0.05		9.08	-9.8	10
	$ML_2/M.L^2$	16.0 ±0.0 [b] 16.4[b] ±0.1		15.9	-19.3	10
	$M(OH)_2L/ML.(OH)^2$	8.31[b]			-7.4	13
	$M_2(OH)_2L_2/(ML)^2.(OH)^2$	15.02[b]			-11.2	31
Zn^{2+}	ML/M.L $ML_2/M.L^2$	4.47 7.90				

[a] 25°, 0.2; [b] 25°, 0.5

Bibliography:

H^+	72CV,76SG	Cu^{2+}	76GS
Ni^{2+}	75OT,79SG	Zn^{2+}	79SG

$$CH_3CH_2 \diagdown$$
$$\qquad NCH_2CH_2NH_2$$
$$CH_3CH_2 \diagup$$

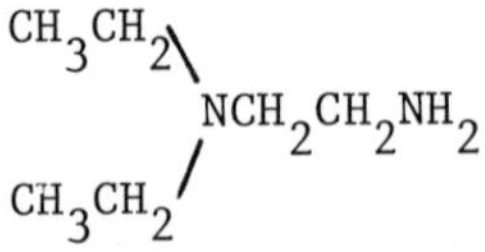

$C_6H_{16}N_2$		N,N-Diethylethylenediamine (Other values on Vol.2, p.121)				L
Metal ion	Equilibrium	Log K 25°, 0.1	Log K 25°, 1.0	Log K 25°, 0	ΔH 25°, 0.5	ΔS 25°, 0.5
H^+	HL/H.L	9.94 ±0.04 [b] 10.11[b] ±0.04	10.30 ±0.08	9.93	-10.1	12
	$H_2L/HL.H$	6.81 ±0.04 [b] 7.05[b] ±0.03	7.25 ±0.05	6.51	-10.0	-1

[b] 25°, 0.5

N,N-Diethylethylenediamine (continued)

Metal ion	Equilibrium	Log K 25°, 0.1	Log K 25°, 1.0	Log K 25°, 0	ΔH 25°, 0.5	ΔS 25°, 0.5
Cu^{2+}	ML/M.L	8.14 ±0.03 8.31[b]	8.53	8.05	-7.0	15
	$ML_2/M.L^2$	13.68 ±0.05 14.09[b]	14.59	13.52	-13.2	20
	MOHL/ML.OH	4.7[b]			-6	1
	$M_2(OH)_2L_2/(ML)^2.(OH)^2$ 4.0[b]				-13	-25

[b]25°, 0.5

Bibliography: 75BF

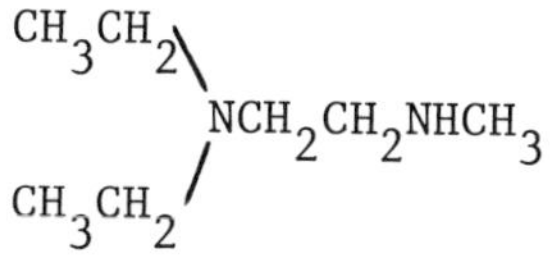

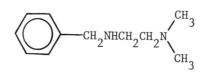

$C_7H_{18}N_2$ N,N-Diethyl-N'-methylethylenediamine   L
(Other values on Vol.2, p.123)

Metal ion	Equilibrium	Log K 25°, 0.1	Log K 25°, 0.5	Log K 25°, 0
H^+	HL/H.L	9.96	10.12	9.92
	$H_2L/HL.H$	6.76	7.07	6.48
Ni^{2+}	ML/M.L	3.7	3.8	3.5

Bibliography: 75OT

$C_{11}H_{18}N_2$ N-Benzyl-N',N'-dimethylethylenediamine   L
(Other values on Vol.2, p.123)

Metal ion	Equilibrium	Log K 25°, 0.1	Log K 25°, 0.5	Log K 25°, 0
H^+	HL/H.L	9.12	9.32	9.07
	$H_2L/HL.H$	5.83	6.11	5.57
Ni^{2+}	ML/M.L	4.34	4.50	4.50
	$ML_2/M.L^2$			6.4

Bibliography: 75OT

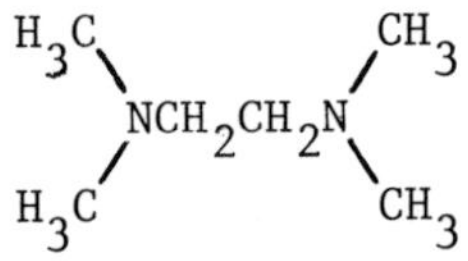

$$H_3C \quad\quad CH_3$$
$$NCH_2CH_2N$$
$$H_3C \quad\quad CH_3$$

$C_6H_{16}N_2$	N,N,N',N'-Tetramethylethylenediamine (Other values on Vol.2, p.125)				L

Metal ion	Equilibrium	Log K 25°, 0.1	Log K 25°, 1.0	Log K 25°, 0	ΔH 25°, 0.5	ΔS 25°, 0.5
H^+	HL/H.L	9.15 ±0.01	9.45	9.15	-7.40	17.7
		9.28[b]			-7.47[c]	18.2[c]
	H_2L/HL.H	5.91 ±0.03	6.36	5.58	-6.64	5.8
		6.13[b]			-7.15[c]	5.1[c]
Cu^{2+}	ML/M.L	7.30 ±0.1			-6.2	13
		7.38[b]				
	MOHL/ML.OH	5.78			-4.0	13
		5.70[b]				
	$M(OH)_2L/ML.(OH)^2$	9.84			-5.9	22
		9.19[b]				
	$M_2(OH)_2L/M(OH)_2L.M$	(7.26)[b]				
	$M_2(OH)_2L_2/(ML)^2.(OH)^2$					
		15.43			-10.1	36
		15.32[b]				
	$M_3(OH)_4L_2/(M(OH)_2L)^2.M$					
		(13.67)[b]			-12.3	21

---
[b]25°, 0.5; [c]25°, 1.0

Bibliography:

H^+	72CV,76SG	Other reference: 77TJ
Cu^{2+}	76SG	

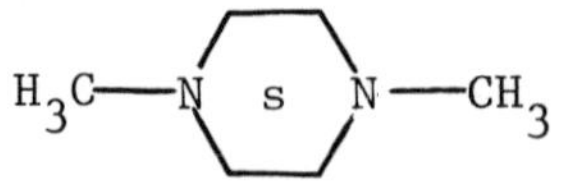

$$H_3C{-}N \quad s \quad N{-}CH_3$$

$C_6H_{14}N_2$	N,N'-Dimethylpiperazine (Other reference on Vol.2, p.338)				L

Metal ion	Equilibrium	Log K 25°, 0.1	Log K 25°, 1.0	ΔH 25°, 0.1	ΔS 25°, 0.1
H^+	HL/H.L	8.13	8.54	-6.66	15.4
				-6.21[c]	18.2[c]
	H_2L/HL.H	4.18	4.63	-3.76	6.5
				-4.42[c]	6.3[c]
Ag^+	ML/M.L	2.20			
	$ML_2/M.L^2$	3.46			

---
[c]25°, 1.0

Bibliography: H^+ 72CV; Ag^+ 74HB

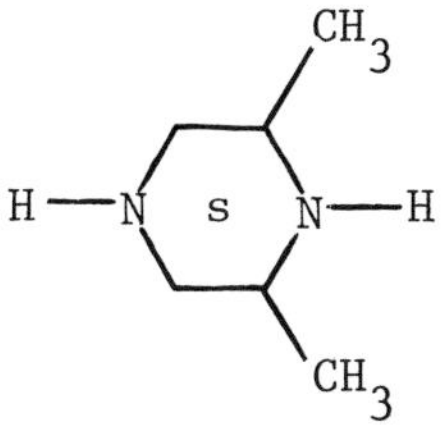

$C_6H_{14}N_2$		2,6-Dimethylpiperazine (Other reference on Vol.2, p.336)			L

Metal ion	Equilibrium	Log K 25°, 0.1	ΔH 25°, 0.1	ΔS 25°, 0.1
H^+	HL/H.L	9.57	-10.47	9.2
	$H_2L/HL.H$	5.40	-6.65	2.9
Ag^+	$ML/M.L$	3.30		
	$ML_2/M.L^2$	6.66		

Bibliography: 74HB

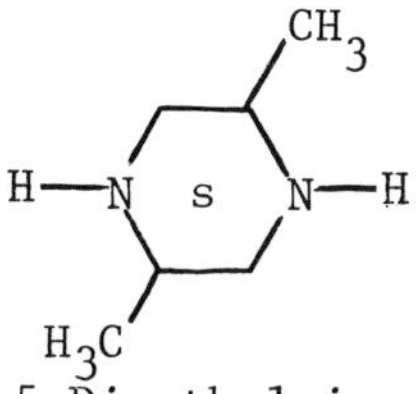

$C_6H_{14}N_2$		cis-2,5-Dimethylpiperazine (Other reference on Vol.2, p.336)			L

Metal ion	Equilibrium	Log K 25°, 0.1	ΔH 25°, 0.1	ΔS 25°, 0.1
H^+	HL/H.L	9.76	-10.76	9.1
	$H_2L/HL.H$	5.32	-7.40	0.0
Ag^+	$ML/M.L$	3.60		
	$ML_2/M.L^2$	5.98		

Bibliography: 74HB

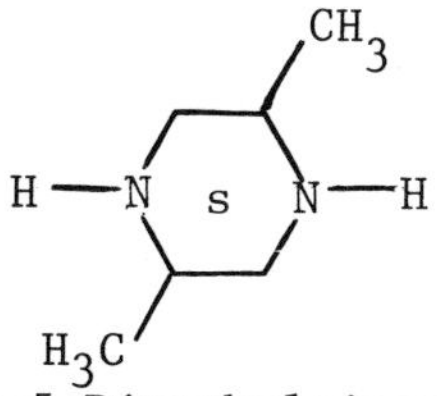

$C_6H_{14}N_2$		trans-2,5-Dimethylpiperazine (Other reference on Vol,2, p.336)			L

Metal ion	Equilibrium	Log K 25°, 0.1	ΔH 25°, 0.1	ΔS 25°, 0.1
H^+	HL/H.L	9.58	-10.37	9.6
	$H_2L/HL.H$	5.37	-7.32	0.5
Ag^+	$ML/M.L$	3.48		
	$ML_2/M.L^2$	6.43		

Bibliography: 74HB

$$HONHCCH_2 - N \quad s \quad N - CH_2CNHOH$$

(with O double bonds above each C)

| $C_8H_{16}O_4N_4$ | Piperazine-N,N'-diacethydroxamic acid | H_2L |

Metal ion	Equilibrium	Log K 20°, 0.1
H^+	$HL/H.L$	9.80
	$H_2L/HL.H$	8.40
	$H_3L/H_2L.H$	5.52
	$H_4L/H_3L.H$	1.5
Cu^{2+}	$M_2L/M^2.L$	17.12
	$M_2L_2/M^2.L^2$	20.40
Fe^{3+}	$ML/M.L$	17.56
	$MHL/M.HL$	12.18
	$MH_2L/M.H_2L$	9.63
	$M_2H_2L_4/M^2.(HL)^2.L^2$	48.8
	$M_2L_4/M^2.L^4$	50.9

Bibliography: 75Kf

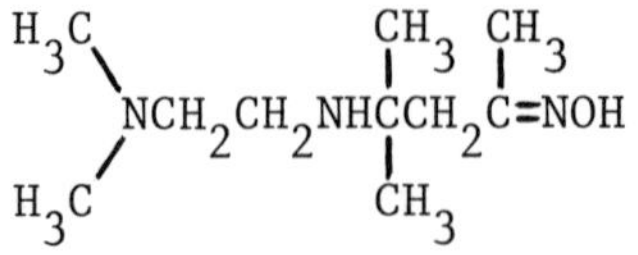

$$\begin{array}{c} H_3C \\ \quad\quad NCH_2CH_2NHCCH_2C=NOH \\ H_3C \end{array}$$

with CH_3 groups: on the NHCCH carbon CH_3 and CH_3 below CH_3

| $C_{10}H_{23}ON_3$ | 2,6,6-Trimethyl-2,5-diazanonan-8-one oxime | L |

Metal ion	Equilibrium	Log K 25°, 0.1
H^+	$HL/H.L$	9.36
	$H_2L/HL.H$	6.13
Cu^{2+}	$ML/M.L$	11.29
	$ML/M(H_{-1}L).H$	5.9

Bibliography: 77PR

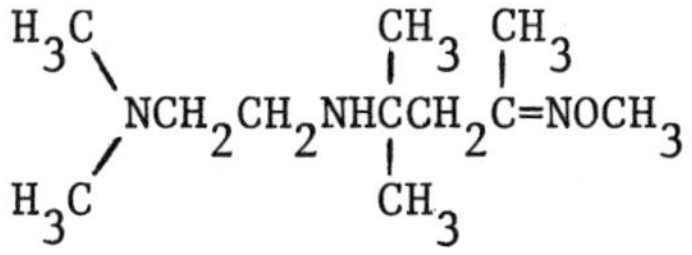

$$H_3C$$
$$\diagdown$$
$$NCH_2CH_2NHCCH_2C=NOCH_3$$
$$\diagup$$
$$H_3C$$

with CH_3 CH_3 on the upper chain and CH_3 below

$C_{11}H_{25}ON_3$ <u>2,6,6-Trimethyl-2,5-diazanonan-8-one oxime methylether</u>   L

Metal ion	Equilibrium	Log K 25°, 0.1
H[+]	HL/H.L	9.23
	H_2L/HL.H	5.95
Cu[2+]	ML/M.L	8.85
	MOHL/ML.OH	5.6

Bibliography: 77PR

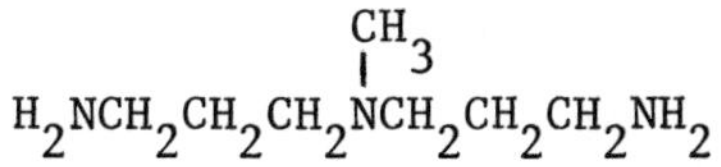

$$CH_3$$
$$|$$
$$H_2NCH_2CH_2CH_2NCH_2CH_2CH_2NH_2$$

$C_7H_{19}N_3$ 5-Methyl-1,5,9-triazanonane
(Other values on Vol.2, p.132)   L

Metal ion	Equilibium	Log K 25°, 0.1
H[+]	HL/H.L	10.50
	H_2L/HL.H	9.60
	H_3L/H_2L.H	6.98
Cu[2+]	ML/M.L	12.86
	MOHL/ML.OH	4.32

Bibliography: 75KH

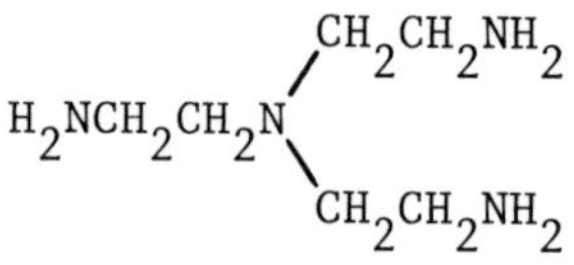

$$\begin{array}{c}
\text{CH}_2\text{CH}_2\text{NH}_2 \\
| \\
\text{H}_2\text{NCH}_2\text{CH}_2\text{N} \\
| \\
\text{CH}_2\text{CH}_2\text{NH}_2
\end{array}$$

$C_6H_{18}N_4$ <u>Nitrilotris(ethyleneamine)</u>   <u>(tris(2-aminoethyl)amine, tren)</u>   L
(Other values on Vol.2, p.137)

Metal ion	Equilibrium	Log K 25°, 0.1	Log K 25°, 1.0	Log K 25°, 0	ΔH 20°, 0.1	ΔS 25°, 0.1
H^+	HL/H.L	10.12 ±0.02	10.39	10.03	-11.7	7
		10.14[b]	9.83[n]			
	$H_2L/HL.H$	9.43 ±0.02	9.81	9.13	-12.8	0
		9.68[b]	9.12[n]			
	$H_3L/H_2L.H$	8.41 ±0.04	8.89	7.85	-12.2	-2
		8.64[b]	8.10[n]			
	$H_4L/H_3L.H$	2.60				
Mn^{2+}	ML/M.L	5.8 ±0.0			-3.0	16
Co^{2+}	ML/M.L	12.6 ±0.2			-10.7	22
Ni^{2+}	ML/M.L	14.5 ±0.1			-15.2	15
Cu^{2+}	ML/M.L	18.8 ±0.1	19.3	18.4	-20.4	18
			18.4[n]			
	$ML_2/M.L^2$		21.7[n]			
	ML/MOHL.H	9.13 -0.1	9.51		-9.8	9
			9.09[n]			
	MHL/ML.H		4.16			
			3.63[n]			
	$M(HL)L/ML_2.H$		9.59[n]			
	$M(HL)_2/M(HL)L.H$		8.5[n]			
Zn^{2+}	ML/M.L	14.4 +0.1			-13.9	19
		14.5[b]				
	ML/MOHL.H	10.84[b]				
Cd^{2+}	ML/M.L	11.7				
		11.8[h]				
Hg^{2+}	ML/M.L	22.8[i]				
	MHL/ML.H	4.5[h]				

[b]25°, 0.5;   [h]20°, 0.1;   [i]20°, 0.5;   [n]37°, 0.15

Bibliography:

H^+ 75APB,75JT,75MMc,76SA,79SPT Ni^{2+} 75APB,75JT
Mn^{2+},Cd^{2+} 75JT Cu^{2+} 71AW,75APB,75JT,76SA,78WNA,79SPT
Co^{2+} 75JT,75MMc Zn^{2+} 73RB,75JT

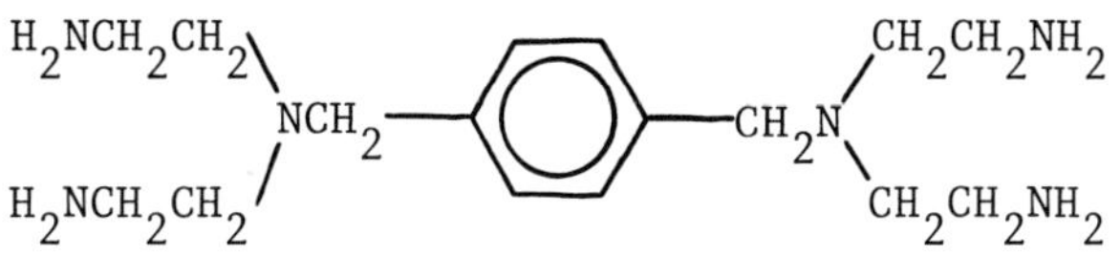

$$H_2NCH_2CH_2 \diagdown$$
$$NCH_2 \text{—benzene—} CH_2N$$
$$H_2NCH_2CH_2 \diagup \quad\quad\quad \diagup CH_2CH_2NH_2 \text{ / } CH_2CH_2NH_2$$

$C_{16}H_{32}N_6$ 1,4-Bis(bis(2-aminoethyl)aminomethyl)benzene   (PXBDE)   L

Metal ion	Equilibrium	Log K 25°, 0.1
H^+	$HL/H.L$	10.06
	$H_2L/HL.H$	9.71
	$H_3L/H_2L.H$	9.11
	$H_4L/H_3L.H$	8.57
	$H_5L/H_4L.H$	1.8
	$H_6L/H_5L.H$	1.2
Co^{2+}	$MHL/M.HL$	8.82
	$MH_2L/M.H_2L$	6.84
	$M_2L/M^2.L$	14.58
Ni^{2+}	$MHL/M.HL$	10.21
	$MH_2L/M.H_2L$	9.20
	$M_2L/M^2.L$	19.77
Cu^{2+}	$ML/M.L$	15.53
	$MHL/M.HL$	14.46
	$MH_2L/M.H_2L$	13.21
	$M_2L/M^2.L$	27.78
	$M_2OHL/M_2L.OH$	5.44
	$M_2(OH)_2L/M_2OHL.OH$	4.64
Zn^{2+}	$ML/M.L$	10.05
	$MHL/M.HL$	9.12
	$MH_2L/M.H_2L$	7.19
	$M_2L/M^2.L$	16.14

Bibliography: 79NM

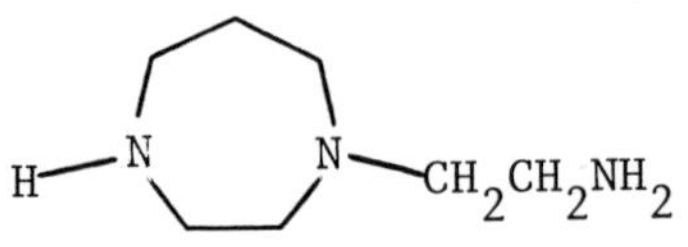

H—N（   ）N—CH₂CH₂NH₂

C₇H₁₇N₃		N-(2-Aminoethyl)-1,4-diazacycloheptane	L

Metal ion	Equilibrium	Log K 25°, 0.1
H⁺	HL/H.L	10.09
	H₂L/HL.H	9.22
	H₃L/H₂L.H	2.8
Ni²⁺	ML/M.L	6.50
	MHL/ML.H	8.84
Cu²⁺	ML/M.L	14.31
	MHL/ML.H	8.82

Bibliography: 77PB

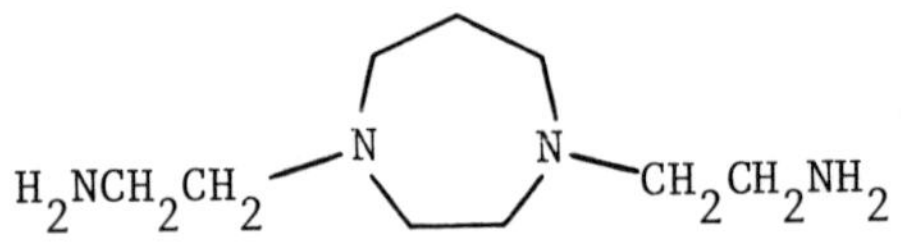

H₂NCH₂CH₂—N（   ）N—CH₂CH₂NH₂

C₉H₂₂N₄		N,N'-Bis(2-aminoethyl)-1,4-diazacycloheptane	L

Metal ion	Equilibrium	Log K 25°, 0.1
H⁺	HL/H.L	9.76
	H₂L/HL.H	9.13
	H₃L/H₂L.H	6.14
Co²⁺	ML/M.L	8.37
	MHL/ML.H	9.88
Ni²⁺	ML/M.L	13.5
Cu²⁺	ML/M.L	20.3
	MHL/ML.H	13.4
Zn²⁺	ML/M.L	10.18
	MHL/ML.H	8.38

Bibliography: 77PB

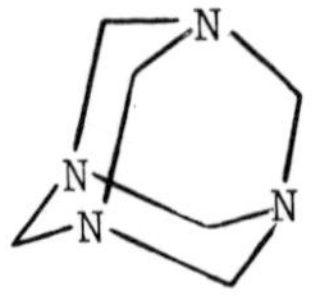

C₆H₁₂N₄	1,3,5,7-Tetrazatricyclo[3.3.1.1³,⁷]decane (hexamethylenetetramine)			L
	(Other references on Vol.2, p.348)			

Metal ion	Equilibrium	Log K 25°, 0.1	Log K 25°, 0.5	Log K 25°, 1.0
H⁺	HL/H.L	4.89	5.05	
Cu²⁺	ML/M.L₂			(-0.34)
Ag⁺	ML₂/M.L²	3.49		
Cd²⁺	ML₂/ M.L²	0.11		

Bibliography: H⁺ 75BN,75IH; Cu²⁺ 70GH; Ag⁺,Cd²⁺ 75BN; Other reference : 78PSc

$C_6H_{15}N_3$ 1,4,7-Triazacyclononane   ([9]aneN$_3$)   L

Metal ion	Equilibrium	Log K 25°, 0.1	Log K 25°, 0.5	ΔH 25°, 0.1	ΔS 25°, 0.1
H$^+$	HL/H.L	10.44 ±0.03	(12.5)	−10.4	13
	H$_2$L/HL.H	6.81 ±0.05	7.09	−9.9	−2
Ni^{2+}	ML/M.L	16.2		−3	
Cu^{2+}	ML/M.L$_2$	15.5 ±0.5	(17.5)	−14.2	23
	ML$_2$/M.L^2		(31.5)		
	MOHL/ML.OH	8.25			
	(MOHL)$_2$/(ML)^2.(OH)^2		15.0		
Zn^{2+}	ML/M.L	11.6 ±0.4		−11.9	13
Cd^{2+}	ML/M.L	9.4 ±0.2		(−8)^s	(16)*
Pb^{2+}	ML/M.L	11.0		(−8)^s	(23)*

^{s}15-35°, 0.2;   *assuming ΔH for 0.2=ΔH for 0.1

Bibliography:

H$^+$,Cu^{2+} 73AH,76YZ,77FZ,77KK,79RK   Cd^{2+} 73AH,78KKa

Ni^{2+} 73AH,76YZ   Pb^{2+} 78KKa

Zn^{2+} 73AH,76YZ,77FZ,78KKa

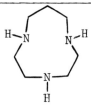

$C_7H_{17}N_3$ 1,4,7-Triazacyclodecane   ([10]aneN$_3$)   L

Metal ion	Equilibrium	Log K 25°, 0.1	Log K 25°, 0.5
H$^+$	HL/M.L	12.0	(12.6)
	H$_2$L/HL.H	6.61 ±0.03	6.71
Ni^{2+}	ML/M.L	14.6	
Cu^{2+}	ML/M.L$_2$	15.5	(16.1)
	ML$_2$/M.L^2		(26.4)
	(MOHL)$_2$/(ML)^2.(OH)^2		14.5
Zn^{2+}	ML/M.L	11.3	
Cd^{2+}	ML/M.L	8.8	
Pb^{2+}	ML/M.L	9.8	

Bibliography: H$^+$ 77RK,78KKa,78Z;   Ni^{2+},Zn^{2+} 78Z;   Cu^{2+} 77RK,78Z,79ZK;   Cd^{2+},Pb^{2+} 78KKa

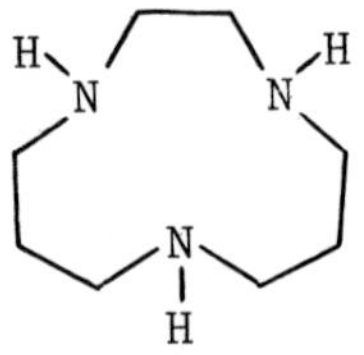

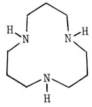

$C_8H_{19}N_3$ 1,4,8-Triazacycloundecane   ([11])aneN$_3$)   L

Metal ion	Equilibrium	Log K 25°, 0.1
H$^+$	HL/H.L	12.0
	H$_2$L/HL.H	7.61
Ni^{2+}	ML/M.L	12.9
Cu^{2+}	ML/M.L	14.4
Zn^{2+}	ML/M.L	10.4
	MOHL/ML.OH	5.60

Bibliography: 78Z

_____

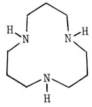

$C_9H_{21}N_3$ 1,5,9-Triazacyclododecane   ([12])aneN$_3$)   L

Metal ion	Equilibrium	Log K 25°, 0.1	Log K 25°, 0.5
H$^+$	HL/H.L	(12.6)	(13.0)
	H$_2$L/HL.H	7.57	7.82
	H$_3$L/H$_2$L.H	2.41	
Ni^{2+}	ML/M.L	10.9	
Cu^{2+}	ML/M.L	12.6	(13.2)
	ML$_2$/M.L^2		(20.8)
	MOHL/ML.OH	5.64	
	(MOHL)$_2$/(ML)^2.(OH)^2		12.23
Zn^{2+}	ML/M.L	8.8	
	MOHL/ML.OH	6.29	

Bibliography:
H$^+$,Cu^{2+} 77RK,78Z   Ni^{2+},Zn^{2+} 78Z

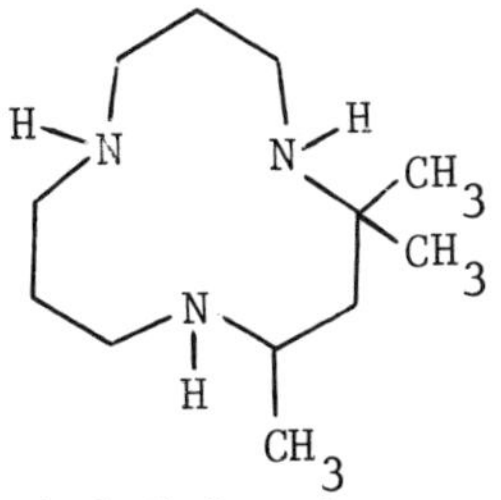

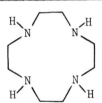

$C_{12}H_{27}N_3$ DL-2,2,4-Trimethyl-1,5,9-triazacyclododecane   L

Metal ion	Equilibrium	Log K 25°, 0.1
H^+	HL/H.L	(12.3)
	$H_2L/HL.H$	7.34
	$H_3L/H_2L.H$	2.51
Ni^{2+}	ML/MOHL.H	9.8
	$M_2(OH)_2L_2/(MOHL)^2$	$(2.4)^w$
Cu^{2+}	ML/M.L	11.6
	ML/MOHL.H	8.48
	$MOHL/M(OH)_2L.H$	11.9
	$M_2(OH)_2L_2/(MOHL)^2$	$(2.00)^w$
Zn^{2+}	ML/M.L	7.7
	ML/MOHL.H	9.6

^{w}Optical isomerism not stated.
Bibliography: 79RJ

$C_8H_{20}N_4$ 1,4,7,10-Tetraazacyclododecane   ([12]aneN$_4$)   L

Metal ion	Equilibrium	Log K 25°, 0.2	Log K 25°, 0.5	Log K 25°, 0	ΔH 25°, 0.2	ΔS 25°, 0.2
H^+	HL/H.L	10.6	10.82	10.53	$(-8)^s$	(22)
	$H_2L/HL.H$	9.6	9.72	9.60	$(-8)^s$	(17)
	$H_3L/H_2L.H$		1.15^q			
Cu^{2+}	ML/M.L	24.8			-22.7^c	37*
Zn^{2+}	ML/M.L	16.2			-14.5^c	25*
Cd^{2+}	ML/M.L	14.3			$(-8)^s$	(39)
Hg^{2+}	ML/M.L	25.5			$(-24)^s$	(36)
Pb^{2+}	ML/M.L	15.9			$(-7)^s$	(49)

^{c}25°, 1.0; ^{q}25°, 0.6; ^{s}10-40°, 0.2;   *assuming ΔH for 1.0=ΔH for 0.2

Bibliography:

H^+ 76KKb,77CGR,78LH Cd^{2+},Pb^{2+} 77KKa

Cu^{2+} 76KKb,78AF Hg^{2+} 76KKd

Zn^{2+} 77KKa,78AF

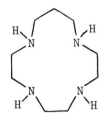

$C_9H_{22}N_4$ 1,4,7,10-Tetraazacyclotridecane   ([13]aneN$_4$)   L

Metal ion	Equilibrium	Log K 25°, 0.1	Log K 25°, 0.5	ΔH 25°, 0.1	ΔS 25°, 0.1
H^+	HL/H.L	11.0	11.0	(−8)^s	(24)
	H_2L/HL.H	10.0	9.97	(−8)^s	(19)
	H_3L/H_2L.H	1.6			
Ni^{2+}	ML(yellow)/ML(blue)	0.84		(7)^t	(30)
Cu^{2+}	ML/M.L	29.1		−25.6^c	47*
Zn^{2+}	ML/M.L	15.6		(−8)^s	(45)
Hg^{2+}	ML/M.L	25.3		(−25)^s	(32)

^{c}25°, 1.0; ^{s}10-40°, 0.2; ^{t}21-48°, 0.1; *assuming ΔH for 1.0=ΔH for 0.1

Bibliography: Ni^{2+} 77AF

H^+ 76KKc,78LH Zn^{2+} 77KKa

Cu^{2+} 76KKc,79FM Hg^{2+} 76KKd

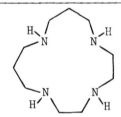

$C_{10}H_{24}N_4$ 1,4,7,11-Tetraazacyclotetradecane   (isocyclam)   L

Metal ion	Equilibrium	Log K 25°, 0.1	Log K 25°, 0.5	ΔH 25°, 0.1	ΔS 25°, 0.1
H^+	HL/H.L		11.1		
	H_2L/HL.H		10.04		
	H_3L/H_2L.H		4.17		
Ni^{2+}	ML(yellow)/ML(blue)	0.19		(5)^t	(19)
Cu^{2+}	ML/M.L			−27.8^c	

^{c}25°, 1.0; ^{t}21-45°, 0.1

Bibliography:

H^+ 78LH Cu^{2+} 79FM

Ni^{2+} 79SF

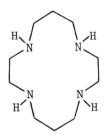

$C_{10}H_{24}N_4$ 1,4,8,11-Tetraazacyclotetradecane ([14]aneN$_4$, <u>cyclam</u>)   L

Metal ion	Equilibrium	Log K 25°, 0.1	Log K 25°, 0.5	ΔH 25°, 0.5	ΔS 25°, 0.1
H$^+$	HL/H.L	11.5 -0.1	11.6 +0.1	-12.3	11
	H$_2$L/HL.H	10.2	10.62-0.01	-12.8	4
Ni^{2+}	ML/M.L	22.2		-20.3^c	33*
	ML(yellow)/ML(blue)	0.39		(5)^r	(20)
	MOHL/ML.H	13.0			
Cu^{2+}	ML/M.L	27.2		-32.4^c	16*
Zn^{2+}	ML/M.L	15.5		-14.8^c	21*
Hg^{2+}	ML/M.L	23.0		(-33)^s	(-3)

^{c}25°, 1.0; ^{r}19-49°, 0.1; ^{s}10-40°, 0.2; *assuming ΔH for 1.0=ΔH for 0.1

Bibliography:

H$^+$ 74HM,77KKa,78LH,78MPV,78MSP   Zn^{2+} 77KKa,78AF

Ni^{2+} 74HM,77AF,78FP   Hg^{2+} 76KKd

Cu^{2+} 77KK,78AF   Other references: 68PT,70FT

$C_{12}H_{28}N_4$ <u>meso-5,12-Dimethyl-1,4,8,11-tetraazacyclotetradecane</u>   L

Metal ion	Equilibrium	Log K 25°, 0.1	ΔH 25°, 0.1	ΔS 25°, 0.1
H$^+$	HL/H.L	11.7		
Ni^{2+}	ML/M.L	21.9	(-28)^s	(6)
	ML(yellow)/ML(blue)	0.37	(5)^t	(20)
	ML/MOHL.H	12.7		

^{s}10-40°, 0.1; ^{t}21-48°, 0.1

Bibliography:

H$^+$ 74HM   Ni^{2+} 74HM,77AF

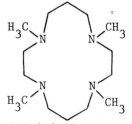

C$_{16}$H$_{36}$N$_4$ 5,5,7,12,12,14-Hexamethyl-1,4,8,11-tetraazacyclotetradecane   L

Isomer	Metal ion	Equilibrium	Log K 25°, 0.1
meso-	H$^+$	HL/H.L	(12.6)
		H$_2$L/HL.H	10.4
		H$_3$L/H$_2$L.H	0.8
	Cu^{2+}	ML/M.L(blue)	20
		ML/M.L(red)	28

DL-	Ni^{2+}	ML/M.L	18.2
		ML/MOHL.H	12.9

Bibliography:

H$^+$,Cu^{2+} 69CM   Other referenes: 79CK,79CM

Ni^{2+} 74HM

C$_{14}$H$_{32}$N$_4$ 1,4,8,11-Tetramethyl-1,4,8,11-tetraazacyclotetradecane   L

Metal ion	Equilibrium	Log K 25°, 0.5	ΔH 25°, 0.5	ΔS 25°, 0.5
H$^+$	HL/H.L	9.70	-5.1	28
	H$_2$L/HL.H	9.31	-10.3	8
	H$_3$L/H$_2$L.H	3.09	-3.6	2
	H$_4$L/H$_3$L.H	2.64	-6.9	-11
Co^{2+}	ML/M.L	10.9		
	ML/MOHL.H	8.31		
Ni^{2+}	ML/M.L	11.8		
Cu^{2+}	ML/M.L	17.7		
Zn^{2+}	ML/M.L	12.2		

Bibliography:

H$^+$ 78MPV,78MSP   Ni^{2+}-Zn^{2+} 74HK

Co^{2+} 74BSK,74HK

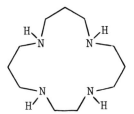

| $C_{11}H_{26}N_4$ | | | 1,4,8,12-Tetraazacyclopentadecane | ([15]aneN$_4$) | | L |

Metal ion	Equilibrium	Log K 25°, 0.2	Log K 25°, 0.5	ΔH 25°, 0.5	ΔS 25°, 0.5
H$^+$	HL/H.L	11.1	11.1 ±0.0	-10.8	15
	H$_2$L/HL.H	10.0	10.3 ±0.2	-12.3	6
	H$_3$L/H$_2$L.H		5.23 ±0.06	-7.2	0
	H$_4$L/H$_3$L.H		3.62 ±0.02	-7.7	-9
Cu^{2+}	ML/M.L	24.4		-26.5[c]	23[a*]
Zn^{2+}	ML/M.L	15.0		-16.5[c]	13[a*]
Hg^{2+}	ML/M.L	23.7		(-25)[s]	(25)[a]

[a] 25°, 0.2; [c] 25°, 1.0; [s] 10-40°, 0.2; *assuming ΔH for 1.0=ΔH for 0.2

Bibliography:

H$^+$ 76KKe,78AF,78LH,78MSP,78MPV   Zn^{2+} 77KKa,78AF

Cu^{2+} 76KKe,78AF   Hg^{2+} 76KKd

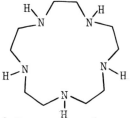

| $C_{10}H_{25}N_5$ | | 1,4,7,10,13-Pentaazacyclopentadecane | | | L |

Metal ion	Equilibrium	Log K 25°, 0.2	ΔH 25°, 0.2	ΔS 25°, 0.2
H$^+$	HL/H.L	10.73	(-7)[s]	(26)
	H$_2$L/HL.H	9.53	(-7)[s]	(20)
	H$_3$L/H$_2$L.H	5.88	(-8)[s]	(0)
	H$_4$L/H$_3$L.H	1.6		
	H$_5$L/H$_4$L.H	1.0		
Cu^{2+}	ML/M.L	28.3	(-33)[s]	(19)
Zn^{2+}	ML/M.L	19.1	(-14)[s]	(40)
Cd^{2+}	ML/M.L	19.2	(-13)[s]	(44)
Hg^{2+}	ML/M.L	28.5	(-33)[s]	(20)
Pb^{2+}	ML/M.L	17.3	(-10)[s]	(46)

[s] 15-35°, 0.2

Bibliography:

H$^+$,Cu^{2+} 78KK   Zn^{2+}-Pb^{2+} 78KKa

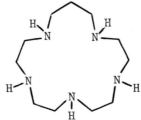

$C_{11}H_{27}N_5$ 1,4,7,10,13-Pentaazacyclohexadecane   L

Metal ion	Equilibrium	Log K 25°, 0.2		ΔH 25°, 0.2	ΔS 25°, 0.2
H^+	HL/H.L	10.52		$(-9)^s$	(18)
	$H_2L/HL.H$	9.37		$(-9)^s$	(13)
	$H_3L/H_2L.H$	7.16		$(-9)^s$	(3)
	$H_4L/H_3L.H$	1.6			
	$H_5L/H_4L.H$	1.3			
Cu^{2+}	ML/M.L	27.1		$(-33)^s$	(13)
Zn^{2+}	ML/M.L	17.9		$(-14)^s$	(35)
Cd^{2+}	ML/M.L	18.1		$(-13)^s$	(39)
Hg^{2+}	ML/M.L	27.4		$(-34)^s$	(11)
Pb^{2+}	ML/M.L	14.3		$(-11)^s$	(29)

^{s}15-35°, 0.2

Bibliography:
H^+, Cu^{2+} 78KK $Zn^{2+}-Pb^{2+}$ 78KKa

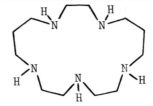

$C_{12}H_{29}N_5$ 1,4,7,11,14-Pentaazacycloheptadecane   L

Metal ion	Equilibrium	Log K 25°, 0.2		ΔH 25°, 0.2	ΔS 25°, 0.2
H^+	HL/H.L	10.20		$(-9)^s$	(16)
	$H_2L/HL.H$	9.50		$(-9)^s$	(13)
	$H_3L/H_2L.H$	7.24		$(-9)^s$	(3)
	$H_4L/H_3L.H$	3.98			
	$H_5L/H_4L.H$	2.26			
Cu^{2+}	ML/M.L	23.8		$(-27)^s$	(18)
Zn^{2+}	ML/M.L	15.8		$(-13)^s$	(29)
Cd^{2+}	ML/M.L	15.5		$(-13)^s$	(27)
Hg^{2+}	ML/M.L	26.5		$(-33)^s$	(10)
Pb^{2+}	ML/M.L	11.6		$(-10)^s$	(20)

^{s}15-35°, 0.2

Bibliography: H^+, Cu^{2+} 78KK $Zn^{2+}-Pb^{2+}$ 78KKa

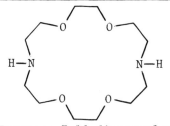

| $C_{10}H_{22}O_3N_2$ | 1,4,10-Trioxa-7,13-diazacyclopentadecane | | L |

Metal ion	Equilibrium	Log K 25°, 0.1
H^+	HL/H.L	8.65^z
	$H_2L/HL.H$	7.93^z
Co^{2+}	ML/M.L	5.0^z
Ni^{2+}	ML/M.L	3.6^z
Cu^{2+}	ML/M.L	7.1^z
Ag^+	ML/M.L	5.8^z
Zn^{2+}	M1/M.L	5.1^z
Cd^{2+}	ML/M.L	6.4^z
Pb^{2+}	ML/M.L	5.8^z

$^z(C_2H_5)_4NClO_4$ used as background electrolyte.

Bibliography: 77ASS

| $C_{12}H_{24}O_4N_2$ | 1,4,10,13-Tetraoxa-7,16-diazacyclooctadecane | | | L |

Metal ion	Equilibrium	Log K 25°, 0.1	ΔH 25°, 0.1	ΔS 25°, 0.1
H^+	HL/H.L	$9.08^z+0.01$	-8.6	13
	$H_2L/HL.H$	$7.94^z-0.03$	-9.5	4
Sr^{2+}	ML/M.L	2.6^z	-2.6	3
Ba^{2+}	ML/M.L	2.97^z	-3.0	4
Cu^{2+}	ML/M.L	6.1^z		
Ag^+	ML/M.L	7.8^z	-9.2	5
Zn^{2+}	ML/M.L	3.1^z		
Cd^{2+}	ML/M.L	5.25^z	-0.7	22
Hg^{2+}	ML/M.L	17.85^z	-17.2	24
Pb^{2+}	ML/M.L	6.8^z		

$^zR_4N^+$ salt used as background electrolyte.

Bibliography:

H^+,Ag^+,Cd^{2+} 75A,77ASS Cu^{2+},Zn^{2+},Pb^{2+} 77ASS

Sr^{2+},Ba^{2+},Hg^{2+} 75A

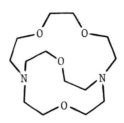

$C_{14}H_{28}O_4N_2$ 4,7,13,18-Tetraoxa-1,10-diazabicyclo[8.5.5]cosane   ([2.1.1]cryptand)   L

Metal ion	Equilibrium	Log K 25°, 0.05	ΔH 25°, 0.06	ΔS 25°, 0.05
H^+	HL/H.L	10.64		
	H_2L/HL.H	7.85		
Li^+	ML/M.L	5.5	-5.1	8
Na^+	ML/M.L	3.2	-5.4	-3
Mg^{2+}	ML/M.L	2.5		
Ca^{2+}	ML/M.L	2.4 ±0.1	-0.1	10

Bibliography:

H^+,Mg^{2+} 75LS Ca^{2+} 75LS,76KLa,77LPW

Li^+,Na^+ 75LS,76KLa   Other reference: 77ASS

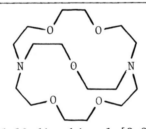

$C_{16}H_{32}O_5N_2$ 4,7,13,16,21-Pentaoxa-1,10-diazabicyclo[8.8.5]tricosane   ([2.2.1]cryptand)   L

Metal ion	Equilibrium	Log K 25°, 0.05	ΔH 25°, 0.06	ΔS 25°, 0.05
H^+	HL/H.L	10.53		
	H_2L/HL.H	7.50		
Li^+	ML/M.L	2.50	0.0	11
Na^+	ML/M.L	5.40	-5.4	6
K^+	ML/M.L	3.95	-6.8	-5
Rb^{2+}	ML/M.L	2.55	-5.4	-7
Ca^{2+}	ML/M.L	6.90 ±0.06	-2.9	22
Sr^{2+}	ML/M.L	7.35	-6.1	13
Ba^{2+}	ML/M.L	6.30	-6.3	8
Ag^+	ML/M.L	10.6		

Bibliography:

H^+,Ag^+ 75LS Ca^{2+} 76KLa,77LPW

Li^+-Rb^{2+},Sr^{2+},Ba^{2+} 75LS,76KLa   Other references: 77ASS

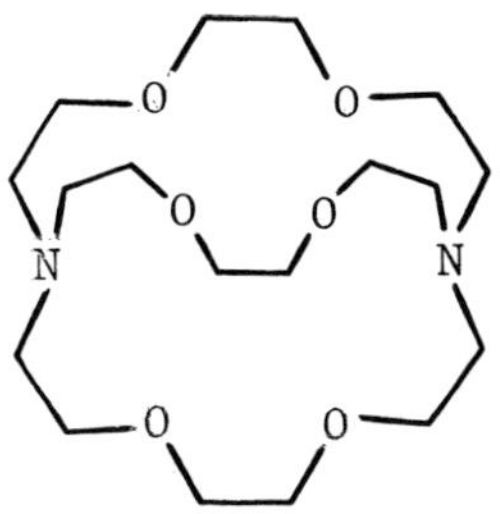

$C_{18}H_{36}O_6N_2$ 4,7,13,16,21,24-Hexaoxa-1,10-diazabicyclo[8.8.8]hexacosane   L
                           ([2.2.2]cryptand)

Metal ion	Equilibrium	Log K 25°, 0.1	ΔH 25°, 0.1	ΔS 25°, 0.1
H^+	HL/H.L	9.60^a 9.71^z	-10.8	8
	$H_2L/HL.H$	7.28^a 7.31^z	-4.5	18
Na^+	ML/M.L	3.9^a 4.11^z	-7.4 ±0.0	-6
K^+	ML/M.L	5.4^a 5.58^z	-11.2 ±0.2	-12
Rb^+	ML/M.L	4.35^a 4.06^z	-11.8 ±0.0	-21
Ca^{2+}	ML/M.L	4.45 ±0.0 4.57^z	-0.2 ±0.0	20
Sr^{2+}	ML/M.L	8.0^a 8.26^z	-10.6 ±0.0	2
Ba^{2+}	ML/M.L	9.5^a 9.7^z	-14.2 ±0.1	-4
Cu^{2+}	ML/M.L	6.5^z		
Ag^+	ML/M.L	9.6^a 9.6^z	-12.8	1
Tl^+	ML/M.L	6.3^a 5.5^z	-13.2	-15
Cd^{2+}	ML/M.L	6.8^z	0.5	33
Hg^{2+}	ML/M.L	18.2^z	-16.0	30
Pb^{2+}	ML/M.L	12.0 12.36^z	-13.8	10

---

[a] 25°, 0.05;   [z] R_4N^+ salt used as background electrolyte.

Bibliography:
H^+, Cu^{2+}-Pb^{2+} 75A,75LS,77ASS
Na^+-Ba^{2+} 75A,75LS,76KLa,77LPW

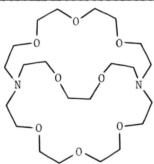

$C_{20}H_{40}O_7N_2$ 4,7,10,16,19,24,27-Heptaoxa-1,13-diazabicyclo[12.8.8]nonacosane   L
                      ([3.2.2]cryptand)

Metal ion	Equilibrium	Log K 25°, 0.05	ΔH 25°, 0.06	ΔS 25°, 0.05
H$^+$	HL/H.L	8.50		
	H$_2$L/HL.H	7.33		
Na$^+$	ML/M.L	1.65		
K$^+$	ML/M.L	2.2	-3	0
Rb$^+$	ML/M.L	2.05	-4.2	-5
Cs$^+$	ML/M.L	2.0	-5.4	-10
Ca^{2+}	ML/M.L	2	-0.2	10
Sr^{2+}	ML/M.L	3.4	-3.3	4
Ba^{2+}	ML/M.L	6.0	-6.2	7

Bibliography:
H$^+$,Na$^+$ 75LS   K$^+$-Ba^{2+} 75LS,76KLa

$C_{22}H_{44}O_8N_2$ 4,7,10,16,19,22,27,30-Octaoxa-1,13-diazabicyclo[11.11.8]dotriacontane   L
                      ([3.3.2]cryptand)

Metal ion	Equilibrium	Log K 25°, 0.05
H$^+$	HL/H.L	8.16
	H$_2$L/HL.H	7.31
Ca^{2+}	ML/M.L	2
Sr^{2+}	ML/M.L	2
Ba^{2+}	ML/M.L	3.65

Bibliography: 75LS

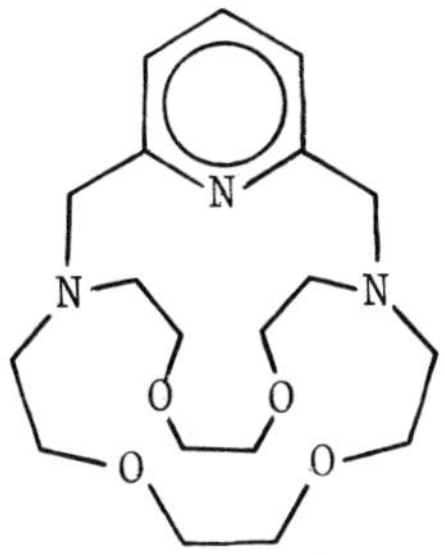

$C_{19}H_{31}O_4N_3$ 4,7,13,16-Tetraoxa-1,10,26-triazatricyclo[8.8.7.1^{20,24}]-   L
hexacosane-21,23,26-triene

Metal ion	Equilibrium	Log K 25°, 0
Na$^+$	ML/M.L	4.89
K$^+$	ML/M.L	4.78

Bibliography: 77TMW

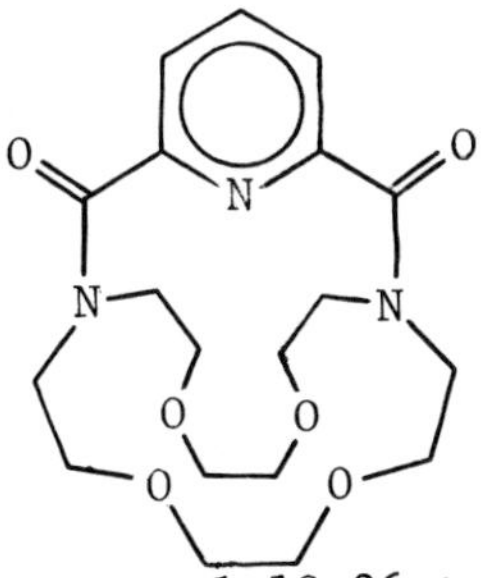

$C_{19}H_{27}O_6N_3$ 19,25-Dioxo-4,7,13,16-tetraoxa-1,10,26-triazatricyclo[8.8.7.1^{20,24}]-   L
hexacosane-21,23,26-triene

Metal ion	Equilibrium	Log K 25°, 0
Na$^+$	ML/M.L	4.58
K$^+$	ML/M.L	5.25

Bibliography: 77TMW

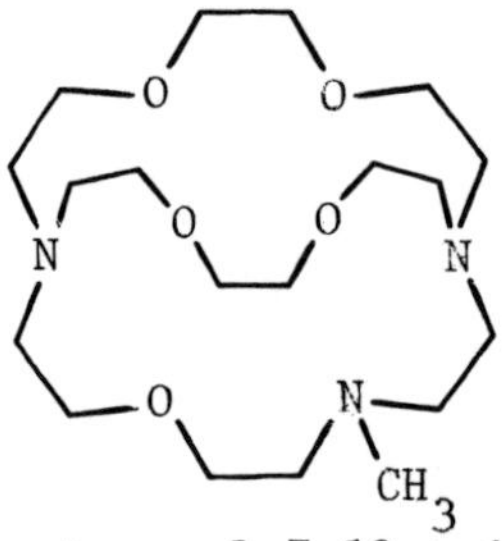

$C_{19}H_{39}O_5N_3$ 7-Methyl-4,13,16,21,24-pentaoxa-1,7,10-triazabicyclo[8.8.8]hexacosane   L

Metal ion	Equilibrium	Log K 25°, 0.1
H^+	HL/H.L	10.55
	$H_2L/HL.H$	8.57
	$H_3L/H_2L.H$	2.55
Li^+	ML/M.L	1.5
Na^+	ML/M.L	3.2
K^+	ML/M.L	4.2
Rb^+	ML/M.L	3.0
Mg^{2+}	ML/M.L	1.9
Ca^{2+}	ML/M.L	4.6
Sr^{2+}	ML/M.L	7.4
Ba^{2+}	ML/M.L	9.0
	MHL/ML.H	4.5
Co^{2+}	ML/M.L	5.2
Ni^{2+}	ML/M.L	5.0
Cu^{2+}	ML/M.L	9.7
	MHL/ML.H	5.4
Ag^+	ML/M.L	10.8
	MHL/ML.H	5.4
Tl^+	ML/M.L	6.3
Zn^{2+}	ML/M.L	6.3
Cd^{2+}	ML/M.L	9.7
	MHL/ML.H	4.7
Hg^{2+}	ML/M.L	21.7
	MHL/ML.H	5.6
Pb^{2+}	ML/M.L	14.1
	MHL/ML.H	2.7

Bibliography: 78LM

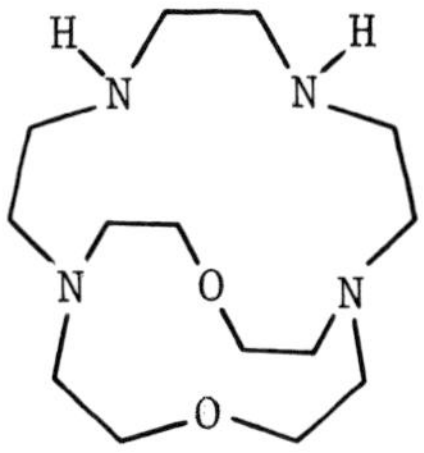

D. CYCLIC AMINES

$C_{14}H_{30}O_2N_4$ <u>13,18-Dioxa-1,4,7,10-tetraazabicyclo[8.5.5]cosane</u>   L

Metal ion	Equilibrium	Log K 25°, 0.1
H^+	HL/H.L	10.25
	H_2L/HL.H	9.55
Li^+	ML/M.L	1.6
Mg^{2+}	ML/M.L	1.9
Ni^{2+}	ML/M.L	7.8
Cu^{2+}	ML/M.L	19.7
Ag^+	ML/M.L	11.5
Zn^{2+}	ML/M.L	11.3
Cd^{2+}	ML/M.L	16.3

Bibliography: 78LM

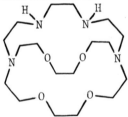

$C_{16}H_{34}O_2N_4$ 4,7-Dimethyl-13,18-dioxa-1,4,7,10-tetraazabicyclo[8.5.5]cosane   L

Metal ion	Equilibrium	Log K 25°, 0.1
H^+	HL/H.L	11.18
	$H_2L/HL.H$	9.75
	$H_3L/H_2L.H$	2.42
Li^+	ML/M.L	3.8
Mg^{2+}	ML/M.L	2.4
Ca^{2+}	ML/M.L	2.2
Co^{2+}	ML/M.L	9.9
Ni^{2+}	ML/M.L	10.0
Cu^{2+}	ML/M.L	16.0
Ag^+	ML/M.L	12.7
	MHL/ML.H	4.8
Tl^+	ML/M.L	3.9
Zn^{2+}	ML/M.L	11.2
Cd^{2+}	ML/M.L	12.4
Hg^{2+}	ML/M.L	26.6
	MHL/ML.H	3.9

Bibliography: 78LM

$C_{18}H_{38}O_4N_4$ 4,7,13,16-Tetraoxa-1,10,21,24-tetraazabicyclo[8.8.8]hexacosane   L

Metal ion	Equilibrium	Log K 25°, 0.1
H^+	HL/H.L	10.19
	$H_2L/HL.H$	8.08
	$H_3L/H_2L.H$	3.76
K^+	ML/M.L	1.5
Ag^+	ML/M.L	8.7
Tl^+	ML/M.L	4.2
Cd^{2+}	ML/M.L	12.7

Bibliography: 78LM

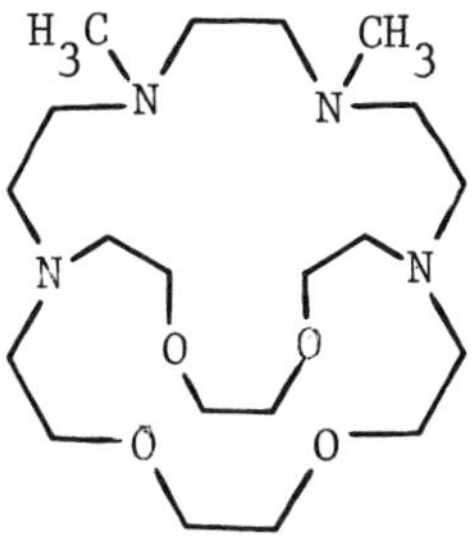

$C_{20}H_{42}O_4N_4$ 21,24-Dimethyl-4,7,13,16-tetraoxa-1,10,21,24-tetraazabicyclo-   L
                         [8.8.8]hexacosane

Metal ion	Equilibrium	Log K 25°, 0.1
H^+	HL/H.L	10.01
	H_2L/HL.H	8.92
	H_3L/H_2L.H	2.75
Li^+	ML/M.L	2.4
Na^+	ML/M.L	2.5
K^+	ML/M.L	2.7
Rb^+	ML/M.L	2.3
Mg^{2+}	ML/M.L	2.6
Ca^{2+}	ML/M.L	4.3
	MHL/ML.H	7.3
Sr^{2+}	ML/M.L	6.1
	MHL/ML.H	6.5
Ba^{2+}	ML/M.L	6.7
	MHL/ML.H	6.0
Co^{2+}	ML/M.L	4.9
Ni^{2+}	ML/M.L	5.1
Cu^{2+}	ML/M.L	12.7
	MHL/ML.H	3.8
Ag^+	ML/M.L	11.5
	MHL/ML.H	5.3
Tl^+	ML/M.L	5.5
	MHL/ML.H	6.4
Zn^{2+}	ML/M.L	6.0
Cd^{2+}	ML/M.L	12.0
	MHL/ML.H	3.8
Hg^{2+}	ML/M.L	24.9
	MHL/ML.H	3.7
Pb^{2+}	ML/M.L	15.3
	MHL/ML.H	2.7

Bibliography: 78LM

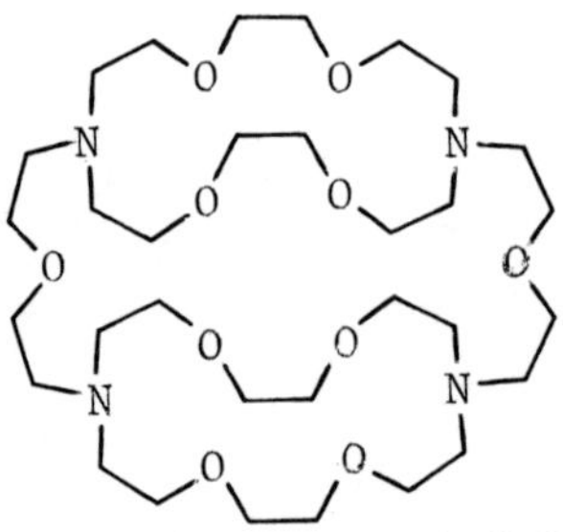

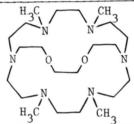

$C_{32}H_{64}O_{10}N_4$ 4,10,13,19,25,28,33,36,41,44-Decaoxa-1,7,16,22-tetraazatricyclo-   L
[20.8.8.8^{7,16}]hexatetracontane

Metal ion	Equilibrium	Log K 25°, 0.05
K^+	ML/M.L	1.5
Rb^+	ML/M.L	1.5
Sr^{2+}	ML/M.L	3.5
Ba^{2+}	ML/M.L	4.4

Bibliography: 77LSa

$C_{22}H_{48}O_2N_6$ 13,16,21,24-Tetramethyl-4,7-dioxa-1,10,13,16,21,24-hexaabicyclo-   L
[8.8.8]hexacosane

Metal ion	Equilibrium	Log K 25°, 0.1
H^+	HL/H.L	9.68
	$H_2L/HL.H$	9.37
	$H_3L/H_2L.H$	5.65
	$H_4L/H_3L.H$	2.26
K^+	ML/M.L	1.7
Ca^{2+}	ML/M.L	1.5
Sr^{2+}	ML/M.L	1.5
Ba^{2+}	ML/M.L	3.7
	MHL/ML.H	7.2
Co^{2+}	ML/M.L	5.2
Ni^{2+}	ML/M.L	5.7
Cu^{2+}	ML/M.L	12.5
	MHL/ML.H	8.3
Ag^+	ML/M.L	13.0
	MHL/ML.H	6.4
Tl^+	ML/M.L	4.1
	MHL/ML.H	7.5
Zn^{2+}	ML/M.L	6.8
	MHL/ML.H	6.6
Cd^{2+}	ML/M.L	10.7
	MHL/ML.H	5.0
Hg^{2+}	ML/M.L	26.1
	MHL/ML.H	4.2
Pb^{2+}	ML/M.L	15.5
	MHL/ML.H	3.4

Bibliography: 78LM

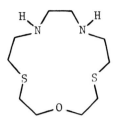

| $C_{10}H_{22}ON_2S_2$ | | 1-Oxa-4,13-dithia-7,10-diazacyclopentadecane | | | L |

Metal ion	Equilibrium	Log K 25°, 0.1	ΔH 25°, 0.1	ΔS 25°, 0.1
H^+	HL/H.L	8.9	-8.1	14
	H_2L/HL.H	5.21	-9.8	-9
Co^{2+}	ML/M.L	5.2		
Ni^{2+}	ML/M.L	8.1		
Cu^{2+}	ML/M.L	13.3	-12	21
Ag^+	ML/M.L	9.9		
	MHL/ML.H	5.31		
	ML_2/ML.M	3		
Zn^{2+}	ML/M.L	4.4		
	ML_2/M.L^2	7.9		
Cd^{2+}	ML/M.L	7.1		
Pb^{2+}	ML/M.L	6.8	-10	-2

Bibliography: 77LA,78ASJ,79AS

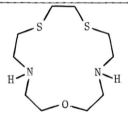

| $C_{10}H_{22}ON_2S_2$ | | 1-Oxa-7,10-dithia-4,13-diazacyclopentadecane | | L |

Metal ion	Equilibrium	Log K 25°, 0.1
H^+	HL/H.L	8.6
	H_2L/HL.H	7.55
Co^{2+}	ML/M.L	5.4
Ni^{2+}	ML/M.L	8.0
Cu^{2+}	ML/M.L	11.6
Ag^+	ML/M.L	9.0
	MHL/ML.H	5.2
	M_2L/ML.M	3.2
Zn^{2+}	ML/M.L	5.1
Cd^{2+}	ML/M.L	6.5
Pb^{2+}	ML/M.L	5.7

Bibliography: 75AS,75ASa,79AS

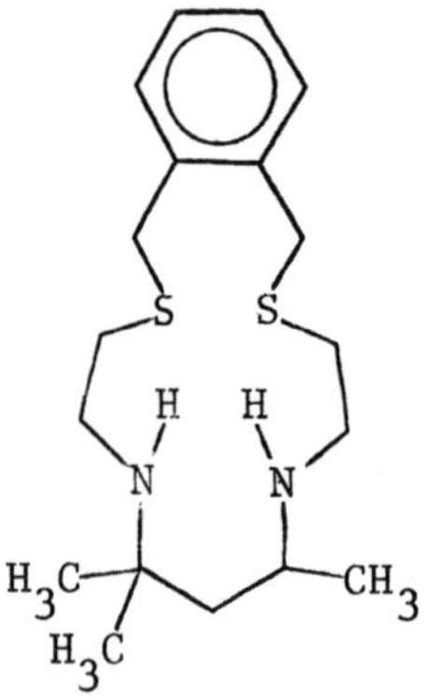

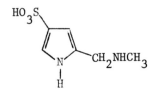

$C_{18}H_{30}O_2S_2$ 10,10,12-Trimethyl-3,4-benzo-1,6-dithia-9,13-diazacyclopentadecane   L

Metal ion	Equilibrium	Log K 25°, 0.5
H^+	HL/H.L	11.03
	H_2L/HL.H	5.09
Cu^{2+}	ML/M.L	18.42

Bibliography: 79KKb

$C_6H_9O_3N_2S$ 2-(Methylaminomethyl)pyrrole-4-sulfonic acid   HL

Metal ion	Equilibrium	Log K 25°, 0.5
H^+	HL/H.L	9.58
Ag^+	ML/M.L	2.6
	$ML_2/M.L^2$	5.6

Bibliography:

H^+ 76SA

Ag^+ 78SA

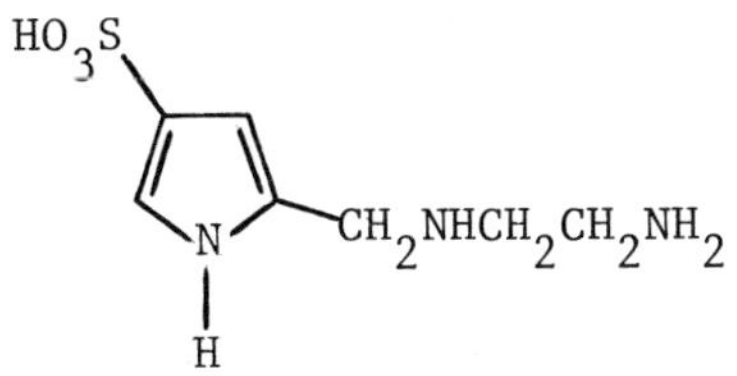

$C_7H_{13}O_3N_3S$ 2-(2-Aminoethylaminomethyl)azole-4-sulfonic acid   HL
(N-(4-sulfonyl-2-pyrrylmethyl)ethylenediamine)

Metal ion	Equilibrium	Log K 25°, 1.0
H^+	HL/H.L	9.74
	$H_2L/HL.H$	6.56
Ni^{2+}	$ML/M.L$	6.52
	$ML_2/M.L^2$	11.58
	ML/MOHL.H	9.19
	$MOHL/M(OH)_2L.H$	10.5
	$ML_2/MOHL_2.H$	9.14
	$MOHL_2/M(OH)_2L_2.H$	11.2
Cu^{2+}	$ML/M.L$	9.49
	$ML_2/M.L^2$	17.33
	$ML/MH_{-1}L.H$	6.48
	$MH_{-1}L/MOH(H_{-1}L).H$	9.86
	$ML_2/MOHL_2.H$	9.46
	$MOHL_2/M(OH)_2L_2.H$	12.3
	$M_2OH(H_{-1}L)_2/M(H_{-1}L).MOH(H_{-1}L)$	1.8

Bibliography: 76SA

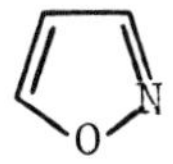

C_3H_3ON 1,2-Oxazole (isoxazole)   L

Metal ion	Equilibrium	Log K 25°, 0.5
Ag^+	$ML/M.L$	1.21
	$ML_2/M.L^2$	2.59

Bibliography: 79BB

Other reference: 78LK

C_3H_3NS		1,2-Thiazole (isothiazole)	L

Metal ion	Equilibrium	Log K 25°, 0.5
Co^{2+}	$ML/M.L$	(0.34)
	$ML_2/M.L^2$	1.0
Ni^{2+}	$ML/M.L$	(0.65)
	$ML_2/M.L^2$	1.6
Cu^{2+}	$ML/M.L$	(0.53)
	$ML_2/M.L^2$	1.8
Ag^+	$ML/M.L$	1.44 −0.3
	$ML_2/M.L^2$	2.69 ±0.0
Zn^{2+}	$ML/M.L$	(0.36)

Bibliography:

$Co^{2+}-Cu^{2+}, Zn^{2+}$ 78KL Ag^+ 78KL,79BB

C_3H_3NS		1,3-Thiazole	L

Metal ion	Equilibrium	Log K 25°, 0.5	ΔH 25°, 0.5	ΔS 25°, 0.5
H^+	$HL/H.L$	2.80 −0.1	−2.27	5.2
Co^{2+}	$ML/M.L$	1.43		
	$ML_2/M.L^2$	2.4		
	$ML_3/M.L^3$	3.0		
Ni^{2+}	$ML/M.L$	1.96		
	$ML_2/M.L^2$	3.5		
	$ML_3/M.L^3$	4.7		
Ag^+	$ML/M.L$	1.95		
	$ML_2/M.L^2$	3.92		
Zn^{2+}	$ML/M.L$	1.20		
	$ML_2/M.L^2$	2.0		
	$ML_3/M.L^3$	2.4		

Bibliography:

H^+ 74LKB,77BE Ag^+ 79BB

$Co^{2+}, Ni^{2+}, Zn^{2+}$ 74LKB

| C$_4$H$_5$NS | 4-Methyl-1,3-thiazole | | L |
</br>

Metal ion	Equilibrium	Log K 25°, 0.5
H$^+$	HL/H.L	3.27
Co^{2+}	ML/M.L	0.54
	ML$_2$/M.L^2	0.8
Ni^{2+}	ML/M.L	0.59
	ML$_2$/M.L^2	1.0
Cu^{2+}	ML/M.L	1.39
	ML$_2$/M.L^2	2.3
	ML$_3$/M.L^3	2.9
Zn^{2+}	ML/M.L	0.55
	ML$_2$/M.L^2	0.8

Bibliography: 76LK

| C$_3$H$_4$N$_2$ | 1,2-Diazole (pyrazole) | | L |

(Other references on Vol.2, p.349)

Metal ion	Equilibrium	Log K 25°, 0.1	Log K 25°, 0.5	Log K 25°, 1.0	ΔH 25°, 0.5	ΔS 25°, 0.5
H$^+$	HL/H.L	2.58	2.61 ±0.05	2.78	-3.65	-0.3
Mn^{2+}	ML/M.L	0.3				
	ML$_2$/M.L^2	0.3				
Fe^{2+}	ML/M.L	0.8				
	ML$_2$/M.L^2	1.1				
	ML$_3$/M.L^3	1.3				
	ML$_4$/M.L^4	1.5				
Co^{2+}	ML/M.L		1.3 ±0.1			
	ML$_2$/M.L^2		2.3 ±0.1			
	ML$_3$/M.L^3		2.9			
	ML$_4$/M.L^4		3.2			
Ni^{2+}	ML/M.L	1.8 ±0.0	1.9 ±0.1	1.9		
	ML$_2$/M.L^2	3.3	3.5 ±0.2			
	ML$_3$/M.L^3	4.2	4.6 ±0.6			
	ML$_4$/M.L^4	4.6	5.3 ±0.8			
Cu^{2+}	ML/M.L	2.33	2.38	2.42		
	ML$_2$/M.L^2	4.2	4.3			
	ML$_3$/M.L^3	5.7	5.7			
	ML$_4$/M.L^4	6.6	6.6			

Pyrazole (continued)

Metal ion	Equilibrium	Log K 25°, 0.1	Log K 25°, 0.5	Log K 25°, 1.0	ΔH 25°, 0.5	ΔS 25°, 0.5
Ag^+	$ML/M.L$	2.05	2.11			
	$ML_2/M.L^2$	(4.37)	4.24			
Zn^{2+}	$ML/M.L$	1.0	1.1 ±0.1			
	$ML_2/M.L^2$	1.4	1.9 ±0.4			
	$ML_3/M.L^3$	1.6	2.5			
	$ML_4/M.L^4$		2.7			
Cd^{2+}	$ML/M.L$	1.1	1.2			
	$ML_2/M.L^2$	1.6	2.2			
	$ML_3/M.L^3$	1.8				
Pb^{2+}	$ML/M.L$	-0.4				

Bibliography:

H^+ 70MH,77BE,77LN Ag^+ 77BB,77PM

Mn^{2+}-Cu^{2+},Zn^{2+}-Pb^{2+} 70MH,74MMK,77BB,77LN,
                                          78LN

4-Methylpyrazole   L

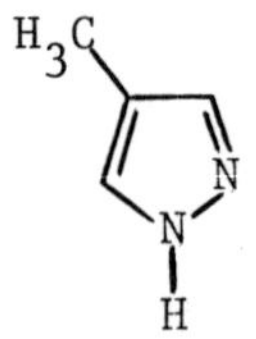

$C_4H_6N_2$

Metal ion	Equilibrium	Log K 25°, 0.5
H^+	$HL/H.L$	3.21
Co^{2+}	$ML/M.L$	1.54
	$ML_2/M.L^2$	2.7
	$ML_3/M.L^3$	3.6
	$ML_4/M.L^4$	4.2
	$ML_5/M.L^5$	4.4
Ni^{2+}	$ML/M.L$	2.01
	$ML_2/M.L^2$	3.6
	$ML_3/M.L^3$	4.9
	$ML_4/M.L^4$	5.9
	$ML_5/M.L^5$	6.6
	$ML_6/M.L^6$	6.9
Cu^{2+}	$ML/M.L$	2.70
	$ML_2/M.L^2$	5.0
	$ML_3/M.L^3$	6.8
	$ML_4/M.L^4$	8.2
Zn^{2+}	$ML/M.L$	1.30
	$ML_2/M.L^2$	2.3
	$ML_3/M.L^3$	2.9
	$ML_4/M.L^4$	3.2

Bibliography: 78LKa

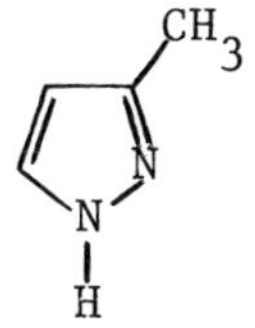

$$CH_3$$

(pyrazole structure)

$C_4H_6N_2$	3-Methylpyrazole	L

(Other reference on Vol.2, p.349)

Metal ion	Equilibrium	Log K 25°, 0.5
H^+	$HL/H.L$	3.62
Co^{2+}	$ML/M.L$	1.43
	$ML_2/M.L^2$	2.5
	$ML_3/M.L^3$	3.3
Ni^{2+}	$ML/M.L$	1.80
	$ML_2/M.L^2$	3.2
	$ML_3/M.L^3$	4.4
	$ML_4/M.L^4$	5.3
Cu^{2+}	$ML/M.L$	2.44
	$ML_2/M.L^2$	4.5
	$ML_3/M.L^3$	6.2
	$ML_4/M.L^4$	7.4
Zn^{2+}	$ML/M.L$	1.23
	$ML_2/M.L^2$	2.1
	$ML_3/M.L^3$	2.8

Bibliography:

$H^+,Co^{2+},Ni^{2+},Zn^{2+}$ 75LWa; Cu^{2+} 78LN   Other reference: 76BMc

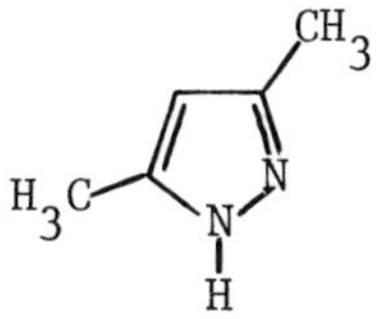

$$H_3C \quad CH_3$$

(dimethylpyrazole structure)

$C_5H_8N_2$	3,5-Dimethylpyrazole	L

Metal ion	Equilibrium	Log K 25°, 0.5
H^+	$HL/H.L$	4.10
Co^{2+}	$ML/M.L$	0.62
	$ML_2/M.L^2$	1.0
	$ML_3/M.L^3$	1.1
Ni^{2+}	$ML/M.L$	0.92
	$ML_2/M.L^2$	1.6
	$ML_3/M.L^3$	2.0
Cu^{2+}	$ML/M.L$	1.91
	$ML_2/M.L^2$	3.5
	$ML_3/M.L^3$	4.9
	$ML_4/M.L^4$	6.0

3,5-Dimethylpyrazole (continued)

Metal ion	Equilibrium	Log K 25°, 0.5
Zn^{2+}	$ML/M.L$	(-0.15)
	$ML_2/M.L^2$	(0.6)
	$ML_3/M.L^3$	(2.1)

Bibliography:

H^+-Ni^{2+},Zn^{2+} 77LGa

Cu^{2+} 78LN

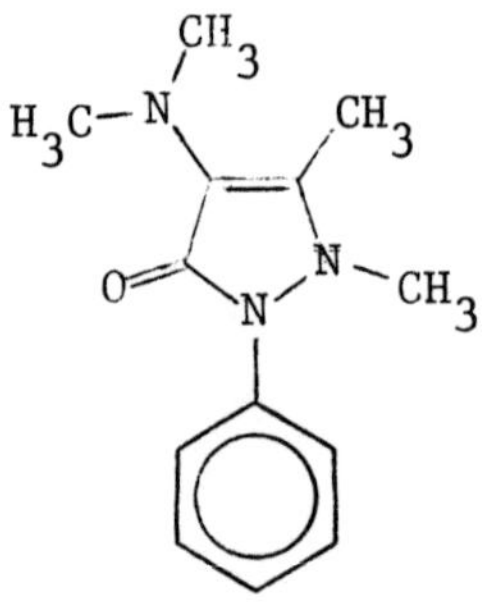

$C_{13}H_{17}ON_3$ 2,3-Dimethyl-4-dimethylamino-1-phenylpyrazol-5-one   (pyramidone)   L
(Other reference on Vol.2, 349)

Metal ion	Equilibrium	Log K 25°, 0.5
H^+	$HL/H.L$	5.21
Co^{2+}	$ML/M.L$	0.78
	$ML_2/M.L^2$	0.8
Ni^{2+}	$ML/M.L$	1.07
	$ML_2/M.L^2$	1.4
Cu^{2+}	$ML/M.L$	1.91
	$ML_2/M.L^2$	3.1
	$ML_3/M.L^3$	3.5
Zn^{2+}	$ML/M.L$	1.00
	$ML_2/M.L^2$	1.3

Bibliography: 78LW   Other reference: 76GJ

C$_3$H$_4$N$_2$ <u>1,3-Diazole</u> (imidazole)   L
                              (Other values on Vol.2, p.144)

Metal ion	Equilibrium	Log K 25°, 0.16	Log K 25°, 1.0	Log K 25°, 0	ΔH 25°, 0.16	ΔS 25°, 0.16
H$^+$	HL/H.L	7.01 ±0.03	7.31 ±0.03	6.993	−8.80 ±0.01	2.5
		7.17[b] ±0.05	7.58[d]	7.90[e] ±0.02	−9.10[b] ±0.02	2.3[b]
Co^{2+}	ML/M.L	2.40 ±0.07	2.50		(−4)[s]	(−2)
		2.49[b]	2.53[d]			
	ML$_2$/M.L^2	4.39 ±0.02			(−8)[s]	(−7)
	ML$_3$/M.L^3	5.92 ±0.03			(−12)[s]	(−13)
	ML$_4$/M.L^4	7.0 ±0.2			(−15)[s]	(−18)
	ML$_5$/M.L^5	7.4			(−18)[s]	(−27)
	ML$_6$/M.L^6	7.4			(−22)[s]	(−40)
Ni^{2+}	ML/M.L	2.98 ±0.04	3.10 +0.01	3.34[e]	−5.8	−6
		3.03[b]	3.24[d]	3.45[f]		
	ML$_2$/M.L^2	5.45 ±0.1	5.56 ±0.2	6.09[e]	(−11)[s]	(−12)
		5.54[b]	5.63[d]			
	ML$_3$/M.L^3	7.5	7.6 ±0.2	8.31[e]	(−15)[s]	(−16)
			8.0[d]			
	ML$_4$/M.L^4	9.1	9.4 ±0.6	9.92[e]	−18.4	−20
	ML$_5$/M.L^5	10.2			(−21)[s]	(−24)
	ML/MOHL.H			2.51[e]		
Cu^{2+}	ML/M.L	4.18 ±0.02	4.30	4.66[e]	−7.6	−6
		4.21[b]	4.47[d]			
	ML$_2$/M.L^2	7.66 ±0.04	7.93	8.64[e]	(−14)[s]	(−12)
		7.84[b]	8.11[d]			
	ML$_3$/M.L^3	10.50 −0.01	11.18	11.94[e]	(−19)[s]	(−16)
		10.72[b]	11.64[d]			
	ML$_4$/M.L^4	12.6 ±0.1	13.9	14.6[e]	−23.0	−20
	ML$_6$/M.L^6			17.5[e]		
	M$_2$(OH)$_2$L$_3$.H^2/M^2.L^3			2.52[e]		
Ag$^+$	ML/M.L	3.17[h]	3.08 ±0.03	3.34[e]	−7.3[c]	−10[c]
	ML$_2$/M.L^2	6.94[h]	6.90 ±0.05	7.55[e]	−15.7[c]	−21[c]
	ML/MOHL.H			8.12[e]		
CH$_3$Hg$^+$	ML/M.L	6.93				
		7.14[h]				
	ML/MOHL.H	9.65[h]				
	M$_2$OHL/M.MOHL	8.18[h]				

---

[b]25°, 0.5; [c]25°, 1.0; [d]25°, 2.0; [e]25°, 3.0; [f]25°, 4.0; [h]20°, 0.1; [s]10−50°, 0.16

Imidazole (continued)

Metal ion	Equilibrium	Log K 25°, 0.16	Log K 25°, 1.0	Log K 25°, 0	ΔH 25°, 0.16	ΔS 25°, 0.16
Zn^+	ML/M.L	2.56 ±0.04	2.66	2.92[e]	-3.8	-1
		2.61[b]	2.75[d]	3.44[f]		
	$ML_2/M.L^2$	4.89 ±0.05		(4.93)[e]		
	$ML_3/M.L^3$	7.16 -0.01		8.77[e]		
	$ML_4/M.L^4$	9.19 ±0.03		(11.41)[e]		
Cd^{2+}	ML/M.L	2.77 ±0.03	2.74 ±0.04			
		2.72[b]±0.05	2.89[d]	3.09[f]	-4.9[b]	-4[b]
	$ML_2/M.L^2$	4.9 ±0.2	5.0 ±0.2			
		4.8[b] ±0.1	5.0[d]		-10.[b]	-12[b]
	$ML_3/M.L^3$	6.3 ±0.2	6.6			
		6.2[b] ±0.2	7.0[d]			
	$ML_4/M.L^4$	6.9	7.6			
		7.3[b] ±0.2				
Hg^{2+}	ML/M.L			9.18[e]	-18[b]	(-18)[e*]
	$ML_2/M.L^2$	16.87		18.19[e]	-26[b]	(-4)[e*]
	ML/MOHL.H			(1.54)[p]	(-4)[b]	(-6)[e*]

[b]25°, 0.5; [d]25°, 2.0; [e]25°, 3.0; [f]25°, 4.0; *assuming ΔH for 0.5=ΔH for 3.0

Bibliography:

H[+] 68IS,76EW,76PJB,77BE,77ERG,78MH,79Ab,79MB	CH₃Hg[+] 77ERG
Co[2+] 68IS,74LV	Zn[2+] 68IS,74LV,77F
Ni[2+] 68IS,74LV,75FS,78F,79F,79SKb	Cd[2+] 74LV,75J,78MH
Cu[2+] 68IS,72MPa,74LV,77Sb	Hg[2+] 77Sa,78MH
Ag[+] 79GS	Other references: 72CM,72S,77HMa,78AMW, 78SPK

	$C_4H_6N_2$	1-Methylimidazole				L

Metal ion	Equilibrium	Log K 25°, 0.1		ΔH 25°, 0.1	ΔS 25°, 0.1
H[+]	HL/H.L	7.05 ±0.02		(-7.73)	(6.3)
Co[2+]	ML/M.L	2.29			
	$ML_2/M.L^2$	4.25			
	$ML_3/M.L^3$	5.32			
	$ML_4/M.L^4$	6.70			
Cu[+]	$ML_2/M.L^2$	11.45[h]			
CH₃Hg[+]	ML/M.L	6.96			

[h]20°, 0.1

Bibliography:

H[+] 76EW,76PJB,77ASa,77ERG	Cu[+] 69Z
Co[2+] 77ASa	CH₃Hg[+] 77ERG

| $C_5H_8N_2$ | | 1-Ethylimidazole | L |

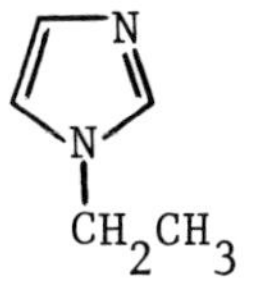

Metal ion	Equilibrium	Log K 25°, 0.5
H^+	HL/H.L	7.21
Co^{2+}	ML/M.L	2.32
	$ML_2/M.L^2$	4.2
	$ML_3/M.L^3$	5.4
	$ML_4/M.L^4$	7.0
	$ML_5/M.L^5$	7.4
Ni^{2+}	ML/M.L	3.04
	$ML_2/M.L^2$	5.5
	$ML_3/M.L^3$	7.5
	$ML_4/M.L^4$	9.0
	$ML_5/M.L^5$	9.8
	$ML_6/M.L^6$	10.2
Cu^{2+}	ML/M.L	4.40
	$ML_2/M.L^2$	8.0
	$ML_3/M.L^3$	11.0
	$ML_4/M.L^4$	13.2
	$ML_5/M.L^5$	(14.2)
Zn^{2+}	ML/M.L	2.50
	$ML_2/M.L^2$	4.8
	$ML_3/M.L^3$	7.4
	$ML_4/M.L^4$	9.3
	$ML_3/M.L^5$	10.1

Bibliography: 79LB

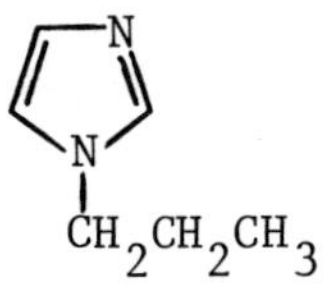

| $C_6H_{10}N_2$ | | 1-Propylimidazole | L |

Metal ion	Equilibrium	Log K 25°, 0.5
H^+	HL/H.L	7.22
Co^{2+}	ML/M.L	2.38
	$ML_2/M.L^2$	4.2
	$ML_3/M.L^3$	5.4
	$ML_4/M.L^4$	6.9
	$ML_5/M.L^5$	7.9
	$ML_6/M.L^6$	8.4

1-Propylimidazole (continued)

Metal ion	Equilibrium	Log K 25°, 0.5
Ni^{2+}	$ML/M.L$	3.06
	$ML_2/M.L^2$	5.6
	$ML_3/M.L^3$	7.6
	$ML_4/M.L^4$	9.2
	$ML_5/M.L^5$	10.3
	$ML_6/M.L^6$	11.0
Cu^{2+}	$ML/M.L$	4.25
	$ML_2/M.L^2$	7.8
	$ML_3/M.L^3$	10.7
	$ML_4/M.L^4$	13.1
	$ML_5/M.L^5$	(14.2)
Zn^{2+}	$ML/M.L$	2.62
	$ML_2/M.L^2$	4.7
	$ML_3/M.L^3$	7.2
	$ML_4/M.L^4$	9.2
	$ML_5/M.L^5$	10.0

Bibliography: 79LB

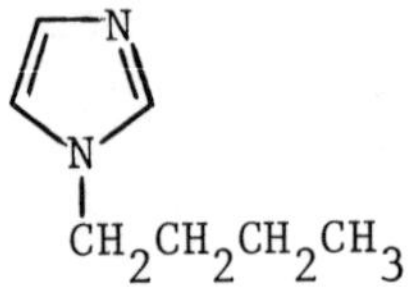

$CH_2CH_2CH_2CH_3$

$C_7H_{12}N_2$ 1-Butylimidazole   L

Metal ion	Equilibrium	Log K 25°, 0.5
H^+	$HL/H.L$	7.22
Co^{2+}	$ML/M.L$	2.75
	$ML_2/M.L^2$	4.8
	$ML_3/M.L^3$	6.0
	$ML_4/M.L^4$	6.5
Ni^{2+}	$ML/M.L$	3.30
	$ML_2/M.L^2$	5.9
	$ML_3/M.L^3$	8.0
Cu^{2+}	$ML/M.L$	4.40
	$ML_2/M.L^2$	8.1
	$ML_3/M.L^3$	11.2
	$ML_4/M.L^4$	13.6
Zn^{2+}	$ML/M.L$	2.57
	$ML_2/M.L^2$	5.0
	$ML_3/M.L^3$	7.2
	$ML_4/M.L^4$	9.3
	$ML_5/M.L^5$	10.8

Bibliography: 77LBK

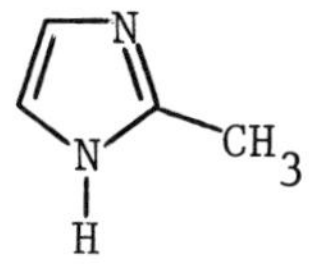

$C_4H_6N_2$ 2-Methylimidazole   L
                                   (Other values on Vol.2, p.148)

Metal ion	Equilibrium	Log K 25°, 0.1	Log K 25°, 1.0	Log K 25°, 0	ΔH 25°, 0.1	ΔS 25°, 0.1
H^+	HL/H.L	7.88 -0.01	8.13	7.86	(-9.26)	(5.0)
Co^{2+}	ML/M.L	8.05[b]				
	$ML_2/M.L^2$	1.73[b]				
	$ML_3/M.L^3$	3.0[b]				
		3.8[b]				
Cu^{2+}	ML/M.L	3.35[b]				
	$ML_2/M.L^2$	6.4[b]				
	$ML_3/M.L^3$	9.2[b]				
	$ML_4/M.L^4$	11.9[b]				
Zn^{2+}	ML/M.L	1.88[b]				

[b]25°, 0.5

Bibliography:
H^+ 74LKL,76EW,76PJB
Co^{2+}-Zn^{2+} 74LKL

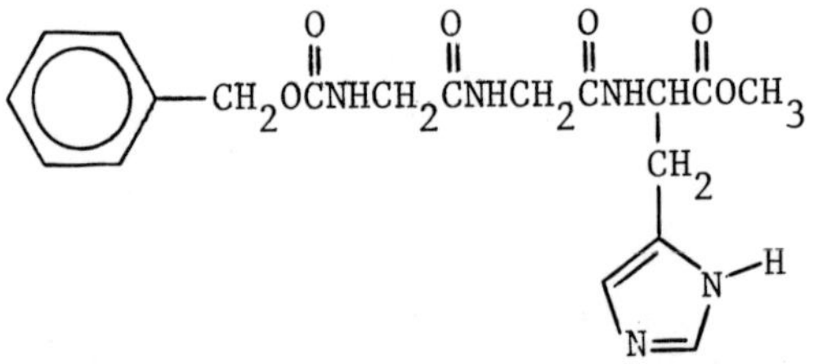

$C_{19}H_{23}O_6N_5$ <u>Phenylmethyloxycarbamylglycylglycyl-L-histidine methylester</u>   L
<u>(carbobenzoxyglyclglycyl-L-histidine methylester)</u>

Metal ion	Equilibrium	Log K 37°, 0.15
H^+	HL/H.L	6.11
Cu^{2+}	ML/M.L	3.49
	$ML/M(H_{-2}L).H^2$	11.78
	$M(H_{-2}L)/M(H_{-3}L).H$	7.14
Zn^{2+}	ML/M.L	1.81

Bibliography: 77AP

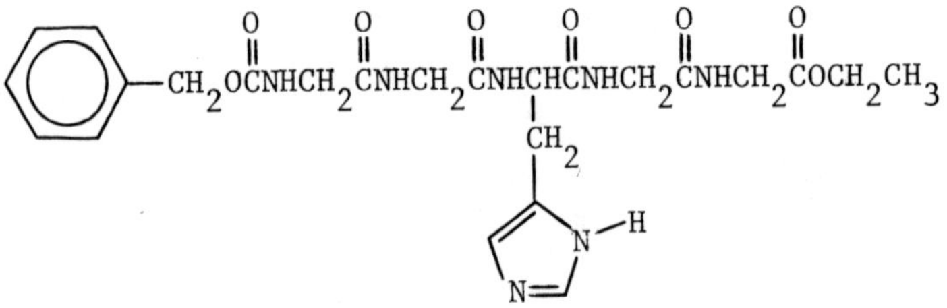

$C_{24}H_{31}O_8N_7$ <u>Phenylmethyloxycarbamylglycylglycyl-L-histidineglycylglycine ethylester</u>   L
<u>(carbobenzoxyglycylglycyl-L-histidineglycylglycine ethylester)</u>

Metal ion	Equilibrium	Log K 37°, 0.15
H^+	HL/H.L	6.08
Cu^{2+}	ML/M.L	3.60
	$ML/M(H_{-2}L).H^2$	11.48
	$M(H_{-2}L)/M(H_{-3}L).H$	7.24

Bibliography: 77AP

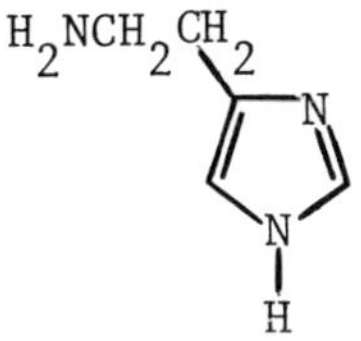

$H_2NCH_2CH_2$ — imidazole

$C_5H_9N_3$	4-(2-Aminoethyl)imidazole (histamine)					L
	(Other values on Vol.2, p.156)					

Metal ion	Equilibrium	Log K 25°, 0.1	Log K 30°, 1.1	Log K 25°, 0	ΔH 25°, 0.1	ΔS 25°, 0.1
H^+	HL/H.L	9.83 ±0.03	9.88 / 9.45[n]	9.80 ±0.05	-12.6 ±0.1	3
	$H_2L/HL.H$	6.07 ±0.05	6.34 / 5.89[n]	5.93 ±0.04	-7.6 ±0.1	2
Co^{2+}	ML/M.L	5.09 ±0.07	5.34 / 4.89[n]	5.12	$(-7)^r$	$(0)^*$
	$ML_2/M.L^2$	8.78 ±0.02	9.09 / 8.43[n]	8.89	$(-12)^r$	$(0)^*$
	$ML_3/M.L^3$		11.0			
Ni^{2+}	ML/M.L	6.80 ±0.04	6.84 / 6.60[n]	6.90	-12.2	-10
	$ML_2/M.L^2$	11.85 ±0.07	11.79 / 11.44[n]	12.02	-19.0	-10
	$ML_3/M.L^3$	15.0 ±0.1	14.9 / 14.4[n]	15.3		
	MHL/M.HL	4.71				
Cu^{2+}	ML/M.L	9.53 ±0.05	9.60 / 9.28[n]	9.64	-13.0	-4
	$ML_2/M.L^2$	16.0 ±0.1	16.1 / 15.6[n]	16.2	-22.8	-3
	MHL/M.HL	3.07				
	$MHL_2/ML_2.H$	5.71 ±0.02			-8.0	-1
	ML/MOHL.H	8.4				
	$ML_2/M(OH)L_2.H$	10.74				
	$M_2(OH)_2L_2.H^2/M^2.L^2$	7.2 ±0.2			-10.9	-4
Zn^{2+}	ML/M.L	5.25 ±0.1	(5.77) / 5.03[n]			
	$ML_2/M.L^2$	10.20 -0.2	10.50 / 9.81[n]			
	$ML_3/M.L^3$		12.1[n]			
	MHL/M.HL	1.65				

[n] 37°, 0.15; [r] 10-40°, 0; *assuming ΔH for 0.0=ΔH for 0.1

Bibliography:

H^+ 67PSS,76DO,76GS,77ST,79MBb Cu^{2+} 67PSS,76DO,76GS,79MBb

Co^{2+} 69PS Zn^{2+} 69PS,77DO

Ni^{2+} 69PS,76DOb,76GS   Other reference: 74PS

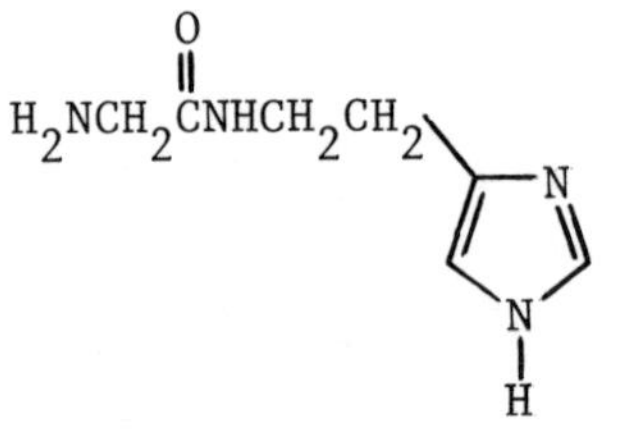

$$H_2NCH_2\overset{\displaystyle O}{\overset{\displaystyle \|}{C}}NHCH_2CH_2 \text{—imidazole ring}$$

$C_7H_{12}ON_4$ Glycylhistamine   L

Metal ion	Equilibrium	Log K 25°, 0.1
H^+	HL/H.L	8.04
	$H_2L/HL.H$	6.78
Co^{2+}	ML/M.L	2.94
	$ML/M(H_{-1}L).H$	7.93

Bibliography: 77HM

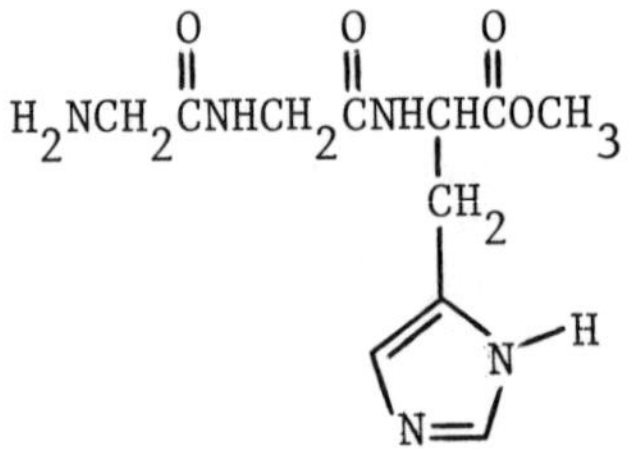

$$H_2NCH_2\overset{\displaystyle O}{\overset{\displaystyle \|}{C}}NHCH_2\overset{\displaystyle O}{\overset{\displaystyle \|}{C}}NHCHCOCH_3$$
$$CH_2\text{—imidazole ring}$$

$C_{11}H_{17}O_4N_5$ Glycylglycyl-L-histidine methylester   L

Metal ion	Equilibrium	Log K 37°, 0.15
H^+	HL/H.L	7.47
	$H_2L/HL.H$	6.12
Cu^{2+}	ML/M.L	6.87
	$ML_2/M(H_{-2}L).H^2$	8.01
	$M(H_{-1}L)L.H/M.L^2$	9.21
	$M(H_{-1}L)L/M(H_{-1}L)_2.H$	6.82
Zn^{2+}	ML/M.L	3.00
	$ML/M(H_{-1}L).H$	7.12

Bibliography: 77AP

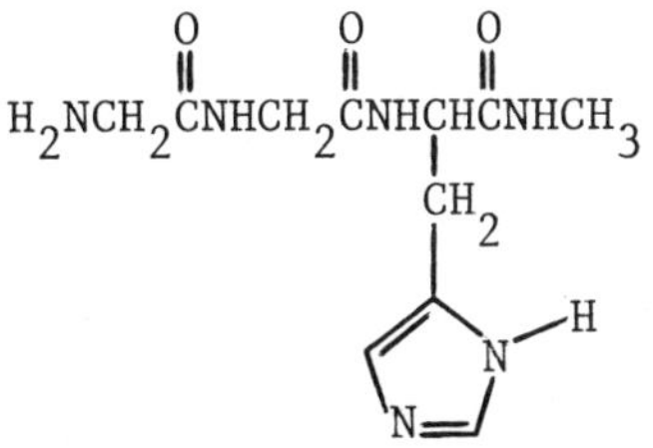

$$H_2NCH_2\overset{\displaystyle O}{\overset{\|}{C}}NHCH_2\overset{\displaystyle O}{\overset{\|}{C}}NHCHCH\overset{\displaystyle O}{\overset{\|}{C}}NHCH_3$$

(imidazole ring structure: CH_2 attached to imidazole N-H)

| $C_{11}H_{17}O_3N_6$ | Glycylglycyl-L-histidine methylamide | | | | L |

Metal ion	Equilibrium	Log K 25°, 0.15	ΔH 25°, 0.15	ΔS 25°, 0.15
H^+	HL/H.L	7.87 +0.01	-9.4	4
	$H_2L/HL.H$	6.33 ±0.02	-6.4	7
Co^{2+}	ML/M.L	5.01		
	MHL/ML.H	6.21		
	$ML/M(H_{-1}L).H$	6.39		
	$M(H_{-1}L)/M(H_{-2}L).H$	8.14		
Cu^{2+}	$M(H_{-2}L).H^2/M.L$	0.65	-1.3	-6
Zn^{2+}	$ML_2/M.L^2$	6.58		
	MHL/M.HL	2.33		
	$MHL/M(H_{-1}L).H^2$	15.22		
	$M(H_{-1}L)/M(H_{-2}L).H$	7.68		
	$ML_2/M(H_{-2}L).H^2$	18.36		

Bibliography:

H^+ 76KLb,79AR,79LS Cu^{2+} 79AR

Co^{2+},Zn^{2+} 79LS

$$H_2NCH_2\overset{\displaystyle O}{\overset{\|}{C}}NHCH_2\overset{\displaystyle O}{\overset{\|}{C}}NHCHCH\overset{\displaystyle O}{\overset{\|}{C}}NHCH_2\overset{\displaystyle O}{\overset{\|}{C}}NHCH_2\overset{\displaystyle O}{\overset{\|}{C}}OCH_2CH_3$$

(imidazole ring structure: CH_2 attached to imidazole N-H)

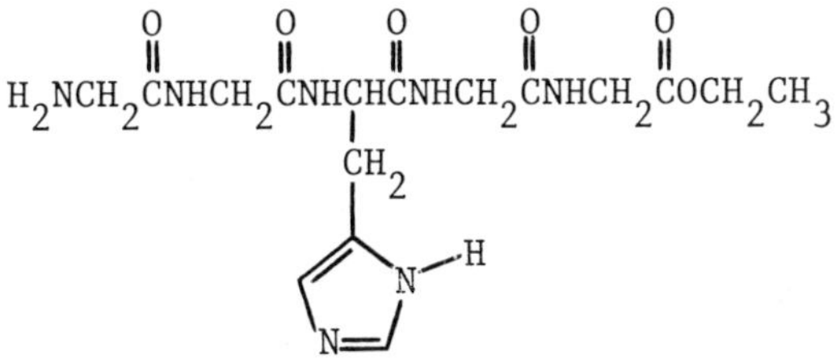

| $C_{16}H_{25}O_6N_7$ | Glycylglycyl-L-histidylglycylglycine ethylester | | L |

Metal ion	Equilibrium	Log K 37°, 0.15
H^+	HL/H.L	7.36
	$H_2L/HL.H$	6.01
Cu^{2+}	ML/M.L	6.65
	$ML/M(H_{-2}L).H^2$	7.51
Zn^{2+}	ML/M.L	3.04
	$ML/M(H_{-1}L).H$	7.08

Bibliography: 77AP

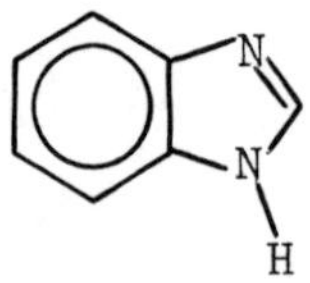

216VIII. AZOLES

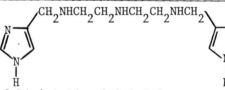

$C_7H_6N_2$	Benzo-1,3-diazole (benzimidazole) (Other values on Vol.2, p.163)				L

Metal ion	Equilibrium	Log K 25°, 0.15	Log K 25°, 0.5	ΔH 25°, 0.5	ΔS 25°, 0.5
H^+	HL/H.L	5.47 -0.01	5.77	-7.17	2.3
Ag^+	ML/M.L		3.1		
	$ML_2/M.L^2$		6.25		

Bibliography:

H^+ 77BE Ag^+ 79BB

CH$_2$NHCH$_2$CH$_2$NHCH$_2$CH$_2$NHCH$_2$

$C_{12}H_{21}N_7$	1,9-Di-4-imidazolyl-2,5,8-triazanonane (N,N''-bis(4-imidazolylmethyl)diethylenetriamine, 4-IMDIEN)	L

Metal ion	Equilibrium	Log K 25°, 0.1
H^+	HL/H.L	9.22
	$H_2L/HL.H$	8.18
	$H_3L/H_2L.H$	4.91
	$H_4L/H_3L.H$	3.90
	$H_5L/H_4L.H$	2.92
Co^{2+}	ML/M.L	13.84
	MHL/ML.H	3.3
Ni^{2+}	ML/M.L	17.35
	MHL/ML.H	2.01
Cu^{2+}	ML/M.L	20.41
	MHL/ML.H	3.35
Zn^{2+}	ML/M.L	13.30
	MHL/ML.H	2.69

Bibliography: 78THM

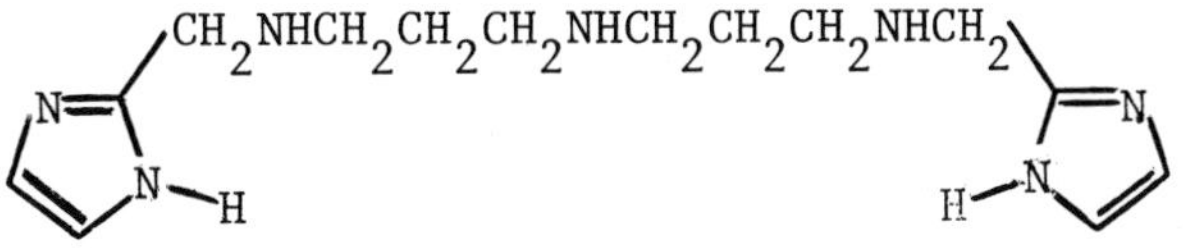

| $C_{14}H_{25}N_7$ | 1,11-Di-2-imidazolyl-2,6,10-triazaundecane | L |
| | (N,N''-bis(2-imidazolylmethyl)di-1,3-propylenetriamine, 2-IMDPT) | |

Metal ion	Equilibrium	Log K 25°, 0.1
H^+	HL/H.L	9.76
	$H_2L/HL.H$	7.22
	$H_3L/H_2L.H$	6.51
	$H_4L/H_3L.H$	3.92
	$H_5L/H_4L.H$	3.29
Co^{2+}	ML/M.L	11.55
Ni^{2+}	ML/M.L	15.01
	MHL/ML.H	2.59
Cu^{2+}	ML/M.L	18.46
	MHL/ML.H	2.93
Zn^{2+}	ML/M.L	11.90
	MHL/ML.H	3.0

Bibliography: 78THM

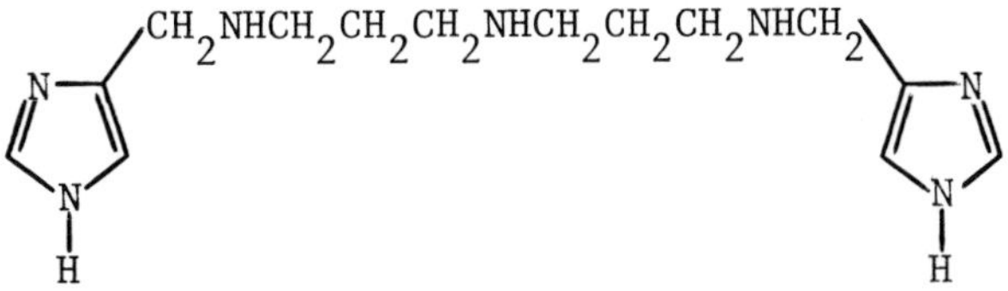

| $C_{14}H_{25}N_7$ | 1,11-Di-4-imidazolyl-2,6,10-triazaundecane | L |
| | (N,N''-bis(4-imidazolylmethyl)di-1,3-propylenetriamine, 4-IMDPT) | |

Metal ion	Equilibrium	Log K 25°, 0.1
H^+	HL/H.L	10.12
	$H_2L/HL.H$	8.62
	$H_3L/H_2L.H$	7.36
	$H_4L/H_3L.H$	4.51
	$H_5L/H_4L.H$	3.82
Co^{2+}	ML/M.L	11.36
	MHL/ML.H	3.99
Ni^{2+}	ML/M.L	14.93
	MHL/ML.H	3.23
Cu^{2+}	ML/M.L	18.97
	MHL/ML.H	3.62
Zn^{2+}	ML/M.L	11.83
	MHL/ML.H	4.52

Bibliography: 78THM

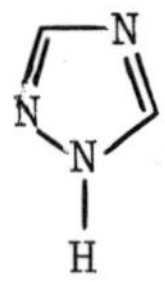

$C_2H_3N_3$		1,2,4-Triazole			L

Metal ion	Equilibrium	Log K 25°, 0.5	Log K 25°, 0	ΔH 25°, 0	ΔS 25°, 0
H^+	HL/H.L	9.99	9.97	-8.8 -9.42[b]	16 14.1[b]
	$H_2L/HL.H$	2.42	2.39	-2.6 -2.43[b]	2 2.9[b]
Ag^+	ML/M.L $ML_2/M.L^2$	2.60 4.38			

[b]25°, 0.5

Bibliography:

H^+ 74LPa,77BE Ag^+ 79BB

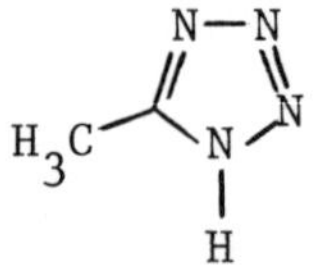

H_3C—[ring]—H

$C_2H_4N_4$		5-Methyltetrazole	L

Metal ion	Equilibrium	Log K 25°, 0.5
H^+	HL/H.L	5.32
Co^{2+}	ML/M.L	3.10
	$ML_2/M.L^2$	5.9
	$ML_3/M.L^3$	8.6
	$ML_4/M.L^4$	10.8
	$ML_5/M.L^5$	12.5
Ni^{2+}	ML/M.L	5.02
	$ML_2/M.L^2$	9.8
	$ML_3/M.L^3$	13.5
	$ML_4/M.L^4$	15.8
	$ML_5/M.L^5$	17.6
	$ML_6/M.L^6$	19.1
Zn^{2+}	ML/M.L	(1.72)
	$ML_2/M.L^2$	4.0
	$ML_4/M.L^4$	7.7
	$ML_6/M.L^6$	11.9

Bibliography: 71LB

C_5H_5N Pyridine (azine)   L
                                (Other values on Vol.2, p.165)

Metal ion	Equilibrium	Log K 25°, 0.1	Log K 25°, 1.0	Log K 25°, 0	ΔH 25°, 0	ΔS 25°, 0
H^+	HL/H.L	5.24 ± 0.02	5.39 ± 0.02	5.229	-4.80 ± 0.1	7.8
		$5.31^b \pm 0.02$			-5.41^b	6.1^b
Mn^{2+}	ML/M.L	0.24^b			-2^c	-6^c
	$ML_2/M.L^2$	-0.4^b				
Fe^{2+}	ML/M.L	0.7^b				
	$ML_2/M.L^2$	0.9^b				
Co^{2+}	ML/M.L	1.20 ± 0.0	1.35^j			
		$1.24^b \pm 0.01$				
	$ML_2/M.L^2$	$1.75^b \pm 0.05$	1.9^j			
	$ML_3/M.L^3$	$1.8^b \pm 0.2$				
	$ML_4/M.L^4$	$1.6^b \pm 0.2$				
Ni^{2+}	ML/M.L	$1.85 +0.2$	1.94		-3^c	-1^c
		$1.88^b \pm 0.02$				
	$ML_2/M.L^2$	$3.10^b \pm 0.02$				
	$ML_3/M.L^3$	$3.6^b \pm 0.2$				
	$ML_4/M.L^4$	$3.4^b \pm 0.2$				
Cu^{2+}	ML/M.L	2.56 ± 0.02	2.59 ± 0.2	2.50	-4.0	-2
		$2.56^b \pm 0.04$				
	$ML_2/M.L^2$	$4.45^b \pm 0.07$	4.5 ± 0.1	4.30	-8.9	-10^b
	$ML_3/M.L^3$	$5.7^b \pm 0.2$	5.9 ± 0.1	(5.2)	-16	-28^b
	$ML_4/M.L^4$	$6.4^b \pm 0.2$	6.6 ± 0.1	6.0	-22	-45^b
Ag^+	ML/M.L	1.93 ± 0.03		2.01 ± 0.04	-4.6	-6
		$2.03^b \pm 0.03$			$-4.78^b \pm 0.05$	-6.6^b
	$ML_2/M.L^2$	4.22 ± 0.03		4.13 ± 0.03	-11.2	-19
		$4.16^b \pm 0.06$			$-11.26^b \pm 0.08$	-18.7^b
CH_3Hg^+	ML/M.L	4.72^h				
Zn^{2+}	ML/M.L	1.04 ± 0.10	1.2 ± 0.2		-3^c	-5^c
		$1.08^b -0.10$				
	$ML_2/M.L^2$	$1.4^b \pm 0.3$	$1.8^j \pm 0.2$			
		$1.5^b \pm 0.1$				
	$ML_3/M.L^3$	$1.6^b \pm 0.1$				
	$ML_4/M.L^4$	1.4^b				
Cd^{2+}	ML/M.L	1.31 ± 0.04	1.51^j			
		$1.38^b \pm 0.02$				
	$ML_2/M.L^2$	2.03 ± 0.01	2.46^j			
		$2.18^b \pm 0.06$				
	$ML_3/M.L^3$	$2.2^p \pm 0.2$				
		$2.5^b \pm 0.1$				
	$ML_4/M.L^4$	2.3^b				

$^b 25°, 0.5; \quad ^c 25°, 1.0; \quad ^h 20°, 0.1; \quad ^j 20°, 1.0; \quad ^p 30°, 0.1$

Pyridine   (continued)

Metal ion	Equilibrium	Log K 25°, 0.1	Log K 25°, 1.0	Log K 25°, 0	ΔH 25°, 0	ΔS 25°, 0.1
Hg^{2+}	$ML/M.L$	5.2^b			-9^b	-6^b
	$ML_2/M.L^2$	10.0^b			-17^b	-11^b
	$ML_3/M.L^3$	10.3^b				
	$ML_4/M.L^4$	10.6^b				

_____

^{b}25°, 0.5

Bibliography:

H^+ 59ML,72TP,74IL,74LP Zn^{2+} 53Nb,67FL,76GS,74IL,75SSa

Co^{2+} 74IL Cd^{2+} 67FL,67GS,74IL

Ni^{2+} 70DT,72TP,74IL Hg^{2+} 78MH

Cu^{2+} 57LH,73CP,74IL   Other references: 62HP,68TR,69RB,
                                                         74SS,75D,77HF,79DP
Ag^+ 55MB,74BE,74IL,77PM

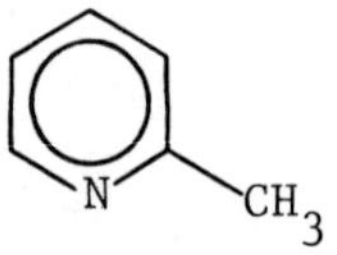

CH_3Hg^+73GE

_____

C_6H_7N		2-Methylpyridine (2-picoline) (Other values on Vol.2, p.167)				L

Metal ion	Equilibrium	Log K 25°, 0.1	Log K 25°, 0.5	Log K 25°, 0	ΔH 25°, ∿0	ΔS 25°, 0
H^+	$HL/H.L$	5.95 +0.01	6.02 ±0.04	5.95 ±0.02	-6.1 ±0.1	7
					-6.55^b	5.6^b
Ni^{2+}	$ML/M.L$	0.4				
Ag^+	$ML/M.L$		2.32 ±0.05		-5.8^b	-9^b
	$ML_2/M.L^2$		4.68 ±0.03		-10.2^b	-13^b

_____

^{b}25°, 0.5

Bibliography:

H^+ 59ML,72TP Ag^+ 74BE

Ni^{2+} 72TP   Other references: 69RB,76AS

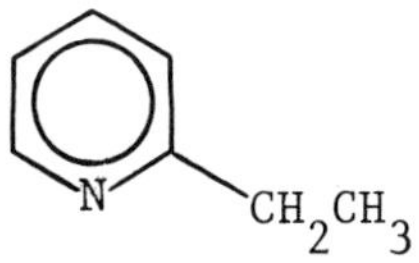

A. PYRIDINES

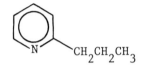

| C₇H₉N | | 2-Ethylpyridine
(Other references on Vol.2, p.340) | | | | L |

C_7H_9N

<u>2-Ethylpyridine</u>
(Other references on Vol.2, p.340)   L

Metal ion	Equilibrium	Log K 25°, 0.1	Log K 25°, 0	ΔH 25°, 0.5	ΔS 25°, 0.5
H^+	HL/H.L	6.07	5.89	-6.48	6.0
Ag^+	ML/M.L	2.29		-5.7	-9
	$ML_2/M.L^2$	4.54		-9.9	-12

Bibliography: 73MB,76BEF

$C_8H_{11}N$

<u>2-Propylpyridine</u>
(Other reference on Vol.2, p.340)   L

Metal ion	Equilibrium	Log K 25°, 0.1	Log K 25°, 0.5	ΔH 25°, 0.5	ΔS 25°, 0.5
H^+	HL/H.L	6.12	6.14	-6.83	5.2
Co^{2+}	ML/M.L	1.4			
Ni^{2+}	ML/M.L	1.6			
Cu^{2+}	ML/M.L	2.2			
Ag^+	ML/M.L	2.15	2.12	-5.6	-9
	$ML_2/M.L^2$	4.45	4.10	-9.7	-14
Zn^{2+}	ML/M.L	1.3			
Cd^{2+}	ML/M.L	1.1			

Bibliography:

H^+ 73MB,74IL

Co^{2+}-Cu^{2+},Zn^{2+},Cd^{2+} 74IL

Ag^+ 73MB,74IL,76BEF

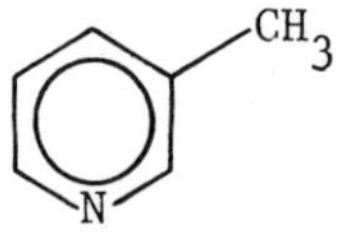

$$\text{3-methylpyridine structure with } CH_3 \text{ and } N$$

| C$_6$H$_7$N | 3-Methylpyridine (3-picoline) | L |
| | (Other values on Vol.2, p.168) | |

Metal ion	Equilibrium	Log K 25°, 0.1	Log K 25°, 0.5	Log K 25°, 0	ΔH 25°, 0	ΔS 25°, 0
H$^+$	HL/H.L	5.73 ±0.04	5.80 ±0.07	5.70 ±0.02	-5.7 ±0.1	7
					-6.04[b]	6.3[b]
Co^{2+}	ML/M.L		1.40			
	ML$_2$/M.L^2		2.2			
	ML$_3$/M.L^3		2.5			
Ni^{2+}	ML/M.L	1.87 ±0.02	2.02 ±0.05			
	ML$_2$/M.L^2		3.3 ±0.1			
	ML$_3$/M.L^3		4.1 ±0.2			
	ML$_4$/M.L^4		4.6			
Cu^{2+}	ML/M.L	2.77	2.74 ±0.04	2.76[c]		
	ML$_2$/M.L^2		4.8 ±0.1			
	ML$_3$/M.L^3		6.3 ±0.2			
	ML$_4$/M.L^4		7.2 ±0.2			
Ag$^+$	ML/M.L		2.20 ±0.05		-5.2[b]	-7[b]
	ML$_2$/M.L^2		4.46 ±0.02		-11.9[b]	-20[b]
Zn^{2+}	ML/M.L		1.00			
	ML$_2$/M.L^2		2.1			
	ML$_3$/M.L^3		2.6			
	ML$_4$/M.L^4		(3.7)			
Cd^{2+}	ML/M.L	1.34[P] ±0.07	1.62			
	ML$_2$/M.L^2	2.3[P] ±0.1	2.8			
	ML$_3$/M.L^3	2.5[P]	3.6			
	ML$_4$/M.L^4		4.0			

[b]25°, 0.5;   [c]25°, 1.0;   [P]30°, 0.1

Bibliography:

H$^+$ 59ML,72TP,74LP   Ag$^+$ 74BE

Co^{2+},Cu^{2+},Zn^{2+} 78LR   Cd^{2+} 79LR

Ni^{2+} 72TP,78LR   Other references: 69RB,76AS,79DP

$$\text{(structure: pyridine with } CH_2CH_3 \text{ at 3-position, N at bottom)}$$

C$_7$H$_9$N   3-Ethylpyridine   L
                        (Other references on Vol.2, p.340)

Metal ion	Equilibrium	Log K 25°, 0.5	Log K 25°, 0	ΔH 25°, 0.5	ΔS 25°, 0.5
H$^+$	HL/H.L	5.82	(5.56)	-6.10	-6.2
Ag$^+$	ML/M.L	2.23		-5.2	-7
	ML$_2$/M.L^2	4.53		-12.0	-19

Bibliography:

H$^+$ 76FE   Ag$^+$ 76BEF

$$\text{(structure: pyridine with } CH_3 \text{ at 4-position, N at bottom)}$$

C$_6$H$_7$N   4-Methylpyridine   (4-picoline)   L
                        (Other values on Vol.2, p.169)

Metal ion	Equilibrium	Log K 25°, 0.1	Log K 25°, 1.0	Log K 25°, 0	ΔH 25°, 0	ΔS 25°, 0
H$^+$	HL/H.L	6.04 ±0.06	6.08 ±0.02	6.00 ±0.02	-6.05 ±0.1	7
		6.07^b±0.05			-6.44^b	6.2^b
Ni^{2+}	ML/M.L	2.11 ±0.00	2.15			
		2.09^b				
	ML$_2$/M.L^2	3.59^b	2.83			
	ML$_3$/M.L^3	4.34^b	4.81			
	ML$_4$/M.L^4	4.70^b				
Cu^{2+}	ML/M.L	2.88	2.93 ±0.07			
		2.93^b				
	ML$_2$/M.L^2	5.16^b	5.17 ±0.03			
	ML$_3$/M.L^3	6.77^b	6.83 ±0.01			
	ML$_4$/M.L^4	8.08^b	7.92 ±0.05			
	ML$_5$/M.L^5		8.3			
Ag$^+$	ML/M.L	2.21^b±0.03		2.03	-6.1^b	-11^b
	ML$_2$/M.L^2	4.67^b±0.03		4.39	-12.8^b	-22^b
Cd^{2+}	ML/M.L	1.51^P±0.01	1.62			
	ML$_2$/M.L^2	2.5^P -0.3	2.8			
	ML$_3$/M.L^3	2.9^P ±0.1	3.6			
	ML$_4$/M.L^4		4.0			

---

^{b}25°, 0.5; ^{P}30°, 0.1

Bibliography: Cu^{2+},Cd^{2+} 79LR
H$^+$ 59ML,72TP,74LP,79LR   Ag$^+$ 74BE
Ni^{2+} 72TP   Other references: 69RB,76AS,79DP

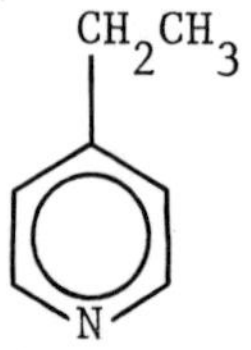

C$_7$H$_9$N		4-Ethylpyridine				L

(Other values on Vol.2, p.170)

Metal ion	Equilibrium	Log K 25°, 0.1	Log K 25°, 1.0	Log K 25°, 0	ΔH 25°, 0.5	ΔS 25°, 0.5
H$^+$	HL/H.L	6.06 (6.33)^b	6.14	5.95 ±0.08	−6.67	(6.6)
Ni^{2+}	ML/M.L	2.30	(1.91)			
	ML$_2$/M.L^2		3.34			
	ML$_3$/M.L^3		4.02			
Cu^{2+}	ML/M.L		2.77			
	ML$_2$/M.L^2		5.07			
	ML$_3$/M.L^3		6.87			
	ML$_4$/M.L^4		(8.2)			
	ML$_5$/M.L^5		(9.1)			
Ag$^+$	ML/M.L	2.34^b			−6.1	−10
	ML$_2$/M.L^2	4.67^b			−12.7	−21
Cd^{2+}	ML/M.L		1.56			
	ML$_2$/M.L^2		2.73			
	ML$_3$/M.L^3		3.53			
	ML$_4$/M.L^4		3.95			

^{b}25°, 0.5

Bibliography:

H$^+$	72TP,73MB,79LR	Cu^{2+},Cd^{2+}	79LR
Ni^{2+}	72TP	Ag$^+$	73MB,76BEF

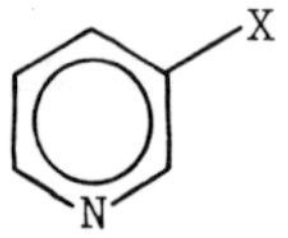

C$_5$H$_4$NX			3-Haligenopyridine				L

(Other references on Vol.2, p.340)

X=	Metal ion	Equilibrium	Log K 25°, 0.5	Log K 25°, 0	ΔH 25°, 0	ΔS 25°, 0
Cl 3-Chloro	H$^+$	HL/H.L	3.01	2.83	−2.60 −2.89^b	4.2 4.1^b
	Ag$^+$	ML/M.L	1.59		−3.9^b	−6^b
		ML$_2$/M.L^2	3.02		−9.0^b	−16^b
Br 3-Bromo	H$^+$	HL/H.L	3.08	2.88	−2.78 −3.00^b	3.8 4.0^b
	Ag$^+$	ML/M.L	1.66		−4.1^b	−6^b
		ML$_2$/M.L^2	3.31		−9.4^b	−16^b

^{b}25°, 0.5

3-Haligenopyridine   (continued)

X=	Metal ion	Equilibrium	Log K 25°, 0.5	Log K 25°, 0	ΔH 25°, 0	ΔS 25°, 0
I	H^+	HL/H.L	3.46		-3.54^b	4.6^b
3-Iodo	Ag^+	ML/M.L	1.80			
		$ML_2/M.L^2$	4.12			

---

^{b}25°, 0.5

Bibliography:

H^+ 74LP,76FE Ag^+ 76BEF

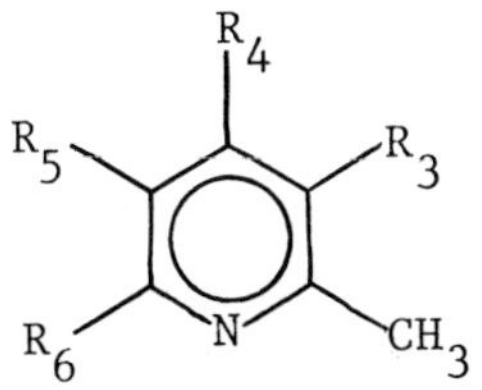

C_7H_9N 2,X-Dimethylpyridine   (2,X-lutidine)   L
                     (Other values on Vol.2, p.171 and p.172)

R=	Metal ion	Equilibrium	Log K 25°, 0.5	Log K 25°, 0	ΔH 25°, 0	ΔS 25°, 0
$R_3=CH_3$	H^+	HL/H.L	6.75	6.75	-7.46^b	5.8^b
$R_4=R_5=R_6=H$	Ag^+	ML/M.L	2.46		-6.2^b	-10^b
2,3-Dimethyl		$ML_2/M.L^2$	4.79		-11.6^b	-17^b
$R_4=CH_3$	H^+	HL/H.L	6.83 ±0.01	6.63	-7.2	6
$R_3=R_5=R_6=H$					$(-8.26)^b$	$(3.5)^b$
2,4-Dimethyl	Ag^+	ML/M.L	2.51 ±0.04		-7.1^b	-12^b
		$ML_2/M.L^2$	5.12 ±0.06		-12.8^b	-20^b
$R_5=CH_3$	H^+	HL/H.L	6.58 -0.1	6.40	-6.8	7
$R_3=R_5=R_6=H$					-7.12^b	6.2^b
2,5-Dimethyl	Ag^+	ML/M.L	2.50 ±0.08		-6.3^b	-9^b
		$ML_2/M.L^2$	4.93 ±0.02		-11.5^b	-16^b
$R_6=CH_3$	H^+	HL/H.L	6.90 -0.2	6.72 +0.03	-7.2	7
$R_3=R_4=R_6=H$					$(-8.36)^b$	$(3.5)^b$
2,6-Dimethyl	Ag^+	ML/M.L	2.68		-7.2^b	-12^b
		$ML_2/M.L^2$	5.06		-12.2^b	-18^b

---

^{b}25°, 0.5

Bibliography: 74BE

Other references: 59ML,69RB

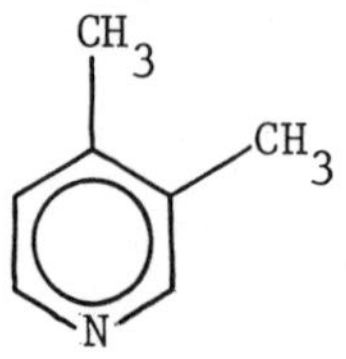

$$\text{3,4-Dimethylpyridine structure with CH}_3\text{ groups}$$

C$_7$H$_9$N		3,4-Dimethylpyridine (3,4-lutidine)			L

(Other references on Vol.2, p.173)

Metal ion	Equilibrium	Log K 25°, 0.5	Log K 25°, 0	ΔH 25°, 0.5	ΔS 25°, 0.5
H$^+$	HL/H.L	6.57 ±0.08	6.46	−7.48	5.0
Co^{2+}	ML/M.L	1.43			
	ML$_2$/M.L^2	2.4			
	ML$_3$/M.L^3	3.1			
Ni^{2+}	ML/M.L	2.18 ±0.08			
	ML$_2$/M.L^2	3.5 ±0.3			
	ML$_3$/M.L^3	5.0 ±0.2			
	ML$_4$/M.L^4	5.5			
	ML$_5$/M.L^5	5.7			
Cu^{2+}	ML/M.L	3.01 ±0.10			
	ML$_2$/M.L^2	5.38 ±0.00			
	ML$_3$/M.L^3	7.42 ±0.04			
	ML$_4$/M.L^4	8.8 ±0.1			
	ML$_5$/M.L^5	(10.1)			
Ag$^+$	ML/M.L	2.43		−6.0	−9
	ML$_2$/M.L^2	4.85		−13.3	−22
Zn^{2+}	ML/M.L	1.43			
	ML$_2$/M.L^2	2.5			
	ML$_3$/M.L^3	3.3			
Cd^{2+}	ML/M.L	1.65			
	ML$_2$/M.L^2	2.9			
	ML$_3$/M.L^3	3.8			
	ML$_4$/M.L^4	4.2			

Bibliography:

H$^+$ 74BE,79LR   Ag$^+$ 74BE

Co^{2+}-Cu^{2+},Zn^{2+},Cd^{2+} 79LR

$$\text{5-Ethyl-2-methylpyridine structure with CH}_3\text{CH}_2\text{ and CH}_3\text{ groups}$$

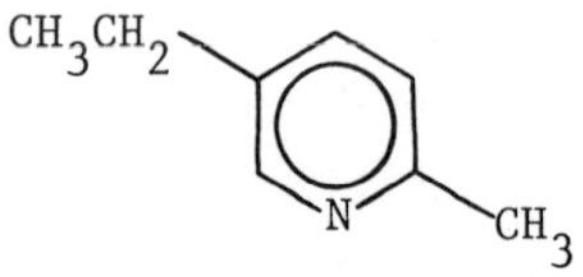

C$_8$H$_{11}$N		5-Ethyl-2-methylpyridine	L

Metal ion	Equilibrium	Log K 25°, 0.1
H$^+$	HL/H.L	6.43
Ni^{2+}	ML/M.L	0.83

Bibliography.   72TP

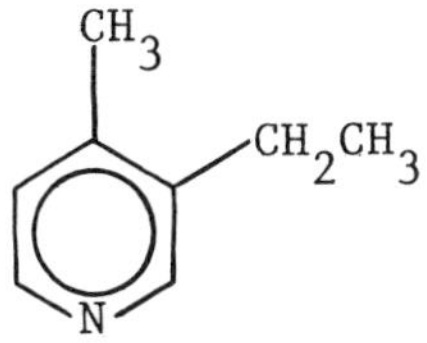

$C_8H_{11}N$ 3-Ethyl-4-methylpyridine   L

Metal ion	Equilibrium	Log K 25°, 1.0
H^+	ML/M.L	6.35
Co^{2+}	ML/M.L	1.28
	$ML_2/M.L^2$	2.15
Ni^{2+}	ML/M.L	2.02
	$ML_2/M.L^2$	3.60
Zn^{2+}	ML/M.L	1.58
	$ML_2/M.L^2$	2.71
	$ML_3/M.L^3$	3.36
Cd^{2+}	ML/M.L	1.07
	$ML_2/M.L^2$	2.73
	$ML_3/M.L^3$	3.50

Bibliography:
H^+-Zn^{2+} 75LP Cd^{2+} 79LR

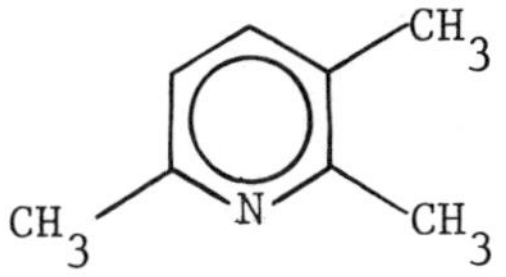

$C_8H_{11}N$ 2,3,6-Trimethylpyridine   (2,3,6-collidine)   L
                  (Other reference on Vol.2, p.340)

Metal ion	Equilibrium	Log K 25°, 0.5	ΔH 25°, 0.5	ΔS 25°, 0.5
H^+	HL/H.L	7.44	−8.33	6.1
Ag^+	ML/M.L	2.70		
	$ML_2/M.L^2$	5.00		

Bibliography: 73MB

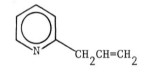

C$_7$H$_7$N		2-Vinylpyridine		L

Metal ion	Equilibrium	Log K 25°, 0.1	Log K 25°, 0.23
H$^+$	HL/H.L	5.03	5.05
Co^{2+}	ML/M.L	0.8	
Ni^{2+}	ML/M.L	1.2	
Cu^{2+}	ML/M.L	(1.1)	1.72
Ag$^+$	ML/M.L ML$_2$/M.L^2	1.75 3.55	
Zn^{2+}	ML/M.L	0.9	
Cd^{2+}	ML/M.L	1.1	

Bibliography: H$^+$,Cu^{2+} 74IL,78HH; Co^{2+},Ni^{2+},Ag$^+$-Cd^{2+} 74IL

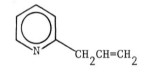

C$_8$H$_9$N		2-Prop-2-enylpyridine (2-allylpyridine)	L

Metal ion	Equilibrium	Log K 25°, 0.1
H$^+$	HL/H.L	5.34
Co^{2+}	ML/M.L	1.5
Ni^{2+}	ML/M.L	1.7
Cu^{2+}	ML/M.L	1.8
Ag$^+$	ML/M.L ML$_2$/M.L^2	2.97 4.8
Zn^{2+}	ML/M.L	1.6
Cd^{2+}	ML/M.L	1.3

Bibliography: 74IL

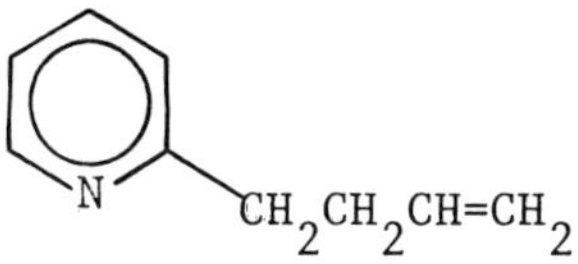

$$\text{pyridine—CH}_2\text{CH}_2\text{CH=CH}_2$$

C$_9$H$_{11}$N   2-But-3-enylpyridine   L

Metal ion	Equilibrium	Log K 25°, 0.1
H$^+$	HL/H.L	5.88
Co^{2+}	ML/M.L	1.2
Ni^{2+}	ML/M.L	1.7
Cu^{2+}	ML/M.L	2.10
Ag$^+$	ML/M.L	2.72
	ML$_2$/M.L^2	4.5
Zn^{2+}	ML/M.L	1.4
Cd^{2+}	ML/M.L	1.3

Bibliography: 74IL

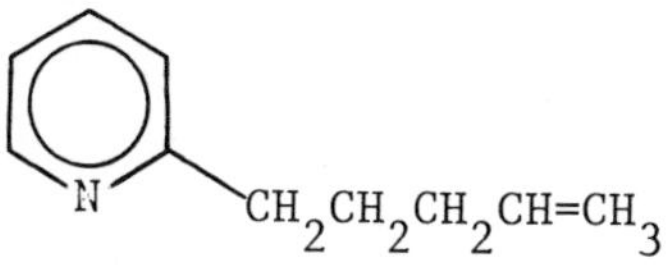

$$\text{pyridine—CH}_2\text{CH}_2\text{CH}_2\text{CH=CH}_3$$

C$_{10}$H$_{13}$N   2-Pent-4-enylpyridine   L

Metal ion	Equilibrium	Log K 25°, 0.1
H$^+$	HL/H.L	5.97
Co^{2+}	ML/M.L	1.2
Ni^{2+}	ML/M.L	1.6
Cu^{2+}	ML/M.L	2.0
Ag$^+$	ML/M.L	2.27
	ML$_2$/M.L^2	4.37
Zn^{2+}	ML/M.L	1.5
Cd^{2+}	ML/M.L	1.0

Bibliography: 74IL

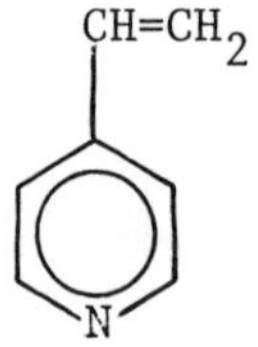

CH=CH$_2$

C$_7$H$_7$N   4-Vinylpyridine   L

Metal ion	Equilibrium	Log K 25°, 0.1
H$^+$	HL/H.L	5.40
Co^{2+}	ML/M.L	1.6
Ni^{2+}	ML/M.L	2.09
Cu^{2+}	ML/M.L	2.73
Ag$^+$	ML/M.L	1.98
	ML$_2$/M.L^2	4.08
Zn^{2+}	ML/M.L	1.6
Cd^{2+}	ML/M.L	1.6

Bibliography: 74IL

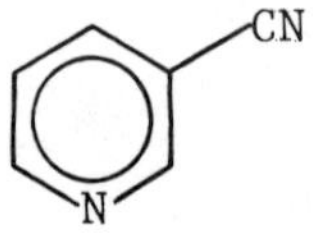

CN

C$_6$H$_4$N$_2$ 3-Cyanopyridine   L
                           (Other value on Vol.2, p.177)

Metal ion	Equilibrium	Log K 25°, 0.1	ΔH 25°, 0.1	ΔS 25°, 0.1
H$^+$	HL/H.L	1.64	-1.0 +0.1	4
Ni^{2+}	ML/M.L	1.23	-1.1	2

Bibliography:
H$^+$ 74LP,79HM
Ni^{2+} 79HM

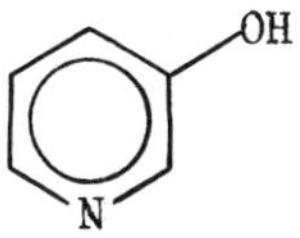

C_5H_5ON		3-Hydroxypyridine			L
		(Other reference on Vol.2, p.340)			

Metal ion	Equilibrium	Log K 25°, 0.5	Log K 25°, 0	ΔH 25°, 0	ΔS 25°, 0
H^+	HL/H.L	5.00	4.77	-4.01	8.4
Co^{2+}	$ML/M.L$	0.98			
	$ML_2/M.L^2$	1.6			
	$ML_3/M.L^3$	1.9			
Ni^{2+}	$ML/M.L$	1.44			
	$ML_2/M.L^2$	2.5			
	$ML_3/M.L^3$	3.1			
	$ML_4/M.L^4$	3.3			
Cu^{2+}	$ML/M.L$	2.03			
	$ML_2/M.L^2$	3.6			
	$ML_3/M.L^3$	4.8			
	$ML_4/M.L^4$	5.6			
Zn^{2+}	$ML/M.L$	0.70			
	$ML_2/M.L^2$	1.6			
	$ML_3/M.L^3$	1.8			
Cd^{2+}	$ML/M.L$	0.81			
	$ML_2/M.L^2$	2.1			
	$ML_3/M.L^3$	2.2			

Bibliography: 78LRa

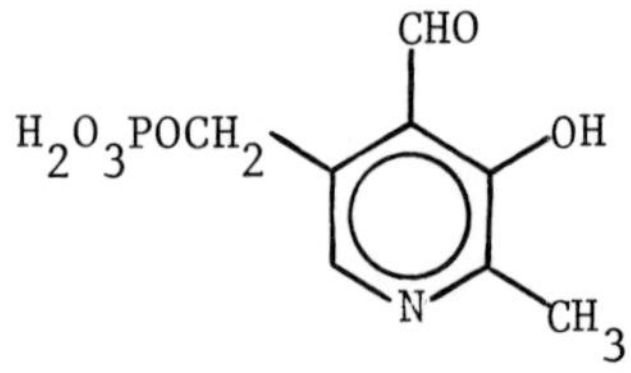

$$\text{H}_2\text{O}_3\text{POCH}_2 \underset{\underset{\text{CH}_3}{\underset{|}{N}}}{\overset{\overset{\text{CHO}}{|}}{\bigcirc}} \text{OH}$$

		Pyridoxal-5-(dihydrogenphosphate)		H_3L
$\text{C}_8\text{H}_{10}\text{O}_6\text{NP}$		(Other values on Vol.2, p.182)		

Metal ion	Equilibrium	Log K 25°, 0.1	Log K 25°, 0.5	Log K 25°, 2.0
H^+	HL/H.L	8.45	(7.99) +0.01	8.17
	$\text{H}_2\text{L}/\text{HL.H}$	6.01	5.84 ±0.01	5.75
	$\text{H}_3\text{L}/\text{H}_2\text{L.H}$	3.44	3.52 ±0.04	3.58
	$\text{H}_4\text{L}/\text{H}_3\text{L.H}$	1.4		1.64
Pr^{3+}	$\text{MH}_2\text{L}/\text{M.H}_2\text{L}$		2.16	
	$\text{MH}_3\text{L}/\text{M}_2\text{L.H}$		7.31	
Dy^{3+}	$\text{MH}_2\text{L}/\text{M.H}_2\text{L}$		2.10	
	$\text{MH}_3\text{L}/\text{MH}_2\text{L.H}$		7.46	

Bibliography: 78AA

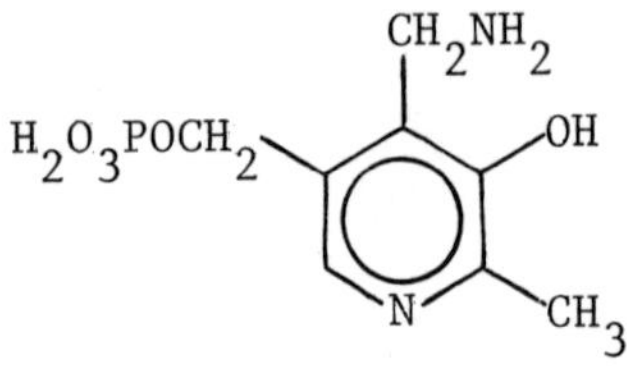

$$\text{H}_2\text{O}_3\text{POCH}_2 \underset{\underset{\text{CH}_3}{\underset{|}{N}}}{\overset{\overset{\text{CH}_2\text{NH}_2}{|}}{\bigcirc}} \text{OH}$$

		Pyridoxamine-5-(dihydrogenphosphate)	H_3L
$\text{C}_8\text{H}_{13}\text{O}_5\text{N}_2\text{P}$			

Metal ion	Equilibrium	Log K 25°, 0.5
H^+	HL/H.L	10.87 −0.4
	$\text{H}_2\text{L}/\text{HL.H}$	8.28 −0.04
	$\text{H}_3\text{L}/\text{H}_2\text{L.H}$	5.51 ±0.00
	$\text{H}_4\text{L}/\text{H}_3\text{L.H}$	3.29 +0.1
Zn^{2+}	ML/M.L	6.7
	$\text{ML}_2/\text{ML.L}$	4.09
	$\text{ML}_3/\text{ML}_2\text{L}$	2.25
	MHL/ML.H	7.5

Bibliography:

H^+ 74FL,78AA

Zn^{2+} 74FL

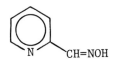

Metal ion	Equilibrium	Log K 25°, 0.1	Log K 25°, 0.5	Log K 25°, 0	ΔH 25°, 0	ΔS 25°, 0
$C_6H_6ON_2$		Pyridine-2-carboxaldehyde oxime (Other values on Vol.2, p.184)				HL
H^+	HL/H.L	9.90 ±0.04	9.96 ±0.07	10.18 −0.01	−2.1	39
	H_2L/HL.H	3.55 ±0.05	3.62 ±0.02	3.55 ±0.05	−4.8	0
Cr^{3+}	ML/M.L		9.6			
	$ML_2/M.L_2$		17.7			
	$ML_3/M.L_3$		24.9			
Fe^{3+}	ML/M.L		11.9			
	$ML_2/M.L_2$		23.3			
	$ML_3/M.L_3$		32.6			

Bibliography:

H^+ 70KMB,75CPM,75STa   Other reference: 79SV

Cr^{3+},Fe^{3+} 76CPM

Metal ion	Equilibrium	Log K 25°, 0.05	Log K 20°, 0.5	Log K 25°, 0	ΔH 25°, 0.05	ΔS 25°, 0.05
$C_7H_7O_2N_3$		Pyridine-2,6-dicarboxaldehyde dioxime (Other values on Vol.2, p.185)				H_2L
H^+	HL/H.L	10.54	10.21	10.88	(−7)[s]	(25)
	H_2L/HL.H	9.91	9.47	10.08	(−5)[s]	(29)
	$H_3L/H_2L.H$		2.18			
Cu^{2+}	$ML_2/M.L^2$		21.75			
	MHL/M.HL		13.93			
	$M(HL)_2/M.(HL)^2$		20.19			
	MOHL/M.OH.L		23.3			

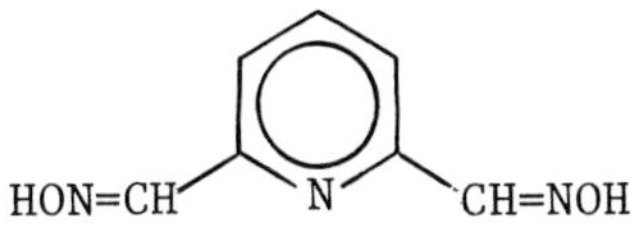

[s]20-35°, 0.05

Bibliography: 74PPF

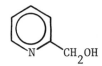

| C$_6$H$_7$ON | 2-(Hydroxymethyl)pyridine (Other values on Vol.2, p.186) | | | L |

Metal ion	Equilibrium	Log K 25°, 0.1	Log K 25°, 0.6	Log K 25°, 0
H$^+$	L/(H$_{-1}$L).H		(13.7)[i]	(13.9)[j]
	HL/H.L	4.89	4.95 5.09[i]	4.86 5.15[j]
Cu^{2+}	ML/M.L	3.41	3.56 3.51[i]	3.75[j]
	ML$_2$/M.L^2	6.22	(6.23) (6.90)[i]	6.68[j]
	ML$_2$/M(H$_{-1}$L)$_2$.H^2		(11.9)[i]	(11.1)[j]
	ML$_2$/M(H$_{-1}$L)L.H		5.2[i]	
Cd^{2+}	ML/M.L		1.7[i]	
	ML$_2$/M.L^2		2.2[i]	
	ML$_3$/M.L^3		3.0[i]	
	ML$_2$/M(H$_{-1}$L)$_2$.H^2		(19.9)[i]	
	ML$_3$/M(H$_{-1}$L)L.H^2		(20.0)[i]	
	M(H$_{-1}$L)$_2$L/M(H$_{-1}$L)$_3$.H		(12.1)[i]	
	ML/MOH(H$_{-1}$L).H^2		(15.8)[i]	

[i] 20°, 0.5; [j] 20°, 1.0

Bibliography:

H$^+$ 75PPF,76CPF   Cd^{2+} 76CPF

Cu^{2+} 75PPF

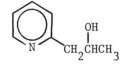

| C$_8$H$_{11}$ON | 2-(2-Hydroxypropyl)pyridine | | L |

Metal ion	Equilibrium	Log K 25°, 0.1
H$^+$	HL/H.L	5.48
Co^{2+}	ML/M.L	1.55
Ni^{2+}	ML/M.L	1.7
Cu^{2+}	ML/M.L	2.25
Zn^{2+}	ML/M.L	1.45
Cd^{2+}	ML/M.L	1.6

Bibliography: 74IL

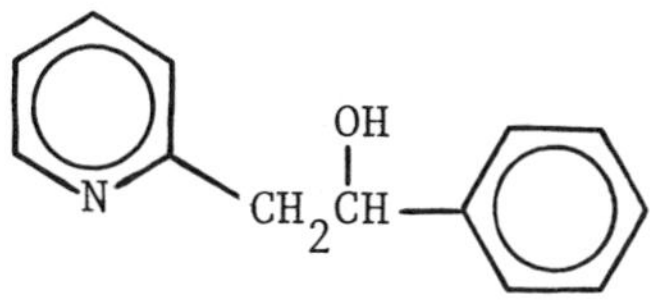

$C_{13}H_{13}ON$ 2-(2-Hydroxy-2-phenylethyl)pyridine   L

Metal ion	Equilibrium	Log K 25°, 0.1
H^+	HL/H.L	5.20
Co^{2+}	ML/M.L	1.40
Ni^{2+}	ML/M.L	1.65
Cu^{2+}	ML/M.L	2.10
Zn^{2+}	ML/M.L	1.2
Cd^{2+}	ML/M.L	1.35

Bibliography: 74IL

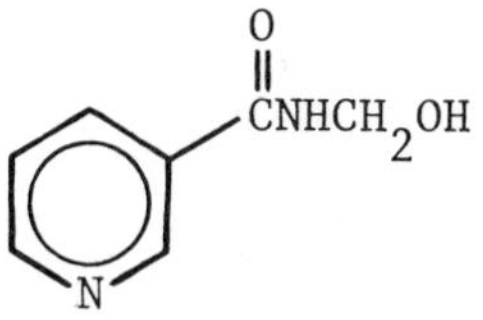

$C_7H_8O_2N_2$ Pyridine-3- carboxylic acid hydroxymethylamide (N-hydroxymethylnicotinamide) L

Metal ion	Equilibrium	Log K 25°, 0.5
H^+	HL/H.L	3.41
Co^{2+}	ML/M.L	0.98
	$ML_2/M.L^2$	1.49
Ni^{2+}	ML/M.L	1.41
	$ML_2/M.L^2$	2.34
Cu^{2+}	ML/M.L	1.77
	$ML_2/M.L^2$	2.73
Zn^{2+}	ML/M.L	0.78
	$ML_2/M.L^2$	1.18

Bibliography: 74Wc

		Log K			

C$_7$H$_9$O$_2$N   2,6-Bis(hydroxymethyl)pyridine   L
                              (Other references on Vol.2, p.341)

Metal ion	Equilibrium	Log K 20°, 0.5
H$^+$	(H$_{-1}$L)/(H$_{-2}$L).H	(14.6)
	L/(H$_{-1}$L).H	(13.5)
	HL/H.L	4.39
Cu^{2+}	ML/M.L	3.86
	ML$_2$/M.L^2	6.82
	ML/M(H$_{-1}$L).H	5.3
	M(H$_{-1}$L)/M(H$_{-2}$L).H	8.2
	M(H$_{-2}$L)/MOH(H$_{-2}$L).H	10.7
	M$_2$(H$_{-2}$L)$_2$/(M(H$_{-2}$L))^2	2.4
Cd^{2+}	ML/M.L	1.84
	ML$_2$/M.L^2	2.6
	ML/M(H$_{-1}$L).H	9.30
	M(H$_{-1}$L)/M(H$_{-2}$L).H	10.3
	ML$_2$/M(H$_{-1}$L)$_2$.H^2	20.0

Bibliography:

H$^+$,Cu^{2+} 72PP   Cd^{2+} 74CP

C$_6$H$_5$ON   Pyridine-2-carboxaldehyde   L
                            (Other values on Vol.2, p.191)

Metal ion	Equilibrium	Log K 20°, 0.5	Log K 25°, 0	ΔH 25°, 0	ΔS 25°, 0
H$^+$	L(H$_{-1}$L).H	(11.92)			
	HL/H.L	3.92	3.84	-6.5	-4
Cu^{2+}	ML/M.L	3.29			
	ML$_2$/M.L^2	6.58			
	ML/M(H$_{-1}$L).H	4.18			
	ML$_2$/ML(H$_{-1}$L).H	4.14			
	MOHL$_2$/M(H$_{-1}$L)$_2$.H	5.07			

Bibliography: 76PPR   Other references: 76EE,77EE,79EE

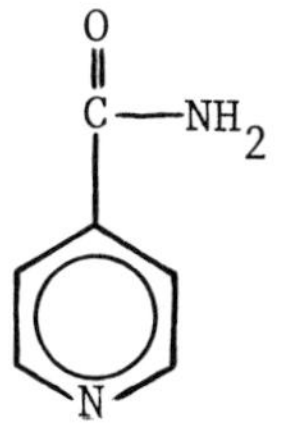

$C_6H_6ON_2$ <u>Pyridine-4-carboxylic acid amide</u> (isonicotinamide)   L
(Other value on Vol.2, p.198)

Metal ion	Equilibrium	Log K 25°, 0.5	Log K 25°, 1.0	Log K 20°, 0
H^+	HL/H.L	3.67	3.68	3.61
Co^{2+}	ML/M.L	1.04		
	$ML_2/M.L^2$	1.60		
Ni^{2+}	ML/M.L	1.48		
	$ML_2/M.L^2$	2.48		
Cu^{2+}	ML/M.L	1.83	2.33	
	$ML_2/M.L^2$	3.08	3.38	
Zn^{2+}	ML/M.L	1.00		
	$ML_2/M.L^2$	1.40		

Bibliography: 74Wc

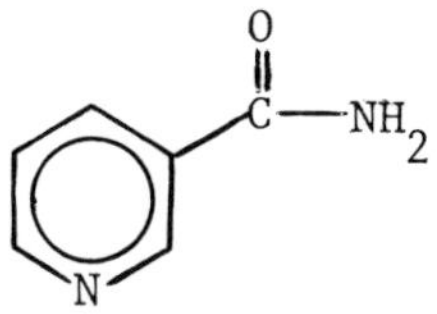

$C_6H_6ON_2$ <u>Pyridine-3-carboxylic acid amide</u> (nicotinamide)   L
(Other values on Vol.2, p.197)

Metal ion	Equilibrium	Log K 25°, 0.5	Log K 20°, 0
H^+	HL/H.L	3.47	3.35
Co^{2+}	ML/M.L	0.9 ±0.2	
	$ML_2/M.L^2$	1.6	
Ni^{2+}	ML/M.L	1.41 ±0.08	
	$ML_2/M.L^2$	2.2 ±0.3	
	$ML_3/M.L^3$	3.0	
Cd^{2+}	ML/M.L	0.9	
	$ML_2/M.L^2$	1.1	

Bibliography: 77BN   Other reference: 76BMc

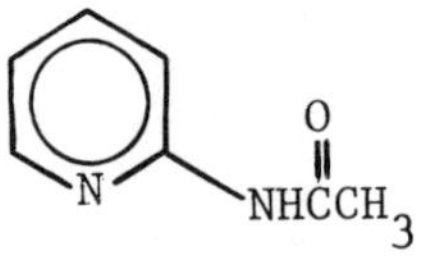

2-Acetamidopyridine

$C_7H_8ON_2$ L

Metal ion	Equilibrium	Log K 25°, 0.5
H^+	HL/H.L	4.15
Co^{2+}	ML/M.L	3.15
Ni^{2+}	ML/M.L	5.10
Cu^{2+}	ML/M.L	5.40
Zn^{2+}	ML/M.L	2.67

Bibliography: 76Wa

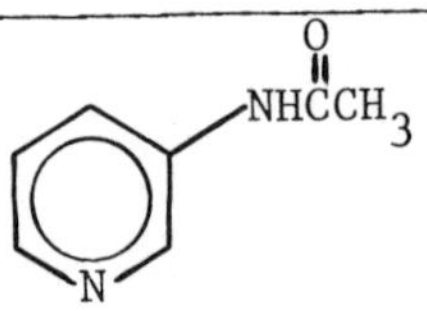

3-Acetamidopyridine

$C_7H_8ON_2$ L

Metal ion	Equilibrium	Log K 25°, 0.5
H^+	HL/H.L	4.55
Co^{2+}	ML/M.L $ML_2/M.L^2$	1.22 1.56
Ni^{2+}	ML/M.L $ML_2/M.L^2$	1.72 2.87
Cu^{2+}	ML/M.L $ML_2/M.L^2$	2.10 3.68
Zn^{2+}	ML/M.L $ML_2/M.L^2$	1.45 1.70

Bibliography: 74Wc

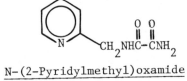

N-(2-Pyridylmethyl)oxamide

$C_8H_9O_2N_3$ L

Metal ion	Equilibrium	Log K 20°, 0.25
H^+	HL/H.L	4.20
Co^{2+}	$M.L/M(H_{-1}L).H$	(6.01)
	$M(H_{-1}L).L/M(H_{-1}L).H$	4.84

Bibliography: 70DG

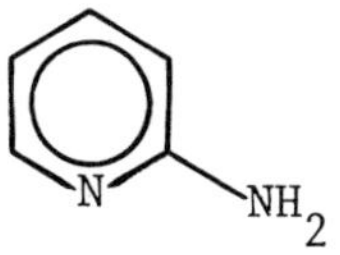

$C_5H_6N_2$			2-Aminopyridine (Other values on Vol.2, p.205)				L

Metal ion	Equilibrium	Log K 25°, 0.1	Log K 25°, 1.0	Log K 25°, 0	ΔH 25°, 0	ΔS 25°, 0
H^+	HL/H.L	6.78	6.93	6.71	-8.40	2.5
		$6.85^b \pm 0.04$			-8.39^b	3.2^b
Ag^+	$ML/M.L$	$2.38^b \pm 0.00$				
	$ML_2/M.L^2$	4.85				
		$4.76^b \pm 0.03$				

^{b}25°, 0.5

Bibliography: H^+ 73MB,76BBc; Ag^+ 73MB

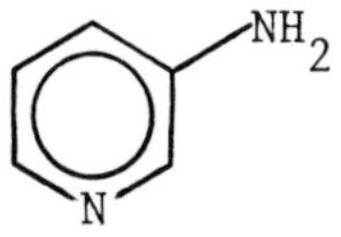

$C_5H_6N_2$			3-Aminopyridine (Other values on Vol.2, p.207)				L

Metal ion	Equilibrium	Log K 25°, 0.1	Log K 25°, 0.5	Log K 25°, 0	ΔH 25°, 0	ΔS 25°, 0
H^+	HL/H.L	6.06	6.16 ±0.05	6.03	-6.42 ±0.01	6.0
					6.66^b	5.8^b
Co^{2+}	$ML/M.L$		1.23			
	$ML_2/M.L^2$		2.1			
	$ML_3/M.L^3$		2.5			
Ni^{2+}	$ML/M.L$		1.90 ±0.07			
	$ML_2/M.L^2$		3.2 ±0.0			
	$ML_3/M.L^3$		4.2 -0.1			
	$ML_4/M.L^4$		4.7			
Ag^+	$ML/M.L$		2.23 ±0.02		-5.5^b	-8^b
	$ML_2/M.L^2$		4.41 ±0.01		-12.4^b	-21^b
Zn^{2+}	$ML/M.L$		1.23			
	$ML_2/M.L^2$		2.1			
	$ML_3/M.L^3$		2.5			
Cd^{2+}	$ML/M.L$		1.50			
	$ML_2/M.L^2$		2.6			
	$ML_3/M.L^3$		3.3			
	$ML_4/M.L^4$		3.6			

^{b}25°, 0.5

Bibliography:
H^+ 73MB,74LP Ag^+ 73MB,76BEF
Co^{2+},Ni^{2+},Zn^{2+},Cd^{2+} 78LRa

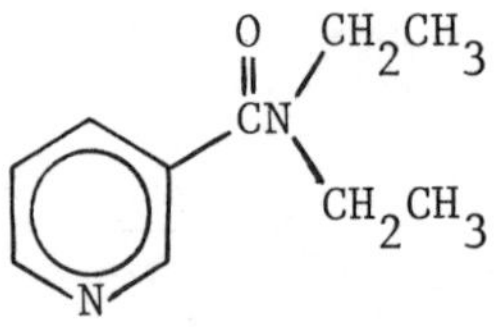

$C_{10}H_{14}ON_2$ <u>Pyridine-3-carboxylic acid diethylamide</u>   (<u>N,N-diethylnicotinamide</u>)   L

Metal ion	Equilibrium	Log K 25°, 0.5
H^+	HL/H.L	3.42
Co^{2+}	ML/M.L	0.85
	$ML_2/M.L^2$	1.15
Ni^{2+}	ML/M.L	1.39
	$ML_2/M.L^2$	2.45
Cu^{2+}	ML/M.L	1.78
	$ML_2/M.L^2$	2.72
	$ML_3/M.L^3$	3.26
Zn^{2+}	ML/M.L	(0.48)
	$ML_2/M.L^2$	1.28

Bibliography: 74Wb

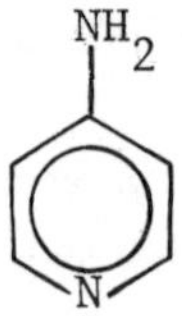

$C_5H_6N_2$ <u>4-Aminopyridine</u>   L
                        (Other values on Vol.2, p.208)

Metal ion	Equilibrium	Log K 25°, 0.1	Log K 25°, 1.0	Log K 25°, 0	ΔH 25°, 0	ΔS 25°, 0
H^+	HL/H.L	9.17 ±0.03	9.33 ±0.06	9.114	-11.28 ±0.03	4.0
		9.25[b]±0.01			-11.34[b]	4.3[b]
Ag^+	ML/M.L	2.90[b]				
	$ML_2/M.L^2$	6.04[b]				
		5.99[b]				

[b] 25°, 0.5

Bibliography:
H^+ 73MB,74LP,74PC,77VJ
Ag^+ 73MB

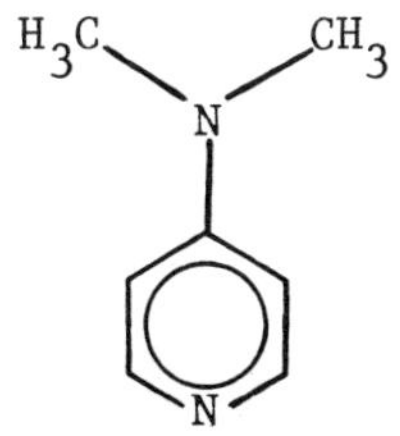

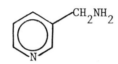

| $C_7H_{10}N_2$ | 4-(Dimethylamino)pyridine | L |

Metal ion	Equilibrium	Log K 25°, 0.1	ΔH 25°, 0.1	ΔS 25°, 0.1
H$^+$	HL/H.L	9.55	-11.4	5
	H$_2$L/HL.H	1.0		
Ni^{2+}	ML/M.L	2.70	-6.0	-8

Bibliography: 79HM

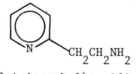

| $C_6H_8N_2$ | 3-(Aminomethyl)pyridine   (3-picolylamine) | L |

Metal ion	Equilibrium	Log K 20°, 1.0
H$^+$	HL/H.L	9.12
	H$_2$L/HL.H	4.06
Cu^{2+}	MHL/M.HL	1.97
	M(HL)$_2$/M.(HL)^2	3.32
Cu$^+$	ML$_2$/M.L^2	9.45
	M(HL)$_2$/M.(HL)^2	6.75
	MOHL/M.OH.L	10.05

Bibliography: 78CP

| $C_7H_{10}N_2$ | 2-(2-Aminoethyl)pyridine (Other values on Vol.2, p.213) | L |

Metal ion	Equilibrium	Log K 25°, 0.1	Log K 25°, 1.0	Log K 25°, 0	ΔH 25°, 0.5	ΔS 25°, 0.5
H$^+$	HL/H.L	9.59 9.70[b]	9.84	9.60	-12.12	3.7
	H$_2$L/HL.H	3.92 4.16[b]	4.37	3.75	-4.60	3.6

[b]25°, 0.5

2-(2-Aminoethyl)pyridine (continued)

Metal ion	Equilibrium	Log K 25°, 0.1	Log K 25°, 1.0	Log K 25°, 0	ΔH 25°, 0.5	ΔS 25°, 0.5
Cu^{2+}	ML/M.L	7.3 7.64[b]	7.71	7.48	−9.4	4
	$ML_2/M.L^2$	12.9 13.23[b]	13.34		−17.7	1
	$M(OH)_2L_2/ML_2.(OH)^2$		4.79[j]			
	$M_2(OH)_2L_2/M(OH)_2L_2.M$		(13.4)[j]			

[b]25°, 0.5;   [j]20°, 1.0

Bibliography: 75CP

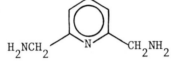

$C_7H_{11}N_3$		2,6-Bis(aminomethyl)pyridine	L

Metal ion	Equilibrium	Log K 20°, 1.0
H^+	HL/H.L	9.53
	$H_2L/HL.H$	9.15
Cu^{2+}	$ML/M.L$	15.7
	$ML_2/M.L^2$	21.1
	MOHL/ML.OH	5.0
	$MHL_2/ML_2.H$	1.6

Bibliography: 75CPa,78CPa

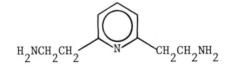

$C_9H_{15}N_3$		2,6-Bis(2-aminoethyl)pyridine	L

Metal ion	Equilibrium	Log K 20°, 1.0
H^+	HL/H.L	10.34
	$H_2L/HL.H$	9.71
	$H_3L/H_2L.H$	3.43
Cu^{2+}	$ML/M.L$	13.2
	$ML_2/M.L^2$	15.7
	MOHL/ML.OH	4.2
	$MHL_2/ML_2.H$	0.7
Cu^+	ML/M.L	11.34

Bibliography:

H^+ 76CF Cu^+ 76CF,78CPa

Cu^{2+} 76CF,76CP

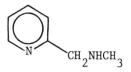

| $C_7H_{10}N_2$ | | | 2-(Methylaminomethyl)pyridine (Other values on Vol.2, p.214) | | | | L |

(Other values on Vol.2, p.214)

Metal ion	Equilibrium	Log K 25°, 0.1	Log K 25°, 0.5	Log K 25°, 0	ΔH 25°, 0.1	ΔS 25°, 0
H^+	HL/H.L	8.91	9.01	8.96	-9.86[b] / -9.88[b]	7.7[a] / 8.1[b]
	$H_2L/HL.H$		1.91		-2.34[b]	0.9[b]
Co^{2+}	ML/M.L	5.10	5.22	5.14	-6.6[b] / -6.8[b]	1[a] / 1[b]
	$ML_2/M.L^2$	8.94	9.20	8.98	-12.9[b] / -12.7[b]	-3[a] / -1[b]
	$ML_3/M.L^3$			11.5	(-15)[r]	(2)
Ni^{2+}	ML/M.L	6.71	6.91	6.82	-9.0[b] / -8.5[b]	0[a] / 3[b]
	$ML_2/M.L^2$	12.0	12.4	12.2	-17.2[b] / -17.0[b]	-3[a] / 0[b]
	$ML_3/M.L^3$	14.7		15.1	-22.0	-7[a]
Cu^{2+}	ML/M.L	9.12	9.07	9.09	-11.1[b] / -10.7[b]	4[a] / 6[b]
	$ML_2/M.L^2$	15.7	15.8	15.8	-20.3[b] / -18.8[b]	4[a] / 9[b]
	$ML_3/M.L^3$			18.5		
Zn^{2+}	ML/M.L	4.83	4.96	4.88	-5.8[b] / -5.8[b]	2[a] / 3[b]
	$ML_2/M.L^2$	8.41	8.58		-10.5[b] / -10.2[b]	3[a] / 5[b]
Cd^{2+}	ML/M.L	4.50	4.49	4.55	-5.3[b] / -5.2[b]	3[a] / 3[b]
	$ML_2/M.L^2$	8.1	7.84	8.02	-9.9[b] / -10.0[b]	4[a] / 2[b]
	$ML_3/M.L^3$			10.7		
Hg^{2+}	$ML_2/M.L^2$	18.8			-22.2	11[a]

[a] 25°, 0.1;   [b] 25°, 0.5;   [r] 10-40°, 0

Bibliography: 75APB

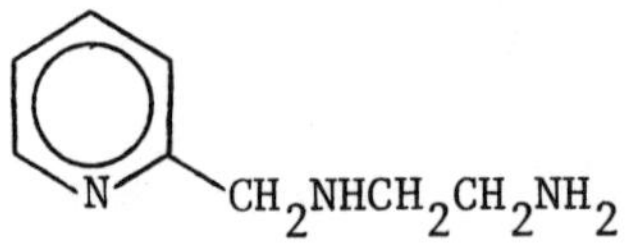

$$CH_2NHCH_2CH_2NH_2$$

$C_8H_{13}N_2$ 2-(2-Aminoethylaminomethyl)pyridine   (N-2-picolylethylenediamine)   L

Metal ion	Equilibrium	Log K 25°, 0.1
H$^+$	HL/H.L	9.46
	H$_2$L/HL.H	5.91
	H$_3$L/H$_2$L.H	2.73
Co^{2+}	ML/M.L	8.49
	ML$_2$/M.L^2	16.1

Bibliography: 70DG

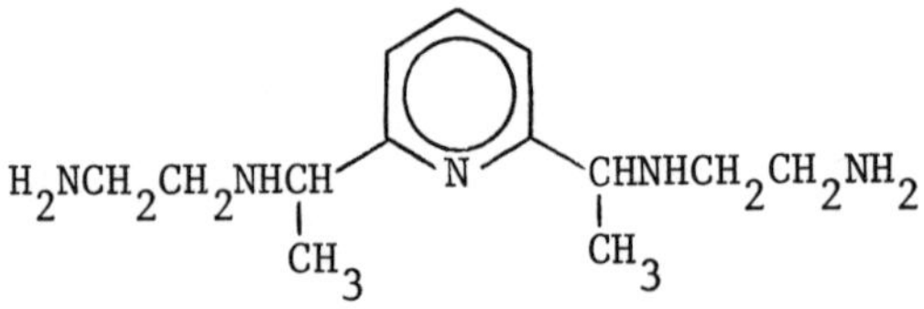

$$H_2NCH_2CH_2NHCH \quad CHNHCH_2CH_2NH_2$$
$$\qquad\qquad CH_3 \qquad\qquad\qquad CH_3$$

$C_{13}H_{25}N_5$ 2,6-Bis[1-(2-aminoethylamino)ethyl]pyridine   (epyden)   L

Metal ion	Equilibrium	Log K 25°, 0.1
H$^+$	HL/H.L	9.75
	H$_2$L/HL.H	9.05
	H$_3$L/H$_2$L.H	6.32
	H$_4$L/H$_3$L.H	5.47
Co^{2+}	ML/M.L	13.99
	MHL/ML.H	4.35
Ni^{2+}	ML/M.L	17.78
	MHL/ML.H	3.50
Cu^{2+}	ML/M.L	21.22
	MHL/ML.H	4.95
Zn^{2+}	ML/M.L	15.73
	MHL/ML.H	2.74

Bibliography: 78HM

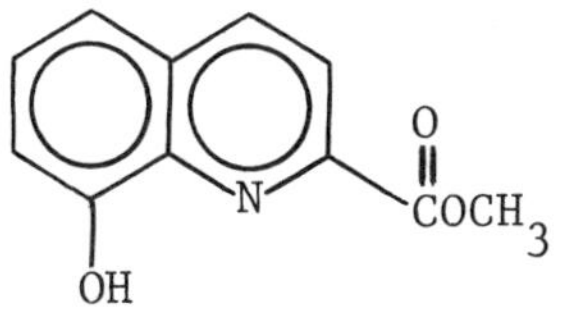

| $C_{11}H_9O_3N$ | 8-Hydroxyquinoline-2-carboxylic acid methylester | | HL |

Metal ion	Equilibrium	Log K 25°, 0.1	Log K 25°, 0
H^+	HL/H.L	9.14	9.36
Mn^{2+}	ML/M.L	4.69	
Co^{2+}	ML/M.L	5.53	
Ni^{2+}	ML/M.L	5.75	
Zn^{2+}	ML/M.L	6.58	

Bibliography: 77HCa

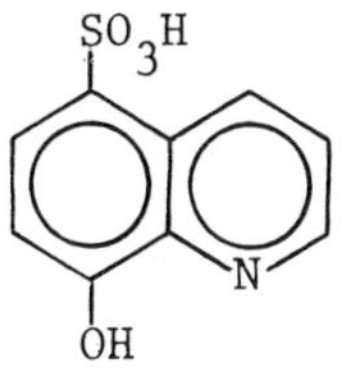

| $C_9H_7O_4NS$ | 8-Hydroxyquinoline-5-sulfonic acid (sulfoxine) | | H_2L |

(Other values on Vol.2, p.227)

Metal ion	Equilibrium	Log K 25°, 0.1	Log K 25°, 1.0	Log K 25°, 0	ΔH 25°, 0.1	ΔS 25°, 0.1
H^+	HL/H.L	8.42 ±0.07 8.23[b] ±0.01	8.22 ±0.1	8.757	-4.0 -4.8[b]	25 21[b]
	H_2L/HL.H	3.94 ±0.07 3.89[b] ±0.06	3.97 ±0.1	4.112	-4.4 -4.4[b]	3 3[b]
Mn^{2+}	ML/M.L	5.67	5.47	6.94		
	$ML_2/M.L^2$	10.72	10.36			
	$ML_3/M.L^3$		14.30			

[b] 25°, 0.5

Bibliography: H^+ 76SG,77ML; Mn^{2+} 75SG

Other references: 64CL,68BBa,69GT,73PM,70KC, 73PM,75AM,75ST,76ABc,76B,77SM,78MM,79SV

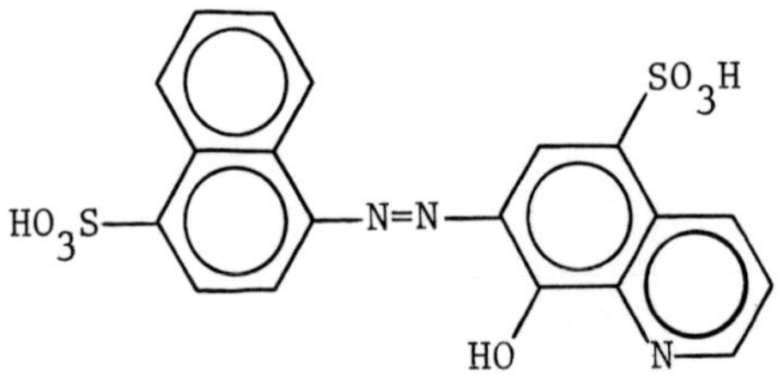

$C_{19}H_{13}O_7N_3S_2$ 7-(4-Sulfo-1-naphthylazo)-8-hydroxyquinoline-5-sulfonic acid (SNAZOXS) H_3L

Metal ion	Equilibrium	Log K 25°, 0.1
H^+	HL/H.L	6.86
	$H_2L/HL.H$	2.88
	$H_3L/H_2L.H$	(0.1)
Co^{2+}	$ML/M.L$	(6.97)
	$ML_2/M.L^2$	(14.82)
	MHL/M.HL	(3.29)
	$M(HL)_2/M.(HL)^2$	(8.88)
Ni^{2+}	$ML/M.L$	(7.57)
	$ML_2/M.L^2$	(15.42)
	MHL/M.HL	(3.29)
	$M(HL)_2/M.(HL)^2$	(8.81)
Cu^{2+}	$ML/M.L$	(9.90)
	$ML_2/M.L^2$	(19.99)
	MHL/M.HL	(5.05)
	$M(HL)_2/M.(HL)^2$	(9.75)
Zn^{2+}	$ML/M.L$	(6.14)
	$ML_2/M.L^2$	(13.91)
	MHL/M.HL	(2.10)
	$M(HL)_2/M.(HL)^2$	(5.83)
Pb^{2+}	$ML/M.L$	(7.24)
	$ML_2/M.L^2$	(14.63)
	MHL/M.HL	(3.20)
	$M(HL)_2/M.(HL)^2$	(6.55)

Bibliography:

H^+ 76MPb

Co^{2+},Ni^{2+} 78MC

$Cu^{2+}-Pb^{2+}$ 79MPa

Other reference: 69GT

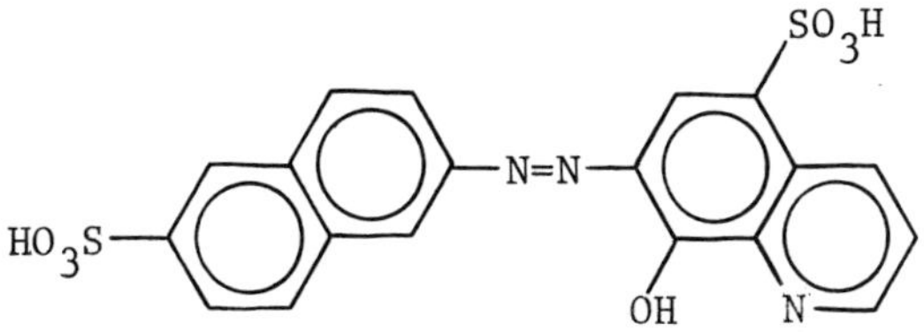

C$_{19}$H$_{13}$O$_7$N$_3$S$_2$ 7-(6-Sulfo-2-naphthylazo)-8-hydroxyquinoline-5-sulfonic acid   H$_3$L
(Naphthylazoxine 6S)

Metal ion	Equilibrium	Log K 25°, 0.1
H$^+$	HL/H.L	7.23
	H$_2$L/HL.H	2.95
	H$_3$L/H$_2$L.H	(0.2)
Cu^{2+}	ML/M.L	(10.38)
	ML$_2$/M.L^2	(20.91)
	MHL/M.HL	(5.03)
	M(HL)$_2$/M.(HL)^2	(9.95)
Zn^{2+}	ML/M.L	(5.87)
	ML$_2$/M.L^2	(13.83)
	MHL/M.HL	(1.65)
	M(HL)$_2$/M.(HL)^2	(5.42)

Bibliography:

H$^+$ 76MPb

Cu^{2+},Zn^{2+} 79MPa

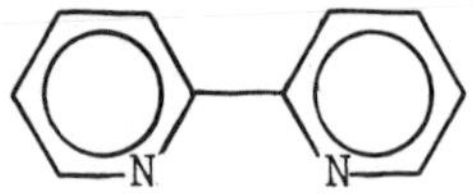

$C_{10}H_8N_2$	2-(2-Pyridyl)pyridine (2,2'-bipyridyl)					L
	(Other values on Vol.2, p.235)					

Metal ion	Equilibrium	Log K 25°, 0.1	Log K 25°, 1.0	Log K 25°, 0	ΔH 25°, 0.1	ΔS 25°, 0.1
H^+	HL/H.L	4.43 ±0.04		4.35	-3.6 ±0.1	8
		4.54[b] ±0.03	4.65 ±0.02		-4.0[c] ±0.0	8[c]
	H_2L/HL.H	1.5				
Fe^{2+}	ML/M.L	4.20	4.65[t]	4.36		
	$ML_2/M.L^2$	7.90				
	$ML_3/M.L^3$	17.2 ±0.2		17.4	-31.4[h]	-27
			17.5		-28.0[t]	-14[c]
Co^{2+}	ML/M.L	5.8 ±0.1			-8.2[h]	-1
			5.81		-7.2[t]	2[c]
	$ML_2/M.L^2$	11.3 ±0.1			-15.2[h]	0
			11.3		-14.4[t]	3[c]
	$ML_3/M.L^3$	16.0 ±0.2			-21.3[h]	1
			16.2		-19.7[t]	8[c]
Ni^{2+}	ML/M.L	7.04 ±0.03			-9.6[h]	0
			7.06		-8.9[t]	2[c]
	$ML_2/M.L^2$	13.9 ±0.1			-19.0[h]	0
			14.0		-17.8[t]	4[c]
	$ML_3/M.L^3$	20.2 ±0.0			-28.2[h]	-2
			20.5		-26.7[t]	4[c]
Cu^{2+}	ML/M.L	8.0 ±0.1			-11.3 ±0.6	-1
			8.52		-10.2[t]	5[c]
	$ML_2/M.L^2$	13.5 ±0.1			-17.0 ±0.3	5
			14.3		-19.0[t]	1[c]
	$ML_3/M.L^3$	16.9 ±0.1			-24.2 ±0.4	-4
			17.9		-21.6[t]	9[c]
	ML/MOHL.H	7.8 ±0.1			-4	22
	MOHL/M(OH)$_2$L.H	9.8 ±0.0			-11	8
	$M_2(OH)_2L_2.H^2/(ML)^2$	10.7 ±0.1			-16	-5
Zn^{2+}	ML/M.L	5.12 ±0.09			-7.1[h]	0
			5.34		-6.3[t]	3[c]
	$ML_2/M.L^2$	9.6 ±0.2			-12.5[h]	2
			9.96		-11.8[h]	6[c]
	$ML_3/M.L^3$	13.2 ±0.2			-17.5[h]	2
			14.0		-15.9[t]	11[c]

[b] 25°, 0.5;   [c] 25°, 1.0;   [h] 20°, 0.1;   [t] 30°, 1.0

2,2'-Bipyridyl (continued)

Metal ion	Equilibrium	Log K 25°, 0.1	Log K 25°, 1.0	Log K 25°, 0	ΔH 25°, 0.1	ΔS 25°, 0.1
Cd^{2+}	$ML/M.L$	4.18 ±0.06			-5.1^h	2
	$ML_2/M.L^2$	7.8 ±0.2			-9.4^h	4
	$ML_3/M.L^3$	10.4 ±0.2			-14.0^h	0

^{h}20°, 0.1

Bibliography:

H^+ 62KM,76AC,76MMR,77BP,79HJ,79MBa

Fe^{2+} 78ABD

Co^{2+},Zn^{2+} 75DOb

Ni^{2+} 76DOC

Cu^{2+} 62IM,63A,63Aa,65DD, 75NW,75OD,76AC,76MMR,77BP,79MB

Other references: 50DL,61CM,69S,72MR,75HL, 76ABc,77BA,77TJ,78ABS,78M,78MAS,78RM, 79S

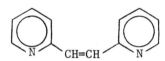

| $C_{12}H_{10}N_2$ | 1,2-Di-2-pyridylethene (vinylene-2,2'-dipyridine) | | L |

Metal ion	Equilibrium	Log K 25°, 0.1
H^+	$HL/H.L$	4.95
	$H_2L/H.L^2$	2.81
Cu^{2+}	$ML/M.L$	1.0

Bibliography: 73IL

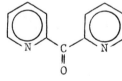

| $C_{11}H_8ON_2$ | 2-(2-Pyridyloxomethyl)pyridine (di-2-pyridylcarbone) | | L |

Metal ion	Equilibrium	Log K 25°, 0.1	Log K 25°, 0.5
H^+	$L/H_{-1}L.H$		13.6
	$HL/H.L$	2.95	
Mn^{2+}	$ML/M.L$		1.0
	$ML/MOHL.H$	7.7	
Co^{2+}	$ML/M.L$	2.6	
	$ML/MOHL.H$	5.6	
Ni^{2+}	$ML/M.L$	3.98	
	$ML/MOHL.H$	6.23	
Cu^{2+}	$ML/M.L$	5.1	
	$ML/MOHL.H$	5.1	
Zn^{2+}	$ML/M.L$	2.1	
	$ML/MOHL.H$	5.2	

Bibliography: 74FSa

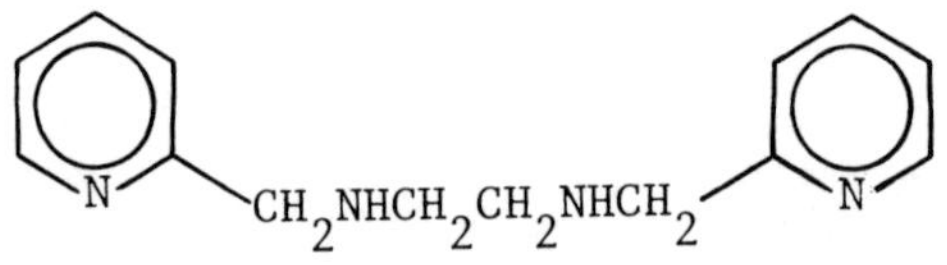

$C_{14}H_{18}N_2$		N,N'-Bis(2-pyridylmethyl)-1,2-diaminoethane (1,6-di(2-pyridyl)-2,5-diazahexane)		L

Metal ion	Equilibrium	Log K 25°, 0.1	ΔH 25°, 0.1	ΔS 25°, 0.1
H[+]	HL/H.L	8.16	-7.7	12
	H₂L/HL.H	5.33	-6.7	2
Mn^{2+}	ML/M.L	5.60	-4.5	10
Co^{2+}	ML/M.L	11.96	-14.2	7
Ni^{2+}	ML/M.L	14.48	-17.4	8
Cu^{2+}	ML/M.L	17.03	-18.2	17
Ag^{+}	ML/M.L	6.15		
	$M_2L_2/M^2.L^2$	15.9	-26.3	-15
Zn^{2+}	ML/M.L	11.13	-11.3	13
Cd^{2+}	ML/M.L	9.66	-9.7	12
Hg^{2+}	ML/M.L	19.1	-22.0	14
	$ML_2/M.L^2$	24.7	-27.5	21

Bibliography: 75APB

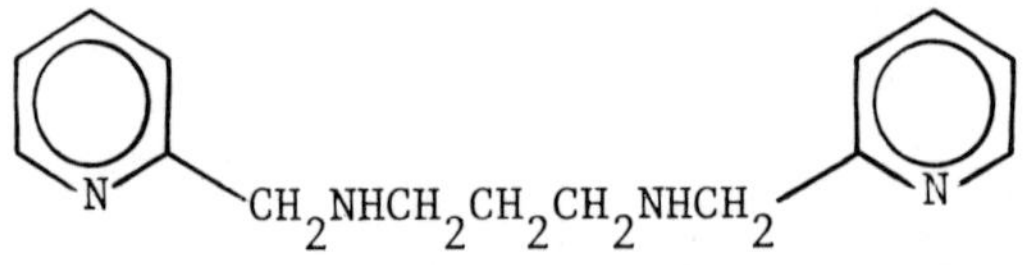

$C_{15}H_{20}N_2$		N,N'-Bis(2-pyridylmethyl)-1,3-diaminopropane (1,7-di(2-pyridyl)-2,6-diazaheptane)		L

Metal ion	Equilibrium	Log K 25°, 0.1	ΔH 25°, 0.1	ΔS 25°, 0.1
H[+]	HL/H.L	8.33	-8.9	10
	H₂L/HL.H	7.40	-8.8	2
Mn^{2+}	ML/M.L	4.45	-3.2	10
Co^{2+}	ML/M.L	11.18	-13.2	7
Ni^{2+}	ML/M.L	14.2	-16.3	10
Cu^{2+}	ML/M.L	18.35	-20.1	16
Ag^{+}	ML/M.L	6.12	-10.9	-8
Zn^{2+}	ML/M.L	10.33	-9.7	15
Cd^{2+}	ML/M.L	8.58	-7.8	13
Hg^{2+}	ML/M.L	18.8	-22.6	10

Bibliography: 75APB

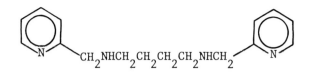

| $C_{16}H_{22}N_2$ | | N,N'-Bis(2-pyridymethyl)-1,4-diaminobutane (1,8-di(2-pyridyl)-2,7-diazaoctane) | | | L |

Metal ion	Equilibrium	Log K 25°, 0.1	ΔH 25°, 0.1	ΔS 25°, 0.1
H^+	HL/H.L	9.06	-9.6	9
	$H_2L/HL.H$	7.56	-9.6	2
Mn^{2+}	ML/M.L	2.57	-0.2	11
Co^{2+}	ML/M.L	7.95	-8.4	8
Ni^{2+}	ML/M.L	11.13	-12.4	9
Cu^{2+}	ML/M.L	15.54	-16.8	15
Ag^+	ML/M.L	5.82	-11.5	-12
Zn^{2+}	ML/M.L	7.68	-5.9	15
Cd^{2+}	ML/M.L	7.04	-6.7	10
Hg^{2+}	ML/M.L	17.52	-20.8	10

Bibliography: 75APB

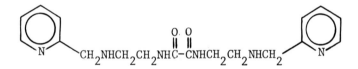

| $C_{18}H_{24}O_2N_6$ | | N,N'-Bis[2-(2-pyridylmethylamino)ethyl]oxamide | L |

Metal ion	Equilibrium	Log K 25°, 0.1
H^+	HL/H.L	7.79
	$H_2L/HL.H$	6.83
	$H_3L/H_2L.H$	1.85
	$H_4L/H_3L.H$	1.25
Co^{2+}	ML/M.L	4.8
	MHL/ML.H	7.2
	$ML/M(H_{-1}L).H$	6.13
	$M(H_{-1}L)/M(H_{-2}L).H$	9.31
Ni^{2+}	ML/M.L	7.48
	$ML/M(H_{-1}L).H$	5.53
	$M(H_{-1}L)/M(H_{-2}L).H$	10.58
Cu^{2+}	ML/M.L	11.10
	MHL/ML.H	4.34
	$ML/M(H_{-1}L).H$	(7.08)
	$M(H_{-1}L)/M(H_{-2}L).H$	9.40
	$M_2L/ML.M$	3.88
	$M_2(H_{-2}L)/M(H_{-2}L).M$	13.90

Bibliography: 73BZ

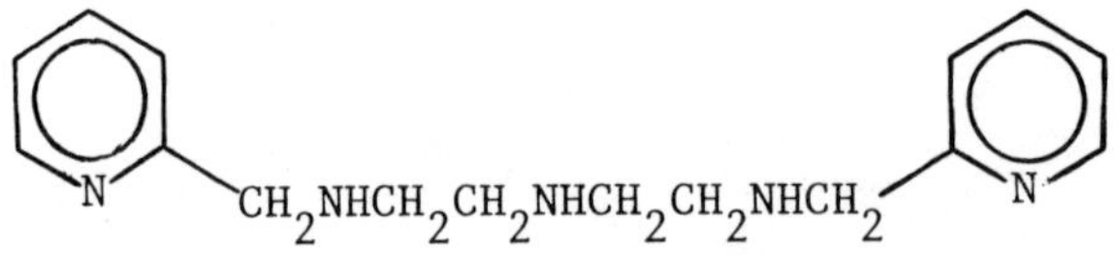

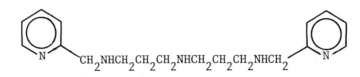

C$_{16}$H$_{23}$N$_5$

1,9-Di(2-pyridyl)-2,5,8-triazanonane
(N,N''-bis(2-pyridylmethyl)diethylenetriamine, pydien)

L

Metal ion	Equilibrium	Log K 25°, 0.1
H$^+$	HL/H.L	8.88
	H$_2$L/HL.H	7.04
	H$_3$L/H$_2$L.H	3.82
	H$_4$L/H$_3$L.H	1.44
Co^{2+}	ML/M.L	14.73
	MHL/ML.H	2.28
Ni^{2+}	ML/M.L	19.2
Cu^{2+}	ML/M.L	20.9
Zn^{2+}	ML/M.L	13.71
	MHL/ML.H	1.83

Bibliography: 78HM

C$_{18}$H$_{27}$N$_5$

1,11-Di(2-pyridyl)-2,6,10-triazaundecane
(N,N''-bis(2-pyridylmethyl)di-1,3-propylenetriamine, pydpt)

L

Metal ion	Equilibrium	Log K 25°, 0.1
H$^+$	HL/H.L	9.92
	H$_2$L/HL.H	7.63
	H$_3$L/H$_2$L.H	6.76
	H$_4$L/H$_3$L.H	1.79
Co^{2+}	ML/M.L	11.47
	MHL/ML.H	4.42
Ni^{2+}	ML/M.L	15.38
	MHL/ML.H	3.35
Cu^{2+}	ML/M.L	18.85
	MHL/ML.H	2.46
Zn^{2+}	ML/M.L	11.18
	MHL/ML.H	4.03

Bibliography: 78HM

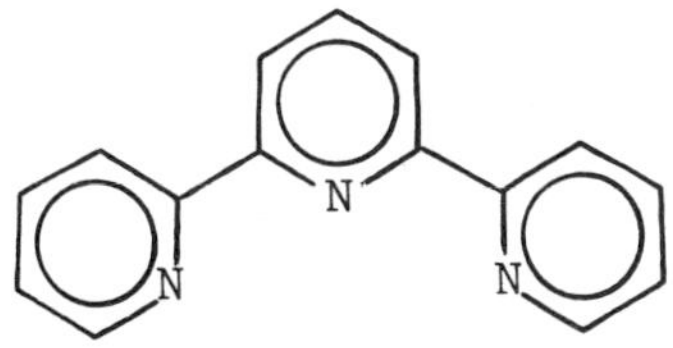

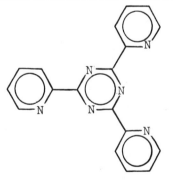

$C_{15}H_{11}N_3$ <u>2,6-Di(2-pyridyl)pyridine   (2,2',2"-terpyridine)</u>   L
                      (Other references on Vol.2, p.352)

Metal ion	Equilibrium	Log K 25°, 0.1	ΔH 25°, 0.1	ΔS 25°, 0.1
H^+	HL/H.L	4.64	-4.11	7.4
	$H_2L/HL.H$	3.42	-5.93	-4.3
Mn^{2+}	ML/M.L	4.4^t		
Fe^{2+}	ML/M.L	7.1^t		
	$ML_2/M.L^2$	20.9^t		
Co^{2+}	ML/M.L	9.5	-10.7	8
	$ML_2/M.L^2$	18.6	-22.4	10
Ni^{2+}	$ML/M.L_2$	10.7^t	-12.9	6
	ML_2/ML^2	21.8^t	-27.2	8
Cu^{2+}	ML/M.L	13.2 ±0.2	-15.8	7
	MOHL/ML.OH	5.74 ±0.07		
Cu^+	ML/M.L	9.3^q		
Zn^{2+}	ML/M.L	6.0^t		
Cd^{2+}	ML/M.L	5.1^t		

^{q}25°, 0.3; ^{t}25°, 0.002-0.1

Bibliography:

H^+ 77KN

$Mn^{2+},Fe^{2+},Zn^{2+},Cd^{2+}$ 66HH

Co^{2+} 69PPa,77KN

Ni^{2+} 66HH,77KN

Cu^{2+} 61JW,73YB,77KN,78WNA

Cu^+ 61JW

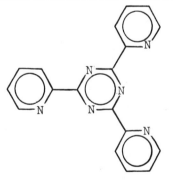

$C_{18}H_{12}N_6$ <u>2,4,6-Tri(2-pyridyl)-1,3,5-triazine</u>   L
                      (Other references on Vol.2, p.353)

Metal ion	Equilibrium	Log K 25°, 0.1
H^+	HL/H.L	3.53
	$H_2L/HL.H$	2.73
Fe^{2+}	$ML_2/M.L^2$	12.7

Bibliography: 68PM   Other references: 71PP,72FE

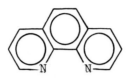

$C_{12}H_8N_2$		1,10-Phenanthroline (Other values on Vol.2, p.251)				L

Metal ion	Equilibrium	Log K 25°, 0.1	Log K 25°, 0.5	Log K 25°, 0	ΔH 25°, 0.1	ΔS 25°, 0.1
H^+	HL/H.L	4.93 ±0.04	5.02[c] ±0.01 5.15[c] ±0.03	4.86	-3.8 ±0.2 -4.6[c]	10 8[c]
	$H_2L/HL.H$	1.5 ±0.4				
Mg^{2+}	ML/M.L	1.5 1.2[h]				
Ca^{2+}	ML/M.L	1.1 0.7[h]				
Mn^{2+}	ML/M.L $ML_2/M.L_2$ $ML_3/M.L_3$	4.10 ±0.09 7.3 ±0.2 10.3 ±0.2			-3.5[h] -7.0[h] -9.0[h]	7 10 17
Ni^{2+}	ML/M.L $ML_2/M.L_2$ $ML_3/M.L_3$	8.6 +0.1 16.7 +0.1 24.3 +0.1	8.65 17.1 24.9		-11.2[h] -20.5[h] -30.0[h]	2 8 11
Cu^{2+}	ML/M.L $ML_2/M.L_2$ $ML_3/M.L_3$ $ML/M(OH)_2L.H^2$ $M_2(OH)_2L_2.H^2/(HL)^2$	9.08 ±0.08 15.8 ±0.1 21.0 ±0.0 17.3 10.69	9.16 16.1		-11.7[h] -18.2[h] -26.4[h]	2 11 7
Zn^{2+}	ML/M.L $ML_2/M.L_2$ $ML_3/M.L_3$	6.4 ±0.1 12.2 ±0.2 17.1 ±0.2	6.52[c]	6.2 12.1 17.3	-7.5[h] ±0.0 -15.0[h] -19.3[h]	4 5 14

[c] 25°, 1.0;   [h] 20°, 0.1

Bibliography:

H^+	62KM,68WF,76OD,78YO,79MBa	Zn^{2+}	78YO

Mg^{2+}-Mn^{2+} 78MSa

Fe^{2+} 78ABD

Ni^{2+} 59BB

Cu^{2+} 59A,59BB,63A,63Aa,67SPa,79OD

Other references: 50DL,68AJ,68Sa,69TN, 74KZ,75NW,75PJ,77TJ,78M,79MT

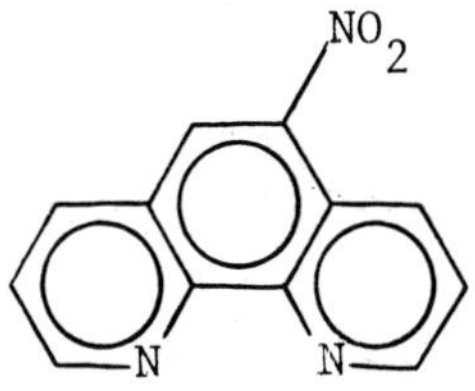

| $C_{12}H_7O_2N_3$ | | 5-Nitro-1,10-phenanthroline (Other values on Vol.2, p.257) | | | | L |

Metal ion	Equilibrium	Log K 25°, 0.1	Log K 25°, 0.3	Log K 25°, 0	ΔH 25°, 0	ΔS 25°, 0
H^+	HL/H.L	3.22	3.25	3.23	$(-2)^r$	(8)
Fe^{2+}	ML/M.L	5.06				
	$ML_3/M.L^3$			17.39	$(-33)^s$	(-31)

^{a}25°, 0.1; ^{r}25°-45°, 0; ^{s}25-82°, 0

Bibliography: 78ABD   Other reference: 68AJ

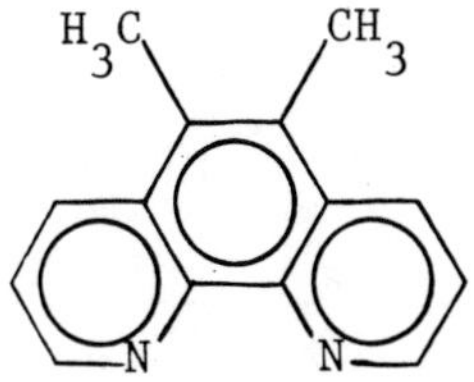

| $C_{14}H_{12}N_2$ | | 5,6-Dimethyl-1,10-phenanthroline (Other values on Vol.2, p.262) | | | | L |

Metal ion	Equilibrium	Log K 25°, 0.1	Log K 25°, 0	ΔH 25°, 0	ΔS 25°, 0
H^+	HL/H.L	5.60	5.44	$(-5)^r$	(8)
Fe^{2+}	ML/M.L	6.37			
	$ML_3/M.L^3$		21.97	$(-31)^r$	(-4)
Fe^{3+}	$ML_3/M.L^3$		18.55		

^{a}25°, 0.1; ^{r}20-40°, 0

Bibliography: 68LA

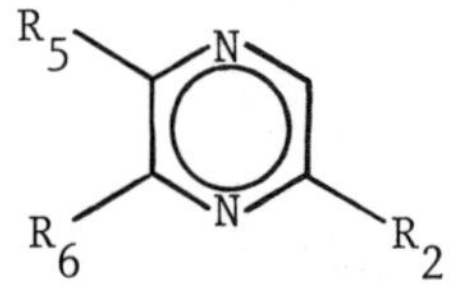

$C_x H_y N_z Cl_q$ Substituted-1,4-diazine   (Substituded-pyrazine)   L

R=	Metal ion	Equilibrium	Log K 25°, 0.1	ΔH 25°, 0.1	ΔS 25°, 0.1
$R_2=CH_3, R_5=R_6=H$	Ag^+	ML/M.L	1.65	-4.8	-9
2-Methyl		$ML_2/M.L^2$	2.76	-8.8	-17
$(C_5H_6N_2)$					
$R_2=R_5=CH_3, R_6=H$	Ag^+	ML/M.L	1.96	-5.2	-8
2,5-Dimethyl		$ML_2/M.L^2$	3.13	-9.6	-16
$(C_6H_8N_2)$					
$R_2=R_6=CH_3, R_5=H$	Ag^+	ML/M.L	1.95	-5.0	-8
2,6-Dimethyl		$ML_2/M.L^2$	3.46	-7.6	-11
$(C_6H_8N_2)$					
$R_2=Cl, R_5=R_6=H$	Ag^+	ML/M.L	0.96	-3.8	-8
2-Chloro		$ML_2/M.L^2$	1.53	-7.7	-18
$(C_4H_3N_2Cl)$					
$R_2=NH_2, R_5=R_6=H$	Ag^+	ML/M.L	1.81		
2-Amino		$ML_2/M.L^2$	3.50		
$(C_4H_5N_3)$					

Bibliography: 74HEB

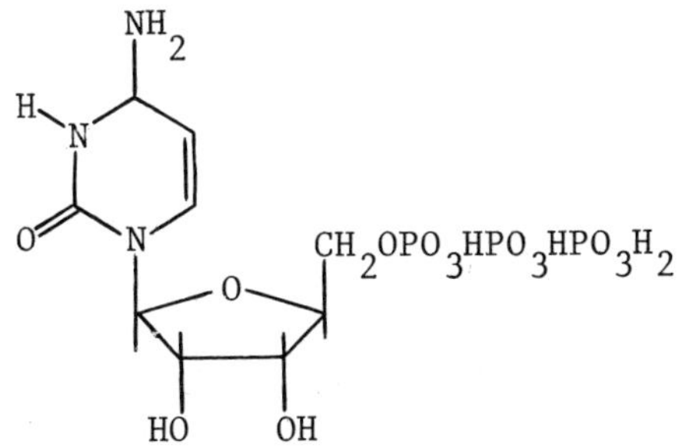

$C_9H_{16}O_{14}N_3P_3$		Cytidine-5'-(tetrahydrogentriphosphate (CTP) (Other values on Vol.2, p.268)				H_4L
Metal ion	Equilibrium	Log K 25°, 0.1	Log K 25°, 0	ΔH 25°, 0	ΔS 25°, 0	
H^+	HL/H.L	6.6 ±0.1	7.65	$(2)^r$	(42)	
	H_2L/HL.L	4.8 ±0.0				
Mg^{2+}	ML/M.L	4.04 ±0.04				
Ca^{2+}	ML/M.L	3.77 ±0.05				
Mn^{2+}	ML/M.L	4.76 ±0.02				
Cu^{2+}	ML/M.L	5.7				
	ML/MOHL.H	7.6				
Zn^{2+}	ML/M.L	4.79				

^{r}10-37°, 0

Bibliography: 77Sg   Other references: 75TR,78FS

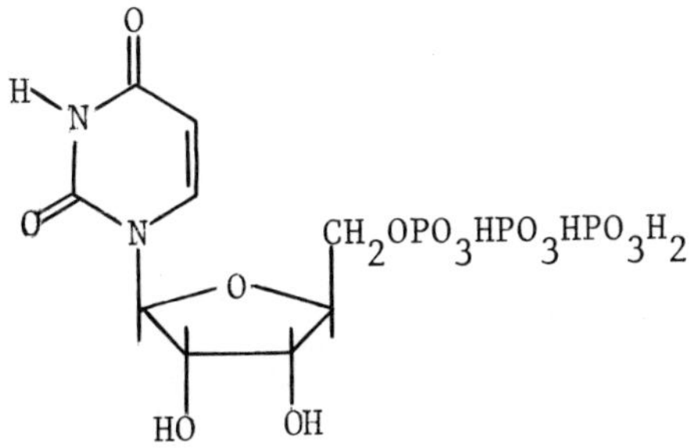

$C_9H_{15}O_{15}N_2P_3$		Uridine-5'-(tetrahydrogentriphosphate) (UTP) (Other references on Vol.2, p.270)				H_4L
Metal ion	Equilibrium	Log K 25°, 0.1	Log K 25°, 0	ΔH 25°, 0	ΔS 25°, 0	
H^+	L/$(H_{-1}L)$.H	9.5 ±0.1				
	HL/H.L	6.5 ±0.1	7.58	$(2)^r$	(36)	
Mg^{2+}	ML/M.L	4.01 ±0.01				
Ca^{2+}	ML/M.L	3.69 ±0.03				
Mn^{2+}	ML/M.L	4.68 ±0.10				
	ML/M$(H_{-1}L)$.H	9.34				
Co^{2+}	ML/M.L	4.54 ±0.01				
Ni^{2+}	ML/M.L	4.29				
	ML/M$(H_{-1}L)$.H	8.99				

^{r}10-37°, 0

Uridine-5'-(tetrahydrogentriphosphate) (continued)

Metal ion	Equilibrium	Log K 25°, 0.1	Log K 25°, 0	ΔH 25°, 0	ΔS 25°, 0
Cu^{2+}	ML/M.L	5.53			
	ML/M(H_{-1}L).H	7.8			
	H(H_{-1}L)/MOH(H_{-1}L).H	8.4			
Zn^{2+}	ML/M.L	4.75			
	ML/M(H_{-1}L).H	8.60			
	M(H_{-1}L)/MOH(H_{-1}L).H	9.13			

Bibliography:

H^+ 75Sc,77Sg Mn^{2+}-Zn^{2+} 75Sc,78FMS,77Sg

Mg^{2+},Ca^{2+} 77Sg   Other reference: 76TR

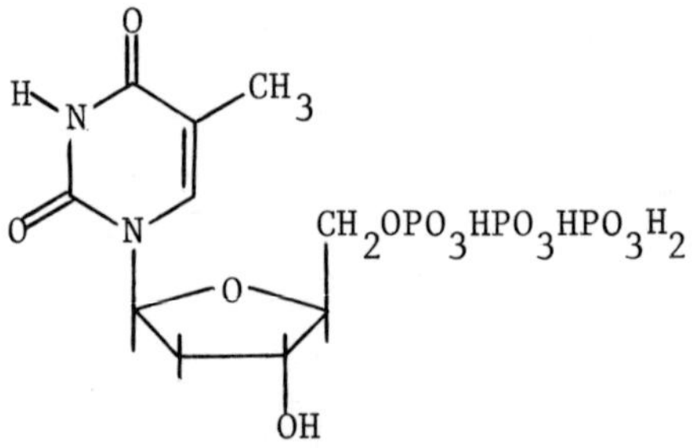

$C_{10}H_{17}O_{14}N_2P_3$ Thymidine-5'-(tetrahydrogentriphosphate) (TTP) H_4L
                              (Other reference on Vol.2, p.273)

Metal ion	Equilibrium	Log K 25°, 0.1
H^+	L/(H_{-1}L).H	9.78
	HL/H.L	6.50
Mg^{2+}	ML/M.L	4.18
Ca^{2+}	ML/M.L	3.78
Mn^{2+}	ML/M(H_{-1}L).H	9.56
Ni^{2+}	ML/M(H_{-1}L).H	8.97
Cu^{2+}	ML/M.L	5.70
	ML/M(H_{-1}L).H	7.6
	M(H_{-1}L)/MOH(H_{-1}L).H	8.2
Zn^{2+}	ML/M.L	4.89
	ML/M(H_{-1}L).H	8.24

Bibliography:

H^+ 75Sc,77Sg Mn^{2+}-Zn^{2+} 75Sc,77Sg

Mg^{2+},Ca^{2+} 77Sg

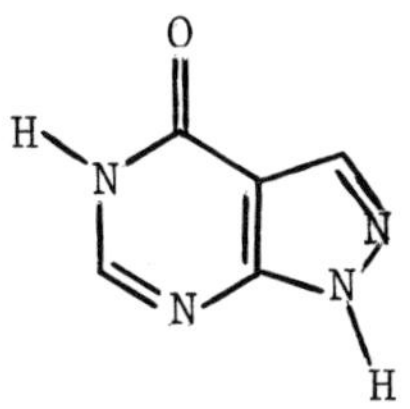

| $C_5H_4ON_4$ | 1,2-Diazolo[3,4-d]-1,3-diazine (allopurinol) | | HL |

Metal ion	Equilibrium	Log K 25°, 0.1
H^+	$L/(H_{-1}L).H$	11.79
	HL/H.L	9.11
	$H_2L/HL.H$	1.35
Ni^{2+}	ML/M.L	5.73
	$ML/M(H_{-1}L).H$	6.81

Bibliography: 76LS

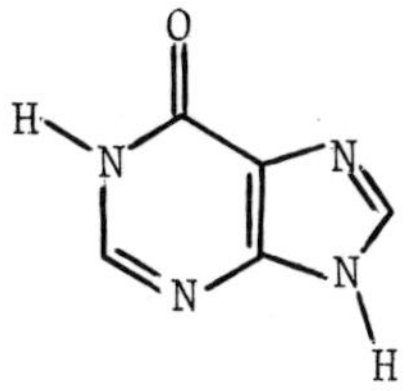

| $C_5H_4ON_4$ | 1,3-Diazolo[4,5-d]-1,3-diazine (hypoxanthine)
(Other references on Vol.2, p.345) | | | | HL |

Metal ion	Equilibrium	Log K 25°, 0.1	Log K 25°, 0	ΔH 25°, 0	ΔS 25°, 0
H^+	$L/(H_{-1}L).H$	11.35	12.03 ±0.04	-9.8 ±0.3	22
	HL/H.L	8.71 ±0.08	8.85 ±0.06	-7.9 ±0.1	14
	$H_2L/HL.H$	2.04 ±0.0	1.85 ±0.06	-2.7 ±0.2	-1
Ni^{2+}	ML/M.L	4.10 ±0.00			
	$ML/M(H_{-1}L).H$	8.11			

Bibliography:

H^+ 76LS,79RP   Other references: 79RPa

Ni^{2+} 64R,76LS

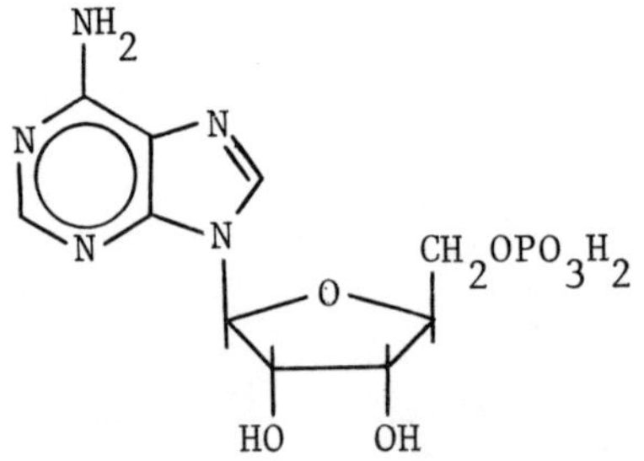

| C$_{10}$H$_{14}$O$_7$N$_5$P | Adenosine-5'-(dihydrogenphosphate)   (AMP-5)
(Other values on Vol.2, p.280) | | | | H$_2$L |

Metal ion	Equilibrium	Log K 25°, 0.1	Log K 25°, 0	ΔH 25°, 0	ΔS 25°, 0
H$^+$	L/(H$_{-1}$L).H		13.06	-10.9	23
	HL/H.L	6.19 ±0.05 6.28^w±0.01	6.67	1.8	34^{a*}
	H$_2$L/HL.H	3.80 ±0.04		-4.2	4^{a*}
Mn^{2+}	ML/M.L	2.35 ±0.05		2.2^a	18^a
Ni^{2+}	ML/M.L	2.49 ±0.04		-2.4^a	3^a
	MHL/M.HL	1.04			

^{a}25°, 0.1; ^{w}(CH$_3$)$_4$N$^+$ salt used as background electrolyte; *assuming ΔH for 0.0=ΔH for 0.1

Bibliography:

H$^+$	74BS,77RSB	Other references: 68K,68RRM,75Kc,77PD,78GB, 78Ka,79TP
Mn^{2+}	77RSB	
Ni^{2+}	74BS,76TD,79MG	

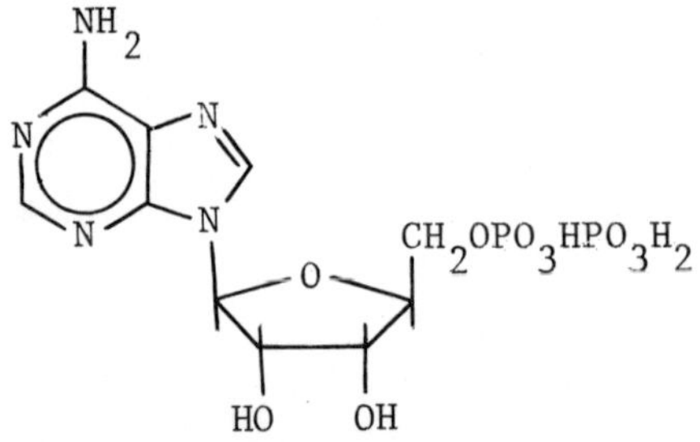

| C$_{10}$H$_{15}$O$_{10}$N$_5$P$_2$ | Adenosine-5'-(trihydrogendiphosphate)   (ADP)
(Other values on Vol.2, p.281) | | | | H$_3$L |

Metal ion	Equilibrium	Log K 25°, 0.1	Log K 25°, 0	ΔH 25°, 0	ΔS 25°, 0.1
H$^+$	HL/H.L	6.40 ±0.05 6.51^w±0.02	7.20	1.3	34*
	H$_2$L/HL.H	3.96 ±0.03		-4.1	4*
Mn^{2+}	ML/M.L	4.1 ±0.2		3.3^a	30
Ni^{2+}	ML/M.L	4.2 ±0.3		1.5^a	24

^{a}25°, 0.2; ^{w}(CH$_3$)$_4$N$^+$ salt used as background electrolyte; *assuming ΔH for 0.0=ΔH for 0.1

Bibliography:

H$^+$,Mn^{2+}	77RSB	Other references: 68K,71KY,75Kc,78DMY, 78FSa
Ni^{2+}	79MG	

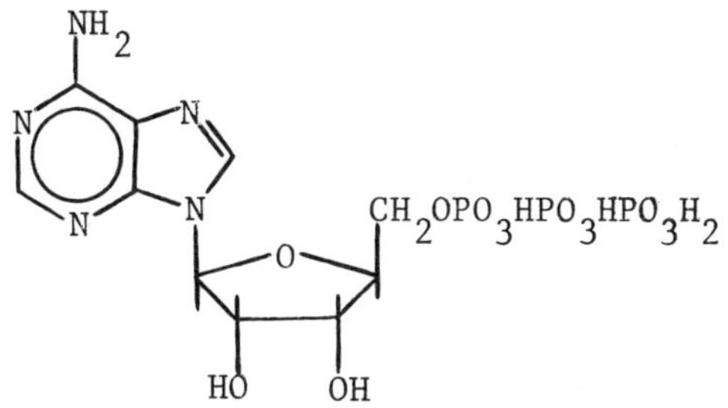

$C_{10}H_{16}O_{13}N_5P_3$ Adenosine-5'-(tetrahydrogentriphosphate) (ATP) H_4L
                         (Other values on Vol.2, p.283)

Metal ion	Equilibrium	Log K 25°, 0.1	Log K 25°, 1.0	Log K 25°, 0	ΔH 25°, 0	ΔS 25°, 0.1
H^+	HL/H.L	6.51 ± 0.03 $6.80^u \pm 0.05$	6.09	7.68	1.2	34^*
	H_2L/HL.H	4.05 ± 0.05	3.97		-3.7	6^*
Mg^{2+}	ML/M.L	4.1 ± 0.2	3.22	5.83	4.5^P	23
	MHL/ML.H	$4.55 +0.3$		5.44	$(0)^r$ -1	(21)
Mn^{2+}	ML/M.L	4.74 ± 0.04			$(-3)^r$	(12)
	MHL/ML.H	$4.14 +0.4$			$(1)^r$	(27)
Cu^{2+}	ML/M.L	6.0 ± 0.2	5.17		$(-4)^r$	(14)
	MHL/ML.H	$3.52 +0.4$			$(1)^r$	(19)
	ML/MOHL.H	$6.47 +1$			$(-8)^r$	(3)
	$MOHL/M(OH)_2L.H$	7.45			$(-4)^r$	(19)
	$(ML)^2/(MOHL)_2.H^2$	10.35			$(-10)^r$	(14)
	$(MOHL)_2/(MOHL)^2$	2.59			$(-6)^r$	(-10)
Zn^{2+}	ML/M.L	4.80 ± 0.05			$(-3)^r$	(13)
	MHL/ML.H	$4.35 +0.4$			$(0)^r$	(20)
	ML/MOHL.H	8.76				

^{r}0-40°, 0.1; $^u(CH_3)_4N^+$ salt used as background electrolyte; *assuming ΔH for 0.0=ΔH for 0.1

Bibliography:

H^+ 76RM,77RSB,78RMa,78MSa

Mg^{2+} 76RM,78RMa,78MSa

Mn^{2+}-Zn^{2+} 75Sc,76RM,77SF,78RMa,78MSa

Other references: 68K,71KY,71RD,75Kc,76RD,
    77GF,78FSa,78GFB,78Ka,79MS,79MT

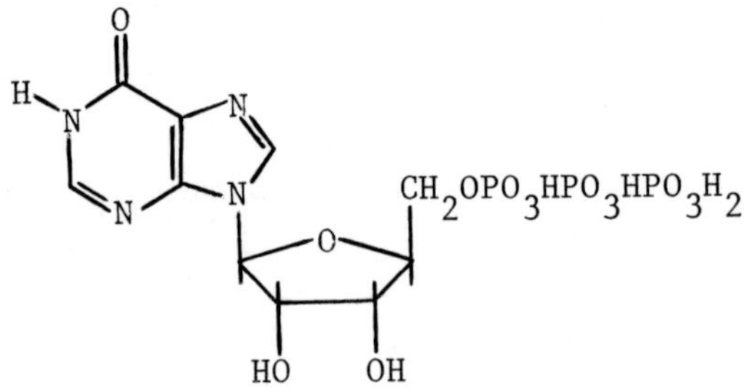

The molecular structure shows a purine ring with CH$_2$OPO$_3$HPO$_3$HPO$_3$H$_2$ substituent, and HO, OH groups on the ribose.

$C_{10}H_{15}O_{14}N_4P_3$ Inosine-5'-(tetrahydrogentriphosphate)   (ITP) H_4L
                                  (Other references on Vol.2, p.286)

Metal ion	Equilibrium	Log K 25°, 0.1	Log K 25°, 0	ΔH 25°, 0	ΔS 25°, 0.1
H^+	$L/(H_{-1}L).H$	9.15			
	$HL/H.L$	6.48 ±0.03	7.68	$(2)^r$	$(36)^*$
Mg^{2+}	$ML/M.L$	4.06 ±0.02		4.5^p	34
Ca^{2+}	$ML/M.L$	3.75 ±0.02			
Mn^{2+}	$ML/M.L$	4.62 ±0.05			
	$ML/M(H_{-1}L).H$	8.82			
Co^{2+}	$ML/M.L$	4.78 ±0.04			
Ni^{2+}	$ML/M.L$	(4.73)			
	$ML/M(H_{-1}L).H$	8.25			
Cu^{2+}	$ML/M.L$	5.99			
	$ML/M(H_{-1}L).H$	7.4			
Zn^{2+}	$ML/M.L$	5.02			
	$ML/M(H_{-1}L).H$	8.20			
	$M(H_{-1}L)/MOH(H_{-1}L).H$	9.3			

^{r}10-37°, 0; ^{p}30°, 0.2; *assuming ΔH for 0.0=ΔH for 0.1

Bibliography:

H^+ 75Sc,77CS,77Sg

Mg^{2+},Ca^{2+} 77Sg

Mn^{2+}-Zn^{2+} 75Sc,77CS,77Sg,77SF

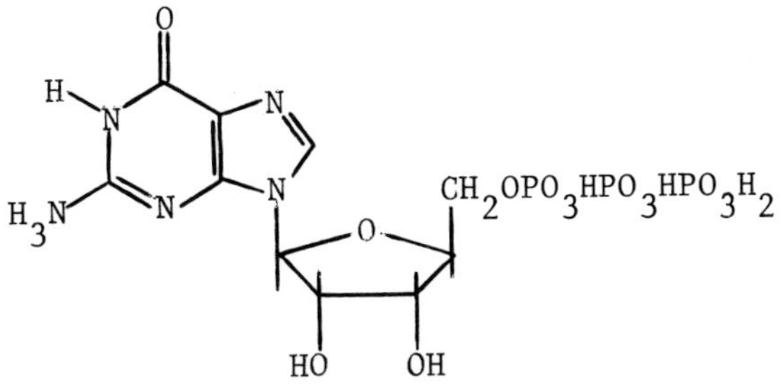

$C_{10}H_{15}O_{14}N_5P_3$ <u>Guanosine-5'-(tetrahydrogentriphosphate)</u> <u>(GTP)</u> H_4L
                                 (Other value on Vol.2, p.290)

Metal ion	Equilibrium	Log K 25°, 0.1	Log K 25°, 0	ΔH 25°, 0	ΔS 25°, 0.1
H^+	$L/(H_{-1}L).H$	9.5 ±0.2			
	$HL/H.L$	6.45 ±0.0	7.65	$(2)^r$	(37)
	$H_2L/HL.H$	3.3			
Mg^{2+}	$ML/M.L$	4.08 ±0.05		4.3^P	33
Ca^{2+}	$ML/M.L$	3.66 ±0.08			
Mn^{2+}	$ML/M.L$	4.69 ±0.05			
	$ML/M(H_{-1}L).H$	9.25			
	$M(H_{-1}L)/MOH(H_{-1}L).H$	11.2			
Ni^{2+}	$ML/M(H_{-1}L).H$	8.53			
	$M(H_{-1}L)/MOH(H_{-1}L).H$	10.46			
Cu^{2+}	$ML/M.L$	5.93			
	$ML/M(H_{-1}L).H$	7.6			
	$M(H_{-1}L)/MOH(H_{-1}L).H$	9.3			
Zn^{2+}	$ML/M.L$	4.96			
	$ML/M(H_{-1}L).H$	8.28			
	$M(H_{-1}L)/MOH(H_{-1}L).H$	9.37			

---

^{P}30°, 0.2; ^{r}10-37°, 0

Bibliography:

H^+ 75Sc,77Sg

Mg^{2+},Ca^{2+} 77Sg

Mn^{2+}-Zn^{2+} 75Sc,77Sg

$$H_2NCH_2PO_3H_2$$

CH_6O_3NP __Aminomethylphosphonic acid__ H_2L
(Other references on Vol.2, p.297)

Metal ion	Equilibrium	Log K 25°, 0.1	Log K 25°, 0.5
H^+	HL/H.L	10.05	9.97
	H_2L/HL.H	5.39	5.32
	H_3L/H_2L.H	0.4	
Mg^{2+}	ML/M.L	2.03	
	MHL/ML.H	9.35	
Ca^{2+}	ML/M.L	1.71	
	MHL/ML.H	9.43	
Co^{2+}	ML/M.L	4.45	
	$ML_2/M.L^2$	8.09	
	MHL/ML.H	7.34	
	$M(HL)L/ML_2$.H	8.7	
	$M(HL)_2/M(HL)L$.H	6.2	
Ni^{2+}	ML/M.L	5.29	4.94
	$ML_2/M.L^2$	8.98	8.5
	MHL/ML.H	6.40	
	$M(HL)L/ML_2$.H	7.4	
	$M(HL)_2/M(HL)L$.H	6.2	
Cu^{2+}	ML/M.L	8.12	7.77
	$ML_2/M.L^2$	14.65	14.1
	ML/MOHL.H	8.5	
	MHL/ML.H	4.44	
	$M(HL)L/ML_2$.H	5.55	
	$M(HL)_2/M(HL)L$.H	4.6	
Zn^{2+}	ML/M.L	5.00	
	ML/MOHL.H	6.9	
	MHL/ML.H	6.72	
	$M(HL)_2/M.(HL)^2$	3.5	

Bibliography:

H^+ 78WN,78WNb   Other reference: 76SO

Mg^{2+}-Zn^{2+} 79WNa

$$\begin{array}{c} NH_2 \\ | \\ CH_3CHPO_3H_2 \end{array}$$

C_2H_8O_3NP	DL-1-Aminoethylphosphonic acid	H_2L

<u>DL-1-Aminoethylphosphonic acid</u>
(Other reference on Vol.2, p.298)

Metal ion	Equilibrium	Log K 25°, 0.1
H[+]	HL/H.L	10.20 −0.08
	H_2L/HL.H	5.59 −0.06
	H_3L/H_2L.H	0.5
Mg[2+]	ML/M.L	2.00
	MHL/ML.H	9.54
Cu[2+]	ML/M.L	8.50
	ML_2/M.L_2	(15.40)[w]
	ML/MOHL.H	8.6
	MHL/ML.H	4.32
	M(HL)L/ML_2.H	(5.6)[w]
	M(HL)_2/M(HL)L.H	(4.9)[w]

[w]Optical isomerism not stated.

Bibliography:

H[+]   78MAb,78WN,78WNb   Mg[2+],Cu[2+] 79WNa

$$\begin{array}{c} NH_2 \\ | \\ CH_3CH_2CH_2CH_2CHPO_3H_2 \end{array}$$

C_5H_14O_3NP	DL-1-Aminopentylphosphonic acid	H_2L

<u>DL-1-Aminopentylphosphonic acid</u>
(Other reference on Vol.2, p.346)

Metal ion	Equilibrium	Log K 25°, 0.1
H[+]	HL/H.L	10.29
	H_2L/HL.H	5.70
	H_3L/H_2L.H	0.6
Mg[2+]	ML/M.L	2.03
	MHL/ML.H	9.56
Cu[2+]	ML/M.L	8.97
	ML_2/M.L_2	(16.27)[w]
	ML/MOHL.H	8.2
	MHL/ML.H	4.28
	M(HL)L/ML_2.H	(5.33)[w]
	M(HL)_2/M(HL).H	(4.3)[w]

[w]Optical isomerism not stated.

Bibliography:

H[+]   78WN,78WNb   Mg[2+],Cu[2+] 79WNa

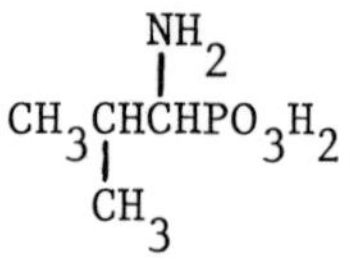

$$\begin{array}{c} NH_2 \\ | \\ CH_3CHCHPO_3H_2 \\ | \\ CH_3 \end{array}$$

$C_4H_{12}O_3NP$	DL-1-Amino-2-methylpropylphosphonic acid	H_2L
	(Other reference on Vol.2, p.299)	

Metal ion	Equilibrium	Log K 25°, 0.1
H^+	HL/H.L	10.36
	$H_2L/HL.H$	5.80
	$H_3L/H_2L.H$	0.6
Mg^{2+}	ML/M.L	2.15
	MHL/ML.H	9.58
Cu^{2+}	$ML/M.L$	9.47
	$ML_2/M.L^2$	$(17.32)^w$
	ML/MOHL.H	7.8
	MHL/ML.H	4.24
	$M(HL)L/ML_2.H$	$(5.21)^w$
	$M(HL)_2/M(HL)L.H$	$(4.2)^w$

---

^{w}Optical isomerism not stated.

Bibliography:

H^+	78WN,78WNb	Mg^{2+},Cu^{2+}	79WNa

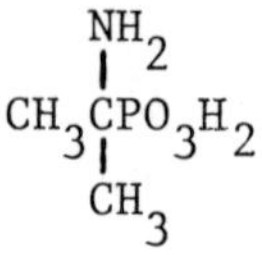

$$\begin{array}{c} NH_2 \\ | \\ CH_3CPO_3H_2 \\ | \\ CH_3 \end{array}$$

$C_3H_{10}O_3NP$	2-Amino-2-propylphosphonic acid	H_2L
	(Other values on Vol.2, p.300)	

Metal ion	Equilibrium	Log K 25°, 0.1
H^+	HL/H.L	10.29 ±0.02
	$H_2L/HL.H$	5.83 ±0.02
	$H_3L/H_2L.H$	0.6 +1
Mg^{2+}	ML/M.L	2.01
	MHL/ML.H	9.61
Cu^{2+}	$ML/M.L$	9.13 -0.7
	$ML_2/M.L^2$	16.64 -1
	MHL/ML.H	5.55 +0.4
	$M(HL)L/ML_2.H$	5.68
	$M(HL)_2/M(HL)L.H$	4.9

Bibliography:

H^+	79WN,78WNb	Mg^{2+},Cu^{2+}	79WNa

A. PRIMARY AMINES

$$H_2NCH_2CH_2PO_3H_2$$

$C_2H_8O_3NP$ 2-Aminoethylphosphonic acid H_2L
                              (Other reference on Vol.2, p.346)

Metal ion	Equilibrium	Log K 25°, 0.1	Log K 25°, 0	ΔH 25°, 0	ΔS 25°, 0
H^+	HL/H.L	11.02 ±0.04	11.499	−14.0	6
	H_2L/HL.L	6.23 ±0.02	6.505	0.3	31
	H_3L/H_2L.H	1.1	1.30	2.0	13
Mg^{2+}	ML/M.L	2.13			
	MHL/ML.H	10.35			
Ca^{2+}	ML/M.L	1.74			
	MHL/ML.H	10.48			
Co^{2+}	ML/M.L	4.67			
	ML/MOHL.H	9.37			
	MHL/ML.H	8.07			
Ni^{2+}	ML/M.L	5.20			
	$ML_2/M.L^2$	10.1			
	MHL/ML.H	7.60			
	$M(HL)L/ML_2$.H	8.7			
	$M(HL)_2/M(HL)L$.H	6.8			
Cu^{2+}	ML/M.L	8.50			
	$ML_2/M.L^2$	14.3			
	ML/MOHL.H	7.46			
	MHL/ML.H	5.25			
	$M(HL)L/ML_2$.H	7.1			
	$M(HL)_2/M(HL)L$.H	5.7			
Zn^{2+}	ML/M.L	6.16			
	ML/MOHL.H	7.86			
	MHL/ML.H	6.83			
	$M(HL)_2/M.(HL)^2$	4.1			

Bibliography:
H^+ 78MAb,78WN,78WNb   Other references: 76SO,77MSW
Mg^{2+}-Zn^{2+} 79WNa

$$H_2NCH_2CH_2CH_2PO_3H_2$$

$C_3H_{10}O_3NP$ 3-Aminopropylphosphonic acid H_2L
                              (Other reference on Vol.2, p.301)

Metal ion	Equilibrium	Log K 25°, 0.1
H^+	HL/H.L	11.07
	H_2L/HL.H	6.89
	H_3L/H_2L.H	1.6
Mg^{2+}	ML/M.L	2.01
	MHL/ML.H	10.56
Ca^{2+}	ML/M.L	1.68
	MHL/ML.H	10.69

3-Aminopropylphosphonic acid (continued)

Metal ion	Equilibrium	Log K 25°, 0.1
Cu^{2+}	ML/M.L	7.2
	ML/MOHL.H	7.1
	MHL/ML.H	6.82

Bibliography:

H^+ 78WN,78WNb   Other reference: 76SO

$Mg^{2+}-Cu^{2+}$ 79WNa

$$\begin{array}{c} CH_2CH_2NH_2 \\ | \\ H_2O_3PCPO_3H_2 \\ | \\ OH \end{array}$$

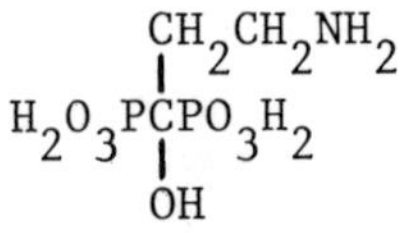

$C_3H_{11}O_7NP_2$ 3-Amino-1-hydroxypropane-1,1-diphosphonic acid H_4L

Metal ion	Equilibrium	Log K 25°, 0.1
H^+	HL/H.L	10.8
	$H_2L/HL.H$	9.9
	$H_3L/H_2L.H$	5.83
	$H_4L/H_3L.H$	2.55
Ca^{2+}	$MH_2L/M.H_2L$	2.85

Bibliography: 78KMD

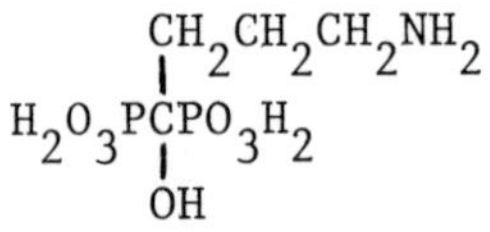

$$\begin{array}{c} CH_2CH_2CH_2NH_2 \\ | \\ H_2O_3PCPO_3H_2 \\ | \\ OH \end{array}$$

$C_4H_{13}O_7NP_2$ 4-Amino-1-hydroxybutane-1,1-diphosphonic acid H_4L

Metal ion	Equilibrium	Log K 25°, 0.1
H^+	HL/H.L	11.6
	$H_2L/HL.H$	10.5
	$H_3L/H_2L.H$	8.73
	$H_4L/H_3L.H$	2.72
Ca^{2+}	ML/M.L	6.1
	$MH_2L/MHL.H$	9.86
	MHL/ML.H	11.5
Cu^{2+}	ML/M.L	12.9
	$MH_2L/MHL.H$	6.92
	MHL/ML.H	9.99

Bibliography: 78KMD

$$\overset{\overset{\text{O}}{\|}}{H_2NCH_2CNHCH_2PO_3H_2}$$

| $C_3H_9O_4N_2P$ | | N-Glycylaminomethylphosphonic acid | H_2L |

Metal ion	Equilibrium	Log K 25°, 0.1
H^+	HL/H.L	8.34
	H_2L/HL.H	6.19
Co^{2+}	ML/M.L	3.68
	MHL/ML.H	6.29
Ni^{2+}	ML/M.L	4.75
	MHL/ML.H	5.79
Cu^{2+}	ML/M.L	6.86
	MHL/ML.H	5.19
	ML/M(H_{-1}L).H	5.17

Bibliography: 75HMa

$$\overset{\overset{\text{O}}{\|}}{H_2NCH_2CNHCH_2CH_2PO_3H_2}$$

| $C_4H_{11}O_4N_2P$ | | N-Glycyl-2-aminoethylphosphonic acid | H_2L |

Metal ion	Equilibrium	Log K 25°, 0.1
H^+	HL/H.L	8.32
	H_2L/HL.H	6.84
Co^{2+}	ML/M.L	3.53
	MHL/ML.H	6.87
Ni^{2+}	ML/M.L	4.44
	MHL/ML.H	6.64
Cu^{2+}	ML/M.L	7.55
	MHL/ML.H	5.21
	ML/M(H_{-1}L).H	5.26

Bibliography: 75HMa

$$CH_3NHCH_2PO_3H_2$$

| $C_2H_8O_3NP$ | | N-Methylaminomethylphosphonic acid | H_2L |

Metal ion	Equilibrium	Log K 25°, 0.1
H^+	HL/H.L	10.91
	H_2L/HL.H	5.31
	H_3L/H_2L.H	0.6

N—Methylaminomethylphosphonic acid   (continued)

Metal ion	Equilibrium	Log K 25°, 0.1
Cu^{2+}	ML/M.L	8.29
	$ML_2/M.L^2$	14.59
	ML/MOHL.H	8.20
	MHL/ML.H	5.03
	$M(HL)L/ML_2.H$	6.4
	$M(HL)_2/M(HL)L.H$	4.8

Bibliography:

H^+ 78WNb,79WNa Cu^{2+} 79WNa

$$CH_3CH_2NHCH_2PO_3H_2$$

$C_3H_{10}O_3NP$ N—Ethylaminomethylphosphonic acid H_2L

Metal ion	Equilibrium	Log K 25°, 0.1
H^+	HL/H.L	11.01
	$H_2L/HL.H$	5.33
	$H_3L/H_2L.H$	0.5
Cu^{2+}	ML/M.L	7.72
	$ML_2/M.L^2$	13.0
	ML/MOHL.H	7.57
	MHL/ML.H	5.70
	$M(HL)L/ML_2.H$	7.7
	$M(HL)_2/M(HL)L.H$	5.58

Bibliography: 79WNa

$$\overset{\displaystyle NHCH_3}{\overset{|}{H_2O_3PCHPO_3H_2}}$$

$C_2H_9O_6NP_2$ N—Methylaminomethylenediphosphonic acid H_4L

Metal ion	Equilibrium	Log K 25°, 0.1
H^+	HL/H.L	10.78
	$H_2L/HL.H$	8.27
	$H_3L/H_2L.H$	5.10
	$H_4L/H_3L.H$	2.00
Ca^{2+}	ML/M.L	4.66
	MHL/ML.H	10.77
Co^{2+}	ML/M.L	7.87
	MHL/ML.H	9.28

Bibliography: 78GMC

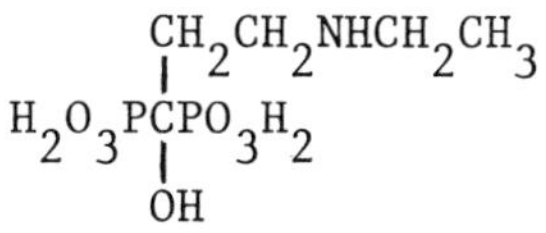

$$CH_2CH_2NHCH_2CH_3$$
$$|$$
$$H_2O_3PCPO_3H_2$$
$$|$$
$$OH$$

$C_5H_{15}O_7NP_2$	1-Hydroxy-3-(ethylamino)propane-1,1-diphosphonic acid	H_4L

Metal ion	Equilibrium	Log K 25°, 0.1
H^+	HL/H.L	11.5
	$H_2L/HL.H$	10.4
	$H_3L/H_2L.H$	6.70
	$H_4L/H_3L.H$	2.49
Ca^{2+}	ML/M.L	6.2
	MHL/ML.H	11.4
	$MH_2L/MHL.H$	9.62
Co^{2+}	ML/M.L	10.2
	MHL/ML.H	10.23
	$MH_2L/MHL.H$	7.85
Cu^{2+}	ML/M.L	13.7
	MHL/ML.H	9.44
	$MH_2L/MHL.H$	6.48

Bibliography: 78KMD

$$H_2O_3PCH_2CH_2NHCH_2CH_2NHCH_2CH_2PO_3H_2$$

$C_6H_{18}O_6N_2P_2$	Ethylenebis(iminoethylenephosphonic acid)	H_4L

Metal ion	Equilibrium	Log K 25°, 0.1
H^+	HL/H.L	11.12
	$H_2L/HL.H$	8.43
	$H_3L/H_2L.H$	6.59
	$H_4L/H_3L.H$	5.58
Ni^{2+}	ML/M.L	12.99
	MHL/ML.H	8.77
	$MH_2L/MHL.H$	3.27
Cu^{2+}	ML/M.L	18.13
	MHL/ML.H	6.15
	$MH_2L/MHL.H$	4.84

Bibliography: 76MD

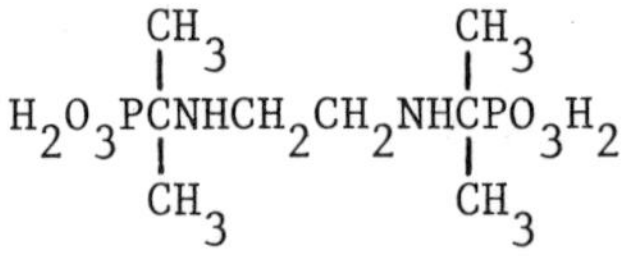

$$\underset{\underset{CH_3}{|}}{\overset{\overset{CH_3}{|}}{H_2O_3PCNHCH_2}}CH_2\underset{\underset{CH_3}{|}}{\overset{\overset{CH_3}{|}}{NHCPO_3H_2}}$$

$C_8H_{22}O_6N_2P_2$	Ethylenebis[imino(dimethyl)methylenephosphonic acid]	H_4L
	(Other values on Vol.2, p.307)	

Metal ion	Equilibrium	Log K 25°, 0.1
H^+	HL/H.L	11.68
	$H_2L/HL.H$	8.55
	$H_3L/H_2L.H$	6.00
	$H_4L/H_3L.H$	4.95
Dy^{3+}	ML/M.L	12.89
	$MH_2L/M.H_2L$	6.20
UO_2^{2+}	ML/M.L	15.84
	$MH_2L/M.H_2L$	8.52
Ni^{2+}	ML/M.L	11.23
	MHL/M.HL	7.35
	$MH_2L/M.H_2L$	3.84
Cu^{2+}	ML/M.L	20.35
	MHL/M.HL	12.38
	$MH_2L/M.H_2L$	8.83

Bibliography: 76MD   Other reference: 79ZKT

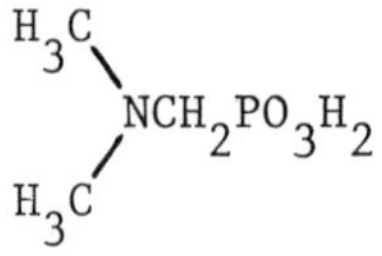

$$\begin{array}{c} H_3C \\ \searrow \\ NCH_2PO_3H_2 \\ \nearrow \\ H_3C \end{array}$$

$C_3H_{10}O_3NP$	N,N-Dimethylaminomethylphosphonic acid	H_2L

Metal ion	Equilibrium	Log K 25°, 0.1
H^+	HL/H.L	11.06
	$H_2L/HL.H$	5.19
	$H_3L/H_2L.H$	0.4
Cu^{2+}	$ML/M.L_2$	7.99
	$ML_2/M.L^2$	13.84
	ML/MOHL.H	7.79
	$M(HL)L/ML_2.H$	7.0
	$M(HL)_2/M(HL)L.H$	5.2

Bibliography:

H^+ 78WNa,78WNb,79WNa Cu^{2+} 78WNa,79WNa

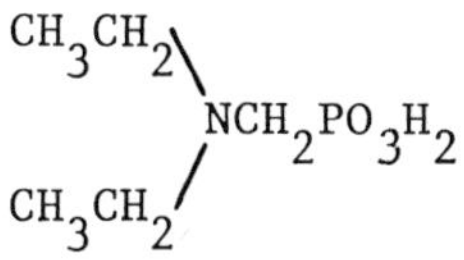

$$CH_3CH_2 \diagdown$$
$$NCH_2PO_3H_2$$
$$CH_3CH_2 \diagup$$

$C_5H_{14}O_3NP$		N,N-Diethylaminomethylphosphonic acid (Other values on Vol.2, p.317)	H_2L

Metal ion	Equilibrium	Log K 25°, 0.1	Log K 25°, 1.0
H^+	HL/H.L	11.81	12.3
	$H_2L/HL.H$	5.28	5.8
Cu^{2+}	ML/M.L	7.46	
	ML/MOHL.H	7.48	
	MHL/ML.H	6.66	
	$M(HL)_2/M.(HL)^2$	3.94	

Bibliography: 79WNa

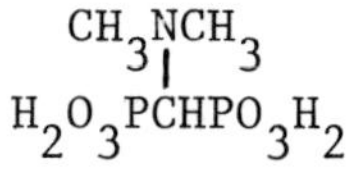

$$CH_3NCH_3$$
$$H_2O_3PCHPO_3H_2$$

$C_3H_{11}O_6NP_2$	N,N-Dimethylaminomethylenediphosphonic acid	H_4L

Metal ion	Equilibrium	Log K 25°, 0.1
H^+	HL/H.L	11.04
	$H_2L/HL.H$	8.71
	$H_3L/H_2L.H$	4.90
	$H_4L/H_3L.H$	2.05
Ca^{2+}	ML/M.L	4.78
	MHL/ML.H	10.82
Mn^{2+}	ML/M.L	7.26
	MHL/ML.H	10.49
Co^{2+}	ML/M.L	7.29
	MHL/ML.H	9.84
Cu^{2+}	ML/M.L	11.92
	MHL/ML.H	8.61
Zn^{2+}	ML/M.L	9.27
	MHL/ML.H	8.96

Bibliography: 78GMC

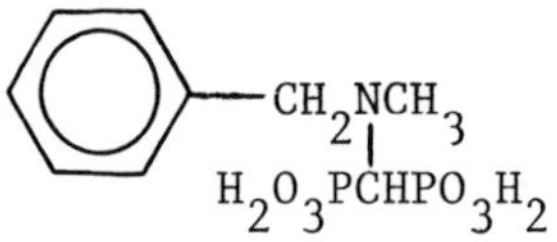

CH₂NCH₃
H₂O₃PCHPO₃H₂

C₉H₁₅O₆NP₂   N-Benzyl-N-methylaminomethylenediphosphonic acid   H₄L

Metal ion	Equilibrium	Log K 25°, 0.1
H⁺	HL/H.L	10.82
	H₂L/HL.H	8.53
	H₃L/H₂L.H	4.91
	H₄L/H₃L.H	2.10
Ca²⁺	ML/M.L	4.20
	MHL/ML.H	10.65
Mn²⁺	ML/M.L	7.03
	MHL/ML.H	10.17
Co²⁺	ML/M.L	7.04
	MHL/ML.H	9.67

Bibliography: 78GMC

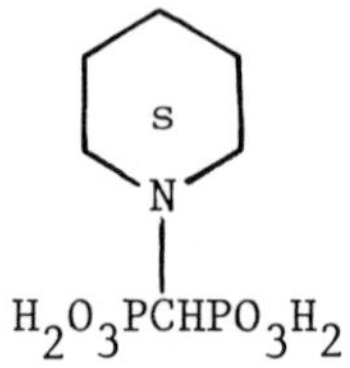

H₂O₃PCHPO₃H₂

C₆H₁₅O₆NP   Pentamethyleniminomethylenediphosphonic acid   H₄L
                 (N-(diphosphonomethyl)piperidine)

Metal ion	Equilibrium	Log K 25°, 0.1
H⁺	HL/H.L	10.72
	H₂L/HL.H	8.08
	H₃L/H₂L.H	4.55
	H₄L/H₃L.H	2.03
Ca²⁺	ML/M.L	4.53
	MHL/ML.H	9.83
Mn²⁺	ML/M.L	7.75
	MHL/ML.H	9.24
Co²⁺	ML/M.L	7.64
	MHL/ML.H	8.80
Cu²⁺	ML/M.L	12.21
	MHL/ML.H	7.73
Zn²⁺	ML/M.L	9.47
	MHL/ML.H	8.05

Bibliography: 78GMC

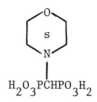

$H_2O_3PCHPO_3H_2$

$C_5H_{13}O_7NP_2$		1,4-Oxazin-4-ylmethylenediphosphonic acid	H_4L

Metal ion	Equilibrium	Log K 25°, 0.1
H^+	HL/H.L	11.17
	$H_2L/HL.H$	8.70
	$H_3L/H_2L.H$	4.80
	$H_4L/H_3L.H$	2.05
Ca^{2+}	ML/M.L	4.18
	MHL/ML.H	11.05
Mn^{2+}	ML/M.L	7.05
	MHL/ML.H	10.70
Co^{2+}	ML/M.L	7.12
	MHL/ML.H	10.15
Cu^{2+}	ML/M.L	11.18
	MHL/ML.H	9.11
Fe^{3+}	ML/M.L	20
Zn^{2+}	ML/M.L	9.46
	MHL/ML.H	(8.96)

Bibliography: 78GMC

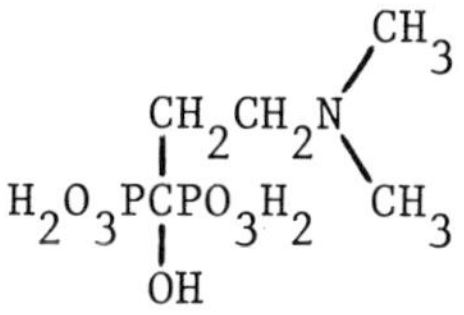

$$H_2O_3PCPO_3H_2 \quad CH_2CH_2N \overset{CH_3}{\underset{CH_3}{}}$$
$$\overset{|}{OH}$$

$C_5H_{15}O_7NP_2$		1-Hydroxy-3-(dimethylamino)propane-1,1-diphosphonic acid	H_4L

Metal ion	Equilibrium	Log K 25°, 0.1
H^+	HL/H.L	10.8
	$H_2L/HL.H$	9.7
	$H_3L/H_2L.H$	5.89
	$H_4L/H_3L.H$	2.35
Mg^{2+}	ML/M.L	6.6
	MHL/ML.H	10.6
Ca^{2+}	ML/M.L	5.7
	MHL/ML.H	10.7
	$MH_2L/MHL.H$	7.51
Sr^{2+}	ML/M.L	4.9
	MHL/ML.H	10.7
Mn^{2+}	ML/M.L	8.3
	MHL/ML.H	10.6

1-Hydroxy-3-(dimethylamino)propane-1,1-diphosphonic acid   (continued)

Metal ion	Equilibrium	Log K 25°, 0.1
Co^{2+}	ML/M.L	**9.0**
	MHL/ML.H	9.68
	MH_2L/MHL.H	6.55
Ni^{2+}	**ML/M.L**	(8.7)
	MHL/ML.H	9.59
Cu^{2+}	ML/M.L	13.0
	MHL/ML.H	8.61
	MH_2L/MHL.H	4.94
Zn^{2+}	ML/M.L	**10.2**
	MHL/ML.H	9.46

Bibliography:

$H^+, Ca^{2+}, Co^{2+}, Cu^{2+}$ 78KMD $Mg^{2+}, Sr^{2+}, Mn^{2+}, Ni^{2+}, Zn^{2+}$ 79KBb

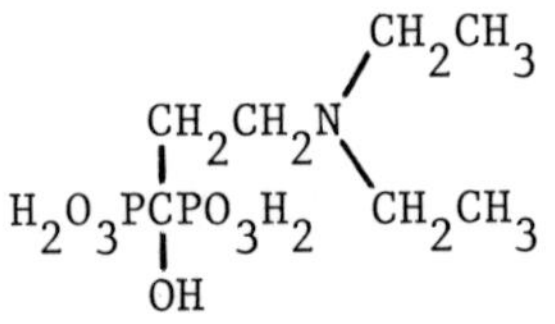

$$H_2O_3PCPO_3H_2 \begin{array}{c} CH_2CH_2N \begin{array}{c} CH_2CH_3 \\ CH_2CH_3 \end{array} \\ OH \end{array}$$

$C_7H_{19}O_6NP_2$	1-Hydroxy-3-(diethylamino)propane-1,1-diphosphonic acid	H_4L

Metal ion	Equilibrium	Log K 25°, 0.1
H^+	HL/H.L	11.2
	H_2L/HL.H	10.0
	H_3L/H_2L.H	5.92
	H_4L/H_3L.H	2.50
Ca^{2+}	ML/M.L	5.7
	MHL/ML.H	9.89
	MH_2L/MHL.H	8.95
Co^{2+}	ML/M.L	7.7
	MHL/ML.H	9.23
	MH_2L/MHL.H	8.46
Cu^{2+}	ML/M.L	12.7
	MHL/ML.H	8.40
	MH_2L/MHL.H	6.11

Bibliography: 78KMD

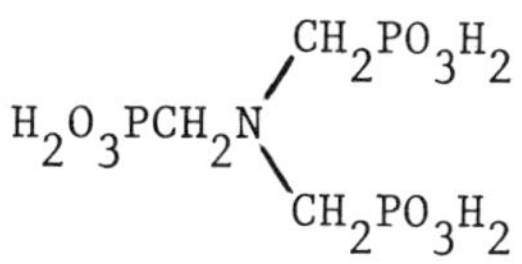

C. TERTIARY AMINES

$$H_2O_3PCH_2N \begin{array}{c} CH_2PO_3H_2 \\ \\ CH_2PO_3H_2 \end{array}$$

$C_3H_{12}O_9NP_3$ **Nitrilotris(methylenephosphonic acid)** H_6L
(Other reference on Vol.2, p.319)

Metal ion	Equilibrium	Log K 25°, 0.1	Log K 25°, 1.0
H^+	HL/H.L	(12.1)	12.3
	$H_2L/HL.H$	7.30	6.66
	$H_3L/H_2L.H$	5.86	5.46
	$H_4L/H_3L.H$	4.64	4.30
	$H_5L/H_4L.H$	1.5	
	$H_6L/H_5L.H$	0.3	
Mg^{2+}	ML/M.L	7.2	6.49
	MHL/ML.H	9.5	9.1
	$MH_2L/MHL.H$	6.6	6.2
	$MH_3L/MH_2L.H$	5.4	4.6
Ca^{2+}	ML/M.L	7.5	6.68
	MHL/ML.H	9.0	8.5
	$MH_2L/MHL.H$	6.6	6.1
	$MH_3L/MH_2L.H$	5.5	4.9
Sr^{2+}	ML/M.L	6.5	
	MHL/ML.H	9.6	
	$MH_2L/MHL.H$	6.9	
	$MH_3L/MH_2L.H$	5.6	
Ba^{2+}	ML/M.L	6.5	
	MHL/ML.H	9.6	
	$MH_2L/MHL.H$	6.8	
	$MH_3L/MH_2L.H$	5.7	
Mn^{2+}	ML/M.L	10.2	
	MHL/ML.H	7.5	
	$MH_2L/MHL.H$	6.1	
	$MH_3L/MH_2L.H$	5.0	
Co^{2+}	ML/M.L	(14.4)	
	MHL/ML.H	6.2	
	$MH_2L/MHL.H$	5.2	
	$MH_3L/MH_2L.H$	4.3	
Ni^{2+}	ML/M.L	(11.1)	
	MHL/ML.H	8.3	
	$MH_2L/MHL.H$	5.8	
	$MH_3L/MH_2L.H$	4.5	
Cu^{2+}	ML/M.L	17.4	
	MHL/ML.H	6.4	
	$MH_2L/MHL.H$	4.7	
	$MH_3L/MH_2L.H$	3.5	
	$MH_4L/MH_3L.H$	0.0	

Nitrilotris(methylenephosphonic acid)   (continued)

Metal ion	Equilibrium	Log K 25°, 0.1	Log K 25°, 1.0
Zn^{2+}	ML/M.L	16.4	
	MHL/ML.H	6.1	
	$MH_2L/MHL.H$	5.1	
	$MH_3L/MH_2L.H$	4.1	
	$MH_4L/MH_3L.H$	1.7	
Cd^{2+}	ML/M.L	11.6	
	MHL/ML.H	7.0	
	$MH_2L/MHL.H$	5.7	
	$MH_3L/MH_2L.H$	4.8	

Bibliography:

H^+	74NG	$Mn^{2+}-Cd^{2+}$	75MN
$Mg^{2+}-Ba^{2+}$	74NGa	Other references:	67H,78MZ,79EFa,79TK

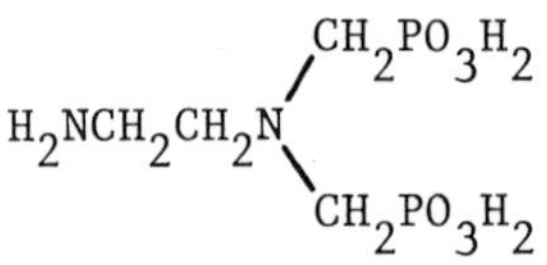

$$H_2NCH_2CH_2N \begin{matrix} CH_2PO_3H_2 \\ \\ CH_2PO_3H_2 \end{matrix}$$

$C_4H_{14}O_6N_2P_2$ N-2-Aminoethyliminobis(methylenephosphonic acid) H_4L

Metal ion	Equilibrium	Log K 25°, 0.1
H^+	HL/H.L	10.85
	$H_2L/HL.H$	8.41
	$H_3L/H_2L.H$	6.34
	$H_4L/H_3L.H$	4.10
Co^{2+}	ML/M.L	10.79
	$MH_2L/ML.H^2$	(11.94)
Ni^{2+}	ML/M.L	11.52
	$MH_2L/ML.H^2$	(12.06)
Cu^{2+}	ML/M.L	14.25
	$MH_2L/ML.H^2$	11.06

Bibliography: 76TI

$$H_2O_3PCH_2NCH_2CH_2NCH_2PO_3H_2$$
$$\underset{CH_3}{|} \qquad \underset{CH_3}{|}$$

$C_6H_{18}O_6N_2P_2$ N,N'-Dimethylethylenebis(nitrilomethylenephosphonic acid) H_4L

Metal ion	Equilibrium	Log K 25°, 0.1
H^+	HL/H.L	11.21
	H_2L/HL.H	7.92
	H_3L/H_2L.H	5.92
	H_4L/H_3L.H	4.00
Dy^{3+}	ML/M.L	15.17
	MHL/ML.H	(8.02)
	MH_2L/MHL.H	4.06
UO_2^{2+}	ML/M.L	14.9
	MHL/ML.H	(6.2)
Ni^{2+}	ML/M.L	14.67
	MHL/ML.H	5.73
	MH_2L/MHL.H	4.69
Cu^{2+}	ML/M.L	19.64
	MHL/ML.H	4.56
	MH_2L/MHL.H	3.66
Fe^{3+}	ML/M.L	22.5

Bibliography: 76MD

$$H_2O_3PCH_2CH_2NCH_2CH_2NCH_2CH_2PO_3H_2$$
$$\underset{CH_3}{|} \qquad \underset{CH_3}{|}$$

$C_8H_{22}O_6N_2P_2$ N,N'-Dimethylethylenebis(nitriloethylenephosphonic acid) H_4L

Metal ion	Equilibrium	Log K 25°, 0.1
H^+	HL/H.L	11.16
	H_2L/HL.H	8.12
	H_3L/H_2L.H	6.38
	H_4L/H_3L.H	4.91
Dy^{3+}	MHL/M.HL	10.4
Ni^{2+}	ML/M.L	8.70
	MHL/ML.H	7.89
	MH_2L/MHL.H	6.24
Cu^{2+}	ML/M.L	15.92
	MHL/ML.H	7.61
	MH_2L/MHL.H	4.08

Bibliography: 76MD

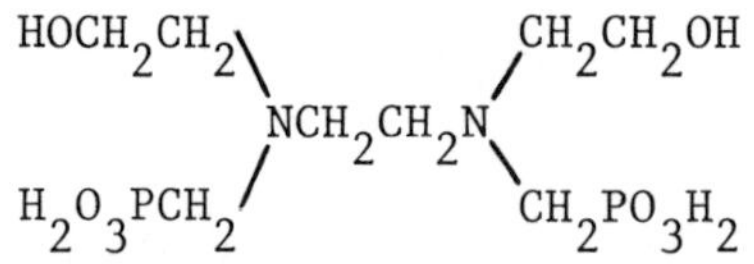

$$\begin{array}{ccc}
\text{HOCH}_2\text{CH}_2 & & \text{CH}_2\text{CH}_2\text{OH} \\
& \text{NCH}_2\text{CH}_2\text{N} & \\
\text{H}_2\text{O}_3\text{PCH}_2 & & \text{CH}_2\text{PO}_3\text{H}_2
\end{array}$$

$C_8H_{22}O_8N_2P_2$	N,N'-Bis(2-hydroxyethyl)ethylenedinitrilo-N,N'-bis(methylenephosphonic acid)	H_4L

Metal ion	Equilibrium	Log K 25°, 0.1
H^+	HL/H.L	11.53
	H_2L/HL.H	9.17
	H_3L/H_2L.H	6.17
	H_4L/H_3L.H	3.68
Dy^{3+}	ML/M.L	11.95
	MHL/ML.H	8.47
	MH_2L/MHL.H	5.57
UO_2^{2+}	ML/M.L	13.04
	MHL/ML.H	8.00
	MH_2L/MHL.H	6.06
Ni^{2+}	ML/M.L	9.60
	MHL/ML.H	10.14
	MH_2L/MHL.H	5.81
Cu^{2+}	ML/M.L	12.61
	MHL/ML.H	9.70
	MH_2L/MHL.H	5.39
Fe^{3+}	ML/M.L	22.1

Bibliography: 76MR

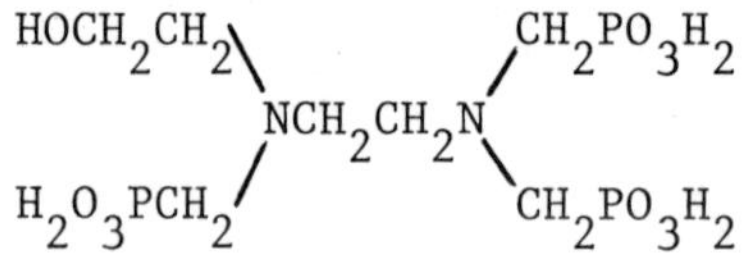

$$\begin{array}{ccc}
\text{HOCH}_2\text{CH}_2 & & \text{CH}_2\text{PO}_3\text{H}_2 \\
& \text{NCH}_2\text{CH}_2\text{N} & \\
\text{H}_2\text{O}_3\text{PCH}_2 & & \text{CH}_2\text{PO}_3\text{H}_2
\end{array}$$

$C_7H_{21}O_{10}N_2P_3$	N-(2-hydroxyethyl)ethylenedinitrilotris(methylenephosphonic acid)	H_6L

Metal ion	Equilibrium	Log K 25°, 0.1
H^+	HL/H.L	10.9
	H_2L/HL.H	9.10
	H_3L/H_2L.H	6.85
	H_4L/H_3L.H	5.50
	H_5L/H_4L.H	2.81
	H_6L/H_5L.H	1.7
Ca^{2+}	ML/M.L	5.6
	MHL/ML.H	9.06
Dy^{3+}	ML/M.L	11.7
	MHL/ML.H	7.59
	MH_2L/MHL.H	6.36
UO_2^{2+}	ML/M.L	10.0
	MHL/ML.H	8.51

N-(2-hydroxyethyl)ethylenedinitrilotris(methylenephosphonic acid) (continued)

Metal ion	Equilibrium	Log K 25°, 0.1
Fe^{2+}	ML/M.L	12.8
	MHL/ML.H	7.10
	$MH_2L/MHL.H$	4.55
Cu^{2+}	ML/M.L	16.3
	MHL/ML.H	7.16
	$MH_2L/MHL.H$	5.58
Fe^{3+}	ML/M.L	19.6
	MHL/ML.H	4.0

Bibliography: 74KR

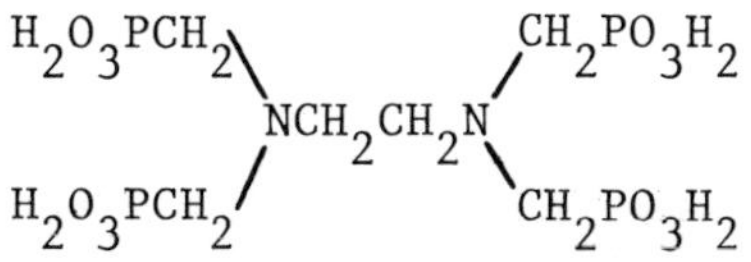

$$H_2O_3PCH_2 \diagdown \qquad\qquad CH_2PO_3H_2$$
$$NCH_2CH_2N$$
$$H_2O_3PCH_2 \diagup \qquad\qquad CH_2PO_3H_2$$

$C_6H_{10}O_{12}N_2P_4$	Ethylenedinitrilotetrakis(methylenephosphonic acid) (Other references on Vol.2, p.323)		H_8L

Metal ion	Equilibrium	Log K 25°, 0.1	Log K 25°, 3.0
H^+	HL/H.L	$(12.99)^z$	12.01
	$H_2L/HL.H$	9.78 +0.04	9.02
	$H_3L/H_2L.H$	7.94 −0.09	7.42
	$H_4L/H_3L.H$	6.42 −0.02	5.88
	$H_5L/H_4L.H$	5.17 +0.05	4.77
	$H_6L/H_5L.H$	3.02 ±0.00	2.82
	$H_7L/H_6L.H$	$(1.33)^z$ +0.2	1.24
Mg^{2+}	ML/M.L	8.43 +0.2	
	MHL/ML.H	9.95 +0.1	
	$MH_2L/MHL.H$	8.79 −0.2	
	$MH_3L/MH_2L.H$	6.96 +0.2	
	$MH_4L/MH_3L.H$	4.97 −0.05	
Ca^{2+}	ML/M.L	9.36 −0.03	
	MHL/ML.H	9.42 +0.4	
	$MH_2L/MHL.H$	8.44 −0.4	
	$MH_3L/MH_2L.H$	6.59 +0.4	
	$MH_4L/MH_3L.H$	5.25 −0.4	
Co^{2+}	ML/M.L	17.11 −2	
	MHL/ML.H	8.31 +0.09	
	$MH_2L/MHL.H$	6.49 +0.4	
	$MH_3L/MH_2L.H$	5.29 +0.4	
	$MH_4L/MH_3L.H$	4.30 +0.4	
Ni^{2+}	ML/M.L	16.38 −1	
	MHL/ML.H	8.94 −0.1	
	$MH_2L/MHL.H$	7.40 −0.1	
	$MH_3L/MH_2L.H$	5.48 +0.2	
	$MH_4L/MH_3L.H$	4.33 +0.2	

^{z}Estimated value calculated from values at 3.0 ionic strength.

Ethylenedinitrilotetrakis(methylenephosphonic acid)   (continued)

Metal ion	Equilibrium	Log K 25°, 0.1	Log K 25°, 3.0
Cu^{2+}	ML/M.L	23.21 −4	
	MHL/ML.H	7.56 +0.4	
	$MH_2L/MHL.H$	5.99 +0.5	
	$MH_3L/MH_2L.H$	4.62 +0.6	
	$MH_4L/MH_3L.H$	3.74 +0.2	
Zn^{2+}	ML/M.L	18.76 −2	
	MHL/ML.H	8.31 +0.2	
	$MH_2L/MHL.H$	6.06 +0.5	
	$MH_3L/MH_2L.H$	4.99 +0.1	
	$MH_4L/MH_3L.H$	3.10 +1	

Bibliography: 76MMM   Other references: 69KT,74KPM,77MRR,78KPS, 79RZ

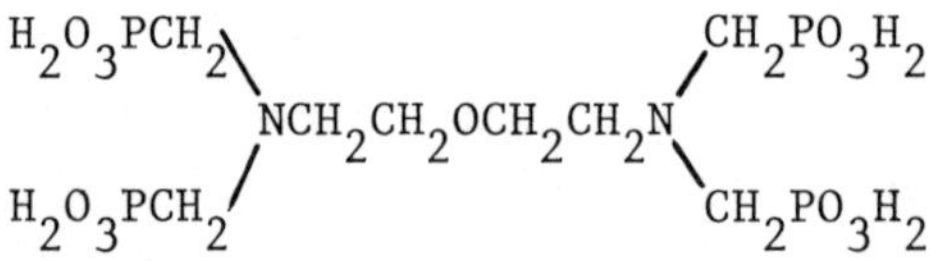

$$H_2O_3PCH_2 \diagdown \qquad \qquad \diagup CH_2PO_3H_2$$
$$NCH_2CH_2OCH_2CH_2N$$
$$H_2O_3PCH_2 \diagup \qquad \qquad \diagdown CH_2PO_3H_2$$

$C_8H_{24}O_{13}N_2P_4$ Oxybis[ethylenenitrilobis(methylenephosphonic acid)] H_8L

Metal ion	Equilibrium	Log K 20°, 0.1
H^+	HL/H.L	10.72
	$H_2L/HL.H$	10.10
	$H_3L/H_2L.H$	7.52
	$H_4L/H_3L.H$	6.41
	$H_5L/H_4L.H$	5.66
	$H_6L/H_5L.H$	4.73
	$H_7L/H_6L.H$	2.72
	$H_8L/H_7L.H$	1.5
Ca^{2+}	ML/M.L	6.59
	MHL/ML.H	8.83
	$MH_2L/MHL.H$	8.36
	$M_2L/ML.M$	2.96
Sr^{2+}	ML/M.L	6.03
	MHL/ML.H	9.33
	$MH_2L/MHL.H$	8.03
	$M_2L/ML.M$	2.49
Y^{3+}	ML/M.L	15.2
Ce^{3+}	ML/M.L	15.6

Bibliography: 69Tb

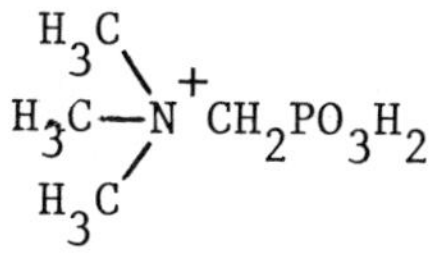

$$H_3C-\overset{\overset{\displaystyle H_3C}{|}}{\underset{\underset{\displaystyle H_3C}{|}}{N^+}}CH_2PO_3H_2$$

$C_4H_{13}O_3NP^+$		Trimethyl(phosponomethyl)ammonium iodide	H_2L^+

Metal ion	Equilibrium	Log K 25°, 0.1
H^+	HL/H.L	5.10
Ca^{2+}	ML/M.L	0.93
Cu^{2+}	ML/M.L	2.18
	ML/MOHL.H	6.9

Bibliography: 79WN

$$HCO_2H$$

CH_2O_2		Methanoic acid (formic acid) (Other values on Vol.3, p.1)				HL
Metal ion	Equilibrium	Log K 25°, 0.1	Log K 25°, 1.0	Log K 25°, 0	ΔH 25°, 0	ΔS 25°, 0
H^+	HL/H.L	3.55 −0.01 3.49[b]	3.53 ±0.00 3.904[e]+0.004	3.745 ±0.007 4.39[g]	0.04 ±0.04	17.3
Hg^+	$ML/M.L$ $ML_2/M.L^2$		2.94[e] 5.45[e]			
CH_3Hg^+	ML/M.L	2.67[v]	2.68			
$(CH_3)_3Pb^+$	ML/M.L	0.86[u]				
Hg^{2+}	$ML/M.L$ $ML_2/M.L^2$		3.66[e] 7.10[e]			
Al^{3+}	$ML/M.L$ $ML_2/M.L^2$ ML/MOHL.H		1.36 2.02 3.43			

[b]25°, 0.5;   [e]25°, 3.0;   [g]25°, 5.0;   [u]25°, 0.3;   [v]25°, 0.4

Bibliography:

H^+,Hg^{2+} 77RW $(CH_3)_3Pb^+$ 77SB

Hg^+ 77RWa Al^{3+} 76KIa

CH_3Hg^+ 78JIa

Other references: 68RS,69KP,70WC,71MT, 73PZ,75M,75SA,76BD,76FK,76FKa,76KF, 77AMa,77FK,77FKa,77YO

$$CH_3CO_2H$$

$C_2H_4O_2$		Ethanoic acid (acetic acid) (Other values on Vol.3, p.3)				HL
Metal ion	Equilibrium	Log K 25°, 0.1	Log K 25°, 1.0	Log K 25°, 0	ΔH 25°, 0	ΔS 25°, 0
H^+	HL/H.L	4.56 ±0.03 4.50[b]±0.02	4.57 ±0.04 4.80[d]±0.00	4.757 ±0.002 5.015[e]±0.005	0.10 ±0.01 −0.75[e]−0.03	22.1 20.5[e]
Fe^{2+}	ML/M.L			1.40 0.54[e]		
Hg^+	$ML/M.L$ $ML_2/M.L^2$			3.57[e] 6.63[e]		
CH_3Hg^+	ML/M.L $M_2L/ML.M$	3.18[v]	3.20 2.08			
$(CH_3)_3Pb^+$	ML/M.L	0.97[u]				
Hg^{2+}	$ML/M.L$ $ML_2/M.L^2$		3.32[p] 7.01[p]	4.22[e] 8.45[e]		
Sn^{2+}	$ML/M.L$ $ML_2/M.L^2$ $ML_3/M.L^3$			3.47[e] 6.04[e] 7.27[e]		

[b]25°, 0.5;   [d]25°, 2.0;   [e]25°, 3.0;   [p]30°, 1.0;   [u]25°, 0.3;   [v]25°, 0.4

Acetic acid   (continued)

Metal ion	Equilibrium	Log K 25°, 0.1	Log K 25°, 1.0	Log K 25°, 0	ΔH 25°, 0	ΔS 25°, 0
Al^{3+}	ML/M.L		1.51			
	ML/MOHL.H		4.54			

Bibliography:

H^+ 77RW,78Da Hg^{2+} 77RW

Fe^{2+} 68CN Sn^{2+} 76Ga

Cu^{2+} 74Ad,79EFa Al^{3+} 76KIa

Hg^+ 77RWa,78THH   Other references: 54CV,69Ma,70WC,71MT,72NP,
                                                 73AB,74DN,74Wa,75CPb,75M,75SA,76BEK,
CH_3Hg^+ 78JIa,   76FK,76FKa,76GC,76KF,76MKD,76NF,77BMO,
                                                 77FK,77FKa,77VV,77YO,78MPa,79ASN,
$(CH_3)_3Pb^+$ 77SB   79EFb,79EFc,79FS

$$CH_3CH_2CO_2H$$

$C_3H_6O_2$	Propanoic acid (Other values on Vol.2, p.8)					HL

Metal ion	Equilibrium	Log K 25°, 0.1	Log K 25°, 1.0	Log K 25°, 0	ΔH 25°, 0	ΔS 25°, 0
H^+	HL/H.L	4.67 ±0.02	4.67 −0.01	4.874 +0.001	0.20 ±0.03	23.0
		4.63[b]	4.89[d]	5.16[e]		
Co^{2+}	ML/M.L		(0.5) 0.7[d]		1.1[c]	6[c]
	$ML_2/M.L^2$		(−0.1) 0.6[d]		4[c]	13[c]
Ni^{2+}	ML/M.L		(0.5) 0.73[d]		1.1[c]	6[c]
	$ML_2/M.L^2$		(0.0) 0.8[d]		3[c]	10[c]
Cu^{2+}	ML/M.L		1.66	2.22[e] 1.86[e]	1.0[c]	11[c]
	$ML_2/M.L^2$		2.62	3.00[e]		
Hg^+	ML/M.L			3.72[e]		
	$ML_2/M.L^2$			6.99[e]		
$(CH_3)_3Pb^+$	ML/M.L	1.08[u]				
Hg^{2+}	ML/M.L			4.33[e]		
	$ML_2/M.L^2$			8.80[e]		

[b] 25°, 0.5;   [c] 25°, 1.0;   [d] 25°, 2.0;   [e] 25°, 3.0;   [u] 25°, 0.3

Bibliography:

H^+,Hg^{2+} 77RW $(CH_3)_3Pb^+$ 77SB,

Co^{2+}-Cu^{2+} 75Aa   Other references: 68RS,69KF,71MC,71TDa,
                                                 71TDG,72TA,75SA,76FK,76FKa,76KF,76PM,
Hg^+ 77RWa   77FK,77FKa,77VPK,77YO,79FS

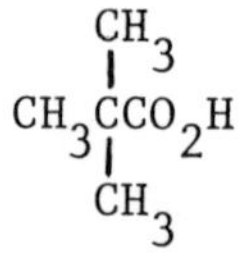

$$\begin{array}{c} CH_3 \\ | \\ CH_3CCO_2H \\ | \\ CH_3 \end{array}$$

$C_5H_{10}O_2$ 2,2-Dimethylpropanoic acid (pivalic acid)   HL
(Other values on Vol.3, p.13)

Metal ion	Equilibrium	Log K 25°, 0.1	Log K 25°, 1.0	Log K 25°, 0	ΔH 25°, 0	ΔS 25°, 0
H^+	HL/H.L	4.83 +0.01	4.79	5.032	0.72 ±0.04	25.5
		4.77[b]	5.33[e]			
$(CH_3)_3Pb^+$	ML/M.L	1.22[u]				

[b] 25°, 0.5;   [e] 25°, 3.0;   [u] 25°, 0.3

Bibliography: 77SB

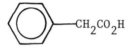

$C_8H_8O_2$ Phenylacetic acid   HL
(Other values on Vol.3, p.13)

Metal ion	Equilibrium	Log K 25°, 0.1	Log K 25°, 1.0	Log K 25°, 0
H^+	HL/H.L	4.10 ±0.02	4.09	4.310 ±0.003
		4.07[b]	4.34[d]	4.56[e]
Co^{2+}	ML/M.L		0.62[d]	
	$ML_2/M.L^2$		0.52[d]	
Ni^{2+}	ML/M.L		0.65[d]	
	$ML_2/M.L^2$		0.99[d]	
Cd^{2+}	ML/M.L		1.15[d]	
	$ML_2/M.L^2$		1.92[d]	

[b] 25°, 0.5;   [d] 25°, 2.0;   [e] 25°, 3.0

Bibliography: 79NT

Other references: 57CR,68RS,74DN

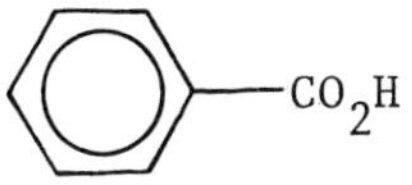

$-CO_2H$

$C_7H_6O_2$		Benzenecarboxylic acid (benzoic acid)				HL
		(Other values on Vol.3, p.16)				

Metal ion	Equilibrium	Log K 25°, 0.1	Log K 25°, 1.0	Log K 25°, 0	ΔH 25°, 0	ΔS 25°, 0
H^+	HL/H.L	4.00[b] ±0.01 3.96[b] ±0.03	4.00 ±0.03	4.202 ±0.002	-0.11 ±0.01 0.46[o]	18.9
Cd^{2+}	ML/M.L	1.4 1.15[t]		1.01 0.99[p]		
	$ML_2/M.L^2$			1.65 1.76[p]		
Pb^{2+}	ML/M.L	2.0 1.99[t]		1.87		
	$ML_2/M.L^2$			2.89		

[b]25°, 0.5; [p]30°, 1.0; [t]30°, 0.4; [o]40°, 0

Bibliography:

H^+ 78La,78OS,78PB,78PM,79SKF   Other references: 69Ta,76AN,76DR

Cd^{2+},Pb^{2+} 78OS

$ClCH_2CO_2H$

$C_2H_3O_2Cl$		Chloroacetic acid				HL
		(Other values on Vol.3, p.16)				

Metal ion	Equilibrium	Log K 25°, 0.1	Log K 25°, 1.0	Log K 25°, 0	ΔH 25°, 0	ΔS 25°, 0
H^+	HL/H.L	2.68 ±0.06 2.60[b]	2.64 ±0.04	2.865 ±0.004 3.023[e] ±0.005	1.07 ±0.08 0.75[c]	16.7
Th^{4+}	ML/M.L	2.75			2.9	22
	$ML_2/M.L^2$	4.63			3.1	19
	$ML_3/M.L^3$	5.79			2.6	14
	$ML_4/M.L^4$	6.53			1.9	10
Hg^+	ML/M.L		2.40[e]			
	$ML_2/M.L^2$		4.4[e]			
$(CH_3)_3Pb^+$	ML/M.L	0.52[u]				
Hg^{2+}	ML/M.L		2.95[e]			
	$ML_2/M.L^2$		5.61[e]			

[b]25°, 0.5; [c]25°, 1.0; [e]25°, 3.0; [u]25°, 0.3

Bibliography:

H^+,Hg^{2+} 77RW $(CH_3)_3Pb^+$ 77SB

Th^{4+} 78DZ   Other references: 69DD,73VBa,74Wa,77YO,79FS

Hg^+ 77RWa

$$H_2O_3PCH_2CO_2H$$

$C_2H_5O_5P$ Phosphonoacetic acid H_3L
                          (Other reference on Vol.3, p.359)

Metal ion	Equilibrium	Log K 37°, 0.15	Log K 25°, 0	ΔH 25°, 0	ΔS 25°, 0
H^+	HL/H.L	7.93	8.69	(-1)[r]	(36)
	H_2L/HL.H	4.80	5.11	(0)[r]	(23)
	H_3L/H_2L.H	1.2	2.0		
Mg^{2+}	ML/M.L	3.90			
	ML_2/M.L^2	5.2			
	MHL/M.HL	1.84			
Ca^{2+}	ML/M.L	3.18			
	ML_2/M.L^2	4.4			
Cu^{2+}	ML/M.L	7.14			
	ML_2/M.L^2	10.99			
	MHL/M.HL	3.83			
	ML/MOHL.H	8.8			
	MHL_2/ML.H	5.4			
	$M(HL)_2$/MHL_2.H	5.1			
	M_2L/ML.M	1.9			
Zn^{2+}	ML/M.L	5.35			
	ML_2/M.L^2	8.79			
	ML_3/M.L^3	10.6			
	MHL/M.HL	2.63			
	ML/MOHL.H	8.8			
	MHL_2/ML.H	5.8			
	$M(HL)_2$/MHL_2.H	5.1			

[r] 1-60°, 0

Bibliography:

H^+ 79HP,79SP Mg^{2+}-Zn^{2+} 79SP

$$\overset{\displaystyle OH}{\underset{\displaystyle CH_2CO_2H}{|}}$$

$C_2H_4O_3$ Hydroxyacetic acid   (glycolic acid)   HL
                          (Other values on Vol.3, p.24)

Metal ion	Equilibrium	Log K 25°, 0.1	Log K 25°, 1.0	Log K 25°, 0	ΔH 25°, 0	ΔS 25°, 0
H^+	HL/H.L	3.62 ±0.03	3.62 ±0.01	3.831 +0.001	-0.16 ±0.05	17.0
		3.57[b]±0.01	3.74[d]±0.03	3.91[e] ±0.02	-0.39[c]	15.3[c]
La^{3+}	ML/M.L	2.55	2.18[d]-0.04		-0.6[d] -0.2	8[d]
		2.27[b]				
	ML_2/M.L^2	4.24	3.75[d]+0.2		-1.1[d] -0.1	14[d]
		3.86[b]				
	ML_3/M.L^3	5.0	4.8[d]		-1.7[d]	16[d]
		(5.2)[b]				
	ML_4/M.L^4		5.1[d]		-2.2[d]	16[d]
	ML_5/M.L^5		4.8[k]			

[b]25°, 0.5;   [c]25°, 1.0;   [d]25°, 2.0;   [e]25°, 3.0;   [k]20°, 2.0

Glycolic acid (continued)

Metal ion	Equilibrium	Log K 25°, 0.1	Log K 25°, 1.0	Log K 25°, 0	ΔH 25°, 0	ΔS 25°, 0
Pr^{3+}	$ML/M.L$	2.78 [b] 2.49	2.43 [k]			
	$ML_2/M.L^2$	4.68 [b] 4.37	4.19 [k]			
	$ML_3/M.L^3$	5.9 [b] 5.5	5.4 [k]			
	$ML_4/M.L^4$ $ML_5/M.L^5$		5.9 [k] 5.7 [k]			
Nd^{3+}	$ML/M.L$	2.89 [b] 2.55	2.50 [d] -0.10		-1.2 [d] +0.2	7 [d]
	$ML_2/M.L^2$	4.85 [b] 4.39	4.31 [d] -0.05		-2.2 [d] +0.8	12 [d]
	$ML_3/M.L^3$	6.1 [b] 5.8	5.6 [d] -0.3		-3.5 [d]	14 [d]
	$ML_4/M.L^4$ $ML_5/M.L^5$		6.0 [d] 5.7 [k]		-4.0 [d]	14 [d]
Sm^{3+}	$ML/M.L$	2.91 [b] 2.54	2.55 [d] -0.09		-1.0 [d] +0.1	8 [d]
	$ML_2/M.L^2$	5.01 [b] 4.58	4.50 [d] -0.10		-2.4 [d] +0.7	14 [d]
	$ML_3/M.L^3$	6.6 (5.7) [b]	5.9 [d] ±0.0		-3.7 [d]	13 [d]
	$ML_4/M.L^4$ $ML_5/M.L^5$		6.4 [d] 6.0 [k]		-4.9 [d]	14 [d]
Gd^{3+}	$ML/M.L$	2.79 2.52 [b]	2.54 [j] 2.47 [d] -0.01	2.59 [l]	-0.6 [d] ±0.0	9 [d]
	$ML_2/M.L^2$	4.85 4.62 [b]	4.48 [j] 4.41 [d] -0.08	4.60 [l]	-1.7 [d] -0.1	14 [d]
	$ML_3/M.L^3$	6.0 5.8 [b]	5.8 [j] 5.7 [d] +0.2	6.1 [l]	-3.5 [d]	15 [d]
	$ML_4/M.L^4$ $ML_5/M.L^5$ $ML_6/M.L^6$		6.4 [d] 6.0 [k]	6.8 [l] 6.8 [l] 6.6 [l]	-4.3 [d]	15 [d]
Er^{3+}	$ML/M.L$	3.00 [b] 2.61	2.60 [d] +0.04		-0.2 [d] +0.1	11 [d]
	$ML_2/M.L^2$	5.19 [b] 4.69	4.57 [d] -0.05		-0.6 [d] +0.1	19 [d]
	$ML_3/M.L^3$	6.8 [b] 6.1	6.0 [d]		-1.3 [d]	23 [d]
	$ML_4/M.L^4$ $ML_5/M.L^5$		6.5 [d] 6.5 [k]		-1.4 [d]	25 [d]
Yb^{3+}	$ML/M.L$	3.13 [b] 2.71	2.72 [d] -0.10		-0.3 [d] -0.1	11 [d]
	$ML_2/M.L^2$	5.37 [b] 4.98	4.8 [d] +0.01		-0.8 [d] -0.4	19 [d]
	$ML_3/M.L^3$	7.1 (6.1) [b]	6.3 [d] -0.1		-1.7 [d] -1	23 [d]
	$ML_4/M.L^4$ $ML_5/M.L^5$		6.8 [d] 7.0 [k]		-0.6 [d]	29 [d]

[b] 25°, 0.5; [d] 25°, 2.0; [j] 20°, 1.0; [k] 20°, 2.0; [l] 20°, 3.0

Glycolic acid   (continued)

Metal ion	Equilibrium	Log K 25°, 0.1	Log K 25°, 1.0	Log K 25°, 0	ΔH 25°, 0	ΔS 25°, 0
Th^{4+}	$ML/M.L$		4.50 ± 0.06		0.5^c	20^c
	$ML_2/M.L^2$		7.40 ± 0.05		-0.2^c	33^c
	$ML_3/M.L^3$		10.1 ± 0.1		-0.7^c	44^c
	$ML_4/M.L^4$		12.0 ± 0.0		-0.9^c	52^c
	$ML_5/M.L^5$		13.4		-0.6^c	59^c
UO_2^{2+}	$ML/M.L$		2.40 ± 0.05		1.3^c	15^c
	$ML_2/M.L^2$		3.99 ± 0.02		3.1^c	29^c
	$ML_3/M.L^3$		5.21 ± 0.03		2.9^c	34^c
Cu^{2+}	$ML/M.L$		2.31 ± 0.05	2.90 ± 0.02	-0.4^c	9^c
			2.40^d	2.50^e		
	$ML_2/M.L^2$		3.72 ± 0.02	4.66^e	-0.7^c	15^c
				4.20^e		
	$ML_3/M.L^3$			4.27^e		
Hg^+	$ML/M.L$			3.01^e		
	$ML_2/M.L^2$			5.71^e		
Hg^{2+}	$ML/M.L$			3.60^e		
	$ML_2/M.L^2$			7.05^e		

$^c 25°, 1.0$; $^d 25°, 2.0$; $^e 25°, 3.0$

Bibliography:

H^+ 76BBP,77CM,77RW,79ZKV Hg^+ 77RWa

$La^{3+}-Yb^{3+}$ 77CM Hg^{2+} 77RW

Th^{4+} 78BR   Other references: 57CR,68T,69Ma,70GF,71AM,
                                                           71Ma,74Wa,75CS,75CSa,75JB,76AMa,76SC,
UO_2^{2+} 76BBP   77SS,77BMO,77RB,78PSa,79Mb

Cu^{2+} 74Ad

$$\begin{array}{c} OH \\ | \\ CH_3CHCO_2H \end{array}$$

$C_3H_6O_3$	D-2-Hydroxypropanoic acid (lactic acid)		HL
	(Other values on Vol.3, p.28)		

Metal ion	Equilibrium	Log K 25°, 0.1	Log K 25°, 1.0	Log K 25°, 0	ΔH 25°, 0	ΔS 25°, 0
H^+	$HL/H.L$	3.66 ± 0.03	3.64 ± 0.01	3.860 ± 0.002	0.08 ± 0.02	17.9
		3.61^b	$3.81^d \pm 0.01$		-1.7^d	
Co^{2+}	$ML/M.L$		1.38 ± 0.03	1.90	0.0^c	6^c
			1.39^d			
	$ML_2/M.L^2$		$(2.37)^w \pm 0.07$		0.1^c	11^c
			2.36^d			
	$ML_3/M.L^3$		$(2.7)^w \pm 0.2$			
			2.7^k			

$^b 25°, 0.5$; $^c 25°, 1.0$; $^d 25°, 2.0$; $^k 20°, 2.0$; ^{w}optical isomerism not stated.

Lactic acid   (continued)

Metal ion	Equilibrium	Log K 25°, 0.1	Log K 25°, 1.0	Log K 25°, 0	ΔH 25°, 0	ΔS 25°, 0
Ni^{2+}	$ML/M.L$		1.64 ± 0.05	2.22 ± 0.0	-0.2^c	7^c
	$ML_2/M.L^2$		1.57^d $(2.76)^w \pm 0.09$		-0.3^c	12^c
	$ML_3/M.L^3$		$(2.94)^{d,w}$ $(3.1)^w \pm 0.1$			
Cu^{2+}	$ML/M.L$	2.54	2.45 ± 0.05 $2.43^d \pm 0.01$	3.02	0.2^c -0.6^d	12^c 9^d
	$ML_2/M.L^2$		$(4.08)^w \pm 0.1$ $(4.08)^{d,w} \pm 0.04$	$(4.84)^w$	0.1^c $(0.4)^d$	19^c 20^d
	$ML_3/M.L^3$		$(4.7)^w \pm 0.4$ $(4.5)^{d,w}$		-0.1^c	21^c
Pb^{2+}	$ML/M.L$		$1.99 -0.01$ $2.16^d -0.01$	2.78^e 2.26	-1.8^d	4^d
	$ML_2/M.L^2$		$(2.88)^w \pm 0.10$ $(3.19)^{d,w} \pm 0.04$	$(3.30)^{e,w}$	2.6^d	23^d
	$ML_3/M.L^3$		$(4.0)^{d,w} \pm 0.3$			

---

^{c}25°, 1.0; ^{d}25°, 2.0; ^{e}25°, 3.0; ^{w}optical isomerism not stated.

Bibliography:

H^+,Pb^{2+}	78FB		Cu^{2+}	75Aa,78FB
Co^{2+}	75Aa		Other references:	68T,70GN,70LN,71ZP,
Ni^{2+}	73HT,75Aa			72SN,74DN,75CS,75JB,75KKa,76AMa,76SC,
				78NP,78PSa,79Mb

OH
|
$CH_3CH_2CHCO_2H$

$C_4H_8O_3$ DL-2-Hydroxybutanoic acid   HL
                              (Other values on Vol.3, p.31)

Metal ion	Equilibrium	Log K 25°, 0.1	Log K 25°, 2.0	Log K 25°, 3.0
H^+	$HL/H.L$	3.68 ± 0.01	3.83 ± 0.02	
Co^{2+}	$ML/M.L$		1.46 ± 0.03	
	$ML_2/M.L^2$		$(2.38)^w$	
	$ML_3/M.L^3$		$(3.04)^w$	
Ni^{2+}	$ML/M.L$		1.72 ± 0.00	
	$ML_2/M.L^2$		$(2.90)^w \pm 0.01$	
	$ML_3/M.L^3$		$(3.63)^w$	
Cd^{2+}	$ML/M.L$		1.26 ± 0.03	
	$ML_2/M.L^2$		$(2.11)^w \pm 0.04$	
	$ML_3/M.L^3$		$(3.4)^w \pm 0.4$	
	$ML_4/M.L^4$		$(4.2)^w$	

---

^{w}Optical isomerism not stated.

DL-2-Hydroxybutanoic acid   (continued)

Metal ion	Equilibrium	Log K 25°, 0.1	Log K 25°, 2.0	Log K 25°, 3.0
Pb^{2+}	$ML/M.L$		2.12 ± 0.04	2.04
	$ML_2/M.L^2$		$(3.0)^W \pm 0.3$	$(2.88)^W$
	$ML_3/M.L^3$		$(4.0)^W$	

---

^{W}Optical isomerism not stated.

Bibliography: 78MMG   Other reference: 76SC

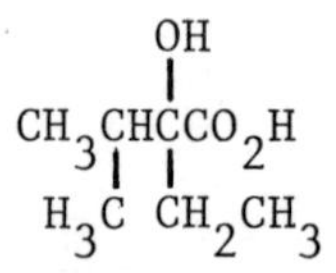

$$\underset{\underset{H_3C \quad CH_2CH_3}{|}}{\overset{\overset{OH}{|}}{CH_3CHCCO_2H}}$$

$C_7H_{14}O_3$	DL-2-Hydroxy-2-ethyl-3-methylbutanoic acid	HL
	(Other reference on Vol.3, p.45)	

Metal ion	Equilibrium	Log K 25°, 0.1
H^+	$HL/H.L$	3.59
Y^{3+}	$ML/M.L$	2.78
	$ML_2/M.L^2$	$(4.52)^W$
La^{3+}	$ML/M.L$	1.75
	$ML_2/M.L^2$	$(2.97)^W$
Ce^{3+}	$ML/M.L$	1.91
	$ML_2/M.L^2$	$(3.02)^W$
Pr^{3+}	$ML/M.L$	1.97
	$ML_2/M.L^2$	$(3.17)^W$
Nd^{3+}	$ML/M.L$	2.08
	$ML_2/M.L^2$	$(3.54)^W$
Sm^{3+}	$ML/M.L$	2.44
	$ML_2/M.L^2$	$(4.17)^W$
Eu^{3+}	$ML/M.L$	2.62
	$ML_2/M.L^2$	$(4.27)^W$
Gd^{3+}	$ML/M.L$	2.68
	$ML_2/M.L^2$	$(4.38)^W$
Tb^{3+}	$ML/M.L$	2.81
	$ML_2/M.L^2$	$(4.60)^W$
Dy^{3+}	$ML/M.L$	2.86
	$ML_2/M.L^2$	$(4.68)^W$
Ho^{3+}	$ML/M.L$	2.87
	$ML_2/M.L^2$	$(4.73)^W$
Er^{3+}	$ML/M.L$	2.95
	$ML_2/M.L^2$	$(4.81)^W$
Tm^{3+}	$ML/M.L$	2.99
	$ML_2/M.L^2$	$(4.88)^W$
Yb^{3+}	$ML/M.L$	3.02
	$ML_2/M.L^2$	$(4.92)^W$

---

^{W}Optical isomerism not stated.

DL-2-Hydroxy-2-ethyl-3-methylbutanoic acid   (continued)

Metal ion	Equilibrium	Log K 25°, 0.1
Lu^{3+}	$ML/M.L$ $ML_2/M.L^2$	3.03 $(4.86)^w$

---

^{w}Optical isomerism not stated.

Bibliography: 76SPa

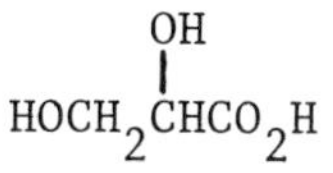

$$
\begin{array}{c}
OH \\
|
\end{array}
$$
$$HOCH_2CHCO_2H$$

$C_3H_6O_4$	DL-2,3-Dihydroxypropanoic acid (glyceric acid) (Other values on Vol.3, p.53)		HL

Metal ion	Equilibrium	Log K 25°, 2.0
H^+	$HL/H.L$	3.54
Zn^{2+}	$ML/M.L$ $ML_2/M.L^2$ $ML_3/M.L^3$	1.46 $(2.36)^w$ $(2.7)^w$
Cd^{2+}	$ML/M.L$ $ML_2/M.L^2$ $ML_3/M.L^3$	1.25 $(2.17)^w$ $(2.7)^w$
Pb^{2+}	$ML/M.L$ $ML_2/M.L^2$ $ML_3/M.L^3$	2.10 $(3.26)^w$ $(3.5)^w$

---

^{w}Optical isomerism not stated.

Bibliography: Other reference: 71AP

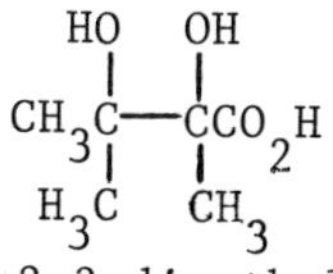

$$
\begin{array}{cc}
HO & OH \\
| & | \\
CH_3C\!-\!CCO_2H \\
| & | \\
H_3C & CH_3
\end{array}
$$

$C_6H_{12}O_4$	DL-2,3-Dihydroxy-2,3-dimethylbutanoic acid	HL

Metal ion	Equilibrium	Log K 25°, 0.1
H^+	$HL/H.L$	3.39
Y^{3+}	$ML/M.L$ $ML_2/M.L^2$ $ML_3/M.L^3$	3.18 $(5.47)^w$ $(8.2)^w$
La^{3+}	$ML/M.L$ $ML_2/M.L^2$ $ML_3/M.L^3$	2.81 $(4.66)^w$ $(5.1)^w$
Ce^{3+}	$ML/M.L$ $ML_2/M.L^2$ $ML_3/M.L^3$	3.08 $(5.02)^w$ $(6.2)^w$

---

^{w}Optical isomerism not stated.

DL-2,3-Dihydroxy-2,3-dimethylbutanoic acid   (continued)

Metal ion	Equilibrium	Log K 25°, 0.1
Pr^{3+}	$ML/M.L$	3.24
	$ML_2/M.L^2$	$(5.34)^W$
	$ML_3/M.L^3$	$(6.5)^W$
Nd^{3+}	$ML/M.L$	3.37
	$ML_2/M.L^2$	$(5.57)^W$
	$ML_3/M.L^3$	$(6.9)^W$
Sm^{3+}	$ML/M.L$	3.52
	$ML_2/M.L^2$	$(5.95)^W$
	$ML_3/M.L^3$	$(7.3)^W$
Eu^{3+}	$ML/M.L$	3.49
	$ML_2/M.L^2$	$(5.95)^W$
	$ML_3/M.L^3$	$(7.7)^W$
Gd^{3+}	$ML/M.L$	3.42
	$ML_2/M.L^2$	$(5.96)^W$
	$ML_3/M.L^3$	$(7.7)^W$
Tb^{3+}	$ML/M.L$	3.31
	$ML_2/M.L^2$	$(5.88)^W$
	$ML_3/M.L^3$	$(7.4)^W$
Dy^{3+}	$ML/M.L$	3.30
	$ML_2/M.L^2$	$(5.89)^W$
	$ML_3/M.L^3$	$(7.5)^W$
Ho^{3+}	$ML/M.L$	3.33
	$ML_2/M.L^2$	(5.83)
	$ML_3/M.L^3$	(7.6)
Er^{3+}	$ML/M.L$	3.38
	$ML_2/M.L^2$	$(5.82)^W$
	$ML_3/M.L^3$	(7.6)
Tm^{3+}	$ML/M.L$	3.44
	$ML_2/M.L^2$	$(5.75)^W$
	$ML_3/M.L^3$	$(7.6)^W$
Yb^{3+}	$ML/M.L$	3.50
	$ML_2/M.L^2$	$(5.91)^W$
	$ML_3/M.L^3$	$(7.1)^W$
Lu^{3+}	$ML/M.L$	3.58
	$ML_2/M.L^2$	$(6.05)^W$
	$ML_3/M.L^3$	$(7.4)^W$

[W] Optical isomerism not stated.

Bibliography: 79PPB

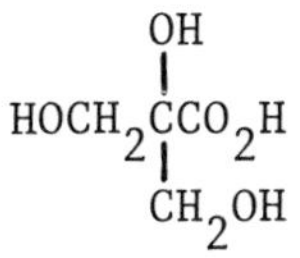

$$OH$$
$$|$$
$$HOCH_2CCO_2H$$
$$|$$
$$CH_2OH$$

| C$_4$H$_8$O$_5$ | | 2,3-Dihydroxy-2-(hydroxymethyl)propanoic acid | | HL |
| | | (Other values on Vol.3, p.57) | | |

Metal ion	Equilibrium	Log K 25°, 0.1	Log K 25°, 0.5
H$^+$	HL/H.L	3.39	3.29
Y^{3+}	ML/M.L	2.94	2.65
	ML$_2$/M.L^2	5.32	4.67
	ML$_3$/M.L^3	7.1	(5.3)
La^{3+}	ML/M.L	2.69	2.40
	ML$_2$/M.L^2	4.69	(3.88)
	ML$_3$/M.L^3	6.3	(4.9)
Pr^{3+}	ML/M.L	2.97	2.75
	ML$_2$/M.L^2	5.20	4.69
	ML$_3$/M.L^3	7.0	6.2
Nd^{3+}	ML/M.L	3.01	(2.81)
	ML$_2$/M.L^2	5.35	(4.62)
	ML$_3$/M.L^3	(7.0)	6.4
Sm^{3+}	ML/M.L	3.09	2.86
	ML$_2$/M.L^2	5.45	(5.07)
	ML$_3$/M.L^3	7.5	6.5
Eu^{3+}	ML/M.L	3.11	2.80
	ML$_2$/M.L^2	5.54	5.00
	ML$_3$/M.L^3	7.4	6.5
Gd^{3+}	ML/M.L	(3.09)	2.69
	ML$_2$/M.L^2	5.51	4.99
	ML$_3$/M.L^3	7.4	6.4
Tb^{3+}	ML/M.L	3.02	2.71
	ML$_2$/M.L^2	5.45	4.88
	ML$_3$/M.L^3	7.4	(6.6)
Dy^{3+}	ML/M.L	3.03	(2.66)
	ML$_2$/M.L^2	5.48	4.87
	ML$_3$/M.L^3	7.5	6.4
Ho^{3+}	ML/M.L	3.04	(2.71)
	ML$_2$/M.L^2	5.49	4.89
	ML$_3$/M.L^3	7.5	(6.2)
Er^{3+}	ML/M.L	3.05	2.79
	ML$_2$/M.L^2	5.53	(4.83)
	ML$_3$/M.L^3	(7.3)	6.6
Tm^{3+}	ML/M.L	3.08	2.85
	ML$_2$/M.L^2	5.60	4.97
	ML$_3$/M.L^3	7.6	6.5
Yb^{3+}	ML/M.L	3.13	2.90
	ML$_2$/M.L^2	5.69	5.07
	ML$_3$/M.L^3	7.7	6.5
Lu^{3+}	ML/M.L	3.16	2.94
	ML$_2$/M.L^2	5.74	5.19
	ML$_3$/M.L^3	7.8	6.9

Bibliography: 76PK

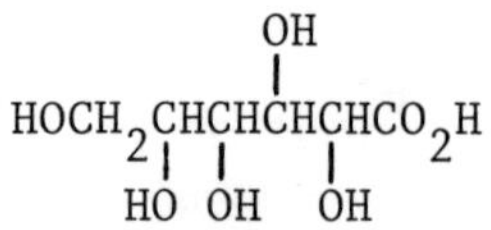

$$\underset{\underset{\text{HO}\quad\text{OH}\quad\text{OH}}{|\quad\;\;|\quad\;\;|}}{\overset{\overset{\text{OH}}{|}}{\text{HOCH}_2\text{CHCHCHCHCO}_2\text{H}}}$$

$C_6H_{12}O_7$	D-2,3,4,5,6-Pentahydroxyhexanoic acid (D-gluconic acid)		HL
	(Other values on Vol.3, p.59)		

Metal ion	Equilibrium	Log K 25°, 0.1	Log K 25°, 1.0
H^+	HL/H.L		3.48
Pb^{2+}	ML/M.L	2.6	2.13
	$ML_2/M.L^2$		3.35

Bibliography: 78CV

Other references: 67TK,70CMa,76SGa,76PP.
    76PPb,76RP,77PP,77RWb,78PB,78PBa,79BR,
    79JA,79PPb,79PPc

$$\text{HOCH}_2\text{CH}_2\text{CO}_2\text{H}$$

$C_3H_6O_3$	3-Hydroxypropanoic acid		HL
	(Other values on Vol.3, p.61)		

Metal ion	Equilibrium	Log K 30°, 0.1	Log K 25°, 1.0	ΔH 25°, 1.0	ΔS 25°, 1.0
H^+	HL/H.L	4.33 ±0.08	4.32		
			4.56[d]	-0.3[d]	20[d]
Co^{2+}	ML/M.L		(0.3)	1.8	7
			0.49[d]		
Ni^{2+}	ML/M.L		(0.5)	1.2	7
			0.78[d]		
	$ML_2/M.L^2$		(1.0)	2.6	13
			1.32[d]		
Cu^{2+}	ML/M.L	2.05	(1.5)	0.9	10
			1.76[d]		
	$ML_2/M.L^2$		(2.9)	1.6	18
			3.21[d]		
Pb^{2+}	ML/M.L		1.95		
			2.13[d]±0.02	0.1[d]	10[d]
	$ML_2/M.L^2$		2.94		
			3.14[d]±0.03	0.6[d]	16[d]
	$ML_3/M.L^3$		3.59[d]±0.03		

[d] 25°, 2.0

Bibliography:
H^+,Pb^{2+} 78FB
$Co^{2+}-Cu^{2+}$ 75Aa

$$\overset{\text{OH}}{\underset{|}{CH_3CHCH_2CO_2H}}$$

$C_4H_8O_3$		DL-3-Hydroxybutanoic acid		HL
		(Other values on Vol.3, p.62)		

Metal ion	Equilibrium	Log K 25°, 0.2	Log K 20°, 1.0	Log K 25°, 3.0
H^+	HL/H.L	4.28 ±0.02	4.35[d] ±0.00 4.55[d] ±0.02	4.76
Co^{2+}	ML/M.L		0.83[d] ±0.08	
	$ML_2/M.L^2$		$(1.12)^{d,w}$ ±0.04	
	$ML_3/M.L^3$		$(1.26)^{d,w}$	
Ni^{2+}	ML/M.L		1.00[d] ±0.00	
	$ML_2/M.L^2$		$(1.35)^{d,w}$ ±0.01	
	$ML_3/M.L^3$		$(1.71)^{d,w}$	
Hg^+	ML/M.L			3.3
Cd^{2+}	ML/M.L		1.20[d] ±0.09	
	$ML_2/M.L^2$		$(2.1)^{d,w}$ ±0.1	
	$ML_3/M.L^3$		$(2.5)^{d,w}$	
Hg^{2+}	ML/M.L			4.26
	$ML_2/M.L^2$			$(8.36)^w$
Pb^{2+}	ML/M.L		2.13[d] ±0.05	
	$ML_2/M.L^2$		$(3.2)^{d,w}$ ±0.2	
	$ML_3/M.L^3$		$(3.7)^{d,w}$ +0.1	

[d] 25°, 2.0; [w] Optical isomerism not stated.

Bibliography:

H^+ 77RW,78MMG	Hg^+ 77RWa
Co^{2+},$Ni^{2+}$$Cd^{2+}$,$Pb^{2+}$ 78MMG	Hg^{2+} 77RW

$$HOCH_2CH_2CH_2CO_2H$$

$C_4H_8O_3$		4-Hydroxybutanoic acid		HL
		(Other values on Vol.3, p.64)		

Metal ion	Equilibrium	Log K 20°, 1.0	Log K 25°, 2.0	Log K 25°, 3.0
H^+	HL/H.L	4.57 +0.01	4.85	4.99
Co^{2+}	ML/M.L		0.47 ±0.02	
Ni^{2+}	ML/M.L		0.58 ±0.06	
Cd^{2+}	ML/M.L		1.39 ±0.06	
	$ML_2/M.L^2$		2.1 ±0.1	
	$ML_3/M.L^3$		2.7	
Hg^{2+}	ML/M.L			4.34
	$ML_2/M.L^2$			8.45
Pb^{2+}	ML/M.L		2.18 ±0.10	
	$ML_2/M.L^2$		3.3 ±0.2	
	$ML_3/M.L^3$		3.7 ±0.1	

Bibliography: H^+,Hg   77RW; Co^{2+},Cd^{2+},Pb^{2+} 78MMG

$$\overset{\overset{\text{O}}{\|}}{\text{CHCO}_2\text{H}}$$

$C_2H_2O_3$		Oxoacetic acid (glyoxylic acid)				HL
		(Other values on Vol.3, p.65)				

Metal ion	Equilibrium	Log K 25°, 0.1	Log K 25°, 1.0	Log K 25°, 0	ΔH 25°, 0	ΔS 25°, 0
H^+	HL/H.L	3.05 +0.01 2.83[b]	2.91	3.46	−0.53 −0.64[a]	14.1 11.8[a]
Cu^{2+}	ML/M.L	2.15[P]				

[a]25°, 0.05; [b]25°, 0.5; [P]30°, 0.1

Bibliography: 78MAa   Other reference: 76CJG

$$\overset{\overset{\text{O}}{\|}}{\text{CH}_3\text{CCO}_2\text{H}}$$

$C_3H_4O_3$		2-Oxopropanoic acid (pyruvic acid)				HL
		(Other values on Vol.3, p.66)				

Metal ion	Equilibrium	Log K 25°, 0.1	Log K 25°, 1.0	Log K 25°, 0	ΔH 25°, 0	ΔS 25°, 0
H^+	HL/H.L	2.26 2.20[b]	(2.45)[d]	2.55 ±0.06 2.40[e]	−2.90 −3.01[t]	1.9
Ca^{2+}	ML/M.L	0.8 0.56[u]		1.08 0.59[e]		
Fe^{2+}	ML/M.L			0.69[e]		
Cu^{2+}	ML/M.L ML$_2$/M.L^2	1.4[b] 2.6[b]	(1.6) (4.0)	(2.2) (4.9)	−0.1[c] −1.2[c]	7[c] 14[c]
Cd^{2+}	ML/M.L			0.98[e]		

[b]25°, 0.5; [c]25°, 1.0; [d]25°, 2.0; [e]25°, 3.0; [t]25°, 0.05; [u]25°, 0.3

Bibliography:
$H^+,Ca^{2+},Fe^{2+},Cd^{2+}$ 78FG   Other reference: 68RS
Cu^{2+} 74Ad,77RL

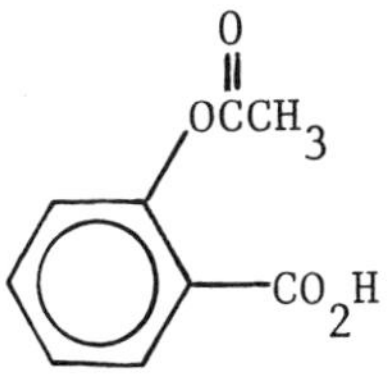

$$\begin{array}{c} O \\ \parallel \\ OCCH_3 \end{array}$$

CO_2H

| $C_9H_8O_4$ | 2-Acetoxybenzoic acid | (acetylsalicylic acid) | HL |
| | (Other reference on Vol.3, p.360) | | |

Metal ion	Equilibrium	Log K 37°, 0.15
H^+	HL/H.L	3.44
Mg^{2+}	ML/M.L	(2.29)
Ca^{2+}	ML/M.L	(2.95)
Cu^{2+}	ML/M.L	(1.26)
	$ML_2/M.L^2$	(3.0)

Bibliography: 78AK

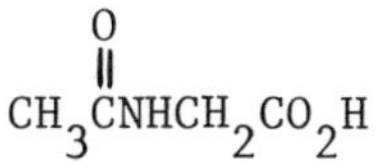

$$\begin{array}{c} O \\ \parallel \\ CH_3CNHCH_2CO_2H \end{array}$$

| $C_4H_7O_3N$ | N-Acetylglycine | HL |
| | (Other values on Vol.3, p.68) | |

Metal ion	Equilibrium	Log K 25°, 0.1	Log K 25°, 1.0	Log K 25°, 0	ΔH 25°, 0	ΔS 25°, 0
H^+	HL/H.L	3.47[b] 3.44[b]	3.46	3.670	0.18 ±0.03	17.4
$(CH_3)_3Pb^+$	ML/M.L	0.82[u]				

[b]25°, 0.5; [u]25°, 0.3

Bibliography: 77SB   Other reference: 77BBG

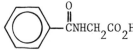

$$\begin{array}{c} O \\ \parallel \\ CNHCH_2CO_2H \end{array}$$

| $C_9H_9O_3N$ | N-Benzoylglycine (hippuric acid) | HL |
| | (Other values on Vol.3, p.69) | |

Metal ion	Equilibrium	Log K 25°, 0.15
H^+	HL/H.L	3.48
Ni^{2+}	ML/M.L	1.25
Cu^{2+}	ML/M.L	2.03

Bibliography: 76FJ

Other references: 67K,75STa,79BBG

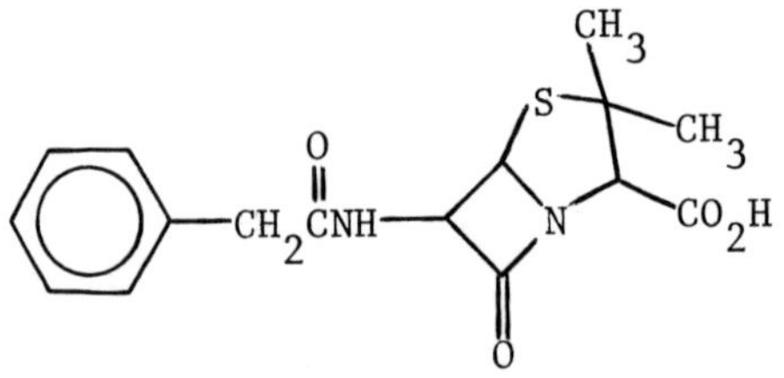

C$_{16}$H$_{18}$O$_4$N$_2$S   <u>DL-Benzylpenicillin</u>   HL

Metal ion	Equilibrium	Log K 25°, 0.15
H$^+$	HL/H.L	2.73
Ni^{2+}	ML/M.L	1.74

Bibliography: 76FJ

CH$_3$OCH$_2$CO$_2$H

C$_3$H$_6$O$_3$ <u>Methoxyacetic acid</u>   HL
                              (Other values on Vol.3, p.69)

Metal ion	Equilibrium	Log K 25°, 0.1	Log K 25°, 1.0	Log K 25°, 0	ΔH 25°, 0	ΔS 25°, 0
H$^+$	HL/H.L	3.32	3.38	3.570 3.74^e	0.95 ±0.01	19.6
Cu^{2+}	ML/M.L		1.83 1.81^d	2.01^e	0.7^c	11^c
	ML$_2$/M.L^2		2.84 3.04^d	3.34^e	2.1^c	20^c
	ML$_3$/M.L^3		3.11^j			
	ML$_4$/M.L^4		2.8^j			
Hg$^+$	ML/M.L			2.98^e		
	ML$_2$/M.L^2			5.41^e		
Hg^{2+}	ML/M.L			3.54^e		
	ML$_2$/M.L^2			6.91^e		

^{c}25°, 1.0; ^{d}25°, 2.0; ^{e}25°, 3.0; ^{j}20°, 1.0

Bibliography:
H$^+$,Hg^{2+} 77RW   Hg$^+$ 77RWa
Cu^{2+} 74Ad

$$CH_3OCH_2CH_2CO_2H$$

$C_4H_8O_3$ 3-Methoxypropanoic acid   HL

Metal ion	Equilibrium	Log K 25°, 3.0
H^+	HL/H.L	4.71
Hg^+	ML/M.L	3.50
	$ML_2/M.L^2$	6.63
Hg^{2+}	ML/M.L	4.24
	$ML_2/M.L^2$	8.45

Bibliography:

H$^+$,Hg^{2+} 77RW   Hg$^+$ 77RWa

$$CH_3CH_2OCH_2CO_2H$$

$C_4H_8O_3$ Ethoxyacetic acid   HL
(Other values on Vol.3, p.71)

Metal ion	Equilibrium	Log K 25°, 1.0	Log K 18°, 0	ΔH 25°, 1.0	ΔS 25°, 1.0
H^+	HL/H.L	3.51	3.65	(-0.9)	(13)
Ni^{2+}	ML/M.L	1.02	1.17[d]	1.0	8
	$ML_2/M.L^2$	1.51	1.91[d]	2.2	14
	$ML_3/M.L^3$	1.23			
Cu^{2+}	ML/M.L	1.80	1.74[d]		
	$ML_2/M.L^2$	2.89	3.23[d]	0.5	10
	$ML_3/M.L^3$	3.21		1.7	19
	$ML_4/M.L^4$	2.8[j]		3	20
Zn^{2+}	ML/M.L	1.13		1.2	9
	$ML_2/M.L^2$	1.87		5	20
	$ML_3/M.L^3$	1.78		(6)	(30)
Cd^{2+}	ML/M.L	1.07		1.1	9
	$ML_2/M.L^2$	1.69		1	10
	$ML_3/M.L^3$	1.54		(10)	(40)

[d] 25°, 2.0; [j] 20°, 1.0

Bibliography: 77A

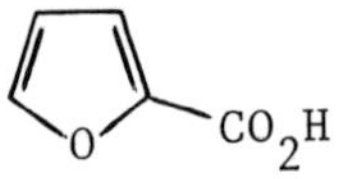

$$\text{(furan ring)}-CO_2H$$

$C_5H_4O_3$	Oxole-2-carboxylic acid (2-furoic acid) (Other values on Vol.3, p.72)					HL
Metal ion	Equilibrium	Log K 25°, 0.1	Log K 25°, 1.0	Log K 25°, 0	ΔH 25°, 0	ΔS 25°, 0
H^+	HL/H.L	2.97 ± 0.02	3.06 3.19^d	3.162 ± 0.007	$(2)^r$ 0.49^d	(20) 15.9^d
La^{3+}	ML/M.L $ML_2/M.L^2$	1.79 2.91	1.48^d		1.5^d	12^d
Pr^{3+}	ML/M.L		1.59^d		1.6^d	13^d
Nd^{3+}	ML/M.L $ML_2/M.L^2$	1.85 3.02	1.60^d		1.7^d	13^d
Sm^{3+}	ML/M.L $ML_2/M.L^2$	1.91 3.12	1.67^d		1.5^d	13^d
Eu^{3+}	ML/M.L		1.67^d		1.5^d	13^d
Gd^{3+}	ML/M.L $ML_2/M.L^2$	1.84 3.03	1.64^d		1.5^d 1.5^d	13^d 13^d
Tb^{3+}	ML/M.L		1.60^d		1.8^d	13^d
Dy^{3+}	ML/M.L		1.51^d		2.1^d	14^d
Ho^{3+}	ML/M.L		1.50^d		2.1^d	14^d
Er^{3+}	ML/M.L $ML_2/M.L^2$	1.74 2.89	1.45^d		2.3^d	14^d
Tm^{3+}	ML/M.L		1.48^d		2.1^d	14^d
Yb^{3+}	ML/M.L		1.47^d		2.1^d	14^d
Lu^{3+}	ML/M.L		1.45^d		2.3^d	14^d

^{d}25°, 2.0; ^{r}25-40°, 0

Bibliography: 76YCB

$$\underset{CH_2CO_2H}{\overset{SH}{|}}$$

$C_2H_4O_2S$	Mercaptoacetic acid (thioglycolic acid) (Other values on Vol.3, p.73)					H_2L
Metal ion	Equilibrium	Log K 25°, 0.1	Log K 25°, 1.0	Log K 25°, 0	ΔH 25°, 0	ΔS 25°, 0
H^+	HL/H.L	10.11 ± 0.1 $9.95^b \pm 0.05$		10.61 ± 0.07	-6.2 ± 0.0 -6.2^b	28 24^b
	$H_2L/HL.H$	3.43 ± 0.04 $3.42^b \pm 0.03$	3.53	3.64 ± 0.04	0.1^b $0.0^c \pm 0.1$	16^b

^{b}25°, 0.5; ^{c}25°, 1.0

Mercaptoacetic acid (continued)

Metal ion	Equilibrium	Log K 25°, 0.1	Log K 25°, 1.0	Log K 25°, 0	ΔH 25°, 0	ΔS 25°, 0
Th^{4+}	MHL/M.HL	10.11 ±0.1	3.22		2.4[c]	23[c]
	$M(HL)_2/M.(HL)^2$		5.69		1.9[c]	18[c]
	$M(HL)_3/M.(HL)^3$		7.20		2.6[c]	16[c]
	$M(HL)_4/M.(HL)^4$		8.54		0.8[c]	9[c]
UO_2^{2+}	MHL/M.HL		1.89		2.1[c]	16[c]
	$M(HL)_2/M.(HL)^2$		3.20		2.5[c]	15[c]
	$M(HL)_3/M.(HL)^3$		4.51		0.0[c]	6[c]
In^{3+}	$ML_3/M.L^3$	30.59[i]				
	$M_2L_3/M^2.L^3$	37.95[i]				
	$M_3L_6/M^3.L^3$	74.10[i]				

---

[b]25°, 0.5;   [c]25°, 1.0;   [i]20°, 0.5

Bibliography:

H^+ 77Ad,78BR,78KSa In^{3+} 78KSa

Th^{4+},UO_2^{2+} 78BR   Other references: 74Wa,76AMa,76L,76SS,77EL

$$HSCH_2CH_2CO_2H$$

3-Mercaptopropanoic acid
(Other values on Vol.3, p.76)

$C_3H_6O_2S$ H_2L

Metal ion	Equilibrium	Log K 25°, 0.1	Log K 25°, 0.5	Log K 25°, 0	ΔH 25°, 0.05	ΔS 25°, 0
H^+	HL/H.L		10.10 +0.01	10.84	-6.10 -6.72[c]	29
	$H_2L/HL.H$	4.16	4.12 ±0.04	4.34	-0.28[c]	20[a*]
Ni^{2+}	$M_5L_{10}/M^5.L^{10}$		64.49			
	$M_6L_{12}/M^6.L^{12}$		78.89			
	$M_6L_9/M^6.L^9$		63.58			
	$M_6L_{11}/M^6.L^{11}$		73.60			
Zn^{2+}	$ML_2/M.L^2$		12.18			
	$M_3L_4/M^3.L^4$		30.40			
	$M_4L_6/M^4.L^6$		45.59			
In^{3+}	$ML_2/M.L^2$		19.92[i]			
	$ML_3/M.L^3$		26.67[i]			
	$ML_4/M.L^4$		30.53[i]			
	$M_2L_2/M^2.L^2$		25.77[i]			
	$M_3L_4/M^3.L^4$		48.61[i]			
	$M_3L_3/M_3OHL_3.H$		4.6[i]			

---

[a]25°, 0.1;   [c]25°, 1.0;   [i]20°, 0.5;   *assuming ΔH for 1.0=ΔH for 0.1

Bibliography:

H^+ 77Ad,78KSa In^{3+} 78KSa

Ni^{2+},Zn^{2+} 76DCG   Other references: 73RS,76SS

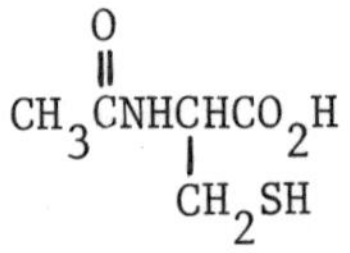

$$\text{CH}_3\overset{\overset{\displaystyle O}{\|}}{\text{C}}\text{NHCHCO}_2\text{H}$$
$$\underset{\text{CH}_2\text{SH}}{|}$$

$C_5H_9O_3NS$ L-N-Acetylcysteine H_2L
(Other references on Vol.3, p.337)

Metal ion	Equilibrium	Log K 20°, 0.1
H^+	HL/H.L	9.71
	$H_2L/HL.H$	3.17
In^{3+}	ML/M.L	10.24

Bibliography: 78KST

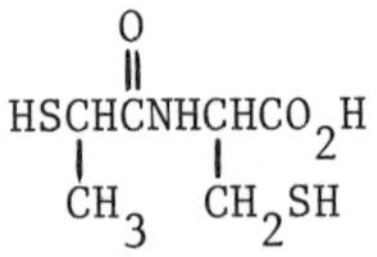

$$\text{HSCHCNHCHCO}_2\text{H}$$
$$\underset{\text{CH}_3}{|}\quad\underset{\text{CH}_2\text{SH}}{|}$$

$C_6H_{11}O_3NS_2$ DL-2-Mercaptopropionyl-L-cysteine H_3L

Metal ion	Equilibrium	Log K 20°, 0.1
H^+	HL/H.L	10.17
	$H_2L/HL.H$	8.46
	$H_3L/H_2L.H$	3.26
In^{3+}	$ML/M.L$	16.45
	$ML_2/M.L^2$	$(29.26)^w$
	MHL/ML.H	2.96
	$MHL_2/ML_2.H$	$(4.52)^w$

[w]Optical activity not stated.
Bibliography: H^+ 76SH; In^{3+} 78KST;   Other reference: 77SH

$$\overset{\text{HS}}{\underset{}{}}\ \overset{\overset{\displaystyle O}{\|}}{}$$
$$\text{HSCH}_2\overset{|}{\text{C}}\text{HCNHCH}_2\text{CO}_2\text{H}$$

$C_5H_9O_3NS_2$ DL-2,3-Dimercaptopropionylglycine H_3L

Metal ion	Equilibrium	Log K 20°, 0.1
H^+	HL/H.L	10.66
	$H_2L/HL.H$	7.63
	$H_3L/H_2L.H$	3.63
Ga^{3+}	$ML/M.L$	17.24
	$ML_2/M.L^2$	$(31.91)^w$
	ML/MOHL.H	3.14
In^{3+}	$ML/M.L$	17.25
	$ML_2/M.L^2$	$(31.46)^w$
	MHL/ML.H	2.44
	$MHL_2/ML_2.H$	$(4.08)^w$

[w]Optical activity not stated.
Bibliograpy: H^+ 77SH; Ga^{3+},In^{3+} 78KST

$$CH_3CH_2SCH_2CO_2H$$

$C_4H_8O_2S$		(Other values on Vol.3, p.77)	Ethylthioacetic acid			HL

Metal ion	Equilibrium	Log K 25°, 0.1	Log K 25°, 1.0	Log K 31°, 2.0	ΔH 25°, 1.0	ΔS 25°, 1.0
H^+	HL/H.L	3.60	3.65	(3.62)	−0.26	15.9
Ni^{2+}	$ML/M.L$		1.04		0.3	6
	$ML_2/M.L^2$		1.81		0.6	10
	$ML_3/M.L^3$		2.28		0	10
Cu^{2+}	$ML/M.L$		2.57		0.6	14
	$ML_2/M.L^2$		4.77		0.6	24
	$ML_3/M.L^3$		4.83		−1	20
Zn^{2+}	$ML/M.L$		0.74		1.2	7
	$ML_2/M.L^2$		1.20		1.2	10
	$ML_3/M.L^3$		1.14		(11)	(40)
Cd^{2+}	$ML/M.L$		1.27		0.7	8
	$ML_2/M.L^2$		2.12		1.0	13
	$ML_3/M.L^3$		2.50		1	15
	$ML_4/M.L^4$		2.7			

Bibliography:

H^+ 77Ad

Ni^{2+}-Cd^{2+} 77A

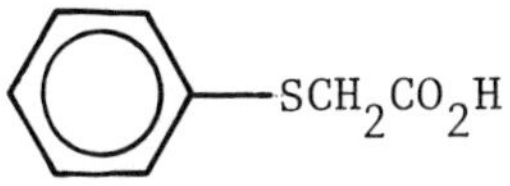

$$-SCH_2CO_2H$$

$C_8H_8O_2S$		(Phenylthio)acetic acid (Other values on Vol.3, p.80)		HL

Metal ion	Equilibrium	Log K 25°, 0.1	Log K 25°, 1.0	ΔH 25°, 0.1	ΔS 25°, 0.1
H^+	HL/H.L	3.33 ±0.05	(3.29)	0.80	18.2
Ni^{2+}	ML/M.L	0.7 −0.4	(0.5)	0.7[c]	5[c]
Cu^{2+}	ML/M.L	1.43 +0.1	(1.3)	1.6[c]	11[c]
Cd^{2+}	ML/M.L	1.2 −0.5	(1.0)	1.0[c]	8[c]

[c] 25°, 1.0

Bibliography: 78Ac

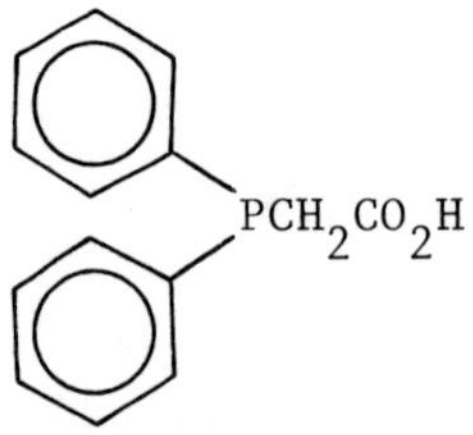

$C_{14}H_{13}O_2P$ Diphenylphosphinoacetic acid   HL
                                  (Carboxymethyldiphenylphosphine)

Metal ion	Equilibrium	Log K 25°, 0.1
Cu^+	$ML/M.L$	11.3
	$ML_2/M.L^2$	16.6
	$ML_3/M.L^3$	19.5
	$ML_4/M.L^4$	21.6
Hg^{2+}	$ML/M.L$	18.7
	$ML_2/M.L^2$	28.9
	$ML_3/M.L^3$	35.4
	$ML_4/M.L^4$	37.2

Bibliography:

Cu^+ 79PP

Hg^{2+} 79PPa

$$HO_2CCO_2H$$

$C_2H_2O_4$	Ethanedioic acid (oxalic acid) (Other values on Vol.3, p.92)					H_2L

Metal ion	Equilibrium	Log K 25°, 0.1	Log K 25°, 1.0	Log K 25°, 0	ΔH 25°, 0	ΔS 25°, 0
H^+	HL/H.L	3.81 ±0.04	3.55 ±0.02	4.266 ±0.001	1.60 ±0.06	24.9
		3.55[b] ±0.06		3.80[e]	0.96[c]	19.5[c]
	H_2L/HL.H	1.04 ±0.10	1.04 ±0.04	1.252	0.9 ±0.1	9
		1.02[b] ±0.03		1.26[e]	0.2[c]	5[c]
Be^{2+}	ML/M.L	4.08[h] 3.52[b]	3.55			
	ML_2/M.L^2	5.38[h] 5.57[b]	5.40			
	M^3.L/$M_3(OH)_3$L.H^3	3.84[b]				
	$M^3.L^3$/$M_3(OH)_3L_3$.H^3	0.59[b]				
Co^{2+}	ML/M.L	3.85 ±0.03	3.29 ±0.04	4.72 ±0.08	0.6	24
	ML_2/M.L^2		5.9 ±0.3	6.9 ±0.2		
	MHL/M.HL	1.61				
	$M(HL)_2$/M.$(HL)^2$	2.89				
Ni^{2+}	ML/M.L		3.7	5.16	0.2	24
	ML_2/M.L^2		6.6			
Cu^{2+}	ML/M.L		5.53 ±1	(6.23)	-0.1	(28)
	ML_2/M.L^2		9.54 ±0.5	(10.27)		
Cr^{3+}	ML_2/MOHL$_2$.H(cis)	7.34[u]				
	MOHL$_2$/M(OH)$_2L_2$.H(cis)	9.55[u]				
Fe^{3+}	ML/M.L		7.58 ±0.02		1.3[c]	39[c]
	ML_2/M.L^2		13.81		0.7[c]	70[c]
	ML_3/M.L^3		18.6		0.1[c]	90[c]
Cd^{2+}	ML/M.L		2.73 ±0.03	3.89[o]		
	ML_2/M.L^2		4.1 +0.1			
	ML_3/M.L^3		5.1 ±0.0			
Pb^{2+}	ML/M.L		4.16 -0.6			
	ML_2/M.L^2		6.33 -0.6			
	MHL/M.HL		1.43			

[b] 25°, 0.5;　[c] 25°, 1.0;　[e] 25°, 3.0;　[h] 20°, 0.1;　[o] 18°, 0;　[u] 4°, 0.1

Bibliography:

H^+　76VSK,77D,77HO,77BP,77DB

Co^{2+}　71MSa

Ni^{2+}　76MKS

Cu^{2+}　62MD,77BP

Cr^{3+}　57H

Fe^{3+}　70GMa,77D

Cd^{2+}　62MD,68VB,76JB,77BD
Pb^{2+}　77HO

Other references:　61CM,63PK,64NN,66S,67AB,
67JK,76KC,76OM,68BM,68VB,69MBJ,70AL,
70MK,70WC,73Ab,74Ad,74NF,74NFa,74SGP,
75BDR,75STb,76BR,76BRa,76BV,76GMb,
76MKD,76MSa,77BL,77CD,77FK,77JKS,77RB,
77N,77VPS,78JB,78KN,78MB,78MPa,78NP,
78PSa,78SK,78TSG,79FS,79KCa,79NMa,
79PZ,79TG

$$HO_2CCH_2CO_2H$$

$C_3H_4O_4$ <u>Propanedioic acid</u> <u>(malonic acid)</u> H_2L
(Other values on Vol.3, p.95)

Metal ion	Equilibrium	Log K 25°, 0.1	Log K 25°, 1.0	Log K 25°, 0	ΔH 25°, 0	ΔS 25°, 0
H^+	HL/H.L	5.28[b]±0.05	5.07[d]±0.05	5.696 ±0.000	1.15 −0.01	30.0[c]
		5.08[b]±0.05	5.14[d]±0.03	5.26[e]	0.48[c]−0.01	24.8[c]
	H_2L/HL.H	2.65[b]±0.05	2.60[d]±0.01	2.847 ±0.00	−0.04 ±0.02	12.9[c]
		2.57[b]±0.05	2.68[d]+0.01	2.81[e] +0.01	−0.36[c]±0.01	10.7[c]
Be^{2+}	ML/M.L	5.30[h]				
		5.10[b]				
	ML_2/M.L^2	8.56[h]				
		8.85[b]				
	$M_3(OH)_3L_3$/$M_3(OH)_3.L^3$					
		0.89[b]				
Ca^{2+}	ML/M.L	1.47 ±0.05		2.35		
	MHL/M.HL	0.47[x]				
Th^{4+}	ML/M.L	7.47			2.8[c]	44[c]
	ML_2/M.L^2	12.79			4.8[c]	75[c]
	ML_3/M.L^3	16.3			6[c]	90[c]
UO_2^{2+}	ML/M.L	(5.43)[b]	(5.42)+0.2		2.1[c]	32[c]
	ML_2/M.L^2	(9.31)[b]	(9.48)+0.2		2.8[c]	21[c]
	ML/MOHL.H	5.02[b]				
Mn^{2+}	ML/M.L	2.5 ±0.2		3.28 ±0.01	3.7	27
Fe^{2+}	ML/M.L		2.17			
	ML_2/M.L^2		3.21			
Co^{2+}	ML/M.L	2.93 ±0.06		3.74 ±0.03	2.9	27
	ML_2/M.L^2	4.4		5.1 ±0.0		
	MHL/M.HL	0.82				
Cu^{2+}	ML/M.L	5.03 ±0.07	4.63	5.70 ±0.1	2.9	36
					1.4[a]	
	ML_2/M.L^2	7.7 ±0.1	7.5 ±0.1	8.1		
	MHL/M.HL	2.15				
Fe^{3+}	ML/M.L	7.57[b]	7.50		2.5[c]	43[c]
	ML_2/M.L^2		13.04		3.4[c]	71[c]
	ML_3/M.L^3		16.6		2[c]	80[c]
Zn^{2+}	ML/M.L	2.91 ±0.06	2.47	3.84 ±0.02	3.1 −0.1	28
	ML_2/M.L^2	4.4	3.8	5.4		
	MHL/M.HL	0.99				
Pb^{2+}	ML/M.L	3.1	2.79			
	ML_2/M.L^2		4.20			
	ML_3/M.L^3		4.16			
	MHL/M.HL		1.13			
	$M(HL)_2$/M.$(HL)^2$		1.40			
	MHL_2/ML_2.H		4.08			
	MHL_3/ML_3.H		4.87			

---

[a] 25°, 0.1;   [b] 25°, 0.5;   [c] 25°, 1.0;   [d] 25°, 2.0;   [e] 25°, 3.0;   [h] 20°, 0.1;   [x] temperature not stated, 0.2

Malonic acid   (continued)

Bibliography:

H^+ 70GS,73OD,76DGN,77BN,77DBa,77HOa,
            79MBb,79PM,79ZKV

Be^{2+} 77DBa

Ca^{2+} 76KM

Th^{4+}, UO_2^{2+} 77BNC

Mn^{2+} 77SF

Fe^{2+}, Fe^{3+} 77D

Co^{2+}, Zn^{2+} 75DOb

Cu^{2+} 70GS,73OD,78AC

Pb^{2+} 77HOa

Other references: 68RS,72HA,74MJ,76KD,
           76MSa,76SJ,76SJa,76SS,77AM,77SJ,
           77SMa,78JB,78SGd,79ED,79KCa,79KN,
           79SJ,79SJa,79SJc

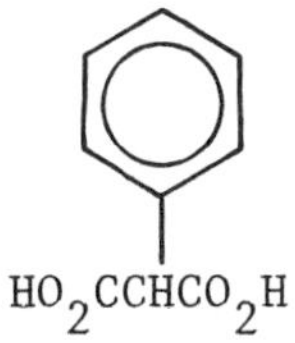

$$HO_2CCHCO_2H$$

$C_9H_8O_4$ Phenylpropanedioic acid (phenylmalonic acid) H_2L
(Other values on Vol.3, p.102)

Metal ion	Equilibrium	Log K 25°, 0.1
H^+	HL/H.L	5.10 ±0.02
	H_2L/HL.H	2.34 ±0.06
Cu^{2+}	ML/M.L	4.54 ±0.03
	ML_2/M.L^2	7.28 ±0.05
	MHL/M.HL	1.8

Bibliography: 73OD   Other references: 74MJ,76DGN,79PZa

$$H_2CCHCH_3$$
$$|$$
$$HO_2CCHCO_2H$$

$C_6H_{10}O_4$ (1-Methylethyl)propanedioic acid (isopropylmalonic acid) H_2L
(Other values on Vol.3, p.100)

Metal ion	Equilibrium	Log K 25°, 0.1
H^+	HL/H.L	5.50 ±0.02
	H_2L/HL.H	2.78 ±0.03
Cu^{2+}	ML/M.L	5.40 ±0.03
	ML_2/M.L^2	9.02 ±0.07
	MHL/M.HL	1.6

Bibliography: 73OD

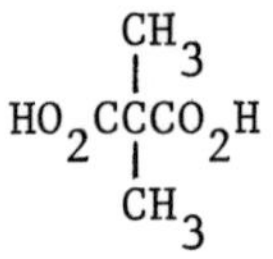

$$\begin{array}{c} CH_3 \\ | \\ HO_2CCCO_2H \\ | \\ CH_3 \end{array}$$

$C_5H_8O_4$	Dimethylpropanedioic acid (dimethylmalonic acid)			H_2L
	(Other values on Vol.3, p.102)			

Metal ion	Equilibrium	Log K 25°, 0.1	Log K 25°, 0.5	Log K 25°, 0
H^+	HL/H.L	5.68	5.42 ±0.04	6.06
	H_2L/HL.H	3.01	2.87 ±0.03	3.17
Fe^{3+}	ML/M.L		7.16	

Bibliography: 77CC   Other references: 74MJ,79ASJ

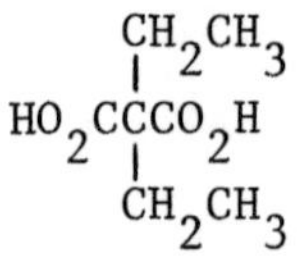

$$\begin{array}{c} CH_2CH_3 \\ | \\ HO_2CCCO_2H \\ | \\ CH_2CH_3 \end{array}$$

$C_7H_{12}O_4$	Diethylpropanedioic acid (diethylmalonic acid)					H_2L
	(Other values on Vol.3, p.103)					

Metal ion	Equilibrium	Log K 25°, 0.1	Log K 25°, 0.5	Log K 25°, 0	ΔH 25°, 0	ΔS 25°, 0
H^+	HL/H.L	6.96 ±0.02	6.71 ±0.09	7.417	0.83 ±0.01	36.8
	H_2L/HL.H	2.00 ±0.04	1.90 ±0.05	2.151	0.71	12.2
Fe^{3+}	ML/M.L		7.86			

Bibliography: 77CC

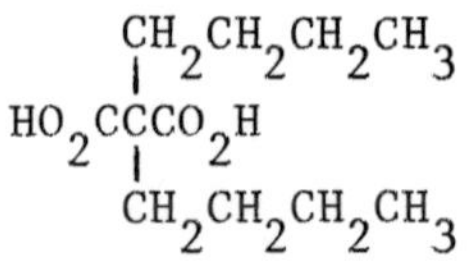

$$\begin{array}{c} CH_2CH_2CH_2CH_3 \\ | \\ HO_2CCCO_2H \\ | \\ CH_2CH_2CH_2CH_3 \end{array}$$

$C_{11}H_{20}O_4$	Dibutylpropanedioic acid (dibutylmalonic acid)	H_2L
	(Other values on Vol.3, p.105)	

Metal ion	Equilibrium	Log K 25°, 0.1
H^+	HL/H.L	7.22 ±0.03
	H_2L/HL.L	1.95 ±0.06
Cu^{2+}	ML/M.L	5.07 ±0.03
	$ML_2/M.L^2$	8.14 ±0.01

Bibliography: 73OD

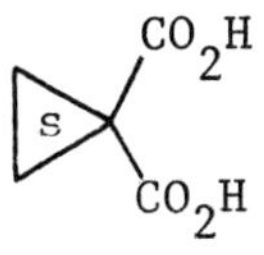

$C_5H_6O_4$		Cyclopropane–1,1-dicarboxylic acid (Other values on Vol.3, p.106)				H_2L
Metal ion	Equilibrium	Log K 25°, 0.1	Log K 25°, 0.5		ΔH 25°, 0.1	ΔS 25°, 0.1
H^+	HL/H.L	7.19 ±0.03	6.90		−0.39	31.6
	H_2L/HL.H	1.64 ±0.04	1.49		0.31	8.5
Co^{2+}	ML/M.L	3.50				
Ni^{2+}	ML/M.L	3.82 ±0.07				
Cu^{2+}	ML/M.L	5.99			0.9	30
	ML_2/M.L^2	9.61			1.0	47
Fe^{3+}	ML/M.L		7.74			
Zn^{2+}	ML/M.L	3.50				

Bibliography:

H^+ 72RV,77CC Cu^{2+} 72RV,77AC

Co^{2+}-Ni^{2+},Zn^{2+} 72RV Fe^{3+} 77CC

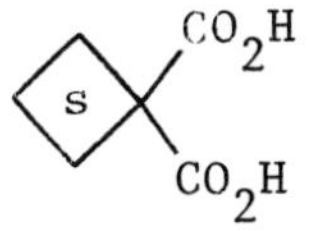

$C_6H_8O_4$		Cyclobutane–1,1-dicarboxylic acid (Other values on Vol.3, p.106)				H_2L
Metal ion	Equilibrium	Log K 25°, 0.1	Log K 25°, 0.5		ΔH 25°, 0.1	ΔS 25°, 0.1
H^+	HL/H.L	5.45 ±0.00	5.22		0.81	27.7
	H_2L/HL.H	2.94 ±0.02	2.86		0.26	14.3
Cu^{2+}	ML/M.L	5.01 ±0.02			2.6	32
	ML_2/M.L^2	8.45 ±0.04			3.3	50
	MHL/M.HL	1.37				

Bibliography: 77AC

Metal		Log K	Log K		ΔH	ΔS
ion	Equilibrium	25°, 0.1	25°, 0.5		25°, 0.1	25°, 0.1
H^+	HL/H.L	5.79 ±0.02	5.49		0.90	26.2
	H_2L/HL.H	3.07 ±0.02	2.96		0.06	16.9
Co^{2+}	ML/M.L	1.92				
Ni^{2+}	ML/M.L	2.08				
Cu^{2+}	ML/M.L	4.89			3.3	33
	ML_2/M.L^2	7.69			4.9	52
	MHL/M.HL	1.30				
Fe^{3+}	ML/M.L		7.62			
Zn^{2+}	ML/M.L	2.38				

$C_7H_{10}O_4$ — Cyclopentane-1,1-dicarboxylic acid (Other values on Vol.3, p.108) — H_2L

Bibliography:

H^+	72RV,77CC		Cu^{2+}	72RV,77AC
Co^{2+},Ni^{2+},Zn^{2+}	72RV		Fe^{3+}	77CC

$C_8H_{12}O_4$ — Cyclohexane-1,1-dicarboxylic acid — H_2L

Metal		Log K		ΔH	ΔS
ion	Equilibrium	25°, 0.1		25°, 0.1	25°, 0.1
H^+	HL/H.L	5.72 ±0.02			
	H_2L/HL.H	3.31 ±0.02			
Co^{2+}	ML/M.L	1.96			
Ni^{2+}	ML/M.L	1.99			
Cu^{2+}	ML/M.L	4.62		3.4	33
	ML_2/M.L	7.24		5.1	50
	MHL/M.HL	1.57			
Zn^{2+}	ML/M.L	2.26			

Bibliography:

H^+	72RV,76MMR		Cu^{2+}	72RV,77AC
Ni^{2+},Zn^{2+}	72RV			

$$HO_2CCH_2CH_2CO_2H$$

$C_4H_6O_4$		Butanedioic acid (succinic acid)				H_2L
		(Other values on Vol.3, p.108)				

Metal ion	Equilibrium	Log K 25°, 0.1	Log K 25°, 1.0	Log K 25°, 0	ΔH 25°, 0	ΔS 25°, 0
H^+	HL/H.L	5.24 ±0.04	5.12 ±0.02	5.636	0.04 ±0.08	26.0
		5.12[b]±0.02	5.21[d]	5.49[e]		
	$H_2L/HL.H$	4.00 ±0.02	3.96 ±0.03	4.207	-0.68 ±0.08	17.0
		3.92[b]±0.03	4.07[d]	4.32[e]		
Be^{2+}	$ML/M.L$	2.74[b]	3.13			
	$ML_2/M.L^2$	4.36[b]				
	ML/MOHL.H		5.59			
	MHL/ML.H		3.41			
	$M(HL)_2/ML_2.H^2$	9.05[b]				
	$M_3(OH)_3HL/M_3(OH)_3.HL^3$	2.00[b]				
	$M_3(OH)_3L_3/M_3(OH)_3.L^3$	5.07[b]				
Ca^{2+}	$ML/M.L$	1.16 ±0.05		2.00		
	MHL/M.HL	0.54[h]				
Cu^{2+}	$ML/M.L$	2.6 ±0.0		3.28 ±0.06	4.6 2.7[a]	30
Cd^{2+}	$ML/M.L$	2.1	1.67[j] 1.47[u]	2.72 ±0.05	4.4	26
	$ML_2/M.L^2$		2.79[j] 2.29[u]			
	$ML_3/M.L^3$		2.74[u]			
	MHL/M.HL		0.99[j]			
Pb^{2+}	$ML/M.L$	2.8	2.68	2.96[e]		
	$ML_2/M.L^2$		3.99	4.44[e]		
	$ML_3/M.L^3$		3.89	4.53[e]		
	MHL/M.HL		1.86	2.06[e]		
	$M(HL)_2/M.(HL)^2$		2.76	3.18[e]		
	$M(HL)L/ML_2.H$		4.85	5.10[e]		
	$M(HL)L_2/ML_3.H$		5.31	5.42[e]		
	$M(HL)_2L/M(HL)L_2.H$			5.16[e]		

[a]25°, 0.1;   [b]25°, 0.5;   [d]25°, 2.0;   [e]25°, 3.0;   [h]20°, 0.1;   [j]20°, 1.0;   [u]27°, 2.1

Bibliography:

H^+ 75DOc 77DBa,77HOa,77RWb   Other references: 54CV,68RS,72TN,73H,73SD,

Be^{2+} 77DBa   76KD,76SJ,77BC,77DS,77SJ,78JB,78KCT,

Ca^{2+} 76KM   78SGe,79Sd,79SJa,79SJc

Cu^{2+} 78AC

Cd^{2+} 72KGa

Pb^{2+} 77HOa

$$CHCO_2H \\ \| \\ CHCO_2H$$

$C_4H_4O_4$	cis-Butenedioic acid (maleic acid) (Other values on Vol.3, p.112)				H_2L
Metal ion	Equilibrium	Log K 25°, 0.1	Log K 25°, 1.0	Log K 25°, 0	ΔH 25°, 0	ΔS 25°, 0

Metal ion	Equilibrium	Log K 25°, 0.1	Log K 25°, 1.0	Log K 25°, 0	ΔH 25°, 0	ΔS 25°, 0
H^+	HL/H.L	5.82 ±0.03	5.62 ±0.02	6.332	0.8 / 0.2[c]	32 / 26[c]
	H_2L/HL.H	1.75 ±0.05 / 1.64[b]	1.63 ±0.03 / 1.71[b]	1.910 / 1.82[e]	-0.1 / 0.1[c]	8 / 8[c]
Ca^{2+}	ML/M.L	1.48 ±0.06		2.43		
Cu^{2+}	ML/M.L / ML_2/M.L^2	3.40 / 5.48			3.4[a]	27[a]
Cd^{2+}	ML/M.L / ML_2/M.L^2 / ML_3/M.L^3	2.3 ±0.1 / 3.6 / 3.8	1.74[u] / 2.66[u]			

[a] 25°, 0.1; [b] 25°, 0.5; [c] 25°, 1.0; [d] 25°, 2.0; [e] 25°, 3.0; [u] 27°, 2.1

Bibliography:

H^+ 76BMb Cd^{2+} 73KG

Ca^{2+} 76KM   Other references: 68RS,76KD,76SS,77JB,
                                                                                     78MAc,79ASJ,79JBa
Cu^{2+} 76BMb,78AC

$$HO_2CCH_2CH_2CH_2CO_2H$$

$C_5H_8O_4$	Pentanedioic acid (glutaric acid) (Other values on Vol.3, p.117)				H_2L

Metal ion	Equilibrium	Log K 25°, 0.1	Log K 25°, 1.0	Log K 25°, 0	ΔH 25°, 0	ΔS 25°, 0
H^+	HL/H.L	5.01 ±0.02 / 4.89[b]±0.00	4.91 ±0.03 / 4.99[d]	5.42 ±0.01 / 5.16[e]	0.6	27
	H_2L/HL.H	4.13 ±0.02 / 4.11[b]±0.02	4.15 ±0.04 / 4.25[d]	4.33 ±0.01 / 4.43[e]	0.1	20
NpO_2^+	ML/M.L / MHL/M.HL	1.43 / 0.9				
Ni^{2+}	ML/M.L	1.6	(1.3)		1.2	10
Cu^{2+}	ML/M.L	2.4	(2.1)	3.16	2.6	19
Zn^{2+}	ML/M.L	1.6	1.25	2.45	3.2	17
Cd^{2+}	ML/M.L	2.0	(1.7) / 1.60[t]		2.6	17
Pb^{2+}	ML/M.L / ML_2/M.L^2	2.80[b]	2.51 / 3.77			
	MHL/M.HL / M(HL)$_2$/M.(HL)^2		2.00 / 3.05			
	MHL_2/ML_2.H		4.95			

[b] 25°, 0.5; [d] 25°, 2.0; [e] 25°, 3.0; [t] 30°, 2.0

Bibliography. H⁺ 77HOa; Ni²⁺-Cd²⁺ 77Ac; NpO_2^{2+} 69EW; Pb^{2+} 72Nc,77HOa
Other references: 76KD,77BC,79SJa,79SJc

$$HO_2CCH_2CH_2CH_2CH_2CO_2H$$

| $C_6H_{10}O_4$ | | Hexanedioic acid (adipic acid) (Other values on Vol.3, p.118) | | | | H_2L |

Metal ion	Equilibrium	Log K 25°, 0.1	Log K 25°, 1.0	Log K 25°, 0	ΔH 25°, 0	ΔS 25°, 0
H^+	HL/H.L	5.03 ±0.03	4.94 ±0.03	5.42 ±0.01	0.6	27
		4.93[b]±0.00	5.05[d]	5.36[e]		
	H_2L/HL.H	4.26 ±0.03	4.26 ±0.05	4.42 ±0.01	0.3	21
		4.22[b]±0.04	4.34[d]	4.76[e]		
Cd^{2+}	ML/M.L	2.1		2.00[e]		
	MHL/M.HL			1.37[e]		
	$M(HL)_2$/M.$(HL)^2$			3.15[e]		
Pb^{2+}	ML/M.L	2.8	2.47			
	ML_2/M.L^2		3.77			
	MHL/M.HL		2.05			
	$M(HL)_2$/M.$(HL)^2$		3.17			
	MHL_2/ML_2.H		(4.99)			

[b]25°, 0.5;   [d]25°, 2.0;   [e]25°, 3.0

Bibliography:

		Other references: 68RS,75JBa,76KD,76SJ,
H^+	77HOa,79N,79PM	77DS,77SJ,79GBS,79JB,79JBa,79SJa,
Cd^{2+}	79N	79SJc
Pb^{2+}	77HOa	

| $C_8H_6O_4$ | | Benzene-1,2-dicarboxylic acid (phthalic acid) (Other values on Vol.3, p.120) | | | | H_2L |

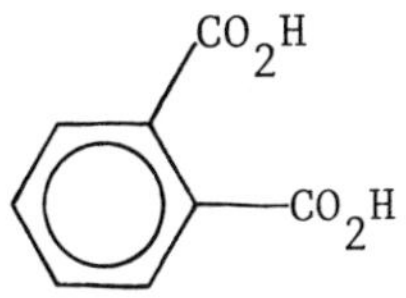

Metal ion	Equilibrium	Log K 25°, 0.1	Log K 25°, 1.0	Log K 25°, 0	ΔH 25°, 0	ΔS 25°, 0
H^+	HL/H.L	4.92 ±0.01	4.73 ±0.05	5.408	0.52 ±0.02	26.5
		4.73[b]±0.01	4.73[d]	4.87[e] ±0.03		
	H_2L/HL.H	2.75 ±0.02	2.66 ±0.03	2.950	0.64 +0.01	15.7
		2.66[b]+0.01	2.80[d]	2.99[e] ±0.07		
Cu^{2+}	ML/M.L	3.22 ±0.03	2.69	4.04 ±0.03	2.0	25
		2.81[b]	2.64[d]		2.5[a]	
	ML_2/M.L^2	5.46	3.73[d] 4.14[d]	(5.3)		
	MHL/M.HL	1.33				
Cd^{2+}	ML/M.L	2.5	1.86 ±0.0			
	ML_2/M.L^2		2.88			
	MHL/M.HL		0.48			
	MHL_2/ML_2.H		3.60			

[a]25°, 0.1;   [b]25°, 0.5;   [d]25°, 2.0;   [e]25°, 3.0

316   XI. CARBOXYLIC ACIDS

Phthalic acid (continued)

Metal ion	Equilibrium	Log K 25°, 0.1	Log K 25°, 1.0	Log K 25°, 0	ΔH 25°, 0	ΔS 25°, 0
Pb^{2+}	$ML/M.L$		2.78			
	$ML_2/M.L^2$		4.01			
	$MHL/M.HL$		1.16			
	$MHL_2/ML_2.H$		3.77			

Bibliography:

H^+ 75DOc,76BMb,78OS Pb^{2+} 78OS

Cu^{2+} 75DOc,76BMb,78AC   Other references: 75PJa,76PA,76PJ,76SJ,
                                                77DS,77JKS,77Pa,77SJ.77SMa,79JB,
Cd^{2+} 77BD,78OS   79PZ

HO₂C—[benzene ring]—CO₂H

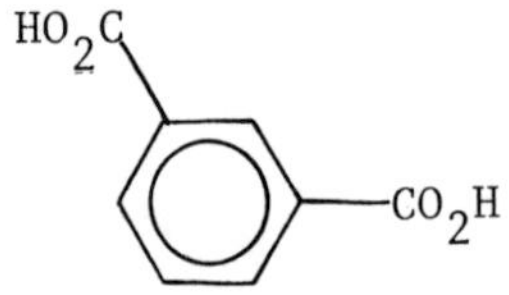

$C_8H_6O_4$	Benzene-1,3-dicarboxylic acid (isophthalic acid)				H_2L
	(Other values on Vol.3, p.122)				

Metal ion	Equilibrium	Log K 25°, 1.0	Log K 25°, 0	ΔH 25°, 0	ΔS 25°, 0
H^+	$HL/H.L$	4.16	4.50	0.4	22
	$H_2L/HL.H$	3.41	3.50	0.0	16
Cd^{2+}	$ML/M.L$	1.33			
	$ML_2/M.L^2$	2.17			
	$MHL/M.HL$	0.82			
Pb^{2+}	$ML/M.L$	2.17			
	$ML_2/M.L^2$	3.36			
	$MHL/M.HL$	1.78			
	$MHL_2/ML_2.H$	3.87			

Bibliography: 78OS

$$\overset{OH}{\underset{|}{HO_2CCH_2CHCO_2H}}$$

$C_4H_6O_5$ L–Hydroxybutanedioic acid (malic acid) H_2L
(Other values on Vol.3, p.124)

Metal ion	Equilibrium	Log K 25°, 0.1	Log K 25°, 1.0	Log K 25°, 0	ΔH 25°, 0	ΔS 25°, 0
H^+	HL/H.L	4.68 ±0.04	4.48 ±0.05	5.097	0.28	24.3
		4.50^b				
	$H_2L/HL.H$	3.20 ±0.08	3.16 ±0.05	3.459	-0.71	13.5
		3.12^b				
Ca^{2+}	ML/M.L	2.02 ±0.06		2.66		
	MHL/M.HL	1.06^h				
Y^{3+}	ML/M.L	4.91				
	$ML_2/M.L^2$	$(8.18)^w$				
La^{3+}	ML/M.L	4.37 +0.01				
	$ML_2/M.L^2$	$(7.16)^w$				
Pr^{3+}	ML/M.L	4.65				
	$ML_2/M.L^2$	$(7.74)^w$				
Nd^{3+}	ML/M.L	4.77				
	$ML_2/M.L^2$	$(7.94)^w$				
Sm^{3+}	ML/M.L	4.89				
	$ML_2/M.L^2$	$(8.16)^w$				
Eu^{3+}	ML/M.L	4.85				
	$ML_2/M.L^2$	$(8.11)^w$				
Gd^{3+}	ML/M.L	4.76				
	$ML_2/M.L^2$	$(8.00)^w$				
Tb^{3+}	ML/M.L	4.77				
	$ML_2/M.L^2$	$(8.03)^w$				
Dy^{3+}	ML/M.L	4.78				
	$ML_2/M.L^2$	$(8.10)^w$				
Ho^{3+}	ML/M.L	4.90				
	$ML_2/M.L^2$	$(8.25)^w$				
Er^{3+}	ML/M.L	4.96				
	$ML_2/M.L^2$	$(8.35)^w$				
Tm^{3+}	ML/M.L	5.00				
	$ML_2/M.L^2$	$(8.50)^w$				
Yb^{3+}	ML/M.L	5.05 -0.1				
	$ML_3/M.L^3$	$(8.58)^w$				
Lu^{3+}	ML/M.L	5.08				
	$ML_2/M.L^2$	$(8.67)^w$				
Cu^{2+}	ML/M.L	3.60				
	MHL/M.HL	2.09				
	$M_2(H_{-1}L)L.H/M^2.L^2$	$(5.06)^x$				
	$M_2(H_{-1}L)L/M_2(H_{-1}L)_2.H$	$(4.24)^x$				
	$M_2L_2/M^2.L^2$		$(8.0)^x$			
	$M_2L_2/M_2(H_{-1}L)_2.H^2$		$(7.8)^x$			

^{b}25°, 0.5; ^{h}20°, 0.1; ^{w}optical isomerism not stated; ^{x}DL–mixture

Malic acid (continued)

Bibliography:

H^+ 75OD,77KP,79PM,79ZKV   Other references: 54CV,69JP,69PV,71PV,

Ca^{2+} 76KM   72SMa,72T,72Tb,73ZGK,75JB,76AMa,76HB,

                                                                   76MP,76MPa,77JB,77MPa,78JS,78KCT,78KN,

Cu^{2+} 75OD   79ASJ,79Mb,79CB

$$\begin{array}{c} \text{OH} \quad \text{O} \\ | \quad\quad \| \\ \text{HO}_2\text{CCCH}_2\text{CCO}_2\text{H} \\ | \\ \text{CH}_3 \end{array}$$

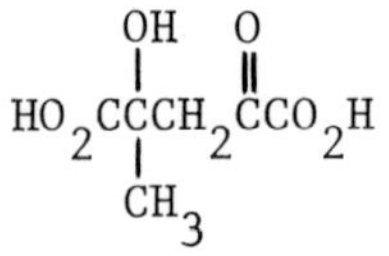

$C_6H_8O_6$	DL-2-Hydroxy-2-methyl-4-oxobutanedioic acid (parapyruvic acid)		H_2L

Metal ion	Equilibrium	Log K 25°, 0.5
H^+	HL/H.L	3.54
	H_2L/HL.H	1.63
Cu^{2+}	ML/M.L	2.76
	M^2.L/$M_2(H_{-1}L)$.H	0.9
	M^2.L^2/$M_2(H_{-1}L)_2$.H^2	$(1.4)^w$

---

[w] Optical isomerism not stated.

Bibliography: 77RL

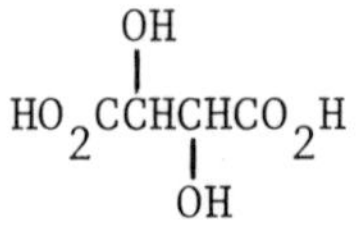

$$\begin{array}{c} \text{OH} \\ | \\ \text{HO}_2\text{CCHCHCO}_2\text{H} \\ | \\ \text{OH} \end{array}$$

$C_4H_6O_6$	D-2,3-Dihydroxybutanedioic acid (D-tartaric acid) (Other values on Vol.3, p.127)		H_2L

Metal ion	Equilibrium	Log K 25°, 0.1	Log K 25°, 1.0	Log K 25°, 0	ΔH 25°, 0	ΔS 25°, 0
H^+	HL/H.L	3.96 ±0.03	3.69 ±0.02	4.366	-0.20	19.3
		3.67^b	3.81^d	3.93^e	-0.36^a	
	H_2L/H.L	2.82 ±0.04	2.74 ±0.05	3.036	-0.74	11.4
		$(2.62)^b$	2.83^d	2.98^e	-0.79^a	
Ca^{2+}	ML/M.L	1.94 ±0.06		2.80		
Co^{2+}	ML/M.L	2.19	1.88	3.05 ±0.03		
	ML_2/M.L^2			$(4.0)^w$ ±0.2		
	MHL/M.HL		0.98			
Ni^{2+}	ML/M.L		2.20			
	MHL/M.HL		1.01			

---

[a] 25°, 0.1; [b] 25°, 0.5; [d] 25°, 2.0; [e] 25°, 3.0; [w] optical isomerism not stated.

D-Tartaric acid   (continued)

Metal ion	Equilibrium	Log K 25°, 0.1	Log K 25°, 1.0	Log K 25°, 0	ΔH 25°, 0	ΔS 25°, 0
Cu^{2+}	ML/M.L	3.47 [b] / 3.39 [b]	2.90 ±0.3			
	$M_2L_2/M^2.L^2$		8.24 ±0.0			
	$M_2L_2/M_2(H_{-1}L)L.H$		4.42			
	$M_2(H_{-1}L)L/M_2(H_{-1}L)_2.H$		4.14			
VO^{2+}	ML/M.L	3.9				
	$ML/M(H_{-1}L).H$	5.2				
	$M_2(H_{-1}L)L.H/M^2.L^2$	9.78				
	$M_2(H_{-1}L)L/M_2(H_{-1}L)_2.H$	3.80				
	$M_2(H_{-1}L)_2/M_2(H_{-2}L)(H_{-1}L).H$	(6.9)				
	$M_2(H_{-2}L)(H_{-1}L)/M_2(H_{-2}L)_2.H$	(5.4)				
Zn^{2+}	ML/M.L	2.65 ±0.04 / 2.2 [b]		3.82		
Cd^{2+}	ML/M.L		1.70			
	MHL/M.HL		0.96			

[b] 25°, 0.5

Bibliography:

H^+	76CG,78PS,78RM,79PM,79ZKV	Zn^{2+}	78RM
Ca^{2+}	76KM		
Co^{2+},Ni^{2+},Cd^{2+}	76CG		
Cu^{2+}	69SL,78RM		
VO^{2+}	78PS		

Other references: 54CV,66S,68PP,69DM,69JP,
69KPa,69MBJ,71BB,71BV,71FL,71LFG,
73ZGa,74VP,75RM,76NCW,77BBA,77RB,
78KCT,78KKA,78KN,78PSa,78SK,79SDD

HO$_2$CCHCHCO$_2$H
            | |
           HO OH

$C_4H_6O_6$	meso-2,3-Dihydroxybutanedioic acid (meso-tartaric acid) H_2L
	(Other values on Vol.3, p.129)

Metal ion	Equilibrium	Log K 25°, 0.1	Log K 25°, 1.0	Log K 25°, 0	ΔH 25°, 0	ΔS 25°, 0
H^+	HL/H.L	4.44 ±0.2	4.10	4.91	-1.5 / -1.4 [a]	17
	$H_2L/HL.H$	2.99 ±0.2	2.86	3.17	-0.8 / -1.0 [a]	12
VO^{2+}	ML/M.L	4.42				
	$M_2(H_{-1}L)L.H/M^2.L^2$	7.75				
	$M_2(H_{-1}L)L/M_2(H_{-1}L)_2.H$	3.29				
	$M_2(H_{-1}L)_2/M_2(H_{-2}L)(H_{-1}L).H$	4.87				

[a] 25°, 0.1

Bibliography: 78PS   Other references: 54CV,78KKa

$$\overset{\displaystyle O}{\overset{\displaystyle \|}{HO_2CCH_2CCO_2H}}$$

$C_4H_4O_5$	Oxobutanedioic acid (oxaloacetic acid) (Other values on Vol.3, p.131)					H_2L

Metal ion	Equilibrium	Log K 25°, 0.1	Log K 25°, 0.2	Log K 25°, 0	ΔH 25°, 0	ΔS 25°, 0
H^+	HL/H.L	3.89 ±0.00	2.78	4.37	(0)[r]	(20)
	H_2L/HL.H	2.27 ±0.05	2.25	2.56	(-4)[r]	(0)
Cu^{2+}	ML/M.L	4.16 4.10[n]	3.9	4.9		
	$M_2(H_{-1}L).H/M^2.L$	2.55				
	$M_2(H_{-2}L).H^2/M^2.L^2$	1.43				
Zn^{2+}	ML/M.L	2.41		3.2		
	M_2L/ML.M			2.3		
	$M_2(H_{-1}L).H/M^2.L$	-1.13				

[r] 25-37°, 0;   [n] 37°, 0.1

Bibliography: 76RL

$$\overset{\displaystyle O\ \ O}{\overset{\displaystyle \|\ \ \|}{HO_2CCH_2CH_2NHC-CNHCH_2CH_2CO_2H}}$$

$C_8H_{12}O_6N_2$	Oxamide-N,N'-di-3-propanoic acid	H_2L

Metal ion	Equilibrium	Log K 20°, 0.1
H^+	HL/H.L	4.57
	H_2L/HL.H	3.81
Cu^{2+}	$M(H_{-1}L).H/M.L$	0.9
	$M_2(H_{-2}L).H/M(H_{-1}L).M$	1.8

Bibliography: 75VH

$$HO_2CCH_2OCH_2CO_2H$$

$C_4H_6O_5$		Oxydiacetic acid (diglycolic acid)				H_2L
		(Other values on Vol.3, p.133)				

Metal ion	Equilibrium	Log K 25°, 0.1	Log K 25°, 1.0	Log K 25°, 0	ΔH 25°, 1.0	ΔS 25°, 1.0
H^+	HL/H.L	3.92 ±0.03	3.76 ±0.00	4.37	0.7 ±0.1	19
		3.76[b]	3.96[d]			
	H_2L/HL.H	2.80 ±0.03	2.75 ±0.06	2.97	-0.6 ±0.2	11
		2.80[b]	2.91[d]			
Ni^{2+}	ML/M.L	3.80 ±0.02	(2.3)		2.8	20
	$ML_2/M.L^2$		2.25[d]			
			3.45[d]			
Cu^{2+}	ML/M.L	3.95 ±0.02	(3.6)		1.9	23
	MHL/M.HL	1.42				
Zn^{2+}	ML/M.L	3.61 ±0.04	(3.3)		1.9	22
	MHL/M.HL	(2.08)				
Cd^{2+}	ML/M.L	3.21 ±0.1	(3.0)		-0.1	14
	MHL/M.HL	1.02				
Pb^{2+}	ML/M.L	4.41 +0.1				
		4.19[b]				
	MHL/M.HL	1.92[b]				
		2.43[b]				

[b] 25°, 0.5; [d] 25°, 2.0

Bibliography:

H^+ 76NC,77Ad,78AM,79HN Pb^{2+} 76NC

Ni^{2+}-Cd^{2+} 78A   Other references: 68RS,76SS,78NS,79SDK

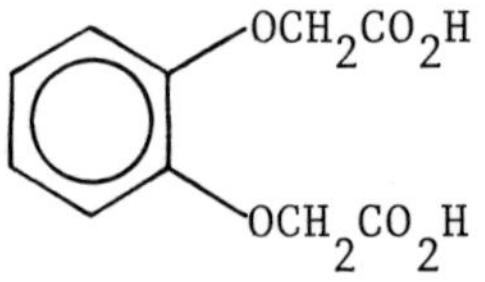

$C_{10}H_{10}O_6$		Benzene-1,2-dioxydiacetic acid		H_2L
Metal ion	Equilibrium	Log K 25°, 0.1	ΔH 25°, 0.1	ΔS 25°, 0.1
H^+	HL/H.L	3.42 ±0.03	1.95	22.2
	H_2L/HL.H	2.6 ±0.2	1.05	15.4
Ca^{2+}	ML/M.L	3.1		
Sr^{2+}	ML/M.L	2.3		
Ba^{2+}	ML/M.L	2.0		
Y^{3+}	ML/M.L	3.78	5.8	37
La^{3+}	ML/M.L	3.81	3.9	31
	ML_2/M.L^2	7.30	6.3	55
Ce^{3+}	ML/M.L	4.16		
	ML_2/M.L^2	7.58		
Pr^{3+}	ML/M.L	4.36	3.4	31
	ML_2/M.L^2	7.69	4.9	52
Nd^{3+}	ML/M.L	4.45	4.1	34
	ML_2/M.L^2	7.65		
Sm^{3+}	ML/M.L	4.57	2.7	30
	ML_2/M.L^2	7.48	3.6	46
Eu^{3+}	ML/M.L	4.60	3.5	33
	ML_2/M.L^2	7.45	3.9	47
Gd^{3+}	ML/M.L	4.48	3.3	32
	ML_2/M.L^2	7.07		
Tb^{3+}	ML/M.L	4.30	4.7	35
	ML_2/M.L^2	6.65		
Dy^{3+}	ML/M.L	4.26	4.8	36
Ho^{3+}	ML/M.L	4.11	5.7	38
Er^{3+}	ML/M.L	3.92	6.4	39
Tm^{3+}	ML/M.L	3.94	8.4	46
Yb^{3+}	ML/M.L	4.02		
Lu^{3+}	ML/M.L	4.04	7.7	44
Mn^{2+}	ML/M.L	(2.8)		
Co^{2+}	ML/M.L	1.1		
Ni^{2+}	ML/M.L	1.6		
Cu^{2+}	ML/M.L	3.3		
Zn^{2+}	ML/M.L	2.0		
Cd^{2+}	ML/M.L	3.8		

Bibliography:

H^+ 68SH,77HC; Ca^{2+}-Ba^{2+},Mn^{2+}-Cd^{2+} 68SH; Y^{3+}-Lu^{3+} 77HC

$$HO_2CCH_2SCH_2CO_2H$$

$C_4H_6O_4S$		Thiodiacetic acid (Other values on Vol.3, p.142)				H_2L
Metal ion	Equilibrium	Log K 25°, 0.1	Log K 25°, 1.0	Log K 25°, 0	ΔH 25°, 1.0	ΔS 25°, 1.0
H^+	HL/H.L	4.13 ±0.04 4.01[b] ±0.03	4.01[d] ±0.04 4.12[d] ±0.02	4.54 ±0.03 4.35[e]	0.43	19.8
	H_2L/HL.H	3.14 ±0.02 3.08[b] ±0.03	3.11[d] ±0.04 3.21[d] ±0.02	3.27 ±0.03 3.46[e]	−0.11	14.0
Sc^{3+}	ML/M.L		3.93			
	ML_2/M.L^2		5.74			
	ML_3/M.L^3		6.18			
Mn^{2+}	ML/M.L	1.73 ±0.0				
	MHL/M.HL	0.6				
Ni^{2+}	ML/M.L	4.20 −0.1	(3.7)[d] 3.93[d]		0.4	18
	ML_2/M.L^2	7.01 −0.3	7.03[d]			
	ML_3/M.L^3		8.55[d]			
	MHL/M.HL	2.15	1.70[d]			
Cu^{2+}	ML/M.L	4.65 −0.1	4.19 −0.2	4.63[e]	0.3	19
	ML_2/M.L^2	7.50 −0.2	7.08[j]	7.86[e]		
	MHL/M.HL	2.62 ±0.02		2.0[e]		
Zn^{2+}	ML/M.L	3.30 −0.3	(2.6)		1.5	17
	ML_2/M.L^2	5.85				
Cd^{2+}	ML/M.L	3.14 −0.5	(2.3)		1.6	16
	ML_2/M.L^2	5.57				
Pb^{2+}	ML/M.L	3.6 3.36[b]				
	MHL/M.HL	1.71[b]				

[b] 25°, 0.5; [d] 25°, 2.0; [e] 25°, 3.0; [j] 20°, 1.0

Bibliography:

H^+ 78AM,79RW

Sc^{3+} 76D

Mn^{2+} 66SY

Ni^{2+} 78A,78JS

Cu^{2+} 78A,78AM,79RW

Zn^{2+},Cd^{2+} 78A

Pb^{2+} 76NC

Other references: 76MG,78SJ,79DS,79S,79SJb, 79SJc,79SSc

$$HO_2CCH_2SSCH_2CO_2H$$

$C_4H_6O_4S_2$ Dithiodiacetic acid H_2L
(Other values on Vol.3, p.144)

Metal ion	Equilibrium	Log K 25°, 0.1	Log K 25°, 1.0	Log K 25°, 0	ΔH 25°, 1.0	ΔS 25°, 1.0
H^+	HL/H.L	3.81 -0.01	3.67	4.21	-0.3	16
		3.66[b]	3.76[d]+0.1	3.93[e]		
	H_2L/HL.H	2.91 ±0.03	2.88	3.07	0.8	15
		2.87[b]	3.00[d]+0.01	3.17[e]		
Ni^{2+}	ML/M.L	1.8	(1.5)		0.3	8
Zn^{2+}	ML/M.L	1.6	(1.3)		0.7	8
Cd^{2+}	ML/M.L	1.9	(1.6)		0.9	10

[b]25°, 0.5; [d]25°, 2.0; [e]25°, 3.0

Bibliography: H^+ 79Ab; Ni^{2+}-Cd^{2+} 79Aa; Other references: 79DS

$$HO_2CCH_2CH_2SCH_2CH_2CO_2H$$

$C_6H_{10}O_4S$ 3,3'-Thiodipropanoic acid H_2L
(Other values on Vol.3, p.146)

Metal ion	Equilibrium	Log K 25°, 0.1	Log K 25°, 1.0	Log K 20°, 0	ΔH 25°, 1.0	ΔS 25°, 1.0
H^+	HL/H.L	4.68 ±0.02	4.54	5.08	-0.8	18
		4.55[b]	4.81[d]	4.98[e]		
	H_2L/HL.H	3.87 ±0.03	3.87	4.08	0.0	17
		3.83[b]	4.15[d]	4.36[e]		
Mn^{2+}	ML/M.L	0.5				
Co^{2+}	ML/M.L	1.6				
Ni^{2+}	ML/M.L	1.6 ±0.0	(1.3)		0.5	8
			1.2[d]			
	MHL/M.HL		0.67[d]			
Cu^{2+}	ML/M.L	2.97 ±0.0	(2.7)	3.64[e]	3	22
	MHL/M.HL	1.59		1.83[e]		
Zn^{2+}	ML/M.L	1.72 -0.1	1.4 ±0.1		2	11
	MHL/M.HL	1.28				
Cd^{2+}	ML/M.L	2.31 -0.3	1.8 -0.1		2	13
	MHL/M.HL	1.77				

[b]25°, 0.5; [d]25°, 2.0; [e]25°, 3.0

Bibliography:

H^+ 79Ab
Mn^{2+},Co^{2+} 68SKM
Ni^{2+} 78JS,79Aa

Cu^{2+} 79Aa,79RW
Zn^{2+},Cd^{2+} 79Aa

Other references: 79DS,79S,79SJ,79SJc

$$HO_2CCH_2CH_2CH_2SCH_2CH_2CH_2CO_2H$$

| $C_8H_{14}O_4S$ | | 4,4'-Thiodibutanoic acid | | H_2L |

(Other reference on Vol.3, p.342)

Metal ion	Equilibrium	Log K 18°, 0.1	Log K 18°, 1.0	Log K 18°, 0
H^+	HL/H.L	4.89[i] 4.79[l]	4.80[k] 4.94[k]	5.26 5.31[e]
	H_2L/HL.H	4.20[i] 4.13[l]	4.17[k] 4.31[k]	4.35 4.70[e]
Cu^{2+}	ML/M.L			2.50[e]
	MHL/M.HL			1.80[e]

[e]25°, 3.0; [i]18°, 0.5; [k]18°, 2.0

Bibliography: 79RW

$$HO_2CCH_2SCH_2CH_2SCH_2CO_2H$$

| $C_6H_{10}O_4S_2$ | | (Ethylenedithio)diacetic acid | | H_2L |

(Other values on Vol.3, p.148)

Metal ion	Equilibrium	Log K 25°, 0.1	Log K 25°, 1.0	Log K 18°, 0
H^+	HL/H.L	4.00 ±0.1 3.83[i]	3.85 ±0.02 3.94[k]+0.2	4.35
	H_2L/HL.H	3.20 ±6.04 3.15[i]	3.21 ±0.05 3.29[k]+0.2	3.38
Cu^{2+}	$ML_2/M.L^2$	11.16		
	$M(HL)L/ML_2.H$	4.20		
	$M(HL)_2/M(HL)L.H$	3.19		
	$M(H_2L)(HL)/M(HL)_2.H$	2.95		

[i]18°, 0.5; [k]18°, 2.0

Bibliography: 76P

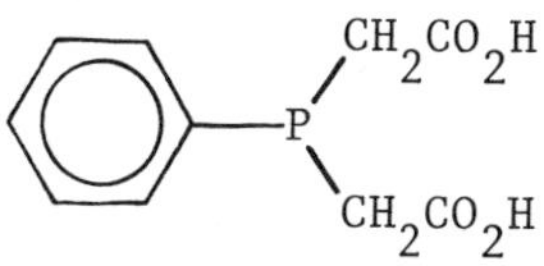

| $C_{10}H_{11}O_4P$ | | P-Phenylphosphinodiacetic acid | H_2L |

[bis(carboxymethyl)phenylphosphine]

Metal ion	Equilibrium	Log K 25°, 0.1
H^+	HL/H.L	4.76
	H_2L/HL.H	3.36
Ni^{2+}	ML/M.L	3.68
	$ML_2/M.L^2$	6.49
Cu^+	ML/M.L	11.0
	$ML_2/M.L^2$	15.9
	$ML_3/M.L^3$	18.9
	$ML_4/M.L^4$	20.9

Bis(carboxymethyl)phenylphosphine (continued)

Metal ion	Equilibrium	Log K 25°, 0.1
Hg^{2+}	$ML/M.L$	19.7
	$ML_2/M.L^2$	31.92
	$ML_3/M.L^3$	34.21

Bibliography:

H^+,Ni^{2+} 78P Hg^{2+} 79PPa

Cu^+ 79PP

$$CO_2H$$
$$|$$
$$HO_2CCH_2CHCH_2CO_2H$$

$C_6H_8O_6$ Propane-1,2,3-tricarboxylic acid (tricarballylic acid) H_3L
(Other values on Vol.3, p.158)

Metal ion	Equilibrium	Log K 25°, 0.1	Log K 25°, 1.0	Log K 25°, 0	ΔH 25°, 0	ΔS 25°, 0
H^+	$HL/H.L$	5.82 ±0.05	5.52 ±0.02	6.38	−1.81	35.2
	$H_2L/HL.H$	4.50 ±0.03	4.38 ±0.02	4.87	−1.49	25.3
	$H_3L/H_2L.H$	3.48 ±0.02	3.43 ±0.03	3.67	−0.56	14.9
Be^{2+}	$ML/M.L$		3.75			
	$MHL/M.HL$		2.47			
Pb^{2+}	$ML/M.L$		3.17			
	$ML_2/M.L^2$		4.70			
	$MHL/ML.H$		4.74			
	$MH_2L/MHL.H$		3.68			
	$M(HL)L/ML_2.H$		5.26			
	$M(HL)_2/M(HL)L.H$		4.74			
	$M(H_2L)(HL)/M(HL)_2.H$		4.1			
	$M_2L_2/(ML)^2$		2.34			

Bibliography:

H^+ 74VG,79AO

Be^{2+} 74VG

Pb^{2+} 79AO

Other reference: 69L

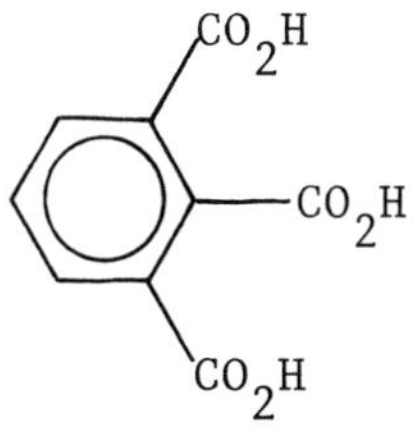

$C_9H_6O_6$ Benzene-1,2,3-tricarboxylic acid (hemimellitic acid) H_3L
(Other values on Vol.3, p.159)

Metal ion	Equilibrium	Log K 25°, 0.1	Log K 25°, 1.0	Log K 25°, 0	ΔH 25°, 0	ΔS 25°, 0
H^+	HL/H.L	5.51	4.96	7.13	-0.37	31.4
	$H_2L/HL.H$	3.82	3.69	4.75	-0.21	21.4
	$H_3L/H_2L.H$	2.62	2.51	2.88	1.06	16.7
Cd^{2+}	$ML/M.L$		2.39			
	$ML_2/M.L^2$		3.79			
	MHL/M.HL		1.66			
	$M_2L_2/M^2.L^2$		6.36			

Bibliography: 79AOa

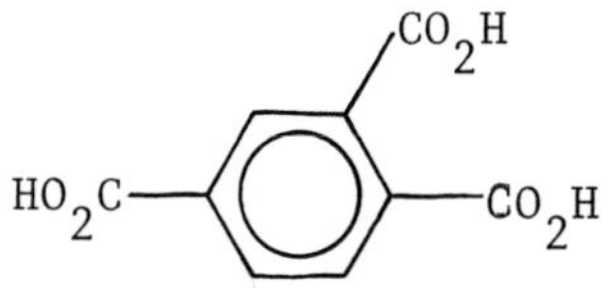

$C_9H_6O_6$ Benzene-1,2,4-tricarboxylic acid (trimellitic acid) H_3L
(Other values on Vol.3, p. 160)

Metal ion	Equilibrium	Log K 25°, 0.1	Log K 25°, 1.0	Log K 25°, 0	ΔH 25°, 0	ΔS 25°, 0
H^+	HL/H.L	5.01	4.53	5.54	0.95	28.6
	$H_2L/HL.H$	3.71	3.51	4.04	0.09	18.5
	$H_3L/H_2L.H$	2.4	2.27	2.48	1.24	25.5
Cd^{2+}	$ML/M.L$		1.88			
	$ML_2/M.L^2$		2.96			
	MHL/M.HL		1.34			
	$M_2L_2/M^2.L^2$		5.11			

Bibliography: 79AOa

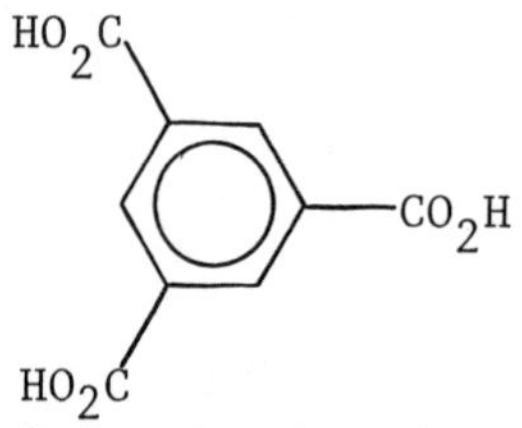

HO_2C, / CO_2H / HO_2C (benzene ring structure)

$C_9H_6O_6$	Benzene-1,3,5-tricarboxylic acid					H_3L
	(Other references on Vol.3, p.343)					

Metal ion	Equilibrium	Log K 25°, 0.1	Log K 25°, 1.0	Log K 25°, 0	ΔH 25°, 0	ΔS 25°, 0
H^+	HL/H.L	4.49	4.22	5.18	1.17	27.7
	$H_2L/HL.H$	3.71	3.57	4.10	0.49	20.4
	$H_3L/H_2L.H$	3.01	3.00	3.12	-0.87	11.4
Cd^{2+}	$ML/M.L$		1.52			
	$ML_2/M.L^2$		2.58			
	MHL/M.HL		1.26			

Bibliography: 79AOa

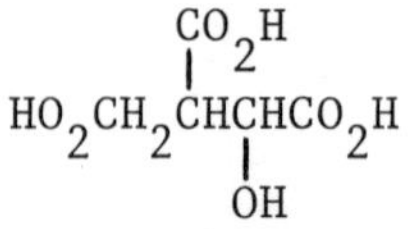

CO_2H / HO_2CH_2CHCHCO_2H / OH

$C_6H_8O_7$	DL-1-Hydroxypropane-1,2,3-tricarboxylic acid (isocitric acid)		H_3L
	(Other values on Vol.3, p.160)		

Metal ion	Equilibrium	Log K 25°, 0.1	Log K 25°, 1.0
H^+	HL/H.L	5.75^w	5.06
	$H_2L/HL.H$	4.28^w	3.73
	$H_3L/H_2L.H$	3.02^w	2.29
Mg^{2+}	$ML/M.L$	2.32 2.72^w	1.94
	MHL/M.HL	1.43^w	0.9
Mn^{2+}	$ML/M.L$	2.55 3.06^w	2.27
	MHL/M.HL	1.76^w	1.20
Co^{2+}	$ML/M.L$		2.73

$^w (CH_3)_4NCl$ used as background electrolyte and corrected for Cl^-.

Bibliography: 76PCB

$$\begin{array}{c} CO_2H \\ | \\ HO_2CCH_2CCH_2CO_2H \\ | \\ OH \end{array}$$

$C_6H_8O_7$ 2-Hydroxypropane-1,2,3-tricarboxylic acid (citric acid) H_3L
(Other values on Vol.3, p.161)

Metal ion	Equilibrium	Log K 25°, 0.1	Log K 25°, 1.0	Log K 25°, 0	ΔH 25°, 0	ΔS 25°, 0
H^+	HL/H.L	5.66 ±0.05	5.34[u] -0.01	6.396 +0.004	0.80	32.0
		5.83[w] ±0.01	5.15[v] ±0.03	5.18[d,v]		
			5.62[n]			
	$H_2L/HL.H$	4.34 ±0.04	4.11 ±0.02	4.761 -0.002	-0.58	19.8
			4.29[n]	4.16[d]		
	$H_3L/H_2L.H$	2.90 ±0.05	2.80 ±0.05	3.128 -0.07	-1.00	11.0
			2.87[n]	2.90[d]		
Be^{2+}	ML/M.L	4.5[p]	4.31[u]			
	MHL/M.HL	2.2[p]	2.23			
	$MH_2L/M.H_2L$	1.4[p]				
	$M_2L_2/M^2.L^2$		13.10[u]			
	$M_2L_2/M_2OHL_2.H$		4.87[u]			
	$M_2OHL_2/M_2(OH)_2L_2.H$		5.08[u]			
	$M_4(OH)_2L_2/M_2(OH)_2L_2.M^2$		7.63[u]			
	$M_4(OH)_2L_2/M_4(OH)_3L_2.H$		3.70[u]			
	$M_4(OH)_3L_2/M_4(OH)_4L_2.H$		5.11[u]			
Mn^{2+}	ML/M.L	3.70 ±0.03	3.80[n] ±0.02			
		4.15[w]				
	MHL/M.HL	2.08	2.22[n]			
		2.16[w]				
	$M^2.L^2/M_2(OH)_2L_2.H^2$		5.73[n]			
Fe^{2+}	ML/M.L	(4.8)	4.56[n]			
		4.4[h]				
	MHL/M.HL	2.9	3.10[n]			
		2.65[h]				
	$MHL_2/ML.HL$		1.73[n]			
	$M^2.L^2/M_2(OH)_2L_2.H^2$		5.4[n]			
Ni^{2+}	ML/M.L	5.35 ±0.05				
		5.40[h]				
	MHL/M.HL	3.25 ±0.02				
		3.30[h]				
	$MH_2L/M.H_2L$	1.75[h]				
	$M^2.L^2/M_2(OH)_2L_2.H^2$	4.71				

---

[d] 25°, 2.0; [h] 20°. 0.1; [n] 37°, 0.15; [p] 34°, 0.15; [u] K^+ salt used as background electrolyte; [v] Na^+ salt used as background electrolyte; [w] $(CH_3)_4N^+$ salt used as background electrolyte or a correction was made for the background electrolyte.

Citric acid   (continued)

Metal ion	Equilibrium	Log K 25°, 0.1	Log K 25°, 1.0	Log K 25°, 0	ΔH 25°, 0	ΔS 25°, 0
Cu^{2+}	ML/M.L	$(5.90)^h$				
	MHL/M.HL	3.7 ±0.1				
		3.42^h				
	$MH_2L/M.H_2L$	2.26^h				
	MOHL.H/M.L	1.63				
		1.56^h				
	$M_2L/ML.M$	2.20^h				
	$M_2L_2/M^2.L^2$	14.8 ±0.2	13.2			
	$M_2L_2/M_2OHL_2.H$	3.85				
	$M_2L_2/M_2(OH)_2L_2.H^2$	8.60	8.03			
Fe^{3+}	ML/M.L	11.2				
		11.4^h				
	MHL/M.HL	6.7				
	MOHL.H/M.L	8.5				
	$M_2(OH)_2L_2.H^2/M^2.L^2$	21.2^h				
Zn^{2+}	ML/M.L	4.86 ±0.04	5.37^n			
		4.94^h ±0.04				
		4.27^b				
	$ML_2/M.L^2$	5.90^b				
	MHL/M.HL	3.87 ±0.09				
		2.98^h				
		2.94^b				
	$MH_2L/M.H_2L$	1.25^h				
	$M_2(OH)_2L_2.H^2/M^2.L^2$	2.94				
Pb^{2+}	ML/M.L		4.44^v	$4.08^{d,v}$		
	$ML_2/M.L^2$		5.92^v	$6.1^{d,v}$		
	MHL/M.HL		2.98	2.97^d		
	$MH_2L/M.H_2L$		1.70	1.51^d		
	$MHL_2/ML_2.H$		4.69^v			
	$M_2L_2/M^2.L^2$		10.70^v			
	$M_2OHL_2/M_2L_2.OH$		6.06^v			
	$M_2(OH)_2L_2/M_2OHL_2.OH$		7.45^v			
Bi^{3+}	ML/M.L		$10.78^{n,v}$			
	$ML_2/M.L^2$		$15.8^{n,v}$			

---

^{b}25°, 0.5; ^{d}25°, 2.0; ^{h}20°, 0.1; ^{n}37°, 0.15; ^{v}Na$^+$ salt used as background electrolyte

Bibliography:

H^+	74VG,75DOa,78EO,78RM,79AD,79HP	Fe^{3+}	74FM
Be^{2+}	74VG	Zn^{2+}	77DO,77RWb,78RM
$Y^{3+}-Cf^{3+}$	71Sf	Pb^{2+}	78EO
Mn^{2+}	77RWb,79AD	Bi^{3+}	77W
Fe^{2+}	74FM,79AD		
Ni^{2+}	75FC,76DOb		
Cu^{2+}	74FM,75DOa		

Other references: 61E,69L,70AM,71Z,72OO,
73H,73KP,73ZG,74RM,75KBR,75RM,76MKD,
76NCW,76VK,77GD,77KP,77LK,78BH,78KC,
78NP,78SK,78TGY,78TSG,78VK,79SFK

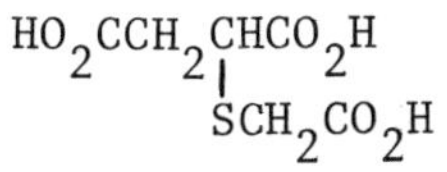

$$HO_2CCH_2CHCO_2H$$
$$|$$
$$SCH_2CO_2H$$

$C_6H_8O_6S$	DL-Carboxymethylthiobutanedioic acid (Other reference on Vol.3, p.165)		H_3L

Metal ion	Equilibrium	Log K 20°, 0.1
H^+	HL/H.L	5.12
	$H_2L/HL.H$	3.79
	$H_3L/H_2L.H$	3.26
Mn^{2+}	ML/M.L	2.11
Co^{2+}	ML/M.L	3.45
	MHL/ML.H	4.49
	$MH_2L/MHL.H$	3.41
Ni^{2+}	ML/M.L	4.32
	MHL/ML.H	3.92
	$MH_2L/MHL.H$	3.18
Cu^{2+}	ML/M.L	4.80
	MHL/ML.H	3.64
	$MH_2L/MHL.H$	2.98
Ag^+	ML/M.L	3.52
Zn^{2+}	ML/M.L	3.19
	MHL/ML.H	4.60
	$MH_2L/MHL.H$	3.52
Cd^{2+}	ML/M.L	2.71
	MHL/ML.H	4.72
	$MH_2L/MHL.H$	3.71

Bibliography: 77CA

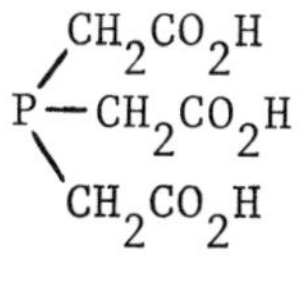

$$P\begin{array}{l} \diagup CH_2CO_2H \\ -CH_2CO_2H \\ \diagdown CH_2CO_2H \end{array}$$

$C_6H_9O_6P$	Phosphinotriacetic acid (tris(carboxymethyl)phosphine)		H_3L

Metal ion	Equilibrium	Log K 25°, 0.1
H^+	HL/H.L	5.43
	$H_2L/HL.H$	3.77
	$H_3L/H_2L.H$	2.73
Cu^+	ML/M.L	10.2
	$ML_2/M.L^2$	14.9
	$ML_3/M.L^3$	18.0
	$ML_4/M.L^4$	19.7
Hg^{2+}	ML/M.L	19.93
	$ML_2/M.L^2$	32.27
	$ML_3/M.L^3$	34.16

Bibliography:

H^+ 78Pa Hg^{2+} 79PPa

Cu^+ 79PP

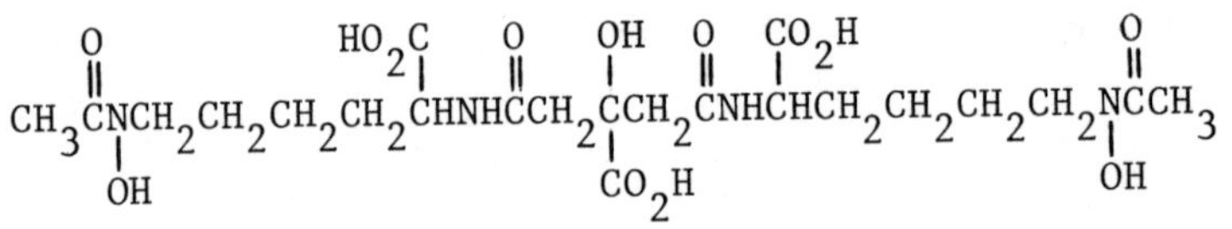

$$\underset{\text{CH}_3\underset{\underset{\text{OH}}{|}}{\overset{\overset{\text{O}}{\|}}{\text{C}}}\text{NCH}_2\text{CH}_2\text{CH}_2\text{CH}_2\text{CHNHCCH}_2\text{CCH}_2\text{CNHCHCH}_2\text{CH}_2\text{CH}_2\text{CH}_2\text{NCCH}_3}{}$$

Aerobactin

$C_{20}H_{34}O_{13}N_4$ H_5L

Metal ion	Equilibrium	Log K 25°, 0.1
H^+	HL/H.L	9.44
	$H_2L/HL.H$	8.93
	$H_3L/H_2L.H$	4.31
	$H_4L/H_3L.H$	3.48
	$H_5L/H_4L.H$	3.11
Fe^{3+}	ML/M.L	22.5
	ML/MOHL.H	4.4
	MHL/ML.H	3.6
	$MH_2L/MHL.H$	3.1
	$MH_3L/MH_2L.H$	2.4

Bibliography: 79HCR

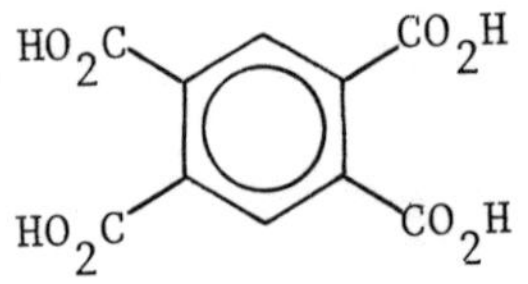

Benzene-1,2,4,5-tetracarboxylic acid (pyromellitic acid)

$C_{10}H_6O_8$ H_4L
(Other references on Vol.3, p.343)

Metal ion	Equilibrium	Log K 25°, 1.0	Log K 25°, 0	ΔH 25°, 0	ΔS 25°, 0
H^+	HL/H.L	4.71	6.23	1.60	33.9
	$H_2L/HL.H$	3.79	4.92	0.79	25.2
	$H_3L/H_2L.H$	2.46	3.12	1.57	19.6
	$H_4L/H_3L.H$	1.67	1.70	3.11	18.2
Cd^{2+}	ML/M.L	2.13			
	$ML_2/M.L^2$	5.28			
	MHL/M.HL	1.74			
	$M_2L_2/M^2.L^2$	6.95			

Bibliography: 79GO

$$CH_3PO_3H_2$$

| CH$_5$O$_3$P | | Methylphosphonic acid | | H$_2$L |

Methylphosphonic acid
(Other references on Vol.3, p. 360)

Metal ion	Equilibrium	Log K 25°, 0.1	Log K 25°, 0
H$^+$	HL/H.L	7.55	8.00
	H$_2$L/HL.H	2.19	
Ca^{2+}	ML/M.L	1.51	
Cu^{2+}	ML/M.L	3.52	
	ML/MOHL.H	6.9	

Bibliography:

H$^+$ 77KT,79WN   Ca^{2+},Cu^{2+} 78WN.

$$CH_3CH_2PO_3H_2$$

Ethylphosphonic acid

C$_2$H$_7$O$_3$P   H$_2$L

Metal ion	Equilibrium	Log K 25°, 0.1
H$^+$	HL/H.L	7.79
	H$_2$L/HL.H	2.29
Ca^{2+}	ML/M.L	1.54
Cu^{2+}	ML/M.L	3.59
	ML/MOHL.H	6.56

Bibliography: 79WN

$$ClCH_2PO_3H_2$$

Chloromethylphosphonic acid

CH$_4$O$_3$ClP   H$_2$L

Metal ion	Equilibrium	Log K 25°, 0.1	Log K 25°, 0
H$^+$	HL/H.L	6.19 ±0.02	6.59
	H$_2$L/HL.H	1.04	
Mg^{2+}	ML/M.L	1.59	
Ca^{2+}	ML/M.L	1.46 ±0.09	
Sr^{2+}	ML/M.L	1.18	
Ba^{2+}	ML/M.L	1.11	
Co^{2+}	ML/M.L	(1.89)	
Ni^{2+}	ML/M.L	1.81	
Cu^{2+}	ML/M.L	2.88 ±0.01	
	ML$_2$/M.L^2	5.32	
	ML/MOHL.H	6.62	
Zn^{2+}	ML/M.L	2.11	
Cd^{2+}	ML/M.L	2.43	

Bibliography:

H$^+$ 70TN,77KT,78WNb,79WN   Ca^{2+},Cu^{2+} 70TN,79WN

Mg^{2+},Sr^{2+}-Ni^{2+},Zn^{2+},Cd^{2+} 70TN

$$Cl_2CHPO_3H_2$$

$CH_3O_3Cl_2P$ Dichloromethylphosphonic acid H_2L

Metal ion	Equilibrium	Log K 25°, 0.1	Log K 25°, 0
H^+	HL/H.L	5.21	5.60
	H_2L/HL.H	(0.7)	
Ca^{2+}	ML/M.L	1.26	
Cu^{2+}	ML/M.L	2.49	
	ML/MOHL.H	7.20	

Bibliography:

H^+ 77KT,78WNb,79WN Ca^{2+},Cu^{2+} 79WN

$$Cl_3CPO_3H_2$$

$CH_2O_3Cl_3P$ Trichloromethylphosphonic acid H_2L

Metal ion	Equilibrium	Log K 25°, 0.1	Log K 25°, 0
H^+	HL/H.L	4.48	4.93
	H_2L/HL.H	(0.8)	
Ca^{2+}	ML/M.L	1.25	
Cu^{2+}	ML/M.L	2.17	

Bibliography:

H^+ 77KT,78WNb,79WN Ca^{2+},Cu^{2+} 79WN

$$BrCH_2PO_3H_2$$

CH_4O_3BrP Bromomethylphosphonic acid H_2L

Metal ion	Equilibrium	Log K 25°, 0.1
H^+	HL/H.L	6.24
	H_2L/HL.H	1.15
Ca^{2+}	ML/M.L	1.34
Cu^{2+}	ML/M.L	2.95
	ML/MOHL.H	6.86

Bibliography:

H^+ 78WNb,79WN Ca^{2+},Cu^{2+} 79WN

$$ICH_2PO_3H_2$$

CH_4O_3IP		Iodomethylphosphonic acid	H_2L

Metal ion	Equilibrium	Log K 25°, 0.1
H^+	HL/H.L	6.44
	H_2L/HL.H	1.27
Ca^{2+}	ML/M.L	1.37
Cu^{2+}	ML/M.L	3.04
	ML/MOHL.H	6.76

Bibliography:

H^+ 78WNb,79WN Ca^{2+},Cu^{2+} 79WN

$$HOCH_2PO_3H_2$$

CH_5O_4P		Hydroxymethylphosphonic acid (Other values on Vol.3, p.175)	H_2L

Metal ion	Equilibrium	Log K 25°, 0.1	Log K 25°, 0
H^+	HL/H.L	6.97 7.01^w	7.36
	H_2L/HL.H	1.70 ±0.0	
Ca^{2+}	ML/M.L	1.68 1.87^w	
Cu^{2+}	ML/M.L	3.53	
	ML/MOHL.H	6.18	

$^w(CH_3)_4NCl$ used as background electrolyte.

Bibliography:

H^+ 77KT,78WNb,79WN Ca^{2+},Cu^{2+} 79WN

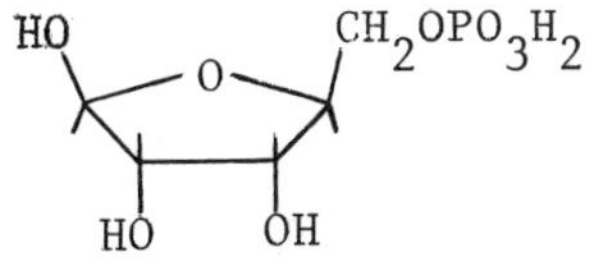

$C_5H_{11}O_8P$		D(-)-Ribose-5-dihydrogenphosphate (Other references on Vol.3, p.346)	H_2L

Metal ion	Equilibrium	Log K 25°, 0.1	Log K 25°, 0	ΔH 25°, 0	ΔS 25°, 0
H^+	L/H$_{-1}$L.H		(13.05)	(-6.1)	(39)
	HL/H.L	6.28	6.70	2.7	40
	H_2L/HL.H	2.0			
Ni^{2+}	ML/M.L	1.99			
	MHL/M.HL	0.8			

Bibliography: 76TD

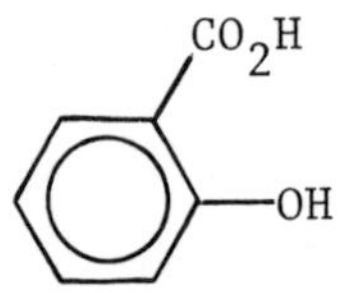

		2-Hydroxybenzoic acid (salicylic acid)				H_2L

(Other values on Vol.3, p.186)

Metal ion	Equilibrium	Log K 25°, 0.1	Log K 25°, 1.0	Log K 25°, 0	ΔH 25°, 0	ΔS 25°, 0
H^+	$HL/H.L$	(13.4) ±0.0	13.2	(13.66)±0.07	-8.56 ±0.05	33.8
		(13.0)[n] ±0.0	13.1[e]			
	$H_2L/HL.H$	2.81 ±0.01	2.78 ±0.02	2.98 ±0.00	-0.7 ±0.1	11
		2.78[b] ±0.04	2.79[n] ±0.02	3.16[e] ±0.02	-0.9[a]	
Mg^{2+}	$ML/M.L$		5.2[n]			
Ca^{2+}	$ML/M.L$		4.3[n]			
	$MHL/M.HL$	0.15 -0.01		0.36		
Eu^{3+}	$MHL/M.HL$	2.59[t]				
	$M(HL)_2/M.(HL)^2$	4.21[t]				
	$M(HL)_3/M.(HL)^3$	4.86[t]				
Cu^{2+}	$ML/M.L$	10.6 ±0.1	10.1[n] ±0.0		-4.4	34
	$ML_2/M.L^2$	18.5	18.2[n] -1			
V^{3+}	$ML.H/M.HL$		2.34[u]		(-1)[s]	(7)[u]
	$MHL/M.HL$		3.32[u]		(4)[s]	(29)[u]
Ga^{3+}	$ML/M.L$	14.5				

[a] 25°, 0.1; [b] 25°, 0.5; [e] 25°, 3.0; [n] 37°, 0.15; [s] 7-45°, 1.0; [t] 22°, 0.1; [u] 27°, 1.0

Bibliography:

H^+ 72LL,77AR,78AK,78Pb

Mg^{2+},Ca^{2+},Cu^{2+} 78AK

Eu^{3+} 70IS

V^{3+} 79PT

Ga^{3+} 77PST

Other references: 61CM,74J,76ABc,76ABd, 76PC,76TM,77JKS,78ABS,79AV,79BK,79FK, 79MPS,79PTa,79ZKV

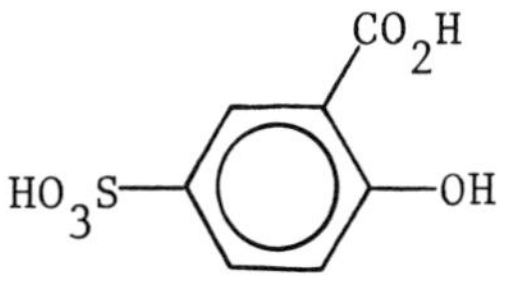

| $C_7H_6O_6S$ | 2-Hydroxy-5-sulfobenzoic acid (5-sulfosalicylic acid) (Other values on Vol.3, p.190) | | | | | H_3L |

Metal ion	Equilibrium	Log K 25°, 0.1	Log K 25°, 1.0	Log K 25°, 0	ΔH 25°, 0	ΔS 25°, 0
H^+	HL/H.L	11.75 ±0.1	11.40[b] ±0.08	12.53	-9.3[b]	26[b]
			11.53[b] ±0.06	11.74[e]	(-7)[s]	(30)[e]
	H_2L/HL.H	2.49 ±0.02	2.32[b] ±0.02	2.84	-0.8	10
			2.35[b] ±0.05	2.67[e]		
Be^{2+}	ML/M.L	11.62[h]	11.0[b]			
		11.32[h]				
	ML_2/M.L^2	20.4[h]	19.8[b]			
		20.0[h]				
UO_2^{2+}	ML/M.L	11.16 ±0.01	10.51			
		11.15[h]	10.52[j]			
	ML_2/M.L^2	19.2[h]				
		18.6[h]				
PuO_2^{2+}	ML/M.L		9.46[j]			
Cu^{2+}	ML/M.L	9.50 ±0.07	8.84[b]	10.74	-4.0	30
			8.97[b]			
	ML_2/M.L^2	16.3 ±0.1	15.7[b]	17.2		
			15.9[b]			

[b] 25°, 0.5;   [e] 25°, 3.0;   [h] 20°, 0.1;   [j] 20°, 1.0;   [s] 15-35°, 3.0

Bibliography:

H^+	77AR,79ZKV	Mn^{2+}	75SG
Be^{2+}	74SRR	Cu^{2+}	69CM,74SRR,78RM,79MB
UO_2^{2+}	60BS,64RM,72BP		
PuO_2^{2+}	72BP		

Other references: 68GI,68MN,69MD,69PP,69Sb, 72PD,74J,76ABd,76CS,76SJ,77DS,77JKS, 77KTS,77SJ,77TJ,77UB,79CP,79SJ,75PM

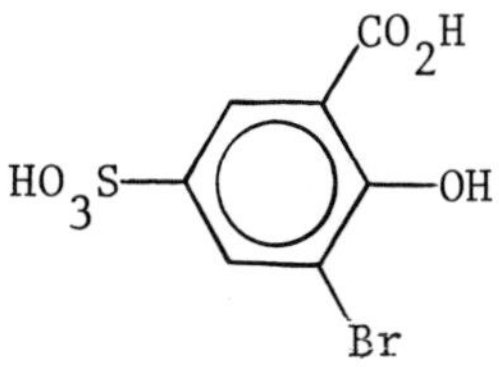

| $C_7H_6O_6BrS$ | 3-Bromo-2-hydroxy-5-sulfobenzoic acid (Other reference on Vol.3, p.349) | H_3L |

Metal ion	Equilibrium	Log K 25°, 3.0
H^+	HL/H.L	10.47
	H_2L/HL.H	2.03
Pb^{2+}	MHL/M.HL	1.12
	$M(HL)_2$/M.$(HL)^2$	1.93
	MHL/ML.H	5.99
	$M(HL)_2$/ML_2.H^2	13.12

Bibliography: H^+ 76La; Pb^{2+} 76Lb

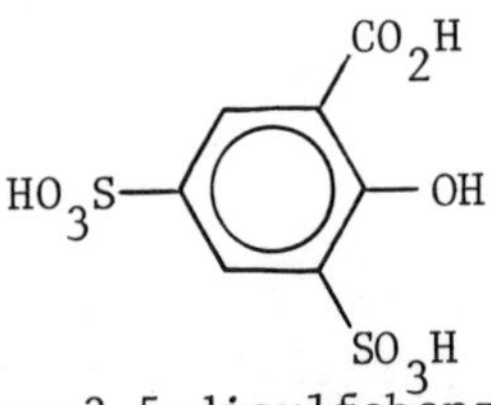

$$CO_2H$$
$$HO_3S \quad OH$$
$$SO_3H$$

$C_7H_6O_9S_2$ 2-Hydroxy-3,5-disulfobenzoic acid H_4L
                              (Other values on Vol.3, p.192)

Metal ion	Equilibrium	Log K 25°, 0.1	Log K 25°, 1.0	Log K 25°, 0
H^+	HL/H.L	11.55 11.07[b]	10.95	12.50
	$H_2L/HL.H$	2.03 1.70[b]	1.71	2.69
Be^{2+}	$ML/M.L$	10.50[b]		
	$ML_2/M.L^2$	18.69[b]		
Y^{3+}	$ML/M.L$	8.64[b]		
	$ML_2/M.L$	14.38[b]		
	MHL/ML.H	4.1[b]		
Pr^{3+}	$ML/M.L$	7.66[b]		
	$ML_2/M.L^2$	12.72[b]		
Nd^{3+}	$ML/M.L$	7.77[b]		
	$ML_2/M.L^2$	12.88[b]		
Sm^{3+}	$ML/M.L$	8.20[b]		
	$ML_2/M.L^2$	13.54[b]		
Eu^{3+}	$ML/M.L$	8.35[b]		
	$ML_2/M.L^2$	13.76[b]		
Gd^{3+}	$ML/M.L$	8.59[b]		
	$ML_2/M.L^2$	14.12[b]		
Tb^{3+}	$ML/M.L$	8.74[b]		
	$ML_2/M.L^2$	14.42[b]		
Dy^{3+}	$ML/M.L$	8.82[b]		
	$ML_2/M.L^2$	14.53[b]		
Ho^{3+}	$ML/M.L$	8.77[b]		
	$ML_2/M.L^2$	14.59[b]		
Er^{3+}	$ML/M.L$	8.81[b]		
	$ML_2/M.L^2$	14.64[b]		
Tm^{3+}	$ML/M.L$	8.85[b]		
	$ML_2/M.L^2$	14.70[b]		
Yb^{3+}	$ML/M.L$	8.90[b]		
	$ML_2/M.L^2$	14.89[b]		
Lu^{3+}	$ML/M.L$	8.86[b]		
	$ML_2/M.L^2$	14.85[b]		
UO_2^{2+}	$ML/M.L$	10.77[b]		
	$ML_2/M.L^2$	18.45[b]		
	ML/MOHL.H	4.21[b]		
	$ML_2/MOHL_2.H$	11.37[b]		
	$M_2L/ML.M$	2.30[b]		

---

[b] 25°, 0.5

Bibliography: H^+,Be^{2+} 74SRL; Y^{3+} 76Lg; Pr^{3+}-Lu^{3+} 76Ld; UO_2^{3+} 79LPa

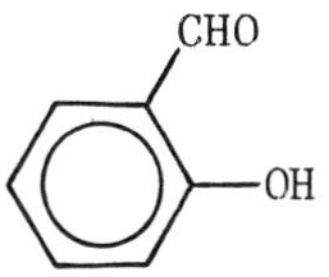

| $C_7H_6O_2$ | 2-Hydroxybenzaldehyde (salicylaldehyde) | | | | | HL |

(Other values on Vol.3, p.194)

Metal ion	Equilibrium	Log K 25°, 0.1	] Log K 25°, 0.5	Log K 25°, 0	ΔH 25°, 0	ΔS 25°, 0
H^+	HL/H.L	8.13 ±0.01	8.07	8.37 ±0.00	-5.2	22
		8.22[t]	8.80[e]		(-6)[s]	(20)[o]
Mg^{2+}	ML/M.L		1.72			
Mn^{2+}	ML/M.L		2.13 ±0.03			
	$ML_2/M.L^2$		4.0			
Co^{2+}	ML/M.L		3.22			
Cd^{2+}	ML/M.L		1.60			
Pb^{2+}	ML/M.L		3.04			

[e] 25°, 3.0;   [s] 15-35°, 3.0;   [t] 20°, 0.15

Bibliography: H^+ 64MAH; Mg^{2+}-Pb^{2+} 69HL

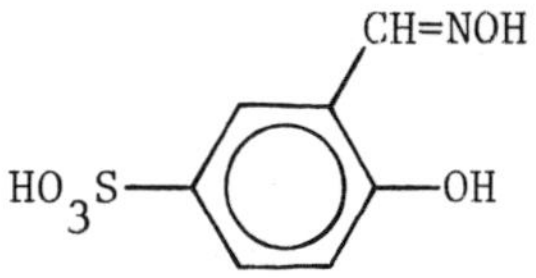

| $C_7H_7O_2N$ | 2-Hydroxy-5-sulfobenzaldehyde oxime (salicylaldoxime-5-sulfonic acid) | | | H_3L |

Metal ion	Equilibrium	Log K 25°, 0.1	Log K 25°, 1.0	Log K 25°, 0
H^+	HL/H.L	11.50	11.17	12.00
			11.25[d]	
	H_2L/HL.H	8.03	7.98	8.42
			8.20[d]	
Ni^{2+}	MHL/M.HL	5.11	4.84	5.83
	$M(HL)_2/M.(HL)^2$	10.04	9.80	10.83
	$M(HL)_2/MHL_2$.H	7.34	7.09	7.94

[d] 25°, 2.0

Bibliography: 72Sa

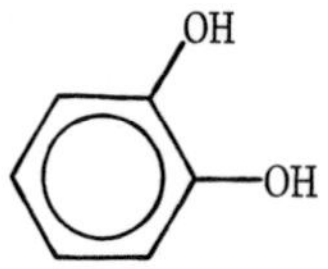

| $C_6H_6O_2$ | | 1,2-Dihydroxybenzene (catechol) (Other values on Vol.3, p.200) | | | | H_2L |

Metal ion	Equilibrium	Log K 25°, 0.1	Log K 25°, 1.0	Log K 25°, 0	ΔH 25°, 0.1	ΔS 25°, 0.1
H^+	HL/H.L	(13.0) ±0.1	13.0	(12.8)	−5	40
	H_2L/HL.H	9.24±0.04	9.23	9.40 ±0.05	−6.0	22
Cu^{2+}	$ML/M.L$	13.0 ±0.1	13.6 ±0.0	(14.1) ±0.0	−10	30
	$ML_2/M.L^2$	24.9 ±0.1	24.9 ±0.0	(24.6) −0.1		
Fe^{3+}	$ML/M.L$	20.0				
	$ML_2/M.L^2$	34.7				
	$ML_3/M.L^3$	43.8				
	$ML.H_2/M.H_2L$		−1.35			
As(V)	$M(OH)_2L_2/H_2MO_4.(H_2L)^2$	6.53[h]				
	$ML_3/H_2MO_4.(H_2L)^3$	8.64[h]				

[h]20°, 0.1

Bibliography:

H^+ 76GS,78AS

Cu^{2+} 70GS,76GK,79MBb

Fe^{3+} 73MP,78AS

As(V)   77VBa

Other references: 58P,68AM,69SZ,70B,70CB, 70GO,76MMa,76SSb,77DS,77KVc,77SF, 77TS,78Ke,78MSb,79GKD,79ND

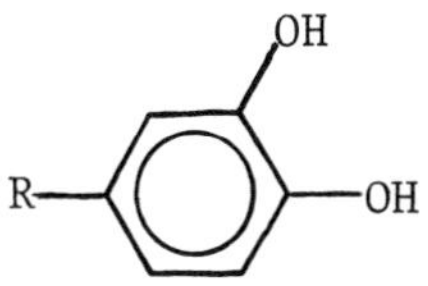

$C_xH_yO_2$		4-Substitueded-1,2-dihydroxybenzene		H_2L

R	Metal ion	Equilibrium	Log K 25°, 1.0
R=CH$_3$ (4-Methyl-) (C$_7$H$_8$O$_2$)	Fe^{3+}	ML.H^2/M.H$_2$L	-1.55
R=C(CH$_3$)$_3$ (4-[1,1-Dimethylethyl]-) (C$_{10}$H$_{14}$O$_2$)	Fe^{3+}	ML.H^2/M.H$_2$L	-1.65
R=CN (4-Cyano-) (C$_7$H$_5$O$_2$N)	Fe^{3+}	ML.H^2/M.H$_2$L	-0.69

Bibliography: 76MPS

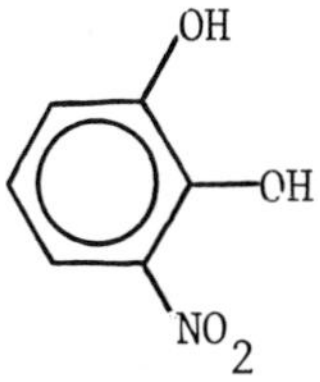

$C_6H_5O_4N$	3-Nitro-1,2-dihydroxybenzene (Other references on Vol.3, p.350)	H_2L

Metal ion	Equilibrium	Log K 25°, 0.1
H$^+$	HL/H.L	11.83
	H$_2$L/HL.H	6.49
B(III)	M(OH)$_2$L.H/M(OH)$_3$.H$_2$L	-3.54

Bibliography:

B(III)   72HK

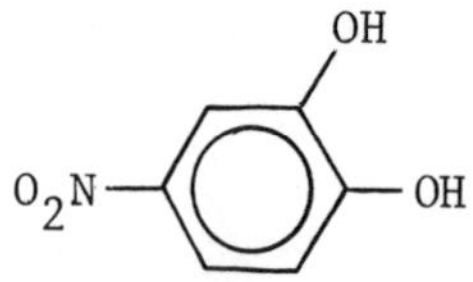

$C_6H_5O_4N$		4-Nitro-1,2-dihydroxybenzene				H_2L
		(Other values on Vol.3, p.203)				

Metal ion	Equilibrium	Log K 25°, 0.1	Log K 20°, 0.1	Log K 30°, 0.1	ΔH 25°, 0.1	ΔS 25°, 0.1
H^+	HL/H.L	10.85 ±0.05	11.0	10.75		
	H_2L/HL.H	6.69 ±0.03	6.76 ±0.03	6.59	-5.7	11
Fe^{3+}	ML/M.L	17.1				
	$ML_2/M.L^2$	30.5				
	$ML_3/M.L^3$	40.0				
B(III)	$M(OH)_2L.H/M(OH)_3.H_2L$					
		-3.95 ±0.05	-4.0			
As(III)	$M(OH)_2L.H/M(OH)_3.H_2L$		-9.49[t]			

---

[t] 22°, 0.1

Bibliography:

H^+,Fe^{3+}	78AS	As(III)	75VHa
B(III)	72HK	Other reference;	76NV

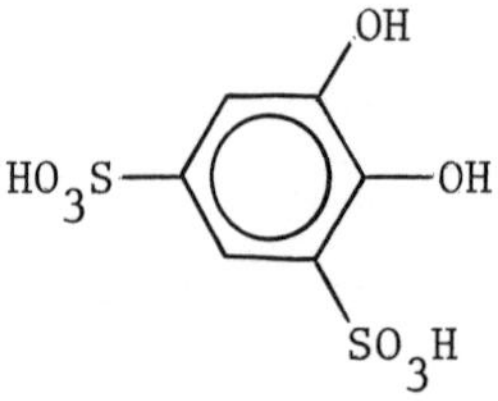

$C_6H_6O_8S_2$		1,2-Dihydroxybenzene-3,5-disulfonic acid (tiron)				H_4L
		(Other values on Vol.3, p.205)				

Metal ion	Equilibrium	Log K 25°, 0.1	Log K 25°, 1.0	Log K 25°, 0	ΔH 25°, 0.1	ΔS 25°, 0.1
H^+	HL/H.L	12.5 ±0.1	11.8 ±0.1	13.3	(-20)[r]	(-10)
		12.6[h]	12.0[b] ±0.1			
	H_2L/HL.H	7.62 ±0.08	7.20 ±0.05			
		7.66[h]	7.28[b] ±0.03			
Be^{2+}	ML/M.L	12.88[h]	12.57[b]			
	$ML_2/M.L^2$	22.25[h]	22.12[b]			
	MHL/M.HL	4.24[h]	4.47[b]			
	$MHL_2/ML_2.H$	5.54[h]	4.9[b]			
Y^{3+}	ML/M.L	13.72	13.33[b]			
	$ML_2/M.L^2$		22.50[b]			
	MHL/ML.H	3.89				
	$MHL_2/ML_2.H$		7.61[b]			
Pr^{3+}	ML/M.L	13.47	12.6[b]			
	$ML_2/M.L^2$		20.4[b]			
	$MHL_2/ML_2.H$		8.4[b]			

---

[b] 25°, 0.5,   [h] 20°, 0.1,   [r] 25-35°, 0.1

Tiron (continued)

Metal ion	Equilibrium	Log K 25°, 0.1	Log K 25°, 1.0	Log K 25°, 0	ΔH 25°, 0.1	ΔS 25°, 0.1
Nd^{3+}	$ML/M.L$	13.69	12.67^b			
	$ML_2/M.L^2$		20.42^b			
	$MHL/ML.H$	4.40				
	$MHL_2/ML_2.H$		8.36^b			
Sm^{3+}	$ML/M.L$	13.92	13.16^b			
	$ML_2/M.L^2$		21.50^b			
	$MHL/ML.H$	4.28				
	$MHL_2/ML_2.H$		8.04^b			
Eu^{3+}	$ML/M.L$		13.33^b			
	$ML_2/M.L^2$		21.71^b			
	$MHL_2/ML_2.H$		7.88^b			
Gd^{3+}	$ML/M.L$	14.10	13.29^b			
	$ML_2/M.L^2$		21.93^b			
	$MHL/ML.H$	4.30				
	$MHL_2/ML_2.H$		7.87^b			
Tb^{3+}	$ML/M.L$	14.14	13.48^b			
	$ML_2/M.L^2$		22.48^b			
	$MHL/ML.H$	4.05				
	$MHL_2/ML_2.H$		7.75^b			
Dy^{3+}	$ML/M.L$	14.36	13.58^b			
	$ML_2/M.L^2$		22.75^b			
	$MHL/ML.H$	3.71				
	$MHL_2/ML_2.H$		7.73^b			
Ho^{3+}	$ML/M.L$	14.39	13.67^b			
	$ML_2/M.L^2$		23.15^b			
	$MHL/ML.H$	3.50				
	$MHL_2/ML_2.H$		7.64^b			
Er^{3+}	$ML/M.L$	14.48	13.68^b			
	$ML_2/M.L^2$		23.19^b			
	$MHL/ML.H$	3.45				
	$MHL_2/ML_2.H$		7.61^b			
Tm^{3+}	$ML/M.L$	14.36	13.75^b			
	$ML_2/M.L^2$		23.29^b			
	$MHL/ML.H$	3.79				
	$MHL_2/ML_2.H$		7.56^b			
Yb^{3+}	$ML/M.L$	14.43	14.04^b			
	$ML_2/M.L^2$		23.55^b			
	$MHL/ML.H$	3.79				
	$MHL_2/ML_2.H$		7.51^b			
Lu^{3+}	$ML/M.L$		14.08^b			
	$ML_2/M.L^2$		23.64^b			
	$MHL_2/ML_2.H$		7.44^b			

---

^{b} 25°, 0.5

Tiron (continued)

Metal ion	Equilibrium	Log K 25°, 0.1	Log K 25°, 1.0	Log K 25°, 0	ΔH 25°, 0.1	ΔS 25°, 0.1
Fe^{3+}	$ML/M.L$	20.2 ± 0.2	19.1 ± 0.1			
		20.7^h	19.5^b			
	$ML_2/M.L^2$	35.3 ± 0.1	33.9			
		35.9^h	34.6^b			
	$ML_3/M.L^3$	45.5 ± 0.3	45.3			
		46.9^h	$(46.0)^b$			
	$MHL/M.HL$	9.7	9.1			
		10.0^h	$(9.1)^b$			
Zn^{2+}	$ML/M.L$	10.41^h	9.93^b	11.68	$(-9)^s$	(20)
		10.19^p				
	$ML_2/M.L^2$	18.52^p	17.69^b			
	$MHL/M.HL$	3.30^h				
Cd^{2+}	$ML/M.L$		8.47^b	10.29		
	$ML_2/M.L^2$		14.07^b			
$As(III)$	$M(OH)_2L.H/M(OH)_3.H_2L$	-7.10^t				
	$ML_2.H/M(OH)_3.(H_2L)^2$			-8.19		

^{b}25°, 0.5; ^{h}20°, 0.1; ^{p}30°, 0.1; ^{s}20-30°, 0.1; ^{t}22°, 0.1

Bibliography:

H^+	75VHa,78AS	Zn^{2+},Cd^{2+}	77LM
Be^{2+}	77SL	$As(III)$	75VHa
Y^{3+}	76Lg,78Kc		
$Pr^{3+}-Lu^{3+}$	76Lc,78Kc		
Fe^{3+}	76MPS,78AS		

Other references: 68AM,68PL,69BF,69SZ,71AO,
74NJ,77MSa,78RMa,78SKG

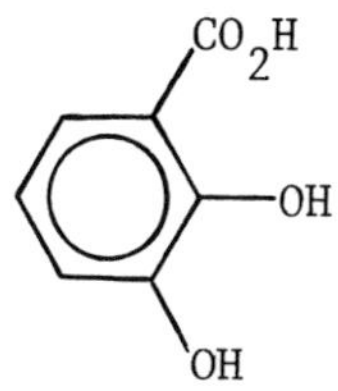

$C_7H_6O_4$ 2,3-Dihydroxybenzoic acid H_3L
                              (Other references on Vol.3, p.208)

Metal ion	Equilibrium	Log K 27°, 0.1
H^+	HL/H.L	(13.1)
	$H_2L/HL.H$	10.06
	$H_3L/H_2L.H$	2.70
Fe^{3+}	$ML/M.L$	21.4
	$ML_2/M.L_2$	28.7
	$ML_3/M.L_3$	33.3
	$M(HL)L/ML_2.H$	8.2

Bibliography: 78AS

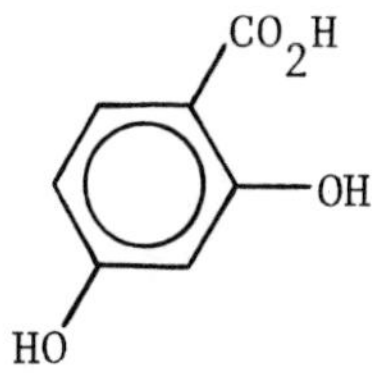

$C_7H_6O_4$ 2,4-Dihydroxybenzoic acid   (β-resorcylic acid) H_3L
                              (Other values on Vol.3, p.209)

Metal ion	Equilibrium	Log K 25°, 0.1	Log K 25°, 0.5	Log K 25°, 1.0	ΔH 25°, 0.5	ΔS 25°, 0.5
H^+	HL/H.L		(13.37)			
	$H_2L/HL.H$	8.60	8.56	8.69	-6.9	16
	$H_3L/H_2L.H$	3.13 ±0.03	3.12	3.19		
Be^{2+}	MHL/M.HL		6.87			
	$M(HL)_2/M.(HL)^2$		11.19			
	$M(HL)_2/M(HL)L.H$		8.91			
	$M(HL)L/ML_2.H$		9.22			

Bibliography:

H^+ 78MGK,79LK

Be^{2+} 79LK

Other references: 70DD,72AK,76SJ,77SJ, 78SD

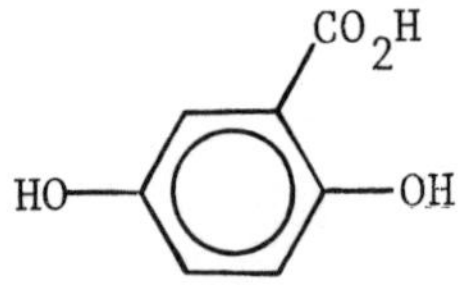

$C_7H_6O_4$ 2,5-Dihydroxybenzoic acid (gentistic acid) H_3L
                               (Other values on Vol.3, p.209)

Metal ion	Equilibrium	Log K 25°, 0.1	Log K 25°, 0.5
H^+	HL/H.L		12.74
	$H_2L/HL.H$		10.00
	$H_3L/H_2L.H$	2.70	2.73
Be^{2+}	MHL/M.HL		9.10
	$M(HL)_2/M.(HL)^2$		15.87
	$M(HL)_2/M(HL)L.H$		9.94
	$M(HL)L/ML_2.H$		10.44

Bibliography: 78LKb   Other reference: 73KJ

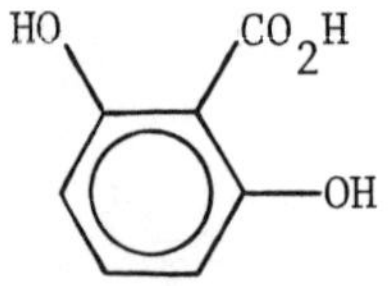

$C_7H_6O_4$ 2,6-Dihydroxybenzoic acid (γ-resorcylic acid) H_3L
                               (Other values on Vol.3, p.209)

Metal ion	Equilibrium	Log K 25°, 0.1	Log K 25°, 0.5
H^+	HL/H.L		(13.28)
	$H_2L/HL.H$		12.57
	$H_3L/H_2L.H$	(1.08)	1.20
Be^{2+}	MHL/M.HL		11.92
	$M(HL)_2/M.(HL)^2$		21.96
	$M(HL)_2/M(HL)L.H$		11.76
	$M_2L/M^2.L$		25.09

Bibliography: 79LK

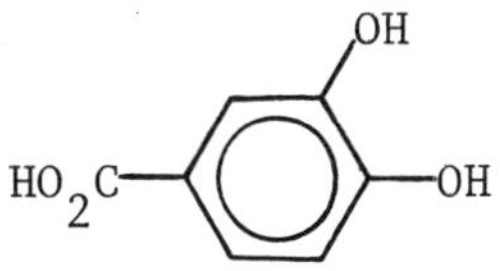

$C_7H_6O_4$ 3,4-Dihydroxybenzoic acid (protocateuchuic acid) H_3L
                    (Other values on Vol.3, p.209)

Metal ion	Equilibrium	Log K 25°, 0.1	Log K 30°, 0.1	Log K 25°, 1.0	ΔH 25°, 0.1	ΔS 25°, 0.1
H^+	HL/H.L	(11.7)	(12.2)±0.3	(12.8)	(-9)[r]	(23)
	H_2L/HL.H	8.83 ±0.01	8.70±0.03	8.68	(-7)[r]	(17)
	$H_3L/H_2L.H$	4.32 ±0.03	4.34±0.05	4.34		
Fe^{3+}	$ML.H^2/M.H_2L$			-0.99		

[r] 1-45°, 0.1

Bibliography: H^+ 77IH; Fe^{3+} 76MPS   Other references: 69SZ,76MMa,76NV

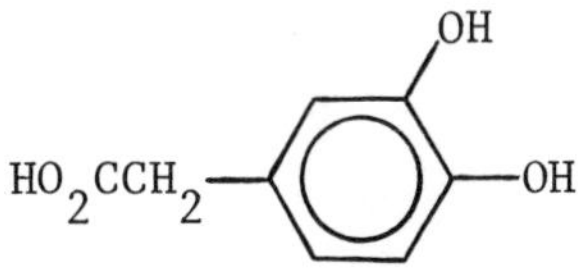

$C_8H_8O_4$ 3,4-Dihydroxyphenylacetic acid H_3L

Metal ion	Equilibrium	Log K 25°, 0.1	ΔH 25°, 0.1	ΔS 25°, 0.1
H^+	HL/H.L	(13.7) −1	(-10)[r]	(29)
	H_2L/HL.H	9.50 ±0.03	(-6)[r]	(23)
	$H_2L/H_3L.H$	4.14 ±0.03		
Fe^{3+}	ML/M.L	20.2		
	$ML_2/M.L^2$	35.1		
	$ML_3/M.L^3$	44.0		

[r] 1-45°, 0.1

Bibliography: H^+ 77IH,78AS; Fe^{3+} 78AS

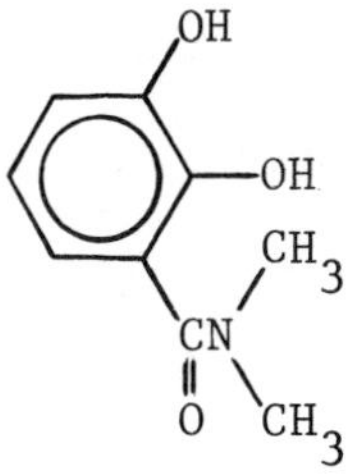

$C_9H_{11}O_3N$ N,N-Dimethyl-2,3-dihydroxybenzamide H_2L

Metal ion	Equilibrium	Log K 25°, 0.1
H^+	HL/H.L	(12.1)
	H_2L/HL.H	8.42
Fe^{3+}	ML/M.L	17.8
	$ML_2/M.L^2$	31.7
	$ML_3/M.L^3$	40.2

Bibliography: 79HCC

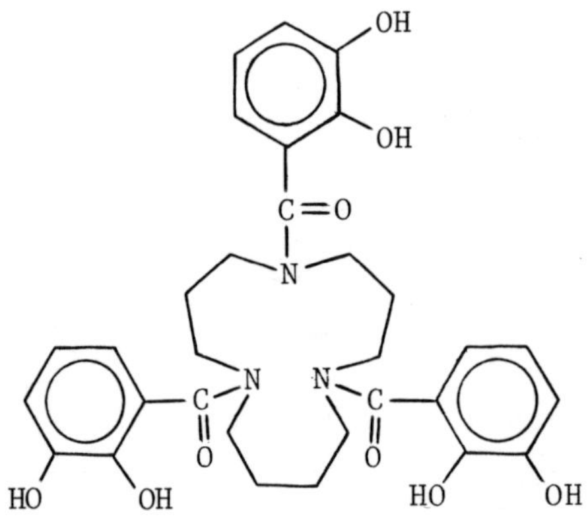

$C_{25}H_{35}O_9N_3$ **1,5,9-Tris(2,3-dihydroxybenzoyl)-1,5,9-triazatridecane** H_6L

Metal ion	Equilibrium	Log K 25°, 0.1
H^+	$H_4L/H_3L.H$	9.3
	$H_5L/H_4L.H$	8.65
	$H_6L/H_5L.H$	7.87
Fe^{3+}	$ML.H^3/M.H_3L$	1.5
	$MHL/ML.H$	9.0
	$MH_2L/MHL.H$	7.6
	$MH_3L/MH_2L.H$	5.9

Bibliography: 79HR

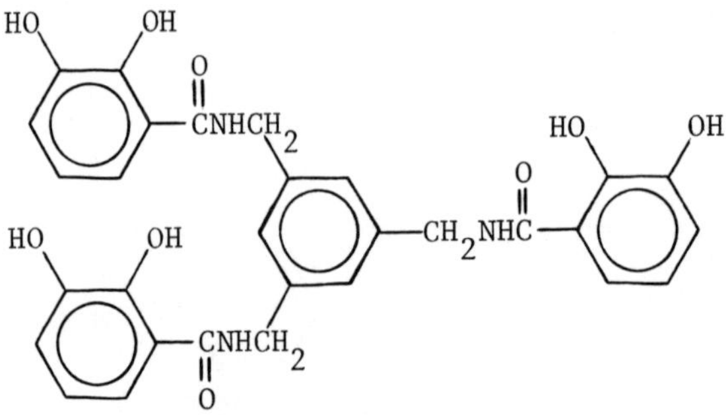

$C_{30}H_{27}O_9N_3$ **1,3,5-Tris(2,3-dihydroxybenzoylaminomethyl)benzene** H_6L

Metal ion	Equilibrium	Log K 25°, 0.1
Fe^{3+}	$ML.H^3/M.H_3L$	9.5
	$MHL/ML.H$	6.9
	$MH_2L/MHL.H$	5.7

Bibliography: 79HR

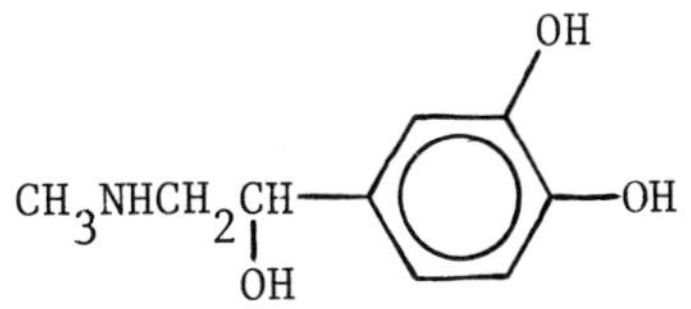

$C_9H_{13}O_3N$ L-2-(Methylamino)-1-(3,4-dihydroxyphenyl)ethanol (adrenaline) H_3L
                              (Other references on Vol.2, p.335)

Metal ion	Equilibrium	Log K 25°, 0.1	Log K 25°, 1.0
H^+	HL/H.L	(13)	
	$H_2L/HL.H$	9.90 ±0.05	
	$H_3L/H_2L.H$	8.65 ±0.01	
Fe^{3+}	$ML.H^2/M.H_2L$		-1.16

Bibliography: 76MPS

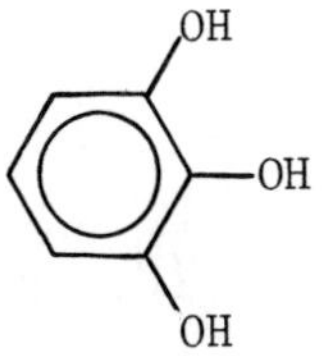

$C_6H_6O_3$ 1,2,3-Trihydroxybenzene (pyrogallol) H_3L
                              (Other values on Vol.3, p.212)

Metal ion	Equilibrium	Log K 25°, 0.1	Log K 20°, 0.1
H^+	HL/H.L		(14)
	$H_2L/HL.H$		11.08
	$H_3L/H_2L.H$	8.98	8.94
As(III)	$M(OH)_2HL.H/M(OH)_3.H_2L$	(-6.32)	(-7.73)[t]
	$M(HL)_2.H/M(OH)_3.(H_2L)^2$	-6.05	
As(V)	$ML_2(OH)_2/H_2MO_4.(H_2L)^2$		10.37
	$ML_3/H_2MO_4.(H_3L)^3$		28.06

[t] 22°, 0.1

Bibliography: As(III) 75VHa;   As(V) 77VBa;   Other references: 76MMa,77TS,78MSb,78SM

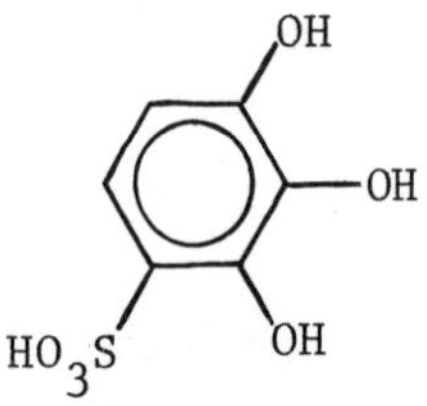

$C_6H_6O_6S$	2,3,4-Trihydroxybenzenesulfonic acid (pyrogallolsulfate)		H_4L
	(Other value on Vol.3, p.213)		

Metal ion	Equilibrium	Log K 20°, 0.1
H^+	$H_2L/HL.H$	11.3
	$H_3L/H_2L.H$	8.28 ±0.01
As(III)	$M(OH)_2L.H/M(OH)_3.H_2L$	-6.64^t

^{t}22°, 0.1

Bibliography: 75VHa

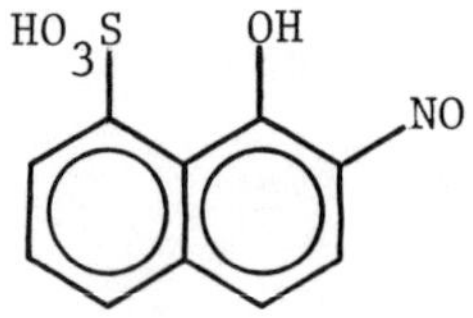

$C_{10}H_7O_8NS$	2-Nitroso-1-naphthol-8-sulfonic acid			H_2L
	(Other values on Vol.3, p.320)			

Metal ion	Equilibrium	Log K 25°, 0.1	Log K 25°, 1.0	Log K 25°, 0
H^+	HL/H.L	7.74	7.50	8.19
		7.50^b		
Y^{3+}	ML/M.L	4.82		
La^{3+}	ML/M.L	4.70		
Ce^{3+}	ML/M.L	5.04		
Pr^{3+}	ML/M.L	5.28		
Nd^{3+}	ML/M.L	5.46		
Sm^{3+}	ML/M.L	5.65		
Eu^{3+}	ML/M.L	5.61		
Gd^{3+}	ML/M.L	5.50		
Tb^{3+}	ML/M.L	5.33		
Dy^{3+}	ML/M.L	5.21		
Ho^{3+}	ML/M.L	5.10		
Er^{3+}	ML/M.L	5.01		
Tm^{3+}	ML/M.L	4.97		
Yb^{3+}	ML/M.L	5.06		
Lu^{3+}	ML/M.L	5.06		

^{b}25°, 0.5

Bibliography: 78PP

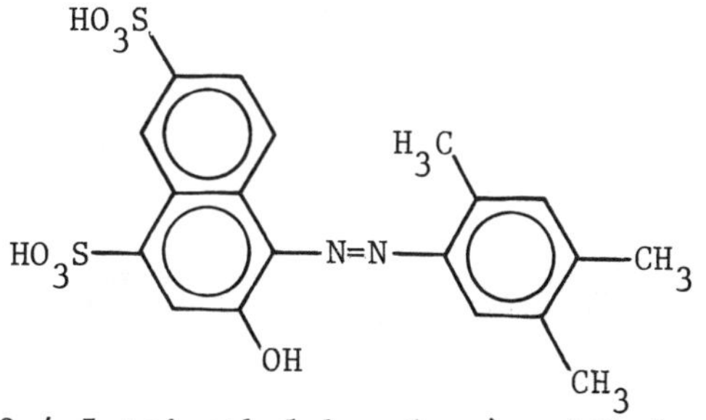

$C_{19}H_{18}O_7N_2S_2$ 3-Hydroxy-4-(2,4,5-trimethylphenylazo)naphthalene-2,7-disulfonic acid H_3L
(Ponceau 3R)

Metal ion	Equilibrium	Log K 25°, 0.1	Log K 25°, 1.0	Log K 25°, 0
H^+	HL/H.L	11.56	10.96[b] 11.12[b]	12.28
Cu^{2+}	ML/M.L	8.90	8.42[b]	10.11
	$ML_2/M.L^2$	16.1	16.0[b]	

[b] 25°, 0.5

Bibliography:

H^+ 73HP

Cu^{2+} 73HKa

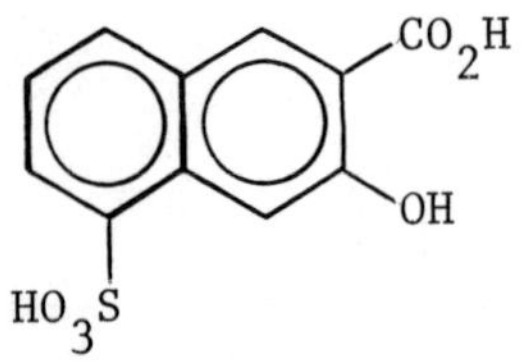

$C_{11}H_8O_6S$ 3-Hydroxy-5-sulfo-2-naphthoic acid H_3L

Metal ion	Equilibrium	Log K 25°, 0.1
H^+	HL/H.L	11.49
	H_2L/HL.H	2.39
Be^{2+}	ML/M.L	11.05
	ML_2/M.L	18.94
Cu^{2+}	ML/M.L	8.88
	ML_2/M.L	14.67

Bibliography: 74SRR

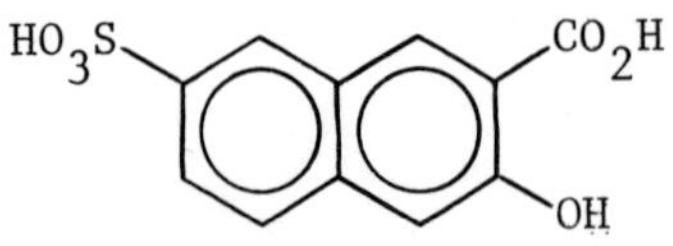

$C_{11}H_8O_6S$ 3-Hydroxy-7-sulfo-2-naphthoic acid H_3L

Metal ion	Equilibrium	Log K 25°, 0.1
H^+	HL/H.L	11.73
	H_2L/HL.H	2.44
Be^{2+}	ML/M.L	11.15
	ML_2/M.L_2	19.56
Cu^{2+}	ML/M.L	8.89
	ML_2/M.L_2	15.22

Bibliography: 74SRR

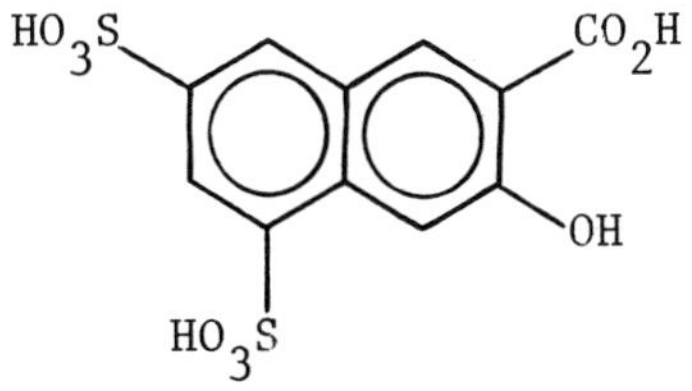

| C$_{11}$H$_8$O$_9$S | 3-Hydroxy-5,7-disulfo-2-naphthoic acid (Other values on Vol.3, p.233) | | | H$_4$L |

Metal ion	Equilibrium	Log K 25°, 0.1	Log K 25°, 1.0	Log K 25°, 0
H$^+$	HL/H.L	10.28[b] 10.91[b]	10.81	12.03
	H$_2$L/HL.H	2.37[b] 2.14[b]	2.18	2.98
Be^{2+}	ML/M.L ML$_2$/M.L^2	10.18[b] 18.17[b]		
UO$_2$$^{2+}$	ML/M.L ML$_2$/M.L^2	9.81[b] 17.40[b]		
Al^{3+}	ML/M.L ML$_2$/M.L^2	10.81[b] 19.26[b]		

[b]25°, 0.5

Bibliography:
H$^+$,Be^{2+} 74SRL   Al^{3+} 78L
UO$_2$$^{2+}$ 78LKc

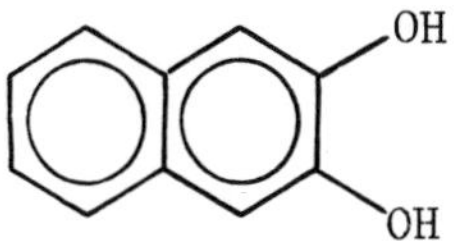

| C$_{10}$H$_8$O$_2$ | 2,3-Dihydroxynaphthalene (Other values on Vol.3, p.233) | | H$_2$L |

Metal ion	Equilibrium	Log K 25°, 0.1	Log K 20°, 0.1
H$^+$	HL/H.L		(12.5)
	H$_2$L/HL.H	8.55	8.68
Fe^{3+}	ML.H^2/M.H$_2$L		−1.15
	ML$_2$.H^2/ML.H$_2$L		−6.5
	ML$_3$.H^2/ML$_2$.H$_2$L		−12.0
As(III)	M(OH)$_2$L.H/M(OH)$_3$.H$_2$L		−6.98[t]

[t]22°, 0.1

Bibliography:
Fe^{3+} 69ZS   As(III)   75VHa

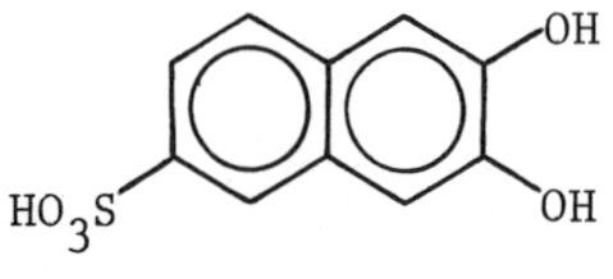

| $C_{10}H_8O_5S$ | 2,3-Dihydroxynaphthalene-6-sulfonic acid | | | H_3L |
| | (Other values on Vol.3, p.234) | | | |

Metal ion	Equilibrium	Log K 25°, 0.1	Log K 25°, 1.0	Log K 25°, 0
H^+	HL/H.L	12.0 ±0.3	11.45	12.7
		12.2 [h]	11.75 [b]	
	H_2L/HL.H	8.13 [h] ±0.04	8.02 [b]	8.55
		8.21 [h] ±0.03	8.03 [b]	
Y^{3+}	$ML/M.L$		10.14 [b]	
	$ML_2/M.L^2$		18.22 [b]	
	$ML_3/M.L^3$		24.1 [b]	
	$MHL_2/ML_2.H$		6.6 [b]	
Pr^{3+}	$ML/M.L$		9.19 [b]	
	$ML_2/M.L^2$		16.27 [b]	
	$MHL_2/ML_2.H$		7.33 [b]	
Nd^{3+}	$ML/M.L$		9.26 [b]	
	$ML_2/M.L^2$		16.40 [b]	
	$MHL_2/ML_2.H$		7.31 [b]	
Sm^{3+}	$ML/M.L$		9.82 [b]	
	$ML_2/M.L^2$		17.15 [b]	
	$MHL/ML.H$		5.6 [b]	
	$MHL_2/ML_2.H$		7.28 [b]	
Eu^{3+}	$ML/M.L$		9.90 [b]	
	$ML_2/M.L^2$		17.25 [b]	
	$MHL/M.HL$	4.10	3.7 [b]	
	$MHL_2/MHL.L$	9.02	9.0 [b]	
	$MHL/ML.H$		5.6 [b]	
	$MHL_2/ML_2.H$		7.26 [b]	
Gd^{3+}	$ML/M.L$		9.92 [b]	
	$ML_2/M.L^2$		17.49 [b]	
	$MHL/ML.H$		4.3 [b]	
	$MHL_2/ML_2.H$		7.30 [b]	
Tb^{3+}	$ML/M.L$		10.20 [b]	
	$ML_2/M.L^2$		18.10 [b]	
	$MHL_2/ML_2.H$		6.94 [b]	
Dy^{3+}	$ML/M.L$		10.25 [b]	
	$ML_2/M.L^2$		18.29 [b]	
	$MHL_2/ML_2.H$		6.88 [b]	
Ho^{3+}	$ML/M.L$		10.26 [b]	
	$ML_2/M.L^2$		18.40 [b]	
	$MHL_2/ML_2.H$		6.83 [b]	

[b]25°, 0.5; [h]20°, 0.1

2,3-Dihydroxynaphthalene-6-sulfonic acid   (continued)

Metal ion	Equilibrium	Log K 25°, 0.1	Log K 25°, 1.0	Log K 25°, 0
Er^{3+}	$ML/M.L$		10.31^b	
	$ML_2/M.L^2$		18.67^b	
	$MHL_2/ML_2.H$		6.82^b	
Tm^{3+}	$ML/M.L$		10.37^b	
	$ML_2/M.L^2$		18.92^b	
	$ML_3/M.L^3$		23.3^b	
	$MHL_2/ML_2.H$		6.81^b	
Yb^{3+}	$ML/M.L$		10.43^b	
	$ML_2/M.L^2$		19.08^b	
	$ML_3/M.L^3$		23.6^b	
	$MHL_2/ML_2.H$		6.77^b	
Lu^{3+}	$ML/M.L$		10.44^b	
	$ML_2/M.L^2$		19.28^b	
	$ML_3/M.L^3$		24.1^b	
	$MHL_2/ML_2.H$		6.76^b	
Ni^{2+}	$ML/M.L$		8.41^b	
	$ML_2/M.L^2$		14.14^b	
	$ML_3/M.L^3$		17.3^b	
Cu^{2+}	$ML/M.L$	13.3 ± 0.2	12.83^b	
	$ML_2/M.L^2$	23.7 ± 0.3	23.34^b	
	$MHL/ML.H$	3.55		
Fe^{3+}	$ML.H^2/M.H_2L$	-0.75^h		
	$ML_2.H^2/ML.H_2L$	-6.05^h		
	$ML_3.H^2/ML_2.H_2L$	-10.5^h		
Zn^{2+}	$ML/M.L$		8.98^b	
	$ML_2/M.L^2$		16.51^b	
Cd^{2+}	$ML/M.L$		7.70^b	
	$ML_2/M.L^2$		13.23^b	
Al^{3+}	$ML/M.L$		15.11^b	
	$ML_2/M.L^2$		27.88^b	
	$ML_3/M.L^3$		37.47^b	
As(III)	$M(OH)_2L.H/M(OH)_3.H_2L$	-6.97^t		

---

^{b}25°, 0.5; ^{h}20°, 0.1; ^{t}22°, 0.1

Bibliography:

$H^+, Pr^{3+}-Lu^{3+}$ 76Le

Y^{3+} 76Lg

$Ni^{2+}, Cu^{2+}, Zn^{2+}-Al^{3+}$ 76Lf

Fe^{3+} 69ZS

As(III)   75VHa

Other references: 75PM,78SM

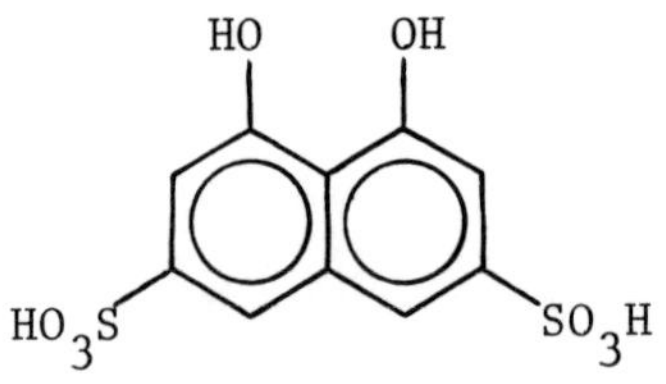

$C_{10}H_8O_8S_2$ 1,8-Dihydroxynaphthalene-3,6-disulfonic acid   (chromotropic acid) H_4L
                       (Other values on Vol.3, p.235)

Metal ion	Equilibrium	Log K 25°, 0.1	Log K 25°, 0.5	Log K 25°, 0	ΔH 25°, 0.1	ΔS 25°, 0.1
H^+	$H_2L/HL.H$	5.35 ±0.05	5.13	5.98 −0.01	−3.3	13
Be^{2+}	$ML.H/M.HL$	0.74[h]	0.63			
	$ML_2.H^2/M.(HL)^2$	−3.01[h]	−2.21			
	$MHL/M.HL$	2.9[h]	3.16			
Sm^{3+}	$MHL/M.HL$	2.26		3.71		
Eu^{3+}	$MHL/M.HL$	2.37		3.87		
Gd^{3+}	$MHL/M.HL$	2.41		3.76		
Tb^{3+}	$MHL/M.HL$	2.41		3.83		
Dy^{3+}	$MHL/M.HL$	2.35		3.43		
Ho^{3+}	$MHL/M.HL$	2.20		3.28		
$Ge(IV)$	$ML_3.H^2/M(OH)_4.(H_2L)^3$	−4.8[h]				

------

[h]20°, 0.1

Bibliography:

H^+ 74ML   Ge(IV)   79MBc

Be^{2+} 77SL   Other reference: 79BG

Sm^{3+}-Ho^{3+} 74ML

$$CH_3\overset{\overset{O}{\|}}{C}CH_2\overset{\overset{O}{\|}}{C}CH_3$$

$C_5H_8O_2$ <u>Pentane-2,4-dione</u>   (acetylacetone)   HL
(Other values on Vol.3, p.244)

Metal ion	Equilibrium	Log K 25°, 0.1	Log K 25°, 1.0	Log K 25°, 0	ΔH 25°, 0	ΔS 25°, 0
H^+	HL/H.L	8.80 ±0.06	8.80 ±0.09	9.02 ±0.03	-3.3[a]±0.1	29[a]
Mg^{2+}	ML/M.L			3.65	-1.0[u]	13
	$ML_2/M.L^2$			6.25	-4.3[u]	14
La^{3+}	ML/M.L	4.94 ±0.02		5.1[t]	-0.2[a]±0.1	22[a]
	$ML_2/M.L^2$	8.42		9.0[t]	-0.7[a]	36[a]
	$ML_3/M.L^3$	10.9		11.9[t]	-1[a]	47[a]
Nd^{3+}	ML/M.L	5.36 ±0.06		5.6[t]	-0.5[a]±0.2	23[a]
	$ML_2/M.L^2$	9.4		9.9[t]	-1.3[a]	39[a]
	$ML_3/M.L^3$	12.6		13.1[t]	-4[a]	44[a]
Sm^{3+}	ML/M.L	5.67 ±0.07		5.9[t]	-0.8[a]±0.0	23[a]
	$ML_2/M.L^2$	10.05		10.4[t]	-2.8[a]	37[a]
	$ML_3/M.L^3$	13.0		13.6[t]	-5[a]	43[a]
Gd^{3+}	ML/M.L	5.90 ±0.01			-1.0[a]±0.0	24[a]
	$ML_2/M.L^2$	10.42			-3.2[a]	37[a]
	$ML_3/M.L^3$	13.9			-6[a]	43[a]
Dy^{3+}	ML/M.L	6.06 ±0.02			-1.2[a]±0.1	24[a]
	$ML_2/M.L^2$	10.74			-3.5[a]	37[a]
	$ML_3/M.L^3$	14.1			-7[a]	41[a]
Ho^{3+}	ML/M.L	6.07 -0.01			-1.1[a]±0.0	24[a]
	$ML_2/M.L^2$	10.77			-3.4[a]	38[a]
	$ML_3/M.L^3$	14.2			-7[a]	42[a]
Yb^{3+}	ML/M.L	6.18 +0.01			-1.0[a]±0.0	25[a]
	$ML_2/M.L^2$	11.08			-3.1[a]	40[a]
	$ML_3/M.L^3$	14.7			-6[a]	47[a]
Mn^{2+}	ML/M.L	4.07	4.09	4.21	-1.4[a]±0.1	15
	$ML_2/M.L^2$		6.98	7.30	-4.4[u]	19
Co^{2+}	ML/M.L	5.18		5.40	-2.8[u]	15
	$ML_2/M.L^2$	9.42		9.54	-8.1[u]	16
Ni^{2+}	ML/M.L	5.72		6.00	-3.7[a]±0.3	15
	$ML_2/M.L^2$	9.66		10.60	-8.4[a]±0.8	20
	$ML_3/M.L^3$			12.8	-14[u]	12
Cu^{2+}	ML/M.L	8.16	8.22	8.25	-4.6[a]±0.2	22
	$ML_2/M.L^2$	14.76	14.81	15.05	-10.3[a]±0.3	33
Zn^{2+}	ML/M.L	4.68		5.06	-1.6[a]±0.2	16
	$ML_2/M.L^2$	7.92		9.00	-4.1[a]±0.7	22
Cd^{2+}	ML/M.L		3.75	3.83	-0.8[u]	15
	$ML_2/M.L^2$		6.49	6.65	-2.8[u]	21
Ga^{3+}	ML/M.L		9.29	9.4[t]		
	$ML_2/M.L^2$		17.27	17.8[t]		
	$ML_3/M.L^3$		23.6	23.7[t]		

---

[a] 25°, 0.1;   [t] 30°, 0;   [u] 25°, 0.05

Acetylacetone (continued)

Bibliography:

H^+ 77BMK Cd^{2+} 77SI

$Mg^{2+}, Cu^{2+}, Zn^{2+}$ 79PK Ga^{3+} 72LV

$La^{3+} - Yb^{3+}$ 76KP   Other references: 68BD,71MS,71RM,71SG,76FA,
                                                        77BMa,77SMa,78MY
Mn^{2+} 77SMT

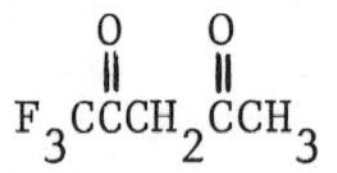

Co^{2+} 69Sa

$$\underset{F_3CCCH_2CCH_3}{\overset{O\quad O}{\overset{\|\quad\|}{}}}$$

		Log K	Log K	
$C_5H_5O_2F_3$		**1,1,1-Trifluoropentane-2,4-dione**		HL
		(Other references on Vol.3, p.363)		
Metal ion	Equilibrium	25°, 0.1	25°, 1.0	
H^+	HL/H.L	6.09	6.09	
Mn^{2+}	ML/M.L		(0.94)	
	$ML_2/M.L^2$		(2.96)	
Co^{2+}	ML/M.L		3.50	
	$ML_2/M.L^2$		5.60	
Zn^{2+}	ML/M.L		2.72	
	$ML_2/M.L^2$		4.48	
Cd^{2+}	ML/M.L		2.08	
	$ML_2/M.L^2$		3.20	

Bibliography:

H^+, Zn^{2+} 71ST Co^{2+} 71MS

Mn^{2+} 77SMT Cd^{2+} 77SI

$$\underset{F_3CCCH_2CCF_3}{\overset{O\quad O}{\overset{\|\quad\|}{}}}$$

		Log K	Log K	
$C_5H_2O_2F_6$		**1,1,1,5,5,5-Hexafluoropentane-2,4-dione**		HL
		(Other values on Vol.3, p.247)		
Metal ion	Equilibrium	20°, 0.1	25°, 1.0	
H^+	HL/H.L	4.35	4.34	
Mn^{2+}	ML/M.L		1.04	
Co^{2+}	ML/M.L		1.56	
	$ML_2/M.L^2$		2.32	
Zn^{2+}	ML/M.L		1.0	

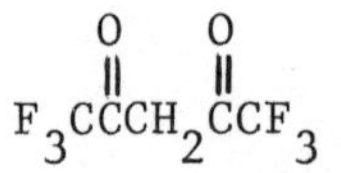

Bibliography:

H^+, Zn^{2+} 71SI Co^{2+} 71MS

Mn^{2+} 77SMT

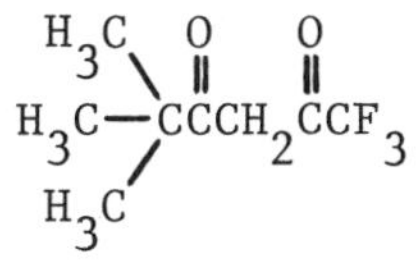

$$H_3C-\underset{\underset{CH_3}{|}}{\overset{\overset{CH_3}{|}}{C}}-\overset{O}{\overset{||}{C}}CH_2\overset{O}{\overset{||}{C}}CF_3$$

| C$_8$H$_{11}$O$_2$F$_3$ | 2,2-Dimethyl-5,5,5-trifluorohexane-3,5-dione | | HL |
| | (pivaloyltrifluoroacetone) | | |

Metal ion	Equilibrium	Log K 25°, 0.1	Log K 25°, 1.0
H$^+$	HL/H.L	7.01	6.90
Co^{2+}	ML/M.L		3.7
	ML$_2$/M.L^2		5.9
Zn^{2+}	ML/M.L		3.3
	ML$_2$/M.L^2		5.6

Bibliography:

H$^+$ 73SHI   Co^{2+},Zn^{2+} 74SMN

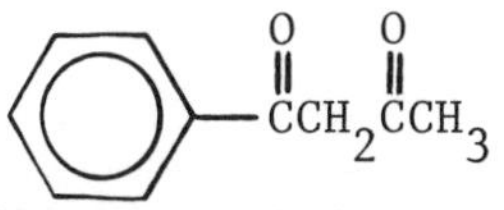

$$\text{(phenyl)}-\overset{O}{\overset{||}{C}}CH_2\overset{O}{\overset{||}{C}}CH_3$$

| C$_{10}$H$_{10}$O$_2$ | 1-Phenylbutane-1,3-dione (benzoylacetone) | | | HL |
| | (Other values on Vol.3, p.247) | | | |

Metal ion	Equilibrium	Log K 25°, 0.1	Log K 25°, 1.0	Log K 25°, 0
H$^+$	HL/H.L	8.41 ±0.02	8.4 ±0.1	8.72
Co^{2+}	ML/M.L		4.40	
	ML$_2$/M.L^2		8.00	
Zn^{2+}	ML/M.L		4.00	
	ML$_2$/M.L^2		7.55	
Cd^{2+}	ML/M.L		3.55	
	ML$_2$/M.L^2		5.85	

Bibliography:

H$^+$ 73SHI, 79MBc   Zn^{2+} 74SMN

Co^{2+} 71MS   Cd^{2+} 77SI

                                                   Other references: 71SG,78MMd

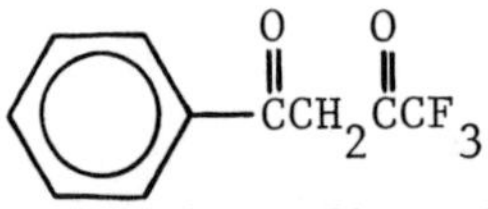

$C_{10}H_7O_2F_3$ 1-Phenyl-4,4,4-trifluorobutane-1,3-dione   HL
                     (Other values on Vol.3, p.248)

Metal ion	Equilibrium	Log K 25°, 0.1	Log K 25°, 1.0
H^+	HL/H.L	6.03 ±0.3	6.01 ±0.2
Mn^{2+}	ML/M.L		(0.80)
	$ML_2/M.L^2$		(2.63)
Co^{2+}	ML/M.L		3.40
	$ML_2/M.L^2$		5.24
Zn^{2+}	ML/M.L		3.23
	$ML_2/M.L^2$		5.49

Bibliography:

H^+ 73SHI Zn^{2+} 74SMN

Mn^{2+} 77SMT   Other references: 71OM,77SPa

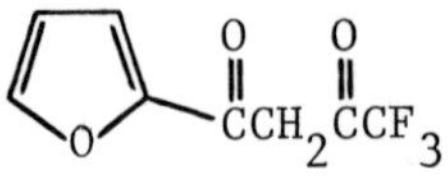

Co^{2+} 71MS

$C_8H_5O_3F_3$ 1-(2-Furanyl)-4,4,4-trifluorobutane-1,3-dione   HL
                   (Other values on Vol.3, p.248)

Metal ion	Equilibrium	Log K 25°, 0.1	Log K 25°, 1.0
H^+	HL/H.L	6.22 -0.3	6.18
Co^{2+}	ML/M.L		3.38
	$ML_2/M.L^2$		5.10
Zn^{2+}	ML/M.L		3.29
	$ML_2/M.L^2$		5.62

Bibliography:

H^+ 73SHI Co^{2+},Zn^{2+} 74SMN

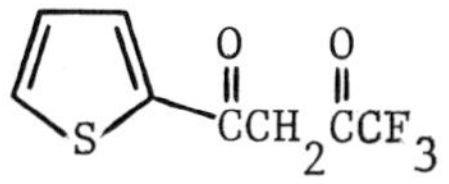

A. DI-OXO LIGANDS

| C$_8$H$_5$O$_2$F$_3$S | 1-(2-Thienyl)-4,4,4-trifluorobutane-1,3-dione (thenoyltrifluoroacetone) (Other values on Vol.3, p.249) | | | HL |

Metal ion	Equilibrium	Log K 25°, 0.1	Log K 25°, 1.0
H$^+$	HL/H.L	6.38 ±0.05	6.28 +0.2
Np^{4+}	ML.H/M.HL		1.68^d
Pu^{4+}	ML.H/M.HL		1.92^d
NpO$_2^+$	ML/M.L	2.89	
	ML$_2$/M.L^2	5.48	
Co^{2+}	ML/M.L		3.36
	ML$_2$/M.L^2		5.16
Ag$^+$	ML/M.L	1.10	
Zn^{2+}	ML/M.L		3.45
	ML$_2$/M.L^2		5.93

^{d}25°, 2.0

Bibliography:

H$^+$ 72GK,73SHI

Np^{4+},Pu^{4+} 76BRb

NpO$_2^+$ 72GK

Co^{2+},Zn^{2+} 74SMN

Ag$^+$ 73ST

Other references: 69LSS,70IK,71SG,78MMd

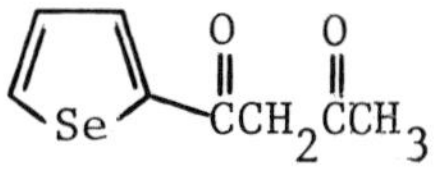

| C$_8$H$_8$O$_2$Se | 1-(2-Selenoyl)butane-1,3-dione (selenoylacetone) (Other values on Vol.3, p.249) | | HL |

Metal ion	Equilibrium	Log K 25°, 0.1
H$^+$	HL/H.L	8.55
Nd^{3+}	ML/M.L	5.62
	ML$_2$/M.L^2	11.04
	ML$_3$/M.L^3	15.5

Bibliography: 66PE

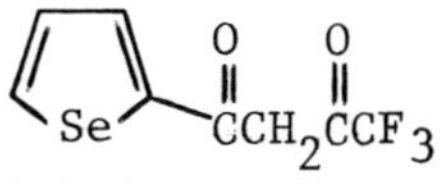

| C$_8$H$_5$O$_2$F$_3$Se | 1-(2-Selenoyl)-4,4,4-trifluorobutane-1,3-dione (Other values on Vol.3, p.250) | | HL |

Metal ion	Equilibrium	Log K 25°, 0.1	Log K 25°, 1.0
H$^+$	HL/H.L	6.33	6.32
Nd^{3+}	ML/M.L	5.04	
	ML$_2$/M.L^2	9.82	
	ML$_3$/M.L^3	13.7	

Bibliography: 66PE

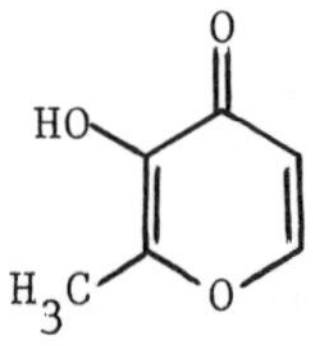

$C_6H_6O_3$ 3-Hydroxy-2-methyl-4-pyrone (maltol)   HL
(Other values on Vol.3, p.256)

Metal ion	Equilibrium	Log K 25°, 0.1	Log K 25°, 0.5	Log K 25°, 2.0	ΔH 25°, 2.0	ΔS 25°, 2.0
H^+	$HL/H.L$	8.48 ± 0.03	8.40 ± 0.05	8.52 ± 0.09	$(-5)^r$	(22)
		8.51^h	8.43^c			
UO_2^{2+}	$ML/M.L$	8.18^h				
	$ML_2/M.L_2$	14.81^h				
	$ML_3/M.L_3$	18.19^h				
	$MHL/ML.H$	1.28^h				
	$MHL_2/ML_2.H$	2.49^h				
	$MHL_3/ML_3.H$	5.8^h				
	$ML/M(OH)_2L.H^2$	11.55^h				
	$M(OH)_2L_4.H^2/M.L^4$	18.6^h				
	$M_2(OH)_3L_3.H^3/M^2.L^3$	8.6^h				
	$M_2(OH)_5L.H^5/M^2.L$	14.6^h				
Zn^{2+}	$ML/M.L$		5.53	5.37	$(-2)^r$	(18)
	$ML_2/M.L_2$		10.20	9.96	$(-4)^r$	(32)
	$ML_3/M.L_3$			12.22	$(-7)^r$	(32)
$Ge(IV)$	$M(OH)_2L_2/M(OH)_4.(HL)^2$	4.2^h	3.90			
	$ML_3/M(OH)_4.(HL)^3.H$	8.3^h				

^{c}25°, 1.0; ^{h}20°, 0.1; ^{r}20-40°, 2.0

Bibliography: H^+,Zn^{2+} 75GD; UO_2^{2+} 72H; $Ge(IV)$ 79MBc

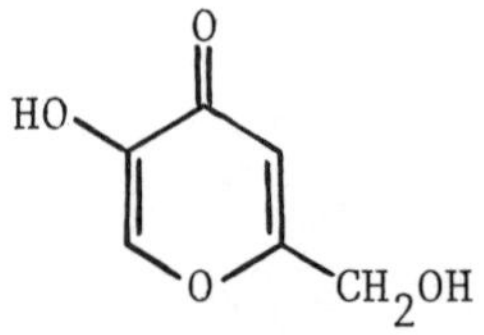

$C_6H_6O_4$ 5-Hydroxy-2-hydroxymethyl-4-pyrone (kojic acid)   HL
(Other values on Vol.3, p.259)

Metal ion	Equilibrium	Log K 25°, 0.1	Log K 25°, 1.0	Log K 25°, 0	ΔH 25°, 0.1	ΔS 25°, 0.1
H^+	$HL/H.L$	7.66 ± 0.05	$7.67 -0.01$	7.88	-3.5	23
		$7.62^b \pm 0.04$	7.88^b		$(-5)^s$	
$Ge(IV)$	$M(OH)_2L_2/M(OH)_4.(HL)^2$	3.2^h				
		2.81^b				
	$ML_3/M(OH)_4.(HL)^3.H$	6.0^h				

^{b}25°, 0.5; ^{h}20°, 0.1; ^{s}20-40°, 2.0

Bibliography: 79MBc

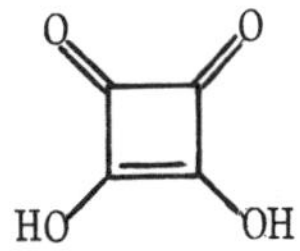

$C_4H_2O_4$	3,4-Dihydroxy-3-cyclobutene-1,2-dione (squaric acid) (Other values on Vol.3, p.263)					H_2L
Metal ion	Equilibrium	Log K 25°, 0.1	Log K 25°, 1.0	Log K 25°, 0	ΔH 25°, 0.1	ΔS 25°, 0.1
H^+	HL/H.L	3.10[b] 2.78[b]	2.8 3.19[e]	3.48 ±0.00	1.75	20.1
	H_2L/HL.H		0.40[e] 0.96[e]	0.55 ±0.05		
Y^{3+}	ML/M.L	2.73			2.4	20
	ML_2/M.L^2	4.25			3.3	31
La^{3+}	ML/M.L	2.71			1.5	18
Ce^{3+}	ML/M.L	2.72			1.8	18
Pr^{3+}	ML/M.L	2.73			1.9	19
Nd^{3+}	ML/M.L	2.74			2.0	19
Sm^{3+}	ML/M.L	2.81			2.0	20
	ML_2/M.L^2	4.08			3.0	29
Eu^{3+}	ML/M.L	2.84			2.0	20
	ML_2/M.L^2	4.12			4.3	33
Gd^{3+}	ML/M.L	2.87			2.0	20
	ML_2/M.L^2	4.16			3.8	32
Tb^{3+}	ML/M.L	2.87			2.2	20
	ML_2/M.L^2	4.21			3.4	31
Dy^{3+}	ML/M.L	2.87			2.2	20
	ML_2/M.L^2	4.25			4.1	33
Ho^{3+}	ML/M.L	2.85			2.1	20
	ML_2/M.L^2	4.27			3.8	31
Er^{3+}	ML/M.L	2.82			2.4	21
	ML_2/M.L^2	4.25			4.6	35
Tm^{3+}	ML/M.L	2.78			2.3	21
	ML_2/M.L^2	4.21			4.0	33
Yb^{3+}	ML/M.L	2.74			2.4	21
	ML_2/M.L^2	4.15			3.8	32
Lu^{3+}	ML/M.L	2.68			2.5	21
	ML_2/M.L^2	4.03			3.9	32

[b]25°, 0.5; [e]25°, 3.0

Bibliography: 760C

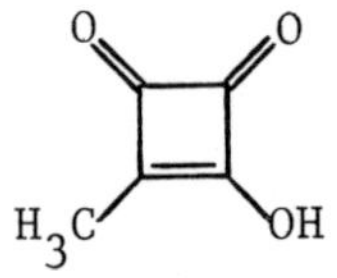

| $C_5H_4O_3$ | Methylhydroxycyclobutene-1,2-dione | | | HL |

Metal ion	Equilibrium	Log K 25°, 1.0	ΔH 25°, 1.0	ΔS 25°, 1.0
Eu^{3+}	ML/M.L	0.65	(-3)[r]	(-6)

[r] 2-51°, 1.0

Bibliography: 76YC

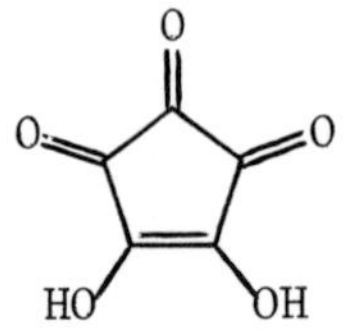

| $C_5H_2O_5$ | 4,5-Dihydroxy-4-cyclopentene-1,2,3-trione (croconic acid) | | | | H_2L |
| | (Other values on Vol.3, p.264) | | | | |

Metal ion	Equilibrium	Log K 25°, 0.1	Log K 25°, 2.0	Log K 25°, 0	ΔH 25°, 0	ΔS 25°, 0
H^+	HL/H.L	1.8	1.51	2.23	3.0	20
	H_2L/HL.H	(0.6)	(0.32)	0.75	-3.9	-10
Y^{3+}	ML/M.L	2.79			2.7[a]	22[a]
	ML_2/M.L^2	4.46			1.2[a]	12[a]
La^{3+}	ML/M.L	3.05			1.0[a]	17[a]
	ML_2/M.L^2	4.50			1.1[a]	10[a]
Ce^{3+}	ML/M.L	3.10			0.8[a]	17[a]
	ML_2/M.L^2	4.40			1.1[a]	10[a]
Pr^{3+}	ML/M.L	3.21			0.7[a]	17[a]
	ML_2/M.L^2	4.45			1.3[a]	10[a]
Nd^{3+}	ML/M.L	3.23			0.6[a]	17[a]
	ML_2/M.L^2	(4.43)			(0.8)[a]	(6)[a]
Sm^{3+}	ML/M.L	3.08			0.9[a]	17[a]
	ML_2/M.L^2	(3.83)			(-0.8)[a]	(-2)[a]
Eu^{3+}	ML/M.L	3.17			1.3[a]	18[a]
	ML_2/M.L^2	(4.18)			(0.0)[a]	(1)[a]
Gd^{3+}	ML/M.L	2.98			1.7[a]	20[a]
	ML_2/M.L^2	(4.21)			(0.6)[a]	(8)[a]
Tb^{3+}	ML/M.L	2.95			2.4[a]	21[a]
	ML_2/M.L^2	4.93			0.4[a]	11[a]
Dy^{3+}	ML/M.L	2.88			2.6[a]	22[a]
	ML_2/M.L^2	4.64			0.9[a]	11[a]

[a] 25°, 0.1

Croconic acid (continued)

Metal ion	Equilibrium	Log K 25°, 0.1	Log K 25°, 2.0	Log K 25°, 0	ΔH 25°, 0	ΔS 25°, 0
Ho^{3+}	ML/M.L	2.90			2.6[a]	22[a]
	ML$_2$/M.L^2	4.68			1.1[a]	12[a]
Tm^{3+}	ML/M.L	2.91			2.6[a]	22[a]
	ML$_2$/M.L^2	4.59			1.7[a]	13[a]
Yb^{3+}	ML/M.L	2.93			2.6[a]	22[a]
	ML$_2$/M.L^2	4.57			1.9[a]	14[a]
Lu^{3+}	ML/M.L	2.92			2.6[a]	22[a]
	ML$_2$/M.L^2	4.45			2.4[a]	15[a]

[a] 25°, 0.1

Bibliography:

H$^+$ 77GSL

Y^{3+}-Lu^{3+} 78CO

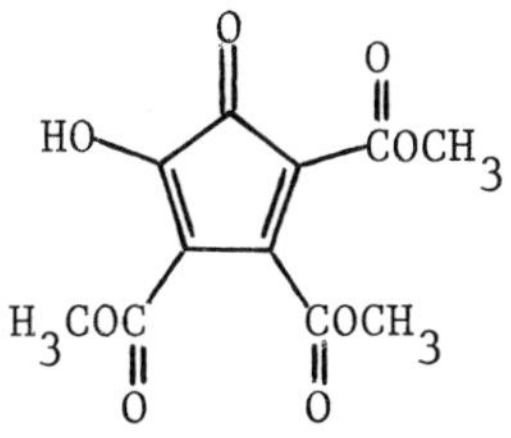

$C_{11}H_{10}O_8$ 4-Hydroxy-5-oxocyclopentadiene-1,2,3-tricarboxylic acid trimethylester   HL

Metal ion	Equilibrium	Log K 25°, 0.1	ΔH 25°, 0.1	ΔS 25°, 0.1
H$^+$	HL/H.L	7.84	(3)[s]	(46)
Ca^{2+}	ML/M.L	3.96	(-3)[s]	(8)
Co^{2+}	ML/M.L	5.30	(-2)[s]	(18)
Ni^{2+}	ML/M.L	5.73	(-2)[s]	(20)
Cu^{2+}	ML/M.L	7.23	(-3)[s]	(23)
Zn^{2+}	ML/M.L	5.84	(-2)[s]	(20)

[s] 15-35°, 0.1

Bibliography: 78MSc

Other references: 73MV,75MSa,77LB,79BL,79BLa

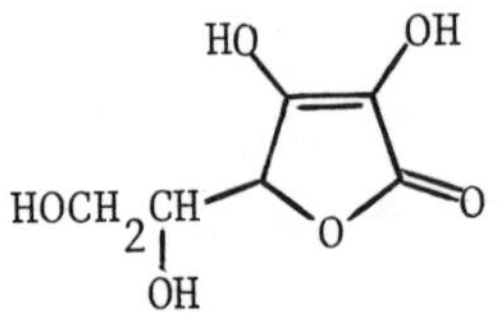

$C_6H_8O_6$

<u>L-Ascorbic acid</u>
(Other values on Vol.3, p.264)

H_2L

Metal ion	Equilibrium	Log K 25°, 0.1	Log K 25°, 3.0	Log K 25°, 0
H^+	$HL/H.L$	11.34	11.34 -0.01	
	$H_2L/HL.H$	4.03 ±0.01	4.37 ±0.01	
Cd^{2+}	$MHL/M.HL$		0.49	1.3
	$M_2(HL)_2/M^2.(HL)^2$		1.3	
	$M_3L_3.H^3/M^3.(HL)^3$		-13.37	
	$M_3L_3/M_3(OH)L_3.H$		7.89	
	$M_5L_6.H^5/M^5.(HL)^6$		-26.20	
	$M_5HL_6/M_5L_6.H$		6.1	

Bibliography:

H^+ 72UW,78La

Cd^{2+} 72UW

Other references: 76KKa,78DP,79FD

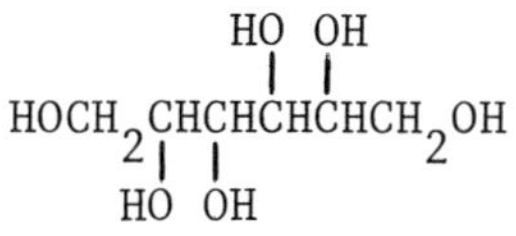

$$\begin{array}{cc} & \text{HO OH} \\ & |\quad| \\ \text{HOCH}_2\text{CHCHCHCHCH}_2\text{OH} \\ & |\quad| \\ & \text{HO OH} \end{array}$$

$C_6H_{14}O_6$ D(-)-Mannitol   L
                              (Other values on Vol.3, p.270)

Metal ion	Equilibrium	Log K 25°, 0.1	Log K 20°, 0.1	Log K 25°, 0.5
Ge(IV)	$HMO_2(H_{-2}L).H/M(OH)_4.L$			−6.43
	$HMO(H_{-2}L)_2.H/M(OH)_4.L^2$	−4.05	−4.0	−3.95
	$MO_2(H_{-2}L)_3.H^2/M(OH)_4.L^3$		−13.7	
	$H_2M_2O_3(H_{-2}L)(H_{-4}L).H^2/[M(OH)_4]^2.L^2$			−10.62

Bibliography: 79MBd

Other references: 76P,77EF,77Pb,78MB

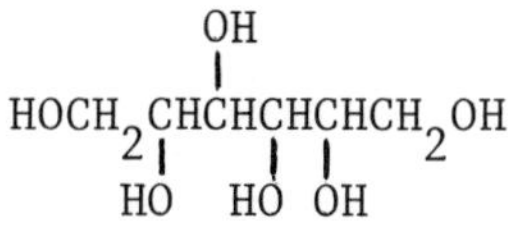

$$\begin{array}{cc} & \text{OH} \\ & | \\ \text{HOCH}_2\text{CHCHCHCHCH}_2\text{OH} \\ & |\quad\ |\quad| \\ & \text{HO HO OH} \end{array}$$

$C_6H_{14}O_6$ D(-)-Sorbitol   L
                              (Other values on Vol.3, p.271)

Metal ion	Equilibrium	Log K 25°, 0.1	Log K 20°, 0.1
Ge(IV)	$HMO(H_{-2}L)_2.H/M(OH)_4.L^2$	−3.49	−3.7
	$MO_2(H_{-2}L)_3.H^2/M(OH)_4.L^3$	−12.3	

Bibliography: 79MBd

Other references: 77NFK,77Pb,78MB

C$_8$H$_{16}$O$_4$ 1,4,7,10-Tetraoxacyclododecane   (12-crown-4-ether)   L

Metal ion	Equilibrium	Log K 25°, 0.1	ΔH 25°, 0.1	ΔS 25°, 0.1
Pb^{2+}	ML/M.L	2.00^w	(−2)^s	(2)

^{s}15-35°, 0.1; ^{w}(CH$_3$CH$_2$)$_3$NClO$_4$ used as background electrolyte

Bibliography: 78KKb   Other reference: 79HRB

CH$_3$OCH$_2$CH$_2$OCH$_2$CH$_2$OCH$_2$CH$_2$OCH$_2$CH$_2$OCH$_3$

C$_{10}$H$_{22}$O$_5$ 2,5,8,11,14-Pentaoxapentadecane   (tetraglyme)   L

Metal ion	Equilibrium	Log K 25°, 0.1	ΔH 25°, 0.1	ΔS 25°, 0.1
Pb^{2+}	ML/M.L	0.5^w	(−3)^s	(−8)
	ML$_2$/M.L^2	1.6^w	(−6)^s	(−13)

^{s}10-35°, 0.1; ^{w}(CH$_3$CH$_2$)$_3$NClO$_4$ used as background electrolyte.

Bibliography: 76KKf

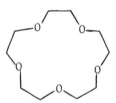

C$_{10}$H$_{20}$O$_5$ 1,4,7,10,13-Pentaoxacyclopentadecane   (15-crown-5-ether)   L

Metal ion	Equilibrium	Log K 25°, 0.1	ΔH 25°, 0.1	ΔS 25°, 0.1
Na$^+$	ML/M.L	0.70	−1.5	−2
K$^+$	ML/M.L	0.74	−4.1	−10
Rb$^+$	ML/M.L	0.62	−1.9	−4
Cs$^+$	ML/M.L	0.8	−1.3	−1
NH$_4^+$	ML/M.L	1.7	−0.2	7
Sr^{2+}	ML/M.L	1.95	−0.9	6
Ba^{2+}	ML/M.L	1.71	−1.1	4
Ag$^+$	ML/M.L	0.94	−3.2	−7
Tl$^+$	ML/M.L	1.23	−4.0	−8
Hg^{2+}	ML/M.L	1.68	−3.6	−4
Pb^{2+}	ML/M.L	1.85 2.05^w	−3.3	−2

^{w}(CH$_3$CH$_2$)$_4$NClO$_4$ used as background electrolyte.

15-Crown-5-ether (continued)

Bibliography:
Na$^+$-Hg^{2+} 76ITH   Other references: 77RLW,79HRB
Pb^{2+} 76ITH,78KKb

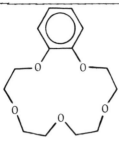

C$_{14}$H$_{20}$O$_5$	Benzo-1,4,7,10,13-Pentaoxacyclopentadecane	(benzo-15-crown-5-ether)			L
Metal ion	Equilibrium	Log K 25°, 0.1		ΔH 25°, 0.1	ΔS 25°,0.1
K$^+$	ML/M.L	0.38		-2.3	-6

Bibliography: 76ITN

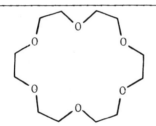

C$_{12}$H$_{24}$O$_6$		1,4,7,10,13,16-Hexaoxacyclooctadecane	(18-crown-6-ether)			L
Metal ion	Equilibrium	Log K 25°, 0.1	Log K 25°, 0	ΔH 25°, 0.1	ΔS 25°,0.1
Na$^+$	ML/M.L	0.80		-2.3	-4
K$^+$	ML/M.L	2.03		-6.2	-11
Rb$^+$	ML/M.L	1.56		-3.8	-6
Cs$^+$	ML/M.L	0.99		-3.8	-8
NH$_4$$^+$	ML/M.L	1.23		-2.3	-2
CH$_3$NH$_3$$^+$	ML/M.L	1.22 1.13			
CH$_3$CH$_2$NH$_3$$^+$	ML/M.L	0.99			
(CH$_2$CH$_2$)$_2$NH$_2$$^+$	ML/M.L	0.9			
(CH$_3$CH$_2$)$_3$NH$^+$	ML/M.L	0.7			
CH$_3$(CH$_2$)$_3$NH$_3$$^+$		0.94			
Sr^{2+}	ML/M.L	2.72		-3.6	0
Ba^{2+}	ML/M.L	3.87		-7.6	-8

18-Crown-6-ether (continued)

Metal ion	Equilibrium	Log K 25°, 0.1	Log K 25°, 0	ΔH 25°, 0.1	ΔS 25°, 0.1
Ag^+	ML/M.L	1.50		-2.2	0
Tl^+	ML/M.L	2.27 -0.1		-4.4	-5
Hg^{2+}	ML/M.L	2.42		-4.7	-5
Pb^{2+}	ML/M.L	4.27 +0.1		-5.2	2

Bibliography:

Na^+-Ag^+,Hg^{2+} 76ITH   Other reference: 79HRB

Tl^+,Pb^{2+} 76ITH,76KKf

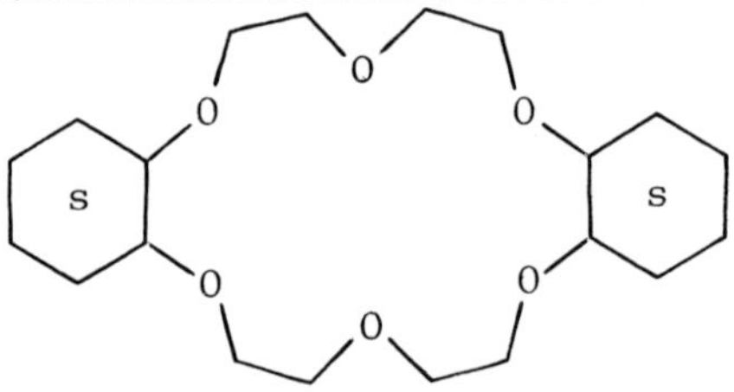

$C_{20}H_{36}O_6$ 2,5,8,15,18,21-Hexaoxatricyclo$[20,4,0,0^{9,14}]$hexacosane   L
                         (dicyclohexyl-18-crown-6-ether)
                         (Other values on Vol.3, p.278)

Isomer	Metal ion	Equilibrium	Log K 25°, 0.1	ΔH 25°, 0.1	ΔS 25°,0.1
cis-syn-cis (A)	Na^+	ML/M.L	1.21	0.2	6
	$CH_3NH_3^+$	ML/M.L	0.8	-0.8	1
	Ag^+	ML/M.L	2.4	0.1	11
	Hg_2^{2+}	ML/M.L ML_2/M.L^2	1.93 3.1	-2.2 3.9	2 28
	Tl^+	ML/M.L	2.44	-3.6	-1
	Hg^{2+}	ML/M.L	2.75	-0.7	10
	Pb^{2+}	ML/M.L	5.29	-5.5	3
cis-anti-cis (B)	Na^+	ML/M.L	0.7	-1.6	-2
	$CH_3NH_3^+$	ML/M.L	0.66	-0.9	0
	Hg_2^{2+}	ML/M.L ML_2/M.L^2	1.57 2.7	-4.3 -10.0	-7 -21
	Tl^+	ML/M.L	1.83	-4.3	-6
	Hg^{2+}	ML/M.L	2.60	-2.6	3
	Pb^{2+}	ML/M.L	4.43	-4.2	6

Bibliography: 76ITH

$$HOCH_2CH_2SCH_2CH_2OH$$

$C_4H_{10}O_2S$ 2,2'-Thiodiethanol (β-thiodiglycol)   L

(Other reference on Vol.3, p.364)

Metal ion	Equilibrium	Log K 25°, 0.5	Log K 25°, 1.0	ΔH 25°, 0.5	ΔS 25°, 0.5
Mg^{2+}	ML/M.L		-0.28		
Ca^{2+}	ML/M.L		-0.09		
Sr^{2+}	ML/M.L		-0.11		
Ba^{2+}	ML/M.L		-0.08		
Mn^{2+}	ML/M.L		-0.22		
Co^{2+}	ML/M.L		-0.20		
Ni^{2+}	ML/M.L		-0.16		
Cu^{2+}	ML/M.L		0.18		
Ag^+	ML/M.L	3.53	3.43	-7.5	-9
	$ML_2/M.L^2$	5.81		-16	-27
	$ML_3/M.L^3$	6.7		-20	-37
Zn^{2+}	ML/M.L		-0.18		
Cd^{2+}	ML/M.L		-0.32		
Hg^{2+}	ML/M.L	6.4		-8.0	2
	$ML_2/M.L^2$	10.47		-17	-9
	$ML_3/M.L^3$	(11.5)		(-10)	(19)
	$ML_4/M.L^4$	14.1		(-20)	(-3)

Bibliography:

Mg^{2+}-Cu^{2+}, Zn^{2+}, Cd^{2+} 79SR Hg^{2+} 76MH

Ag^+ 76MH,79SR

$C_8H_{16}O_3S$ 1,4,7-Trioxa-10-thiacyclododecane   L

Metal ion	Equilibrium	Log K 25°, 0	ΔH 25°, ∿0	ΔS 25°, 0
Ag^+	ML/M.L	2.71	-10.2	-22
Pb^{2+}	ML/M.L	0.94	-5.9	-16

Bibliography: 78IT

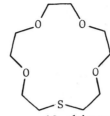

1,4,7,10-Tetraoxa-13-thiacyclopentadecane   L

$C_{10}H_{20}O_4S$

Metal ion	Equilibrium	Log K 25°, 0	ΔH 25°, ∿0	ΔS 25°, 0
Ag$^+$	ML/M.L	5.0	-9.4	-9
	ML$_2$/ML.L	2.45	-3.5	-1
Tl$^+$	ML/M.L	0.80	-7.7	-22
Pb^{2+}	ML/M.L		-16.9	
	ML$_2$/ML.L	1.65	-5.1	-10

Bibliography: 78IT

$CH_3SCH_2CH_2OCH_2CH_2OCH_2CH_2OCH_2CH_2SCH_3$

5,8,11-Trioxa-2,14-dithiapentadecane   L

$C_{10}H_{22}O_3S_2$

Metal ion	Equilibrium	Log K 25°, 0	ΔH 25°, ∿0	ΔS 25°, 0
Ag$^+$	ML/M.L	4.46	-14.4	-28
Hg^{2+}	ML/M.L	5.3	-14.7	-25

Bibliography: 78IT

$CH_3OCH_2CH_2SCH_2CH_2OCH_2CH_2SCH_2CH_2OCH_3$

2,8,14-Trioxa-5,11-dithiapentadecane   L

$C_{10}H_{22}O_3S_2$

Metal ion	Equilibrium	Log K 25°, 0	ΔH 25°, ∿0	ΔS 25°, 0
Ag$^+$	ML/M.L		-14.1	
	ML$_2$/ML.L	3.06	-3.7	2
Hg^{2+}	ML/M.L		-14.0	
	ML$_2$/ML.L	3.22	-7.1	-9

Bibliography: 78IT

$C_{10}H_{20}O_3S_2$	1,4,10-Trioxa-7,13-dithiacyclopentadecane			L

Metal ion	Equilibrium	Log K 25°, 0	ΔH 25°, ∼0	ΔS 25°, 0
Ag^+	ML/M.L		-16.6	
	$ML_2/ML.L$	2.7	-1.0	9
Hg^{2+}	ML/M.L		-16.1	
	$ML_2/ML.L$	2.91	-5.0	-3
Pb^{2+}	ML/M.L	1.62	-8	-19

Bibliography: 78IT

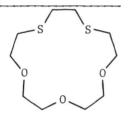

$C_{10}H_{20}O_3S_2$	1,4,7-Trioxa-10,13-dithiacyclopentadecane			L

Metal ion	Equilibrium	Log K 25°, 0	ΔH 25°, ∼0	ΔS 25°, 0
Ag^+	ML/M.L		-12.1	
	$ML_2/ML.L$	3.31	-5.6	-4
Hg^{2+}	ML/M.L		-11.3	
	$ML_2/ML.L$	5.1	-7.8	-3
Pb^{2+}	ML/M.L	1.21	-6	-15

Bibliography: 78IT

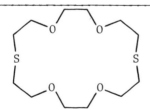

$C_{12}H_{24}O_4S_2$	1,4,10,13-Tetraoxa-7,16-dithiacyclooctadecane			L

Metal ion	Equilibrium	Log K 25°, 0	ΔH 25°, ∼0	ΔS 25°, 0
Ag^+	ML/M.L		-16.7	
Tl^+	ML/M.L	0.9	-11	-33
Hg^{2+}	ML/M.L		-17.8	
Pb^{2+}	ML/M.L	3.13	-21.2	-57

Bibliography: 78IT

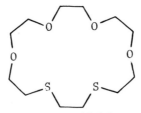

$C_{12}H_{24}O_4S_2$ 1,4,7,10-Tetraoxa-13,16-dithiacyclooctadecane   L

Metal ion	Equilibrium	Log K 25°, 0	ΔH 25°, ~0	ΔS 25°, 0
Ag^+	ML/M.L	3.0	-16	-40
Tl^+	ML/M.L	1.38	-7.3	-18
Hg^{2+}	ML_2/ML.L		-28	
Pb^{2+}	ML/M.L	2.62	-8.8	-18

Bibliography: 78IT

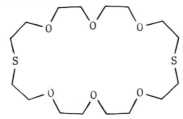

$C_{16}H_{32}O_6S_2$ 1,4,7,13,16,19-Hexaoxa-10,22-dithiacyclotetracosane   L

Metal ion	Equilibrium	Log K 25°, 0	ΔH 25°, ~0	ΔS 25°, 0
Ag^+	ML/M.L		-13.8	
Hg^{2+}	ML/M.L		-14	

Bibliography: 78IT

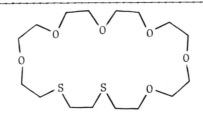

$C_{16}H_{32}O_6S_2$ 1,4,7,10,13,16-Hexaoxa-19,22-dithiacyclotetracosane   L

Metal ion	Equilibrium	Log K 25°, 0	ΔH 25°, ~0	ΔS 25°, 0
Ag^+	ML/M.L	4.5	-14	-26
	ML_2/ML.L	5.0	-2	16
Hg^{2+}	ML/M.L		-14	

Bibliography: 78IT

$$CH_3OCH_2CH_2SCH_2CH_2SCH_2CH_2SCH_2CH_2OCH_3$$

$C_{10}H_{22}O_2S_3$ 2,14-Dioxa-5,8,11-trithiapentadecane   L

Metal ion	Equilibrium	Log K 25°, 0	ΔH 25°, ∿0	ΔS 25°, 0
Ag^+	ML/M.L		-17.6	
	ML_2/ML.L	3.75	-6.9	-3
Hg^{2+}	ML/M.L		-17.1	
	ML_2/ML.L	4.7	-9.4	-10

Bibliography: 78IT

$$HOCH_2CH_2SH$$

C_2H_6OS 2-Mercaptoethanol   HL
                      (Other values on Vol.3, p.280)

Metal ion	Equilibrium	Log K 25°, 0.1	Log K 25°, 0.5	Log K 25°, 0	ΔH 25°, 0	ΔS 25°, 0
H^+	HL/H.L	9.40	9.34	9.72	-6.21	23.7
		$9.48^h\pm0.04$			-6.5^h	21^h
Cd^{2+}	$M_{10}L_{16}/M^{10}.L^{16}$	147^h				

^{h}20°, 0.1

Bibliography: 72SG   Other references: 74DV,76HS,76SK

$$\overset{\displaystyle SH}{\underset{\displaystyle HOCH_2CHCH_2SH}{|}}$$

$C_3H_8OS_2$ 2,3-Dimercaptopropanol   (BAL) H_2L
                      (Other values on Vol.3, p.283)

Metal ion	Equilibrium	Log K 25°, 0.1	Log K 30°, 0.1	Log K 20°, 0.1
H^+	HL/H.L	10.68	10.60 ±0.01	10.78
	H_2L/HL.H	(8.58)	8.60 ±0.02	8.76
Hg^{2+}	$ML/M.L_2$	25.7		
	$ML_2/M.L^2$	34.3		
In^{3+}	ML/M.L			17.14

Bibliography:
H^+,In^{3+} 78KST   Other reference: 76HS
Hg^{2+} 77CJ

$$HOCH_2CH=CH_2$$

C_3H_5O Prop-2-en-1-ol   (allyl alcohol)   L
                  (Other values on Vol.3, p.290)

Metal ion	Equilibrium	Log K 25°, 0.1	Log K 25°, 1.0	Log K 25°, 2.0	ΔH 25°, 1.0	ΔS 25°, 1.0
Ag^+	$ML/M.L$	1.15 ±0.01	1.25	1.36	−4.4	−9
	$ML_2/M.L^2$	0.60	1.0	1.12		

Bibliography: 77HSA

$$HOCH_2CH=CHCH_3$$

C_4H_8O trans-But-2-en-1-ol   (crotyl alcohol)   L
                  (Other value on Vol.3, p.290)

Metal ion	Equilibrium	Log K 25°, 0.1	Log K 25°, 1.0	Log K 25°, 2.0	ΔH 25°, 1.0	ΔS 25°, 1.0
Ag^+	$ML/M.L$	0.59	0.76 ±0.05	0.90	−5.0	−13
	$ML_2/M.L^2$			0.66		

Bibliography: 77HSA

$$HO(CH_2)_nCH=CH_2$$

C_xH_yO ω-Alkene-1-ol   L

n	Metal ion	Equilibrium	Log K 25°, 1.0	ΔH 25°, 1.0	ΔS 25°, 1.0
n=2	Ag^+	$ML/M.L$	1.80	−5.7	−11
but-3-en-1-ol (C_4H_7O)		$ML_2/M.L^2$	2.25		
n=3	Ag^+	$ML/M.L$	1.96	−6.3	−12
but-4-en-1-ol (C_5H_9O)		$ML_2/M.L^2$	2.30		
n=4	Ag^+	$ML/M.L$	2.04	−6.6	−13
hex-5-en-1-ol ($C_6H_{11}O$)		$ML_2/M.L^2$	2.92		

Bibliography: 77HSA

$$CH_3C≡N$$

C_2H_3N Cyanomethane   (acetonitrile)   L
                  (Other values on Vol.3, p.293)

Metal ion	Equilibrium	Log K 25°, 0.1
Cu^+	$ML_2/M.L^2$	4.35

Bibliography: 63HS

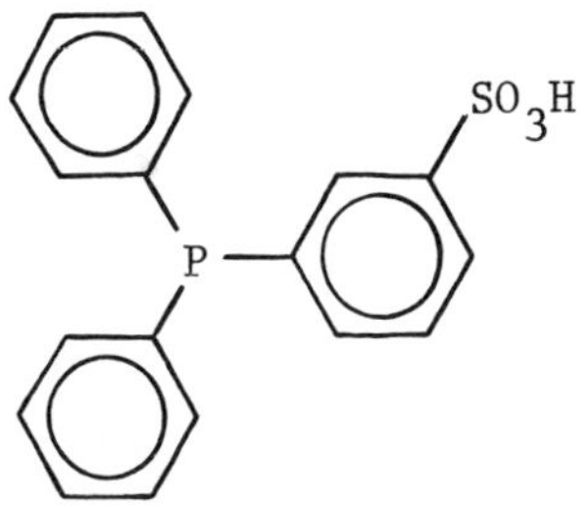

$C_{18}H_{15}O_3SP$ 3-(Diphenylphosphino)benzenesulfonic acid   (3-sulfotriphenylphosphine)   HL
(Other values on Vol.3, p.298)

Metal ion	Equilibrium	Log K 25°, 0.1	Log K 25°, 1.0
H^+	HL/H.L		(0.63)
Pb^{2+}	$ML/M.L$		10.2
	$ML_2/M.L^2$		20.0
	$ML_3/M.L^3$		26.3
	$ML_4/M.L^4$	33.9	31.2

Bibliography: 77HF

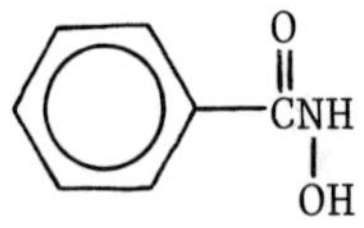

$C_7H_7O_2N$ Benzohydroxamic acid   HL
                              (Other references on Vol.3, p.302)

Metal ion	Equilibrium	Log K 20°, 0.1	Log K 25°, 2.0	Log K 25°, 3.0	ΔH 25°, 2.0	ΔS 25°, 2.0
H^+	HL/H.L	8.79	(8.50)	9.03	(-6)[s]	(19)
Fe^{3+}	ML.H/M.HL	2.27	2.24[c]		(3)[s]	(20)[c*]
	$ML_2/M.L^2$	20.43				
	$ML_3/M.L^3$	27.8				
Bi^{3+}	ML/M.L			9.2		
	$ML_2/M.L^2$			18.0		
	$ML_3/M.L^3$			25.3		

[c]25°, 1.1;   [e]25°, 3.0;   [r]25-35°, 0;   [s]0-50°, 2.0;   *assuming ΔH for 2.0=ΔH for 1.1

Bibliography:

H^+ 77BH,79MC Bi^{3+} 77BH

Fe^{3+} 79MC   Other reference: 76ABc

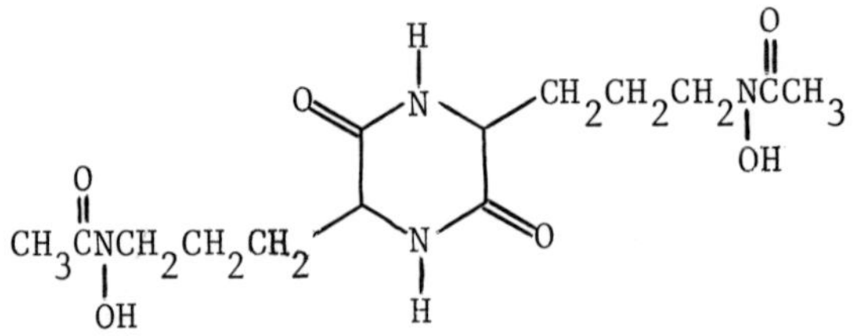

$C_{14}H_{24}O_6N_4$ Rhodotorulic acid H_2L

Metal ion	Equilibrium	Log K 25°, 0.1	Log K 25°, 1.0
H^+	HL/H.L	9.44	
	$H_2L/HL.H$	8.49	
Fe^{3+}	ML/M.L	21.99	21.55
	$ML_2/M.L^2$		40.73
	$M_2L_3/M^2.L^3$	62.2	

Bibliography: 79CC

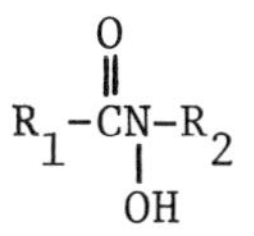

$$O$$
$$\|$$
$$R_1-CN-R_2$$
$$|$$
$$OH$$

$C_xH_yO_2N$		Substitutedhydroxamic acids				HL
R	Metal ion	Equilibrium	Log K 25°, 0.1	Log K 25°, 2.0	ΔH 25°, 2.0	ΔS 25°, 2.0
$R_1=CH_3, R_2=H$	H^+	HL/H.L	9.36^h	9.02	$(-4)^s$	(28)
(aceto-) $(C_2H_5O_2N)$	Fe^{3+}	ML.H/M.HL	2.06^h	2.04^c	$(2)^s$	$(16)^{c*}$
(Other values on Vol.3, p.301)		$ML_2/M.L^2$	21.10^h			
		$ML_3/M.L^3$	28.33^h			
$R_1=R_2=CH_3$	H^+	HL/H.L		8.63	$(-1)^s$	(36)
(N-methylaceto-) $(C_3H_7O_2N)$	Fe^{3+}	ML.H/M.HL		2.50^c	$(1)^s$	$(15)^{c*}$
$R_1=CH_3, R_2=C_6H_5$	H^+	HL/H.L		8.34	$(-11)^s$	(1)
(N-phenylaceto-) $(C_8H_9O_2N)$	Fe^{3+}	ML.H/M.HL		2.35^c	$(1)^s$	$(14)^{c*}$
$R_1=C_6H_5, R_2=CH_3$	H^+	HL/H.L		7.87	$(-4)^s$	(23)
(N-methylbenzo-) $(C_8H_9O_2N)$	Fe^{3+}	ML.H/M.HL		2.74^c	$(1)^s$	$(15)^{c*}$
$R_1=R_2=C_6H_5$	H^+	HL/H.L	8.15	8.00	$(-8)^s$	(10)
(N-phenylbenzo-) $(C_{13}H_{11}O_2N)$	Fe^{3+}	ML.H/M.HL		2.42^c	$(1)^s$	$(14)^{c*}$
(Other values on Vol.3, p.302)						

$^c25°, 1.1;$ $^h20°, 0.1;$ $^s0-50°, 2.0;$ *assuming ΔH for 2.0=ΔH for 1.1

Bibliography: 79MC   Other references: 65HA, 71LFG,78SSR

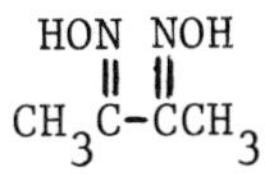

$$HON \quad NOH$$
$$\| \quad \|$$
$$CH_3C-CCH_3$$

$C_4H_8O_2N_2$	Butane-2,3-dione dioxime (dimethylglyoxime) (Other values on Vol.3, p.306)				H_2L
Metal ion	Equilibrium	Log K 25°, 0.1	Log K 25°, 0	ΔH 25°, 0	ΔS 25°, 0
H^+	HL/H.L	11.9	12.0		
	$H_2L/H.HL$	10.45 ±0.03	10.66 ±0.0	$(-2)^r$	(40)
Cu^{2+}	MHL/M.HL	(8.75)			
	$M(HL)_2/M.(HL)^2$	19.3 ±0.1			

$^r25-40°, 0$

Bibliography: Cu^{2+} 76LU,77MT

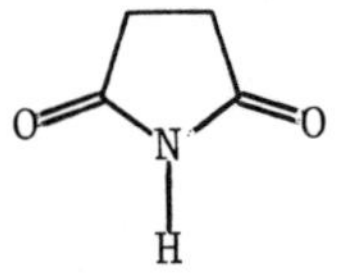

$C_4H_5O_2N$		Pyrrolidine-2,5-dione (succinimide)			HL
		(Other references on Vol.3, p.311)			

Metal ion	Equilibrium	Log K 25°, 0.1	Log K 25°, 0.5	ΔH 25°, 0.5	ΔS 25°, 0.5
H^+	HL/H.L	9.38	9.59	-6.63	21.6
Ni^{2+}	$ML/M.L$		2.93	-2.2	6
	$ML_2/M.L^2$		4.49	-4	6
	$ML_3/M.L^3$		6.63	-6	11
Cu^{2+}	$ML/M.L$		4.61	-3.9	5
	$ML_2/M.L^2$		8.75	-8.0	13
Ag^+	$ML/M.L$	4.45	4.36	-5.6	1
	$ML_2/M.L^2$	9.54	9.64	-12.1	4
Cd^{2+}	$ML/M.L$		3.31	-2.0	8
	$ML_2/M.L^2$		5.50	-3.9	12
	$ML_3/M.L^3$		6.92	-6	13

Bibliography:

H^+ 77BE Ni^{2+}-Cd^{2+} 79BE   Other references: 76SS

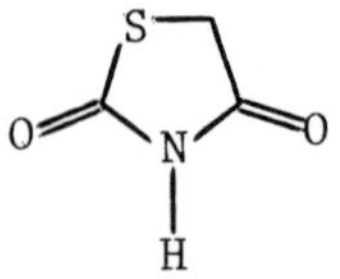

$C_3H_3O_2NS$		1,3-Thiazolidine-2,4-dione		HL

Metal ion	Equilibrium	Log K 25°, 0.5	ΔH 25°, 0.5	ΔS 25°, 0.5
H^+	HL/H.L	6.46	-6.40	8.1
Ag^+	$ML/M.L$	3.5		
	$ML_2/M.L^2$	7.05		

Bibliography:

H^+ 77BE

Ag^+ 79BB

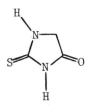

| $C_3H_4O_2N_2$ | 1,3-Diazolidine-2,4-dione (hydantoin) | | | | HL |

(Other references on Vol.3, p.311)

Metal ion	Equilibrium	Log K 25°, 0.1	Log K 25°, 0.5		ΔH 25°, 0.5	ΔS 25°, 0.5
H^+	HL/H.L	8.93	9.10		-7.58	16.2
Co^{2+}	ML/M.L		2.93		-2.3	6
	$ML_2/M.L_2$		4.2		-4	5
	$ML_3/M.L_3$		5.2		-7	0
Ni^{2+}	ML/M.L		3.33		-2.8	6
	$ML_2/M.L_2$		5.11		-7	0
	$ML_3/M.L_3$		6.09		-7	6
Cu^{2+}	ML/M.L		4.73		-4.5	7
	$ML_2/M.L_2$		8.36		-9	10
	$ML_3/M.L_3$		11.8		-14	7
Ag^+	ML/M.L	4.29	4.42		-5.8	1
	$ML_2/M.L_2$	9.20	9.05		-12.7	-1
Zn^{2+}	ML/M.L		3.09		-2.0	8
	$ML_2/M.L_2$		5.6		-4	12
	$ML_3/M.L_3$		8.8		-7	18
Cd^{2+}	ML/M.L		3.33		-2.7	6
	$ML_2/M.L_2$		5.34		-4.9	8
	$ML_3/M.L_3$		6.5		-7	6

Bibliography:

H^+ 77BE $Co^{2+}-Cd^{2+}$ 79BEa

| $C_3H_4ON_2S$ | 1,3-Diazolidine-2-thione-4-one (thiohydantoin) | | | HL |

Metal ion	Equilibrium	Log K 25°, 0.5	ΔH 25°, 0.5	ΔS 25°, 0.5
H^+	HL/H.L	8.51	-7.75	12.9
Ag^+	ML/M.L	4.26		
	$ML_2/M.L_2$	6.87		

Bibliography:

H^+ 77BE Ag^+ 79BB

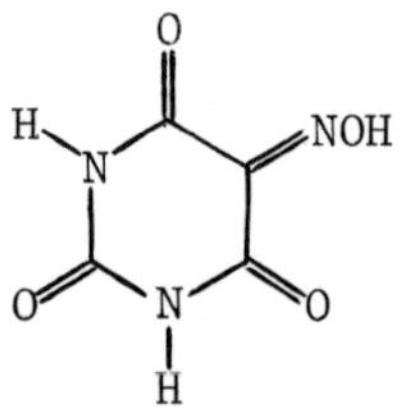

| $C_4H_3O_4N_3$ | 2,4,5,6-Tetraoxopyrimidine 5-oxime (violuric acid)
(Other reference on Vol.2, p.343) | | | H_3L |

Metal ion	Equilibrium	Log K 25°, 0.5	Log K 25°, 1.0	Log K 25°, 0
H^+	HL/H.L		(13.4)	
	$H_2L/HL.H$	9.6		
	$H_3L/H_2L.H$	4.10		4.34
Fe^{2+}	$M(H_2L)_3/M.(H_2L)^3$	8.5		
Co^{2+}	$MH_2L/M.H_2L$	2.4		
	$M(H_2L)_2/M.(H_2L)^2$	4.7		
Ni^{2+}	$MH_2L/M.H_2L$	3.5		
	$M(H_2L)^2/M.(H_2L)^2$	6.3		
Cu^{2+}	$MH_2L/M.H_2L$	4.3		
	$M(H_2L)^2/M.(H_2L)^2$	7.4		

Bibliography: 78DD   Other references: 73IB,74EI,75Ba

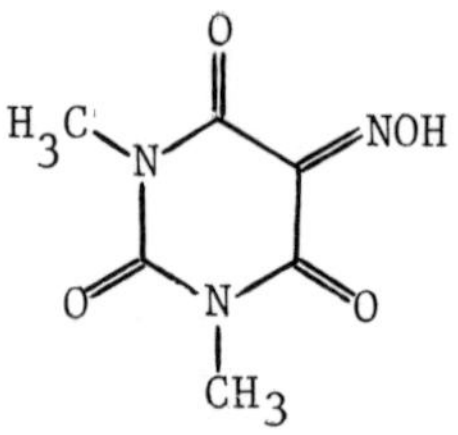

| $C_6H_7O_4N_3$ | 1,3-Dimethylvioluric acid | | | HL |

Metal ion	Equilibrium	Log K 25°, 0.1	Log K 25°, 0.5	Log K 25°, 0
H^+	HL/H.L	4.57	4.42	4.82
Fe^{2+}	$ML_2/M.L^2$		10.25	
Co^{2+}	ML/M.L		(2.34)	
	$ML_3/M.L^3$		5.52	
Ni^{2+}	ML/M.L		3.76	
	$ML_2/M.L^2$		7.12	
Cu^{2+}	ML/M.L		4.67	
	$ML_2/M.L^2$		7.95	
Zn^{2+}	ML/M.L		2.32	

Bibliography: 76VN,77VNa

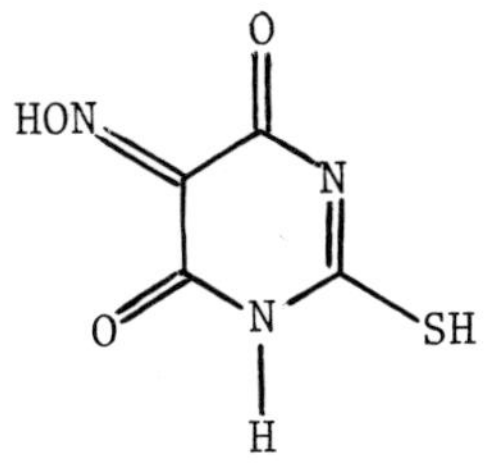

| $C_4H_3O_3N_3S$ | 2-Mercapto-4,5,6-trioxopyrimidine-5-oxime | H_3L |
| | (2-thiovioluric acid) | |

Metal ion	Equilibrium	Log K 25°, 0.1	Log K 25°, 0.5	Log K 25°, 1.0
H^+	$HL/H.L$			(13.3)
	$H_2L/HL.H$	8.71	8.60	
	$H_3L/H_2L.H$	3.84	3.79	
Fe^{2+}	$M(H_2L)_3/M.(H_2L)^3$	9.04		
Co^{2+}	$MH_2L/M.H_2L$	2.61		
	$M(H_2L)_2/M.(H_2L)^2$	4.77		
Ni^{2+}	$MH_2L/M.H_2L$	3.51		
	$M(H_2L)_2/M.(H_2L)^2$	5.89		
Cu^{2+}	$MH_2L/M.H_2L$	3.9		
	$M(H_2L)_2/M.(H_2L)^2$	7		
	$M.H_2L/MH_2L(s)$	-10.19		
Pb^{2+}	$MH_2L_4/M.L^2.(HL)^2$	20.02		

Bibliography: 79DDH   Other references: 71TT,73IB,74EI

$$CH_3CH_2 \quad CH_2CH_3$$

(structure of 5,5-diethylbarbituric acid)

$C_8H_{12}O_3N_2$	5,5-Diethylbarbituric acid (barbitol)			HL
	(Other references on Vol.2, p.342)			

Metal ion	Equilibrium	Log K 25°, 0.1	Log K 25°, 0	ΔH 25°, 0	ΔS 25°, 0
H^+	HL/H.L	7.78	7.980 ±0.02	-5.81	17.0
Ca^{2+}	ML/M.L	0.7			
Sr^{2+}	ML/M.L	0.5			

Bibliography: 52SL   Other reference: 76BMc

$$\underset{H_2NCNH_2}{\overset{\overset{\displaystyle S}{\|}}{}}$$

CH_4N_2S	Thiocarbamide (thiourea)				L
	(Other values on Vol.3, p.313)				

Metal ion	Equilibrium	Log K 25°, 0.1	Log K 25°, 1.0	Log K 25°, 0	ΔH 25°, 0.1	ΔS 25°, 0.1
H^+	HL/H.L	1.18	0.97	1.44		
Co^{2+}	$ML/M.L$	0.7				
	$ML_2/M.L^2$	1.2				
	$ML_3/M.L^3$	1.5				
Cu^+	$ML_2/M.L^2$	12.30				
	$ML_3/M.L^3$	14.30				
	$ML_4/M.L^4$	15.53 +0.1				
		15.4[b]				
Ag^+	$ML/M.L$	7.11[b]±0.07		6.46[t]	-19[t]	-34
	$ML_2/M.L^2$	10.61[b]-0.01		10.90[t]	-27[t]	-41
	$ML_3/M.L^3$	12.73[b]±0.02		12.88[t]	-30[t]	-40
Cd^{2+}	$ML/M.L$	1.3 ±0.1	1.5[d] ±0.1		-4.0	-7
		1.4[b] ±0.1	1.6[d] ±0.1	1.8[e] ±0.1	-4.6[b]	8[b]
	$ML_2/M.L^2$	2.1 ±0.1	2.3[d] ±0.1		-11	-27
		2.2[b] ±0.1	2.5[d] ±0.2	2.7[e] ±0.2	-10[b]	24[b]
	$ML_3/M.L^3$	2.6 ±0.2	2.8[d] ±0.2		-6	-8
		2.7[b] ±0.2	3.0[d] ±0.2	3.2[e] ±0.2		
	$ML_4/M.L^4$	3.1 ±0.2	3.5[d] ±0.3		-8	-10
		3.3[b] ±0.2	4.0[d] ±0.3	4.8[e] ±0.3		
Hg^{2+}	$ML/M.L$	11.4[b]			-17	-5[b]
	$ML_2/M.L^2$	21.7	22.1		-33[p]	-11
		22.1[b]			-34[b]	-13[b]
	$ML_3/M.L^3$	24.6	24.7		(-35)[p]	(-5)
		25.1[b]			(-45)[b]	(-36)[b]
	$ML_4/M.L^4$	26.4	26.5		-45[p]	-34
		27.1[b]			-48[b]	-37[b]
	$M_2L_3/M^2.L^3$	36.0[b]			-50[b]	-3[b]

[b] 25°, 0.5;   [d] 25°, 2.0;   [e] 25°, 3.0;   [p] 30°, 0.1;   [t] 25°, 0.02

Thiourea (continued)

Metal ion	Equilibrium	Log K 25°, 0.1	Log K 25°, 1.0	Log K 25°, 0	ΔH 25°, 0.1	ΔS 25°, 0.1
Pb^{2+}	ML/M.L	0.3 ±0.1	0.6	1.7[e]	-3.9	-12
		0.4[b]	1.1[d]			
	$ML_2/M.L^2$	0.9 ±0.1	1.4		-3	-6
		1.1[b]	1.9[d]	2.5[e]		
	$ML_3/M.L^3$	1.4 ±0.2	1.8		-4	-7
		1.6[b]	2.2[d]	2.9[e]		
	$ML_4/M.L^4$	1.6 ±0.2	2.0		-12	-30
		1.8[b]	2.4[d]	3.1[e]		
Bi^{3+}	ML/M.L		1.2	2.1		

[b]25°, 0.5; [d]25°, 2.0; [e]25°, 3.0; [P]30°, 0.1; [t]25°, 0.02

Bibliography:

H^+	78VSG	Hg^{2+}	76MH
Co^{2+}	66IG	Pb^{2+}	75FFa,75M,77FN,79FFG
Cu^+	57LQ,76FL	Bi^{3+}	78FBG
Ag^+	73EB,76MH		
Cd^{2+}	75FF,75FFa,75M,76BD,76MH,77FN	Other references:	68M,70S,77HF,77MTV, 78DK,78GF,78TL,79VS

$C_nH_mN_2S$

N,N'-Alkylthiourea   L
(Other values on Vol.3, p.314 and 315)

R	Metal ion	Equilibrium	Log K 25°, 0.1	Log K 25°, 1.0	ΔH 25°, 0.1	ΔS 25°, 0.1
$R_1=CH_3$	Cu^+	$ML_2/M.L^2$	13.28			
		$ML_3/M.L^3$	14.46			
$R_2=R_3=R_4=H$		$ML_4/M.L^4$	15.52			
(N-methyl-)	Cd^{2+}	ML/M.L	1.4		(-4)[t]	(7)
		$ML_2/M.L^2$	2.4		(0)[t]	(11)
$(C_2H_6N_2S)$		$ML_3/M.L^3$	2.9		(-7)[t]	(-10)
		$ML_4/M.L^4$	4.1		(-13)[t]	(-25)
	Pb^{2+}	ML/M.L	0.4	0.6		
		$ML_2/M.L^2$	0.6			
		$ML_3/M.L^3$	(1.9)			
		$ML_4/M.L^4$	(1.3)			
	Bi^{3+}	ML/M.L		1.45		
$R_1=CH_3CH_2$	Cu^+	$ML_2/M.L^2$	13.08			
		$ML_3/M.L^3$	14.42			
$R_2=R_3=R_4=H$		$ML_4/M.L^4$	16.23			
(N-ethyl-)	Pb^{2+}	ML/M.L	0.5 +0.1	0.6	(-4)[t]	(-11)
		$ML_2/M.L^2$	0.8 -0.1		(-6)[t]	(-16)
$(C_3H_8N_2S)$		$ML_3/M.L^3$	(2.1) -1.0		(-13)[t]	(-34)
		$ML_4/M.L^4$	(2.4) -0.4		(-10)[t]	(-23)
	Bi^{3+}	ML/M.L		1.46		

[t]7-45°, 0.1

N,N'-Alkylthiourea (continued)

R	Metal ion	Equilibrium	Log K 25°, 0.1	Log K 25°, 1.0	ΔH 25°, 0.1	ΔS 25°, 0.1
$R_1=CH_3CH_2CH_2CH_2$	Pb^{2+}	$ML/M.L$		0.66		
$R_2=R_3=R_4=H$	Bi^{3+}	$ML/M.L$		1.55		
(N-butyl-)						
$(C_5H_{12}N_2S)$						
$R_1=R_3=CH_3$	Cu^+	$ML_2/M.L^2$	12.52			
$R_2=R_4=H$		$ML_3/M.L^3$	13.91			
(N,N'-dimethyl-)		$ML_4/M.L^4$	14.98			
$(C_3H_8N_2S)$						
$R_1=R_3=CH_3CH_2$	Cu^+	$ML_2/M.L^2$	14.04			
$R_2=R_4=H$		$ML_3/M.L^3$	15.00			
(N,N'-diethyl-)		$ML_4/M.L^4$	15.87			
$(C_5H_{12}N_2S)$						
$R_1=R_2=R_3=CH_3,R_4=H$ Bi^{2+}		$ML/M.L$		1.05		
(N,N,N'-trimethyl-)						
$(C_4H_{10}N_2S)$						
$R_1=R_3=CH_3$	Bi^{3+}	$ML/M.L$		1.04		
$R_2=C_6H_5, R_4=H$						
(N,N'-dimethyl-N-phenyl-)						
$(C_9H_{12}N_2S)$						
$R_1=CH_2=CHCH_2$	Cu^+	$ML_2/M.L^2$	13.18			
$R_2=R_3=R_4=H$		$ML_3/M.L^3$	14.58			
(N-allyl-)		$ML_4/M.L^4$	15.90			
$(C_4H_8N_2S)$						
$R_1=R_3=CH_2=CHCH_2$	Cu^{2+}	$ML_2/M.L^2$	13.00			
$R_2=R_4=H$		$ML_3/M.L^3$	14.43			
(N,N'-diallyl-)		$ML_4/M.L^4$	15.58			
$(C_7H_{12}N_2S)$						
$R_1=CH_3C=O$	Cu^{2+}	$ML_2/M.L^2$	11.53			
$R_2=R_3=R_4=H$		$ML_3/M.L^3$	12.75			
(N-acetyl-)		$ML_4/M.L^4$	13.81			
$(C_3H_6ON_2S)$						

Bibliogarphy: Cu^+ 76FL; Cd^{2+} 77FFa; Pb^{2+} 75FFa,76FF,79FFG; Bi^{3+} 79FFG

Other reference: 57LQ

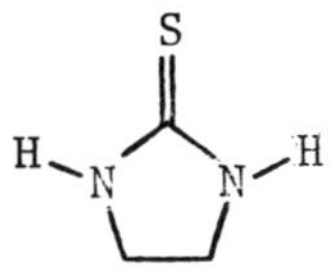

| $C_3H_6N_2S$ | 1,3-Diazolidine-2-thione (ethylenethiourea) | | | L |
| | (Other values on Vol.3, p.316) | | | |

Metal ion	Equilibrium	Log K 25°, 0.1	Log K 25°, 0.5
Cu^+	$ML_2/M.L^2$	11.91	
	$ML_3/M.L^3$	13.52	
	$ML_4/M.L^4$	14.86	
Ag^+	$ML/M.L$		5.97
	$ML_2/M.L^2$		10.2
	$ML_3/M.L^3$		12.3

Bibliography: Cu^+ 76FL; Ag^+ 79BB

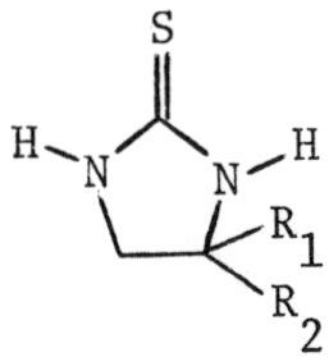

| $C_nH_mN_2S$ | | 4-Substitutedethylenethiourea | | L |

R	Metal ion	Equilibrium	Log K 25°, 0.1
R_1=CH$_3$, R_2=H (4-methyl-) ($C_4H_8N_2S$)	Cu^+	$ML_2/M.L^2$	12.43
		$ML_3/M.L^3$	13.78
		$ML_4/M.L^4$	15.11
R_1=R_2=CH$_3$ (4,4-dimethyl-) ($C_5H_{10}N_2S$)	Cu^+	$ML_2/M.L^2$	12.04
		$ML_3/M.L^3$	13.65
		$ML_4/M.L^4$	15.18

Bibliography: 76FL

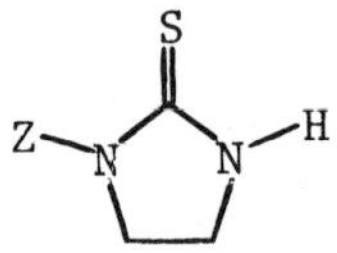

| $C_n H_m O_o N_o S_p$ | | 1-Substitutedethylenethiourea | | L |

Z	Metal ion	Equilibrium	Log K 25°, 0.1	
Z=HOCH$_2$CH$_2$	Cu$^+$	ML$_2$/M.L^2	11.34	
(1-(2-hydroxyethyl)-)		ML$_3$/M.L^3	13.00	
(C$_5$H$_{10}$ON$_2$S)		ML$_4$/M.L^4	14.08	

Z=H$_2$NCH$_2$CH$_2$	Cu$^+$	ML$_2$/M.L^2	10.28	
(1-(2-aminoethyl)-)		ML$_3$/M.L^3	11.66	
(C$_5$H$_{11}$N$_3$S)		ML$_4$/M.L^4	12.86	

Bibliography: 76FL

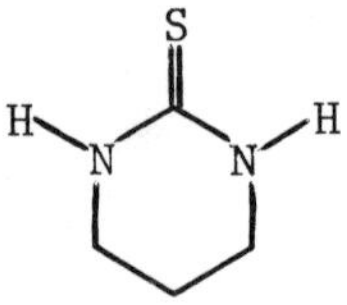

| C$_4$H$_8$N$_2$S | | 1,3-Diazinidine-2-thione   (propylenethiourea) | | L |

Metal ion	Equilibrium	Log K 25°, 0.1
Cu$^+$	ML$_2$/M.L^2	12.79
	ML$_3$/M.L^3	14.98
	ML$_4$/M.L^4	16.43

Bibliography: 76FL

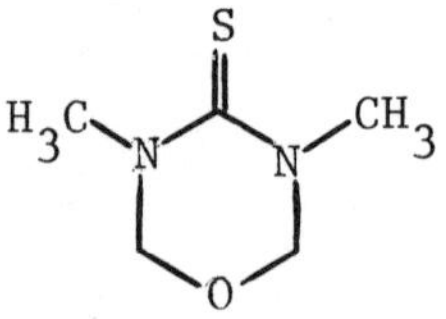

| C$_6$H$_{10}$ON$_2$S | | 3,5-Dimethyl-1,3,5-oxadiazinidine-4-thione | | L |

Metal ion	Equilibrium	Log K 25°, 0.1
Cu$^+$	ML$_2$/M.L^2	11.32
	ML$_3$/M.L^3	12.87
	ML$_4$/M.L^4	14.15

Bibliography: 76FL

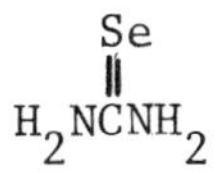

$$\overset{\overset{\text{Se}}{\|}}{\text{H}_2\text{NCNH}_2}$$

CH$_4$N$_2$Se		Selenocarbamide (selenourea)			L

(Other values on Vol.3, p.316)

Metal ion	Equilibrium	Log K 25°, 0.6	ΔH 30°, 0.1	ΔS 25°, 0.1
H$^+$	HL/H.L	(0.6)	0	3
Pd^{2+}	ML$_4$/M.L^4	32.1		

Bibliography: 77HF

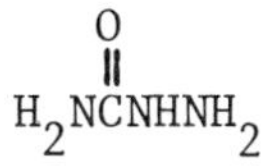

$$\overset{\overset{\text{O}}{\|}}{\text{H}_2\text{NCNHNH}_2}$$

CH$_5$ON$_3$		Semicarbazide			L

(Other values on Vol.3, p.316)

Metal ion	Equilibrium	Log K 30°, 0.1	ΔH 30°, 0.1	ΔS 30°, 0.1
H$^+$	HL/H.L	3.53	-6.1	-4
Ni^{2+}	ML$_2$/M.L^2	4.79	-14.2	25
	ML$_3$/M.L^3	6.65	-16.2	23

Bibliography:
Ni^{2+} 77AG

$$\overset{\overset{\text{S}}{\|}}{\text{H}_2\text{NCNHNH}_2}$$

CH$_5$N$_3$S		Thiosemicarbazide				L

(Other values on Vol.3, p.317)

Metal ion	Equilibrium	Log K 25°, 0.1	Log K 25°, 0.5	Log K 25°, 1.0	ΔH 25°, 0.1	ΔS 25°, 0.1
H$^+$	HL/H.L	1.7 ±0.1	1.87	1.8	-4.5[p]	-7
Co^{2+}	ML/M.L		(1.14)			
	ML$_2$/M.L^2		2.86			
	ML$_3$/M.L^3		4.17			
Ni^{2+}	ML/M.L		(2.04)			
	ML$_2$/M.L^2	4.94	4.92		-10.7[p]	-13
	ML$_3$/M.L^3	7.46	7.17		-20.8[p]	-36
	ML$_4$/M.L^4		7.50			
Cu^{2+}	ML/M.L	6.23	(3.30)		-9.8[p]	-4
	ML$_2$/M.L^2	11.82	(7.87)		-17.8[p]	-6
Ag$^+$	ML/M.L	7.55			-22	-39
	ML$_2$/M.L^2	10.67			-26.1	-39
	ML$_3$/M.L^3	12.33	12.9[q]		-29.0	-41
	ML$_4$/M.L^4	13.61				
	M$_2$L$_3$/M^2.L^3	21.6			-47	-59

[p] 30°, 0.1;   [q] 25°, 0.8

Thiosemicarbazide (continued)

Metal ion	Equilibrium	Log K 25°, 0.1	Log K 25°, 0.5	Log K 25°, 1.0	ΔH 25°, 0.1	ΔS 25°, 0.1
Zn^{2+}	$ML/M.L$		(1.34)			
	$ML_2/M.L^2$	2.8	2.58			
	$ML_3/M.L^3$		3.71			
Cd^{2+}	$ML/M.L$	2.33	(1.81)	2.58 ±0.02	-4.3^P -4.7^c	-4 -4^c
	$ML_2/M.L^2$	4.51	4.50	4.69 ±0.01	-8.9^P -9.3^c	-9 -10^c
	$ML_3/M.L^3$		5.39	5.86 ±0.00	-13.0^c	-17^c

^{c}25°, 1.0; ^{P}30°, 0.1

Bibliography:

$H^+, Co^{2+}, Cu^{2+}, Zn^{2+}$ 79LG Cd^{2+} 76T,79LG,79T

Ni^{2+} 77AG,79LG   Other references: 79AOA,79KKa

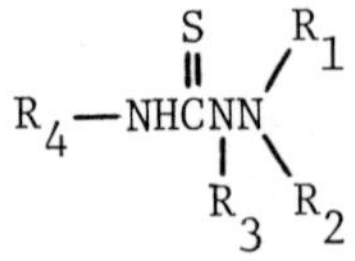

Ag^+ 76BB,76BBa

$C_nH_mN_3S$			N,N',N''-Substitutedthiosemicarbazide			L

R=	Metal ion	Equilibrium	Log K 25°, 1.0	ΔH 25°, 1.0	ΔS 25°, 1.0
$R_1=CH_3, R_2=R_3=R_4=H$ (1-methyl-) ($C_2H_7N_3S$)	Cd^{2+}	$ML/M.L$ $ML_2/M.L^2$ $ML_3/M.L^3$	2.34 4.5 5.8		
$R_3=CH_3, R_1=R_2=R_4=H$ (2-methyl-) ($C_2H_7N_3S$)	Cd^{2+}	$ML/M.L$ $ML_2/M.L^2$ $ML_3/M.L^3$	2.31 4.22 5.53		
$R_4=CH_3, R_1=R_2=R_3=H$ (4-methyl-) ($C_2H_7N_3S$)	Cd^{2+}	$ML/M.L$ $ML_2/M.L^2$ $ML_3/M.L^3$	2.75 5.09 6.30	-4.8 -9.8 -13.5	-4 -10 -17
$R_4=CH_2=CHCH_2$ $R_1=R_2=R_3=H$ (4-allyl-) ($C_4H_9N_3S$)	Cd^{2+}	$ML/M.L$ $ML_2/M.L^2$ $ML_3/M.L^3$	2.60 4.81 5.95		
$R_4=(CH_3)_3C$ $R_1=R_2=R_3=H$ (4-(1,1-dimethylethyl)-) ($C_5H_{12}N_3S$)	Cd^{2+}	$ML/M.L$ $ML_2/M.L^2$	2.86 5.4		

N,N',N'-Substitutedthiosemicarbazide   (continued)

R=	Metal ion	Equilibrium	Log K 25°, 1.0
$R_3=R_4=CH_3$, $R_1=R_2=H$ (2,4-dimethyl-) ($C_3H_9N_3S$)	Cd^{2+}	$ML/M.L$ $ML_2/M.L^2$ $ML_3/M.L^3$	2.18 3.88 4.92
$R_3=(CH_3)_2CH$, $R_4=CH_3$ $R_1=R_2=H$ (4-methyl-2-(1-methylethyl)-) ($C_5H_{13}N_3S$)	Cd^{2+}	$ML/M.L$	1.51
$R_1=R_2=R_4=CH_3$, $R_2=H$ (1,1,4-trimethyl-) ($C_4H_{11}N_3S$)	Cd^{2+}	$ML/M.L$ $ML_2/M.L^2$	1.76 3.17

Bibliography: 76T,79T

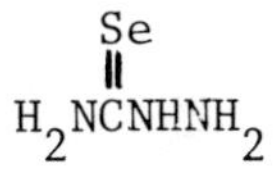

$$\underset{H_2NCNHNH_2}{\overset{\overset{\displaystyle Se}{\|}}{}}$$

CH_5N_3Se	Selenosemicarbazide (Other values on Vol.3, p.317)			L

Metal ion	Equilibrium	Log K 30°, 0.1	ΔH 30°, 0.1	ΔS 30°, 0.1
H^+	$HL/H.L$	0.8	-4.5	-11
Ni^{2+}	$ML_2/M.L^2$ $ML_3/M.L^3$	3.30 5.08	-13 -15	26 25

Bibliography:
Ni^{2+} 77AG

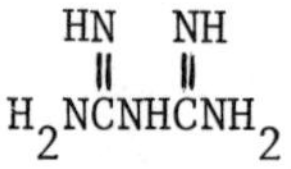

$$\begin{array}{cc} HN & NH \\ \parallel & \parallel \\ H_2NCNHCNH_2 & \end{array}$$

$C_2H_7N_5$ Biguanide   L
                          (Other references on Vol.3, p.327)

Metal ion	Equilibrium	Log K 25°, 0.1			ΔH 25°, 0.1	ΔS 25°, 0.1
H^+	IIL/H.L	(13.0)			(-23)	(-18)
	$H_2L/HL.H$	3.01			-5.0	-3
Ni^{2+}	$ML_2.H^2/M.(HL)$	-9.9			21	25
Cu^{2+}	ML.H/M.HL	-1.31			5.7	13
	$ML_2.H^2/M.(HL)^2$	-4.70			8.4	7

Bibliography:

H^+ 77FM Ni^{2+}, Cu^{2+} 78FMP

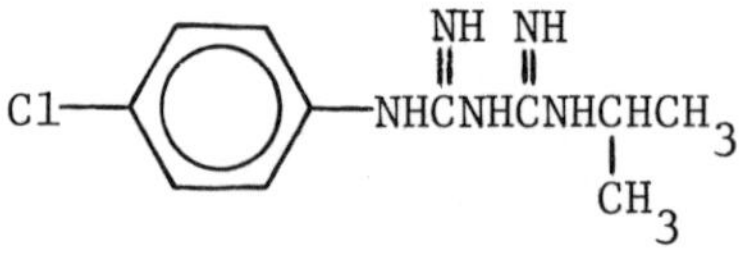

$$\begin{array}{cc} NH & NH \\ \parallel & \parallel \\ Cl-\bigcirc-NHCNHCNHCHCH_3 \\ & CH_3 \end{array}$$

$C_{11}H_{16}N_5Cl$ 1-(4-Chlorophenyl)-5-(1-methylethyl)biguanide   (paludrine)   L

Metal ion	Equilibrium	Log K 25°, 0.5	Log K 25°, 1.0	Log K 25°, 0
H^+	HL/H.L			11.3
	$H_2L/HL.H$	2.44	2.64[d]	2.02
			2.98[d]	
Cu^{2+}	$ML/M.L$			8.7
	$ML_2/M.L^2$			16.5

[d] 25°, 2.0

Bibliography: 72KH

$$OH^-$$

HO$^-$ Hydroxide ion   L$^-$
(Other values on Vol.4, p.1)

Metal ion	Equilibrium	Log K 25°, 0.5	Log K 25°, 1.0	Log K 25°, 0	ΔH 25°, 0	ΔS 25°, 0
H$^+$	HL/H.L	13.74 ±0.02	13.79 ±0.02	13.977 ±0.003	-13.34^b ±0.01	19.3$_b$
		13.78^a±0.01			-13.55^b±0.05	17.7^b
		13.96^d±0.01	14.18^e±0.04		-13.08^e±0.03	21.0^e
		13.95^h		13.544^z	-12.67^z±0.06	
Be^{2+}	ML/M.L	8.3^h		8.6		
	ML$_2$/M.L^2	16.5	17.4^e ±0.2	(14.4)		
		16.2^a				
		16.7^h				
	ML$_3$/M.L^3			18.8		
	ML$_4$/M.L^4			18.6		
	M$_2$L/M^2.L	10.54	10.98^e±0.04	(10.0)	-9.0^e ±0.1	20^e
		10.5^a				
		10.68^d				
	M$_3$L$_3$/M^3.L^3	32.41	33.88^e±0.00	33.1	-24.8^e ±0.0	105^e
		32.53^a				
		32.98^d				
	M$_6$L$_8$/M^6.L^8			(85)	(-58)^r	(200)
Mg^{2+}	ML/M.L		2.2^e	2.80 ±0.0	(-3)^s	(0)^e
	M$_2$L/ML.M		-0.3^e		(0)^s	(-1)^e
	M$_4$L$_4$/M^4.L^4		18.1^e		(0)^s	(83)^e
UO$_2$$^{2+}$	ML/M.L	8.0 ±0.0	8.1^e	8.2	(-2)^t	(30)^b
		8.0^a ±0.3				
	M$_2$L$_2$/M^2.L^2	21.55 ±0.02	21.64 ±0.02	22.4		
		21.70^a±0.03	22.32^e		16.7^e	46^e
	M$_3$L$_4$/M^3.L^4	42.9^a	42.4			
			43.5^e			
	M$_3$L$_5$/M^3.L^5	52.4 ±0.1	52.6 ±0.1	54.4		
		52.5^a	54.4^e		41.1^e	111^e
Fe^{2+}	ML/M.L		4.3	4.7 ±0.2		
	ML$_2$/M.L^2			7.4		
	ML$_3$/M.L^3			9.3		
	ML$_4$/M.L^4			8.9		
	M.L^2/ML$_2$(s)			-14.39		
Co^{2+}	ML/M.L	3.9	4.0 ±0.1	4.3		
			4.4^e			
	ML$_2$/M.L^2		8.5	8.4		
	ML$_3$/M.L^3		9.7	9.7		
	ML$_4$/M.L^4			10.2		
	M$_2$L/M^2.L		4.7^e	(2.7)		
	M$_4$L$_4$/M^4.L^4		27.5^e±0.0	(25.6)		
	M.L^2/ML$_2$(s)		-14.6	-14.9		

---

^{a}25°, 0.1; ^{b}25°, 0.5; ^{d}25°, 2.0; ^{e}25°, 3.0; ^{h}20°, 0.1; ^{r}0-60°, 1.0 molal;
^{s}25-100°, 3.0; ^{t}25-95°, 0.5; ^{z}40°, 0

Hydroxide ion (continued)

Metal ion	Equilibrium	Log K 25°, 0.5	Log K 25°, 1.0	Log K 25°, 0	ΔH 25°, 0	ΔS 25°, 0
Cu^{2+}	$ML/M.L$	$6.1^a \pm 0.0$	6.6^e	6.2	-5.0^a	11^a
	$ML_2/M.L^2$		12.8			
	$ML_3/M.L^3$		14.5			
	$ML_4/M.L^4$		15.6	16.4		
	$M_2L_2/M^2.L^2$	$16.8^a \pm 0.2$	17.0	17.4 ± 0.2	-8.5^a	48^a
			$17.5^e \pm 0.3$		-10.4^e	45^e
	$M_3L_4/M^3.L^4$	$33.7^a -0.1$		35.2	-28^a	60^a
Mn^{3+}	$ML/M.L$		14.6^e	14.4^f	$(-8)^u$	$(40)^f$
	$ML_2/M.L^2$		28.5^e			
Fe^{3+}	$ML/M.L$	11.03 ± 0.09	11.90 ± 0.10	11.81 ± 0.03	-3^a	41^a
		$11.26^a \pm 0.02$	$11.21^e \pm 0.08$		-2^e	45^e
		11.14^d				
	$ML_2/M.L^2$	21.7	21.9^j	22.3		
			22.1^e			
	$ML_4/M.L^4$			34.4		
	$M_2L_2/M^2.L^2$	24.5 ± 0.02	$24.9 -0.1$	25.1		
		24.7^a	$25.6^e \pm 0.2$		-16.2^e	63^e
		25.3^d				
	$M_3L_4/M^3.L^4$		51.0^e	49.7	-38^e	106^e
Hg_2^{2+}	$ML/M.L$	8.7	9.30^e			
	$M_2L/M^2.L$		11.50^e			
	$M_5L_4/M^5.L^4$		48.24^e			
CH_3Hg^+	$ML/M.L$	$9.24^a \pm 0.02$	9.10		-8.5^h	14^a
	$M_2L/ML.M$	2.37^a	2.96			
	$M.L/ML(s)$		-13.66			
$C_6H_5Hg^+$	$ML/M.L$	9.75^a				
$(CH_3)_3Pb^+$	$ML/M.L$	9.03^v				
	$M_2L/ML.M$	1.49^v				
Zn^{2+}	$ML/M.L$			5.0 ± 0.0	$(0)^w$	(25)
			3.8^e		$(-2)^x$	$(12)^e$
	$ML_2/M.L^2$		8.3^e	10.2		
	$ML_3/M.L^3$		13.7^e	13.9		
	$ML_4/M.L^4$		18.0^e	15.6		
	$M_2L/ML.M$		1.7^e	0.0	$(0)^w$	$(8)^e$
Hg^{2+}	$ML/M.L$	10.04	10.14	10.60	-8.6^c	18^c
			$10.67^e \pm 0.07$		-9.0^e	19^e
	$ML_2/M.L^2$	21.2	21.3	21.8	-17.2^c	40^c
		21.2^a				
		21.6^d	$22.13^e \pm 0.02$		-18.0^e	41^e
	$ML_3/M.L^3$			20.9		
	$M_2L/M^2.L$		11.5^e	10.7	-10.0^e	19^e
	$M_3L_3/M^3.L^3$		36.1^e	35.6		
	$M.L^2/MO(s,red)$		-25.6^e	-25.44	7.9^c	-91^e

$^a 25°, 0.1;$ $^c 25°, 1.0;$ $^d 25°, 2.0;$ $^e 25°, 3.0;$ $^f 25°, 4.0;$ $^h 20°, 0.1;$ $^j 20°, 1.0;$

$^u 1-35°, 4.0;$ $^v 25°, 0.3;$ $^w 15-42°, 0;$ $^x 25-100°, 3.0$

Hydroxide ion (continued)

Metal ion	Equilibrium	Log K 25°, 0.5	Log K 25°, 1.0	Log K 25°, 0	ΔH 25°, 0	ΔS 25°, 0
Sn^{2+}	$ML/M.L$		10.48^e			
	$ML_3/M.L^3$		24.58^e			
	$M_2L_2/M^2.L^2$		23.9^e			
	$M_3L_4/M^3.L^4$		$49.93^e \pm 0.03$			
Pb^{2+}	$ML/M.L$	6.0^v	6.3^e -0.1	6.3		
	$ML_2/M.L^2$	10.3^v	10.9^e	10.9		
	$ML_3/M.L^3$	13.3^v	13.7^e	13.9		
	$M_2L/M^2.L$		7.9^e	7.6		
	$M_3L_4/M^3.L^4$	31.7^v	34.0^e ± 0.2	32.1	-25.9^e	69^e
	$M_4L_4/M^4.L^4$	35.1^v	37.6^e ± 0.1	(35.1)	-32.3^e	64^e
	$M_6L_8/M^6.L^8$	67.4^v	71.5^e ± 0.2	68.4	-55.4^e	141^e
In^{3+}	$ML/M.L$	10.5^a	10.4 $9.76^e \pm 0.0$	10.0	-8.2^e	17^e
	$ML_2/M.L^2$	20.3^a	20.0 19.6^e $+0.5$	20.2		
	$ML_3/M.L^3$	29.3^a		29.6		
	$ML_4/M.L^4$			33.9		
	$M_3L_4/M^3.L^4$		47.4^e	50.2		
Tl^{3+}	$ML/M.L$	12.8^a	12.7 $13.02^e \pm 0.02$	13.4	7^e	83^e
	$ML_2/M.L^2$	25.3^a	25.3 $25.75^e \pm 0.02$	26.4	22^e	192^e
	$ML_3/M.L^3$	37.6^a		38.7		
	$ML_4/M.L^4$		40.0^e	41.0		
	$M.L^3/(M_2O_3)^{0.5}(s)$		-45.0^e	-45.2		
As(III)	$ML_3/ML_2.L$		14.5^j			
Bi^{3+}	$ML/M.L$	12.36^a	12.60^e	12.9	4^e	71^e
	$ML_3/M.L^3$	31.9^a		33.1		
	$ML_4/ML_3.L$		0.95	1.1		
	$M_6L_{12}/M^6.L^{12}$		164.95 170.49^e		-34^e	667^e
	$M_9L_{20}/(M_6L_{12})^{1.5}.L^2$	$23.9^a \pm 0.2$				
	$M_9L_{21}/M_9L_{20}.L$	$10.6^a \pm 0.0$				
	$M_9L_{22}/M_9L_{21}.L$	$11.1^a \pm 0.1$				
	$ML_2/(M_2O_3)^{0.5}(s,\alpha)$		-5.34	-5.4		
Te^{4+}	$ML/M.L$		6.51		-6.5^c	8^c
	$ML_2/M.L^2$		7.48		-7.3^c	10^c
	$M_2L_2/M^2.L^2$		14.13		-13.9^c	18^c
	$M_2L/M^2.L$		9.84		$(-18)^y$	$(-15)^c$
	$M_2L_3/M^2.L^3$		18.44		$(-16)^y$	$(31)^c$

^{a}25°, 0.1; ^{c}25°, 1.0; ^{e}25°, 3.0; ^{j}22°, 1.0; ^{v}25°, 0.3; ^{y}25-45°, 1.0

Hydroxide ion   (continued)

Bibliography:

H^+ 64PK,73HW,74VS,75OH,75Q,76ACR,77AR,
          77CF,77OO,77SG,78BL,78BM,78BS,78CKM,
          78VV

Be^{2+} 70KM,75TKM,79IO

Mg^{2+} 78BBG

Ca^{2+} 76KVS

UO_2^{2+} 79SDa

Fe^{2+} 78JBa

Co^{2+} 78GT

Cu^{2+} 72OK,76ACa,79SD

Cr^{3+} 58JB

Mn^{3+} 78BP

Fe^{3+} 75CG,75CPM,76BKa,77CB,78BK,78Da

Hg_2^{2+} 76HHc

CH_3Hg^+ 78JIa

$C_6H_5Hg^+$ 74Aa

$(CH_3)_3Pb^+$ 77SB

Zn^{2+} 75RMa,78BG

Hg^{2+} 75CGP,76HH,77RW,77S

Sn^{2+} 58T,76G,77M

Pb^{2+} 76Lb

Al^{3+} 75T

In^{3+} 79GB

Tl^{3+} 76KY,77YKU,79YR

As(III) 76IBG

Bi^{3+} 75O

Te^{4+} 75KMA,75KMB

Other references: 42M,62VI,63TP,65HA,67MS,
          68SS,69ALb,70D,70IE,70KA,71BS,71KYK,
          71SR,72KA,72KEM,72MA,73PP,74Gb,75AN,
          75BDT,75DOd,75IKV,75KI,75KY,75KZ,75L,
          75LBH,75MH,75Q,75UK,75ZG,76BHS,76BK,
          76BM ,76E,76IM,76LMP,76RP,76SKa,76TZ,
          77AN,77ANa,77BLa,77E,77KH,77KV,77KVb,
          77KY,77Lb,77LPK,77NP,77NS,77UB,77YKP,
          78BH,78Ka,78KD,78KS,78MMc,78SKM,78TE,
          78THH,78WNa,78YK,79CP,79DR,79E,79KD,
          79LP,79Sa

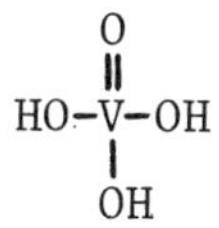

$$\begin{array}{c} O \\ \parallel \\ HO-V-OH \\ \mid \\ OH \end{array}$$

H_3O_4V	Hydrogen vanadate (vanadic acid) (Other values on Vol.4, p.15)				H_3L

Metal ion H^+	Equilibrium	Log K 25°, 0.5	Log K 25°, 1.0	Log K 25°, 0	ΔH 25°, 0	ΔS 25°, 0
	$HL/H.L$	13.0[h] 13.5[h]		14.3 13.2[e]		
	$H_2L/HL.H$	7.85[h] 8.23[h] 8.31[h,w]	8.07 8.04[j]	8.5		
	$H_3L/H_2L.H$	3.78	3.8	4.0	(0)[s]	(17)[c]
	$VO_2/H_3L.H$	3.20	(3.7)	3.3	(0)[s]	(17)[c]
	$V_2O_7/(HL)^2$	0.4		0.6		
	$HV_2O_7/(HL)^2.H$	10.58		11.0 10.93[e]		
	$V_3O_9/(HL)^3.H^3$	30.66		31.8 31.6[e]		
	$V_4O_{12}/(HL)^4.H^4$	41.4				
	$V_4O_{12}/V_4O_{13}.H^2$	8.9				
	$V_3O_9/(H_3L)^3$	7.14[s]				
	$V_4O_{12}/(H_2L)^4$	10.10[s]				
	$HV_6O_{17}/(H_2L)^6.H^3$	33.04[s]				
	$V_{10}O_{27}/(H_2L)^{10}.H^6$	61.8[s]				
	$VO_2/H.(V_2O_5)^{0.5}(s)$			-0.68	4.2	11
	$V_{10}O_{27}.H^{14}/(VO_2)^{10}$		-6.8	-5.5[e]		
	$V_{10}O_{27}/HV_{10}O_{28}.H$	4.34[h] 4.45[h,w] 4.39[s]	3.6 3.6[j] 4.5[j,w]	3.5[e]		
	$HV_{10}O_{28}/V_{10}O_{28}.H$	6.94[h] 7.52[h,w]	5.8 6.06[j] 7.6[j,w]			
Ca^{2+}	$M.(VO_3)^2/M(VO_3)_2(H_2O)_4(s)$	-4.10[t]	-3.66[u]			
	$M^3.(VO_4)^2/M_3(VO_4)_2(H_2O)_4(s)$	-17.48[t]	-17.36[u]			
	$M^2.V_2O_7/M_2V_2O_7(H_2O)_2(s)$	-7.24[t]	-7.11[u]			
	$M^3.V_{10}O_{28}/M_3V_{10}O_{28}(s)$	-5.17[t]				
	$M.V_{12}O_{31}/MV_{12}O_{31}(s)$	-4.00[t]				

---

[c]25°, 1.0; [e]25°, 3.0; [h]20°, 0.1; [j]20°, 1.0; [s]15-35°, 1.0; [t]22°, 1.0; [u]22°, 3.0;
[w]$(CH_3)_4NCl$ used as background electrolyte.

Hydrogen vanadate (continued)

Metal ion	Equilibrium	Log K 25°, 0.5	Log K 25°, 1.0	Log K 25°, 0	ΔH 25°, 0	ΔS 25°, 0
Sr^{2+}	$M.(VO_3)^2/M(VO_3)_2(H_2O)_4(s)$		-9.00^t	-8.22^u		
	$M^3.(VO_4)^2/M_3(VO_4)_2(s)$		-20.60^t	-20.29^u		
	$M^2.V_2O_7/M_2V_2O_7(H_2O)_{1.5}(s)$		-10.28^t	-9.96^u		
	$M.V_6O_{16}/MV_6O_{16}(s)$		-0.07^t			
	$M^3.V_{10}O_{28}/M_3V_{10}O_{28}(s)$		-8.37^t			
	$M.V_{12}O_{31}/MV_{12}O_{31}(s)$		-4.1^t			
Ba^{2+}	$M.(VO_3)^2/M(VO_3)_2(H_2O)(s)$		-11.92^t	-11.00^u		
	$M^3.(VO_4)^2/M_3(VO_4)_2(s)$		-24.40^t	-24.10^u		
	$M^2.V_2O_7/M_2V_2O_7(H_2O)_{0.5}(s)$		-12.92^t	-12.70^u		
	$M.V_6O_{16}/MV_6O_{16}(s)$		-0.7			
	$M^3.V_{10}O_{28}/M_3V_{10}O_{28}(s)$		-13.0^t			
	$M.V_{12}O_{31}/MV_{12}O_{31}(s)$		-4.7^t			

^{t}22°, 1.0; ^{u}22°, 3.0

Bibliography:

H^+ 75YF,79YY   Other references: 75DK,75KIa,78HJ,79IC

$Ca^{2+}-Ba^{2+}$ 76IC,77IC

$$\begin{array}{c} O \\ \| \\ HO-Cr-OH \\ \| \\ O \end{array}$$

| H_2O_4Cr | Hydrogen chromate (chromic acid) (Other values on Vol.4, p.17) | | | | | H_2L |

Metal ion	Equilibrium	Log K 25°, 0.5	Log K 25°, 1.0	Log K 25°, 0	ΔH 25°, 0	ΔS 25°, 0
H^+	HL/H.L	5.81 $6.09^a \pm 0.04$	5.74 5.90^e	6.51 ±0.02	0.7 1.1^e	32 31^e
	$H_2L/HL.H$	-0.7^i -0.6^h	0.7 ±0.1 -0.6^e	-0.2^o	$(9)^s$	$(27)^c$
	$Cr_2O_7/(HL)$	1.84 $1.72^a \pm 0.02$	$1.97^e \pm 0.03$ $2.17^e \pm 0.03$	1.53 ±0.02	-4.7 -4.8^e	-9 -6^e
K^+	ML/M.L			$(0.57)^u$		
	$MCr_2O_7/M.Cr_2O_7$			(0.76)	$(-7)^r$	(-20)
NH_4^+	$MCr_2O_7/M.Cr_2O_7$			(0.88)	$(-5)^r$	(-13)
Pb^{2+}	M.L/ML(s)			-12.60	10.56	-22.2

^{a}25°, 0.1; ^{c}25°, 1.0; ^{e}25°, 3.0; ^{h}20°, 0.1; ^{i}20°, 0.5; ^{r}25-35°, 0; ^{s}15-35°, 1.0; ^{u}18°, 0; ^{o}20°, 0

Bibliography:

K^+,NH_4^+ 77SSc   Other references: 76MPR,77BL,78FU

Pb^{2+} 76DMH

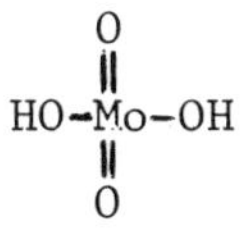

$$
\begin{array}{c}
O \\
\parallel \\
HO-Mo-OH \\
\parallel \\
O
\end{array}
$$

H_2O_4Mo		Hydrogen Molybdate (Molybdic acid)				H_2L

(Other values on Vol.4, p.18)

Metal ion	Equilibrium	Log K 25°, 0.1	Log K 25°, 1.0	Log K 25°, 0	ΔH 25°, 0	ΔS 25°, 0
H^+	HL/H.L	3.74	3.51 ±0.04	4.24[o]	(5)[r]	(34)[a]
			3.89[e]		14[e]	65[e]
	H_2L/HL.H	3.77	3.70 ±0.05	4.00[o]	(-12)[r]	(-23)[a]
			3.61[e]			
Pb^{2+}	M.L/ML(s)			-15.62	12.89	-28.2

[a]25°, 0.1; [e]25°, 3.0; [o]20°, 0; [r]10-40°, 0.1

Bibliography:

		Other references: 72RC,73J,74MP,76CH,76MPR,
H^+	75CR	77CCa
Pb^{2+}	76DMH	

$$
\begin{array}{c}
O \\
\parallel \\
HO-W-OH \\
\parallel \\
O
\end{array}
$$

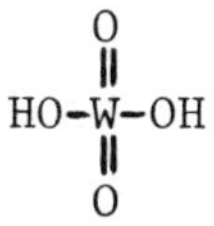

H_2O_4W		Hydrogen wolframate (tungstic acid)				H_2L

(Other values on Vol.4, p.19)

Metal ion	Equilibrium	Log K 20°, 0.1	Log K 25°, 1.0	Log K 25°, 0	ΔH 25°, 0	ΔS 25°, 0
H^+	HL/H.L	(3.5)				
	$H_2L/H^2.L$	(8.1)	11.30[e]			
Pb^{2+}	M.L/ML(s)			-16.07	12.50	-31.6

[e]25°, 3.0

Bibliography:

		Other references: 76MPR,77CCa,79IR
Pb^+	76DMH	

$$MnO_4^{\,-}$$

O_4Mn^-		Manganate(VII) ion (permanganate ion)			L^-

(Other references on Vol.4, p.135)

Metal ion	Equilibrium	Log K 25°, 0	ΔH 25°, 0	ΔS 25°, 0
Ag^+	M.L/ML(s)	-9.88	(7)[s]	(-22)

[s]15-35°, 0

Bibliography: 75DM

$$Fe(CN)_6^{3-}$$

C$_6$N$_6$Fe^{3-} Hexacyanoferrate(III) ion   L^{3-}
                          (Other values on Vol.4, p.22)

Metal ion	Equilibrium	Log K 25°, 0.5	Log K 25°, 1.0	Log K 25°, 0	ΔH 25°, 0	ΔS 25°, 0
Na$^+$	ML/M.L		-0.3^e	1.08		
K$^+$	ML/M.L	-0.24 0.85^a	0.18^d 0.30^e	1.41 ±0.05	0.5	8
Rb$^+$	ML/M.L			1.57		
Cs$^+$	ML/M.L		0.52^e	1.71		
Ag$^+$	M^3.L/M$_3$L(s)	-27.9^h				

^{a}25°, 0.1; ^{d}25°, 2.0; ^{e}25°, 3.0; ^{h}20°, 0.1

Bibliography:

Na$^+$-Cs$^+$ 76LL

Ag$^+$ 76CC

$$B(OH)_3$$

H$_3$O$_3$B   Hydrogen borate (boric acid)   HL
                          (Other values on Vol.4, p.25)

Metal ion	Equilibrium	Log K 25°, 0.5	Log K 25°, 1.0	Log K 25°, 0	ΔH 25°, 0	ΔS 25°, 0
H$^+$	HL/H.L	8.97^a±0.02 8.94^d±0.07	8.85 8.97^e+0.01	9.236 ±0.001	-3.4 ±0.0 -3.9^d-0.1	31 28^d
Ca^{2+}	ML/M.L		0.99			

^{a}25°, 0.1; ^{d}25°, 2.0; ^{e}25°, 3.0

Bibliography:

H$^+$ 75BDR,75PBa,76FSH   Other references: 72DS,76MI

Ca^{2+} 76FSH

$$HCN$$

CHN   Hydrogen cyanide (hydrocyanic acid)   HL
                          (Other values on Vol.4, p.26)

Metal ion	Equilibrium	Log K 25°, 0.5	Log K 25°, 1.0	Log K 25°, 0	ΔH 25°, 0	ΔS 25°, 0
H$^+$	HL/H.L	9.01^a 9.14^h	8.95 9.48^e	9.21 ±0.01	-10.43^h -10.9^h	7.2 5^a
Ag$^+$	ML$_2$/M.L^2 ML$_3$/M.L^3 ML$_4$/M.L^4	20.4^h	20.0 20.3^t 20.8^t	20.48 21.4	-32.9 -33.5	-17 -13
	MOHL/M.OH.L		12.7	13.2		
	M.L/ML(s)	-15.5^h	-15.4	-15.66		
	M^2.L^2/M$_2$L$_2$(s)	-10.9^h				

^{a}25°, 0.1; ^{e}25°, 3.0; ^{h}20°, 0.1; ^{t}30°, 1.0

Bibliography: H$^+$ 76TCI; Ag$^+$ 76CC;   Other references: 69BH,76HF,70MR,78MMb,79BPU,79MMa

HNCS

CHNS		Hydrogen thiocyanate (thiocyanic acid)				HL
		(Other values on Vol.4, p.29)				
Metal ion	Equilibrium	Log K 25°, 0.5	Log K 25°, 1.0	Log K 25°, 0	ΔH 25°, 0	ΔS 25°, 0
H^+	HL/H.L			-1.1		
NpO_2^+	ML/M.L	0.32[d]				
Ni^{2+}	ML/M.L	1.23 ±0.02	1.13 ±0.01	1.76 -0.09	-2.3	0
		(1.29)[a]	1.34[e]		-2.9[c]	-4[c]
	$ML_2/M.L^2$	(2.18)[a]	1.57 ±0.01		-5.2[c]	-10[c]
	$ML_3/M.L^3$		1.5 ±0.2		-8.6[c]	-22[c]
Cu^{2+}	ML/M.L	1.90[a]	1.75 ±0.01	2.33	-3.0	1
			1.91[e]		-3.1[c]	-2[c]
	$ML_2/M.L^2$	3.00[a]	2.72	3.65	-6.2[c]	-8[c]
Fe^{3+}	ML/M.L	2.14 ±0.03	2.10 ±0.03	3.02 ±0.01	-1.3[c]	5[c]
		2.33[a]	2.21[e] ±0.03		-0.6[d]	
	$ML_2/M.L^2$	3.3 ±0.2	3.2 ±0.1	3.64[o]		
			3.64[e] ±0.04			
	$ML_3/M.L^3$		5.0[e]			
	$ML_4/M.L^4$		6.3[e]			
Cu^+	$ML_3/M.L^3$			11.60[g]		
	$ML_4/M.L^4$			12.02[g]		
	$M_2L_6/M^2.L^6$			24.34[g]		
	M.L/ML(s)			-14.77[g]		
Tl^+	M.L/ML(s)			-3.77 ±0.03	(16)[r]	(36)
$(CH_3)_3Pb^+$	ML/M.L		-0.42[t]			
Zn^{2+}	ML/M.L	1.07[a] 0.74[d]	0.71	1.33[f] 1.11[f]	-1.4[c]	-1[c]
	$ML_2/M.L^2$	1.15[d]	1.04	1.91[f] 1.81[f]	-1.8[c]	-1[c]
	$ML_3/M.L^3$	1.3[d]	1.2	2.0[f] 2.8[f]	-2[c]	-1[c]
	$ML_4/M.L^4$	1.7[d]	1.5	1.6[f] 2.8[f]	-4[c]	-6[c]
Cd^{2+}	ML/M.L	1.34 ±0.01 1.58[a] ±0.05	1.32 ±0.02	1.89	-2.3[c]	-2[c]
		1.34[d]	1.41[e] -0.05		-1.9[e]	0[e]
	$ML_2/M.L^2$	2.04 2.32[a] 2.05[d]	1.99 2.24[e] -0.2	2.78	-4.2[c] -3.7[e]	-5[c] -2[e]
	$ML_3/M.L^3$	2.1 2.2[d]	2.0 2.5[e] ±0.1	2.8	-6[c] -5.2[e]	-12[c] -6[e]
	$ML_4/M.L^4$	2.0 2.0[d]	1.9 2.5[e] ±0.1	2.3	-6.2[e]	-10[e]
Hg^{2+}	ML/M.L		9.08		-12.0[c] ±0.2	1[c]
	$ML_2/M.L^2$	16.43[a]	16.86	17.26	-24.2[c] ±0.3	-4[c]
	$ML_3/M.L^3$	19.1[a]	19.7	20.0	-28.9[c] ±0.1	-7[c]
	$ML_4/M.L^4$	21.2[a] ±0.1	21.7	21.8 ±0.1	-33.9[c] ±0.8	-15[c]
	ML/MOHL.H		3.4		-5.3[c]	-2[c]

[a] 25°, 0.1; [c] 25°, 1.0; [d] 25°, 2.0; [e] 25°, 3.0; [f] 25°, 4.0; [g] 25°, 5.0; [o] 20°, 0;
[r] 10-40°, 0; [t] 25°, 1.5

Hydrogen thiocyanate (continued)

Bibliography:

H^+	69Pa,76CM	Tl^+	75PT
NpO_2^+	79VG	Zn^{2+}	71PB
Ni^{2+}	71PB,76KK	Cd^{2+}	71PB,77ARM
Cu^{2+}	76KK	Hg^{2+}	75CGP,76CGP
Fe^{3+}	76VKM,78BW		
Cu^+	77AT		
Ag^+	76CC		
CH_3Hg^+	76RT		

Other references: 68GS,66GB,68TR,70H,71LF,
73JP,74Ba,75ABR,75BS,75LMb,75M,76DMP,
76JB,76LP,76OST,76PPa,77HF,78FN,78VBa,
79JBa

$$NCSe^-$$

$CNSe^-$

Selenocyanate ion
(Other values on Vol.4, p.35)

L^-

Metal ion	Equilibrium	Log K 25°, 0.1	Log K 25°, 1.0	Log K 25°, 0	ΔH 25, 0	ΔS 25°, 0
Co^{2+}	ML/M.L			1.49	-1.5	2
Ni^{2+}	ML/M.L		0.99	1.66	-2.1 -3.1[c]	1 -6[c]
	$ML_2/M.L^2$		1.26		-6[c]	-14[c]
CH_3Hg^+	ML/M.L	6.79				
Cd^{2+}	ML/M.L		1.47	1.98	-2.7 -2.4[c]	0 -1[c]
	$ML_2/M.L^2$		2.30		-6.3[c]	-11[c]
	$ML_3/M.L^3$		2.8		(-1)[c]	(9)[c]
	$ML_4/M.L^4$		4.04		-10[c]	-15[c]

---

[c] 25°, 1.0

Bibliography:

Co^{2+},Ni^{2+},Cd^{2+}	75SSD	CH_3Hg^+	76RT

$$O$$
$$\|$$
$$HO-C-OH$$

CH_2O_3	Hydrogen carbonate (carbonic acid) (Other values on Vol.4, p.37)				H_2L

Metal ion	Equilibrium	Log K 25°, 0.5	Log K 25°, 1.0	Log K 25°, 0	ΔH 25°, 0	ΔS 25°, 0
H^+	HL/H.L	10.00^a	9.57 ± 0.03 $9.56^e +0.01$	10.329 ± 0.01	-3.5 -0.1	36
	$H_2L/HL.H$	6.16^a 5.97^d	6.02 ± 0.03 6.33^e	$6.352 +0.01$	-2.0 ± 0.1	22
	$H_2L/CO_2(g)$		-1.51^e -1.55^e	-1.464 ± 0.01		
Na^+	ML/M.L	0.63^q -0.4		1.27 -0.3		
	MHL/M.HL	$-0.55^q \pm 0.0$		-0.25 ± 0.0		
Mg^{2+}	ML/M.L	2.05^q	1.73^e	2.92 ± 0.07	$(3)^r -1$	(23)
	MHL/M.HL	0.21^q	0.15^e	1.01 ± 0.06	$(1)^s$	(8)
	$M_2L/ML.M$	0.54^q				
Ca^{2+}	ML/M.L	2.21^q		3.15 ± 0.05	$(3)^t +1$	(25)
	MHL/M.HL	0.29^q		1.1 ± 0.2	$(5)^r \pm 1$	(22)
Eu^{3+}	ML/M.L			7.11		
	$ML_2/M.L^2$			10.6		
	$M^2.L^3/M_2L_3(s)$			-32.3		
UO_2^{2+}	ML/M.L		9.02^e			
	$ML_2/M.L^2$	16.22^h				
	$ML_3/M.L^3$	21.54^h				
	$M_3(OH)_3L/M^3.(OH)^3.L$		16.34^e			
	$M_{11}(OH)_{12}L_6/M^{11}.(OH)^{12}.L^6$		72.1^e			
Mn^{2+}	MHL/M.HL		0.45^e	1.27	$(1)^u$	(9)
Fe^{2+}	MHL/M.HL			1.10		
Hg_2^{2+}	$M.CO_2(g)/H^2.ML(s)$		4.19^e			
CH_3Hg^+	ML/M.L		6.10			
$(CH_3)_3Pb^+$	ML/M.L	2.60^v				
Zn^{2+}	MHL/M.HL			1.50	$(1)^w$	(9)
Hg^{2+}	ML/M.L		10.7^e			
	MHL/M.HL		5.47^e			
	$M^3.CO_2(g)/H^6.ML(MO)^2(s)$		7.2^e			

---

^{a}25°, 0.1; ^{e}25°, 3.0; ^{q}25°, 0.7; ^{r}10–50°, 0; ^{s}10–90°, 0; ^{t}temp. var., 0; ^{u}5–55°, 0; ^{v}25°, 0.3; ^{w}10–70°, 0

Bibliography:

H^+	70BHa,75Rc,77RGa,78CKP	Mn^{2+}	78LB
Na^+	61GT,62GT,70BHa,74PH	Fe^{2+}	78JBa
Mg^{2+}	74PH,77RGa,77SHa	Hg_2^{2+}	76HH
Ca^{2+}	68Na,70La,74PH	CH_3Hg^+	76RT
UO_2^{2+}	69T,75OSH,77Se,79CF		

Hydrogen carbonate   (continued)

$(CH_3)_3Pb^+$ 78ME

Zn^{2+} 78RB

Hg^{2+} 76HH,76HHa

Other references: 64Z,76BHS,77BL,78DB,78DJ,
78MPa,79BK,79BKa

$$NH_3$$

H_3N

Ammonia
(Other values on Vol.4, p.40)

L

Metal ion	Equilibrium	Log K 25°, 0.5	Log K 25°, 1.0	Log K 25°, 0	ΔH 25°, 0	ΔS 25°, 0
H^+	$HL/H.L$	9.32 ± 0.03 $9.29^a -0.01$	9.40 ± 0.04	9.244 ± 0.005 9.96^e	-12.45 ± 0.05 -13.9^e	0.5 -1^e
Cr^{3+}	$ML_5/ML_4.L$			1.6^t		
	$ML_6/ML_5.L$			1.5^t		
	$ML_6/M.L^6$			13^t		
	$ML_4/MOHL_4.H$(cis)		4.96			
	$MOHL_4/M(OH)_2L_4.H$		7.53			
	$ML_4/MOHL_4.H$(trans)		4.38			
	$MOHL_4/M(OH)_2L_4.H$		7.78			
Ag^+	$ML/M.L$	3.30		3.31 ± 0.06 3.62^e	-4.9 -5.7^e	-1 -3^c
	$ML_2/M.L^2$			7.22 ± 0.01 7.93^e	-13.4 ± 0.1 -14.1^e	-12 -11

---

^{a}25°, 0.1; ^{e}25°, 3.0; ^{t}24°, 4.5

Bibliography:

H^+ 750a,76HMa,76M,79MA

Ni^{2+} 76KSA

Cr^{3+} 75ABa,76MM

Cu^+ 63HS

Ag^+ 79MA

Cd^{2+} 77SFb

Other references: 66BP,71WN,75APS,75NW,
75SSa,76ES,77SFa,77SFb,78SL,78Y,
79FS,79Ha,79WNb

HN_3

HN_3		Hydrogen azide (hydrazoic acid)				HL
		(Other values on Vol.4, p.45)				

Metal ion	Equilibrium	Log K 25°, 0.5	Log K 25°, 1.0	Log K 25°, 0	ΔH 25°, 0	ΔS 25°, 0
H^+	HL/H.L	4.38	4.44 ±0.00	4.65 ±0.02	-3.6	9
		4.45[a]	4.78[e]	4.99[f]	-3.1[c]	10[c]
Mn^{2+}	ML/M.L		0.64			
Co^{2+}	ML/M.L		0.74 ±0.02			
Cu^{2+}	ML/M.L	2.44[h]	2.21	2.86[o]		
	$ML_2/M.L^2$			2.56[f]		
	$ML_3/M.L^3$			4.48[f]		
	$ML_4/M.L^4$			6.11[f]		
	$M.L^2/ML_2(s)$			7.82[f]-0.01		
					3.6	
V^{3+}	ML/M.L		2.63			
Cr^{3+}	ML/M.L		3.4			

[a]25°, 0.1; [c]25°, 1.0; [e]25°, 3.0; [f]25°, 4.0; [h]20°, 0.1; [o]20°, 0

Bibliography:

H^+ 75AA,76KK Cr^{3+} 76Ta
Mn^{2+},Cu^{2+} 77GA Fe^{3+},Hg^{2+} 76AA
Co^{2+} 78GS Zn^{2+} 75AAa
Ni^{2+} 75AA   Other references: 75EA,76DMN,78FN
V^{3+} 70EP

NO

ON		Nitric Oxide	L
		(Other references on Vol.4, p.135)	

Metal ion	Equilibrium	Log K 22°, 0.1
Fe^{2+}	ML/M.L	2.57

Bibliography: 77ZF

HNO_2

HO$_2$N	Hydrogen nitrite (nitrous acid) (Other values on Vol.4, p.47)					HL
Metal ion	Equilibrium	Log K 25°, 0.5	Log K 25°, 1.0	Log K 25°, 0	ΔH 25°, 0	ΔS 25°, 0
H$^+$	HL/H.L	2.94 / 3.24[d]	3.00 -0.1	3.15	(-2)[r]	(7)
Na$^+$	M.L/ML(s)				3.35	
Am^{3+}	ML/M.L		9.96			
Cm^{3+}	ML/M.L		0.82			
NpO$_2^+$	ML/M.L	-0.05[d]				
Fe^{3+}	ML/M.L			3.15	-1.9	8

[d] 25°, 2.0;   [r] 15-35°, 0

Bibliography:

Na$^+$ 77CP

Am^{3+},Cm^{3+} 78VKP

NpO$_2^+$ 79VG

Fe^{3+} 76ZB

Other references: 70KCa,77HF

NO_3^-

O$_3$N$^-$	Nitrate ion (Other values on Vol.4, p.48)					L$^-$
Metal ion	Equilibrium	Log K 25°, 0.5	Log K 25°, 1.0	Log K 25°, 0	ΔH 25°, 0	ΔS 25°, 0
Pu^{4+}	ML/M.L		0.54	1.8	(2)[s]	(10)[d]
		0.65[d]		0.74[m]		
	ML$_2$/M.L^2	1.12[d]		1.37[m]		
	ML$_3$/M.L^3			1.2		
NpO$_2^+$	ML/M.L	-0.6[d]				
Cu^{2+}	ML/M.L		-0.01 / -0.02[e]	0.5 / 0.11[f]	-0.9[c]	-3[c]
		-0.06[d]				
	ML$_2$/M.L^2		-0.6 / -0.5[e]	-0.4 / -0.4[f]		
		-0.6[d]				
Ag$^+$	M.L/ML(s)				5.43	
Sn^{2+}	ML/M.L		0.44			

[c] 25°, 1.0;   [d] 25°, 2.0;   [e] 25°, 3.0;   [f] 25°, 4.0;   [m] 20°, 4.0;   [s] 10-25°, 2.0

Bibliography:

Pu^{4+} 76BR,77BRP

NpO$_2^+$ 79VG

Cu^{2+} 75Ab

Ag$^+$ 73WK

Sn^{2+} 79ASR

Other references: 67KN,75APS,75BC,75FR, 75MHB,75MSB,76BKa,76MI,77EK,79MMN, 79MA

$$\begin{array}{c} O \\ \| \\ HO-P-OH \\ | \\ OH \end{array}$$

H_3O_4P	Hydrogen phosphate (phosphoric acid)	H_3L
	(Other values on Vol.4, p.56)	

Metal ion	Equilibrium	Log K 25°, 0.5	Log K 25°, 1.0	Log K 25°, 0	ΔH 25°, 0	ΔS 25°, 0
H^+	$HL/H.L$	11.40^q		12.35 ± 0.02	-3.5 ± 0.9	45
		$11.74^a \pm 0.08$	$10.79^e \pm 0.07$			
	$H_2L/HL.H$	6.58 ± 0.04	6.46 ± 0.02	7.199 ± 0.002	-0.9 ± 0.1	30
		$6.73^a \pm 0.03$	6.36^d		-1.2^a	
		$6.79^{t,w}$	$6.26^e \pm 0.02$			
	$H_3L/H_2L.H$	1.85 ± 0.04	1.70 ± 0.02	2.148 ± 0.002	1.9 ± 0.1	16
		$2.0^a \pm 0.1$	$1.86^e \pm 0.03$			
Mg^{2+}	$M.HL/MHL(H_2O)_3(s)$		-4.5^e	-5.82		
Ni^{2+}	$MHL/M.HL$	$2.10^a \pm 0.02$	2.00^p			
	$MH_2L/M.H_2L$	0.5^a				
Cr^{3+}	$MH_2L/M.H_2L$	2.56^t				
VO^{2+}	$MH_2L/M.H_2L$		3.20			
	$M(H_2L)_2/M.(H_2L)^2$		5.15			
	$MH_2L/MHL.H$		3.2			
	$M(H_2L)_2/M(HL)(H_2L).H$		3.5			
VO_2^+	$MHL/M.HL$		4.6			
	$M(HL)_2/M.(HL)^2$		9.31			
Hg_2^{2+}	$M.HL/MHL(s)$		-10.70^e	-12.40		
	$M^3.(HL)^2/M_3L_2(s)$		-21.40^e			
$(CH_3)_3Pb^+$	$MHL/M.HL$	1.88^u				
Hg^{2+}	$ML/M.L$		9.5^e			
	$MHL/M.HL$		8.8^e			
	$M.HL/MHL(s)$		-13.1^e			
	$M^3.(HL)^2/H^2.M_3L_2(s)$		-24.6^e			
	$M^3.HL/H^4.(MOH)_3L(s)$		-9.4^e			
Ga^{3+}	$MHL/M.HL$		7.46			
	$MH_2L/M.H_2L$		1.48			

---

[a] $25°, 0.1$;   [d] $25°, 2.0$;   [e] $25°, 3.0$;   [P] $15°, 0.1$;   [q] $25°, 0.68$;   [t] $25°, 0.2$;   [u] $25°, 0.3$;

[w] $(C_3H_7)_4NCl$ used as background electrolyte.

Bibliography:

H^+ 76ACP,76PSc,77VKa,78MAc

Mg^{2+} 76HHb

Ni^{2+} 76TD

Cr^{3+} 76AMA

VO^{2+} 76CKP

VO_2^+ 70IVb

Hg_2^{2+}, Hg^{2+} 75Q

$(CH_3)_3Pb^+$ 78ME

Ga^{3+} 75MG

In^{3+} 74FKb

Other references: 68RRM,71KY,72DSB,
72PA,73LM,74F,74RM,75IV,75RM,76F,
76MFG,77KG,78MPa,79DRa,79GGP,79JW

```
        O   O
        ‖   ‖
     HO-P-O-P-OH
        |   |
        OH  OH
```

$H_4O_7P_2$	Hydrogen diphosphate (pyrophosphoric acid)					H_4L
	(Other values on Vol.4, p.59)					

Metal ion	Equilibrium	Log K 25°, 0.1	Log K 25°, 1.0	Log K 25°, 0	ΔH 25°, 0	ΔS 25°, 0
H^+	HL/H.L	8.37 ±0.08	7.43 ±0.07	9.40 ±0.1	-0.39 ±0.0	42
		9.00[w]±0.05	8.74[w]±0.04	7.17[e]		
	$H_2L/HL.H$	6.04 ±0.04	5.41 ±0.05	6.70 ±0.1	-0.13 ±0.0	30
		6.19[w]+0.07	5.98[w]±0.00	5.29[e]		
	$H_3L/H_2L.H$	1.8	1.4 +0.1	2.2 ±0.1	1.0	13
		2.0[w] ±0.1	1.7[w] ±0.0	1.4[e]		
	$H_4L/H_3L.H$		0.8	0.8 ±0.1	1.5	9
		0.8[w]	0.7[w] ±0.1			
VO^{2+}	$MH_2L/M.H_2L$	4.20[t]				

---

[e]25°, 3.0; [t]25°, 0.2; [w]$(CH_3)_4N^+$ salt used as background electrolyte.

Bibliography:

H^+ 75VA

VO^{2+} 71KY

Other references: 70SK,72SKT,75SW,77SSM, 77WH,78CI,78TLa

```
        O   O   O
        ‖   ‖   ‖
     HO-P-O-P-O-P-OH
        |   |   |
        OH  OH  OH
```

$H_5O_{10}P_3$	Hydrogen triphosphate (triphosphoric acid)					H_5L
	(Other values on Vol.4, p.63)					

Metal ion	Equilibrium	Log K 25°, 0.1	Log K 25°, 1.0	Log K 25°, 0	ΔH 25°, 0	ΔS 25°, 0
H^+	HL/H.L	8.00 ±0.1		9.25 ±0.01	-0.1	42[h,w]
		8.70[w]±0.1	8.61[w]±0.05		-0.1[h,w]	40[h,w]
	$H_2L/HL.H$	5.50 ±0.1		6.54 ±0.07	1.4 ±0.1	35
		5.90[w]±0.1	5.69[w]-0.01			
	$H_3L/H_2L.H$	(2.6)		2.5 ±0.3		
		2.2[w] -0.1	2.0[w] ±0.0			
	$H_4L/H_3L.H$		1.0[w] ±0.1			
VO^{2+}	$MH_2L/M.H_2L$	4.8[t]				

---

[h]20°, 0.1; [t]25°, 0.2; [w]$(CH_3)_4N^+$ salt used as background electrolyte.

Bibliography:

VO^{2+} 71KY

Other references: 68AS,72SKT,74DS,77RLE,77SF,79KW

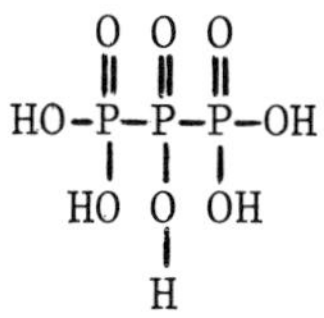

$$
\begin{array}{c}
\text{O O O} \\
\| \| \| \\
\text{HO-P-P-P-OH} \\
| \ | \ | \\
\text{HO O OH} \\
| \\
\text{H}
\end{array}
$$

$H_5O_8P_3$		Hydrogen triphosphate(IV,III,IV)	H_5L

Metal ion	Equilibrium	Log K 25°, 0.1
H^+	HL/H.L	10.28^w
	$H_2L/HL.H$	7.32^w
Li^+	ML/M.L	4.26^w
	MHL/M.HL	1.87^w
Na^+	ML/M.L	2.87^w
	MHL/M.HL	1.51^w
K^+	ML/M.L	2.53^w
	MHL/M.HL	1.36^w

$^w(CH_4)_4NCl$ used as background electrolyte.

Bibliography: 75Sb

$$As(OH)_3$$

H_3O_3As		Hydrogen Arsenite (arsenous acid) (Other references on Vol.4, p.132)				HL

Metal ion	Equilibrium	Log K 25°, 0.5	Log K 25°, 1.0	Log K 25°, 0	ΔH 25°, 0	ΔS 25°, 0
H^+	HL/H.L	9.13 ± 0.04	9.07	9.29	−6.6	20
	$H_2L_2/(HL)^2$	−0.21	0.21	−0.92		
	$H_2L_2/HL_2.H$	8.31	8.44			
	$HL/(As_4O_6)^{0.25}(s)$			−0.69		
Ag^+	$M^3.L/M_3L(s)$	-3.13^h				

^{h}20°, 0.1

Bibliography:

H^+ 59AR,76IV   Other reference :   79IV

Ag^+ 76CC

$$\begin{array}{c} O \\ \parallel \\ HO-As-OH \\ | \\ OH \end{array}$$

H_3O_4As	Hydrogen Arsenate (arsenic acid) (Other references on Vol.4, p.133)					H_3L

Metal ion	Equilibrium	Log K 20°, 0.1	Log K 25°, 1.0	Log K 25°, 0	ΔH 25°, 0	ΔS 25°, 0
H^+	HL/H.L			11.50	-4.4	38
	$H_2L/HL.H$		6.39	6.96 ±0.02	-0.8	29
	$H_3L/H_2L.H$			2.24 ±0.06	1.7	16
Ag^+	$M^3.L/M_3L(s)$	-22.2				

Bibliography: 76CC   Other reference : 75P

$$H_2S$$

H_2S	Hydrogen sulfide (hydrosulfuric acid) (Other values on Vol.4, p.76)					H_2L

Metal ion	Equilibrium	Log K 25°, 0.1	Log K 25°, 1.0	Log K 25°, 0	ΔH 25°, 0	ΔS 25°, 0
H^+	HL/H.L	(13.0) 13.6^q	13.9 ±0.1	(13.9)±0.1	-12	23
	$H_2L/HL.H$	6.82 ±0.03 6.71^q±0.02	6.61	7.02±0.04	-5.3	14
In^{3+}	MHL/M.HL		11^j			
	$M(HL)_2/M.(HL)^2$		17^j			
	$M^2.L^3/M_2L_3(s)$		-77^j	-69.4		
As(III)	$(M_3L_6)^{1/3}.M(OH)_3.H^3.L/M_2L_3(s)$	-36.5^t				

^{j}20°, 1.0; ^{q}25°, 0.7; ^{t}22°, 1.0

Bibliography:

H^+ 75GK,76AD,76TCl   As(III) 79IVa

In^{3+} 70TS   Other reference: 77BSb

$$\begin{array}{c} O \\ \parallel \\ HO-S-OH \end{array}$$

H_2O_3S	Hydrogen sulfite (sulfurous acid) (Other values on Vol.4, p.78)					H_2L

Metal ion	Equilibrium	Log K 25°, 0.5	Log K 25°, 1.0	Log K 25°, 0	ΔH 25°, 0	ΔS 25°, 0
H^+	HL/H.L	6.79^h	6.34 6.36^e	7.18 ±0.03	$(3)^r$	43
	$H_2L/HL.H$	1.62^a	1.37 1.72^e	1.91 ±0.02	4.0	22
	$H_2L/SO_2(g)$		0.03 0.06^e	0.09	$(-6)^s$	(-20)
	$SO_2(g)/HL.H$			1.852±0.005	$(4)^t$	(22)

^{a}25°, 0.1; ^{e}25°, 3.0; ^{h}20°, 0.1; ^{r}10-50°, 0; ^{s}25-50°, 0; ^{t}0-40°, 0

Hydrogen sulfite (continued)

Metal ion	Equilibrium	Log K 25°, 0.5	Log K 25°, 1.0	Log K 25°, 0	ΔH 25°, 0	ΔS 25°, 0
CH_3Hg^+	ML/M.L	8.11[h]	7.96[d]			
$(CH_3)_3Pb^+$	ML/M.L	1.28[u]				
Pd^{2+}	$ML_4/M.L^4$	29.1[q]				
Hg^{2+}	$ML_2/M.L^2$	22.33	22.85[l]	24.07[o]	-20[b]	40[b]
	$ML_3/M.L^3$	24.1		25.96[o]		

[b]25°, 0.5;   [d]25°, 1.7;   [h]20°, 0.1;   [l]18°, 3.0;   [o]18°, 0;   [q]25°, 0.8;   [u]25°, 0.3

Bibliography:

H^+	61EA,77HE	Pd^{2+}	77HF
CH_3Hg^+	76RT	Hg^{2+}	76MH
$(CH_3)_3Pb^+$	78ME	Other reference:	78MPa

$$HSO_4^-$$

HO_4S^-	Hydrogen sulfate ion (sulfuric acid) (Other values on Vol.4, p.79)					HL^-

Metal ion	Equilibrium	Log K 25°, 0.5	Log K 25°, 1.0	Log K 25°, 0	ΔH 25°, 0	ΔS 25°, 0
H^+	HL/H.L	1.32 ±0.06	1.10 ±0.08	1.99 ±0.01	5.4 ±0.2	27
		1.55[a]±0.05			5.6[c]	24[c]
		1.01[d]±0.07	0.91[e]±0.02	1.53[g]		
Li^+	ML/M.L			0.77 -0.1	0	4
Na^+	ML/M.L	0.40	0.33[q]±0.07	0.82 ±0.1	1.1	7
K^+	ML/M.L	0.4[a]	0.33[q]±0.07	0.85 ±0.1	1.0	7
Rb^+	ML/M.L			0.60		
Cs^+	ML/M.L			0.3		
NH_4^+	ML/M.L			0.94 +0.2		
Mg^{2+}	ML/M.L	1.46[a]	0.9[q] ±0.1	2.23 ±0.02	1.4 ±0.1	15
Ca^{2+}	ML/M.L	1.51[a] ±0.02	1.2[q] ±0.2	2.32 ±0.05	1.6 ±0.1	16
	M.L/ML(s)	-3.02	-2.94 ±0.02	-4.62 ±0.02	0±3	-20
		-3.01[d]	-3.21[e]±0.05	-3.55[f]		
Sr^{2+}	ML/M.L	1.41[a]				
	M.L/ML(s)	-5.11	-5.02	-6.50 ±0.05	0.5	-28
		-5.02[d]	-5.19[e]	5.38[f]		
Pu^{3+}	ML/M.L	1.65[d]	1.73			
	$ML_2/M.L^2$	3.29[d]	3.39			
Np^{4+}	ML/M.L	3.52[d] ±0.02	3.41[e]	3.53[m]		
	$ML_2/M.L^2$	6.06[d]	5.42[e]	5.92[m]		
Pu^{4+}	ML/M.L	3.83[d] ±0.01	3.66		(7)[s]	(41)[d]
	$ML_2/M.L^2$	6.69[d]				

---

[a]25°, 0.1;   [c]25°, 1.0;   [d]25°, 2.0:   [e]25°, 3.0;   [f]25°, 4.0;   [g]25°, 5.0;   [q]25°, 0.7;
[m]20°, 4.0 assuming HL/H.L=0.83;   [s]10-25°, 2.0

Hydrogen sulfate ion   (continued)

Metal ion	Equilibrium	Log K 25°, 0.5	Log K 25°, 1.0	Log K 25°, 0	ΔH 25°, 0	ΔS 25°, 0
NpO_2^+	ML/M.L	0.45[d]				
UO_2^{2+}	ML/M.L	1.89[d]	1.81 ±0.0	2.95	5.0 / 4.4[c]	30 / 24[c]
	$ML_2/M.L^2$	3.25[d]	2.5 ±0.2	4.0	8.4[c]	41[c]
	$ML_3/M.L^3$		3.7[j]			
NpO_2^{2+}	ML/M.L	2.07 / 2.20[a] / 2.08[d]	1.82[j]	3.27		
	$ML_2/M.L^2$	3.4 / 3.8[a]	2.62[j]			
PuO_2^{2+}	ML/M.L	2.17[d]				
Mn^{2+}	ML/M.L	0.77 / 0.60[d]	(0.57) / 0.56[e]	2.26 +0.02 / 0.72[f]	2.1 ±0.1	17
Co^{2+}	ML/M.L		0.23[e]	2.34 ±0.07	1.4 ±0.1	16
Ni^{2+}	ML/M.L	1.0[t]	0.57 / 0.26[e]	2.34 ±0.06	1.5 -0.1	16
Cu^{2+}	ML/M.L		0.95 / 0.70[e]	2.36 ±0.03	2.1 ±0.4	18
Cr^{3+}	ML/M.L		2.60[u]		7.6[b]	
	ML/MOHL.H	4.65[a]				
CH_3Hg^+	ML/M.L	0.94[q]				
Tl^+	ML/M.L	0.65 / 0.08[d]	0.40 / -0.12[e]	1.37 ±0.07 / -0.24[f]	-0.2 / 1.6[d]	6
Zn^{2+}	ML/M.L	0.93 / 0.76[d]	0.89 / 0.70[e]	2.34 ±0.04 / 0.61[f]	1.5 ±0.1 / 0.5[d]	16
Cd^{2+}	ML/M.L	1.08 / 0.86[d]	0.95 / 0.71[e] ±0.01	2.46 / 0.95[f]	2.3 ±0.1 / 1.0[d]	19
Al^{3+}	ML/M.L		1.48 / 1.16[e]	3.89 / 1.20[g]	(-5.1)[e]	(-12)[e]

---

[a]25°, 0.1;   [b]25°, 0.5;   [c]25°, 1.0;   [d]25°, 2.0;   [e]25°, 3.0;   [f]25°, 4.0;   [g]25°, 5.0;
[j]20°, 1.0;   [q]25°, 0.7;   [t]25°, 0.2;   [u]48°, 1.0

Bibliography:

H^+ 71KV,74EW,74FP,75RB,76KK,77AH,77PRS, Cu^{2+} 68HP
          78MAc,78YS
                                                    Cr^{3+} 76DH,77Sf
$Li^+-NH_4^+$ 74EW,75EW,75Rb,75SCC
                                                    Fe^{3+} 77AH
$Mg^{2+}-Ba^{2+}$ 69DI,71IS,73K,75LMa,76K,76KF,77FKa, Ag^+ 76MS
          78EF,78EW,79EW,79FF
                                                    CH_3Hg^+ 76RT
Pu^{3+} 76FB
Np^{4+} 76BRa Tl^+ 75FR
Pu^{4+} 74FP,76BR,76BRb,77BRP
                                                    Zn^{2+},Cd^{2+} 75FCK,76FKM,76K,78Ab
$UO_2^{2+}-PuO_2^{2+}$ 76PR
                                                    Al^{3+} 71KV
$Mn^{2+}-Ni^{2+}$ 73K,76K,79FF

Hydrogen sulfate ion (continued)

Other references: 42KP,67PH,69F,69M,70AL,
    71MSa,73AB,75APS,75F,75FBb,75FFb,75MA,
    75MV,75NT,75YY,76MKS,77FF,77KG,77VV,
    78FF,78FKD,78GSM,78KKc.78MPa,78NM,
    78SGc,78VB,79AB,79FK,79Sc,79TZ

$$H_2S_2O_3$$

$H_2O_3S_2$	Hydrogen thiosulfate (thiosulfuric acid) (Other values on Vol.4, p.86)					H_2L
Metal ion	Equilibrium	Log K 25°, 0.5	Log K 25°, 1.0	Log K 25°, 0	ΔH 25°, 0	ΔS 25°, 0
H^+	HL/H.L	1.0 1.3[a]	0.8	1.6 ±0.1		
	H_2L/HL.H			0.6		
Na^+	ML/M.L	(0.04)	0.15	0.53 ±0.05	1.1[b]	4[b]
Tl^+	ML/M.L		1.57	0.86[f]		
	ML_2/M.L^2		1.94	0.72[f]		
	ML_3/M.L^3			0.2[f]		
$(CH_3)_3Pb^+$	ML/M.L	2.14[u]				
Pd^{2+}	ML_4/M.L^4	35.0[u]				

[a] 25°, 0.1; [b] 25°, 0.5; [f] 25°, 4.0; [u] 25°, 0.3

Bibliography:

Na^+	75SPF	Pd^{2+}	77HF
Tl^+	75GF,77PC	Other references:	71MSa,74Gc,75GB,76MKS,
$(CH_3)_3Pb^+$	78ME		77DBb,77JB,77SSb,78JB,79JB

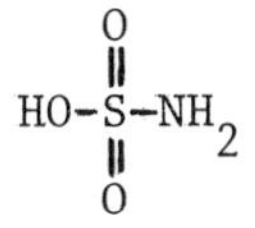

$$\begin{array}{c} O \\ \| \\ HO-S-NH_2 \\ \| \\ O \end{array}$$

H_3O_3NS	Hydrogen amidosulfate (sulfamic acid) (Other values on Vol.4, p.88)					HL
Metal ion	Equilibrium	Log K 25°, 1.0	Log K 25°, 3.0	Log K 25°, 0	ΔH 25°, 0	ΔS 25°, 0
H^+	HL/H.L			0.988	−0.25	3.7
Zn^{2+}	ML/M.L	0.75	0.85	1.60[v]		
	ML_2/M.L^2	0.3	0.3	2.5[v]		
	ML_3/M.L^3		0.1	0.5[v]		
Cd^{2+}	ML/M.L	0.85	1.00	1.80[v]		
	ML_2/M.L^2	0.7	0.9	2.3[v]		
	ML_3/M.L^3	0.2	0.3	1.4[v]		

[v] 25°, 6.0

Bibliography: 78NF   Other reference: 78MAc

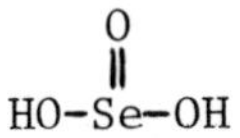

$$\overset{\overset{\textstyle O}{\|}}{\text{HO--Se--OH}}$$

H_2O_3S **Hydrogen selenite (selenous acid)** H_2L

(Other values on Vol.4, p.91)

Metal ion	Equilibrium	Log K 25°, 0.3	Log K 25°, 1.0	Log K 25°, 0	ΔH 25°, 0	ΔS 25°, 0
H^+	HL/H.L	7.94	7.78 8.05[e]		-1.20[c] -1.26[c]	31.6[c] 32.6[e]
	H_2L/HL.H	2.35	2.27 2.61[e]	2.65	1.5[c] 1.3[e]	15[c] 16[e]
Ag^+	M^2.L/M_2L(s)	-15.5[h]		-15.55	10.7	-35
CH_3Hg^+	ML/M.L MHL/M.HL		6.46[q] 2.70[q]			
$(CH_3)_3Pb^+$	ML/M.L	1.95				

[c] 25°, 1.0; [e] 25°, 3.0; [h] 20°, 0.1; [q] 25°, 0.8

Bibliography:

H^+ 75VKb $(CH_3)_3Pb^+$ 78ME

Ag^+ 76CC

CH_3Hg^+ 76RT

$$HSeO_4^-$$

HO_4Se^- **Hydrogen selenate ion (selenic acid)** HL^-

(Other values on Vol.4, p.93)

Metal ion	Equilbrium	Log K 25°, 1.0	Log K 25°, 0	ΔH 25°, 0	ΔS 25°, 0
H^+	HL/H.L		1.70 ±0.05	5.6 ±0.1	27
CH_3Hg^+	ML/M.L	1.12[q]			

[q] 25°, 0.7

Bibliography: 76RT

HF

		HF Hydrogen fluoride (hydrofluoric acid) (Other values on Vol.4, p.96)				HL

Metal ion	Equilibrium	Log K 25°, 0.5	Log K 25°, 1.0	Log K 25°, 0	ΔH 25°, 0	ΔS 25°, 0
H^+	HL/H.L	2.92 ±0.01 2.92[a]±0.03 3.12[d]±0.02	2.97 ±0.02 3.30[e]±0.03	3.165±0.01 3.54[f]	3.20 ±0.02 2.89[c]±0.03 2.37[e]	25.2[c] 23.3[c] 23.1[e]
	HL_2/HL.L	0.59 ±0.02 0.84[d]	0.59 ±0.1 0.94[e]±0.1	0.57 ±0.04 1.05[f]	1.0 0.8[c] ±0.1	6 5[c]
Na^+	ML/M.L	$(-1.4)^q$	-0.79	-0.45	$(-3)^t$	(-16)
Mg^{2+}	ML/M.L	1.32 ±0.02 1.86[w]	1.32 ±0.02		3.2[c]	17[c]
Ca^{2+}	ML/M.L	0.6 ±0.1 1.57[w]	0.60 ±0.03		3.5[c]	15[c]
	$M.L^2/ML_2(s)$			-10.50 ±0.05	$(3)^r$	(-38)
Sr^{2+}	ML/M.L	1.38[w]	0.1 ±0.1		4[c]	13[c]
Ba^{2+}	ML/M.L	1.32[w]	-0.3 ±0.1		$(4)^x$	(12)[c]
Y^{3+}	$M.L^3/ML_3(s)$			-18.3		
La^{3+}	$M.L^3/ML_3(s)$			-18.7		
Ce^{3+}	ML/M.L M.L /ML (s)	3.14 ±0.01	2.81	4.0 -19.1	4.8[c]	29[c]
Nd^{3+}	M.L /ML (s)			-20.3		
Eu^{3+}	ML/M.L $M.L^3/ML_3(s)$	3.40 -0.01	3.26 ±0.07	 -21.9	9.2[c]	46[c]
Er^{3+}	$M.L^3/ML_3(s)$			-18.0		
Am^{3+}	ML/M.L		2.49		$(7)^u$	(35)[c]
Cm^{3+}	ML/M.L		2.61		$(7)^u$	(35)[c]
Bk^{3+}	ML/M.L		2.89		$(7)^u$	(37)[c]
Cf^{3+}	ML/M.L		3.03		$(7)^u$	(37)[c]
Th^{4+}	ML/M.L $ML_2/M.L^2$ $ML_3/M.L^3$ $ML_4/M.L^4$	7.59 ±0.04 7.56[d] 13.44 ±0.02 17.9	7.46 7.80[e] 13.80[e] 18.8[e]	8.44 ±0.2 8.19[m] 15.08 ±0.2 14.67[m] 19.8 ±0.4 23.2	3.6[c] $(-2)^s$ $(-3)^s$ $(-4)^s$	46[c] (70) (90) (100)
U^{4+}	ML/M.L $ML_2/M.L^2$ $ML_3/M.L^3$		7.78	8.91[f] 9.05[m] 15.37[f] 15.74[m] 20.1[f] 21.2[m]	$(3)^v$	(46)[c]

[a] 25°, 0.1; [c] 25°, 1.0; [d] 25°, 2.0; [e] 25°, 3.0; [f] 25°, 4.0; [m] 20°, 4.0 H^+; [q] 27°, 0.7;
[r] 1-40°, 0; [s] 5-45°, 0; [t] 15-35°, 0.7; [v] 3-47°, 1.0·H^+; [w] 25°, 0.05 $(C_2H_5)_4NClO$;
[x] 15-60°, 1.0; [u] 10-55°, 1.0H^+

Hydrogen fluoride (continued)

Metal ion	Equilibrium	Log K 25°, 0.5	Log K 25°, 1.0	Log K 25°, 0	ΔH 25°, 0	ΔS 25°, 0
Np^{4+}	$ML/M.L$	7.83^d	7.56	8.33^m	$(4)^u$	$(48)^c$
	$ML_2/M.L_2$			14.59^m		
	$ML_3/M.L_3$			20.3^m		
	$ML_4/M.L_4$			25.1^m		
Pu^{4+}	$ML/M.L$	7.77^d			$(1)^z$	$(39)^d$
NpO_2^+	$ML/M.L$	0.99^d				
UO_2^{2+}	$ML/M.L$	4.4 4.69^d	4.54 -0.1	5.02^f	0.4^c	22^c
	$ML_2/M.L_2$		7.97 -0.08		0.5^c	28^c
	$ML_3/M.L_3$		10.55 -0.09		0.6^c	51^c
	$ML_4/M.L_4$		12.0 -0.2		0.1^c	55^c
NpO_2^{2+}	$ML/M.L$	4.04 4.12^a 4.25^d	3.85^j	4.6		
	$ML_2/M.L_2$	7.00 7.01^a	6.97^j			
PuO_2^{2+}	$ML/M.L$	4.21^d				
Mn^{2+}	$ML/M.L$		0.7 ±0.1 1.00^e		3.3^b 3.6^e	14^c 11^e
Co^{2+}	$ML/M.L$		0.4 0.64^e		2.5^b 3.3^e	10^c 14^e
Ni^{2+}	$ML/M.L$		0.5 ±0.2 0.76^e		1.9^b 1.4^e	9^c 8^e
Cu^{2+}	$ML/M.L$	0.8 -0.1 0.9^y	0.8 ±0.2 1.06^e		3.6^b 3.2^e	16^c 16^e
Cr^{3+}	$ML/M.L$	4.36		5.2	2.7^b	29^b
	$ML_2/M.L_2$	7.70				
	$ML_3/M.L_3$	10.2				
Fe^{3+}	$ML/M.L$	5.18 ±0.04	5.16 ±0.02	6.0	2.3^b	31^b
	$ML_2/M.L_2$	9.13 ±0.04	9.07		3.5^b	54^b
	$ML_3/M.L_3$	11.9 ±0.1	12.1		4.4^b	69^b
Zr^{4+}	$ML/M.L$	8.94^d		9.8 9.51^f 9.39^m		
	$ML_2/M.L_2$	16.4^d		17.26^m		
	$ML_3/M.L_3$	22.4^d		23.8^m		
	$ML_4/M.L_4$			29.6^m		
	$ML_5/M.L_5$			34.6^m		
	$ML_6/M.L_6$			38.4^m		
Hf^{4+}	$ML/M.L$			9.04^f 9.03^m		
	$ML_2/M.L_2$			16.58^m		
	$ML_3/M.L_3$			23.1^m		
	$ML_4/M.L_4$			28.8^m		
	$ML_5/M.L_5$			34.0^m		
	$ML_6/M.L_6$			38.0^m		

---

^{a}25°, 0.1; ^{b}25°, 0.5; ^{c}25°, 1.0; ^{d}25°, 2.0; ^{e}25°, 3.0; ^{f}25°, 4.0; ^{j}20°, 1.0;

^{m}20°, 4.0 H$^+$; ^{u}10-55°, 1.0 H$^+$; ^{y}25°, 0.05; ^{z}10-25°, 2.0

Hydrogen fluoride (continued)

Metal ion	Equilibrium	Log K 25°, 0.5	Log K 25°, 1.0	Log K 25°, 0	ΔH 25°, 0	ΔS 25°, 0
Zn^{2+}	ML/M.L	0.7 0.8[d]	0.7 ±0.2 0.88[e] -0.1	1.15	3.0[b] 2.2[e]	13[c] 12[e]
Cd^{2+}	ML/M.L	0.5[q]	0.5 0.6[e] -0.1		1.2[c] 1.0[e]	6[c] 6[e]
Pb^{2+}	ML/M.L $ML_2/M.L^2$	(1.26)[d] 2.55[d]	1.44 ±0.04 2.53 ±0.01			
In^{3+}	ML/M.L	3.73 ±0.02 3.74[d]	3.70 ±0.03	4.6	2.6 2.2[c] ±0.0	30 24[c]
	$ML_2/M.L^2$	6.5 ±0.1 6.6[d]	6.4 ±0.1	8.1	5 3[c] ±1	50 40[c]
	$ML_3/M.L^3$	8.6 9.0[d]	8.6 ±0.0	10.3	7 5[c] ±2	70 60[c]
	$ML_4/M.L^4$	9.9 10.3[d]	9.8 +0.1	11.5	9 8[c]	80 70[c]
As(III)	$M(OH)_2L/M(OH)_2.L$		4.48[j]			

[b]25°, 0.5; [c]25°, 1.0; [d]25°, 2.0; [e]25°, 3.0; [j]22°, 1.0; [q]25°, 0.8

Bibliography:

H^+ 70Ka,72CT,73BC,73KC,74GC,75SSa,73BC

Na^+ 70BH,71RDB,76MK

Be^{2+} 69MB

Mg^{2+} 78BB,78EW

Ca^{2+} 77MB,77VKT,78BB

Sr^{2+},Ba^{2+} 78BB

$Y^{3+},La^{3+},Nd^{3+},Er^{3+}$ 77VK

Ce^{3+} 67LN,77VK

Eu^{3+} 66LN,76CU,77VK

Gd^{3+} 67LN

$Am^{3+}-Cf^{3+},U^{4+}$ 76CU

Th^{4+} 75PR,76CU

Np^{4+} 76BRb,76CU

Pu^{4+} 76BR,77BRP

NpO_2^+ 79VG

$UO_2^{2+}-PuO_2^{2+}$ 76PR

$Mn^{2+}-Ni^{2+},Zn^{2+}$ 74Ac,76KBM

Cu^{2+} 73BC,74AC,74GC,76KBM,73BC

Cr^{3+} 76DH

Fe^{3+} 69AL,75Ja

Zr^{4+},Hf^{4+} 73N

Cd^{2+} 75AC,75VK,76BCa

Pb^{2+} 63MH,73BH
Al^{3+} 75VK
In^{3+} 73BC

As(III)   76IBG

Other references: 69B,69Ba,69D,69M,70GMa,
71B,74H,74KI,75MHB,75NA,76APS,76IS,
76KBa,76SB,76VKK,77BL,77IK,77KG,77MK,
77SSa,79BH,80BH

In^{3+} 73BC

$$Cl^-$$

Chloride ion
(Other values on Vol.4, p.104)

$$Cl^- \qquad\qquad\qquad\qquad\qquad\qquad\qquad\qquad\qquad\qquad\qquad\qquad\qquad\qquad L^-$$

Metal ion	Equilibrium	Log K 25°, 0.5	Log K 25°, 1.0	Log K 25°, 0	ΔH 25°, 0	ΔS 25°, 0
Mg^{2+}	$ML/M.L$		-0.10		$(-1)^r$	$(-5)^c$
			-1.0^e			
Ca^{2+}	$ML/M.L$	0.08^q	-0.11		$(-2)^r$	$(-6)^c$
Sr^{2+}	$ML/M.L$		-0.21		$(-2)^r$	$(-9)^c$
Ba^{2+}	$ML/M.L$		-0.42		$(-3)^r$	$(-11)^c$
Th^{4+}	$ML/M.L$	0.30 ± 0.05	0.18	1.38		
		$0.10^d \pm0.02$		$0.17^f \pm0.06$		
	$ML_2/M.L^2$	-1.0^d		$-0.9^f \pm0.0$		
Np^{4+}	$ML/M.L$	0.15^i	-0.04^j			
		-0.05^d				
		-0.04^k				
	$ML_2/M.L^2$	-0.15^d	-0.24^j			
		-0.15^k				
	$ML_3/M.L^3$		-0.5^j			
Pu^{4+}	$ML/M.L$	0.15^d	0.14	0.15^m		
	$ML_2/M.L^2$		-0.17	0.08^m		
	$ML_3/M.L^3$			-1.0^m		
NpO_2^{2+}	$ML/M.L$	$-0.35^d \pm0.07$				
Mn^{2+}	$ML/M.L$		-0.2 ± 0.2			
Fe^{2+}	$M.L^2/ML_2(s)$				-19.8	
Co^{2+}	$ML/M.L$		0.0 ± 0.1			
		-0.2^d	$-0.3^e \;\; +0.1$	-1.1^g	0.5^d	1^d
Ni^{2+}	$ML/M.L$		$0.0 \;\; +0.1$			
		$-0.2^d \pm0.1$	-0.5^e		0.5^d	1^d
Cu^{2+}	$ML/M.L$		0.1 ± 0.0	0.4 ± 0.1		
		$0.0^e \pm0.1$	$0.0^d \;\; -0.1$	0.1^f	1.6^d	10^d
		$0.2^g \pm0.1$		0.4^v		
Cr^{3+}	$ML/M.L$		-0.5		6.3^b	19^c
		-0.4^d		0.05^f	6.6^g	22^f
Zr^{4+}	$M.L^4/ML_4(s)$				-70.6	
Cu^+	$ML_2/M.L^2$	5.19^h		5.5^o		
		6.06^g		5.79^f		
	$ML_3/ML_2.L$			0.2^o		
	$M_2L_4/M^2.L^4$	-0.12^g		-0.28^f	-4.8^g	-17^g
		13.0^g				
	$M.L/ML(s)$	-7.38^g		-6.73		

---

^{b}25°, 0.5; ^{c}25°, 1.0; ^{d}25°, 2.0; ^{e}25°, 30; ^{f}25°, 4.0; ^{g}25°, 5.0; ^{h}20°, 0.1;
^{i}20°, 0.5; ^{j}20°, 1.0; ^{k}20°, 2.0; ^{m}20°, 4.0; ^{o}20°, 0; ^{q}25°, 0.7; ^{r}15-60°, 1.0;
^{v}25°, 6.0

Chloride ion (continued)

Metal ion	Equilibrium	Log K 25°, 0.5	Log K 25°, 1.0	Log K 25°, 0	ΔH 25°, 0	ΔS 25°, 0
Ag^+	ML/M.L	3.08[a]	3.36	3.31 ±0.00		
		3.70[g]		3.45[m]		
	$ML_2/M.L^2$	5.08[a]	5.20	5.25 ±0.01		
		5.62[g]		5.67[m]		
	$ML_3/M.L^3$	6.4[g]		6.0[m]	-9.3[g]	-2[g]
	$ML_4/M.L^4$	6.1[g]		6.0[m]	-14[g]	-19[g]
	M.L/ML(s)	-9.62	-9.74	-9.752±0.002	15.70 ±0.01	8.1
			-10.05[e]	-10.21[f]	15.0[g]	
CH_3Hg^+	ML/M.L	5.18[a]	5.32	5.50[t]	-6.0[h]	4[a]
$(CH_3)_3Pb^+$	ML/M.L	0.08[u]	0.32			
Pd^{2+}	ML/M.L		4.47 -0.5	6.1 -0.1	-3.0[c]	10[c]
	$ML_2/M.L^2$		7.74 -0.3	10.7 -0.1	-5.6[c]	17[c]
	$ML_3/M.L^3$		10.2 -0.4	13.1 ±0.0	-8.2[c]	19[c]
	$ML_4/M.L^4$	9.5	11.5 ±0.6	15.4 ±0.3	-11.6[c]	14[c]
Pt^{2+}	ML/M.L		5.0			
	$ML_2/M.L^2$		9.0			
	$ML_3/M.L^3$		11.9			
	$ML_4/M.L^4$		14.0			
	$ML_2(cis)/ML_2(trans)$	0.08				
Zn^{2+}	ML/M.L		0.0 ±0.1	0.46 ±0.03		
		-0.5[d]	-0.19[e]±0.00	0.10[f]	1.3[e]	4[e]
	$ML_3/M.L^3$			0.6		
		0.0[d]	-0.4[e] ±0.2	0.1[f]	9[e]	28[e]
				0.5		
	$ML_4/M.L^4$	-0.1[d]	0.0[e] ±0.2	0.3[f]	0[e]	5[e]
				0.2		
			-0.7[e]	0.2[f]		
Cd^{2+}	ML/M.L	1.35 ±0.02	1.35 ±0.02	1.98 ±0.03	0.3[b]	7[b]
		1.52[a]±0.07				
		1.44[d]±0.02	1.54[e]±0.05	1.66[f]±0.1	-0.1[e]	7[e]
	$ML_2/M.L^2$	1.7[d] ±0.1	1.7 ±0.1	2.6[f] ±0.1	0.9[b]	11[b]
		1.9[d] ±0.1	2.2[e] ±0.1	2.4[f] ±0.1	0.0[e]	10[e]
	$ML_3/M.L^3$	1.9[d] ±0.1	1.5 ±0.2	2.4[f] ±0.1	2.4[c]	15[c]
			2.3[e] ±0.1	2.8[f] ±0.3	1.9[e]	17[e]
	$ML_4/M.L^4$			1.7		
			1.6[e]	2.2[f] +0.3	6.1[f]	31[f]
Hg^{2+}	ML/M.L	6.74 -0.1	6.72		-5.5[c]	12[c]
			7.15[e]±0.08		-5.8[e]	13[e]
	$ML_2/M.L^2$	13.22 -0.2	13.23		-12.2[c]	20[c]
			13.99[e]±0.01		-12.3[e]	23[e]
	$ML_3/M.L^3$	14.1 ±0.1	14.2 -0.1		-12.4[c]	23[c]
			14.9[e] ±0.2		-13.3[c]	23[e]
	$ML_4/M.L^4$	15.1 ±0.1	15.3 ±0.0		-14.2[c]	22[c]
			16.1[e] ±0.1		-14.8[c]	24[e]
	ML/MOHL.H		3.05		-4.8[c]	-2[c]
			3.05[e]			

[a] 25°, 0.1; [b] 25°, 0.5; [c] 25°, 1.0; [d] 25°, 2.0; [e] 25°, 3.0; [f] 25°, 4.0; [g] 25°, 5.0; [h] 20°, 0.1; [m] 20°, 4.0; [t] 25°, 2.5; [u] 25°, 0.3

Chloride ion (continued)

Metal ion	Equilibrium	Log K 25°, 0.5	Log K 25°, 1.0	Log K 25°, 0	ΔH 25°, 0	ΔS 25°, 0
Sn^{2+}	$ML/M.L$	1.05 ± 0.05	1.05 ± 0.03	1.64	2.2	15
		$1.08^d\pm0.03$	$1.17^e\pm0.02$	$1.40^f\pm0.06$	1.6^e	11^e
				1.80^v		
	$ML_2/M.L^2$	1.42 ± 0.06	1.5 ± 0.3	2.43	$(4)^s$	21^e
		$1.68^d\pm0.06$	$1.75^e\pm0.05$	$2.24^f\pm0.06$		
				3.04^v		
	$ML_3/M.L^3$	$1.3^d \pm0.2$	$1.67^e\pm0.02$	$2.24^f\pm0.06$	$(6)^s$	28^e
				3.30^v		
Bi^{3+}	$ML/M.L$	2.5	2.4 ± 0.2	3.5 ± 0.2	4.0	29
		$2.3^d \pm0.2$	$2.2^e \pm0.2$	$2.3^f \pm0.2$	0.5^f	12^f
	$ML_2/M.L^2$		$4.0 \ -0.3$	$5.5 \ \pm0.0$		
		$3.9^d \pm0.4$	$3.9^e \pm0.4$	$4.2^f +0.3$	$(6.3)^f$	$(40)^f$
	$ML_3/M.L^3$		$5.2 \ -0.3$	$7.1 \ \pm0.2$		
		$5.5^d \pm0.4$	$5.6^e \pm0.2$	$6.0^f +0.1$	$(-4.6)^f$	$(12)^f$
	$ML_4/M.L^4$		$6.4 \ +0.5$	$8.1 \ \pm0.2$		
		$6.7^d \pm0.5$	$6.9^e -0.2$	$7.3^f -0.4$	$(4.2)^f$	$(48)^f$
	$ML_5/M.L^5$	$6.8^d -0.1$	$7.4^e \pm0.2$	$8.3^f +0.3$	$(-7.2)^f$	$(14)^f$
	$ML_6/M.L^6$	6.6^d	7.4^e	$7.9^f \pm0.5$		

$^d 25°. 2.0;$ $^e 25°, 3.0;$ $^f 25°, 4.0;$ $^s 0-45°, 3.0;$ $^v 25°, 6.0$

Bibliography:

Mg^{2+} 75EW,75SCH,78EW

Ca^{2+} 75EW,75SCH

Sr^{2+},Ba^{2+} 75SCH

Th^{4+},Np^{4+} 75PR

Pu^{4+} 76BR

NpO_2^+ 64GS,79VG

Mn^{2+} 75FK

Fe^{2+} 77CH

Co^{2+} 75BHS,78SB

Ni^{2+} 76MKS,78F

Cu^{2+} 75MSB,77BSa,77Sb

Cr^{3+} 76DH

Zr^{4+} 78VL

Cu^+ 77AT,77ATT,78PK

Ag^+ 73WK

CH_3Hg^+ 78JI

$(CH_3)_3Pb^+$ 78ME

Pd^{2+} 77HF

Pt^{2+} 78E

Zn^{2+} 75BA,77F,78FKK

Cd^{2+} 71PB,75KL

Hg^{2+} 75CGP,76Ba,76CGP,77S,78BKP

Sn^{2+} 75FB,76SL,76VKV

Pb^{2+} 69FK,73BH,75M,76FS,77BL

Bi^{3+} 73VG,76VG

Other references: 52HH,54CV,66SM,70H,70L,
    70LT,71MSa,71WB,73BM,73E,73S,74GW,
    74PK,75BA,75Fa,75FBa,75KB,75LBa,75MHB,
    75NF,75VKa,76BKa,76CW,76FRI,76KB,76KSb,
    76Ra,76SF,76SM,76TS,76YK,77FF,77KD,
    77NT,77Sc,77SJH,77SK,78AR,78BZ,78FF,
    78FR,78JP,78PT,78SGc,79BKa,79O,79SFa,
    79SPM

ClO_3^-

| O_3Cl^- | | Chlorate ion
(Other values on Vol.4, p.113) | | | L^- |
</br>

Metal ion	Equilibrium	Log K 25°, 0.1	ΔH 25°, 0.1	ΔS 25°, 0.1
Eu^{3+}	ML/M.L	0.02	-1.5	-5

Bibliography: 77CE

ClO_4^-

| O_4Cl^- | | Perchlorate ion
(Other values on Vol.4, p.114) | | | L^- |

Metal ion	Equilibrium	Log K 25°, 1.0	Log K 25°, 0	ΔH 25°, 1.0	ΔS 25°, 1.0
K^+	M.L/ML(s)		-1.94		
Rb^+	M.L/ML(s)		-2.54		
Cs^+	M.L/ML(s)		-2.38		
Cu^{2+}	ML/M.L	-0.34		-1.6	-7

Bibliography:
K^+-Cs^+ 69G   Other references: 75APS,78GD
Cu^{2+} 75Ab

Br^-

| Br^- | | Bromide ion
(Other values on Vol.4, p.115) | | | L^- |

Metal ion	Equilibrium	Log K 25°, 0.5	Log K 25°, 1.0	Log K 25°, 0	ΔH 25°, 0	ΔS 25°, 0
Th^{4+}	ML/M.L	-0.13[d]				
Np^{4+}	ML/M.L	-0.21[d]				
Pu^{4+}	ML/M.L	0.33[d]				
Mn^{2+}	ML/M.L		-0.4			
Cr^{3+}	ML/M.L				8.9	
Cu^+	$ML_2/M.L^2$	6.28[g]		5.9		
	$ML_3/M.L^3$	7.45[g]				
	M.L/ML(s)	-8.89[g]		-8.3		
Ag^+	ML/M.L	4.30[a]±0.00		4.7 ±0.4		
	$ML_2/M.L^2$	7.11[a]		7.5 ±0.1		
	$ML_3/M.L^3$	8.0[a]		8.4 ±0.1		
		8.9[g]			-13.1[g]	-3[g]
	$ML_4/M.L^4$	9.2[g]		8.5 +0.2	-19.1[g]	-22[g]
	M.L/ML(s)	-12.10[a]	-11.92	-12.30±0.02	20.26 ±0.02	11.7
		-12.66[g]			19.1[g]	

[a]25°, 0.1; [d]25°, 2.0; [g]25°, 5.0

Bromide ion   (continued)

Metal ion	Equilibrium	Log K 25°, 0.5	Log K 25°, 1.0	Log K 25°, 0	ΔH 25°, 0	ΔS 25°, 0
CH_3Hg^+	ML/M.L	6.49[a]	6.37	6.6[t]	-9.9[h]	-4[a]
$(CH_3)_3Pb^+$	ML/M.L	0.30[u]				
Pd^{2+}	$ML/M.L$		5.17		-5.1[c]	7[c]
	$ML_2/M.L_2$		9.42			
	$ML_3/M.L_3$		12.7			
	$ML_4/M.L_4$	14.2	14.9 ±0.1		-13.1[a]	23[a]
		14.7[a]				
Pt^{2+}	$ML/M.L$		5.28			
	$ML_2/M.L_2$		9.7			
	$ML_3/M.L_3$		13.3			
	$ML_4/M.L_4$		16.1			
	$ML_2(cis)/ML_2(trans)$	0.2				
Zn^{2+}	ML/M.L		-0.58[e]±0.03	-0.38[f]	0.4[e]	-1[e]
Cd^{2+}	ML/M.L	1.55 ±0.05	1.55 ±0.03	2.14 ±0.02	-0.8[b]	5[b]
		1.80[a]				
		1.63[d]±0.05	1.74[e]±0.09		-1.0[e]±0.0	5[e]
	$ML_2/M.L^2$	2.1	2.1 ±0.1	3.0 ±0.1	-0.8[b]	7[b]
		2.2[d] ±0.1	2.4[e] ±0.1		-1.6[e]-1	6[e]
	$ML_3/M.L^3$	2.3	2.6 ±0.1	3.0 ±0.1	0.0[b]	12[b]
		2.8[d] ±0.2	3.3[e] ±0.2		0.2[e]-1	16[e]
	$ML_4/M.L^4$		2.6 ±0.2	2.9 0.2		
		3.2[d] ±0.2	3.8[e] ±0.2		0.5[e]-1	19[e]
Hg^{2+}	ML/M.L	9.00 ±0.06	9.40[e]		-10.6	
					-10.1[b]±0.1	7[b]
					-9.6[e]	11[e]
	$ML_2/M.L^2$	17.1 ±0.2	17.98[e]		-20.9	
					-20.8[b]±0.4	8[b]
					-19.2[e]	18[e]
	$ML_3/M.L^3$	19.4 ±0.2	20.7[e]		-23.8[b]	9[b]
					-21.8[e]	22[e]
	$ML_4/M.L^4$	21.0 ±0.2	22.2[e]		-25.9	
					-27.8[b]±0.1	3[b]
					-25.2[e]	17[e]
	ML/MOHL.H	3.37	3.50[e]			

[a] 25°, 0.1;   [b] 25°, 0.5;   [c] 25°, 1.0;   [d] 25°, 2.0;   [e] 25°, 3.0;   [f] 25°, 4.0;   [h] 20°, 0.1;

[t] 25°, 2.5;   [u] 25°, 0.3

Bibliography:

$Th^{4+}-Pu^{4+}$	75RR		Pd^{2+}	76AM,77HF
Cr^{3+}	76DH		Pt^{2+}	78E
Mn^{2+}	75FK		Zn^{2+}	78FKK
Cu^+	77AT		Cd^{2+}	71PB,75FCK,76BHb
Ag^+	73WK,77ATT		Hg^{2+}	72AL,78BKP
CH_3Hg^+	78JI		Pb^{2+}	73BH,76FS,77DG
$(CH_3)_3Pb^+$	78WH		Bi^{3+}	76FKb

Bromide ion   (continued)

Other references: 52HH,62AT,66SM,70H,70IY,
    70L,71FK,73E,74PK,75Fa,75KB,75MHB,75NF,
    75PPa,76CC,76LM,76SSc,77BL,77KSa,
    77MTV,77Na,78JP,78JPa,78LK,79O

$$BrO_3^-$$

O_3Br^- Bromate ion L^-
                              (Other values on Vol.4, p.121)

Metal ion	Equilibrium	Log K 25°, 0.1	ΔH 25°, 0.1	ΔS 25°, 0.1
Eu^{3+}	ML/M.L	0.58	-0.6	1

Bibliography: 77CE

$$I^-$$

I^- Iodide ion L^-
                              (Other values on vol.4, p.122)

Metal ion	Equilibrium	Log K 25°, 0.5	Log K 25°, 1.0	Log K 25°, 0	ΔH 25°, 0	ΔS 25°, 0
Cu^+	$ML_2/M.L^2$	8.7^g				
	$ML_3/M.L^3$	10.43^g				
	$ML_4/M.L^4$	9.4^g				
	$ML_5/M.L^5$	22.0^g				
Ag^+	M.L/ML(s)	-15.7^h		-16.08 ± 0.04	26.54 ± 0.02	15.5
$(CH_3)_3Pb^+$	ML/M.L	0.28^u				
Pd^{2+}	$ML_4/ML_2.L$		2.56			
	$M_2L_6.L^2/(ML_4)^2$		1.32			
Zn^{2+}	ML/M.L		-1.5^e		-1^e	-10^e
Cd^{2+}	ML/M.L	1.86 ± 0.04	1.89 ± 0.02	2.28 ± 0.1	-2.3 -0.1	3
		$2.0^a \pm 0.1$			$-2.5^c \pm 0.0$	0^c
		$1.99^d \pm 0.02$	$2.13^e \pm 0.07$		$-2.2^e \pm 0.1$	2^e
	$ML_2/M.L^2$	3.2 ± 0.1	3.2 ± 0.1	3.92 ± 0.1	-3.0^c	5^c
		$3.4^d \pm 0.1$	$3.6^e \pm 0.1$		-2.5^e	8^e
	$ML_3/M.L^3$	4.4 ± 0.1	4.5 ± 0.1	5.0 ± 0.1	-4.4^c	6^c
		$4.8^d \pm 0.1$	$5.1^e \pm 0.1$		-3.2^e	13^e
	$ML_4/M.L^4$	5.5 ± 0.1	5.6 ± 0.1	6.0 ± 0.1	-8.4^c	-3^c
		$6.1^d \pm 0.1$	$6.6^e \pm 0.1$		-7.0^e	7^e

$^a 25°, 0.1;$ $^c 25°, 1.0;$ $^d 25°, 2.0;$ $^e 25°, 3.0;$ $^g 25°, 5.0;$ $^h 20°, 0.1;$ $^u 25°, 0.3$

Bibliography:

K^+ 76DK

Cu^+ 77AT

Ag^+ 73WK

$(CH_3)_3Pb^+$ 78ME

Pd^{2+} 77EO

Cd^{2+} 71PB,75KL,76BHb,78LP,

Pb^{2+} 73BH,76FS

Other references: 52HH,60CL,70H,70IY,74PK,
    75Fa,75KB,75MHB,75NF,77BL,77HMB,78FKK,
    78PRR,79O

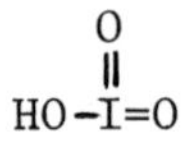

$$\begin{array}{c} O \\ \parallel \\ HO-I=O \end{array}$$

HO_3I		Hydrogen iodate (iodic acid)				HL
		(Other values on Vol.4, p.126)				

Metal ion	Equilibrium	Log K 25°, 0.5	Log K 25°, 1.0	Log K 25°, 0	ΔH 25°, 0	ΔS 25°, 0
H^+	HL/H.L			0.77 ±0.03	0.66	5.7
Eu^{3+}	ML/M.L	1.15[a]-0.2		1.83	2.7[a]	14[a]
NpO_2^+	ML/M.L	0.32[d]				
NpO_2^{2+}	ML/M.L	0.61[u]				
Co^{2+}	$M.L^2/ML_2(s)$			-5.9		
Ni^{2+}	$M.L^2/ML_2(s)$			-5.3		

[a]25°, 0.1; [d]25°, 2.0; [u]25°, 0.3

Bibliography:

H^+	78WH,79BP	NpO_2^{2+}	78PRR
Eu^{3+}	77CE	Co^{2+}-Cu^{2+},Zn^{2+}	76FRS
NpO_2^+	79VG	Other reference : 77BL	

PROTONATION VALUES FOR OTHER LIGANDS

A. Aminoacids

1.   2-Aminocarboxylic acids

Ligand	Equilibrium	Log K 25°, 0	ΔH 25°, 0	ΔS 25°, 0	Bibliography

$$\overset{\displaystyle \overset{NH_2}{|}}{RHO_3PO-R'-CHCO_2H}$$

Ligand	Equilibrium	Log K 25°, 0	ΔH 25°, 0	ΔS 25°, 0	Bibliography
L-2-Amino-3-phosphopropanoic acid (phosphoserine)* $(C_3H_8O_6NP)$, H_3L (Other values on Vol.1, p.29)	HL/H.L	9.71^a+0.01 9.53^n			78MA, Other references: 78MAa,79MAb, 79MBa
	$H_2L/HL.H$	6.19 5.64^a±0.02 5.65^n			
	$H_3L/H_2L.H$	2.09^a±0.02 1.98^n			
DL-2-Amino-3-(phenylphospho)-propanoic acid (O-phenyl-phosphorylserine) $(C_9H_{12}O_4NP)$, H_2L	HL/H.L $H_2L/HL.H$	8.79^a 2.13^a			59FO
L-2-Amino-3-phosphobutanoic acid (phosphothreonine)* $(C_4H_{10}O_6NP)$, H_3L	HL/H.L $H_2L/HL.H$ $H_3L/H_2L.H$	9.67^a 5.83^a 2.25^a			78MA
L-2-Amino-2-methyl-3-phospho-propanoic acid (phospho-α-methylserine)* $(C_4H_{10}O_6NP)$, H_3L	HL/H.L $H_2L/HL.H$ $H_3L/H_2L.H$	10.07^a 5.68^a 2.07^a			78MA

$$\overset{\displaystyle \overset{NH_2}{|}}{H_2O_3P-R-CHCO_2H}$$

Ligand	Equilibrium	Log K 25°, 0	ΔH 25°, 0	ΔS 25°, 0	Bibliography
DL-2-Amino-3-phosphono-propanoic acid* $(C_3H_8O_5NP)$, H_3L	HL/H.L $H_2L/HL.H$ $H_3L/H_2L.H$	10.68^a 6.05^a 2.34^a			78MAb

Ligand	Equilibrium	Log K 25°, 0	ΔH 25°, 0	ΔS 25°, 0	Bibliography
DL-2-Amino-3-(3-hydroxyphenyl)-propanoic acid $(C_9H_{11}O_3N)$, H_2L (Other values on Vol.1, p.30)	HL/H.L $H_2L/HL.H$ $H_3L/H_2L.H$	10.05^a±0.06 8.90^a±0.07 2.33^a	-6.4^a -10.1^a	25^a 7^a	77IH

---

^{a}25°, 0.1; ^{n}37°, 0.15; *metal constants were also reported but not included in the compilation of selected constants.

2-Aminocarboxylic acids (continued)

Ligand	Equilibrium	Log K 25°, 0	ΔH 25°, 0	ΔS 25°, 0	Bibliography

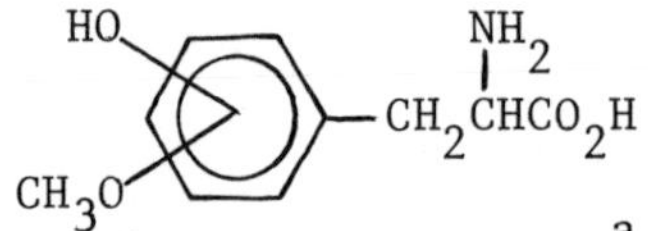

HO—⟨ring⟩(CH$_3$O)—CH$_2$CHCO$_2$H (NH$_2$)

DL-2-Amino-3-(3-hydroxy-4-methoxyphenyl)propanoic acid ($C_{10}H_{13}O_4N$), H_2L	HL/H.L	10.12[a]			77IH
	H_2L/HL.H	8.84[a]			
	H_3L/H_2L.H	2.23[a]			
DL-2-Amino-3-(4-hydroxy-3-methoxyphenyl)propanoic acid ($C_{10}H_{13}O_4N$), H_2L	HL/H.L	10.14[a]			77IH
	H_2L/HL.H	8.78[a]			
	H_3L/H_2L.H	2.13[a]			

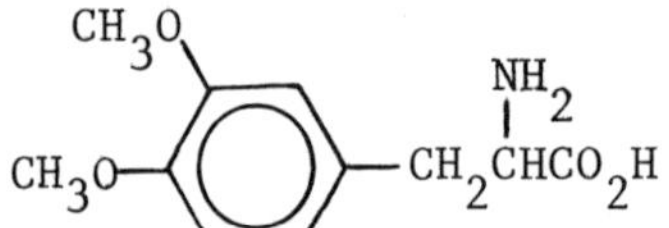

CH$_3$O—(CH$_3$O)⟨ring⟩—CH$_2$CHCO$_2$H (NH$_2$)

| DL-2-Amino-3-(3,4-dimethoxy-phenyl)propanoic acid ($C_{11}H_{15}O_4N$), HL | HL/H.L | 9.02[a] | | | 77IH |
| | H_2L/HL.H | 2.37[a] | | | |

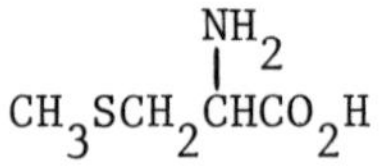

CH$_3$SCH$_2$CHCO$_2$H (NH$_2$)

L-2-Amino-3-(methylthio)-propanoic acid (S-methyl-cysteine) ($C_4H_9O_2NS$), HL (Other values on Vol.1, p.49)	HL/H.L	8.97	-10.1	7	78L, Other reference: 76HS
		8.73[a]			
		8.74[b]			
	H_2L/HL.H	1.99[b]			

H$_2$N-R'—⟨ring⟩—CH$_2$CHCO$_2$H (NH$_2$)

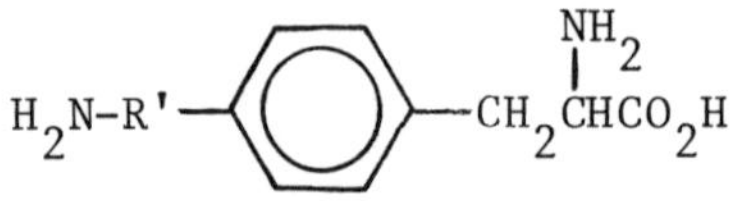

DL-3-(4-Aminophenyl)alanine ($C_9H_{12}O_2N_2$), HL	HL/H.L	8.93[n]			77OK
DL-3-(4-Aminomethylphenyl)-alanine ($C_{10}H_{14}O_2N_2$), HL	HL/H.L	8.42[n]			77OK
	H_2L/HL.H	8.40[n]			
DL-3-[4-(N,N-Dimethylamino)-phenyl]alanine ($C_{10}H_{14}O_2N_2$), HL	HL/H.L	8.86[n]			77OK

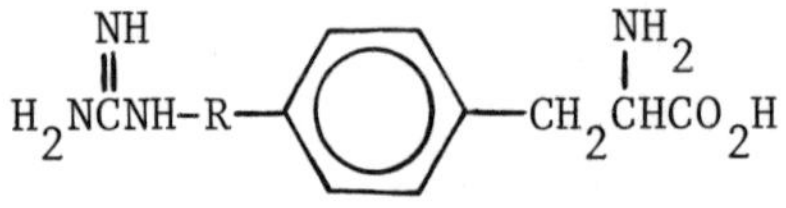

H$_2$NCNH-R—⟨ring⟩—CH$_2$CHCO$_2$H (NH)(NH$_2$)

DL-3-(4-Guanidinophenyl)-alanine ($C_{10}H_{14}O_2N_4$), HL	HL/H.L	10.91[n]			77OK
	H_2L/HL.H	8.44[n]			
DL-3-(4-Guanidinomethylphenyl)-alanine ($C_{11}H_{16}O_2N_4$), HL	H_2L/HL.H	8.83[n]			77OK

[a]25°, 0.1; [b]25°, 0.5; [n]37°, 0.15

2-Aminocarboxylic acids (continued)

$$\begin{array}{c} NHR \\ | \\ R-CHCO_2H \end{array}$$

Ligand	Equilibrium	Log K 25°, 0	Log K 25°, 0	Log K 25°, 0	Bibliography
N-[Tris(hydroxymethyl)methyl]-glycine (tricine) ($C_6H_{15}O_5N$), HL (Other reference on Vol.1, p. 408)	HL/H.L	8.135	-7.52	12.0	73RRB, Other references: 77JKK, 78KJK
	$H_2L/HL.H$	2.023	-1.41	4.6	
L-4-Hydroxypyrrolidine-2-carboxylic acid (hydroxy-proline)*($C_5H_9O_3N$), HL (Other values on Vol.1, p.73)	HL/H.L	9.662 $9.47^a \pm 0.01$ 9.47^b 9.45^c 9.16^n 10.05^e	-9.4	12	73SK,76MT,78VV, Other references: 73RG,73SC,73SCa, 73SKa,74KH,74SK, 75HKa,77KD,78KZa, 79GC,79H
	$H_2L/HL.H$	1.818 1.80^a 1.89^b 1.92^c 1.67^n 2.21^e	-0.9	5	

$$\begin{array}{c} HO_2CCH_2CHNH-R-NHCHCH_2CO_2H \\ \quad\quad | \quad\quad\quad\quad\quad\quad | \\ \quad HO_2C \quad\quad\quad\quad\quad CO_2H \end{array}$$

Ligand	Equilibrium	Log K 25°, 0			Bibliography
Ethylenediiminodibutanedioic acid (ethylenediamine-N,N'-disuccinic acid)*($C_{10}H_{16}O_8N_2$), H_4L (Other values on Vol.1, p.91)	HL/H.L	9.96^a 10.13^h			79GSN, Other references: 73GK, 73GKS,73GSK,79MMb,
	$H_2L/HL.H$	6.83^a 6.93^h			
	$H_3L/H_2L.H$	3.88^a 3.85^h			
	$H_4L/H_3L.H$	2.95^a 3.00^h			
trans-1,2-Cyclohexylene-diiminodibutanedioic acid (trans-1,2-diaminocyclo-hexane-N,N'-disuccinic acid) ($C_{14}H_{22}O_8N_2$), H_4L	HL/H.L	10.95^a			79SGN
	$H_2L/HL.H$	6.91^a			
	$H_3L/H_2L.H$	3.75^a			
	$H_4L/H_3L.H$	2.05^a			
Nitrilobis(ethyleneimino)-tributanedioic acid (diethylenetriamino-trisuccinic acid) ($C_{16}H_{25}O_{12}N_3$), H_6L	HL/H.L	10.35^a			79GSN
	$H_2L/HL.H$	9.62^a			
	$H_3L/H_2L.H$	8.75^a			
	$H_4L/H_3L.H$	5.60^a			
	$H_5L/H_4L.H$	2.55^a			
	$H_6L/H_5L.H$	1.9^a			

---

^{a}25°, 0.1; ^{b}25°, 0.5; ^{c}25°, 1.0; ^{e}25°, 3.0; ^{h}20°, 0.1; ^{n}37°, 0.15; *metal constants were also reported but not included in the compilation of selected constants.

2-Aminocarboxylic acids (continued)

Ligand	Equilibrium	Log K 25°, 0	Log K 25°, 0	Log K 25°, 0	Bibliography

(structure: HS-imidazole with N-H, $^+N(CH_3)_3$ and $-CH_2CHCO_2H$)

| DL-1-Carboxy-2-(2-mercapto-4-imidazolyl)ethyl(trimethyl)-ammonium (chloride)* (ergothioneine) ($C_9H_{16}O_2N_3S^+$), H_2L | HL/H.L | 10.44^a | $(-12)^s$ | $(8)^a$ | 77ST, Other reference: 74MMS |

2.   3-,4-Aminocarboxylic acids $H_2N-R-CO_2H$

DL-3-Amino-3-phenylpropanoic acid (phenyl-β-alanine)* ($C_9H_{11}O_2N$), HL	HL/H.L $H_2L/HL.H$	9.00^b 3.40^b			71KS, Other references: 73BS, 73SKa, 73SK
DL-3-Amino-2-hydroxypropanoic acid (isoserine)* ($C_3H_7O_3N$), HL (Other values on Vol.1, p.39)	HL/H.L $H_2L/HL.H$	$9.14^a-0.01$ 2.66^a	-10.1^a	8^a	76BMD
DL-4-Amino-3-hydroxybutanoic acid*($C_4H_9O_3N$), HL (Other values on Vol.1, p.39)	HL/H.L $H_2L/HL.H$	$9.55^a\pm0.06$ 3.83^a	-10.9^a $(-1)^s$	7^a $(14)^a$	75BM

3.   Iminodiacetic acid derivatives

(structure: $R'N$ with CH_2CO_2H and $CHRCO_2H$)

| L-2-(2-Pyridylmethyl)imino-diacetic acid ($C_{10}H_{11}O_4N_2$), H_2L | HL/H.L $H_2L/HL.H$ | 8.94^a 4.21^a | | | 77BR |
| N-[2-(Carboxymethylthio)-ethyl]iminodiacetic acid ($C_8H_{13}O_6NS$), H_3L (Other reference on Vol.1, p. 409) | HL/H.L $H_2L/HL.H$ $H_3L/H_2L.H$ | 8.82^a 3.53^a 2.0^a | | | 74P |

(structure: HO_2CCH_2 and HO_2CCH_2 — $N(CH_2CH_2N)_nCH_2CH_2N$ with CH_2CO_2H, CH_2CO_2H, CH_2CO_2H, CH_2CO_2H)

| Tetraethylenepentamine-N,N,N',N'',N''',N'''-heptaacetic acid (TPHA) ($C_{22}H_{37}O_{14}N_5$), H_7L (Other references on Vol.1, p.400) | HL/H.L $H_2L/HL.H$ $H_3L/H_2L.H$ $H_4L/H_3L.H$ $H_5L/H_4L.H$ $H_6L/H_5L.H$ $H_7L/H_6L.H$ $H_8L/H_7L.H$ | 10.76^a 9.87^a 8.13^a 4.75^a 2.76^a 2.10^a 1.8^a 1.5^a | | | 79LM, Other references: 71LS, 75HK, 76GA |

---

^{a}25°, 0.1; ^{b}25°, 0.5; ^{s}5-35°, 0.1;   *metal constants were also reported but not included in the compilation of selected constants.

Iminodiacetic acid derivatives   (continued)

Ligand	Equilibrium	Log K 25°, 0	Log K 25°, 0	Log K 25°, 0	Bibliography
Pentaethylenehexamine-N,N,N',	HL/H.L	10.85^a			79LM
N'',N''',N'''',N'''''-octacetic acid	$H_2L/HL.H$	9.96^a			
(PHOA) $(C_{26}H_{44}O_{16}N_6)$, H_8L	$H_3L/H_2L.H$	8.85^a			
	$H_4L/H_3L.H$	6.20^a			
	$H_5L/H_4L.H$	4.48^a			
	$H_6L/H_5L.H$	3.57^a			
	$H_7L/H_6L.H$	2.52^a			
	$H_8L/H_7L.H$	2.02^a			
	$H_9L/H_8L.H$	1.7^a			
	$H_{10}L/H_9L.H$	1.4^a			

4. Peptides

$$H_2N-R-\overset{\overset{O}{\|}}{C}NH-R'-CO_2H$$

Ligand	Equilibrium	Log K 25°, 0	Log K 25°, 0	Log K 25°, 0	Bibliography
Glycyl-DL-2-aminobutanoic acid $(C_6H_{12}O_3N_2)$, HL	HL/H.L	8.331	-10.8	2	75K
	$H_2L/HL.H$	3.155	0.7	17	
(Other reference on Vol.1, p.401)					
Glycyl-L-leucine*$(C_8H_{16}O_3N_2)$,	HL/H.L	8.327 $8.14^a\pm0.02$	-10.6	3	75K, Other references: 75NM,
HL (Other values on Vol.1, p.299)	$H_2L/HL.H$	3.180 $3.09^a\pm0.08$	0.8	17	76PN,79KC
L-Leucylglycine $(C_8H_{16}O_3N_2)$,	HL/H.L	$7.95^a\pm0.04$			74MS
HL (Other values on Vol.1, p. 308)	$H_2L/HL.H$	$3.14^a\pm0.03$			
L-Alanyl-D-alanine $(C_6H_{12}O_3N_2)$,	HL/H.L	$8.19^a\pm0.02$	$(-10)^s$	$(4)^a$	74NA,75KMa
HL (Other values on Vol.1, p.317)	$H_2L/HL.H$	$3.04^a\pm0.03$	$(1)^s$	$(17)^a$	

$$HS-(R-\overset{\overset{O}{\|}}{C}NH)_n\underset{\underset{\underset{\underset{N}{\diagup\diagdown}}{\text{imidazole}}}{CH_2}}{C}HCO_2H$$

Ligand	Equilibrium	Log K 25°, 0	Log K 25°, 0	Log K 25°, 0	Bibliography
N-(Mercaptoacetyl)-L-histidine*	HL/H.L	8.70^h			77SH
$(C_8H_{11}O_3N_2S)$, H_2L	$H_2L/HL.H$	7.14^h			
	$H_3L/H_2L.H$	3.43^h			
N-(3-Mercaptopropionyl)-L-histidine*$(C_9H_{13}O_3N_2S)$, H_2L	HL/H.L	9.86^h			77SH
	$H_2L/HL.H$	7.25^h			
	$H_3L/H_2L.H$	3.48^h			
N-(Mercaptoacetyl)glycyl-L-histidine*$(C_{10}H_{14}O_4N_4S)$, H_2L	HL/HL.H	8.65^h			78S
	$H_2L/HL.H$	7.10^h			
	$H_3L/H_2L.H$	3.47^h			

[a]25°, 0.1; [h]20°, 0.1; [s]9-37°, 0.1; *metal constants were also reported but not included in the compilation of selected constants.

5. <u>Pyridinecarboxylic acids</u>

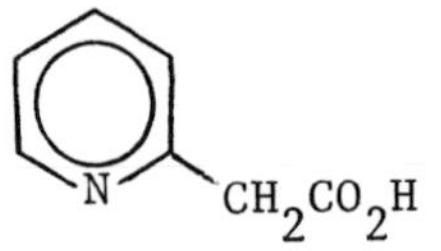

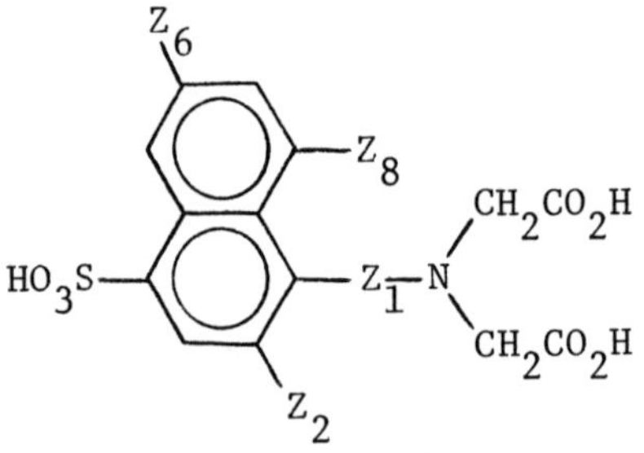

Ligand	Equilibrium	Log K 25°, 0	ΔH 25°, 0	ΔS 25°, 0	Bibliography

| 2-Pyridylacetic acid ($C_7H_7O_2N$), HL (Other values on Vol.1, p.387) | HL/H.L | 5.81^b 5.84^c | -4.3^c | 12^c | 77Ad |
| | $H_2L/HL.H$ | 2.73^b 2.74^c | -1.2^c | 9^c | |

6. <u>Aminonaphthalene carboxylic acids</u>

1-Bis(carboxymethyl)amino- naphthalene-4-sulfonic acid ($C_{14}H_{13}O_7NS$), H_3L	HL/H.L $H_2L/HL.H$	3.92^a 2.76^a			69TD
1-Bis(carboxymethyl)amino- 2-hydroxynaphthalene-4- sulfonic acid* ($C_{14}H_{13}O_8NS$), H_4L	HL/H.L $H_2L/HL.H$ $H_3L/H_2L.H$	8.85^a 3.38^a 2.16^a			69TD
1-Bis(carboxymethyl)amino- 8-hydroxynaphthalene-4,6- disulfonic acid ($C_{14}H_{13}O_{11}NS_2$), H_5L	HL/H.L $H_2L/HL.H$ $H_3L/H_2L.H$	9.48^a 3.87^a 2.58^a			69TD
1-[N-Carboxymetyl-2-(bis- (carboxymethyl)amino)ethyl- amino]naphthalene-4-sulfonic acid*($C_{18}H_{20}O_{10}N_2S$), H_4L	HL/H.L $H_2L/HL.H$ $H_3L/H_2L.H$	9.01^a 4.37^a 2.91^a			69TD

B. <u>Amines</u>

1. <u>Primary amines</u>   R–NH$_2$

Hexylamine ($C_6H_{15}N$), L (Other value on Vol.2, p.4)	HL/H.L	10.630	-14.0 ± 0.1	2	75BO
Cyclohexylamine ($C_6H_{13}N$), L (Other values on Vol.2, p.6)	HL/H.L	10.609+0.03	-14.0 ± 0.2	1	75BO,76LMR
2-Phenylethylamine ($C_8H_{11}N$), L	HL/H.L	$9.97^a \pm 0.09$	$(-9)^s$	$(15)^a$	71RD,78IH, Other reference: 72RD

---

^{a}25°, 0.1; ^{b}25°, 0.5; ^{c}25°, 1.0; ^{s}15-35°, 0.1; *metal constants were also reported but not included in the compilation of selected values.

Primary amines   (continued)

Ligand	Equilibrium	Log K 25°, 0	ΔH 25°, 0	ΔS 25°, 0	Bibliography

Ligand	Equilibrium	Log K 25°, 0	ΔH 25°, 0	ΔS 25°, 0	Bibliography
2-(2-Hydroxyphenyl)ethylamine ($C_8H_{11}ON$), HL	$H_2L/HL.H$	9.27^a			71RD
2-(3-Hydroxyphenyl)ethylamine ($C_8H_{11}ON$), HL	$HL/H.L$ $H_2L/HL.H$	$10.48^a \pm 0.09$ $(-9)^s$ $9.31^a \pm 0.03$ $(-6)^s$		$(18)^a$ $(22)^a$	71RD,78IH
2-(4-Hydroxyphenyl)ethylamine (tyramine)* ($C_8H_{11}ON$), HL (Other references on Vol.2, p.332)	$HL/H.L$ $H_2L/HL.H$	$10.56^a \pm 0.02$ $(-9)^s$ $9.47^a \pm 0.03$ $(-5)^s$		$(18)^a$ $(27)^a$	54L,75KA,71RD, 78IH
2-(3-Hydroxy-4-methoxy-phenyl)ethylamine ($C_9H_{13}O_2N$), HL	$HL/H.L$ $H_2L/HL.H$	10.38^a 9.23^a			78IH
2-(4-Hydroxy-3-methoxyphenyl)-ethylamine ($C_9H_{13}O_2N$), HL	$HL/H.L$ $H_2L/HL.H$	10.47^a 9.34^a			78IH
DL-2-Hydroxy-2-(3,4-di-hydroxyphenyl)ethylamine (norepinephrine)* ($C_8H_{11}O_3N$), H_3L	$HL/H.L$ $H_2L/HL.H$ $H_3L/H_2L.H$	$(11.13)^a$ 9.73^a 8.57^a			71RD

Ligand	Equilibrium	Log K 25°, 0	ΔH 25°, 0	ΔS 25°, 0	Bibliography
2-Aminobenzenearsonic acid ($C_6H_8O_3NAs$), H_2L	$HL/H.L$ $H_2L/HL.H$ $H_3L/H_2L.H$	8.93 3.79 0.69			7300
4-Aminobenzenearsonic acid ($C_6H_8O_3NAs$), H_2L	$HL/H.L$ $H_2L/HL.H$ $H_3L/H_2L.H$	9.19 4.13 1.91			7300

$$HO-R-NH_2$$

Ligand	Equilibrium	Log K 25°, 0	ΔH 25°, 0	ΔS 25°, 0	Bibliography
DL-2-Aminobutanol ($C_4H_{11}ON$), L (Other references on Vol.2, p.322)	$HL/H.L$	9.516 9.55^a 9.67^b 9.81^c 10.06^d	-12.46	1.8	75NL
2-Amino-2-methylpropanol ($C_4H_{11}ON$), L (Other references on Vol.2, p.322)	$HL/H.L$	9.694 $9.72^a \pm 0.00$ 9.81^b 9.93^c 10.16^d	-12.91 ± 0.02	1.1	75NL

---

[a] 25°, 0.1; [b] 25°, 0.5; [c] 25°, 1.0; [d] 25°, 2.0; [e] 25°, 3.0; [P] 30°, 0.1; [q] 25°, 0.7; [s] 15-35°,0.1
*metal constants were also reproted but not included in the compilation of selected values.

Primary amines   (continued)

Ligand	Equilibrium	Log K 25°, 0	ΔH 25°, 0	ΔS 25°, 0	Bibliography

Ligand	Equilibrium	Log K 25°, 0	Bibliography
2-Amino-2-deoxy-D-gluco-pyranose (D-glucosamine)* $(C_6H_{13}O_5N)$, L	HL/H.L	$7.42^P \pm 0.05$	65TMS,79MN, Other reference: 68TM
1-Methoxy-D-glucosamine* $(C_7H_{15}O_5N)$, L	HL/H.L	7.15^P	65TMS, Other reference: 65TMa
3-Methoxy-D-glucosamine $(C_7H_{15}O_5N)$, L	HL/H.L	7.10^P	65TMa
3,4,6-Trimethoxy-D-glucosamine* $(C_9H_{19}O_5N)$, L	HL/H.L	6.92^P	65TMS, Other reference: 65TMa
N-Methyl-D-glucosamine* $(C_7H_{15}O_5N)$, L	HL/H.L	7.86^P	79MN
D-Galactosamine* $(C_6H_{13}O_5N)$, L	HL/H.L	7.61^P	79MN
D-Mannosamine* $(C_6H_{13}O_5N)$, L	HL/H.L	7.50^P	79MN
D-Talosamine* $(C_6H_{13}O_5N)$, L	HL/H.L	7.98^P	79MN

$$CH_3O-R-NH_2$$

Ligand	Equilibrium	Log K 25°, 0	Bibliography
Methoxyamine (CH_5ON), L	HL/H.L	4.62^a	77PR
2-(3,4-Dimethoxyphenyl)ethyl-amine $(C_{10}H_{15}O_2N)$, L	HL/H.L	9.78^a	78IH

$$HSe-R-NH_2$$

Ligand	Equilibrium	Log K 25°, 0	Bibliography
2-Aminoethylselenol* (C_2H_7NSe), HL	HL/H.L H_2L/HL.H	10.88^t 4.90^t	70TSY, Other reference: 70YS
2-Amino-2-propylselenol* (C_3H_9NSe), HL	HL/H.L H_2L/HL.H	10.74^t 5.08^t	71TS
2-Amino-2-methyl-2-propyl-selenol* $(C_4H_{11}NSe)$, HL	HL/H.L H_2L/HL.H	10.84^t 5.10^t	71TS
3-Aminopropylselenol* (C_3H_9NSe), HL	HL/H.L H_2L/HL.H	11.2^t 6.16^t	71TS

$$CH_3SeCH_2CH_2NH_2$$

Ligand	Equilibrium	Log K 25°, 0	Bibliography
2-(Methylseleno)ethylamine* (C_3H_9NSe), L	HL/H.L	9.42^t	70TSY,71YS

---

^{a}25°, 0.1; ^{P}30°, 0.1; ^{t}22°, 0.1; *metal constants were also reported but not included in the compilation of selected values.

Primary amines   (continued)

Ligand	Equilibrium	Log K 25°, 0	ΔH 25°, 0	ΔS 25°, 0	Bibliography

$$H_2NCH_2CHNH_2$$
$$\underset{\underset{CH_3}{|}}{H_3CCCH_3}$$

Ligand	Equilibrium	Log K 25°, 0	ΔH 25°, 0	ΔS 25°, 0	Bibliography
(1,1-Dimethylethyl)ethylene-diamine ($C_6H_{16}N_2$), L	HL/H.L	9.78[a]			72YY
		6.26[a]			

$$H_2NCH_2CH_2SeCH_2CH_2NH_2$$

4-Selena-1,7-diazaheptane* (selenobis(ethyleneamine)) ($C_4H_{12}N_2Se$), L	HL/H.L	9.85[t]			71YS
	$H_2L/HL.H$	8.52[t]			

$$\underset{H_2N-R-S}{\overset{H_2N-R-S}{>}}CH-R'-CH\underset{S-R-NH_2}{\overset{S-R-NH_2}{<}}$$

Ligand	Equilibrium	Log K 25°, 0	Bibliography
Ethanediylidenetetrakis-(thiotrimethyleneamine) ($C_{14}H_{34}N_4S_4$), L	HL/H.L	10.30[a]	77CJa
	$H_2L/HL.H$	9.86[a]	
	$H_3L/H_2L.H$	9.57[a]	
	$H_4L/H_3L.H$	8.89[a]	
Propane-1,3-diylidene-tetrakis(thiotrimethylene-amine) ($C_{15}H_{36}N_4S_4$), L	HL/H.L	10.36[a]	77CJa
	$H_2L/HL.H$	9.78[a]	
	$H_3L/H_2L.H$	9.66[a]	
	$H_4L/H_3L.H$	8.98[a]	
Butane-1,4-diylidene-tetrakis(thiotrimethylene-amine) ($C_{16}H_{38}N_4S_4$), L	HL/H.L	10.33[a]	77CJa
	$H_2L/HL.H$	9.79[a]	
	$H_3L/H_2L.H$	9.71[a]	
	$H_4L/H_3L.H$	9.05[a]	
Pentane-1,5-diylidene-tetrakis(thiotrimethylene-amine) ($C_{17}H_{40}N_4S_4$), L	HL/H.L	10.44[a]	77CJa
	$H_2L/HL.H$	9.82[a]	
	$H_3L/H_2L.H$	9.68[a]	
	$H_4L/H_3L.H$	9.08[a]	
Phthal-1,2-diylidene-tetrakis(thioethylene-amine) ($C_{16}H_{30}N_4S_4$), L	HL/H.L	9.57[a]	76CJ
	$H_2L/HL.H$	9.10[a]	
	$H_3L/H_2L.H$	8.60[a]	
	$H_4L/H_3L.H$	8.07[a]	
Phthal-1,2-diylidene-tetrakis(thiotrimethylene-amine) ($C_{20}H_{38}N_4S_4$), L	HL/H.L	10.20[a]	77CJa
	$H_2L/HL.H$	9.73[a]	
	$H_3L/H_2L.H$	9.57[a]	
	$H_4L/H_3L.H$	8.92[a]	

2. Secondary amines   R-NH-R'

Ligand	Equilibrium	Log K 25°, 0	ΔH 25°, 0	ΔS 25°, 0	Bibliography
Diethylamine ($C_4H_{11}N$), L (Other values on Vol.2, p.72)	HL/H.L	10.933±0.09	-12.7±0.1	7	76IB,76LMR, Other references:
		10.97[b]			68PMa,78SL,79FS
		11.12[d,w]			

[a]25°, 0.1; [b]25°, 0.5; [d]25°, 2.0; [t]22°, 0.1; [w]$(CH_3CH_2)_2NH_2NO_3$ used as background electrolyte; *metal constants were also reported but not included in the compilation of selected values.

Secondary amines (continued)

Ligand	Equilibrium	Log K 25°, 0	ΔH 25°, 0	ΔS 25°, 0	Bibliography
Dipropylamine ($C_6H_{15}N$), L (Other values on Vol.2, p.335)	HL/H.L	10.96 ±0.04	-13.1±0.1	6	75BO,78BO
Di-2-propylamine ($C_6H_{15}N$), L (Other references on Vo.2, p.335)	HL/H.L	11.15 ±0.04	-13.6±0.0	5	78BO, Other reference: 69MP
Pyrrolidine (tetramethylene-imine) (C_4H_9N), L (Other values on Vol.2, p.73)	HL/H.L	11.305 11.48[b]	-12.9±0.1 -13.04[b]	8 8.8[b]	77BE
Hexamethyleneimine ($C_5H_{11}N$), L (Other references on Vol.2, p.335)	HL/H.L	11.12	-13.0±0.0	7	78BO
N-Methylpentylamine ($C_6H_{15}N$), L	HL/H.L	10.88	-12.6	7	78BO

HO—[benzene ring]—CHCH$_2$NHCH$_3$ with OH below CH

Ligand	Equilibrium	Log K 25°, 0	ΔH 25°, 0	ΔS 25°, 0	Bibliography
DL-2-(Methylamino)-(3-hydroxyphenyl)ethanol (phenylephrin)* ($C_9H_{13}O_2N$), H_2L	$H_3L/H_2L.H$	9.13			71RD, Other reference: 72RD

$HOCH_2$
$HOCH_2CNHCH_2CH_2SO_3H$
$HOCH_2$

Ligand	Equilibrium	Log K 25°, 0	ΔH 25°, 0	ΔS 25°, 0	Bibliography
2-[Tris(hydroxymethyl)-methylamino]ethanesulfonic acid ($C_6H_{15}O_6NS$), HL	HL/H.L	7.550	-7.7	9	76VBc

[ring structure: O, S, R, N—H, R]

Ligand	Equilibrium	Log K 25°, 0	ΔH 25°, 0	ΔS 25°, 0	Bibliography
Perhydro-1,4-oxazine (morpholine) (C_4H_9ON), L (Other values on Vol.2, p.81), L	HL/H.L	8.492 8.58[b]±0.03	-9.3 -9.83[c]	8 6.3[b]	75BE
2,6-Dimethylmorpholine ($C_6H_{13}ON$), L	HL/H.L	8.60[b]	-9.81[b]	6.4[b]	75BE

$HSeCH_2CH_2CH_2NHCH_2CH_3$

Ligand	Equilibrium	Log K 25°, 0	ΔH 25°, 0	ΔS 25°, 0	Bibliography
3-(Ethylamino)propyl-selenol* ($C_4H_{11}NSe$), HL	HL/H.L $H_2L/HL.H$	10.80[t] 4.93[t]			71YS

[b] 25°, 0.5; [c] 25°, 1.0; [t] 22°, 0.1; * metal constants were also reported but not included in the compilation of selected values.

Secondary amines   (continued)

Ligand Ligand	Equilibrium	Log K 25°, 0	ΔH 25°, 0	ΔS 25°, 0	Bibliography

$$CH_3N-R-NH_2$$

Ligand	Equilibrium	Log K 25°, 0	ΔH 25°, 0	ΔS 25°, 0	Bibliography
N-Methylethylenediamine $(C_3H_{10}N_2)$, L (Other values on Vol.2, p.82)	HL/H.L	10.04 ±0.02 10.11[a] 10.21[b]±0.08 10.28[c]	-11.25[b] -11.22[c]	9.0[b] 9.4[c]	72CV
	$H_2L/HL.H$	6.76 ±0.02 7.04[a] 7.21[b]±0.01 7.47[c]	-10.3[b] -10.85[c]	-1[b] -2.2[c]	
N-Methyltrimethylenediamine $(C_4H_{12}N_2)$,L	HL/H.L	10.57 10.62[a] 10.78[b] 10.98[c]			72NT
	$H_2L/HL.H$	8.48 8.74[a] 9.01[b] 9.21[c]			

$$\underset{\displaystyle CH_3NHCH_2CCH_2NHCH_3}{\overset{\displaystyle CH_2 \atop \displaystyle \|}{}}$$

Ligand	Equilibrium	Log K 25°, 0	ΔH 25°, 0	ΔS 25°, 0	Bibliography
N,N'-Dimethyl-2-methylene-trimethylenediamine $(C_6H_{14}N_2)$, L	HL/H.L $H_2L/HL.H$	10.42[b] 8.39[b]			75HS

Ligand	Equilibrium	Log K 25°, 0	ΔH 25°, 0	ΔS 25°, 0	Bibliography
Piperazine (perhydro-1,4-diazine) $(C_6H_{10}N_2)$, L	HL/H.L	9.731 9.71[a] ±0.02 10.01[c]	-10.25 -10.3[a]±0.1 -10.40[c]	10.2 10[a] 10.9[c]	72CV

3.   Tertiary amines

$$\underset{R}{\overset{R'}{\diagdown}} N-R''$$

Ligand	Equilibrium	Log K 25°, 0	ΔH 25°, 0	ΔS 25°, 0	Bibliography
Triethylamine $(C_6H_{15}N)$, L (Other values on Vol.2, p.112)	HL/H.L	10.715 10.75[a] 10.77[t]	-10.4 ±0.2	14	75BO
Butyldimethylamine $(C_6H_{15}N)$,L	HL/H.L	10.06	-9.6	14	78BO
1-Azabicyclo[2.2.2]octane (quinuclidine) $(C_7H_{13}N)$, L	HL/H.L	11.15	-11.14±0.0	13.7	76LMR,78BO
4-[4-(Dimethylamino)phenyl-azo]benzenesulfonic acid (Methyl Orange) $(C_{14}H_{15}O_3N_3S)$, HL	HL/H.L	3.44	-4.3	1.4	73BE, Other reference: 76RC
2-[4-(Dimethylamino)phenyl-azo]benzene sulfonic acid $(C_{14}H_{15}O_3N_3S)$, HL	HL/H.L	3.48	-1.1	12.2	73BE

---
[a]25°, 0.1; [b]25°, 0.5;   [c]25°, 1.0;   [t]25°, 0.4

Tertiary amines   (continued)

Ligand	Equilibrium	Log K 25°, 0	ΔH 25°, 0	ΔS 25°, 0	Bibliography

$$CH_3$$
$$|$$
$$CHCH_3$$
$$HOCH_2CH_2N$$
$$CHCH_3$$
$$|$$
$$CH_3$$

		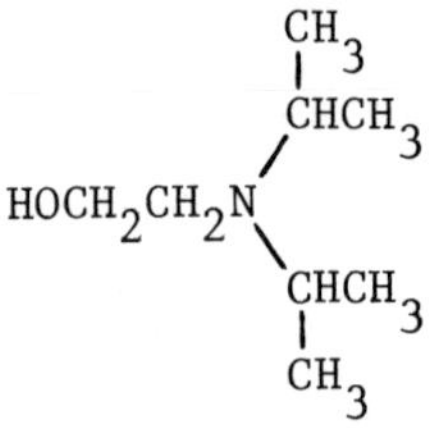			
2-(Di-2-propylamino)ethanol $(C_8H_{19}ON)$, L (Other value on Vol.2, p.116)	HL/H.L	10.03 10.08[a] 10.28[b] 10.51[c] 10.92[d]			75NL

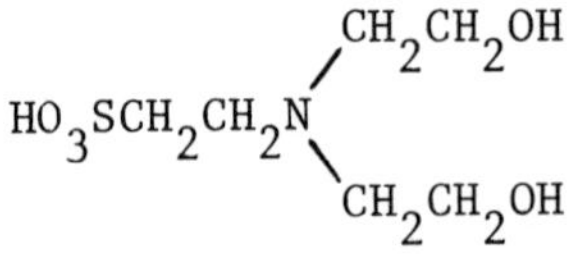

$$CH_2CH_2OH$$
$$HO_3SCH_2CH_2N$$
$$CH_2CH_2OH$$

| 2[Bis(2-hydroxyethyl)amino]-ethanesulfonic acid $(C_6H_{15}O_5NS)$, HL | HL/H.L | 7.187 | −5.8 | 13 | 76VBc |

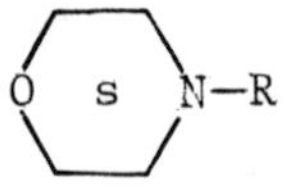

$$O \quad S \quad N-R$$

N-Methylmorpholine $(C_5H_{11}ON)$, L	HL/H.L	7.58[b]	−6.50[b]	12.9[b]	75BE
N-Ethylmorpholine $(C_6H_{13}ON)$, L	HL/H.L	7.62[b]	−6.56[b]	12.9[b]	75BE
N-Aminomorpholine $(C_4H_{10}ON_2)$, L	HL/H.L	4.04[b]	−5.67[b]	−0.5[b]	75BE
2-(N-Morpholino)ethanesul-fonic acid $(C_6H_{13}O_4NS)$, HL	HL/H.L	6.270	−3.5	17	76VBc

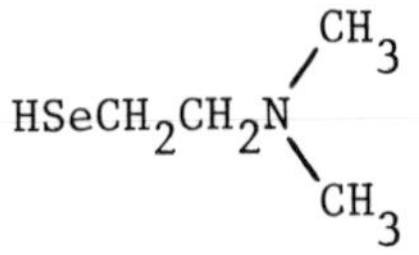

$$CH_3$$
$$HSeCH_2CH_2N$$
$$CH_3$$

| 2-(N,N-Dimethylamino)ethyl-selenol* $(C_4H_{11}NSe)$, HL | HL/H.L H$_2$L/HL.H | 10.70[t] 4.63[t] | | | 71YS |

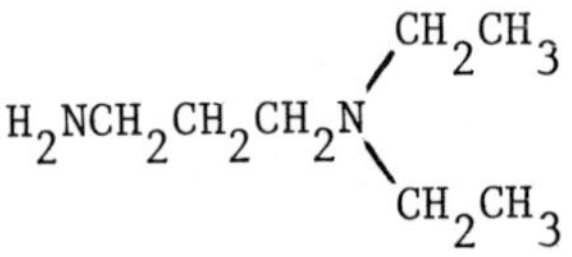

$$CH_2CH_3$$
$$H_2NCH_2CH_2CH_2N$$
$$CH_2CH_3$$

| N,N-Diethyltrimethylene-diamine $(C_7H_{18}N_2)$, L | HL/H.L | 10.48 10.53[a] 10.71[b] 10.94[c] | | | 72NT |
| | H$_2$L/HL.H | 8.30 8.58[a] 8.89[b] 9.13[c] | | | |

[a]25°, 0.1; [b]25°, 0.5; [c]25°, 1.0; [d]25°, 2.0; [t]22°, 0.1

Tertiary amines (continued)

Ligand	Equilibrium	Log K 25°, 0	ΔH 25°, 0	ΔS 25°, 0	Bibliography

$$CH_3NCH_2CH_2N\begin{smallmatrix}CH_3\\CH_3\end{smallmatrix}$$

Ligand	Equilibrium	Log K 25°, 0	ΔH 25°, 0	ΔS 25°, 0	Bibliography
N,N,N'-Trimethylethylene-diamine ($C_5H_{14}N_2$), L (Other values on Vol.2 p. 122)	HL/H.L	9.88[a]	-8.4[a]	13[a]	72CV
	H_2L/HL.H	6.83[c]	-7.6[c]	6[c]	

$$H-N\ \underset{s}{\bigcirc}\ N-CH_3$$

Ligand	Equilibrium	Log K 25°, 0	ΔH 25°, 0	ΔS 25°, 0	Bibliography
1-Methylpiperazine ($C_5H_{12}N_2$), (Other values on Vol.2, p.124), L	HL/H.L	8.98[a]	-8.4[a]	13[a]	72CV
		9.32[c]	-9.09[c]	12.1[c]	

$$\begin{smallmatrix}H_3C\\H_3C\end{smallmatrix}NCH_2CH_2CH_2N\begin{smallmatrix}CH_3\\CH_3\end{smallmatrix}$$

Ligand	Equilibrium	Log K 25°, 0	ΔH 25°, 0	ΔS 25°, 0	Bibliography
N,N,N',N'-Tetramethyltri-methylenediamine ($C_7H_{18}N_2$), L	HL/H.L	9.76 9.81[a] 10.02[b] 10.25[c]			74NT
	H_2L/HL.H	7.53 7.84[a] 8.22[b] 8.53[c]			

$$\begin{smallmatrix}R\\R\end{smallmatrix}NCH_2\overset{\overset{CH_2}{\|}}{C}CH_2N\begin{smallmatrix}R\\R\end{smallmatrix}$$

Ligand	Equilibrium	Log K 25°, 0	ΔH 25°, 0	ΔS 25°, 0	Bibliography
N,N,N',N'-Tetramethyl-2-methylenetrimethylenediamine ($C_8H_{18}N_2$), L	HL/H.L	9.45[b]			75HS
	H_2L/HL.H	7.15[b]			
N,N,N',N'-Tetraethyl-2-methylenetrimethylenediamine ($C_{12}H_{26}N_2$), L	HL/H.L	10.77[b]			75HS
	H_2L/HL.H	7.57[b]			

$$HOCH_2CH_2-N\ \underset{s}{\bigcirc}\ N-CH_2CH_2SO_3H$$

Ligand	Equilibrium	Log K 25°, 0	ΔH 25°, 0	ΔS 25°, 0	Bibliography
N-(2-Hydroxyethyl)piperazine-N'-ethanesulfonic acid ($C_8H_{18}O_4N_2S$), HL	HL/H.L	7.565	-4.9	18	76VBc

---

[a]25°, 0.1; [b]25°, 0.5; [c]25°, 1.0

4. Cyclic polyamines

Ligand	Equilibrium	Log K 25°, 0	ΔH 25°, 0	ΔS 25°, 0	Bibliography

Ligand	Equilibrium	Log K 25°, 0	ΔH 25°, 0	ΔS 25°, 0	Bibliography
1,5,9,13-Tetraazacyclo-hexadecane ($C_{12}H_{28}N_4$), L	$HL/H.L$ $H_2L/HL.H$ $H_3L/H_2L.H$ $H_4L/H_3L.H$	10.70^b 9.65^b 7.06^b 5.54^b			78LH
11,13-Dimethyl-1,4,7,10-tetraazacyclotridecane ($C_{11}H_{26}N_4$), L	$HL/H.L$ $H_2L/HL.H$	11.4^b 9.76^b			75SK
12,12-Dimethyl-1,4,7,10-tetraazacyclotridecane ($C_{11}H_{26}N_4$), L	$HL/H.L$ $H_2L/HL.H$	11.4^b 9.95^b			75SK
11,11,13-Trimethyl-1,4,7,10-tetraazacyclotridecane ($C_{12}H_{28}N_4$), L	$HL/H.L$ $H_2L/HL.H$	11.3^b 9.46^b			75SK
1,4,7,10,12,12-Hexamethyl-1,4,7,10-tetraazacyclotridecane ($C_{15}H_{34}N_4$), L	$HL/H.L$ $H_2L/HL.H$	11.4^b 7.95^b			75SK

Ligand	Equilibrium	Log K 25°, 0	ΔH 25°, 0	ΔS 25°, 0	Bibliography
2,12,-Dimethyl-3,7,11,17-tetraazabicyclo[11.3.1]-heptadeca-14,16,17-triene ($C_{15}H_{26}N_4$), L	$HL/H.L$ $H_2L/HL.H$ $H_3L/H_2L.H$	9.93^b 8.94^b 5.30^b			78SK
2,7,12-Trimethyl-3,7,11,17-tetraazabicyclo[11.3.1]-heptadeca-14,16,17-triene ($C_{16}H_{28}N_4$), L	$HL/H.L$ $H_2L/HL.H$ $H_3L/H_2L.H$	10.06^b 8.77^b 5.06^b			78SK

Ligand	Equilibrium	Log K 25°, 0	ΔH 25°, 0	ΔS 25°, 0	Bibliography
1,4,7,10-Tetraazacyclo-tridecane-11,13-dione* ($C_9H_{18}O_2N_4$), L	$HL/H.L$ $H_2L/HL.H$	8.93^a 3.70^a			79KK

---

^{a}25°, 0.1; ^{b}25°, 0.5; *metal constants were also reported but not included in the compilation of selected values.

Cyclic polyamines (continued)

Ligand	Equilibrium	Log K 25°, 0	ΔH 25°, 0	ΔS 25°, 0	Bibliography
1,4,8,11-Tetraazacyclo-tetradecane-12,14-dione* $(C_{10}H_{20}O_2N_4)$, L	HL/H.L $H_2L/HL.H$	9.45[a] 5.85[a]			79KK
1,4,8,12-Tetraazacyclo-pentadecane-9,10-dione* $(C_{11}H_{22}O_2N_4)$, L	HL/H.L $H_2L/HL.H$	9.28[a] 6.40[a]			79KK

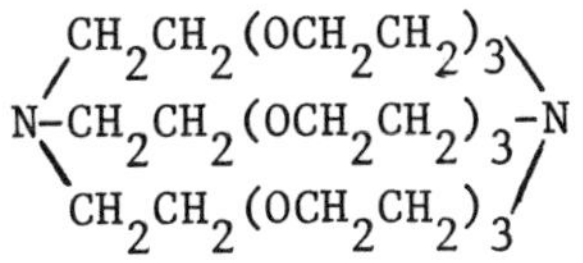

$$\begin{array}{c} CH_2CH_2(OCH_2CH_2)_3 \\ N\text{-}CH_2CH_2(OCH_2CH_2)_3\text{-}N \\ CH_2CH_2(OCH_2CH_2)_3 \end{array}$$

Ligand	Equilibrium	Log K 25°, 0	ΔH 25°, 0	ΔS 25°, 0	Bibliography
4,7,10,16,19,22,27,30,30-Nonaoxa-1,13-diazabicyclo-[11.11.11]pentatriacontane ([3.3.3]cryptand) $(C_{24}H_{48}O_9N_2)$, L	HL/H.L $H_2L/HL.H$	7.70[u] 6.96[u]			75LS

5. Azoles

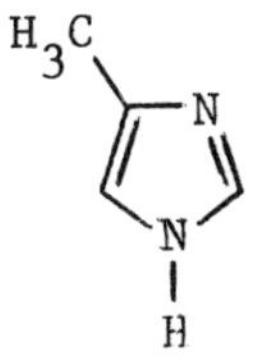

Ligand	Equilibrium	Log K 25°, 0	ΔH 25°, 0	ΔS 25°, 0	Bibliography
4-Methylimidazole $(C_4H_6N_2)$, L	HL/H.L	7.55[a]±0.01	-9.61[a]	2.3[a]	76EW

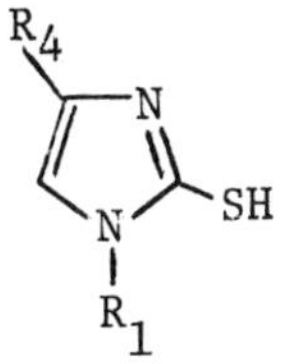

(Other values on Vol.2, p.149)

Ligand	Equilibrium	Log K 25°, 0	ΔH 25°, 0	ΔS 25°, 0	Bibliography
2-Mercaptoimidazole, HL $(C_3H_4N_2S)$	HL/H.L	11.21[a]	(-15)[r]	(1)[a]	77ST
2-Mercapto-1-methylimidazole*, HL $(C_4H_6N_2S)$	HL/H.L	11.64[a]	(-17)[r]	(3)[a]	77ST, Other reference: 77LW
2-Mercapto-4-(2-aminoethyl-imidazole)(2-mercapto-histamine) $(C_5H_9N_3S)$, HL	HL/H.L $H_2L/HL.H$	11.62[a] 9.12[a]	(-7)[r] (-10)[r]	(28)[a] (10)[a]	77ST

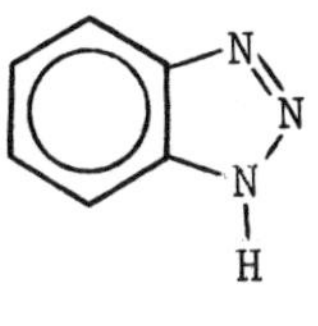

Ligand	Equilibrium	Log K 25°, 0	ΔH 25°, 0	ΔS 25°, 0	Bibliography
Benzotriazole $(C_6H_5N_3)$, L	HL/H.L	8.35[b]	-7.28[b]	13.8[b]	77BE, Other reference: 76CC

[a] 25°, 0.1; [b] 25°, 0.5; [r] 15-35°, 0.1; [u] 25°, 0.05; *metal constants were also reported but not inclued in the compilation of selected values.

6. <u>Azines</u>

Ligand	Equilibrium	Log K 25°, 0	ΔH 25°, 0	ΔS 25°, 0	Bibliography

Ligand	Equilibrium	Log K 25°, 0	ΔH 25°, 0	ΔS 25°, 0	Bibliography
4-Chloropyridine (C_5H_4NCl), L	HL/H.L		-3.6		74LP
4-Bromopyridine (C_5H_4NBr), L	HL/H.L		-3.5		74LP
4-Cyanopyridine ($C_6H_4N_2$), L (Other values on Vol.2, p.177)	HL/H.L		-1.3		74LP
3-Cyanomethylpyridine ($C_7H_6N_2$), L	HL/H.L	4.16^b	-4.51^b	4.1^b	76FE

Ligand	Equilibrium	Log K 25°, 0	ΔH 25°, 0	ΔS 25°, 0	Bibliography
1,3-Dihydroxy-4-(2-pyridyl-azo)benzene (PAR)* ($C_{11}H_9O_2N_3$), H_2L (Other values on Vol.2, p.178)	HL/H.L	12.3^a 12.3^h	-0.3 -0.4		61HS,74PM,78KLH, 79RKS, Other references: 66BBI,68BI,68S, 69HS,70EN,70NE, 70GMT,71BR,71FY, 72BN,72MPV,72NE, 73NE,73NEa,75IYa, 78Kd,78Ke,78TT, 79PGR,79PK
	$H_2L/HL.H$	5.54^a 5.6^h	±0.04 ±0.1		
	$H_3L/H_2L.H$	2.69^a 2.75^h	+0.3 +0.3		

Ligand	Equilibrium	Log K 25°, 0	ΔH 25°, 0	ΔS 25°, 0	Bibliography
Pyridine-2,6-dicarboxaldehyde* ($C_7H_5O_2N$), L	$(H_{-1}L)/(H_{-2}L).H$	$(13.5)^b$			76PPF
	$L/(H_{-1}L).H$	$(11.9)^b$			
	HL/H.L	2.13^b			

Ligand	Equilibrium	Log K 25°, 0	ΔH 25°, 0	ΔS 25°, 0	Bibliography
2-Methoxypyridine (C_6H_7ON), L (Other values on Vol.2, p.190)	HL/H.L	6.47	-6.9	6	74LP

---

^{a}25°, 0.1; ^{b}25°, 0.5; ^{h}20°, 0.1; *metal constants were also reported but not included in the compilation of selected constants.

Azines (continued)

Ligand	Equilibrium	Log K 25°, 0	ΔH 25°, 0	ΔS 25°, 0	Bibliography

Ligand	Equilibrium	Log K 25°, 0	ΔH 25°, 0	ΔS 25°, 0	Bibliography
8-Hydroxyquinoline (oxine)* (C_9H_7ON), HL (Other values on Vol.2, p.223)	HL/H.L	9.77 ± 0.04 $9.65^a \pm 0.03$ $9.65^b \pm 0.04$	$(-7)^r$	(21)	49IE,70NS, Other references: 67SL,74Bc,75BG, 77RR,77Sd,78GMM, 79BK,79YY
	$H_2L/HL.H$	4.95 ± 0.1 $4.99^a \pm 0.04$ $5.14^b \pm 0.05$	$(-5)^s$	$(6)^a$	
Quinoline-8-thiol (C_9H_7NS), HL	HL/H.L	8.36 8.40^a			64SF,70NS, Other reference: 74VSU
Quinoline-8-selenol (C_9H_7NSe), HL	HL/H.L	8.18 $(8.75)^a$			64SF,70NS
7-Iodo-8-hydroxyquinoline-5-sulfonic acid (ferron)* $(C_9H_6O_4NIS)$, H_2L (Other values on Vol.2, p.232)	HL/H.L	7.43 ± 0.01 $7.06^a \pm 0.05$ $6.91^b \pm 0.04$ $6.87^c \pm 0.03$	-3.7^b	19^b	66HC,77ML,78MM, Other references: 68BN,68KB,68RR, 72HKa,73PM,74KJ, 75AM,77KC,77KSK, 77Sd,77SM,78GMM
	$H_2L/HL.H$	2.49 ± 0.03 $2.45^a \pm 0.06$ $2.25^b \pm 0.04$ $2.42^c \pm 0.01$	-2.2^b	3^b	

Ligand	Equilibrium	Log K 25°, 0	ΔH 25°, 0	ΔS 25°, 0	Bibliography
3,3'-Bipyridyl $(C_{10}H_8N_2)$, L	HL/H.L	4.60^h			51K
2,3'-Bipyridyl $(C_{10}H_8N_2)$, L	HL/H.L	4.42^h			51K
2,4'-Bipyridyl $(C_{10}H_8N_2)$, L	HL/H.L	4.77^h			51K
3,4'-Bipyridyl $(C_{10}H_8N_2)$, L	HL/H.L	4.85^h			51K
Vinylene-4,4'-dipyridine $(C_{12}H_{10}O_2)$, L	HL/H.L $H_2L/HL.H$	5.65^a 4.41^a			74IL

Ligand	Equilibrium	Log K 25°, 0	ΔH 25°, 0	ΔS 25°, 0	Bibliography
2,9-Dimethyl-1,10-phenanthroline $(C_{14}H_{12}N_2)$, L	HL/H.L	5.77 $5.85^a + 0.01$ 5.88^t	$(-5)^r$	(10)	68LA,68WF

---

^{a}25°, 0.1; ^{b}25°, 0.5; ^{c}25°, 1.0; ^{h}20°, 0.2; ^{r}20-40°, 0; ^{s}25-45°, 0.1; ^{t}25°, 0.3;
*metal constants were also reported but not included in the compilation of selected constants.

Azines   (continued)

Ligand	Equilibrium	Log K $25°, 0$	ΔH $25°, 0$	ΔS $25°, 0$	Bibliography
1,7-Phenanthroline ($C_{12}H_8N_2$), L	HL/H.L	4.0^h			51K
4,7-Phenanthroline ($C_{12}H_8N_2$), L	HL/H.L	4.0^h			51K

7. Aminophosphonic acids

$$\begin{array}{c} NH_2 \\ | \\ R-CHPO_3H_2 \end{array}$$

Ligand	Equilibrium	Log K			Bibliography
DL-1-Aminopropylphosphonic acid ($C_3H_{10}O_3NP$), H_2L	HL/H.L $H_2L/HL.H$	10.26^a 5.66^a			78WN
DL-1-Aminobutylphosphonic acid ($C_4H_{12}O_3NP$), H_2L	HL/H.L $H_2L/HL.H$	10.29^a 5.68^a			78WN

$$H_2O_3PCH_2CH_2NCH_2CH_2\overset{+}{N}CH_2CH_2PO_3H_2$$
$$\underset{CH_3}{|} \quad \overset{CH_3}{\underset{CH_3}{|}}$$

Ligand	Equilibrium	Log K			Bibliography
Dimethyl(2-phosphoethyl)- (N-methyl-N-phosphoethyl)- 2-aminoethylammonium (iodide) ($C_9H_{25}O_6N_2P_2^+$), H_4L	$H_2L/HL.H$ $H_3L/H_2L.H$ $H_4L/H_3L.H$	6.46^a 5.45^a 1.5^a			76MD

8. Aminophosphates $H_2NCH_2CH_2OPO_3R_2$

Ligand	Equilibrium	Log K			Bibliography
2-Aminoethyl(dihydrogen- phosphate) (O-phosphoryl- ethanolamine)* ($C_2H_8O_4NP$), H_2L (Other values on Vol.2, p.329)	HL/H.L $H_2L/HL.H$	$10.15^a\pm0.05$ $5.56^a\pm0.04$			78MA
2-Aminoethyl(diphenylphosphate) (O-diphenylphosphorylethanolamine) ($C_{14}H_{16}O_4NP$), L	HL/H.L	8.22^a			59FO

C. Other Organic Ligands

1. Carboxylic acids

Ligand	Equilibrium	Log K			Bibliography
1-Naphthoic acid ($C_{11}H_8O_2$), HL (Other reference on Vol.3, p.332)	HL/H.L	3.67 ± 0.03			61CF

$$Cl_3CCO_2H$$

Ligand	Equilibrium	Log K			Bibliography
Trichloroacetic acid ($C_2HO_2Cl_3$), HL (Other values on Vol.3, p.19)	HL/H.L	-0.34			79BP, Other references: 76PT, 76PSd,77YO

---

[a] $25°, 0.1$; [h] $20°, 0.2$; * metal constants were also reported but not included in the compilation of selected values.

Carboxylic acids (continued)

Ligand	Equilibrium	Log K 25°, 0	ΔH 25°, 0	ΔS 25°, 0	Bibliography
	$H_2O_3P-R-CO_2H$				
Phosphonoformic acid (CH_3O_5P), H_3L	$HL/H.L$	7.56^a			79HP
	$H_2L/HL.H$	3.59^a			
	$H_3L/H_2L.H$	1.7^a			
2-Phosphonopropanoic acid ($C_3H_7O_5P$), H_3L	$HL/H.L$	8.54^a			79HP
	$H_2L/HL.H$	5.15^a			
	$H_3L/H_2L.H$	1.8^a			
3-Phosphonopropanoic acid ($C_3H_7O_5P$), H_3L	$HL/H.L$	7.75^a			79HP
	$H_2L/HL.H$	4.63^a			
	$H_3L/H_2L.H$	2.26^a			
	$H_2O_3AsCH_2CO_2H$				
Arsonoacetic acid* ($C_2H_5O_5As$), H_3L	$HL/H.L$	8.83^b			76TN
	$H_2L/HL.H$	3.83^b			
	$H_3L/H_2L.H$	1.86^b			
	$HO-R-CO_2H$				
D-Glucuronic acid* ($C_6H_{10}O_7$), HL	$HL/H.L$	3.231			77HH, Other reference: 77MC
D-Galacturonic acid* ($C_6H_{10}O_7$), HL (Values on Vol.3, p.61)	$L/H_{-1}L.H$	11.42^n			77HH, Other reference: 77MC
	$HL/H.L$	3.471 3.23^n			

Ligand	Equilibrium	Log K 25°, 0	ΔH 25°, 0	ΔS 25°, 0	Bibliography
2,3-Dimethoxybenzoic acid, HL ($C_9H_{10}O_4$)	$HL/H.L$	3.50^a			78PM
	$HS-R-CO_2H$				
3-Mercaptobutanoic acid ($C_4H_8O_2S$), H_2L	$HL/H.L$	10.45^h			75SH
	$H_2L/HL.H$	4.20^h			
DL-(2-Mercaptopropionyl) glycine*, H_2L ($C_5H_9O_3NS$)	$HL/H.L$	8.74^h			75SH,76SH, Other reference: 71FT
	$H_2L/HL.H$	3.60^h			
DL-2-(2-Mercaptopropionyl-amino)propanoic acid*, H_2L ($C_6H_{11}O_3NS$)	$HL/H.L$	8.81^h			76SH
	$H_2L/HL.H$	3.77^h			
(3-Mercaptopropionyl)glycine* H_2L ($C_5H_9O_3NS$)	$HL/H.L$	9.60^h			76SH
	$H_2L/HL.H$	3.71^h			

─────────────

[a] 25°, 0.1; [b] 25°, 0.5; [h] 20, 0.1; [n] 37°, 0.15; *metal constants were also reported but not included in the compilation of selected constants.

Carboxylic acids (continued)

Ligand	Equilibrium	Log K 25°, 0	ΔH 25°, 0	ΔS 25°, 0	Bibliograhy
DL-2-(3-Mercaptopropionyl-amino)propanoic acid*, H_2L ($C_6H_{11}O_3NS$)	HL/H.L $H_2L/HL.H$	9.69^h 3.82^h			76SH
DL-(2-Mercapto-3-methyl-butanoyl)glycine, H_2L ($C_7H_{13}O_3NS$)	HL/H.L $H_2L/HL.H$	9.07^h 3.77^h			76SH
DL-(2-Mercaptopropionyl)-phenylglycine*, H_2L ($C_{11}H_{13}O_3NS$)	HL/H.L $H_2L/HL.H$	8.66^h 3.08^h			76SH
DL-(2-Mercapto-2-phenylacetyl)-glycine, H_2L ($C_{10}H_{11}O_3NS$)	HL/H.L $H_2L/HL.H$	7.80^h 3.20^h			76SH
DL-(2-Mercapto-3-phenyl-propionyl)glycine, H_2L ($C_{11}H_{13}O_3NS$)	HL/H.L $H_2L/HL.H$	8.41^h 3.47^h			76SH

$$HO_2C-R-CO_2H$$

Ligand	Equilibrium	Log K 25°, 0	ΔH 25°, 0	ΔS 25°, 0	Bibliograhy
Butylpropanedioic acid (butylmalonic acid) ($C_7H_{12}O_4$), H_2L (Other values on Vol.3, p.100)	HL/H.L $H_2L/HL.H$	5.96 5.50^a+0.01 5.20^b 3.02 2.86^a±0.05 2.77^b			76BMa, Other reference: 76DGN
Methylenebutanedioic acid (itaconic acid) ($C_5H_6O_4$), H_2L (Other values on Vol.3, p.116)	HL/H.L $H_2L/HL.H$	5.44^o 5.14^a+0.01 4.99^c 3.68^a+0.01 3.63^c	-0.70^c -1.0^c	20.4^c 13^c	77Ad,79PM, Other references: 76Sd, 76SJ,77DS,77SJ, 79JA,79JB,79JBa, 79PBJ,79SJ,79SJa, 79SJc
3,3-Diethylglutaric acid ($C_9H_{16}O_4$), H_2L	HL/H.L $H_2L/HL.H$	(7.13) 3.62			31GI
3,3-Dipropylglutaric acid ($C_{11}H_{20}O_4$), H_2L	HL/H.L $H_2L/HL.H$	(7.31) 3.69			31GI

2. Phosphonic acids $R-PO_3H_2$

Ligand	Equilibrium	Log K 25°, 0	ΔH 25°, 0	ΔS 25°, 0	Bibliograhy
Dibromomethanephosphonic acid ($CH_3O_3Br_2P$), H_2L	HL/H.L $H_2L/HL.H$	5.40^a 0.8^a			78WNb,79WN
2-Bromoethanephosphonic acid ($C_2H_6O_3BrP$), H_2L	HL/H.L $H_2L/HL.H$	6.9^a 1.62^a			79WN
1,1-Dimethylethanephosphonic acid ($C_4H_{11}O_3P$), H_2L	HL/H.L	8.71			77KT

---

^{a}25°, 0.1; ^{b}25°, 0.5; ^{c}25°, 1.0; ^{h}20°, 0.1; ^{o}18°, 0; *metal constants were also reported but not included in the compilation of selected constants.

3. Arsonic acids

Ligand	Equilibrium	Log K 25°, 0	ΔH 25°, 0	ΔS 25°, 0	Bibliography

$R-AsO_3H_2$

Ligand	Equilibrium	Log K 25°, 0	ΔH 25°, 0	ΔS 25°, 0	Bibliography
Methylarsonic acid* (CH_5O_3As), H_2L (Other value on Vol.3, p.180)	HL/H.L	8.77 / 8.66[a]	-1.40	35.4	76LH, Other reference: 76TN
	H_2L/HL.H	4.19 / 3.96[a]	1.65	24.7	
Ethylarsonic acid ($C_2H_7O_3As$), H_2L	HL/H.L	9.19	-1.53	36.9	76LH
	H_2L/HL.H	4.24	1.83	25.5	
Propylarsonic acid ($C_3H_9O_3As$), H_2L	HL/H.L	9.36	-1.93	36.3	76LH
	H_2L/HL.H	4.34	1.53	25.0	
Butylarsonic acid ($C_4H_{11}O_3As$), H_2L	HL/H.L	9.31	-1.88	36.3	76LH
	H_2L/HL.H	4.37	1.56	25.2	
Pentylarsonic acid ($C_5H_{13}O_3As$), H_2L	HL/H.L	9.39	-1.87	36.7	76LH
	H_2L/HL.H	4.34	1.53	25.0	
Hexylarsonic acid ($C_6H_{15}O_3As$), H_2L	HL/H.L	9.41	-1.89	36.7	76LH
	H_2L/HL.H	4.34	1.54	25.0	

R_2AsO_2H

Ligand	Equilibrium	Log K 25°, 0	ΔH 25°, 0	ΔS 25°, 0	Bibliography
Dimethylarsinic acid ($C_2H_7O_2As$), HL	HL/H.L	6.14	0.63	30.2	76LH
	H_2L/HL.H	1.78	0.8	11	
Diethylarsinic acid ($C_4H_{11}O_2As$), HL	HL/H.L	6.43	0.88	32.3	76LH
	H_2L/HL.H	1.53	1.8	13	
Dipropylarsinic acid ($C_6H_{14}O_2As$), HL	HL/HL.H	6.52	0.65	32.0	76LH
	H_2L/HL.H	1.64	1.9	14	
Dibutylarsinic acid ($C_8H_{18}O_2As$), HL	HL/H.L	6.53	0.49	31.5	76LH
	H_2L/HL.H	1.48	1.9	13	
Dipentylarsinic acid ($C_{10}H_{22}O_2As$), HL	HL/H.L	6.54	0.31	30.9	76LH
	H_2L/HL.H	1.69	1.5	13	

4. Phenols

R—C6H4—OH

Ligand	Equilibrium	Log K 25°, 0	ΔH 25°, 0	ΔS 25°, 0	Bibliography
Hydroxybenzene (phenol)* (C_6H_6O), HL (Other values on Vol.3, p.181)	HL/H.L	9.98 ±0.04 / 9.82[a] ±0.04 / 9.64[b] / 9.52[c]	-5.5-0.1 / -6.1[a]	27 / 24[a]	59PC,61BR,66BH, 78Pb, Other references: 76TC, 78Kd,78Ke,78RP
2-Methylphenol(2-cresol)* (C_7H_8O), HL (Other references on Vol.3, p.347)	HL/H.L	10.26 ±0.01	-5.7±0.0	28	77RWc, Other references: 59PC, 78RP

[a]25°, 0.1; [b]25°, 0.5; [c]25°, 1.0; *metal constants were also reported but not included in the compilation of selected constants.

Phenols (continued)

Ligand	Equilibrium	Log K 25°, 0	ΔH 25°, 0	ΔS 25°, 0	Bibliography
3-Methylphenol (3-cresol) (C_7H_8O), HL (Other values on Vol.3, p.181)	HL/H.L	10.99 ±0.01	-5.6 -0.1	27	77RWc, Other reference: 59PC
4-Methylphenol (4-Cresol) (C_7H_8O), HL (Other values on Vol.3, p.181)	HL/H.L	10.26 ±0.01 9,87[b]	-5.6 ±0.1	28	76KSc, Other reference: 59PC
3-Ethylphenol ($C_8H_{10}O$), HL	HL/H.L	10.07	-5.3	28	67BH
2,3-Dimethylphenol ($C_8H_{10}O$), HL	HL/H.L	10.54 -0.01	-5.7	29	57HK,62CL,69P Other reference: 59PC
2,4-Dimethylphenol ($C_8H_{10}O$), HL	HL/H.L	10.59 ±0.01	-5.8	29	57HK,62CL,69P, Other reference: 59PC
2,5-Dimethylphenol ($C_8H_{10}O$), HL	HL/H.L	10.40 +0.01	-5.6	29	57HK,62CL,69P, Other reference: 59PC
2,6-Dimethylphenol ($C_8H_{10}O$), HL (Other reference on Vol.3, p.347)	HL/H.L	10.62 ±0.02 10.58[c]	-5.5 -5.8[c]	30 29[c]	48WB,47HK,62CL, 69P, Other reference: 59PC
3,4-Dimethylphenol ($C_8H_{10}O$), HL	HL/H.L	10.36 -0.01	-5.4	29	57HK,62CL,69P, Other reference: 59PC
3,5-Dimethylphenol ($C_8H_{10}O$), HL (Other reference on Vol.3, p.347)	HL/H.L	10.19 ±0.01 9.87[c]	-5.3 (-7.5)[c]	29 (20)[c]	48WB,57HK,62CL, 69P, Other reference: 59PC

X—⬡—OH

Ligand	Equilibrium	Log K 25°, 0	ΔH 25°, 0	ΔS 25°, 0	Bibliography
2-Chlorophenol (C_6H_5OCl), HL (Other values on Vol.3, p.182)	HL/H.L	8.53 ±0.00 8.29[a]±0.04	-4.5 ±0.3 -4.8[a]	24 22[a]	77RWc
3-Chlorophenol (C_6H_5OCl), HL (Other values on Vol.3, p.182)	HL/H.L	9.13 -0.01 8.78[a]±0.02	-5.3 ±0.2 -6.1[a]	24 20[a]	61OH,66BHa,77RWc
4-Chlorophenol (C_6H_5OCl), HL (Other values on Vol.3, p.182)	HL/H.L	9.43 ±0.01 9.14[a]±0.04	-5.6 -0.1 -6.1[a]	24 21[a]	66BH
3,5-Dichlorophenol ($C_6H_4OCl_2$), HL (Other reference on Vol.3, p.348)	HL/H.L	8.18 ±0.00	-4.9	21	68BHK
3-Bromophenol (C_6H_5OBr), HL (Other values on Vol.3, p.182)	HL/H.L	9.03 ±0.00 8.75[a]	-5.4	23	66BHa
4-Bromophenol* (C_6H_5OBr), HL (Other values on Vol.3, p.182)	HL/H.L	9.35 ±0.01	-5.7	24	66BH, Other reference: 78RP

---

[a]25°. 0.1; [b]25°, 0.5; [c]25°, 1.0; *metal constants were also reported but not included in the compilation of selected constants.

Phenols (continued)

Ligand	Equilibrium	Log K 25°, 0	ΔH 25°, 0	ΔS 25°, 0	Bibliography
3,5-Dibromophenol ($C_6H_4OBr_2$), HL	HL/H.L	8.06	−5.3	19	68BHK
3-Iodophenol (C_6H_5OI), HL (Other values on Vol.3, p.182)	HL/H.L	9.05 ±0.02 8.74[a]	−5.5	23	66BHa
4-Iodophenol (C_6H_5OI), HL (Other values on Vol.3, p.182)	HL/H.L	9.32 ±0.01	−5.4	25	66BH
3,5-Diiodophenol ($C_6H_4OI_2$), HL	HL/H.L	8.10	−5.5	19	68BHK
2-Nitrophenol* ($C_6H_5O_3N$), HL (Other values on Vol.3, p.183)	HL/H.L	7.24 ±0.04 7.05[a]±0.01	−4.5 −0.1	18	74HL,78R, Other references: 78Kd, 78Ke
3-Nitrophenol ($C_6H_5O_3N$), HL (Other values on Vol.3, p.183)	HL/H.L	8.38 ±0.02 8.09[a]±0.05 8.03[b]	−4.9 ±0.3 −4.8[a]	22 21[a]	67BH,74HL,77RWc, 78R
4-Nitrophenol* ($C_6H_5O_3N$), HL (Other values on Vol.3, p.183)	HL/H.L	7.15 ±0.01 6.90[a]−0.01 6.85[b]+0.01	−4.6 ±0.1 −4.8[a]	17 15[a]	74HL,79ZK,78R, Other references: 78Kd,78Ke
2,4-Dinitrophenol* ($C_6H_4O_5N_2$), HL (Other values on Vol.3, p.183)	HL/H.L	4.08 ±0.03 3.93[a]			74HL, Other references: 78Kd, 78Ke
2,5-Dinitrophenol ($C_6H_4O_5N_2$), HL	HL/H.L	5.27			74HL
2,6-Dinitrophenol* ($C_6H_4O_5N_2$), HL (Other references on Vol.3, p.348)	HL/H.L	3.11			74HL, Other references: 78Kd, 78Ke
3,5-Dinitrophenol ($C_6H_4O_5N_2$), HL	HL/H.L	6.73	−4.0 ±0.2	17	68BHK,77RWc
2,4,6-Trinitrophenol (picric acid)* ($C_6H_3O_7N_3$), HL (Other references on Vol.3, p.348)	HL/H.L	0.37 ±0.05			64WMa,74HL, Other reference: 79PS
4-Nitrosophenol ($C_6H_5O_2N$), HL	HL/H.L	6.33			74HL

Ligand	Equilibrium	Log K 25°, 0	ΔH 25°, 0	ΔS 25°, 0	Bibliography
3-Hydroxybenzoic acid ($C_7H_6O_3$), H_2L (Other references on Vol.3, p.349)	HL/H.L	9.96 ±0.04 9.68[a]	(−4)[s]	(31)[a]	77IH, Other reference: 75Jb
	H_2L/HL.H	4.080±0.003 3.96[a]	−0.17+0.01	18.1	

---

[a]25°, 0.1; [b]25°, 0.5; [s]1-45°, 0.1; *metal constants were also reported but not included in the compilation of selected constants.

Phenols (continued)

Ligand	Equilibrium	Log K 25°, 0	ΔH 25°, 0	ΔS 25°, 0	Bibliography
4-Hydroxybenzoic acid ($C_7H_6O_3$), H_2L (Other references on Vol.3, p.349)	HL/H.L	9.46 8.95[a]	-4.3	29	77IH,78Pb, Other reference: 75Jb
	$H_2L/HL.H$	4.58 ±0.01 4.36[a]	-0.37±0.02	19.7	
3-Hydroxyphenylacetic acid ($C_8H_8O_3$), H_2L	HL/H.L $H_2L/HL.H$	9.87[a] 4.17[a]			77IH
4-Hydroxyphenylacetic acid ($C_8H_8O_3$), H_2L	HL/H.L $H_2L/HL.H$	9.88[a] 4.25[a]			77IH
3-(3-Hydroxyphenyl)propanoic acid ($C_9H_{10}O_3$), H_2L	HL/H.L $H_2L/HL.H$	9.91[a] 4.40[a]	(-6)[s]	(25)[a]	77IH
3-(4-Hydroxyphenyl)propanoic acid ($C_9H_{10}O_3$), H_2L	HL/H.L $H_2L/HL.H$	10.03[a] 4.39[a]	(-6)[s]	(26)[a]	77IH

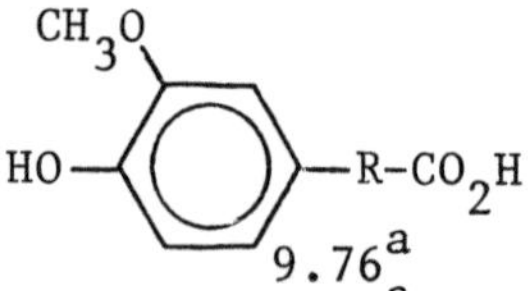

Ligand	Equilibrium	Log K 25°, 0	ΔH 25°, 0	ΔS 25°, 0	Bibliography
4-Hydroxy-3-methoxyphenyl-acetic acid ($C_9H_{10}O_4$), H_2L	HL/H.L $H_2L/HL.H$	9.76[a] 4.31[a]			77IH
3-(4-Hydroxy-3-methoxyphenyl)-propanoic acid ($C_{10}H_{12}O_4$), H_2L	HL/H.L $H_2L/HL.H$	9.87[a] 4.48[a]			77IH

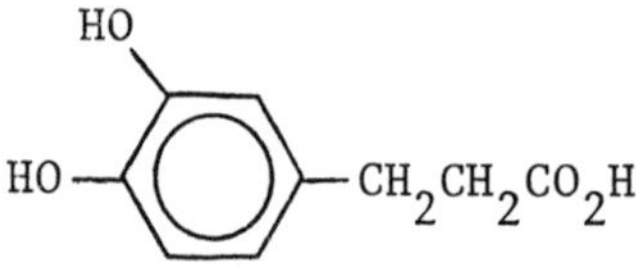

Ligand	Equilibrium	Log K 25°, 0	ΔH 25°, 0	ΔS 25°, 0	Bibliography
3-(3,4-Dihydroxyphenyl)-propanoic acid ($C_9H_{10}O_4$), H_3L	HL/H.L $H_2L/HL.H$ $H_3L/H_2L.H$	(12.0)[a] 9.58[a] 4.36[a]	(-9)[s] (-7)[s]	(25)[a] (20)[a]	77IH

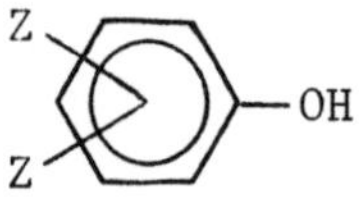

Ligand	Equilibrium	Log K 25°, 0	ΔH 25°, 0	ΔS 25°, 0	Bibliography
3-Methoxyphenol ($C_7H_8O_2$), HL (Other references on Vol.1, p.348)	HL/H.L	9.65 ±0.00	-5.1+0.1	27	67BH
3-Ethoxyphenol ($C_8H_{10}O_2$), HL	HL/H.L	9.65	-5.1	27	68BHK
3,5-Diethoxyphenol ($C_{10}H_{14}O_3$), HL	HL/H.L	9.37	-4.7	27	68BHK
3-Hydroxybenzaldehyde ($C_7H_6O_2$), HL (Other values on Vol.3, p.195)	HL/H.L	9.01 ±0.02	-5.0±0.0	24	64MAH,67BH

[a] 25°, 0.1; [s] 1-45°, 0.1

Phenols (continued)

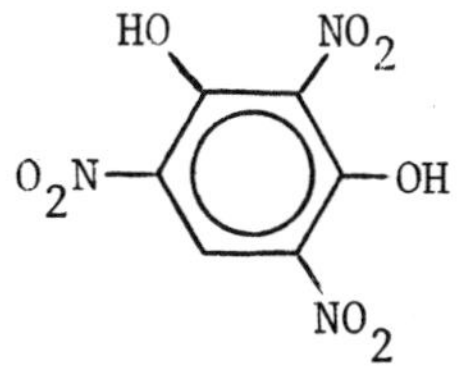

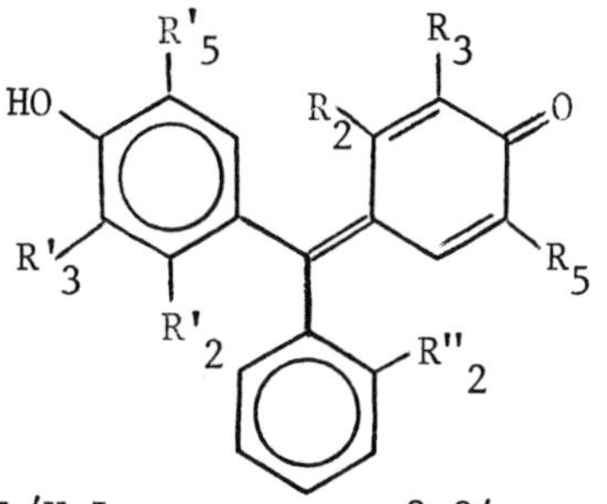

Ligand	Equilibrium	Log K 25°, 0	ΔH 25°, 0	ΔS 25°, 0	Bibliography
4-Hydroxybenzaldehyde $(C_7H_6O_2)$, HL (Other values on Vol.3, p.195)	HL/H.L	7.62 ±0.00 7.41[a]	-4.1 -4.0[a]	21 20[a]	64MAH
2-Hydroxy-3-methoxy-benzaldehyde (O-vanillin) $(C_8H_8O_3)$, HL	HL/H.L	7.91	-3.6	21	55RK,64MAH
2-Hydroxy-4-methoxybenzalde-hyde (isovanillin) $(C_8H_8O_3)$, HL	HL/H.L	8.89	-4.5	25	55RK,64MAH
4-Hydroxy-3-methoxybenzal-dehyde (vanillin) $(C_8H_8O_3)$, HL	HL/H.L	7.45 ±0.05	-4.0	22	55RK,64MAH,74HL

| 2,4,6-Trinitrobenzene-1,3-diol (2,4,6-trinitroresorcinol) $(C_6H_3O_8N_3)$, H_2L | HL/H.L H_2L/HL.H | 4.23 0.1 | | | 64WMa |

4'-Hydroxyfuchson-2"-sulfonic acid (Phenol Red) $(C_{19}H_{14}O_5S)$, H_2L	HL/H.L	8.04			77MKa
4'-Hydroxy-2,2'-dimethyl-5,5'-di(2-propyl)fuchson-2"-sulfonic acid (Thymol Blue) $(C_{23}H_{22}O_5S)$, H_2L	HL/H.L	9.23			77MKa
4'-Hydroxy-3,3'-dibromofuchson-2"-sulfonic acid (Bromophenol Red) $(C_{19}H_{12}O_5Br_2S)$, H_2L	HL/H.L	6.14			77MKa
4'-Hydroxy-3,3'-dibromo-5,5'-dimethylfuchson-2"-sulfonic acid (Bromocresol Purple) $(C_{21}H_{14}O_5Br_2S)$, H_2L	HL/H.L	6.20			77MKa

[a]25°, 0.1

Phenols (continued)

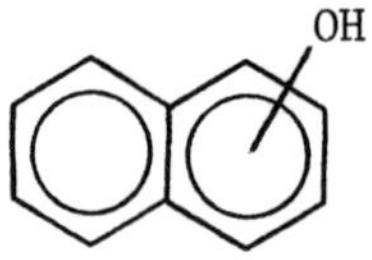

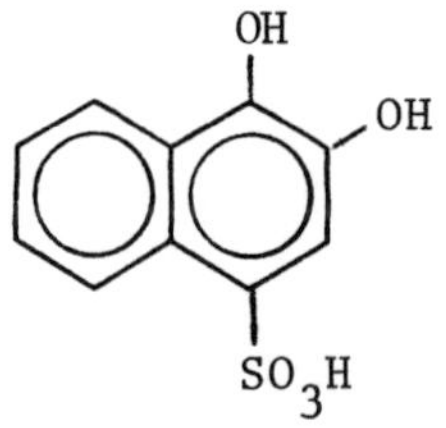

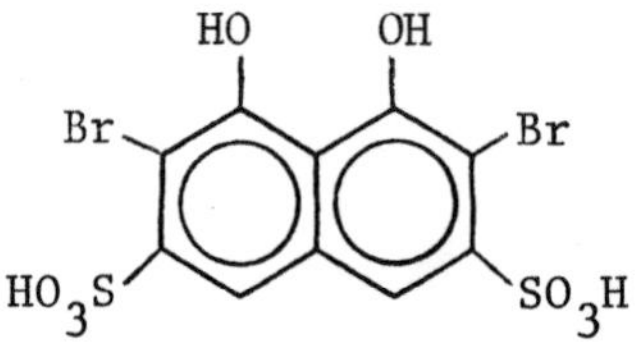

Ligand	Equilibrium	Log K 25°, 0	ΔH 25°, 0	ΔS 25°, 0	Bibliography
4'-Hydroxy-3,3'-dibromo-2,2'-dimethyl-5,5'-di(2-propyl)-fuchson-2''-sulfonic acid (Bromothymol Blue) ($C_{23}H_{20}O_5Br_2S$), H_2L	HL/H.L	7.20			77MKa
4'-Hydroxy-3,3',5,5'-tetra-bromo-2,2'-dimethylfuchson-2''-sulfonic acid (Bromocresol Green) ($C_{21}H_{12}O_5Br_4S$), H_2L	HL/H.L	5.03			77MKa
4'-Hydroxy-2''-carboxyfuchson-2,2'-oxide (Fluorescein) ($C_{20}H_{12}O_5$), H_2L	HL/H.L $H_2L/HL.H$ $H_3L/H_2L.H$	6.50[a] 4.32[a] 2.27[a]			78KA

| 1-Naphthol ($C_{10}H_8O$), HL (Other reference on Vol.3, p.350) | HL/H.L | 9.416 9.14[a] 9.06[b] 9.05[c] | −5.0 | 26 | 61CF,63BM,74HL, 77DF |
| 2-Naphthol ($C_{10}H_8O$), HL (Other reference on Vol.3, p.351) | HL/H.L | 9.573 9.31[a] 9.24[b] 9.25[c] | −4.6 | 28 | 63BM,74HL,77DF |

| 1,2-Dihydroxynaphthalene-4-sulfonic acid ($C_{10}H_8O_5S$), H_3L | HL/H.L $H_2L/HL.H$ | 12.66[a] 8.14[a] | | | 77BM |

| 2,7-Dibromochromotropic acid ($C_{10}H_6O_8Br_2S_2$), H_4L (Other references on Vol.3, p.363) | $H_2L/HL.H$ | 2.34[h] | | | 69BBH |

---

[a]25°, 0.1; [b]25°, 0.5; [c]25°, 1.0; [h]20°, 0.1

5. Carbonyl ligands

Ligand	Equilibrium	Log K 25°, 0	ΔH 25°, 0	ΔS 25°, 0	Bibliography

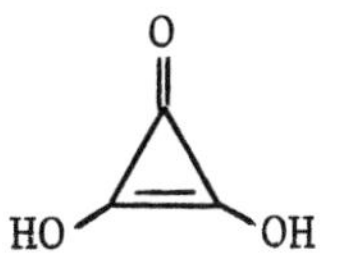

Ligand	Equilibrium	Log K 25°, 0	ΔH 25°, 0	ΔS 25°, 0	Bibliography
Dihydroxycyclopropenone (deltic acid) $(C_3H_2O_3)$, H_2L	HL/H.L $H_2L/HL.H$	6.03 2.57	$(0)^r$ $(0)^r$	(27) (12)	76GSb

$$CX_3CHO$$

| Trifluoroacetaldehyde (fluoral) (C_2HOF_3), HL | HL/H.L | 10.249 | -6.7 | 24 | 76KSc |
| Trichloroacetaldehyde (chloral) (C_2HOCl_3), HL | HL/H.L | 10.07 | -7.1 | 22 | 76KSc |

| 1,2-Naphthoquinone-4-sulfonic acid $(C_{10}H_6O_5S)$, H_2L | HL/H.L | 10.51^a | | | 77BM |

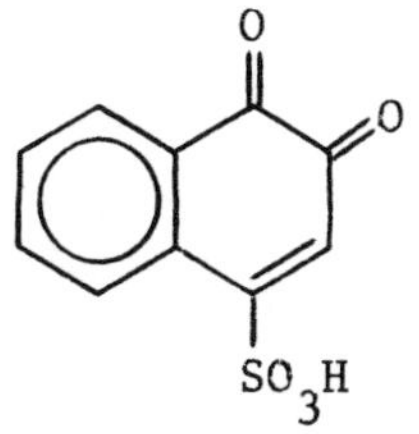

6. Thiols   R-SH

Ethanethiol (C_2H_6S), HL (Other reference on Vol.3, p.355)	HL/H.L	10.61 ±0.00	-6.4	27	76TCI
Naphthalene-1-thiol $(C_{10}H_8S)$, HL	HL/H.L	6.343	(0.1)	(29)	77DFa
Naphthalene-2-thiol $(C_{10}H_8S)$, HL	HL/H.L	6.472	(0.8)	(32)	77DFa

7. Amides

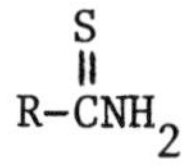

| Thioacetamide (C_2H_5NS), HL | HL/H.L | $(13.22)^c$ | | | 77AHa |
| Thiobenzamide (C_7H_7NS), HL | HL/H.L | 12.56^c | | | 77AHa |

| N^O-Methylthioxamide $(C_3H_6ON_2S)$, HL | HL/H.L | 11.72 11.51^a 11.41^b 11.45^c | | | 77AHa |

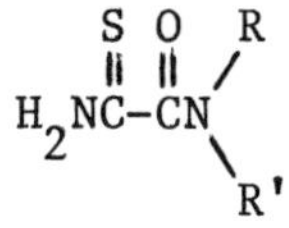

---

^{a}25°, 0.1; ^{b}25°, 0.5; ^{c}25°, 1.0; ^{r}15-35°, 0;   *metal constants were also reported but not included in the compilation of selected constants.

Amides (continued)

Ligand	Equilibrium	Log K 25°, 0	ΔH 25°, 0	ΔS 25°, 0	Bibliography
N^O,N^O-Dimethylthioxamide ($C_4H_8ON_2S$), HL	HL/H.L	10.57 10.37[a] 10.33[b] 10.35[c]			77AHa

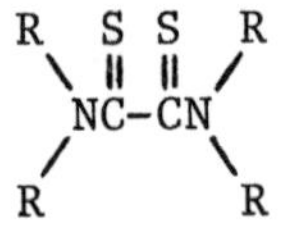

$$\underset{R}{\overset{R}{\diagdown}}N-C\overset{\overset{S}{\parallel}}{{}}\ \ C\overset{\overset{S}{\parallel}}{{}}N\underset{R}{\overset{R}{\diagup}}$$

Ligand	Equilibrium	Log K 25°, 0	ΔH 25°, 0	ΔS 25°, 0	Bibliography
Dithiooxamide ($C_2H_4N_2S_2$), H_2L (Other references on Vol.3, p.366)	HL/H.L H_2L/HL.H	13.11[c] 11.05 10.84[a] 10.78[b] 10.77[c]			77AHa, Other reference: 76AMP
N-Methyldithiooxamide ($C_3H_6N_2S_2$), H_2L	HL/H.L H_2L/HL.H	13.80[c] 11.30 11.09[a] 11.03[b] 11.03[b]			77AHa
N,N'-Dimethyldithiooxamide ($C_4H_8N_2S_2$), H_2L	HL/H.L H_2L/HL.H	14.16[c] 11.67[c]			77AHa, Other reference: 76AMP
N,N-Dimethyldithiooxamide ($C_4H_8N_2S_2$), HL	HL/H.L	10.76 10.56[a] 10.49[b] 10.46[c]			77AHa

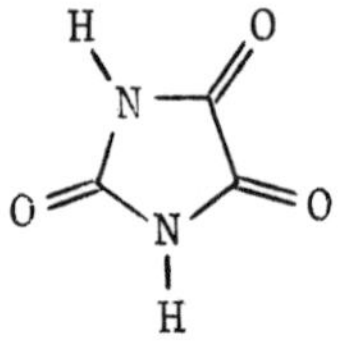

Ligand	Equilibrium	Log K 25°, 0	ΔH 25°, 0	ΔS 25°, 0	Bibliography
1,3-Diazolidine-2,4,5-trione (parabanic acid) ($C_3H_2O_3N_2$) HL	HL/H.L	5.91[b]	-0.07[b]	26.8[b]	77BE
2,4,6-Trioxopyrimidine (barbituric acid) ($C_4H_3O_3N_2$) HL (Other references on Vol.2, p.342)	HL/H.L	4.06 ±0.02 4.06[b] 3.98[c]	-0.06 -0.30[b]	18.4 17.1[b]	77BE

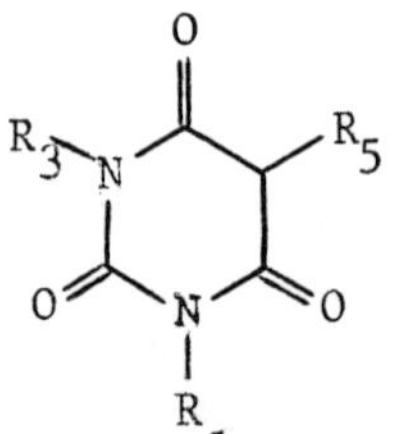

[a] 25°, 0.1; [b] 25°, 0.5; [c] 25°, 1.0

Amides (continued)

Ligand	Equilibrium	Log K 25°, 0	ΔH 25°, 0	ΔS 25°, 0	Bibliography
1,3-Dimethyl-5-nitroso-barbituric acid ($C_6H_7O_4N_3$) HL	HL/H.L	4.60^a			77VN

8. **Amidines**

$$\underset{H_2N\overset{\displaystyle\|}{C}NH_2}{NH}$$

Guanidine (CH_5N_3), L (Other values on Vol.3, p.326)	HL/H.L	$(13.5)^c$	-18^c	1^c	77FM

D. **Inorganic Ligands**

Carbamic acid (H_2NCO_2H), HL	HL/H.L	6.76^b			78CKP
Hydrogen tellurite* (tellurous acid) ($Te(OH)_4$), H_2L (Other value on Vol.4, p.95)	HL/H.L	$9.35^h\pm0.01$ $9.02^i\pm0.02$ 8.91^j			76Ma, Other reference: 76CC
	$H_2L/HL.H$	6.07^h 5.99^i 5.89^j			
Hydrogen silicate (silicic acid), ($Si(OH)_4$), H_2L (Other values on Vol.4, p.39)	HL/H.L	$(13.1)_b$ 12.56^b 12.71^e			77BMb
	$H_2L/HL.H$	9.86 ± 0.05 9.57^a $9.46^b\pm0.00$ 9.47^c 9.43^e			

---

^{a}25°, 0.1; ^{b}25°, 0.5; ^{c}25°, 1.0; ^{h}20°, 0.1; ^{i}20°, 0.5; ^{j}20°, 1.0; ^{e}25°, 3.0; *metal constants were also reported but not included in the compilation of selected constants.

LIGANDS CONSIDERED BUT NOT INCLUDED IN THE COMPILATION OF SELECTED CONSTANTS

A. Aminoacids

1. Aminocarboxylic acids

a. Primary amines

Ligand	Bibliography
2-Amino-2-methylpropanoic acid ($C_4H_9O_2N$) (Values on Vol.1, p.13	77RRa
cis-2-Aminocyclohexanecarboxylic acid ($C_7H_{13}O_2N$)	72KS
trans-2-Aminocyclohexanecarboxylic acid ($C_7H_{13}O_2N$)	72KS
8-Aminooctanoic acid ($C_8H_{17}O_2N$)	74RO
DL-2-Amino-3-sulfopropanoic acid (cysteic acid) ($C_3H_7O_5NS$)	79DZc
DL-2-Aminohexanedioic acid ($C_6H_{11}O_4N$) (H^+ values on Vol.1, p.395)	79FW
DL-2-Aminoheptanedioic acid ($C_7H_{13}O_4N$)	79FW
threo-4-Hydroxy-DL-glutamic acid ($C_5H_9O_5N$)	78KPI
erythro-4-Hydroxy-DL-glutamic acid ($C_5H_9O_5N$)	78KPI
5-Aminoorotic acid ($C_5H_5O_4N_3$)	67TKL
4-Aminobenzoylglycine ($C_9H_{11}O_3N_2$)	77MR
L-Aspartic acid 4-hydrazide ($C_4H_9O_3N_3$)	77MS
L-2-Amino-5-ureidopentanoic acid (citrulline) ($C_6H_{13}O_9N_2$) (Values on Vol.1, p.42)	68TN,79SS
L-2-Amino-4-mercaptobutanoic acid (homocysteine) ($C_4H_9O_2NS$) (H^+ values on Vol.1, p.395)	78BKS
4-Amino-2,2,6,6-tetramethylpiperidine-4-carboxylic acid N-oxide ($C_{10}H_{19}O_3N_2$)	76TC
2,4-Diamino-3-hydroxypentanedioic acid ($C_5H_{10}O_5N_2$)	77ABP
2,3-Bis(2-aminoethylthio)butanedioic acid ($C_8H_{16}O_4N_2S_2$)	78MJ
2,3-Bis(2-amino-2-carboxyethylthio)butanedioic acid ($C_{10}H_{16}O_8N_2S_2$)	78MJ

b. Secondary amines

Ligand	Bibliography
N-Methyl-L-alanine ($C_4H_9O_2N$)	77KDK,78KZa
N-Methyl-L-valine ($C_6H_{13}O_2N$)	77KDK,78KZa
L-Piperidine-2-carboxylic acid (pipecolic acid) ($C_6H_{11}O_2N$) (Values on Vol.1, p.70)	78CU
cis-4-Methylpipecolic acid ($C_7H_{13}O_2N$)	78CU
trans-4-Methylpipecolic acid ($C_7H_{13}O_2N$)	78CU
N-Benzyl-L-alanine ($C_{10}H_{13}O_2N$)	77KDK,78KZa
N-Benzyl-L-valine ($C_{12}H_{17}O_2N$)	77KDK,78KZa
5,5'-Carboxymethylaminomethyl-3,3'-dimethyl-4'-hydroxyfuchson-2"-sulfonic acid (Glycinecresol Red) ($C_{27}H_{28}O_9N_2S$)	73BBC

Aminocarboxylic acids   (continued)

Ligand	Bibliography
5,5'-Carboxymethylaminomethyl-3,3'-bis(1-methylethyl)-6,6-dimethyl-4'-hydroxyfuchson-2"-sulfonic acid (Glycinethymol Blue) $(C_{33}H_{40}O_9N_2S)$	73BBC,75ZL, 76AAA
4-(4-Carboxythiazol-2-ylazo)resorcinol $(C_{27}H_7O_4N_3S)$	71DG
allo-4-Hydroxy-L-proline $(C_5H_9O_3N)$	77KDK,78KZa
DL-N-Benzylalanylglycine $(C_{12}H_{16}O_3N_2)$	67SBG
5-Methylpyrazole-3-carboxylic acid $(C_5H_6O_2N_2)$	79PGS
Ethylenediiminodipropanedioic acid (ethylenediamine-N,N'-dimalonic acid) $(C_8H_{12}O_8N_2)$ (Values on Vol.1, p.105)	73GK,73GKS,73GSK, 73SG,74Ga,74SG
Ethylenediimino-2,2'-bis(3-hydroxypropanoic acid) $(C_8H_{16}O_6N_2)$	75KST
Oxybis(ethyleneimino)diacetic acid $(C_8H_{16}O_5N_2)$	77NF
Tetramethylenebis(2-carboxy-4-piperidine) $(C_{16}H_{28}O_4N_2)$	78CU
Pentamethylenebis(2-carboxy-4-piperidine) $(C_{17}H_{30}O_4N_2)$	78CU

c.   Tertiary amines

N,N-Dimethyl-L-alanine $(C_5H_{11}O_2N)$	77KDK,78KZa
N,N-Dimethyl-L-valine $(C_7H_{15}O_2N)$	77KDK,78KZa
N-Methyl-L-proline $(C_6H_{11}O_2N)$	77KDK,78KZa
Pyrrolidine-N-acetic acid $(C_6H_{11}O_2N)$	74NF
Piperidine-N-acetic acid $(C_7H_{13}O_2N)$ (Values on Vol.1, p.101)	74NF
Hexahydroazepine-N-acetic acid $(C_8H_{15}O_2N)$	74NF
N-Ethylpipecolic acid $(C_8H_{15}O_2N)$	78CU
N-Benzyl-N-methyl-L-alanine $(C_{11}H_{15}O_2N)$	77KDK,78KZa
N-Benzyl-N-methyl-L-valine $(C_{13}H_{19}O_2N)$	77KDK,78KZa
N-Methylpyridylaspartic acid $(C_{10}H_{12}O_4N_2)$	78WNc
N,N-Bis(2-hydroxyethyl)glycine $(C_6H_{13}O_4N)$ (Values on Vol.1, p.105)	62W,68RK,73KA, 74KT,75KK,79JK
allo-N-Benzyl-4-hydroxy-L-proline $(C_{12}H_{15}O_3N)$	77KDK,78KZa
2-Carboxypyrrolidine-N-carbodithioic acid $(C_6H_9O_2NS_2)$	76KNa
1-Ethyl-7-methyl-4-oxo-1,4-diH-1,8-naphthyridine-3-carboxylic acid (nalidixic acid) $(C_{12}H_{12}O_3N_2)$	78TS
1-Ethyl-4-oxo-6,7-bis(oxymethylene)quinoline-3-carboxylic acid (oxolinic acid) $(C_{13}H_{11}O_5N)$	78TS
Folic acid $(C_{17}H_{18}O_6N_6)$	75FL
Ethylenebis(2-carboxy-N-piperidine) $(C_{14}H_{24}O_4N_2)$	78CU
Trimethylenebis(2-carboxy-N-piperidine) $(C_{15}H_{26}O_4N_2)$	78CU
Ethylenedinitrilo-N,N'-bis(ethylenesulfonic)-N-methylenephosphonic N'-acetic acid $(C_{12}H_{24}O_{11}N_2PS_2)$	71SE
Ethylenedinitrilo-N,N'-bis(ethylenesulfonic)-N,N'-diacetic acid $(C_{13}H_{23}O_{10}N_2S_2)$	69TE,71ES, 71SE
Ethylenedinitrilo-N,N'-bis(ethylenesulfonic)-N,N'-di-3-propanoic acid $(C_{15}H_{27}O_{10}N_2S_2)$	71ES

Aminocarboxylic acids (continued)

Ligand	Bibliography
Ethylenedinitrilo-N,N'-diacetic-N,N'-bis(methylenephosphonic) acid ($C_8H_{18}O_{10}N_2P_2$) (H^+ values on Vol.1, p.396)	79ZKT
Ethylenedinitrilo-N,N'-bis(3-carboxy-2-hydroxy-1-naphthylmethyl)-N,N'-diacetic acid ($C_{30}H_{28}O_{10}N_2$)	75TRY

2. Iminodiacetic acid derivatives

N-Ethyliminodiacetic acid ($C_6H_{11}O_4N$)	76JP
N-Propyliminodiacetic acid ($C_7H_{13}O_4N$) (Values on Vol.1, p.127)	76JP
N-Butyliminodiacetic acid ($C_8H_{15}O_4N$)	76JP
N-Pentyliminodiacetic acid ($C_9H_{17}O_4N$)	76JP
N-(2-Propyl)iminodiacetic acid ($C_7H_{13}O_4N$)	76JP
N-(2-Methylpropyl)iminodiacetic acid ($C_8H_{15}O_4N$)	76JP
N-(1,1-Dimethylethyl)iminodiacetic acid ($C_8H_{15}O_4N$)	76JP
N-(3-Methylbutyl)iminodiacetic acid ($C_9H_{17}O_4N$)	76JP
N-(3,3-Dimethylbutyl)iminodiacetic acid ($C_{10}H_{19}O_4N$) (Values on Vol.1, p.128) Hg^{2+} not corrected for Cl^-	
2-Carboxypyrrolidine-N-acetic acid ($C_7H_{11}O_4N$)	74M
N-(Cyanomethyl)iminodiacetic acid ($C_6H_8O_4N_2$) (Values on Vol.1, p.134) Hg^{2+} not corrected for Cl^-	
Hydrazine-N,N-diacetic acid ($C_4H_8O_4N_2$) (Values on Vol.1, p.131)	72K,76BT,77TM, 78BBI,78NB
N-(3-Carboxy-2-hydroxy-1-naphthylmethyl)iminodiacetic acid ($C_{16}H_{15}O_7N$)	75TRY
3-Bis(carboxymethyl)aminomethyl-1,2-dihydroxyanthraquinone (Alizarin Complexone) ($C_{19}H_{15}O_8N$)	75PTa
8-Bis(carboxymethyl)aminomethyl-7-hydroxy-4-methyl-1,2-benzopyranone (Calcein Blue) ($C_{10}H_{15}O_7N$)	74HP
N-(2-Hydroxypropyl)iminodiacetic acid ($C_7H_{13}O_5N$)	76JP
N-(2-Hydroxy-2-methylpropyl)iminodiacetic acid ($C_8H_{15}O_5N$)	76JP
N-(2-Hydroxy-1-methylpropyl)iminodiacetic acid ($C_8H_{15}O_5N$)	76JP
N-(2-Hydroxy-1,1-dimethylpropyl)iminodiacetic acid ($C_9H_{17}O_5N$)	76JP
N-(3-Hydroxypropyl)iminodiacetic acid ($C_7H_{13}O_5N$) (Values on Vol.1, p.167)	76JP
N-(4-Hydroxybutyl)iminodiacetic acid ($C_8H_{15}O_5N$)	76JP
N-(5-Hydroxypentyl)iminodiacetic acid ($C_9H_{17}O_5N$)	76JP
N-(2-Methoxyethyl)iminodiacetic acid ($C_7H_{13}O_5N$) (Values on Vol.1, p.170) Hg^{2+} not corrected for Cl^-	
N-(Carbamylmethyl)iminodiacetic acid ($C_6H_{10}O_5N_2$) (Values on Vol.1, p. 178) Hg^{2+} not corrected for Cl^-	
N,N-Bis(carboxymethyl)glycylglycine ($C_8H_{12}O_7N_2$) (Values on Vol.1, p. 179) same as 73MM.	74MM
N,N-Bis(carboxymethyl)glycylglycylglycine ($C_{10}H_{15}O_8N_3$) (Values on Vol.1, p.180) same as 73MM	74MM

Iminodiacetic acid derivatives   (continued)

<u>Ligand</u>	<u>Bibliography</u>
N,N-Bis(carboxymethyl)glycylglycylglycylglycine ($C_{12}H_{18}O_9N_4$) (Values on Vol.1, p.181) same as 73MM	74MM
N-[2-(Ethoxycarbamyl)ethyl]iminodiacetic acid ($C_9H_{16}O_6N_2$) (Values on Vol.1, p.186) Hg^{2+} not corrected for Cl^-	
N-(2-Mercaptoethyl)iminodiacetic acid ($C_6H_{11}O_4NS$) (Values on Vol.1, p. 188) Hg^{2+} not corrected for Cl^-	
N-[2-(Methylthio)ethyl]iminodiacetic acid ($C_7H_{13}O_4NS$) (Values on Vol.1, p.190) Hg^{2+} not corrected for Cl^-	
N-[Bis(carboxymethyl)amino]ethyltrimethylammonium(perchlorate) ($C_9H_{19}O_4N_2^+$) (Values on Vol.1, p.196) Hg^{2+} not corrected for Cl^-	
2-Methyl-6-bis(carboxymethyl)aminomethyl-4-(4-oxo-3,5-dibromophenyl-imino)phenol (Indoferron) ($C_{18}H_{16}O_6N_2Br_2$)	69SO
N-(2-Hydroxyethyl)diethylenetrinitrilo-N,N'N'',N''-tetraacetic acid ($C_{13}H_{25}O_9N_3$) (Other reference on Vol.1, p.409)	74YKP,76NGT,77TR
Oxomethylenebis[(4-hydroxy-5-methoxy-1,3-phenylene)methylnitrilo]-tetraacetic acid ($C_{24}H_{28}O_{13}N_2$)	73VI
5,5'-Bis[bis(carboxymethyl)aminomethyl]-3,3',6,6'-tetramethyl-4'-hydroxyfuchson-2''-sulfonic acid (Xylenol phthalexone S) ($C_{33}H_{36}O_{13}N_2S$)	75KKP
5,5''-Bis[bis(carboxymethyl)aminomethyl]-3,3'-dibromo-4'-hydroxy-fuchson-2''-sulfonic acid (o-Bromophthalexone S) ($C_{29}H_{26}O_{13}N_2Br_2S$)	71C
DL-(2,3-Dihydroxytetramethylene)dinitrilotetraacetic acid ($C_{12}H_{20}O_{10}N_2$) (Values on Vol.1, p.262)	68EM
Oxybis(ethylenenitrilo)tetraacetic acid (EEDTA) ($C_{12}H_{20}O_9N_2$) (Values on Vol.1, p.263)	73BW
3,3',5,5'-Tetrakis[bis(carboxymethyl)aminomethyl]-6,6'-dimethyl-4'-hydroxyfuchson-2''-sulfonic acid (m-Cresolphthalexone SA) ($C_{41}H_{47}O_{21}N_4S$)	71C

3. Peptides

Glycyl-L-asparagine ($C_6H_{11}O_4N_3$) (H^+ values on Vol.1, p.401)	77BS
DL-Alanyl-DL-alanine ($C_6H_{12}O_3N_2$) (Other references on Vol.1, p.410)	76PN,77GN
DL-Alanyl-DL-valine ($C_8H_{16}O_3N_2$) (Other references on Vol.1, p.410)	76PN
DL-Alanyl-DL-leucine ($C_9H_{18}O_3N_2$)	76PN
L-Alanyl-L-leucine ($C_9H_{18}O_3N_2$)	74KH
L-Alanyl-L-proline ($C_8H_{14}O_3N_2$)	74KH
L-Alanyl-L-arginine ($C_9H_{19}O_3N_5$)	74WK
D-Alanyl-L-arginine ($C_9H_{19}O_3N_5$)	74WK
L-Leucyl-L-alanine ($C_9H_{18}O_3N_2$)	67KKa
L-Leucyl-D-alanine ($C_9H_{18}O_3N_2$)	67KKa
L-Arginyl-L-alanine ($C_9H_{19}O_3N_5$)	74WK

Peptides (continued)

Ligand	Bibliography
L-Arginyl-D-alanine $(C_9H_{19}O_3N_5)$	74WK
DL-Valyl-DL-leucine $(C_{10}H_{20}O_3N_2)$ (Other references on Vol.1, p.410)	76PN
DL-Leucyl-DL-leucine $(C_{11}H_{22}O_3N_2)$	74KH
L-Leucyl-L-phenylalanine $(C_{15}H_{22}O_3N_2)$ also L-Leucyl-D-phenylalanine	67KK
DL-Histidyl-DL-histidine $(C_{12}H_{16}O_3N_6)$ (Other references on Vol.1, p. 410)	76AP
Glycylglycyl-L-proline $(C_9H_{15}O_4N_3)$	74KH
L-Histidylglycylglycine $(C_{10}H_{15}O_4N_5)$	74AY,74YA

4. Anilinecarboxylic acids

3-Aminobenzoic acid $(C_7H_7O_2N)$ (H^+ values on Vol.1, p.405)	74FLa,76KVP
4-Aminobenzoic acid $(C_7H_7O_2N)$ (H^+ values on Vol.1, p.405)	75KKK,77ANY,77EB
2-(Benzoylacetonylamino)benzoic acid $(C_{17}H_{14}O_4N)$	76KMM
N-Phenyliminodiacetic acid $(C_{10}H_{11}O_4N)$ (Values on Vol.1, p.351)	75Kg,75KP
1-Bis(carboxymethyl)aminomethyl-2-hydroxy-3-naphthoic acid $(C_{16}H_{15}O_7N)$	68BW
2-Bis(carboxymethyl)aminomethyl-1,8-dihydroxynaphthalene-3,6-disulfonic acid $(C_{15}H_{15}O_{12}NS_2)$	70EM
1,3-Phenylenedinitrilotetraacetic acid $(C_{14}H_{16}O_8N_2)$ (Values on Vol.1, p.362)	73MR
1,4-Phenylenedinitrilotetraacetic acid $(C_{14}H_{16}O_8N_2)$ (Values on Vol.1, p.364)	73MR

5. Pyridinecarboxylic acids

Pyridine-3-carboxylic acid (nicotinic acid) $(C_6H_5O_2N)$ (Values on Vol.1, p.374)	76BC,79JAa
Pyridine-4-carboxylic acid (isonicotinic acid) $(C_6H_5O_2N)$ (H^+ values on Vol.1, p.407)	76BC
3-Hydroxy-5-hydroxymethyl-2-methylpyridine-4-carboxylic acid (4-pyridoxylic acid) $(C_8H_9O_4N)$	79FL
3-Hydroxy-2-methylpyridine-4,5-dicarboxylic acid (2-methyl-3-hydroxycinchomeronic acid) $(C_8H_7O_5N)$	79FL
Pyridine-2,3-dicarboxylic acid (quinolinic acid) $(C_7H_5O_4N)$ (Values on Vol.1, p.375)	73CB,78BPG,78SJ, 79BPG,79S,79SJa, 79SJb
Pyridine-2,4-dicarboxylic acid (lutidinic acid) $(C_7H_5O_4N)$ (Values on Vol.1, p.375)	70OM
Pyridine-2,5-dicarboxylic acid (isocinchomeronic acid) $(C_7H_5O_4N)$ (Values on Vol.1, p.376)	70OM
Pyridine-2,4-dicarboxylic acid 4-methylester $(C_8H_7O_4N)$	70OM
Pyridine-2,5-dicarboxylic acid 5-methylester $(C_8H_7O_4N)$	70OM

B. Amines

1. Primary amines

Propylamine (C_3H_9N) (Values on Vol.2, p.3)	69PM

Primary amines   (continued)

Ligand	Bibliography

3-Methylaniline (m-toluidine) (C_7H_9N) (H^+ Values on Vol.2, p.331)   72VG

4-Methylaniline (p-toluidine) (C_7H_9N) (Values on Vol.2, p.9)   72VG

3-Bromoaniline (C_6H_6NBr)   72VG

3-Nitroaniline ($C_6H_6O_2N_2$)   72VG

4-Nitroaniline ($C_6H_6O_2N_3$)   72VG

2-Aminoethanesulfonic acid (taurine) ($C_2H_7O_3NS$) (Values on Vol.2,   70FMa,73FA,74FA,
 p.10)   74FAa,74FAb,76KFA

3-Aminobenzenethiosulfonic acid ($C_6H_7O_2NS_2$)   76GSc

3-Amino-4-hydroxybenzenesulfonic acid ($C_6H_7O_4NS$)   73TSc

L-2-Amino-1-(3,4-dihydroxyphenyl)ethanol (noradrenaline) ($C_8H_{11}O_3N$) 74AM
  (H^+ values on Vol.2, p.332)

DL-2-Phenyl-2-propylamine (amphetamine) ($C_9H_{13}N$)   72RD

DL-2-(4-Hydroxyphenyl)-2-propylamine (4-hydroxyamphetamine) ($C_9H_{13}ON$) 76RD

2-(3,4,6-Trihydroxyphenyl)ethylamine (6-hydroxydopamine) ($C_8H_{11}O_3N$) 72RD

DL-2-Amino-1-(4-hydroxyphenyl)ethanol (octopamine) ($C_8H_{11}O_3N$)   72RD,76RD

DL-2-Amino-1-phenylpropanol (norephedrine) ($C_9H_{13}ON$)   72RD,76RD

DL-2-Amino-1-(4-hydroxyphenyl)propanol(4-hydropynorephedrine)   76RD
  ($C_9H_{13}O_2N$)

DL-2-Amino-1-(4-hydroxy-3-methoxyphenyl)propanol(normetanephrine)   76RD
  ($C_{10}H_{15}O_3N$)

DL-2-Amino-1-(3-methoxyphenyl)propanol(3-methoxynorephedrine)   72RD
  ($C_{10}H_{15}O_2N$)

8-Amino-1-naphthol ($C_{10}H_9ON$)   73VJ

8-Amino-1-hydroxynaphthalene-2,5-disulfonic acid ($C_{10}H_9O_7NS_2$)   73VJ

8-Amino-1-hydroxynaphthalene-3,5-disulfonic acid ($C_{10}H_9O_7NS_2$)   73VJ

8-Amino-1-hydroxynaphthalene-3,6-disulfonic acid ($C_{10}H_9O_7NS_2$)   73VJ

1-Amino-4-hydroxyanthraquinone ($C_{14}H_9O_3N$)   72JA

2-Aminoethanol (ethanolamine) (C_2H_7ON) (Values on Vol.2, p.15)   66DP,74MK,74MKa,
                                                                            77GMT,79AP,79TM,
                                                                            79TMa

DL-1-Amino-2-propanol (C_3H_9ON) (Values on Vol.2, p.16)   76RR

2-Methoxyaniline (o-anisidine) (C_7H_9ON) (Values on Vol.2, p.22)   72VG

2-Aminoethylmethacrylate ($C_6H_{11}O_2N$)   70TD

3-Aminopropylmethacrylate ($C_7H_{13}O_2N$)   70TD

4-Aminobenzenesulfacetamide ($C_8H_{10}O_2N_2S$)   76BMc

1-$\frac{1}{-}$(4-Aminophenylsulfonyl)guanidine (sulfaguanidine) ($C_7H_{10}O_2N_4S$)   76CJK

3-Aminophthalhydrazide (luminol) ($C_8H_7O_2N_3$)   72L

2-Aminopropanethiol (C_3H_9NS)   78BKS

3-Aminopropanethiol (C_3H_9NS)   78BKS

4-Aminobutanethiol ($C_4H_{11}NS$)   78BKS

Primary amines (continued)

Ligand	Bibliography
2-Amino-1,1-dimethylethanethiol ($C_4H_{11}NS$)	78BKS
Penicillamine methylester ($C_6H_{13}O_2NS$)	70SY
DL-(trans)-Cyclohexane-1,2-diamine ($C_6H_{14}N_2$) (Values on Vol.2, p.46)	76M
2,2-Dimethyltrimethylenediamine ($C_5H_{14}N_2$) (Values on Vol.2, p.53)	72NB
2,2-Diethyltrimethylenediamine ($C_7H_{18}N_2$)	72NB
1,1-Bis(aminomethyl)cyclopropane ($C_5H_{12}N_2$)	72NB
1,1-Bis(aminomethyl)cyclobutane ($C_6H_{14}N_2$)	72NB
1,1-Bis(aminomethyl)cyclopentane ($C_7H_{16}N_2$)	72NB
1,1,-Bis(aminomethyl)cyclohexane ($C_8H_{18}N_2$)	72NB
Pentamethylenediamine ($C_5H_{14}N_2$) (Values on Vol.2, p.55)	77HS
Hexamethylenediamine ($C_6H_{16}N_2$) (H^+ values on Vol.2, p.334)	76BMc,77HS
Dodecamethylenediamine ($C_{12}H_{26}N_2$)	77HS
1,3-Diaminobenzene ($C_6H_8N_2$)	79AM
1,4-Diaminobenzene ($C_6H_8N_2$)	79AM
2-Aminomethyl-2-methyltrimethylenediamine ($C_5H_{15}N_3$)	70KAT

2. Secondary amines

Methylhydrazine (CH_6N_2)	70BG
Ethylhydrazine ($C_2H_8N_2$)	70BG
DL-Adrenaline (epinephrine) ($C_9H_{13}O_3N$)	72RD
DL-Iminodi-1-(2-propanol) (bis(2-hydroxypropyl)amine) ($C_6H_{15}O_2N$) (Values on Vol.2, p.81)	76RR
DL-2(Methylamino)-1-phenylpropanol (DL-ephedrine) ($C_{10}H_{15}ON$) (H^+ values on Vol.2, p.336)	71RD,72RD
2-(Methylamino)ethanethiol (C_3H_9NS)	78BKS
Streptomycin ($C_{21}H_{39}O_{12}N_7$)	75AV,75AVa
2-(Methylamino)ethylmethacrylate ($C_7H_{13}O_2N$)	70TD
2-(Ethylamino)ethylmethacrylate ($C_8H_{15}O_2N$)	70TD
2-(1-Methylethylamino)ethylmethacrylate ($C_9H_{17}O_2N$)	70TD
2-(Butylamino)ethylmethacrylate ($C_{10}H_{19}O_2N$)	70TD
2-(1-Methylpropylamino)ethylmethacrylate ($C_{10}H_{19}O_2N$)	70TD
2-(2-Methylpropylamino)ethylmethacrylate ($C_{10}H_{19}O_2N$)	70TD
Ethylenediiminodi-2-ethanol (N,N'-bis(2-hydroxyethyl)ethylenediamine) ($C_6H_{16}O_2N_2$) (Values on Vol.2, p.98)	76H
3,7-Diazanonanedioic acid diamide ($C_7H_{16}O_2N_2$) (Other reference on Vol.2, p.348)	74ZK,79ZK
3,7-Diazanonanedioic acid bis(ethylamide) ($C_9H_{20}O_2N_4$) (Other reference on Vol.2, p.348)	74ZK

Secondary amines   (continued)

Ligand	Bibliography
1,5,8,12-Tetraazadodecane (3,2,3-tet) ($C_8H_{22}N_4$) (Values on Vol.2, P. 108)	74SM
1,5,9,13-Tetraazatridecane (3,3,3-tet) ($C_9H_{24}N_4$) (Values on Vol.2, p.109)	76NG

3. Tertiary amines

6-Nitroso-3-(dimethylamino)phenol ($C_8H_{11}O_2N_2$)	71Md
Nitrilotri-2-ethanol (triethanolamine) ($C_6H_{15}O_3N$) (Values on Vol.2, p.118)	68DP,74KS,74UP,75Kb, 74MS,79VKP
N-2-Aminoethyl-N-(2-hydroxyethyl)dithiocarbamic acid ($C_5H_{12}ON_2S_2$)	78SH
10-[3-(Dimethylamino)propyl]-2-(trifluromethyl)phenothiazine (trifluopromazine) ($C_{18}H_{19}N_2F_3S$)	76GTa
2-(Dimethylamino)ethanethiol ($C_4H_{11}NS$) (Other reference on Vol.2, p.348)	72MP,78BKS
2-(Di-2-propylamino)ethanethiol ($C_8H_{19}NS$)	73SCb
2-(Diethylamino)ethanethiol ($C_6H_{15}NS$)	78BKS
2-(Dimethylamino)ethylmethacrylate ($C_8H_{15}O_2N$)	70TD
2-(Diethylamino)ethylmethacrylate ($C_{10}H_{19}O_2N$)	70TD
2-(Butylmethylamino)ethylmethacrylate ($C_{11}H_{21}O_2N$)	70TD
2-[Bis(1-methylethyl)amino]ethylmethacrylate ($C_{12}H_{23}O_2N$)	70TD
2-(Allylmethylamino)ethylmethacrylate ($C_9H_{15}O_2N$)	70TD
2-(Diallylamino)ethylmethacrylate ($C_{10}H_{15}O_2N$)	70TD
Tetracycline ($C_{22}H_{24}O_8N_2$)	75SL
10-[2-(Dimethylamino)propyl]phenothiazine (promethazine) ($C_{17}H_{20}N_2S$)	76GKc
2-Thioxoethyl(dimethyl)ammonium ion ($C_4H_{11}NS^+$)	78BKS

4. Cyclic amines

1-Oxa-4,13-dithia-5,7-diazacyclopentadecane ($C_{10}H_{22}ON_2S_2$)	75ASa

5. Azoles

a. Pyrroles

Tetra(N-methyl-2-pyridyl)porphine ($C_{46}H_{38}N_8$)	77H
Tetraphenylporphyrinsulfonate ($C_{44}H_{30}O_{12}N_4S_4$)	79MBe
Hematoporphrin ($C_{34}H_{38}O_6N_4$)	79MBe

b. Oxazoles

5-Methylisoxazole (C_4H_5ON)	77LKa

c. Thiazoles

4-Chloro-2-(2-thiazolylazo)phenol ($C_9H_6ON_3ClS$)	74KSa

Thiazoles   (continued)

Ligand	Bibliography
4-Methoxy-2-(2-thiazolylazo)phenol ($C_{10}H_9O_2N_3S$)	74KSa
4-(2-Thiazolylazo)resorcinol (TAR) ($C_9H_7O_2N_3S$) (Other references on Vol.2, p.349)	69BIN,69HS,69IBP, 69LS,69MS,69ST, 76BSB
4-(2-Thiazolylazo)-3-methylphenol ($C_{10}H_9O_4N_4S$)	76BSB
2-Nitro-4-(2-thiazolylazo)resorcinol ($C_9H_6O_6N_6S$)	71AG
4-(5-Bromo-2-thiazolylazo)resorcinol ($C_9H_6O_2N_3BrS$)	69BN
4-(5-Sulfo-2-thiazolylazo)resorcinol ($C_9H_7O_5N_3S_2$)	71AG
1-(4-Methyl-2-thiazolylazo)-2-naphthol ($C_{14}H_{11}ON_3S$)	76BSB
1-(2-Thiazolylazo)-2-hydroxynaphalene-3,6-disulfonic acid ($C_{13}H_9O_7N_3S_3$)	75LK
2-(2-Thiazolylazo)chromotropic acid ($C_{13}H_9O_8N_3S_3$)	75RS
4-(2-Benzothiazolylazo)resorcinol ($C_{13}H_9O_2N_3S$)	69IBN
4-(6-Bromo-2-benzothiazolylazo)resorcinol ($C_{13}H_8O_2N_3BrS$)	69IBN
2-Aminothiazole ($C_3H_4N_2S$)	78BBa

d. 1,2-Diazoles

Ligand	Bibliography
4-Acetyl-3-methyl-1-phenylpyrazol-5-one ($C_{12}H_{11}O_2N_2$)	73BK
4-Trichloroacetyl-3-methyl-1-phenylpyrazol-5-one ($C_{12}H_8O_2N_2Cl_3$)	73BK
4-Trifluoroacetyl-3-methyl-1-phenylpyrazol-5-one ($C_{12}H_8O_2N_2F_3$)	73BK
4-Benzoyl-3-methyl-1-phenylpyrazol-5-one ($C_{17}H_{14}O_2N_2$) (Other references on Vol.3, p.366)	69BFa,71OM,72NI, 73BK,73NM,78MMd, 78NMa
Bis(2,3-dimethyl-5-oxo-1-phenylpyrazol-4-yl)methane (diantipyrinylmethane) ($C_{23}H_{24}O_2N_4$)	77PMG
Diantipyrinylpropylmethane ($C_{26}H_{30}O_2N_4$)	79SPa
2,5-Dimethyl-1,3,4-oxadiazole ($C_4H_6ON_2$)	77LG

e. 1,3-Diazoles

Ligand	Bibliography
1-Vinyl-2-(hydroxymethyl)imidazole ($C_6H_8ON_2$)	79DB

f. Triazoles

Ligand	Bibliography
1,2,4-Triazoline-3-thione ($C_2H_3N_3S$)	71RC,72RP,73RR
5-Chlorobenzotriazole ($C_6H_4N_3Cl$)	76CC

g. Tetrazoles

Ligand	Bibliography
1,5-Cyclotrimethylenetetrazole ($C_4H_6N_4$)	69DP
1,5-Cyclotetramethylenetetrazole ($C_5H_8N_4$)	69DP
1,5-Cyclopentamethylenetetrazole ($C_6H_{10}N_4$)	69DP
1,5-Cyclohexamethylenetetrazole ($C_7H_{12}N_4$)	69DP
1,5-Cycloheptamethylenetetrazole ($C_8H_{14}N_4$)	69DP

Tetrazoles (continued)

Ligand	Bibliography
1-(5-Tetrazolylazo)-2-hydroxynaphthalene-3,6-disulfonic acid ($C_{11}H_9O_7N_6S_2$)	77SR
Bis(5-tetrazolylazo)acetic acid ethylester ($C_6H_8O_2N_{12}$)	69FG
Oxybis(ethylene-5-tetrazole) ($C_6H_{10}ON_8$)	79ES
Thiobis(ethylene-5-tetrazole) ($C_6H_{10}N_8S$)	79ES
Iminobis(ethylene-5-tetrazole) ($C_6H_{10}N_9$)	79ES

6. Azines

a. Pyridines

Ligand	Bibliography
3,5-Dimethylpyridine (3,5-lutidine) (C_7H_9N) (Values on Vol.2, p.174)	78LR
2,4,6-Trimethylpyridine (2,4,6-collidine) ($C_8H_{11}N$) (Other references on Vol.2, p.350)	69RB
2-Hydroxypyridine (C_5H_5ON) (H^+ values on Vol.2, p.340)	69PKD
2,3-Dihydroxypyridine ($C_5H_5O_2N$) (Values on vol.2, p.179)	76SPb,77SP
2-Amino-3-hydroxypyridine ($C_5H_6ON_2$)	77SP
1-(2-Pyridylazo)-2-naphthol (PAN) ($C_{15}H_{11}ON_3$) (Other references on Vol.2, p.350)	69BIB,71BR,72BE, 72BN,74PM,76BSB, 79KH
2-(2-Pyridylazo)chromotropic acid ($C_{15}H_{11}O_8N_3S_2$)	74CB
2-[4-(N-Pyrid-2-ylamidosulfonyl)phenylazo[chromotropic acid ($C_{22}H_{17}O_{11}N_4S_3$)	76BBe
1-(2,3-Dihydroxy-4-pyridylazo)benzene-4-sulfonic acid ($C_{11}H_9O_5N_3S$)	77GGS
1-(5-Chloro-2,3-Dihydroxy-4-pyridylazo)benzene-4-sulfonic acid ($C_{11}H_8O_5N_3ClS$)	79GGS
5-Bromo-4-(2-pyridylazo)resorcinol ($C_{11}H_8O_2N_3Br$)	69BN
Pyridoxine ($C_8H_{11}O_3N$) (Values on Vol.2, p.180)	76EEa,77EA,78ZA, 79FL
Pyridoxal ($C_8H_9O_3N$) (Values on Vol.2, p.181)	75EG,76EEa,79FL
Pyridoxamine ($C_8H_{12}O_2N_2$) (Values on Vol.2, p.183)	76EEa,77ERM,78AE, 79EM
6-Methylpyridine-2-carboxaldehyde oxime ($C_7H_8ON_2$)	70DF,70KMB
Pyridine-3-hydroxamic acid (nicotinohydroxamic acid) ($C_6H_5O_2N_2$)	69DS,70Sc
2-(2-Hydroxyethyl)pyridine (C_7H_9ON)	78HH
2-(1,2-Dihydroxyethyl)pyridine ($C_7H_9O_2N$)	78HH
2-(Methoxymethyl)pyridine (C_7H_9ON)	78HH
2-Pyridyloxirane (C_7H_7ON)	78HH
3,3-Diethyl-2,4-dioxotetrahydropyridine ($C_9H_{11}O_2N$)	76BMc
Pyridine-4-carboxylic acid hydrazide (isonicotinic hydrazide) ($C_6H_7ON_3$)	75PSb
2-Mercaptopyridine 1-oxide (C_5H_5ONS)	73EM

Azines (continued)

Ligand	Bibliography
2-(Aminomethyl)pyridine ($C_6H_8N_2$) (Values on Vol.2, p.209)	79SSd
2-[4-(Dimethylamino)phenylazo]pyridine ($C_{13}H_{14}N_4$)	72CG

b. Benzo[b]pyridines

2-Hydroxyquinoline (C_9H_7ON)	72L
8-Hydroxyquinoline-7-sulfonic acid ($C_9H_7O_4NS$) (Other references on Vol.2, p.351)	64CL,68BBa,69BBb, 70BBb,70DB
7-Chloro-8-hydroxyquinoline-5-sulfonic acid ($C_9H_6O_4NClS$) (Other reference on Vol.2, p.351)	64CL,71AB
5-Chloro-8-hydroxyquinoline-7-sulfonic acid ($C_9H_6O_4NClS$)	64CL,68BB,70BB, 70BBa
7-Bromo-8-hydroxyquinoline-5-sulfonic acid ($C_9H_6O_4NBrS$) (Other references on Vol.2, p.351)	64CL,70AB,70ABa, 78KJN
5-Bromo-8-hydroxyquinoline-7-sulfonic acid ($C_9H_6O_4NBrS$)	64CL,69BB,70BB
5-Iodo-8-hydroxyquinoline-7-sulfonic acid ($C_9H_6O_4NIS$)	64CL,69BBa,70BB
7-Nitro-8-hydroxyquinoline-5-sulfonic acid ($C_9H_6O_6N_2S$) (Values on Vol.2, p.231)	72PB,72PS,73PS, 77MO
5-Nitro-8-hydroxyquinoline-7-sulfonic acid ($C_9H_6O_6N_2S$)	77MO
7-Nitroso-8-hydroxyquinoline-5-sulfonic acid ($C_9H_6O_5N_2S$)	77MO
5-Nitroso-8-hydroxyquinoline-7-sulfonic acid ($C_9H_6O_5N_2S$)	77MO
7-Phenylazo-8-hydroxyquinoline-5-sulfonic acid ($C_{15}H_{11}O_4N_3S$)	69GT
7-(4-Sulfophenylazo-8-hydroxyquinoline-5-sulfonic acid ($C_{15}H_{11}O_7N_3S_2$)	69GT
3-(8-Hydroxy-7-quinolylazo)naphthalene-1,5-disulfonic acid ($C_{19}H_{13}O_7N_3S_2$)	76PG
8-(2,4-Dihydroxyphenylazo)quinoline ($C_{15}H_{11}O_2N_3$)	69IBN
8-Mercaptoquinoline-5-sulfonic acid ($C_9H_7O_3NS_2$)	76ABb
8-Mercapto-2-methylquinoline-5-sulfonic acid ($C_{10}H_9O_3NS_2$)	76ABD
2,2'-Diquinolyl ($C_{18}H_{12}N_2$)	71GG
Benzo[c]pyridine (isoquinoline) (C_9H_7N) (Values on Vol.2, p.222)	76CK
6,7-Dimethyl-9-D-1'-ribitylbenzo[b]-1,4-diazino[2,3-d]-1,3-diazin-2,4-dione (riboflavin) ($C_{17}H_{20}O_6N_4$)	61BH,64SM,71GB, 73TM,78FD

c. Dipyridines

Iminodi-2-pyridine (di-2-pyridylamine) ($C_{10}H_9N_3$)	76BBc

d. 1,10-Phenanthrolines

4,7-Dimethyl-1,10-phenanthroline ($C_{14}H_{12}N_2$) (Values on Vol.2, p.261)	68WF
5-Chloro-1,10-phenanthroline ($C_{12}H_7N_2Cl$) (Values on Vol.2, p.255)	68AJ
5-Nitro-1,10-phenanthroline ($C_{12}H_7O_2N_3$) (Values on Vol.2, p.257)	68AJ
4,7-Dihydroxy-1,10-phenanthroline ($C_{12}H_{10}O_2N_2$)	76PD

Azines (continued)

Ligand	Bibliography
e. 1,3-Diazines	
1,3-Diazine (pyrimidine) ($C_4H_4N_2$) (Values on Vol.2, p.264)	76BMc
4-Amino-2-oxopyrimidine (cytosine) ($C_4H_5ON_3$) (Values on Vol.2, p.264)	77TJ,78TJ
Cytidine-5'-(dihydrogenphosphate) (CMP-5) ($C_9H_{14}O_8N_3P$) (Values on Vol.2, p.266)	66RRM,79TP
Cytidine-5'-(trihydrogendiphosphate) (CDP) ($C_9H_{15}O_{11}N_3P_2$) (Values on Vol.2, p.267)	78DMY,78FS
2,4-Dioxopyrimidine (uracil) ($C_4H_4O_2N_2$) (H^+ values on Vol.2, p.342)	61BH,75DW,78SSb
5-Fluorouracil ($C_4H_3O_2N_2F$)	70GKa,70GKb
1-(β-D-Ribofuranosyl)uracil (uridine) ($C_9H_{12}O_6N_2$) (H^+ values on Vol.2, p.342)	70GBa
Uridine-5'-(trihydrogendiphosphate) (UDP) ($C_9H_{14}O_{12}N_2P_2$) (Values on Vol.2, p.269)	78DMY
5-Methyl-2,4-dioxopyrimidine (thymine) ($C_5H_6O_2N_2$) (Values on Vol.2, p.271)	78SSb
Thymidine-5'-(dihydrogenphosphate) (TMP) ($C_{10}H_{16}O_8N_2P$)	68RRM,79TP
4-Methyl-5-methoxyuracil ($C_6H_8O_3N_2$)	76BMc
4-Methyl-5-methoxy-2,6-dioxopyrimidine ($C_6H_6O_3N_2$)	76BMc
Orotic acid methyl ester ($C_6H_6O_4N_2$)	79DZa
5-Ethyl-5-(1-methylbutyl)barbituric acid ($C_{11}H_{18}O_3N_2$)	76BMc
5-Ethyl-5-(3-methylbutyl)barbituric acid ($C_{11}H_{18}O_3N_2$)	76BMc
2,4,5,6-Tetraoxopyrimidine (alloxane) ($C_4H_2O_4N_2$)	76BMc
2-[4-(4,6-Dimethylpyrimidin-2-yl)amidosulfanyl)phenylazo]chromotropic acid ($C_{23}H_{20}O_{11}N_5S_3$)	76BBe
Thiaminephosphate ($C_{12}H_{18}O_4N_4ClPS$)	78TA
Thiaminepyrophosphate ($C_{12}H_{19}O_7N_4ClP_2S$)	78TA
f. Purines	
6-Aminopurine (adenine) ($C_5H_5N_5$) (Values on Vol.2, p.276)	77TJ
1-Methyladenine ($C_6H_7N_5$)	79HMa
9-(β-D-Ribofuranosyl)adenine (adenosine) ($C_{10}H_{13}O_4N_5$) (Values on Vol.2, p.277)	70PM,72MP
Adenosine-3',5'-cyclic(hydrogenphosphate) (cyclic-AMP) ($C_{10}H_{11}O_6N_5P$)	77FS
Adenosine-3'-(dihydrogenphosphate) (AMP-3) ($C_{10}H_{14}O_7N_5P$) (Values on Vol.2, p.279)	76TD
Guanosine-5'-(dihydrogenphosphate) (GMP) ($C_{10}H_{13}O_8N_5P$) (Values on Vol.2, p.289)	68RRM,79TP
Guanosine-5'-(trihydrogendiphosphate) (GDP) ($C_{10}H_{14}O_{11}N_5P_2$) (Values on Vol.2, p.289)	78DMY
1,3-Dimethyl-2,4-dioxopurine (theophylline) ($C_7H_8O_2N_4$)	73KW
3,7-Dimethyl-2,6-dioxopurine (3,7-dimethylxanthine) ($C_7H_8O_2N_4$)	76BMa
9-(β-D-Ribofuranosyl)xanthine (xanthosine) ($C_{10}H_{12}O_6N_4$) (H^+ values on Vol.2, p.345)	76RV,78RR,79RRT

Azines (continued)

Ligand	Bibliography

g. Triazines

2-Amino-1,3-diazolo[4,5-d]1,2,3-triazine (8-azaquanine) ($C_4H_4N_6$)   77ND

7. Aminophosphonic acids

2-Aminopropylphosphonic acid ($C_3H_{10}O_3NP$)   77MSW

(Benzylamino)methylphosphonic acid ($C_8H_{12}O_3NP$)   74SSb,77SSS

1-(Benzylamino)propylphosphonic acid ($C_{10}H_{16}O_3NP$)   74SSb

2-(Benzylamino)propylphosphonic acid ($C_{10}H_{16}O_3NP$)   74SSb

2-(Benzylamino)butylphosphonic acid ($C_{11}H_{18}O_3NP$)   74SSb

2-Amino-1-hydroxy-1-methylethylphosphonic acid ($C_3H_{10}O_4NP$)   75SLB

1-Aminomethyl-1-hydroxybutylphosphonic acid ($C_4H_{12}O_4NP$)   75SLB

1-Aminomethyl-1-hydroxy-3-methylbutylphosphonic acid ($C_5H_{14}O_4NP$)   75SLB

2-Amino-1-hydroxy-2-phenylethylphosphonic acid ($C_8H_{11}O_4NP$)   77MSW

1-Methyl-1-(2-picolylamino)ethylphosphonic acid ($C_9H_{15}O_3N_2P$)   78SSc,79SSe

1-Methyl-1-(3-picolylamino)ethylphosphonic acid ($C_9H_{15}O_3N_2P$)   78SSc,79SSe

1-Methyl-1-(4-picolylamino)ethylphosphonic acid ($C_9H_{15}O_3N_2P$)   78SSc,79SSe

Ethylenebis(iminomethylenephosphonic acid) ($C_4H_{14}O_6N_2P_2$)   73EZ
  (Values on Vol.2, p.305)

Oxybis(ethyleneiminomethylenephosphonic acid) ($C_6H_{18}O_7N_2P_2$)   77NF

Iminobis(methylenephosphonic acid) ($C_2H_9O_6NP_2$)   79ZP

N,N'-Bis(2-hydroxybenzyl)ethylenedinitrilo-N,N'-bis(methylene-   75MMa
  phosphonic acid) ($C_{18}H_{26}O_8N_2P_2$) (Values on Vol.2, p.321)

Diethylenetrinitrilopentakis(methylenephosphonic acid) ($C_9H_{28}O_{15}N_3P_5$)   79ZKT
  (Values on Vol.2, p.325)

C. Other Organic Ligands

1. Carboxylic acids

  a. Mono-carboxylic acids

Butanoic acid ($C_4H_8O_2$) (Values on Vol.3, p.10)   69KP,74DN,75SA,
                                                        76FK,76FKa,76KF,
                                                        77FK,77FKa,77YO,
                                                        78RP

2-Methylpropanoic acid (isobutyric acid) ($C_4H_8O_2$) (Values on Vol.3,   68RSa,77YO,79RRS
  p.11)

Pentanoic acid (valeric acid) ($C_5H_{10}O_2$) (Values on Vol.3, p.12)   69KP,77YO,78GC

3-Methylbutanoic acid (isovaleric acid) ($C_5H_{10}O_2$) (Values on Vol.3,   77YO,78GC
  p.12)

Hexanoic acid (caproic acid) ($C_6H_{12}O_2$) (H^+ values on Vol.3, p.329)   76PM,78RP

Propenoic acid (acrylic acid) ($C_3H_4O_2$) (H^+ values on Vol.3, p.330)   76W,79RS

trans-But-2-enoic acid (crotonic acid) ($C_4H_6O_2$) (Values on Vol.3,   76Sd,76SS,79RS
  p.14)

trans-3-Phenylpropenoic acid (trans-cinnamic acid) ($C_9H_8O_2$)   76Sd,79DD
  (H^+ values on Vol.3, p.330)

Fluoroacetic acid ($C_2H_3O_2F$) (H^+ values on Vol.3, p.333)   70KP

Carboxylic acids (continued)

Ligand	Bibliography
Dichloroacetic acid ($C_2H_2O_2Cl_2$) (Values on Vol.3, p.18)	72PK,76PT,78PT
2-Chloropropanoic acid ($C_3H_5O_2Cl$) (H^+ values on Vol.3, p.333)	73VBa,76W
3-Chloropropanoic acid ($C_3H_5O_2Cl$) (Values on Vol.3, p.19)	73VBa,76W
2-Chlorobenzoic acid ($C_7H_5O_2Cl$) (H^+ values on Vol.3, p.334)	73VBa
Bromoacetic acid ($C_2H_3O_2Br$) (Values on Vol.3, p.19)	76W
2-Bromopropanoic acid ($C_3H_5O_2Br$) (H^+ values on Vol.3, p.333)	76W
3-Bromopropanoic acid ($C_3H_5O_2Br$) (H^+ values on Vol.3, p.333)	76W
Iodoacetic acid ($C_2H_3O_2I$) (Values on Vol.3, p.20)	74Wa,79KA
2-Nitrobenzoic acid ($C_7H_5O_4N$) (Values on Vol.3, p.22)	79R
3-Nitrobenzoic acid ($C_7H_5O_4N$) (Values on Vol.3, p.22)	79R
4-Nitrobenzoic acid ($C_7H_5O_4N$) (Values on Vol.3, p.22)	79R
2-Sulfobenzoic acid ($C_7H_6O_5S$)	76AN
6-Carboxy-6'-hydroxy-3'-methylazobenzene-4-sulfonic acid ($C_{14}H_{12}O_6N_2S$)	75SKa
2-Hydroxy-2-methylpropanoic acid ($C_4H_8O_3$) (Values on Vol.3, p.36)	63DV,76SC
L-Phenylhydroxyacetic acid (mandelic acid) ($C_8H_8O_3$) (Values on Vol.3, p.47)	59CR,75CS,76MPc, 76SC,78KKH,79ZKV
DL-2-Hydroxy-2-phenylpropanoic acid (atrolactic acid) ($C_9H_{10}O_3$) (Values on Vol.3, p.49)	76SC
Diphenylhydroxyacetic acid (benzilic acid) ($C_{14}H_{12}O_3$) (Values on Vol.3, p.50)	79Mb
1,3,4,5-Tetrahydroxycyclohexanecarboxylic acid (quinic acid) ($C_7H_{12}O_6$) (Values on Vol.3, p.58)	77SS
2-Methoxybenzoic acid ($C_8H_8O_3$) (Values on Vol.3, p.73)	78PM
4-Oxopentanoic acid (levulinic acid) ($C_5H_8O_3$) (H^+ values on Vol.3, p.336)	79PZ
2-Benzoylbenzoic acid ($C_{14}H_{10}O_3$)	77OB
Uracil-6-carboxylic acid (orotic acid) ($C_5H_4O_4N_2$) (Values on Vol.3, p.88)	64H,79DZ,79DZb, 79MD,79SB
3-Methylorotic acid ($C_6H_6O_4N_2$)	79DZa,79SB
Thioorotic acid ($C_5H_4O_3N_2S$)	79DZ
Acrylamidoglycolic acid ($C_5H_7O_4N$)	77DP
Benzene-1,2-dicarboxylic acid-4-(acetylamidosulfamyl)-phenylmonoamide (phthalylsulfacetamide) ($C_{16}H_{14}O_6N_2S$)	79AV
DL-2-Mercaptopropanoic acid (thiolactic acid) ($C_3H_6O_2S$) (Values on Vol.3, p.75)	76AMa,76HS,76L, 77Ad,77AA, 77EL,77HSa
N-Acetylpenicillamine ($C_7H_{13}O_3NS$) (H^+ values on Vol. 3, p.337)	70SY
N-Benzoylcysteine ($C_{10}H_{11}O_3NS$)	75ZN,76ZNa,77NZ, 79ZN
Thiole 2 carboxylic acid ($C_5H_4O_2S$) (Values on Vol.3, p.83)	76SKS,76SSa,76SSb, 76SSk

Carboxylic acids (continued)

Ligand	Bibliography
N-(Dithiocarboxy)aminoacetic acid ($C_3H_5O_2NS_2$) (Values on Vol.3, p.91)	66BR,68BR

b. Di-carboxylic acids

Ligand	Bibliography
Methylpropanedioic acid (methylmalonic acid) ($C_4H_6O_4$) (Values on Vol.3, p.99)	76DGN,77ZT,79ASJ
Ethylmalonic acid ($C_5H_8O_4$) (Values on Vol.3, p.99)	79ZT
DL-1,2,2-Trimethylcyclopentane-1,3-dicarboxylic acid (camphoric acid) ($C_{10}H_{16}O_4$) (Other reference on Vol.3, p.359)	78GCa
Nonanedioic acid (azelaic acid) ($C_9H_{16}O_4$) (Values on Vol.3, p.120)	73TP,74TB
cis-Methylbutenedioic acid (citraconic acid) ($C_5H_6O_4$) (Values on Vol.3, p.114)	76Sd,79JA
trans-Butenedioic acid (fumaric acid) ($C_4H_4O_4$) (Values on Vol.3, p.115)	76KD,76KM
Butynedioic acid (acetylenedicarboxylic acid) ($C_4H_2O_4$)	78RSa
Chlorosuccinic acid ($C_4H_5O_4Cl$)	79ASJ
3-Chlorophthalic acid ($C_8H_5O_4Cl$) (Values on Vol.3, p.123)	76PJK,77Pa
3,6-Dichlorophthalic acid ($C_8H_4O_4Cl_2$)	76PJK,77Pa
Bromomalonic acid ($C_3H_3O_4Br$)	76DGN
3-Bromophthalic acid ($C_8H_5O_4Br$) (Values on Vol.3, p.123)	76PJK,77Pa
3-Iodophthalic acid ($C_8H_5O_4I$)	76PJK,77Pa
3-Nitrophthalic acid ($C_8H_5O_6N$) (Values on Vol.3, p.123)	76PJK,77Pa
4-Nitrophthalic acid ($C_8H_5O_6N$) (Values on Vol.3, p.123)	76PJK,77Pa
Hydroxypropanedioic acid (tartronic acid) ($C_3H_4O_5$) (Values on Vol.3, p.124)	72Ta,76MRa,76MRb
2,3,4-Trihydroxypentanedioic acid (trihydroxyglutaric acid) ($C_5H_8O_7$) (Other references on Vol.3, p.359)	68CM,74ZG,75GG, 78KKA,78NP
2,3,4,5-Tetrahydroxyhexanedioic acid (mucic acid) ($C_6H_{10}O_8$) (Values on Vol.3, p.130)	78KKA
D-Saccharic acid ($C_6H_{10}O_8$) (Other references on Vol.3, p.359)	75GO,76GO,77BOS, 79GA
N-Benzoylglutamic acid ($C_{12}H_{13}O_5N$)	52A
Oxamide-N,N'-diacetic acid ($C_6H_8O_6N_2$) (Values on Vol.3, p.133) (reference omitted in bibliography)	74SB
N-Benzenesulfonyl-L(−)-aspartic acid ($C_{10}H_{11}O_6NS$)	76GN
N-Benzenesulfonyl-L(+)-glutamic acid ($C_{11}H_{13}O_6NS$)	70GD,76GN
2-Mercaptobutanedioic acid (thiomalic acid) ($C_4H_6O_4S$) (Values on Vol.3, p.140)	68SG,76AMa,76HS, 76L,77ACM,77CA, 77EL,78SJ,79SJb
2,3-Dimercaptobutanedioic acid (dithiotartaric acid) ($C_4H_6O_4S_2$) (Values on Vol.3, p.141)	79ASJ
2,3-Bis(2-hydroxyethylthio)butanedioic acid ($C_8H_{14}O_6S_2$)	78MJ
Oxybis(ethylenethio)diacetic acid ($C_8H_{14}O_5S_2$) (Values on Vol.3, p.150)	77CA

c. Tri-carboxylic acids

Ligand	Bibliography
4,5-Dihydroxy-3,5-cyclopentadiene-1,2,3-tricarboxylic acid ($C_8H_6O_8$)	77KKV

Carboxylic acids (continued)

Ligand	Bibliography
2,3-Bis(2-carboxyethylthio)butanedioic acid ($C_8H_{10}O_8S_2$)	78MJ

2. Phosphorus acids

a. Hydrogen phosphates

D-3-Phosphoglyceric acid ($C_3H_7O_7P$)	72Lb
5-Phospho-D-ribose-1-(trihydrogendiphosphate) ($C_5H_{13}O_{14}P_3$)	78TLa
Phosphoric acid dimethyl ester ($C_2H_7O_4P$)	66SS
Phosphoric acid diethyl ester ($C_4H_{11}O_4P$) (Other references on Vol.3, p.360)	66SS
Phosphoric acid dipropyl ester ($C_6H_{15}O_4P$)	66SS
Phosphoric acid dibutyl ester ($C_8H_{19}O_4P$) (H^+ values on Vol.3, p.346)	66SS,74GM
Phorphoric acid dipentyl ester ($C_{10}H_{23}O_4P$) (Other reference on Vol.3, p.360)	66SS
Phosphoric acid dioctyl ester ($C_{16}H_{35}O_4P$)	74GM
Phosphoric acid didecyl ester ($C_{20}H_{43}O_4P$)	74GM
Phosphoric acid diphenyl ester ($C_{12}H_{11}O_4P$)	74GM
Phosphoric acid bis(4-methylphenyl) ester ($C_{14}H_{15}O_4P$)	74GM
Phosphoric acid bis(butyloxyethyl) ester ($C_{12}H_{27}O_5P$)	77Nb

b. Phosphonic acids

Benzenephosphonic acid hexyl ester ($C_{12}H_{19}O_3P$)	77Nc
Benzenephosphonic acid octyl ester ($C_{14}H_{23}O_3P$)	77Nc
Benzenephosphonic acid decyl ester ($C_{16}H_{27}O_3P$)	77Nc
trans-Vinylidenediphosphonic acid ($C_2H_6O_6P_2$)	77YKK
1-Hydroxyethane-1,1-diphosphonic acid ($C_2H_8O_7P_2$) (Values on Vol.3, p.177)	78MZa
Ethylenediphosphonic acid ($C_2H_8O_6P_2$) (Values on vol.3, p.178)	77YKK
Methylbis(phosphonomethyl)phosphine oxide ($C_3H_{11}O_7P_3$)	68KD

c. Phosphinic acids

Bis(2-ethylhexyl)phosphinic acid ($C_{16}H_{35}O_2P$)	74GM
Octylphenylphosphinic acid ($C_{14}H_{23}O_2P$)	74GM
2-Ethylhexylphenylphosphinic acid ($C_{14}H_{23}O_2P$)	74GM

d. Phosphoric acid esters

Phenylsulfonylamidophosphoric acid dibutyl ester ($C_{14}H_{24}O_5NPS$)	78SKZ

e. Phosphine oxides

Methylenebis(dihexylphosphine oxide) ($C_{25}H_{54}O_2P_2$)	69GD
Tetramethylenebis(dihexylphosphine oxide) ($C_{28}H_{60}O_2P_2$)	69GD
Pentamethylenebis(dihexylphosphine oxide) ($C_{29}H_{62}O_2P_2$)	69GD

3. Sulfonic acids

Ligand	Bibliography

Dodecane–1–sulfonic acid ($C_{12}H_{26}O_4S$)   79KBc

4. Phenols

   a. Mono–hydroxyphenols

2–Hydroxy–3–methylbenzoic acid (2,3–cresotic acid) ($C_8H_8O_3$)   76ABd
  (Values on Vol.3, p.187)

2–Hydroxy–4–methylbenzoic acid (2,4–cresotic acid) ($C_8H_8O_3$)   76ABd
  (Values on Vol.3, p.188)

2–Hydroxy–5–methylbenzoic acid (2,5–cresotic acid) ($C_8H_8O_3$)   76ABd,78ABS
  (Values on Vol.3, p.188)

2–Hydroxy–5–chlorobenzoic acid (5–chlorosalicylic acid) ($C_7H_5O_3Cl$)   76ABd,76C,78ABS
  (Values on Vol.3, p.188)

5–Bromosalicylic acid ($C_7H_5O_3Br$) (Values on Vol.3, p.188)   76ABd,78ABS

5–Nitrosalicylic acid ($C_7H_5O_5N$) (Values on Vol.3, p.189)   75KA,76ABd,78ABS,
                                                                 79PT

2,3–Dinitrosalicylic acid ($C_7H_4O_7N_2$)   78SM

3,5–Dinitrosalicylic acid ($C_7H_4O_7N_2$) (Values on Vol.3, p.189)   70DDa,75PM,76SJ,
                                                                 77DS,77SJ,79SJ

5–Sulfo–2,3–cresotic acid ($C_8H_8O_6S$)   73CS

6–Sulfo–2,3–cresotic acid ($C_8H_8O_6S$)   76ABd

6–Sulfo–2,4–cresotic acid ($C_8H_8O_6S$)   76ABd

6–Sulfo–2,5–cresotic acid ($C_8H_8O_6S$)   76ABd

2'–Hydroxy–4–methoxy–5'–methylchalkone oxime ($C_{17}H_{17}O_3N$)   77DK

3'–Bromo–2'–hydroxy–4–methoxy–5'–methylchalkone oxime ($C_{17}H_{16}O_3NBr$) 77DK

2–Hydroxybenzaldehyde oxime (salicylaldoxime) ($C_7H_7O_2N$) (H^+ values   72LN
  on Vol.3, p.349)

2–Acetylphenol(2–hydroxyacetophenone) ($C_8H_8O_2$) (Values on Vol.3,   77EP
  p.198)

4–Amino–2–hydroxybenzoic acid ($C_7H_7O_3N$) (Values on Vol.3, p.194)   76TM

3–Carboxy–4–hydroxyazobenzene–4'–sulfonic acid (Solochrome Yellow   78MBB
  2GS) ($C_{13}H_{10}O_6N_2S$)

6'–Carboxy–6–hydroxy–3,5–dichloroazobenzene–4'–sulfonic acid   78GSK
  ($C_{13}H_8O_6N_2Cl_2S$)

4–Hydroxy–3–(oxomethyl)azobenzene–4'–sulfonic acid ($C_{13}H_{10}O_5N_2S$)   74MT,75JS

   b. Di–hydroxyphenols

2,4–Dihydroxy–5–sulfobenzoic acid ($C_7H_6O_7S$)   75SM

4,5–Dibromo–1,2–dihydroxybenzene–3,6–disulfonic acid ($C_6H_4O_8Br_2S_2$)   71AH

2,4–Dihydroxy–1–propionylbenzene (2,4–dihydroxypropiophenone)   68GD
  ($C_9H_{10}O_3$)

7,8–Dihydroxy–2,4–dimethylchromenol ($C_{11}H_{12}O_3$)   70NM

6,7–Dihydroxy–2,4–diphenylbenzopyranol ($C_{21}H_{16}O_4$)   69PSa

Phenols (continnued)

Ligand	Bibliography
2,4-Dihydroxyazobenzene ($C_{12}H_{10}O_2N_2$) (H^+ values on vol.3, p.350)	76PSb
3,4-Dihydroxyazobenzene ($C_{12}H_{10}O_2N_2$)	76Sc
2,4-Dihydroxy-3'-nitroazobenzene ($C_{12}H_9O_4N_3$)	76PSb
2,4-Dihydroxy-4'-nitroazobenzene ($C_{12}H_9O_4N_3$)	76PSb
2,4-Dihydroxyazobenzene-4'-sulfonic acid (tropaeolin O) ($C_{12}H_9O_5N_2S$)	76PSb

c. Poly-hydroxyphenols

1,3,5-Trihydroxybenzene (phloroglucinol) ($C_6H_6O_3$)	77MSb
3,4,5-Trihydroxybenzoic acid (gallic acid) ($C_7H_6O_5$) (Values on Vol.3, p.213)	76ABa,77TS,78MSb
2,2',4-Trihydroxyazobenzene ($C_{12}H_{10}O_3N_2$) (Other reference on Vol.3, p.361)	76Sc
2-(4,5,6-Trihydroxy-3-oxo-3-H-9-xanthenyl)benzene-1-sulfonic acid (Pyrogallol Red) ($C_{19}H_{12}O_8S$) (H^+ values on Vol.3, p.352)	63PLL,63PLW
4,4'-Bis(catecholazo)biphenyl ($C_{24}H_{18}O_2N_2$)	79SO
3,3',4',5,7-Pentahydroxyflavone-5'-sulfonic acid (quercetin-5'-sulfonic acid) ($C_{14}H_{10}O_{10}S$)	76KT,79KNa

d. Fuchsones

5,5'-Dibromo-3,3'-dichloro-4-hydroxyfuchsone-2"-sulfonic acid (Bromochlorophenol Blue) ($C_{19}H_{10}O_5Br_2Cl_2S$)	78SKa
3,3',5,5'-Tetrabromo-4-hydroxyfuchsone-2"-sulfonic acid (Bromophenol Blue) ($C_{19}H_{10}O_5Br_4S$)	78Kd,78Ke,78SKa
4'-Hydroxy-3,3'-dimethyl-2"-sulfofuchsone-5,5'-dicarboxylic acid (Eriochrome Cyanine R) ($C_{23}H_{18}O_9S$) (Values on Vol.3, p.214)	69JM,70SM,73PMa
4',4"-Dihydroxyfuchsone-3,3',3"-tricarboxylic acid (aurintricarboxylic acid) ($C_{22}H_{14}O_9$) (Other references on Vol.3, p.361)	76MSb
3,3',4'-Trihydroxyfuchsone-2"-sulfonic acid (Pyrocatechol Violet) ($C_{19}H_{14}O_7S$) (Values on Vol.3, p.216)	74PM,77KVc,79CSV, 79SSf
5,5'-Dibromo-2,2',3,3',4'-pentahydroxyfuchsone-2"-sulfonic acid (Bromopyrogallol Red) ($C_{19}H_{12}O_9Br_2S$) (Other reference on Vol.3, p.362)	69NM,77KVc
4'-Hydroxydinaphthofuchsone-3,3'-dicarboxylic acid (Naphthochrome Green G) ($C_{29}H_{18}O_6$)	63AM,69AM

e. Naphthols

2-Hydroxy-1-nitrosonaphthalene-3,6-disulfonic acid (nitroso-R acid) ($C_{10}H_7O_8NS_2$) (Values on Vol.3, p.227)	75PM,76GMc,78MMa, 79GBb
7-Hydroxy-8-(phenylazo)naphthalene-1,3-disulfonic acid (Orange G) ($C_{16}H_{12}O_7N_2S_2$)	76DCD
1-(2-Arsonophenylazo)-2-hydroxynaphthalene-3,6-disulfonic acid (Thorin) ($C_{16}H_{13}O_{11}N_2AsS_2$) (Values on Vol.3, p.230)	76GMc,76ND
8-Amino-1-hydroxy-2-(3-chloro-2-hydroxy-5-nitrophenylazo)naphthalene-3,6-disulfonic acid (Gallion) ($C_{16}H_{11}O_{10}N_4ClS_2$)	74PM
2-Hydroxy-1-(2-hydroxy-5-methylphenylazo)naphthalene-4-sulfonic acid (Calmagite) ($C_{17}H_{14}O_5N_2S$) (Other references on Vol.3, p.362)	73NW,76GMc,78MPK, 78SMa
1-Hydroxy-2-(5-chloro-2-hydroxy-3-sulfophenylazo)naphthalene-5-sulfonic acid (Calcichrome Fast Navy 2RD) ($C_{16}H_{11}O_8N_2ClS_2$)	78MBB

Phenols (continued)

Ligand	Bibliography

2,2'-Dihydroxy-1,1'-azonaphthalene (Solochrome Dark Blue) ($C_{20}H_{15}O_2N_2$)   78SMa

2,1'-Dihydroxy-1,2'-azonaphthalene-4-sulfonic acid (Solochrome   76GMc,78MBB,78MPK
Black 6 BR) ($C_{20}H_{14}O_5N_2S$)

2,1'-Dihydroxy-6-nitro-1,2'-azonaphthalene-4-sulfonic acid (Erichrome   78MBB
Black T) (Solochrome Black WDFA) ($C_{20}H_{13}O_7N_3S$) (Values on Vol.3, p.231)

2-(4-Methylphenylazo)chromotropic acid ($C_{17}H_{14}O_8N_2S_2$)   79BG

2-(4-Chlorophenylazo)chromotropic acid ($C_{16}H_{11}O_8N_2ClS_2$)   79BG

2-(4-Nitrophenylazo)chromotropic acid ($C_{16}H_{11}O_{10}N_3S_2$)   79BG
(Values on Vol.3, p.237)

2-(4-Sulfophenylazo)chromotropic acid ($C_{16}H_{12}O_{11}N_2S_3$) (Values on   74PM,78AKD,79BG
Vol.3, p.238)

2,7-Bis(4-chloro-2-phosphonophenylazo)chromotropic acid   77MN
(Chlorophosphonazo III) ($C_{22}H_{16}O_{14}N_4Cl_2P_2S_2$) (Other references on
Vol.3, p.363)

2-(2-Arsonophenylazo)chromotropic acid (Arsenazo I) ($C_{16}H_{13}O_{11}N_2AsS_2$)   76BBe
(Values on Vol.3, p.239)

2-(2-Arsonophenylazo)-7-(3-sulfophenylazo)chromotropic acid   75MB
(Arsenazo M) ($C_{22}H_{17}O_{14}N_4AsS_3$)

2,7-Bis(2-arsonophenylazo)chromotropic acid (Arsenazo III)   76BBe,76Mb
($C_{22}H_{18}O_{14}N_4As_2S_2$) (Other references on Vol.3, p.363)

2,7-Bis(2-arsono-5-carboxyphenylazo)chromotropic acid   76BBd
($C_{24}H_{18}O_{18}N_4As_2S_2$)

2-(4-Acetylamidosulfonylphenylazo)chromotropic acid ($C_{18}H_{15}O_{11}N_3S_3$)   76BBe

2-(4-Guanidinosulfonylphenylazo)chromotropic acid ($C_{17}H_{15}O_{10}N_5S_3$)   76BBe

2-(2-Hydroxy-3,5-dinitrophenylazo)chromotropic acid (Picramine CA)   68GT
($C_{16}H_{10}O_{13}N_4S_2$) (Other reference on Vol.3, p.363)

2,7-Bis(5-chloro-2-hydroxy-3-sulfophenylazo)chromotropic acid   76BBd
(Sulphochlorophenol S) ($C_{22}H_{16}O_{16}N_4Cl_2S_4$) (Other references on Vol.3,
p.363)

3,4,9,10-Tetrahydroxyanthracene-2-sulfonic acid (leucoalizarin)   71NP
($C_{14}H_{10}O_7S$)

f. Hydroxyanthraquinones

1,2-Dihydroxyanthraquinone-3-sulfonic acid (Alizarin Red S) ($C_{14}H_8O_7S$)   74NF,75MSb
(Values on Vol.3, p.242)

1,4-Dihydroxyanthraquinone-2-sulfonic acid) (quinizarin-2-sulfonic   71TA
acid) ($C_{14}H_8O_7S$)

5. Carbonyl ligands

Decan-2,4-dione ($C_{10}H_{18}O_2$)   76JGC

2,5-Dichloro-3,6-dihydroxy-1,4-benzoquinone (chloranilic acid)   78V,79PV
($C_6H_2O_4Cl_2$) (Values on Vol.3, p.262)

2,5-Dibromo-3,6-dihydroxy-1,4-benzoquinone (bromanilic acid)   79PV
($C_6H_2O_4Br_2$)

5,6-Dihydroxy-5-cyclohexene-1,2,3,4-tetrone (rhodizonic acid)   69BM,79MR
($C_6H_2O_6$) (H^+ values on Vol.3, p.354)

Carbonyl ligands   (continued)

Ligand	Bibliography
Dehydroascorbic acid ($C_6H_6O_5$)	76KKa
1,2,4-Trihydroxy-3,4,5-tris(methoxycarbonyl)cyclopentadiene ($C_{11}H_{12}O_9$)	73MV,75MSa,79BL, 79BLa

6. Alcohols

Ligand	Bibliography
Ethane-1,2-diol (ethylene glycol) ($C_2H_6O_2$) (Values on Vol.3, p.266)	75PBa,75PBb,76Pa 77Pb
Propane-1,2-diol (propylene glycol) ($C_3H_8O_2$) (Values on Vol.3, p.266)	76Pa,77Pb
Propane-1,2,3-triol (glycerol) ($C_3H_8O_3$) (Values on Vol.3, p.268)	76Pa,77Pb

7. Polyethers

Ligand	Bibliography
Dicyclohexyl-24-crown-8-ether ($C_{24}H_{44}O_8$)	78KKb

8. Thiols

Ligand	Bibliography
3-Mercaptopropane-1,2-diol (thiovanol) ($C_3H_8O_2S$) (Values on Vol.3, p.281)	74DV,77K,77KPN
Thiodibenzoylmethane ($C_{15}H_{12}OS$)	72YC,74CY
N-Ethyl-3-mercapto-3-phenylpropenoic acid amide ($C_{11}H_{13}ONS$)	72YC,74CY
3-Mercapto-N,3-diphenylpropenoic acid amide ($C_{15}H_{15}ONS$)	72YC,74CY,75CY
3-Mercapto-N-(4-methylphenyl)-3-phenylpropenoic acid amide ($C_{16}H_{16}ONS$)	75CY
3-Mercapto-N-(4-chlorophenyl)-3-phenylpropenoic acid amide ($C_{15}H_{12}ONClS$)	75CY
3-Mercapto-N-(4-methoxyphenyl)-3-phenylpropenoic acid amide ($C_{16}H_{15}O_2NS$)	75CY
3-Mercapto-N-(4-ethoxyphenyl)-3-phenylpropenoic acid amide ($C_{17}H_{17}O_2NS$)	75CY
5-Mercapto-1,3,4-thiadiazolidine-2-thione ($C_2H_3N_2S_3$)	68GK,70GK,70KG,72KG
2,3-Dimercaptopropanesulfonic acid (unithiol) ($C_3H_8O_3S_3$) (Other references on Vol.3, p.364)	72PR,73RP,73RPa, 74PR,74RP,78OSK
2-(2,3-Dimercaptopropoxy)ethanesulfonic acid ($C_5H_{12}O_4S_3$) (Other references on Vol.3, p.364)	72PR,73RP,73RPa, 74PR,74RP
2-(2,3-Dimercaptopropylthio)ethanesulfonic acid ($C_5H_{12}O_3S_4$) (Other references on Vol.3, p.364)	72PR,73RP,73RPa, 74PR,74RP
2-(2,3-Dimercaptopropylsulfonyl)ethanesulfonic acid ($C_5H_{12}O_5S_4$) (Other references on Vol.3, p.364)	72PR,73RP,73RPa, 74PR,74RP
3-Methyl-2,6-dimercaptothiapyran-4-one ($C_6H_6OS_3$)	77CT
3-Ethyl-2,6-dimercaptothiapyran-4-one ($C_7H_8OS_3$)	77CT
3-Propyl-2,6-dimercaptothiapyran-4-one ($C_8H_{10}OS_3$)	77CT
3-Pentyl-2,6-dimercaptothiapyran-4-one ($C_{10}H_{14}OS_3$)	77CT
3-Phenyl-2,6-dimercaptothiapyran-4-one ($C_{11}H_8OS_3$)	77CT
3,5-Dimethyl-2,6-dimercaptothiapyran-4-one ($C_7H_8OS_3$)	77CT
3,5-Diethyl-2,6-dimercaptothiapyran-4-one ($C_9H_{12}OS_3$)	77CT
3,5-Diphenyl-2,6-dimercaptothiapyran-4-one ($C_{17}H_{12}OS_3$)	77CT

9. Thioxo ligands

Ligand	Bibliography
Morpholine-N-dithiocarbamic acid ($C_5H_9ONS_2$)	71GK
Methylthiosulfonic acid (CH_4O_2S)	78GSa
Ethylthiosulfonic acid ($C_2H_6O_2S$)	78GSa
Propylthiosulfonic acid ($C_3H_8O_2S$)	78GSa
4-Thioxopentan-2-one (monothioacetylacetone) (C_5H_8OS)	79LJ
4-Thioxo-1,1,1-trifluoropentan-2-one ($C_5H_5OF_3S$)	79LJ
Monothiodibenzoylmethane ($C_{15}H_{12}OS$)	72YN
Monothiothenoyltrifluoroacetone ($C_8H_5OF_3S$)	72YN

10. Unsaturated hydrocarbons

Ethene (C_2H_4) (Values on Vol.3, p.285)	78VS
Propene (C_3H_6) (Values on Vol.3, p.285)	78VS
But-1-ene (C_4H_8) (Values on Vol.3, p.285)	78VS
cis-But-2-ene (C_4H_8) (Values on Vol.3, p.285)	78VS
Benzene (C_6H_6) (Values on Vol.3, p.294)	78VS
Methylbenzene (toluene) (C_7H_8) (Values on Vol.3, p.294)	78VS
Cycloocta-1,5-diene (C_8H_{12})	77GGR

11. Hydroxamic acids

Chloroacetohydroxamic acid ($C_2H_3O_2NCl$)	78SSR
Hexanohydroxamic acid ($C_7H_{15}O_2N$)	76SR
Heptanohydroxamic acid ($C_8H_{17}O_2N$)	76SR
Octanohydroxamic acid ($C_9H_{19}O_2N$)	76SR
Nonanohydroxamic acid ($C_{10}H_{21}O_2N$)	76SR
2-Phenylacetohydroxamic acid ($C_8H_9O_2N$)	76RSR,77RS,77RSa
3-Phenylpropenohydroxamic acid (cinnamohydroxamic acid) ($C_9H_9O_2N$)	77BKa
N,3-Diphenylpropenohydroxamic acid (N-cinnamoyl-N-phenylhydroxylamine) ($C_{15}H_{13}O_2N$) (Values on Vol.3, p.302)	71LFG
4-Nitrobenzohydroxamic acid ($C_7H_6O_4N_2$)	77LGL
N-Phenyl-4-nitrobenzohydroxamic acid ($C_{13}H_{10}O_4N_2$)	77LGL
N-(4-Sulfophenyl)benzohydroxamic acid ($C_{13}H_{11}O_5NS$)	76GMd
1-Naphthohydroxamic acid ($C_{11}H_9O_2N$)	77AMb
O-Acetylacetohydroxamic acid (acetic acetohydroxamic anhydride) ($C_4H_7O_3N$)	78SSR
N-Phenyl-3-carboxypropanohydroxamic acid (N-phenyl-N-hydroxysuccinamic acid) ($C_{10}H_{11}O_4N$)	76GMe,76GMf

Hydroxamic acids   (continued)

<u>Ligand</u>	<u>Bibliography</u>
N-(2-Methylphenyl)-3-carboxypropanohydroxamic acid (N-2-tolyl-N-hydroxysuccinamic acid) ($C_{11}H_{13}O_4N$)	76GMe,76GMf
N-(3-Methylphenyl)-3-carboxypropanohydroxamic acid (N-3-tolyl-N-hydroxysuccinamic acid) ($C_{11}H_{13}O_4N$)	76GMe,76GMf
N-(4-Methylphenyl)thenohydroxamic acid ($C_{12}H_{11}O_2NS$)	76LMG

12. Oximes

Butane-2,3-dione monoxime ($C_4H_7O_2N$)	73AS
2-Acetyl-4,6-dichlorophenol oxime ($C_8H_6O_2NCl_2$)	75LG
Indoline-2,3-dione 3-oxime ($C_8H_6O_2N_2$)	76PBP
Di-2-oxolylethanedione monoxime (2-furilmonoxime) ($C_{10}H_7O_4N$)	78SSK
N-Benzoylaniline oxime (benzalilidoxime) ($C_{13}H_{12}ON_2$)	69MK
1-Ethoxyethanal oxime (ethylacetohydroxamate) ($C_4H_9O_2N$)	78SSR
Benzamidoxime ($C_7H_8ON_2$) (Other reference on Vol.3, p.366)	78NPM
Pentane-2,3-dione dioxime (ethylmethylglyoxime) ($C_5H_{10}O_2N_2$) (Values on Vol.3, p.307)	72DC
Di-2-oxolylethanedione dioxime (2-furildioxime) ($C_{10}H_8O_4N_2$) (Values on Vol.3, p.309)	78SSK
Acetyldimethylglyoxime ($C_6H_{10}O_4N_2$)	76LU
Methyl(dimethyl) (dimethylglyoxime)ammonium chloride ($C_7H_{20}O_2N_3Cl$)	76LU

13. Amides

Formamide (CH_3ON)	76BMc
N,N-Dimethylacetamide (C_4H_9ON)	70SKS
N,N-Diethylacetamide ($C_6H_{13}ON$)	70SKS
N,N-Dipropylacetamide ($C_8H_{17}ON$)	70SKS
N,N-Dibutylacetamide ($C_{10}H_{21}ON$)	70SKS
N-Acetylpyrrolidine ($C_6H_{11}ON$)	70SKS
N-Acetylpiperidine ($C_7H_{13}ON$)	70SKS
N-Acetylaniline (C_8H_9ON)	70SKS
N-Acetyl-2-methylaniline ($C_9H_{11}ON$)	70SKS
N-Acetyl-3-methylaniline ($C_9H_{11}ON$)	70SKS
N-Acetyl-4-methylaniline ($C_9H_{11}ON$)	70SKS
N-Acetylbenzylamine ($C_9H_{11}ON$)	70SKS
N-Acetylnaphtylamine ($C_{12}H_{11}ON$)	70SKS
N-Acetyl-2,4-dimethylaniline ($C_{10}H_{13}ON$)	70SKS
N-Acetyl-3,5-dimethylaniline ($C_{10}H_{13}ON$)	70SKS
N-Acetyl-2-chloroaniline (C_8H_8ONCl)	70SKS
N-Acetyl-3-chloroaniline (C_8H_8ONCl)	70SKS
N-Acetyl-4-chloroaniline (C_8H_8ONCl)	70SKS
N-Acetyl-3-chloro-2-methylaniline ($C_9H_{10}ONCl$)	70SKS

Amides (continued)

Ligand	Bibliography
N-Acetyl-5-chloro-2-methylaniline ($C_9H_{10}ONCl$)	70SKS
N-Acetyl-2,5-dichloroaniline ($C_8H_7ONCl_2$)	70SKS
N-Acetyl-2-nitroaniline ($C_8H_8O_3N_2$)	70SKS
N-Acetyl-3-nitroaniline ($C_8H_8O_3N_2$)	70SKS
N-Acetyl-4-nitroaniline ($C_8H_8O_3N_2$)	70SKS
N-Acetyl-2-methoxyaniline ($C_9H_{11}O_2N$)	70SKS
N-Acetyl-3-methoxyaniline ($C_9H_{11}O_2N$)	70SKS
N-Acetyl-4-methoxyaniline ($C_9H_{11}O_2N$)	70SKS
N-Acetyl-2,4-dimethoxyaniline ($C_{10}H_{13}O_3N$)	70SKS
N-Acetyl-2,5-dimethoxyaniline ($C_{10}H_{13}O_3N$)	70SKS
Cyclotris(L-prolylglycyl) ($C_{21}H_{36}O_6N_6$)	74MA
Carbamide (urea) (CH_4ON_2) (Values on Vol.3, p.313)	76BMc
Biuret ($C_2H_5O_2N_3$) (Other references on Vol.3, p.366)	79Ab,79SBb,79SBc,
N,N'-Bis(hydroxymethyl)thiourea ($C_3H_8O_2N_2S$)	79TB
N-(Phenylsulfonylimidobenzoyl)thiocarbamic acid ethylester ($C_{16}H_{16}O_3N_2S_2$)	76KU
1,5-Diphenyl-1,2,4,5-tetraazapent-1-en-3-thione (dithizone) ($C_{13}H_{12}N_4S$) (Values on Vol.3, p.319)	72YN
N,N,N',N''-Tetramethyldithiooxamide ($C_6H_{12}N_2S_2$)	76AMP
N,N,N',N''-Tetraethyldithiooxamide ($C_{10}H_{20}N_2S_2$)	76AMP
N,N'-Dicyclohexyldithiooxamide ($C_{14}H_{24}N_2S_2$)	76AMP
N,N'-Dibenzyldithiooxamide ($C_{16}H_{12}N_2S_2$)	76AMP
N,N'-Bis(2-hydroxyethyl)dithiooxamide ($C_6H_{12}O_2N_2S_2$)	76AMP
Purpuric acid (murexide=$L.NH_3$) ($C_8H_5O_6N_5$) (Values on Vol.3, p.322)	78BKG,79FKR

14. Amidines

2-Imino-1-methyl-1,3-diazolidine-4-one (creatinine) ($C_4H_7ON_3$)	76RDa
N',N^3-Bis(benzenesulfonyl)benzamidrazone ($C_{19}H_{17}O_4N_3S_2$)	78KPU

D. Inorganic Ligands

Hydrogen hexacyanoferrate (II) ($H_4Fe(CN)_6$) (Values on Vol.4, p.21)	76CC,77BL
Hydrogen niobate ($H_8Nb_6O_{19}$) (H^+ values on Vol.4, p.131)	76GT
Hydrogen germanate (germanic acid) ($Ge(OH)_4$) (H^+ values on Vol.4, p.131)	57Ae
Hydrazine (H_2NNH_2) (Values on Vol.4, p.43)	70BG,77AGK,78AMK
Hydrogen tetraphosphate (tetraphosphorous acid) ($H_6O_{13}P_4$) (Values on Vol.4, p.66)	64WM,76MTa,79KW
Hydrogen hexaphosphate ($H_8P_6O_{19}$) (Other references on Vol.4, p.135)	77SKa,79KW

Inorganic ligands (continued)

Ligand	Bibliography
Hydrogen octaphosphate ($C_{10}H_8O_{25}$)	77SKa,79KW
Hydrogen hypophosphate (hypophosphoric acid) ($H_4O_6P_2$) (Values on Vol.4, p.73)	77WH
Dodecaoxohexaphosphoric acid ($-(PO_2H)_6-$)	79WNc
Decaoxotetraphosphoric acid ($-(PO_2H-O-PO_2H)_2-$)	79WNc
Hexafluorophosphate ion (PF_6^-) (Values on Vol.4, p.74)	78PSS
Hydrogen peroxide (H_2O_2) (Values on Vol.4, p.75)	77VZ
Hydrogen peroxodisulfate (peroxodisulfuric acid) ($H_2O_8S_2$) (Values on Vol.4, p.89)	77ASb
Hydrogen tellurate (telluric acid) ($Te(OH)_6$) (H^+ values on Vol.4, p.134)	55A,71BG,76CC

BIBLIOGRAPHY

Russian translations have the page of the original in parentheses.

31GI   R. Gane and C.K. Ingold, J. Chem. Soc., 1931, 2153

38PO   J.N. Pearce and L.D. Ough, J. Amer. Chem. Soc., 1938, 60, 80

42KH   F.W. Klemperer, A.B. Hastings, and D.D. Van Slyke, J. Biol. Chem., 1942, 143, 433

42KP   I.M. Kolthoff, R.W. Perlich, and D. Weiblen, J. Phys. Chem., 1942, 46, 561

42M   T. Moeller, J. Amer. Chem. Soc., 1942, 64, 953

48WB   G.W. Wheland, R.M. Brownell, and E.C. Mayo, J. Amer. Chem. Soc., 1948, 70, 2493

49IE   H. Irving, J.A.D. Ewart, and J.T. Wilson, J. Chem. Soc., 1949, 2672

50DL   B.E. Douglas, H.A Laitinen, and J.C. Bailar, Jr., J. Amer. Chem. Soc., 1950, 72,
            2484

51K   P. Krumholz, J. Amer. Chem. Soc., 1951, 73, 3487

52A   A. Albert, Biochem. J., 1952, 50, 690

52BK   M. Blumer and I.M. Kolthoff, Experimentia, 1952, 8, 138

52HH   L.G. Hepler and Z.Z. Hugus, Jr., J. Amer. Chem. Soc., 1952, 74, 6115

52MP   A.E. Martell and R.C. Plumb, J. Phys. Chem., 1952, 56, 993

52SL   J. Schubert and A. Lindenbaum, J. Amer. Chem. Soc., 1952, 74, 3529

53Nb   C.J. Nyman, J. Amer. Chem. Soc., 1953, 75, 3575

53RL   A. Ringbom and E. Linko, Anal. Chim. Acta, 1953, 9, 80

54CV   D. Cozzi and S. Vivarelli, Z. Elektrochem., 1954, 58, 907

55A   P.J. Antikainen, Suomen Kem., 1955, B28, 135

55EH   D. Eckhardt and L. Holleck, Z. Elektrochem., 1955, 59, 202

55MB   R.K. Murmann and F. Basolo, J. Amer. Chem. Soc., 1955, 77, 3484

55MK   W.M. McNevin and O.H. Kriege, J. Amer. Chem. Soc., 1955, 77, 6149

55RK   R.A. Robinson and A.K. Kiang, Trans. Faraday Soc., 1955, 51, 1398

57Ae   P.J. Antikainen, Suomen Kem., 1957, B30, 123

57CR   D. Cozzi and G. Raspi, Ric.Sci., 1957, 27, 2392

57FSa   J.K. Foreman and T.D. Smith, J. Chem. Soc., 1957, 1758

57H   R. Hofstetter, Diss., Univ. Zurich, 1957

57HK   E.F.G. Herington and W. Kynaston, Trans. Faraday Soc., 1957, 53, 138

57J   O. Jantti, Suomen Kem., 1957, B30, 136

57LH   D.L. Leussing and R.C. Hansen, J. Amer. Chem. Soc., 1957, 79, 4270

57LQ   T.J. Lane, J.V. Quagliano, and Ernest Bertin, Anal. Chem., 1957, 29, 481

58CL   C.M. Cook and F.A. Long, J. Amer. Chem. Soc., 1958, 80, 33

58JB   E. Jorgensen and J. Bjerrum, Acta Chem. Scand., 1958, 12, 1047; J. Inorg. Nucl.
            Chem., 1958, 8, 313

58P   D.D. Perrin, Nature, 1958, 182, 741

58T   R.S. Tobias, Acta Chem. Scand., 1958, 12, 198

59A   G. Anderegg, Helv. Chim. Acta., 1959, 42, 344

59AR   P.J. Antikainen and V.M.K. Rossi, Suomen Kem., 1959, B32, 185

59BB   C.V. Banks and R.C. Bystroff, J. Amer. Chem. Soc., 1959, 81, 6153

59FO   G. Folsch and R. Osterberg, J. Biol. Chem., 1959, 234, 2298

59GM   A.D. Gelman and M.P. Mefodeva, Proc. Acad. Sci. USSR, 1959, 124, 69 (815)

59GMA   A.D. Gelman, A.I. Moskvin, and P.I. Artyukhin, Russ. J. Inorg. Chem., 1959, 4,
            599 (1332)

59MA   A.I. Moskvin and P.I Artyukhin, Russ. J. Inorg. Chem., 1959, 4, 269 (591)

59MK   A.I. Moskvin, G.U. Khalturin, and A.D. Gelman, Radiokhim., 1959, 1, 141

59ML   C.T. Mortimer and K.J. Laidler, Trans. Faraday Soc., 1959, 55, 1731

59NT   B.P. Nikolskii, A.M. Trofimov, and N.B. Vysokovstrovskaya, Radiokhim., 1959, 1,
            141

59PC   H.M. Papee, W.J. Canady, T.W. Zawidzki, and K.J. Laidler, Trans. Faraday Soc.,
            1959, 55, 1734

59ZP   T.W. Zawidzki, H.M. Papee, W.J. Canady, and K.J. Laidler, Trans. Faraday Soc.,
            1959, 55, 1738

60BS   C.V. Banks and R.S. Singh, J. Inorg. Nucl. Chem., 1960, 15, 125

60CL   C.P. Chang and L.S. Liu, Acta Chim. Sinica, 1960, 26, 148

60M   R.B. Martin, J. Amer. Chem. Soc., 1960, 82, 6053

60MS   N.N. Matorina and N.D. Safonova, Russ. J. Inorg. Chem., 1960, 5, 151 (313)

60SS   V.P. Shuedov and A.V. Stepanov, Soviet Radiochem., 1960, 1, 77 (162)

61BH   P. Bamberg and P. Hemmerich, Helv. Chim. Acta, 1961, 44, 1001

61BR   A.I. Biggs and R.A. Robinson, J. Chem. Soc., 1961, 388, 2572

61CF   L.K. Creamer, A. Fisher, B.R. Mann, J. Packer, R.B. Richards, and J. Vaughan,
            J. Org. Chem., 1961, 26, 3148

61CM   J.M. Crabtree, D.W. Marsh, J.C. Tomkinson, R.J.P. Williams, and W.C. Fernelius,
            Proc. Chem. Soc., 1961, 336

61DR   L. Davis, F. Roddy, and D.E. Metzler, J. Amer. Chem. Soc., 1961, 83, 127

61E   N. Elenkova, Ann. Inst. Chim. Techn. (Sofia), 1961, 8, 125

61EA   A.J. Ellis and D.W. Anderson, J. Chem. Soc., 1961, 1765

61GT   R.M. Garrels, M.E. Thompson, and R. Siever, Amer. J. Sci., 1961, 259, 24

61HS   M. Hnilickova and L. Sommer, Coll. Czech. Chem. Comm., 1961, 26, 2189

61JW   B.R. James and R.J.P. Williams, J. Chem. Soc., 1961, 1007

61NM   R. Nasanen and P. Merilainen, Suomen Kem., 1961, B34, 47

61OH   W.F. O'Hara and L.G. Hepler, J. Phys. Chem., 1961, 65, 2107

62AT   T. Anderson and A.B. Knutsen, Acta Chem. Scand., 1962, 16, 875

62CL   D.T. Chen and K.J. Laidler, Trans. Faraday Soc., 1962, 58, 480

62GT   R.M. Garrels and M.E. Thompson, Amer. J. Sci., 1962, 260, 57

62HP   C.J. Hawkins and D.D. Perrin, J. Chem. Soc., 1962, 1351

62IM   H. Irving and D.H. Miller, J. Chem. Soc., 1962, 5222

62IN   T. Ishimori and E. Nakamura, Radiochem. Acta, 1962, 1, 6

62KM   F. Ya. Kulba and Yu. A. Makashev, J. Gen. Chem. USSR, 1962, 32, 1709 (1724)

62KP   R.S. Kolat and J.E. Powell, Inorg. Chem., 1962, 1, 485

62LS   P. Lumme and N. Seppalainen, Suomen Kem., 1962, B35, 123

62MD   D.L. McMasters, J.C. DiRaimondo, L.H. Jones, R.P. Lindley and E.W. Zeltmann,
            J. Phys. Chem., 1962, 66, 249

62SY   O.E. Schupp, III, T. Youness, and J.I. Watters, J. Amer. Chem. Soc., 1962, 84
            505

62VI   L.V. Vanyukova, M.M. Isaeva, and B.N. Kabanova, Dokl. Phys. Chem., 1962, 143,
            (377)

62W   P. Wong, J. Hong Kong Baptist Coll., 1962, 1, 49, 85

62YO   Y. Yoshino, A. Ouchi, Y. Tsunoda, and M. Kojima, Canad. J. Chem., 1962, 40, 775

63A   G. Anderegg, Helv. Chim. Acta, 1963, 46, 2397

63Aa   G. Anderegg, Helv. Chim. Acta, 1963, 46, 2813

63AM   L.P. Adamovich, A.P. Mirnaya, and A.V. Starchenko, J. Anal. Chem. USSR, 1963,
            18, 369 (420)

63BM   A. Bryson and R.W. Matthews, Aust. J. Chem., 1963, 16, 401

63DV   H. Deelstra and F. Verbeek, Bull. Soc. Chim. Belg., 1963, 72, 612

63GS   A.M. Golub and V.M. Samoilenko, Ukr. Khim.Zh., 1963, 29, 472

63HS   P. Hemmerich and Ch. Sigwart, Experimentia, 1963, 19, 488

63IDa   H. Irving and J.J.R. Frausto da Silva, J. Chem. Soc., 1963, 458

63MH   S.S. Mesaric and D.N. Hume, Inorg. Chem., 1963, 2, 788

63PK   V.I. Paramonova, A.S. Kereichuk, and A.V. Chizhov, Radiokhim., 1963, 5, 63

63PLL   K. Pan, Z.F. Lin, S.H. Lin, Y.L. Wo, and E.M. Chen, J. Chinese Chem. Soc.
            (Taiwan), 1963, 10, 24

63PLW   K. Pan, Z.F. Lin, and Y.L. Wo, J. Chinese Chem. Soc. (Taiwan), 1963, 10, 175

63TP   L.C.A. Thompson and R. Pacer, J. Inorg. Nucl. Chem., 1963, 25, 1041

64CK   R. Caletka, M. Kirs, and J. Rais, J. Inorg. Nucl. Chem., 1964, 26, 1443

64CL   T.H. Chang, J.T. Lin, T.I. Chow, and S.K. Yang, J. Chinese Chem. Soc. (Taiwan)
            1964, 11, 125

64GS   I. Gainar and K.W. Sykes, J. Chem. Soc., 1964, 4452

64H   A. Haug, J. Amer. Chem. Soc., 1964, 86, 3381

64MAH   F.J. Millero, J.C. Ahluwalia, and L.G. Hepler, J. Chem. Eng. Data, 1964, 9, 319

64NN   G. Nikolov and St. Nacheva, God. Sofii. Univ., Khim. Fak., 1964, 59, 37, 51

64PK   R. Palmaeus and P. Kierkagaard, Acta Chem. Scand., 1964, 18, 2226

64R   H. Reinert, Abhandl. Deut. Akad. Wiss. Berlin, Kl. Med., 1964, 6, 373

64RM   K.S. Rajan and A.E. Martell, J. Inorg. Nucl. Chem., 1964, 26, 789

64SF   E. Sekido, Q. Fernando, and H. Freiser, Anal. Chem., 1964, 36, 1768

64SM   Unpublished values quoted in L.G. Sillen and A.E. Martell, Stability Constants
            of Metal-Ion Complexes, Special Publication No. 17, The Chemical Soc.,
            London, 1964

64WM   J.I. Watters and S. Matsumoto, J. Amer. Chem. Soc., 1964, 86, 3961

64WMa   A.V. Willi and P. Mori, Helv. Chim. Acta, 1964, 47, 155

64Z   S.S. Zavodnov, Zh. Vses. Khim. Obsh. Mendeleeva, 1964, 9, 472

65DD   R.L. Davies and K.W. Dunning, J. Chem. Soc., 1965, 4168

65HA   A. Hamid, I.P. Alimarin, and I.V. Puzdrenkova, Vestnik Moskov. Univ. Khim.,
        1965, 20, 71; Chem. Abs., 1965, 63, 6376d

65KD   R.C. Kapoor, G. Doughty and G. Gorin, Biochem. Biophys. Acta, 1965, 100, 376

65TM   Z. Tamura and M. Miyazaki, Chem. Pharm. Bull. (Japan), 1965, 13, 333, 345

65TMa   Z. Tamura and M. Miyazaki, Chem. Pharm. Bull. (Japan), 1965, 13, 387

65TMS   Z. Tamura, M. Miyazaki, and T. Suzuki, Chem. Pharm. Bull. (Japan), 1965, 13, 330

65UG   E. Uhlig, M. Gentschew, and A. Martin, Chem. Ber., 1965, 98, 983

66BH   P.D. Bolton, F.M. Hall, and I.H. Reece, Spectrochim. Acta, 1966, 22, 1149

66BHa   P.D. Bolton, F.M. Hall, and I.H. Reece, Spectrochim. Acta, 1966, 22, 1825

66BP   M. Bonnet and R.A. Paris, Bull. Soc. Chim. France, 1966, 747

66BR   O. Budevsky and E. Ruseva, Bulg. Akad. Wissen., Inst. Allg. Anorg. Chem., 1966
        4, 5

66DM   C.D. Dwivedi, K.N. Munshi, and A.K. Dey, J. Indian Chem. Soc., 1966, 43, 301

66EA   S.H. Eberle and S.A.H. Ali, Radiochim. Acta, 1966, 5, 58

66GB   E.A. Gyunner and N.D. Belykh, Soviet Progr. Chem. (Ukr. Khim. Zh.), 1966, 32,
        962 (1270)

66GJ   J.N. Gaur and D.S. Jain, Trans. Soc. Adv. Electrochem. Sci. Tech. (Karaikudi),
        1966, 1, 38

66HC   T.M. Hseu and L.S. Chen, J. Chinese Chem. Soc. (Taiwan), 1966, 13, 150

66HH   R.H. Holyer, C.D. Hubbard, S.F.A. Kettle, and R.G. Wilkins, Inorg. Chem.,
        1966, 5, 622

66IG   I.M. Ivanov and L.M. Grindin, Izv. Sibir. Otd. Acad. Nauk. SSSR, 1966, 4, 31

66KF   T. Kaden and S. Fallab, Chimia (Switz.), 1966, 20, 51

66KFa   H. Kock and W.D. Falkenberg, Solv. Extr. Chem., Proc. Int. Conf., Goteborg,
        1966, 26

66KR   R.J. Kula and D.L. Rabenstein, Anal. Chem., 1966, 38, 1934

66KT   N.A. Kostromina, T.B. Ternovaya, E.D. Romanenko, and K.B. Yatsimirskii, Teoret.
        Exp. Khim., 1966, 2, 673

66LN   S.J. Lyle and S.J. Naqvi, J. Inorg. Nucl. Chem., 1966, 28, 2993

66LP   A.V. Lapitsky and L.N. Pankratova, Moscow Univ. Chem. Bull., 1966, 21, 204
        (No. 3, 61)

66PE   V.M. Peshkova, I.P. Efimov, and N.N. Magdesieva, J. Anal. Chem. USSR, 1966,
        21, 446 (499)

66PP   J. Podlahova and J. Podlaha, J. Inorg. Nucl. Chem., 1966, 28, 2267

66S   J. Stary, JINR 2000, 1965; JINR 2001, 1965; JINR 2224, 1965; Talanta, 1966,
        13, 421; Soviet Radiochem., 1966, 8, 467 (504); 471 (509) (Listed as
        65S in Vol. 1)

66SM   M. Shilov and Y. Marcus, J. Inorg. Nucl. Chem., 1966, 28, 2725

66SS   E.I. Sinyavskaya and Z.A. Sheka, Soviet Radiochem., 1966, 8, 380 (410)

66SY   K. Suzuki and K. Yamasaki, J. Inorg. Nucl. Chem., 1966, 28, 473

66ZK   A.I. Zelyanskaya and L. Ya. Kukalo, J. Anal. Chem. USSR, 1966, 21, 1059 (1191)

67AB   Z.F. Andreeva, V.N. Bezuevskaya, I.V. Kolosov, and Z.N. Shevtsova, Izv. Tim. Sel.
          Akad., 1967, 200

67BH   P.D. Bolton, F.M. Hall, and I.H. Reece, J. Chem. Soc.(B), 1967, 709

67BR   T.R. Bhat and T.V. Rao, Z. Anorg. Allg. Chem., 1967, 354, 201

67FL   Ya. D. Fridman and M.G. Levine, Russ. J. Inorg. Chem. 1967, 12, 1425 (2704)

67GS   J.N. Gaur and V.K. Sharma, Rev. Polarog., 1967, 14, 287

67H   H.S. Hendrickson, Anal. Chem., 1967, 39, 998

67JK   D.S. Jain, A. Kumar, and J.N. Gaur, J. Inst. Chem. (India), 1967, 39, 230

67K   F. Karczynski, Zesz. Nauk., Mat. Fiz. Chem., Wyzsza Szk. Pedagog. Gdansku, Wydz,
          Mat., Fiz. Chem., 1967, 7, 149; Chem. Abstr., 1968, 69, 100120g

67KC   V.I. Kravtsov and V.N. Chamaev, Vestn. Leningrad Univ., 1967, 22, 94; Chem.
          Abstr., 1967, 67, 36827w

67KK   F. Karczynski and G. Kupryszewski, Rocz. Chem., 1967, 41, 1019

67KKa   F. Karczynski and G. Kupryszewski, Rocz. Chem., 1967, 41, 1665

67KN   E.E. Kapantsyan and B.I. Nabivanets, Soviet Progr. Chem. (Ukr. Khim. Zh.), 1967,
          33, No. 9, 73 (961)

67LN   S.J. Lyle and S.J. Naqvi, J. Inorg. Nucl. Chem., 1967, 29, 2441

67LP   S. Laxmi and S. Prakash, Z. Phys. Chem. (Frankfurt), 1967, 55, 259

67MP   P.M. Milyukov and N.V. Polenova, Izv. Vyssh. Ucheb. Zaued., Khim., 1967, 10, 277

67MS   S. Mahapatra and R.S. Subrahmanya, Proc. Indian Acad. Sci., 1967, 65A, 283

67OM   T. Oncescu and M. Macorschi, An. Univ. Bucuresti. Ser., Sti. Nat. Chim., 1967,
          16, 77

67PH   S. Petrucci, P. Hemmes, and M. Battistini, J. Amer. Chem. Soc., 1967, 89, 5552

67PS   D.D. Perrin and V.S. Sharma, J. Chem. Soc. (A), 1967, 724

67PSS   D.D. Perrin, I.G. Sayce, and V.S. Sharma, J. Chem. Soc. (A), 1967, 1755

67S   V.S. Sharma, Biochem. Biophys. Acta, 1967, 148, 37

67SBG   U.I. Salakhutdinov, A.P. Borisova, Yu. V. Granoskii, I.A. Savich and
          V.I. Spitsyn, Doklady Chem., 1967, 177, 1039 (365)

67SL   V.M. Savostina, F.I. Lobanov, and V.M. Peshkova, Russ. J. Inorg. Chem., 1967,
          12, 1139 (2162)

67SPa   J.P. Scharff and M.R. Paris, Bull. Soc. Chim. France, 1967, 1782

67TKL   E.R. Tucci, C.H. Ke and N.C. Li, J. Inorg. Nucl. Chem., 1967, 29, 1657

67TKR   T.V. Ternovaya, N.A. Kostromina, and E.D. Romanenko, Soviet Progr. Chem. (Ukr.
          Khim. Zh.), 1967, 33, No. 7, 1 (651)

68AJ   J. Affolter, A. Jacot-Guillarmod, and K. Bernauer, Helv. Chim. Acta, 1968, 51, 293

68AM   V.T. Athavale, N. Mahadevan, and R.M. Sathe, Indian J. Chem., 1968, 6, 660

68AN   V.P. Antonovich and V.A. Nazarenko, J. Anal. Chem. USSR, 1968, 23, 1284 (1460)

68AS   L.P. Andrusenko and I.A. Sheka, Russ. J. Inorg. Chem., 1968, 13, 1363 (2645)

68BB   K. Balachandran and S.K. Banerji, J. Indian Chem. Soc., 1968, 45, 571

68BBa   K. Balachandran and S.K. Banerji, J. Prakt. Chem., 1968, 37, 263

68BBI   A.I. Busev, L.I. Bogdanovich, and V.M. Ivanov, Moscow Univ. Chem. Bull., 1968,
          23, No. 3, 68 (101)

68BD   V.A. Barabanov, S.L. Davydov, and N.A. Plate, Russ. J. Phys. Chem., 1968, 42, 516 (990)

68BHK   P.D. Bolton, F.M. Hall, and J. Kudryuski, Aust. J. Chem., 1968, 21, 1541

68BI   A.I. Busev, V.M. Ivanov, and Zh. I. Nemtseva, Russ. J. Inorg. Chem., 1968, 13, 266 (511)

68BM   A.K. Babko, E.A. Mazurenko, and B.I. Nabivanets, Russ. J. Inorg. Chem., 1968, 13, 375 (718)

68BN   E.A. Biryuk, V.A. Nazarenko, and N.I. Zabolotnaya, J. Anal. Chem. USSR, 1968, 23, 742 (853)

68BR   O. Budevsky, E. Russeva, and T. Sotyrova, Talanta, 1968, 15, 629

68BW   B. Budesinsky and T.S. West, Anal. Chim. Acta, 1968, 42, 455

68CM   E.G. Chikryzova, S.Ya. Mashinskaya, and L.G. Kirnyak, Russ. J. Inorg. Chem., 1968, 13, 1114 (2153)

68CN   L. Ciavatta and G. Nunziata, Ric. Sci., 1968, 38, 109

68CWI   J.J. Christensen, D.P. Wrathall, and R.M. Izatt, Anal. Chem., 1968, 40, 175

68DP   C.W. Davies and B.N. Patel, J. Chem. Soc. (A), 1968, 1824

68EM   A.N. Ermakov, I.N. Marov, N.B. Kalinichenko, and S.S. Travnikov, Russ. J. Inorg. Chem., 1968, 13, 1709 (3316)

68F   W.L. Felty, M.S. Thesis, Ohio State Univ., 1968

68GD   M.H. Gandhi and M.N. Desai, Indian J. Chem., 1968, 6, 371

68GI   P. Gabor-Klatsmanyi, J. Inczedy, and L. Erdy, Acta Chim. Acad. Sci. Hung., 1968, 57, 5

68GK   Z. Gregorowicz and Z. Klima, Coll. Czech. Chem. Comm., 1968, 33, 3380

68GS   L.N. Grigoreva, L.D. Stepin, and T.A. Shurupova, Russ. J. Inorg. Chem., 1968, 13, 1671 (3240)

68GT   S.S. Goyal and J.P. Tandon, Talanta, 1968, 15, 895

68HP   P. Hemmes and S. Petrucci, J. Phys. Chem. 1968, 72, 3986; 1969, 73, 4426 (see also R.A. Matheson, ibid., 1969, 73, 4425)

68IS   J. Israeli and H. Saulnier, Inorg. Chim. Acta, 1968, 2, 482

68K   E.E Kriss, Russ. J. Inorg. Chem., 1968, 13, 1276 (2472)

68KB   W. Kemula and K. Brajter, Chem. Anal. (Warsaw), 1968, 13, 503

68KD   V.V. Medyntsev, Yu. M. Polikarpov, and Z.I. Tsareva, Bull. Acad. Sci. USSR, Chem., 1968, 1955 (2058)

68KR   N.A. Kostromina and E.D. Romanenko, Soviet Progr. Chem. (Ukr. Khim. Zh.), 1968, 34, No. 7, 1 (645)

68KS   J.R. Kuempel and W.B. Schaap, Inorg. Chem., 1968, 7, 2435

68L   P. Lanza, J. Electroanal. Chem., 1968, 19, 289; Ric. Sci., 1968, 38, 1181

68LA   S.C. Lahiri and S. Aditya, J. Inorg. Nucl. Chem., 1968, 30, 2487

68LP   S. Laxmi, S. Prakash, and S. Prakash, Z. Phys. Chem. (Frankfurt), 1968, 61, 247

68M   J. Maslowska, Rocz. Chem., 1968, 42, 1191

68MF   T.V. Malkova and N.A. Fateeva, Russ. J. Inorg. Chem., 1968, 13, 1082 (2094)

68MN   M.G. Mushkina, M.S. Novakovaskii, and Vu Van Lyu, Russ. J. Inorg. Chem., 1968, 13, 1193 (2309)

68MT   Gh. Marcu, M. Tomus, and M. Solea, Stud. Univ. Babes-Bolyai, Chem., 1968, 2, 15

68Na   F.S. Nakayama, Soil Sci., 1968, 106, 429

68PL   G. Popa and C. Lazar, An. Univ. Bucuresti, Ser. Sti., Nat. Chim., 1968, 17, 9

68PM   G.K. Pagenkopf and D.W. Margerum, Inorg. Chem., 1968, 7, 2514

68PMa   Gr. Popa and V. Magearu, An. Univ. Bucuresti, Ser. Sti. Nat. Chim., 1968, 17, 41

68PP   P.K. Patnaik and S. Pani, Indian J. Chem., 1968, 6, 658

68PS   D.D. Perrin and I.G. Sayce, J. Chem. Soc. (A), 1968, 53

68PT   C.K. Poon and M.L. Tobe, Inorg. Chem., 1968, 7, 2398

68RK   E.D. Romanenko and N.A. Kostromina, Russ. J. Inorg. Chem., 1968, 13, 958 (1840)

68RR   V.P.R. Rao and K.V. Rao, J. Inorg. Nucl. Chem., 1968, 30, 2445

68RRM   C. Ropars, M. Rougee, M. Momenteau, and D. Lexa, J. Chim. Phys., 1968, 65, 823

68RS   S. Ramamoorthy and M. Santappa, Curr. Sci. (India), 1968, 37, 403

68RSa   S. Ramamoorthy and M. Santappa, J. Inorg. Nucl. Chem., 1968, 30, 2393

68Sa   M.I. Shtokalo, Soviet Prog. Chem. (Ukr. Khim. Zh.), 1968, 34, No. 11, 70 (1172)

68Sb   M.I. Shtokalo, Russ. J. Inorg. Chem., 1968, 13, 392 (748)

68SG   R.S. Saxena and K.C. Gupta, J. Indian Chem. Soc., 1968, 45, 609

68SH   K. Suzuki, T. Hattori, and K. Yamasaki, J. Inorg. Nucl. Chem., 1968, 30, 161

68SKM   K. Suzuki, C. Karaki, S. Mori, and K. Yamasaki, J. Inorg. Nucl. Chem., 1968, 30, 167

68SP   J.P. Scharff and M.R. Paris, Bull. Soc. Chem. France, 1968, 3184

68SS   N.F. Savenko and I.A. Sheka, Soviet Progr. Chem. (Ukr. Khim. Zh.), 1968, 34, No. 4, 1 (309)

68T   I. Torko, Magyar Kem. Foly., 1968, 74, 590

68TN   K.C. Trikha, B.C. Nair, and R.P. Singh, Indian J. Chem., 1968, 6, 532

68TR   Ya. I. Turyan and O.E. Ruvinskii, Doklady Phys. Chem., 1968, 179, 170 (148)

68VB   E. Verdier and R. Bennes, J. Chim. Phys., 1968, 65, 1465

68WF   C. Woodward and H. Freiser, Anal. Chem., 1968, 40, 345

69AL   A. Aziz and S.J. Lyle, Anal. Chim. Acta, 1969, 47, 49

69ALb   A. Aziz and S.J. Lyle, J. Inorg. Nucl. Chem., 1969, 31, 2531

69AM   L.P. Adamovich, A.P. Mirnaya, and A.K. Khukhryanskaya, J. Anal. Chem. USSR, 1969, 24, 1473 (1816)

69AS   S.S. Arslanova, A.M. Sorochan, M.M. Senyavin, and Kh. R. Rakhimov, Uzbeksh. Khim. Zh., 1969, 4, 32

69B   A.M. Bond, J. Electroanal. Chem., 1969, 20, 109

69Ba   A.M. Bond, J. Electroanal. Chem., 1969, 20, 223

69BB   K. Balachandran and S.K. Banerji, Indian J. Chem., 1969, 7, 185

69BBa   K. Balachandran and S.K. Banerji, Indian J. Chem., 1969, 7, 297

69BBb   K. Balachandran and S.K. Banerji, J. Indian Chem. Soc., 1969, 46, 198

69BBH   M. Bartusek, L. Brchan, and L. Havelkova, Spisy. Prir. Fak. Univ. Purk. Brne, 1969, No. 499, 19

69BF   D. Braun-Steinle and S. Fallab, Chimia (Switz), 1969, 23, 269

69BFa   A.I. Busev and V.Z. Filip, Russ. J. Inorg. Chem., 1969, 14, 1699 (3221)

69BH   G.B. Briscoe and S. Humphries, Talanta, 1969, 16, 1403

69BIB   A.I. Busev, V.M. Ivanov, and L.I. Bogodanovich, Moscow Univ. Chem. Bull., 1966,
         24, No. 3, 63 (86)

69BIN   A.I. Busev, V.M. Ivanov, and Zh. I. Nemtseva, J. Anal. Chem. USSR, 1969, 24,
         299 (414)

69BK   E.A. Biryuk and T.M. Karmalyuk, Russ. J. Inorg. Chem., 1969, 14, 193 (375)

69BL   U.C. Bhattacharyya, S.C. Lahiri, and S. Aditya, J. Indian Chem. Soc., 1969, 46,
         247

69BM   A. Banerjee, S. Mandal, T. Singh, and A.K. Dey, Indian J. Chem., 1969, 7, 733

69BN   A.I. Busev, Zh. I. Nemtseva, and V.M. Ivanov, J. Anal. Chem. USSR, 1969, 24,1111
         (1376)

69CA   T.A. Chernova, K.V. Astakhov, and S.A. Barkov, Russ. J. Phys. Chem., 1969. 43.
         1570 (2796)

69CM   D.K. Cabbiness and D.W. Margerum, J. Amer. Chem. Soc., 1969, 91, 6540

69CMa   G.F. Condike and A.E. Martell, J. Inorg. Nucl. Chem., 1969, 31, 2455

69CP   C.W. Childs and D.D Perrin, J. Chem. Soc. (A), 1969, 1039

69D   H. Diebler, Z. Phys. Chem. (Frankfurt), 1969, 68, 64

69DD   Y. Dartiguenave, M. Dartiguenzve, and J.P. Walter, Bull. Soc. Chim. France, 1969
         2287

69DI   D. Dyrssen, E. Ivanova, and K. Aren, J. Chem. Ed., 1969, 46, 252

69DM   N.A. Dobrynina, L.I. Martynenko, and V.I. Spitsy, Bull. Acad. Sci. USSR, 1969
         912 (1000)

69DP   F.M. D'Itri and A.I. Popov, J. Inorg. Nucl. Chem., 1969, 31, 1069

69DS   N.K. Dutt and T. Seshadri, J. Inorg. Nucl. Chem., 1969, 31, 2153

69EW   S.H. Eberle and U. Wede, Inorg. Nucl. Chem. Letters, 1969, 5, 5

69F   D.P. Fay, Diss., Oklahoma State Univ., 1969; Diss. Abstr. Int. B, 1971, 31,
         4610

69FG   N.S. Frumina, N.N. Goryunova, and I.S. Mustafin, Russ. J. Inorg. Chem., 1969,
         14, 790 (1510)

69FK   V.A. Fedorov, M.Ya. Kutuzova, and V.E. Mironov, Fiz. Khim. Khim. Tekhnol.,
         1969, 326; Chem. Abstr., 1971, 75, 155455z

69G   W.B. Guenther, J. Amer. Chem. Soc., 1969, 91, 7619

69GD   J. Goffart and G. Duychaerts, Anal. Chim. Acta, 1969, 48, 99

69GT   S.S. Goyal and J.P. Tandon, Talanta, 1969, 16, 106

69HL   D. Hopgood and D.L. Leussing, J. Amer. Chem. Soc., 1969, 91, 3740

69HS   M. Hnilickova and L. Sommer, Talanta, 1969, 16, 83

69IBN   V.M. Ivanov, A.I. Busev, and Zh. I. Nemtseva, Moscow Univ. Chem. Bull., 1969,
         24, No. 5, 60 (80)

69IBP   V.M. Ivanov, A.I. Busev, L.V. Popova, and L.I. Bogdanovich, J. Anal. Chem. USSR,
         1969, 24, 850 (1064)

69JM   A.P. Joshi and K.N. Munshi, J. Indian Chem. Soc., 1969, 46, 58

69JP   E. Jercan and G. Popa, An. Univ. Bucuresti, Chim., 1969, 18, 43

69KA   V.K. Khakimova and P.K. Agasyan, Uzb. Khim. Zh., 1969, 13, 6; Chem. Abstr.,
         1970, 72, 93834z

69KK   H. Koch and H. Kupsch, Z. Naturforsch., 1969, 24b, 398

69KP   Ts. B. Konunova, M.S. Popov, and K. Chan, Russ. J. Inorg. Chem., 1969, 14, 1084
            (2066)

69KPa   K. Kustin and R. Pizer, J. Amer. Chem. Soc., 1969, 91, 317

69KT   M. Kodama and Y. Tominaga, Bull. Chem. Soc. Japan, 1969, 42, 2267

69L   T. Lengyel, Acta Chim. Acad. Sci. Hung., 1969, 60, 225

69LA   B.E. Leach and R.J. Angelici, Inorg. Chem., 1969, 8, 907

69LC   T.T. Lai and M.C. Chen, Talanta, 1969, 16, 544

69LS   M. Langova-Hnilickova and L. Sommer, Talanta, 1969, 16, 681

69LSS   F.I. Lobanov, V.M. Savostina, L.V. Serzhenko, and V.M. Peshkova, Russ. J.
            Inorg. Chem., 1969, 14, 562 (1077)

69M   V.A. Milhailov, Russ. J. Inorg. Chem., 1969, 14, 1119 (2133)

69Ma   A.I. Moskvin, Soviet Radiochem., 1969, 11, (458)

69MB   R.E. Mesmer and C.F. Baes, Jr., Inorg. Chem., 1969, 8, 618

69MBJ   G.S. Manku, A.N. Bhat, and B.D. Jain, J. Inorg. Nucl. Chem., 1969, 31, 2533

69MD   D.R. Motiani, S.S. Dube, and M.L. Mittal, Z. Naturforsch., 1969, 24b, 458

69MG   R. Munze, A. Guthert, and H. Matthes, Z. Phys. Chem. (Leipzig), 1969, 241, 240

69MK   G. Manoussakis and Th. Kouimtzis, J. Inorg. Nucl. Chem., 1969, 31, 3851

69MM   M.S. Michailidis and R.B. Martin, J. Amer. Chem. Soc., 1969, 91, 4683

69MP   V. Magearu and Gr. Popa, Rev. Roum. Chim., 1969, 14, 1399

69MS   S.P. Mushran and L. Sommer, Coll. Czech. Chem. Comm., 1969, 34, 3693

69NK   T. Nozaki, K. Koshiba, and Y. Ono, Nippon Kagaku Zasshi, 1969, 90, 1147

69NM   V.A. Nazarenko and N.I. Makrinich, J. Anal. Chem. USSR, 1969, 24, 1373 (1694)

69NN   V.A. Nazarenko and E.M. Nevskaya, J. Anal. Chem. USSR, 1969, 24, 670 (839)

69P   T.N. Pliev, Doklady Chem., 1969, 184, 123 (1113)

69Pa   A.D. Pethybridge, J. Chem. Soc. (A), 1969, 1345

69PKD   S.D. Paul, M.S. Krishnan, and C. Dass, Indian J. Chem., 1969, 7, 299

69PKK   S.A. Popova, B.P. Karadakov, P.N. Kovalenko, and K.N. Bagdasarov, J. Anal. Chem.
            USSR, 1969, 24, 531 (682)

69PM   Gr. Popa and V. Magearu, Rev. Roum. Chim., 1969, 14, 879

69PP   M.M. Petit-Ramel and M.R. Paris, Bull. Soc. Chim. France, 1969, 3070

69PPa   J. Prasad and N.C. Peterson, Inorg. Chem., 1969, 8, 1622

69PS   D.D. Perrin and V.S. Sharma, J. Chem. Soc. (A), 1969, 2060

69PSa   N.S. Pouektov and M.A. Sandu, J. Anal. Chem. USSR, 1969, 24, 1191 (1472)

69PV   A.V. Pavilinova and T.D. Vyotskaya, Soviet Progr. Chem. (Ukr. Khim. Zh.), 1969,
            35, No. 1, 36 (37)

69RB   P.S. Relan and P.K. Bhattacharya, J. Indian Chem. Soc., 1969, 46, 534

69RK   D.L. Rabenstein and R.J. Kula, J. Amer. Chem. Soc., 1969, 91, 2492

69S   M.I. Shtokalo, Soviet Progr. Chem. (Ukr. Khim. Zh.), 1969, 35, No. 4, 51 (393)

69Sa   A.Ya. Sychev, Russ. J. Inorg. Chem., 1969, 14, 506 (971)

69Sb   S.I. Smyshlyaev, Izv. Vyssh. Ucheb. Zaved., Khim., 1969, 12, 384

69SB   K.C. Srivastava and S.K. Banerji, J. Prakt. Chem., 1969, 311, 769

69SF   H.B. Silber, R.D. Farina, and J.H. Swinehart, Inorg. Chem., 1969, 8, 819

69SL   J. Stary and J.O. Liljenzin, Radiochem. Radioanal. Letters, 1969, 1, 273

69SO   T. Shimizu and K. Ogami, Talanta, 1969, 16, 1527

69ST   T. Sakai and K. Tonosaki, Bull. Chem. Soc. Japan, 1969, 42, 2718

69SZ   L.Sommer and M. Zemcikova, Spisy Prir. Fak. Univ. Purk. Brne, 1969, E37, 236

69T   C. Tsymball, CEA-R-3476; Chem. Abstr. 1970, 72, 6684j

69Ta   B. Topuzovski, God. Zb., Prir.-Mat. Fak. Univ., Skopje, Mat., Fiz. Hem., 1969,
        19, 71

69Tb   L.I. Tikhonova, Russ. J. Inorg. Chem., 1969, 14, 1245 (2368)

69TD   V.Ya. Temkin, N.M. Dyatlova, G.F. Yaroshenko, O.Yu.Lavrova, and R.P. Lastovskii,
        J. Anal. Chem. USSR, 1969, 24, 135 (240)

69TE   G.N. Tyurenkova, L.G. Egorova, K.N. Klyachina and I.Ya. Postovskii, J. Gen. Chem.
        USSR, 1969, 39, 2286 (2251)

69TK   N.N. Tananaeva and N.A. Kostromina, Russ. J. Inorg. Chem., 1969, 14, 631 (1205)

69TN   I.A. Tserkovnitskaya and E.I. Novikova, J. Anal. Chem. USSR, 1969, 24, 929
        (1160)

69YH   O. Yamauchi, Y. Hirano, Y. Nakao and A. Nakahara, Canad. J. Chem., 1969, 47,
        3441

69Z   A. Zuberbuhler, Chimia (Switz.), 1969, 23, 416

69ZS   M. Zemcikova and L. Sommer, Spisy Prir. Fak. Univ. Purk. Brne, 1969, E37, 199

70AB   B.K. Avinashi and S.K. Banerji, J. Indian Chem. Soc., 1970, 47, 177

70ABa   B.K. Avinashi and S.K. Banerji, J. Indian Chem. Soc., 1970, 47, 453

70AL   A. Aziz and S.J. Lyle, J. Inorg. Nucl. Chem., 1970, 32, 1925

70AM   G.A. Artyukhina, L.I. Martynenko, and V.I. Spitzin, Bull. Acad. Sci, USSR, 1970,
        477 (522)

70BB   K. Balachandran and S.K. Banerji, J. Inorg. Nucl. Chem., 1970, 32, 3333

70BBa   K. Balachandran and S.K. Banerji, J. Indian Chem. Soc., 1970, 47, 343

70BBb   K. Balachandran and S.K. Banerji, J. Indian Chem. Soc., 1970, 47, 353

70BG   D.W. Bisacchi and H. Goldwhite, J. Inorg. Nucl. Chem., 1970, 32, 961

70BH   J.N. Butler and R. Huston, Anal. Chem., 1970, 42, 1308

70BHa   J.N. Butler and R. Huston, J. Phys. Chem., 1970, 74, 2976

70CB   M.V. Chidambaram and P.K. Bhattacharya, J. Inorg. Nucl. Chem., 1970, 32, 3271

70CBa   M.V. Chidambaram and P.K. Bhattacharya, J. Indian Chem. Soc., 1970, 47, 881

70CM   L.C. Coombs and D.W. Margerum, Inorg. Chem., 1970, 9, 1711

70CMa   L. Czegledi and C. Gh. Macarovici, Rev. Roum. Chim., 1970, 15, 1741

70CS   N.V. Chernaya and S.Ya. Shnaiderman, J. Anal. Chem. USSR, 1970, 24, 424 (495)

70D   B. Desire, NP-18284, 1970

70DB   C.D. Dwivedi and S.K. Banerji, J. Inorg. Nucl. Chem., 1970, 32, 688

70DD   S.S. Dube and S.S Dhindsa, J. Inorg. Nucl. Chem., 1970, 32, 543

70DDa   S.S. Dube and S.S Dhindsa, J. Indian Chem. Soc., 1970, 47, 489

70DF   A. Doadrio, N. Fanny Salas, and A. Perez Villacastin, An. Quim., 1970, 66, 19

70DG   P. Donatsch, K.H. Gerber, A. Zuberbuhler, and S. Fallab, Helv. Chim. Acta, 1970,
          53, 262

70DP   C. Dragulesco, S. Policec, and T. Simonescu, Talanta, 1970, 17, 557

70DPa   C. Dragulescu, S. Polichek, and T. Simonescu, Russ. J. Inorg. Chem., 1970,
          15, 1129 (2192)

70DT   G.S. Dokolina, Ya.I. Turyan, and O.N. Malyavinskaya, Russ. J. Phys. Chem., 1970,
          44, 1679 (2942)

70EM   A.N. Ermakov, I.N. Marov, and N.B. Kalinichenko, Russ. J. Inorg. Chem., 1970,
          15, 879 (1712)

70EN   B. Evtimova and D. Nonova, Compt. Rend. Acad. Bulg. Sci., 1970, 23, 1111

70EP   J.H. Espenson and J.R Pladziewicz, Inorg. Chem., 1970, 9, 1380

70FM   O. Farooq, A.U. Malik, and N. Ahmad, J. Electroanal. Chem., 1970, 24, 233

70FMa   O. Farooq, A.U. Malik, and N. Ahmad, J. Electroanal. Chem., 1970, 26, 411

70FMb   O. Farooq, A.U. Malik, and N. Ahmad, and S.M.F. Rahman, J. Electroanal. Chem.,
          1970, 24, 464

70FT   J. Ferguson and M.L Tobe, Inorg. Chim, Acta, 1970, 4, 109

70GD   N.N. Ghosh and M. Dasgupta, Z. Anorg. Allg. Chem., 1970, 375, 315

70GF   B. Grabaric and I. Filpovic, Croat. Chem. Acta, 1970, 42, 479

70GH   S.L. Grassino and D.N. Hume, J. Inorg. Nucl. Chem., 1970, 32, 3112

70GK   Z. Gregorowicz and Z. Klima, Rocz. Chem., 1970, 44, 503

70GKa   M.I. Gelfman and N.A. Kustova, Russ. J. Inorg. Chem., 1970, 15, 47 (92)

70GKb   M.I. Gelfman and N.A. Kustova, Russ. J. Inorg. Chem., 1970, 15, 1602 (3076)

70GM   A.P. Gerbeleu and P.C. Migal, Proc. 13th I.C.C.C., Poland, 1970, Vol.1, 57 (69)

70GMa   V.I. Gordienko and Yu. I. Mikhailyuk, J. Anal. Chem. USSR, 1970, 25, 1946 (2267)

70GMT   S.S. Goyal, G.J. Misra, and J.P. Tandon, Bull. Acad. Pol. Sci., Ser. Sci, Chem.,
          1970, 18, 425

70GN   R. Ghosh and V.S.K. Nair, J. Inorg. Nucl. Chem., 1970, 32, 3025

70GO   T. Goina, M. Olariu, and L. Bocaniciu, Rev. Roum. Chim., 1970, 15, 1049

70GP   R. Griesser, B. Prijs, H. Sigel, W. Fory, L.D. Wright and D.B. McCormick,
          Biochemistry, 1970, 9, 3285

70GS   R. Griesser and H. Sigel, Inorg. Chem., 1970, 9, 1238

70H   Y. Hasegawa, Bull. Chem. Soc. Japan, 1970, 43, 2665

70IE   B.N. Ivanov-Emin, A.M. Egorov, V.I. Romanyuk, and E.N. Siforova, Russ. J. Inorg.
          Chem., 1970, 15, 628 (1224)

70IK   N. Ikeda, K. Kimura, H. Asai, and N. Oshima, Radioisotopes, 1970, 19, 1

70IS   H.M.N.H. Irving and S.P. Sinha, Anal. Chim. Acta, 1970, 49, 449

70IVb   A.A. Ivakin and E.M. Voronova, Trudy Inst. Khim. Uralsh. Fil. Akad. Nauk SSSR,
          1970, 17, 144

70IY   B.Z. Iofa and A.S. Yushchenko, Soviet Radiochem, 1970, 12, (65)

70JP   T.J. Janic, L.B. Pfendt, and M.B. Celap, Z. Anorg. Allg. Chem., 1970, 373, 83

70K   M. Koskinen, Ann. Acad. Sci. Fenn., 1970, AII, No. 155

70Ka   K. Kleboth, Monat. Chem., 1970, 101, 767

70KA   H. Kakinhana, T. Amaya, and M. Maeda, Bull. Chem. Soc. Japan, 1970, 43, 3155

70KAT   N. Kitajiri, T. Arishima, and S. Takamoto, Nippon Kagaku Zasshi, 1970, 91, 240

70KB   B. Karadakov, D. Bodkova, and A. Alexieva, Compt. Rend. Acad. Bulg. Sci. 1970,
          11,1385

70KC   G.C. Kugler and G.H. Carey, Talanta, 1970, 10, 907

70KCa   B. Kuznik and D.M. Czakis-Sulikowska, Rocz. Chem., 1970, 44, 1155

70KG   Z. Klima and Z. Gregorowicz, Pr. Nauk. Univ. Slack. Katowicach, 1970, 9, 15;
          Chem. Abstr., 1972, 76, 145467m

70KM   H. Kakihana and M. Maeda, Bull. Chem. Soc. Japan, 1970, 43, 109

70KMa   G.N. Kupriyanova and L.I. Martynenko, Russ. J. Inorg. Chem., 1970, 15, 1024
          (1991)

70KMB   N. Kumar, G.S. Manku, A.N. Bhat, and B.D. Jain, J. Less-Common Metals, 1970, 21,
          23

70KN   B.P. Karadakov, P.P. Nenova, and D. St. Kyncheva, Russ. J. Inorg. Chem., 1970,
          15, 216 (417)

70KP   S. Krzewska and L. Pajdowski, Rocz. Chem., 1970, 44, 249

70L   T. Lengyel, Acta Chim. Acad. Sci, Hung., 1970, 64, 331

70La   G.M. Lafon, Geochim. Cosmochim. Acta, 1970, 34, 935

70LN   R. Larsson and G. Nunziata, Acta Chem. Scand., 1970, 24, 2156

70LT   T. Lengyel and J. Torko, Acta Chim. Acad. Sci. Hung., 1970, 62, 151

70MA   L.E. Martynenko, G.A. Artyukhina, N.G. Bogdanovich, and N.I. Pechurova, Russ. J
          Inorg. Chem., 1970, 15, 924 (1799)

70MH   T.R. Musgrave and E.R. Humburg. Jr., J. Inorg. Nucl. Chem., 1970, 32, 2229

70MKS   S.A. Merkusheva, V.N. Kumok, N.A. Skorik, and V.V. Serebrennikov, Soviet
          Radiochem., 1970, 12, 155 (175)

70MS   H. Mizuochi, S. Shirakata, E. Kyuno and R. Tsuchiya, Bull. Chem. Soc. Japan,
          1970, 43, 397

70MU   A. Martin and E. Uhlig. Z. Anorg. Allg. Chem., 1970, 375, 166

70MUa   A. Martin and E. Uhlig, Z. Anorg. Allg. Chem., 1970, 376, 282

70NE   D. Nonova and B. Evtimova, Anal. Chim. Acta, 1970, 49, 103

70NM   V.A. Nazarenko and N.I. Makrinich, J. Anal. Chem. USSR, 1970, 25, 620 (719)

70NS   N. Nakamura and E. Sekido, Talanta, 1970, 17, 515

70OM   G.K.S. Ooi and R.J. Magee, J. Inorg. Nucl. Chem., 1970, 32, 3315

70PL   G. Popa, R. Lereh, and I.C. Cazacu, An. Univ. Bucuresti, Chim., 1970, 19, 9

70PM   J. Pradel and R.P. Martin, Compt. Rend. Acad. Sci. Paris, Ser. C, 1970, 270,
          1863

70PMa   Gr. Popa and V. Magearu, An. Univ. Bucuresti, Chim., 1970, 19, 45

70PR   E.M. Piskunov and A.G. Rykov, Sci. Res. At. Reactor. Inst., Rep., 1970, SRARI-
          p-92

70S   P. Stantcheva, Trav. Sci. Ecol. Norm. Sup. Plovdiv.(Bulg.) 1970, 8, 103

70Sc   T. Seshadri, Indian J. Chem., 1970, 8, 282

70SKS   G. Stockelmann, A. Kettrup, and H. Specker, Z. Anorg. Allg. Chem., 1970, 372,
          134, 144

70SB   K.C. Srivastava and S.K. Banerji, J. Indian Chem. Soc., 1970, 47, 225

70SBa   K.C. Srivastava and S.K. Banerji, Chim. Anal., 1970, 52, 973

70SK   G.A. Selivanova and N.T. Kudryavtsev. Tr. Mosk. Khim. Tekhnol. Inst., 1970,
          No. 67, 79

70SM   S.C. Shrivastawa, K.N. Munshi, and A.K. Dey, J. Indian. Chem. Soc., 1970, 47,
            1013

70SY   Y. Sugiura, A. Yokoyama, and H. Tanaka, Chem. Pharm. Bull. (Japan), 1970, 18, 693

70TD   Ya. I. Turyan, G.S. Dokolina, and M.A. Korshunov, J. Gen. Chem. USSR, 1970, 40,
            1874 (1894)

70TN   G. Tridot, J. Nicole, and M. Wozniak, Chim. Anal. (Paris), 1970, 52, 265

70TS   K. Tunaboylu and G. Schwarzenbach, Chimia (Switz.), 1970, 24, 424

70TSY   H. Tanaka, H. Sakurai, and A. Yokoyama, Chem. Pharm. Bull. (Japan), 1970, 18,
            1015

70VM   G.L. Varlamova, L.I. Martynenko, N.I. Pechurova, and V.I. Spitsin, Russ. J.
            Inorg. Chem., 1970, 15, 1108 (2151)

70WC   R. Wojtas and D.M. Czakis-Sulikowska, Rocz. Chem., 1970, 44, 981

70YS   A. Yokoyama, H. Sakurai, and H. Tanaka, Chem. Pharm. Bull. (Japan), 1970, 18,
            1021

71AB   B.K. Avinashi and S.K. Banerji, J. Indian Chem. Soc., 1971, 48, 174

71AG   L.P. Adamovich, A.P. Gershuns, A.A. Oleinik, and N.M. Shkabara, J. Anal. Chem.
            USSR, 1971, 26, 471 (548)

71AH   G. Ackermann, D. Hesse, and E.H. Muller, Z. Anorg. Allg. Chem., 1971, 382, 157

71AM   Y. Ayabe and H. Matsuda, Denki Kagaku, 1971, 39, 635

71AO   S. Akalin and U.Y. Ozer, J. Inorg. Nucl. Chem., 1971, 33, 4171

71AP   L.A. Albota, A.V. Pavlinova, and R.N. Khomitskaya, Izv. Vys. Uch. Zav. Khim.,
            1971, 14, 675

71AW   G. Anderegg and F. Wenk, Helv. Chim. Acta, 1971, 54, 216

71B   A.M. Bond, Coord. Chem. Rev., 1971, 6, 377

71BB   H. Bednar, V. Bednar, L. Oprei, and M. Suciu, Z. Chem., 1971, 11, 112

71BG   V.P. Biryukov and E. Sh. Ganelina, Russ. J. Inorg. Chem., 1971, 16, 320 (600)

71BR   E.A. Biryuk and R.V. Ravitskaya, J. Anal. Chem. USSR, 1971, 26, 637 (735), 1576
            (1767)

71BS   G.A. Bhat and R.S. Subrahmanya, Proc. Indian Acad. Sci., 1971, 73A, 157

71BV   E. Bottari and M. Vicedomini, Gazz. Chim. Ital., 1971, 101, 661

71C   A.A. Cherkesov, Russ. J. Inorg. Chem., 1971, 16, 952 (1794)

71CL   C.T. Chang and C.F. Liaw, J. Inorg. Nucl. Chem., 1971, 33, 2623, 2717

71CW   R.F. Childers, R.A.D. Wentworth, and L.J. Zompa, Inorg. Chem., 1971, 10, 302

71DG   S.N. Drozdova, L.I. Gen, A.P. Monsenko, and M.A Yampolskii, Russ. J. Inorg.
            Chem., 1971, 16, 1111 (2082)

71ES   L.G. Egorova, N.V. Serebryakova, and G.N. Tyurenkova, J. Gen. Chem. USSR, 1971,
            41, 1816 (1807)

71EV   V.A. Ermakov, V.V. Vorobieva, A.A. Zaitsev, and G.N. Yakovlev, Soviet Radiochem.,
            1971, 13, (840)

71EZ   A.A. Elesin and A.A. Zaitsev, Soviet Radiochem., 1971, 13, (775)

71FK   V.A. Fedorov, T.N. Kalosh, and V.E. Mironov, Russ. J. Inorg. Chem., 1971, 16,
            1596 (3006)

71FT   Y. Funae, N. Toshioka, I. Mita, T. Sugihara, T. Ogura, Y. Nakamura, and
            S. Kawaguchi, Chem. Pharm. Bull. (Japan), 1971, 19, 1618

71FY   S. Funahashi, S. Yamada, and M. Tanaka, Inorg. Chem., 1971, 10, 257

71GB   P.K. Govil and S.K. Banerji, J. Indian Chem. Soc., 1971, 48, 1095

71GBG   R.I. Gorelova, V.A. Babich and I.P. Gorelov, Russ. J. Inorg. Chem., 1971, 16, 995
        (1873)

71GG   A.L. Gershuns and L.G. Grineva, J. Anal. Chem. USSR, 1971, 26, 1327 (1485)

71GK   Z. Gregorowicz, G. Kwapulinska, and Z. Klima, Rocz. Chem., 1971, 45, 1159

71GN   A. Gergely, I. Nagypal, and I. Sovago, Acta Chim. Acad. Sci. Hung., 1971, 67,
        241

71HP   P.S. Hallman, D.D. Perrin, and A.E. Watt, Biochem. J., 1971, 121, 549

71IK   N. Ivicic, B. Kuznar, and V. Simeon, Croat. Chem. Acta, 1971, 43, 237

71IS   E. Inada, K. Shimizu, and J. Osugi, Nippon Kagaku Zasshi, 1971, 92, 1096

71KA   B. Karadakov, A. Aleksieva, D. Venkova, and D. Boikova, God. Vissh.
        Khimikotekhnol. Inst., Sofia, 1971, 18, 125, 185

71KN   B.P. Karadakov, P. Nenova, and K.S. Kuncheva, Khim. Ind. (Sofia), 1971, 43, 163

71KP   J. Kloubek and J. Podlaha, J. Inorg. Nucl. Chem., 1971, 33, 2981

71KS   I.A. Korshunov and G.M. Sergeev, Soviet Rodiochem., 1971, 13, 929 (901)

71KT   K. Kina and K. Toei, Bull. Chem. Soc. Japan, 1971, 44, 1289

71KTa   N.A. Kostromina and N.N. Tananaeva, Russ. J. Inorg. Chem., 1971, 16, 1256 (2356)

71KV   M.M. Kryzhanovskii, Yu. A. Volokhov, L.N. Pavlov, N.I. Eremin, and V.E. Mironov,
        J. Appl. Chem. USSR, 1971, 44, 484 (476)

71KY   E.E. Kriss and K.B. Yatsimirskii, Russ. J. Inorg. Chem., 1971, 16, 202 (386)

71KYK   F.Ya. Kulba, Yu. B. Yakovlev, and E.A. Kopylov, Russ. J. Phys. Chem., 1971, 45,
        408 (727)

71BL   B. Lenarcik, M. Badyoczek-Grzonka, and Z. Grzonka, Rocz. Chem., 1971, 46, 2023

71LF   A.E. Laubscher and K.F. Fouche, J. Inorg. Nucl. Chem., 1971, 33, 3521

71LFG   F.I. Lobanov, V.M. Feskova, and I.M. Gibalo, Russ. J. Inorg. Chem., 1971, 16,
        414 (776)

71LS   P. Letkeman and D.T. Sawyer, Canad. J. Chem., 1971, 49, 2096

71LW   P. Letkeman and J.B. Westmore, Canad. J. Chem., 1971, 49, 2073

71LWa   P. Letkeman and J.B. Westmore, Canad. J. Chem., 1971, 49, 2086

71M   R.B. Martin, J. Phys. Chem., 1971, 75, 2657

71Ma   A.I. Moskvin, Soviet Radiochem., 1971, 13, (582)

71Mb   A.I. Moskvin, Soviet Radiochem., 1971, 13, (575)

71Mc   A.I. Moskvin, Soviet Radiochem., 1971, 13, (641)

71Md   S. Motomizu, Anal. Chim. Acta, 1971, 56, 415

71MA   T.V. Mikhailova, K.V. Astakhov, and N.M. Zhirnova, Russ. J. Phys. Chem., 1971,
        45, 618 (1106)

71MC   P.K. Migal, N.G. Chebotar, and A.M. Sorochinskaya, Russ. J. Inorg. Chem., 1971,
        16, 968 (1823)

71MM   P.J. Morris and R.B. Martin, Inorg. Chem., 1971, 10, 964

71MMW   J.P. Manners, K.G. Morallee, and R.J.P. Williams, J. Inorg. Nucl. Chem., 1971,
        33, 2085

71MS   R. Murai, T. Sekine, and S. Iwahori, Nippon Kagaku Zasshi, 1971, 92, 967

71MSa   R. Murai, T. Sekine, and M. Iguchi, Nippon Kagaku Zasshi, 1971, 92, 1019

71MT   T. Mikami and S. Takei, J. Inorg. Nucl. Chem., 1971, 33, 4283

71NP   B.P. Nikolskii, V.V. Palchevskii, and N.M. Okun, Proc. Acad. Sci. USSR, Doklady
          Chem., 1971, 198, 470 (851)

71O   G. Olofsson, J. Chem. Thermodyn., 1971, 3, 217

71OM   G.P. Ozerova, N.V Melchakova, and V.M. Peshkova, Moscow Univ. Chem. Bull., 1971,
          26, No. 3, 77 (367)

71P   S. Petri, Rocz. Chem., 1971, 45, 529

71PB   M. Polasek and M. Bartusek, Scripta Fac. Sci. Nat. Univ. Purk. Brun., 1971, 1,
          109

71PP   J. Prasad and N.C. Peterson, Inorg. Chem., 1971, 10, 88

71PR   E.M. Psikunov and A.G. Rykov, Soviet Radiochem., 1971, 13, (62)

71PV   A.V. Pavlinova and T.D. Vysotskaya, Izv. Vyssh. Ucheb. Zaved, Khim., 1971, 14,
          1517

71RC   A.V. Radushev, A.N. Chechneva, and I.I. Mudretsova, J. Anal. Chem. USSR, 1971,
          26, 694 (796)

71RD   K.S. Rajan, J.M. Davis, and R.W. Colburn, J. Neurochem., 1971, 18, 345

71RDB   R.A. Robinson, W.C. Duer, and R.G. Bates, Anal. Chem., 1971, 43, 1862

71RM   B. Rao and H.B. Mathur, J. Inorg. Nucl. Chem., 1971, 33, 2919

71RN   E.D. Romanenko, L.V. Novikova, and N.A. Kostromina, Russ. J. Inorg. Chem., 1971,
          16, 821 (1554)

71S   A.B. Shalinets, Soviet Radiochem., 1971, 13, (566)

71Sf   A.V. Stepanov, Russ. J. Inorg. Chem., 1971, 16, 1583 (2981)

71SE   N.V. Serebryakova, L.G. Egorova and G.N. Tyurenkova, J. Gen. Chem. USSR, 1971,
          41, 1821 (1812)

71SG   A.I. Sevastyanov, I.L. Gorodetskaya, and N.P. Rudenko, Moscow Univ. Chem. Bull.,
          1971, 26, No. 3, 45 (328)

71SI   T. Sekine and N. Ihara, Bull. Chem. Soc. Japan, 1971, 44, 2942

71SSa   K. Srinivasan and R.S. Subrahmanya, J. Electroanal. Chem., 1971, 31, 257

71SR   M.A. Salam and M.A. Raza, Chem. Ind. (London), 1971, 601

71TA   J.A. Thomson, and G.F. Atkinson, Talanta, 1971, 18, 935; (see also
          B.W. Budesinsky, Z. Anal. Chem., 1973, 267, 43)

71TDa   P.H. Tedesco and V.B. DeRumi, J. Inorg. Nucl. Chem., 1971, 33, 3833

71TDG   P.H. Tedesco, V.B. DeRumi, and J.A. Gonzalez Quintana, J. Inorg. Nucl. Chem.,
          1971, 33, 3839

71TT   V.F. Toropova, O. Yu. Timofeeva, and E.P. Evteeva, J. Anal. Chem. USSR, 1971,
          26, 1381 (1545)

71WB   E. Wendling, O. Benali-Baitich, and G. Yaker, Rev. Chim. Minerale, 1971, 8, 559

71WN   D. Waysbort and G. Navon, Chem. Comm., 1971, 1410

71YS   A. Yokoyama, H. Sakurai, and H. Tanaka, Chem. Pharm. Bull. (Japan), 1971, 19,
          1089

71Z   A.V. Zholnin, Russ. J. Inorg. Chem., 1971, 16, 537 (1010)

71ZP   A.V. Zholnin and V.N. Podchainova, Russ. J. Inorg. Chem., 1971, 16, 616 (1162)

72AK   A.I. Astakhov, E.N. Knyazeva, and S. Ya. Shnaiderman, J. Gen. Chem. USSR, 1972, 2496 (2505)

72AL   I. Ahlberg and I. Leden, Trans. Royal Inst. Tech. Stockholm, 1972, No. 249

72AP   R.P. Agarwal and D.D. Perrin, Trans. Royal Inst. Tech. Stockholm, 1972, No. 278

72B   J. Bjerrum, Acta Chem. Scand., 1972, 26, 2634

72BB   N.L. Babenko, A.I. Busev, and I.N. Chistyachenko, Russ. J. Inorg. Chem., 1972, 17, 966 (1864)

72BE   A.I. Busev, N.S. Ershova, and V.M. Ivanov, Russ. J. Inorg. Chem., 1972, 17, 538 (1036)

72BN   E.A. Biryuk, V.A. Nazarenko, and R.V. Ravitskaya, J. Anal. Chem. USSR, 1972, 27, 1755 (1934)

72BP   A. Bismondo, R. Portanova, P. DiBernardo, and L. Magon, Energ. Nucl. (Milan), 1972, 19, 402

72CG   E.F. Caldin, M.W. Grant, and B.B. Hasinoff, J. Chem. Soc. Faraday I, 1972, 68, 2247

72CM   S.C. Chang, J.K.H. Ma, J.T. Wang, and N.C. Li, J. Coord. Chem., 1972, 2, 31

72CP   E. Chiacchierini, V. Petrone, A.L. Magri, and F. Balestrieri, Gazz. Chim. Ital., 1972, 102, 911

72CT   A.K. Covington and J.M. Thain, J. Chem. Educ., 1972, 49, 554

72CV   H.S. Creyf and L.C. van Poucke, Thermochim. Acta, 1972, 4, 485

72DC   R. Dias Cadavieco, M.P. Collados de Dias, D. Dyrssen, and B. Egneus, Trans. Royal Inst. Tech. Stockholm, 1972, No. 276

72DS   I. Danielsson and P. Stenius, Trans. Royal Inst. Tech. Stockholm, 1972, No. 254

72DSB   S.F. Deryugina, I.A. Sheka, and L.P. Barchuk, Soviet Progr. Chem. (Ukr. Khim. Zh.), 1972, 38, No. 10, 1 (967)

72FA   J.J.R. Frausto da Silva and M.C.T. Abreu Vaz, Rev. Port. Quim., 1972, 14, 102

72FE   F.H. Fraser, P. Epstein, and D.J. Macero, Inorg. Chem., 1972, 11, 2031

72GK   J. Gross and C. Keller, J. Inorg. Nucl. Chem., 1972, 34, 725

72GS   S.A. Grachev, L.I. Shchelkunova, and Yu. A. Makashev, Russ. J. Inorg. Chem., 1972, 17, 706 (1364)

72H   J. Havel, Trans. Royal Inst. Tech. Stockholm, 1972, No. 277

72HA   S. Harada, K. Amidaiji, and T. Yasunaga, Bull. Chem. Soc. Japan, 1972, 45, 1752

72HK   E.J. Hakoila and J.J. Kankare, Suomen Kem., 1972, B45, 179

72HKa   T.M. Hseu and K.N. Kuo, J. Chinese Chem. Soc. (Taiwan), 1972, 19, 161

72HM   R.W. Hay and P.J. Morris, J. Chem. Soc. Perkin II, 1972, 1021

72JA   A.K. Jain, V.P. Aggarwala, P. Chand, and S.P. Garg, Talanta, 1972, 19, 1481

72K   V.I. Kornev, Russ. J. Phys. Chem., 1972, 46, 484 (834)

72KA   H. Kakihana, T. Amaya, and M. Maeda, Trans. Royal Inst. Tech. Stockholm, 1972, No. 251

72KEM   N.A. Krasnyanskaya, I.I. Eventova, N.V. Melchakova, and V.M. Peshkova, J. Anal. Chem. USSR, 1972, 27, 1672 (1842)

72KG   Z. Klima and Z. Gregorowicz, Pr. Nauk. Univ. Slask. Katowicach, 1972, 27, 21; Chem. Abstr., 1973, 78, 102635p

72KGa   S.C. Khurana and C.M. Gupta, J. Inorg. Nucl. Chem., 1972, 34, 2557

72KH   M. Koskinen and J. Hamalainen, Suomen Kem., 1972, B45, 191

72KI   O.K. Kudra, O.V. Izbekova, and V.V. Chelikidi, Izv. Vyssh. Ucheb. Zaved., Khim.,
         1972, 15, 667

72KNK   D. Kantcheva, P. Nenova, and B. Karadakov, Talanta, 1972, 19, 1450

72KNT   N.A. Kostromina, L.B. Novikova, and R.V. Tikhonova, Soviet Progr. Chem. (Ukr.
         Khim. Zh.), 1972, 38, No. 9, 5 (859)

72KS   Y. Kidani, R. Saito, and H. Koike, Chem. Letters (Japan), 1972, 729

72KT   N.A. Kostromina and T.V. Ternovaya, Russ. J. Inorg. Chem., 1972, 17, 825 (1596)

72L   N.M. Lukovskaya, Soviet Progr. Chem. (Ukr. Khim. Zh.), 1972, 38, No. 5, 73 (485)

72La   S. Lal, Aust. J. Chem., 1972, 25, 1571

72Lb   M. Larsson-Raznikiewicz, Eur. J. Biochem., 1972, 30, 579

72LH   G. Lesgards and J. Haladjian, J. Chim. Phys., 1972, 69, 1183

72LL   Y.H. Lee and G. Lundgren, Trans. Royal Inst. Tech. Stockholm, 1972, No. 267

72LN   P. Lumme and K. Nieminen, Suomen Kem., 1972, B45, 214

72LPa   P. Lumme, K. Ponkala and K. Nieminen, Suomen Kem., 1972, 45, 105

72LV   J.O. Liljenzin, K. Vadasdi, and J. Rydberg, Trans. Royal Inst. Tech. Stockholm,
         1972, No. 280

72LW   P. Letkeman and J.B. Westmore, Canad. J. Chem., 1972, 50, 3821

72MA   M. Maeda, T. Amaya, H. Ohtaki, and H. Kakihana, Bull. Chem. Soc. Japan, 1972, 45,
         2464

72MB   L. Magon, A. Bismondo, G. Tomat, and A. Cassol, Radiochim. Acta, 1972, 17, 164

72MP   M.L. Mittal and A.V. Pandey, J. Inorg. Nucl. Chem., 1972, 34, 2962

72MPa   R.P. Martin and J. Pradel, Trans. Royal Inst. Tech. Stockholm, 1972, No. 272

72MPV   S.P. Mushran, O. Prakash, and J.R. Verma, Bull. Chem. Soc. Japan, 1972, 45, 1709

72MR   S. Musumeci, E. Rizzarelli, I. Fragela, and S. Sammartano, Boll. Sedute Accad.
         Gioenica Sci. Natur. Catania, 1972, 11, 15; Chem. Abstr., 1976, 81, 6761t

72MSS   T.P. Makarova, G.S. Sinitsyna, I.A. Shestakova, A.V. Stepanov, and
         B.I. Shestakov, Soviet Radiochem., 1972, 14, (822)

72Nb   A. Napoli, J. Inorg. Nucl. Chem., 1972, 34, 1347

72Nc   A. Napoli, Gazz. Chim. Ital., 1972, 102, 724

72NB   M.S. Newman, D.H. Busch, G.E. Cheney, and C.R. Gustafson, Inorg. Chem., 1972, 11,
         2890

72NE   D. Nonova and B. Evtimova, Anal. Chim. Acta, 1972, 62, 456

72NI   J.S. Narayan, E.K. Ivanova, and V.M. Peshkova, Moscow Univ. Chem. Bull., 1972,
         27, No. 6, 62 (707)

72NP   N.M. Nikolaeva, A.V. Pirozhkov, and V.A. Antipina, Izv. Sibirsk. Otdel. Acad.
         Nauk. SSSR, 1972, 5, 143

72NT   R. Nasanen, P. Tilus, and E. Eskolin, Trans. Royal Inst. Tech. Stockholm, 1972,
         No. 273

72OK   H. Ohtaki and T. Kawai, Bull. Chem. Soc. Japan, 1972, 45, 1735

72OO   A. Ohyoshi, E. Ohyoshi, H. Ono, and S. Yamakawa, J. Inorg. Nucl. Chem., 1972, 34,
         1955

72OS   R. Osterberg and B. Sjoberg, Trans. Royal Inst. Tech. Stockholm, 1972, No. 275

720T   R. Osterberg and B. Toftgard, Bioinorg. Chem., 1972, 1, 295

72PA   L. Pettersson, I. Andersson, L. Lyhamn, and N. Ingri, Trans. Royal Inst. Tech. Stockholm, 1972, No. 256

72PB   S.K. Patel, I.M. Bhatt, and K.P. Soni, J. Inst. Chem. Calcutta, 1972, 44, 73

72PD   G. Popa and E. Dascalescu, An. Univ. Bucuresti, Chim., 1972, 21, 113

72PK   L. Pajdowski, S. Krzewska, and Z. Pruchnik, Trans. Royal Inst. Tech. Stockholm, 1972, No. 269

72PP   J. Pinart, C. Petitfaux, and J. Faucherre, Bull. Soc. Chim. France, 1972, 4534

72PR   A.T. Pilipenko, O.P. Ryabushko, N.L. Emchenko, L.A. Krivokhizhina, and L.D. Chukhno, Soviet Progr. Chem. (Ukr. Khim. Zh.), 1972, 38, 69, (1269)

72PS   S.K. Patel, K.P. Soni, and I.M. Bhatt, J. Inst. Chem. Calcutta, 1972, 44, 43

72RC   E.F.C.H. Rohwer, J.J. Cruywagen, and H.G. Raubenheimer, J. South African Chem. Inst., 1972, 25, 338

72RD   K.S. Rajan, J.M. Daivs, R.W. Colburn, and F.H. Jarke, J. Neurochem., 1972, 19,1099

72RH   A. Ringbom and L. Harju, Anal. Chim. Acta, 1972, 59, 49

72RK   A. Ringbom and B. Kyrklund, Trans. Royal Inst. Tech. Stockholm, 1972, No. 257

72RP   A.V. Radushev and E.N. Prokhorenko, J. Anal. Chem. USSR, 1972, 27, 2008 (2209)

72RV   E. Roletto, A. Vanni, and G. Ostacoli, J. Inorg. Nucl. Chem., 1972, 34, 2817

72S   S. Sjoberg, Acta Chem. Scand., 1972, 26, 3400

72Sa   H. Saarinen, Suomen Kem., 1972, B45, 219

72SG   G. Schwarzenbach, K. Gautschi, J. Peter, and K. Tunaboylu, Trans. Royal Inst. Tech. Stockholm, 1972, No. 271

72SKT   T. Sugano, T. Kitagawa, Y. Tsuda, T. Shibutani, and K. Kubo, Nippon Kagaku Kaishi, 1972, 93, 734

72SMa   R. Sarin and K.N. Munshi, J. Inorg. Nucl. Chem., 1972, 34, 581

72SN   M. Sakanoue and M. Nakatani, Bull. Chem. Soc, Japan, 1972, 45, 3429

72SS   M.K. Singh and M.N. Srivastava, J. Inorg. Nucl. Chem., 1972, 34, 2067

72SSa   M.K. Singh and M.N. Srivastava, J. Inorg. Nucl. Chem., 1972, 34, 2081

72SSb   M.K. Singh and M.N. Srivastava, Talanta, 1972, 19, 699

72SSc   M.K. Singh and M.N. Srviastava, Vijn. Par. Anus. Patr., 1972, 15, 63

72T   B. Topuzovski, God. Zb., Prir.-Mat. Fak. Univ., Skopje, Mat. Fiz. Hem., 1972, 22, 185; Chem. Abstr., 1973, 78, 165002a

72Ta   B. Topuzovski, God. Zb., Prir.-Mat. Fak. Univ., Skopje, Mat. Fiz. Hem., 1972, 22, 199; Chem. Abstr., 1973, 78, 164988q

72Tb   B. Topuzovski, God. Zb., Prir.-Mat. Fak. Univ., Skopje, Mat. Fiz. Hem., 1972, 22, 237; Chem. Abstr., 1973, 78, 164992m

72TA   P.H. Tedesco and M.C. Anon, J. Inorg. Nucl. Chem., 1972, 34, 2271

72TB   N.N. Tananaeva, E. Bryukher, and N.A. Kostromina, Russ. J. Inorg. Chem., 1972, 17, 1683 (3199)

72TK   T.V. Ternovaya and N.A. Kostromina, Soviet Progr. Chem. (Ukr. Khim. Zh.), 1972, 38, 14 (1211)

72TN   B. Topuzovski, D. Nikolovska, and K. Risteska, God. Zb., Prir.-Mat. Fak. Univ., Skopje, Mat. Fiz. Hem., 1972, 22, 193; Chem. Abstr., 1973, 78, 164988q

72TP   Ya. I. Turiyan, N.I. Pershakova, and O.E. Ruvinskii, J. Gen. Chem. USSR, 1972, 42, 1194 (1198)

72UT   H. Uchiyama and S. Takamoto, Nippon Kagaku Kaishi, 1972, 1084

72UW   P. Ulmgren and O. Wahlberg, Trans. Royal Inst. Tech. Stockholm, 1972, No. 274

72V   V.F. Vetere, Lab. Ensayo Mat. Inv. Tech. An., Ser. 2, 1972, 3, 79

72VG   M.N. Vargaftik, E.D. German, R.R. Dogonadze, and Ya. K. Syrkin, Dokl. Phys.
        Chem., 1972, 206, 769 (370)

72YC   A. Yokoyama, M. Chikuma, and H. Tanaka, Chem. Pharm. Bull. (Japan), 1972, 20,
        2000

72YN   A. Yokoyama, N. Nakanishi, and H. Tanaka, Chem. Pharm. Bull. (Japan), 1972, 20,
        1856

72WN   M. Wozniak, J. Nicole, and G. Tridot, Chim. Anal. (Paris), 1972, 54, 147

72YY   Y. Yanagihara, T. Yano, H. Kobayashi, and K. Ueno, Bull. Chem. Soc. Japan, 1972,
        45, 554

73Ab   N.I. Ampelogova, Soviet Radiochem., 1973, 15, 823 (813)

73AB   B.M. Antipenko, I.M. Batyaev, and T.A. Privalova, Russ. J. Inorg. Chem., 1973,
        18, 318 (607)

73AH   T. Arishima, K. Hamada, and S. Takamoto, Nippon Kagaku Kaishi, 1973, 1119

73AS   M. Atchayya and P.R. Subbaraman, Indian J. Chem., 1973, 11, 1065

73AV   A. Aldaz, J.L. Vazquez, and J. Sancho, An. Quim., 1973, 69, 423

73BBB   B. Borkowski, K. Banczyk and A. Borkowska, Z. Phys. Chem. (Leipziq), 1973, 253,
        379

73BBC   N.L. Babenko, A.I. Busev, and I.N. Chistyachenko, Russ. J. Inorg. Chem., 1973,
        18, 958 (1813)

73BC   N. Bottazzini, V. Crespi, and F. Perini, Chim. Ind. (Milan), 1973, 55, 697

73BE   P.D. Bolton, J. Ellis, K.A. Fleming, and I.R. Lautzke, Aust. J. Chem., 1973,
        26, 1005

73BF   R.I. Burdykina and A.I. Falicheva, Izv. Vyssh. Ucheb. Zaved., Khim. 1973, 16,
        476

73BFP   R. Barbucci, L. Fabbrizzi, and P. Paoletti, Inorg. Chem. Acta, 1973, 7, 157

73BH   A.M. Bond and G. Hefter, J. Electroanal. Chem., 1973, 42, 1

73BK   W. Bacher and C. Keller, J. Inorg. Nucl. Chem., 1973, 35, 2945

73BM   Eu.A. Barbanel and L.P. Mureaveva, Soviet Radiochem., 1973, 15, 221 (227)

73BS   L.P. Berezina, V.G. Samoilenko and A.I. Pozigun, Russ. J. Inorg. Chem., 1973,
        18, 205 (393)

73BSa   G.A. Bhat and R.S. Subrahmanya, Indian J. Chem., 1973, 11, 584

73BW   C.I. Balcombe and B. Wiseall, J. Inorg. Nucl. Chem., 1973, 35, 2859

73BZ   U.P. Buxtorf and A. Zuberbuhler, Helv. Chim. Acta, 1973, 56, 524

73CB   A.K. Chakrabarti and S.P. Bag, Z. Anal. Chem., 1973, 265, 269

73CBP   A. Cassol, P. DiBernardo, R. Portanova, and L. Magon, Inorg. Chim. Acta, 1973,
        7, 353

73CC   C.T. Chang, M.M. Chang, and C.F. Liaw, J. Inorg. Nucl. Chem., 1973, 35, 261

73CG   P. Cauchetier and C. Guichard, Radiochim. Acta, 1973, 19, 137

73CP   Y. Couturier and C. Petitfaux, Bull. Soc. Chim. France, 1973, 439

73CPM   E. Chiacchierini, V. Petrone, A. Magri, and A. Stacchini, Gazz. Chim. Ital.,
        1973, 103, 501

73CS   M.C. Chattopadhyaya and R.S. Singh, Indian J. Chem., 1973, 11, 593

73CT   A.M. Corrie, M.L.D. Touche, and D.R. Williams, J. Chem. Soc. Dalton, 1973, 2561

73DR   P.C. Das and G.S. Rao, J. Indian Chem. Soc., 1973, 50, 172

73E   L.I. Elding, Inorg. Chem. Acta, 1973, 7, 581

73EB   O. Enea and G. Berthon, Thermochim. Acta, 1973, 6, 47

73EZ   A.A. Elesin, A.A. Zaitsev, G.M. Sergeev, and I.I. Nazarova, Soviet Radiochem.,
        1973, 15, 62 (64)

73FA   O. Farooq, N. Ahmad and A.U. Malik, J. Electroanal. Chem., 1973, 48, 475

73GE   G. Geier and I.W. Erni, Chimia (Switz.), 1973, 27, 635

73GK   I.P. Gorelov and M. Kh. Kolosova, J. Anal. Chem. USSR, 1973, 28, 433 (489)

73GKS   I.P. Gorelov, M.Kh. Kolosova, and A.P. Samsonov, J. Anal. Chem. USSR, 1973, 28,
        961 (1080)

73GM   P. Grenouillet, R.P. Martin, A. Rossi, and M. Ptak, Biochim. Biophys. Acta, 1973,
        322, 185

73GS   A. Gergely and I. Sovago, J. Inorg. Nucl. Chem., 1973, 35, 4355

73GSK   I.P. Gorelov, A.P. Samsonov and M.Kh. Kolosova, Russ. J. Inorg. Chem., 1973, 18,
        934 (1767)

73H   J.A. Happe, J. Amer. Chem. Soc., 1973, 95, 6232

73Ha   L. Harju, Suomen Kem., 1973, B46, 199

73HK   M.S. Hague and M. Kopanica, Bull. Chem. Soc. Japan, 1973, 46, 3072

73HKa   E.J. Hakoila and P. Kiilholma, Suomen Kem., 1973, B46, 143

73HP   E.J. Hakoila and S. Piepponen, Suomen Kem., 1973, B46, 86

73HR   E.H. Hansen and J. Ruzicka, Talanta, 1973, 20, 1105

73HT   S. Harada, H. Tanabe, and T. Yasunaga, Bull. Chem. Soc. Japan, 1973, 46, 3125

73HW   L.G. Hepler and E.M. Wooley, in Water, A. Comprehensive treatise, ed. F. Franks,
        Plenum Press, New York, Vol.3, 1973, p.145

73IB   V.M. Ivanov, A.I. Busev, and L.V. Ershova, J. Anal. Chem. USSR, 1973, 28, 395
        (442)

73IY   H. Ishizuka, T. Yamamoto, Y. Arata, and S. Fujiwara, Bull. Chem. Soc. Japan,
        1973, 46, 468

73J   N.D. Jespersen, J. Inorg. Nucl. Chem., 1973, 35, 3873

73JP   S.B. Joshi, M.D. Pundalik, and B.N. Mattoo, Indian J. Chem., 1973, 11, 1297

73K   S. Katayama, Bull. Chem. Soc. Japan, 1973, 46, 106

73Ka   M. Kodama, Bull. Chem. Soc. Japan, 1973, 46, 3422

73KA   R.C. Kapoor and B.S. Aggarwal, Indian J. Chem., 1973, 11, 71

73KB   N.A. Kostromina, N.V. Beloshitskii and I.A. Sheka, Russ. J. Inorg. Chem., 1973,
        18, 823 (1563)

73KBa   N.A. Kostromina, N.V. Beloshitskii and I.A. Sheka, Russ. J. Inorg. Chem., 1973,
        18, 1420 (2675)

73KC   A.J. Kresge and Y. Chiang, J. Phys. Chem., 1973, 77, 822

73KG   S.C. Khurana and C.M. Gupta, J. Inorg. Nucl. Chem., 1973, 35, 209

73KI   B.P. Karadakov and Kh.R. Ivanova, J. Anal. Chem. USSR, 1973, 28, 463, (525)

73KJ   V.D. Khanolkar, D.V. Jahagirdar, and D.D. Khanolkar, Indian J. Chem., 1973, 11,
        286

73KK   H. Kaneko and K. Kaneko, Nippon Kagaku Kaishi, 1973, 2127

73KL   K. Kustin and S.T. Liu, Inorg. Chem., 1973, 12, 2362

73KP   I. Khalil and M.M. Petit-Ramel, Bull. Soc. Chim. France, 1973, 1908

73KS   T.P.A. Kruck and B. Sarkar, Canad. J. Chem., 1973, 51, 3549

73KSa   T.P.A. Kruck and B. Sarkar, Canad. J. Chem., 1973, 51, 3555

73KSD   A.Yu. Kireeva, N.F. Shugal, and N.M. Dyatlova, Russ. J. Inorg. Chem., 1973, 18, 1426 (2685)

73KU   R. Krannich and E. Uhlig, Z. Anorg. Allg. Chem., 1973, 402, 285

73KW   K. Kustin and M.A. Wolff, J. Chem. Soc. Dalton, 1973, 1031

73KZ   A.Yu. Kireeva, B.V. Zhadanov, V.V. Siderenko, and N.M. Dyatlova, J. Gen. Chem. USSR, 1973, 43, 2494 (2508)

73LM   J.E. Land, W.R. Mountcastle, H.T. Peters, and D.R. Holt, Auburn Univ. WRRI Bull., No. 17, Sept. 1973

73MB   E. M'Foundou and G. Berthon, Analusis, 1973-1974, 2, 658

73MMC   W.A.E. McBryde, J.L. McCourt and V. Cheam, J. Inorg. Nucl. Chem., 1973, 35, 4193

73MP   E. Mentasti and E. Pelizzetti, J. Chem. Soc. Dalton, 1973, 2605

73MR   A. Mederos, A. Rodriquer Gonzales and B. Rodriquez Rios, An. Quim., 1973, 69, 731

73MS   M. Morin and J.P. Scharff, Bull. Soc. Chim. France, 1973, 2198

73MV   Yu.A. Maletin, I.M. Vasilkevich, G.F. Dvorko, and V.P. Tikhonov, J. Gen. Chem. USSR, 1973, 43, 2718 (2743), 2721 (2746)

73N   B. Noren, Acta Chem. Scand., 1973, 27, 1369

73ND   R. Nayan and A.K. Dey, J. Indian Chem. Soc., 1973, 50, 98

73NE   D. Nonova and B. Evtimova, Talanta, 1973, 20, 1347

73NEa   D. Nonova and B. Evtimova, J. Inorg. Nucl. Chem., 1973, 35, 3581

73NH   T. Nozaki and T. Hashimoto, Nippon Kagaku Kaishi, 1973, 1794

73NM   O. Navratil and Z. Mikulec, Coll. Czech. Chem. Comm., 1973, 38, 2430

73NO   T. Nozaki and Y. Ohno, Nippon Kagaku Kaishi, 1973, 1455

73NP   A. Napoli and L. Pontelli, Gazz. Chim. Ital., 1973, 103, 1219

73NW   G. Nakagawa, H. Wada, and Y. Fujita, Bull. Chem. Soc. Japan, 1973, 46, 489

73O   H. Ots, Acta Chem. Scand., 1973, 27, 2344

73OD   G. Ostacoli, P.G. Daniele, and A. Vanni, Ann. Chim. (Rome), 1973, 63, 815

73OI   H. Ohtaki and Y. Ito, J. Coord. Chem., 1973, 3, 131

73OO   C. O'Nuallain and S. O'Cinneide, J. Inorg. Nucl. Chem., 1973, 35, 2871

73PJ   J.E. Powell, D.A. Johnson, H.R. Burkholder, and S.C. Vick, J. Chromatogr., 1973, 87, 437

73PM   S.P. Pande and K.N. Munshi, Indian J. Chem., 1973, 11, 1322

73PMa   S.P. Pande and K.N. Munshi, J. Indian Chem. Soc., 1973, 50, 649

73PP   B. Pokric and Z. Pucar, J. Inorg. Nucl. Chem., 1973, 35, 1987

73PS   S.K. Patel, K.P. Soni, and I.M. Bhatt, Indian J. Chem., 1973, 11, 1324

73PZ   V.I. Paramonova, V.Ya. Zamanskii, and V.B. Kolychev, Russ. J. Inorg. Chem., 1973, 18, 1132 (2139)

73RB   D.L. Rabenstein and G. Blakney, Inorg. Chem., 1973, 12, 128

73RD   G. Reinhard, R. Dreyer and R. Munze, Z. Phys. Chem. (Leipzig), 1973, 254, 226

73RG   P.C. Rawat and C.M. Gupta, Bull. Chem. Soc. Japan, 1973, 46, 3079

73RGa   P.C. Rawat and C.M. Gupta, Indian J. Chem., 1973, 11, 186

73RK   E.M. Rogozina, L.F. Konkina, and D.K. Popov, Soviet Radiochem., 1973, 15, 59 (61)

73RP   O.P. Ryabushko, A.T. Pilipenko, L.A. Krivokhizhina, and N.L. Emchenko, Soviet
          Prog. Chem. (Ukr. Khim Zh.), 1973, 39, No. 10, 76 (1051)

73RPa   O.P. Ryabushko, A.T. Pilipenko, and L.A. Krivokhizhina, Soviet Prog. Chem.,
          (Ukr. Khim. Zh.), 1973, 39, No. 12, 93 (1293)

73RR   L.P. Romanenko and A.V. Radushev, J. Anal. Chem. USSR, 1973, 28, 1695 (1908)

73RRB   R.N. Roy, R.A. Robinson and R.G. Bates, J. Amer. Chem. Soc., 1973, 95, 8231

73RS   A. Raghavan and M. Santappa, J. Inorg. Nucl. Chem., 1973, 35, 3363

73S   M.J. Schwing-Weill, Bull. Soc. Chim. France, 1973, 823

73SC   B.S. Sekhon and S.L. Chopra, Thermochim. Acta, 1973, 7, 151

73SCa   B.S. Sekhon and S.L. Chopra, Thermochim. Acta, 1973, 7, 311

73SCb   R.S. Saxena and U.S. Chaturvedi, J. Indian Chem. Soc., 1973, 50, 756

73SD   D. Shishkov and H.I. Doichinova, Dokl. Bolg. Akad. Nauk., 1973, 26, 927

73SG   A.P. Samsonov and I.P. Gorelov, Russ. J. Inorg. Chem., 1973, 18, 1053 (1988)

73SHI   T. Sekine, Y. Hasegawa, and N. Ihara, J. Inorg. Nucl. Chem., 1973, 35, 3968

73SK   G.M. Sergeev and I.A. Korshunov, Soviet Radiochem., 1973, 15, 619 (618)

73SKa   G.M. Sergeev and I.A. Korshunov, Soviet Radiochem., 1973, 15, 623 (621)

73SM   S.K. Srivastava and H.B. Mathur, Indian J. Chem., 1973, 11, 936

73SMa   S.K. Srivastava and H.B. Mathur, Indian J. Chem., 1973, 11, 1293

73ST   T. Sekine and Y. Takahashi, Bull. Chem. Soc. Japan, 1973, 46, 1183

73T   T.V. Ternovaya, Soviet Prog. Chem. (Ukr. Khim. Zh.), 1972, 39, No. 2, 12 (125)

73TG   H. Tomiyasu and G. Gordon, J. Coord. Chem., 1973, 3, 47

73TK   N.N. Tananaeva and N.A. Kostromina, Russ. J. Inorg. Chem., 1973, 18, 1246 (2354)

73TKa   N.N. Tananaeva and N.A. Kostromina, Russ. J. Inorg. Chem., 1973, 18, 352 (674)

73TM   M.M. Taqui Khan and M.S. Mohan, J. Inorg. Nucl. Chem., 1973, 35, 1749

73TP   S.C. Tripathi and S. Paul, Indian J. Chem., 1973, 11, 1042

73TS   R.C. Tewari and M.N. Srivastava, J. Inorg. Nucl. Chem., 1973, 35, 3044

73TSa   R.C. Tewari and M.N. Srivastava, Talanta, 1973, 20, 133

73TSb   R.C. Tewari and M.N. Srivastava, Indian J. Chem., 1973, 11, 700

73TSc   S.C. Tripathi and M. Singh, Indian J. Chem., 1973, 11, 817

73UW   E. Uhlig and D. Walter, Z. Anorg. Allg. Chem., 1973, 397, 187

73VB   P. Vieles and A. Bonniol, Compt. Rend. Acad. Sci. Paris, Ser. C, 1973, 276
          1769

73VBa   R.S. Vaidya and S.N. Banerji, Indian J. Chem., 1973, 11, 1200

73VG   V.P. Vasilev, N.K. Grechina, A.A. Ikonnikov, and A.V. Kuranov, Izv. Vyssh.
          Zaved., Khim., 1973, 16, 702

73VI   N.M. Voronina, A.A. Ivakin, I.V. Podgornaya, I.A. Yalovets and K.N. Klyachkina,
          J. Gen. Chem. USSR, 1973, 43, 629 (632)

73VJ   D.G. Vartak and C.J. Jose, Indian J. Chem., 1973, 11, 1306

73W   D.R. Williams, J. Chem. Soc. Dalton, 1973, 1064

73WK   D.D. Wagman and M.V. Kilday, J. Res. Nat. Bur. Stand., 1973, 77A, 569

73WU   D. Walther and E. Uhlig, Z. Anorg. Allg. Chem., 1973, 400, 189

73YB   O. Yamauchi, H. Benno, and A. Nakahara, Bull. Chem. Soc. Japan, 1973, 46, 3458

73YP   E.G. Yakovleva, N.I. Pechurova, L.I. Martyuenko, and V.I. Spitsyn, Russ. J.
          Inorg. Chem., 1973, 18, 800 (1519)

73YPa   E.G. Yakovleva, N.I. Pechurova, L.I. Martynenko, and V.I. Spitsyn, Bull. Acad.
          Sci. USSR, 1973, 22, 1655 (1699)

73ZG   V.K. Zolotukhin and O.M. Gnatyshin, Russ. J. Inorg. Chem., 1973, 18, 1467 (2761)

73ZGa   V.K. Zolotukhin and O.M. Gnatyshin, Soviet Prog. Chem., (Ukr. Khim. Zh.), 1973,
          39, No. 12, 43 (1243)

73ZGK   V.K. Zolotukhin, Z.G. Galanets, and V.I. Korotya, Soviet Prog. Chem., (Ukr. Khim.
          Zh.), 1973, 39, No. 10, 84 (1059)

74Aa   G. Anderegg, Helv. Chim. Acta, 1974, 57, 1340

74Ab   R.C. Agarwal, J. Indian Chem. Soc., 1974, 51, 772

74Ac   R. Aruga, Ann. Chim. (Rome), 1974, 64, 439

74Ad   R. Aruga, Ann. Chim. (Rome), 1974, 64, 659

74AM   R. Abu-Eittah, Z. Mobarak, and S. El-Lathy, J. Prakt. Chem., 1974, 316, 235

74AY   H. Aiba, A. Yokoyama, and H. Tanaka, Bull. Chem. Soc. Japan, 1974, 47, 136

74AYa   H. Aiba, A. Yokoyama, and H. Tanaka, Bull. Chem. Soc. Japan, 1974, 47, 1437

74B   E.W. Baumann, J. Inorg. Nucl. Chem., 1974, 36, 1827

74Ba   N. Bertazzi, Z. Anorg. Allg. Chem., 1974, 410, 316

74Bb   B.W. Budesinsky, Anal. Chim. Acta, 1974, 71, 333

74Bc   E. Bottari, Monat. Chem., 1974, 105, 187

74BE   G. Berthon and O. Enea, Bull. Soc. Chim. France, 1974, 2793

74BF   I.M. Batyaev and R.S. Fogileva, Russ. J. Inorg. Chem., 1974, 19, 363 (670)

74BFP   R. Barbucci, L. Fabbrizzi, and P. Paoletti, J. Chem. Soc. Dalton, 1974, 2403

74BK   E. Brucher, Cs.E. Kukri, and L. Zekany, J. Inorg. Nucl. Chem., 1974, 36, 2620

74BS   J.L. Banyasz and J.E. Stuehr, J. Amer. Chem. Soc., 1974, 96, 6481

74BSK   R. Buxtorf, W. Steinmann, and T.A. Kaden, Chimia (Switz.), 1974, 28, 15

74BV   R. Barbucci and A. Vacca, J. Chem. Soc. Dalton, 1974, 2363

74BW   A.C. Baxter and D.R. Williams, J. Chem. Soc. Dalton, 1974, 1117

74BWa   J. Biernat and M. Wilgocki, Rocz. Chem., 1974, 48, 1663

74CB   A.K. Chakrabarti and S.P. Bag, Z. Anal. Chem., 1974, 272, 124

74CM   F. Chastellain and A.E. Merbach, Chimia (Switz.), 1974, 28, 609

74CP   J.L. Colin and J. Pinart, Bull. Soc. Chim. France, 1974, 2756

74CPS   G. Carpeni, S. Poize, N. Sabiani and G. Perinet, J. Chim. Phys., 1974, 71, 311

74CY   M. Chikuma, A. Yokoyama, and H. Tanaka, Chem. Pharm. Bull. (Japan), 1974, 22,
          1378

74DB   G. Duc, F. Bertin, and G. Thomas-David, Bull. Soc. Chim. France, 1974, 793

74DC   G. D'Ascenzo, E. Chiacchierini, A. Marino, A. Magri, and G. DeAngelis, Gazz. Chim. Ital., 1974, 104, 607

74DF   E. Dazzi and M.T. Falqui, Gazz. Chim. Ital., 1974, 104, 589

74DN   A.C. Dash and R.K. Nanda, J. Indian Chem. Soc., 1974, 51, 733

74DS   S.F. Deryugina, I.A. Sheka, and L.P. Barchuk, Soviet Progr. Chem., (Ukr. Khim. Zh.), 1974, 40, No. 9, 15 (916)

74DV   H.F. DeBrabander and L.C. Van Poucke, J. Coord. Chem., 1974, 3, 301

74EI   L.V. Ershova, V.M. Ivanov, and A.I. Busev, J. Anal. Chem. USSR, 1974, 29, 1180 (1367)

74EW   B. Elgquist and M. Wedborg, Marine Chem., 1974, 2, 1

74F   L.N. Filatova, Russ. J. Inorg. Chem., 1974, 19, 1827 (3335)

74FA   O. Farooq and N. Ahmad, J. Electroanal. Chem., 1974, 53, 457

74FAa   O. Farooq and N. Ahmad, J. Electroanal. Chem., 1974, 53, 461

74FAb   O. Farooq and N. Ahmad, J. Electroanal. Chem., 1974, 57, 121

74FKb   L.N. Filatova and T.N. Kurdyumova, Russ. J. Inorg. Chem., 1974, 19, 1746 (3190)

74FL   W.L. Felty and D.L. Leussing, J. Inorg. Nucl. Chem., 1974, 36, 617

74FLa   Ya.D. Fridman and M.G. Levina, Russ. J. Inorg. Chem., 1974, 19, 1324 (2422)

74FM   T.B. Field, J.L. McCourt, and W.A.E. McBryde, Canad. J. Chem., 1974, 52, 3119

74FP   J.J. Fardy and J.M. Pearson, J. Inorg. Nucl. Chem., 1974, 36, 671

74FSD   Ya.D. Fridman, O.P. Svanidze, N.V. Kolgashova, and P.V. Gogorishvili, Russ. J. Inorg. Chem., 1974, 19, 1809 (3304)

74G   I.P. Gorelov, J. Anal. Chem. USSR, 1974, 49, 906 (1057)

74Ga   I.P. Gorelov, J. Anal. Chem. USSR, 1974, 49, 1344 (1554)

74Gb   V.I. Gordienko, J. Gen. Chem. USSR, 1974, 44, 853 (885)

74Gc   J.C. Ghosh, J. Indian Chem. Soc., 1974, 51, 361

74GC   S. Gifford, W. Cherry, J. Jecnen, and M. Readnour, Inorg. Chem., 1974, 13, 1434

74GG   I. Grenthe and G. Gardhammer, Acta Chem. Scand., 1974, A28, 125

74GM   N.N. Guseva, A.I. Mikhailichenko, M.K. Karapetyants, and E.V. Sklenskaya, Russ. J. Inorg. Chem., 1974, 19, 1637 (2994)

74GNF   A. Gergely, I. Nagypal, and E. Farkas, Acta Chim. Acad. Sci. Hung., 1974, 82, 43

74GNK   A. Gergely, I. Nagypal, T. Kiss, and R. Kiraly, Acta Chim. Acad. Sci. Hung., 1974, 82, 257

74GS   B. Grgas-Kuznar, V. Simeon, and O.A. Weber, J. Inorg. Nucl. Chem., 1974, 36, 2151

74GW   R.D. Graham and D.R. Williams, J. Chem. Soc. Dalton, 1974, 1123

74H   G. Hefter, Coord. Chem. Rev., 1974, 12, 221

74HB   K. Houngbossa and G. Berthon, Bull. Soc. Chim. France, 1974, 2418

74HEB   K. Houngbossa, O. Enea, and G. Berthon, Thermochim. Acta, 1974, 10, 415

74HF   F. Hogue and H. Frye, Inorg. Nucl. Chem. Letters, 1974, 10, 505

74HK   L. Hertli and T.A. Kaden, Helv. Chim. Acta, 1974, 57, 1328

74HL   S.D. Hamann and M. Linton, J. Chem. Soc. Faraday I, 1974, 70, 2239

74HM   F.P. Hinz and D.W. Margerum, Inorg. Chem., 1974, 13, 2941

74HP   G.M. Huitink, D.P. Poe, and H. Diehl, Talanta, 1974, 21, 1221

74HS   L. Harju and R. Sara, Anal. Chim. Acta, 1974, 73, 129

74IL   M. Israeli, D.K. Laing, and L.D. Pettit, J. Chem. Soc. Dalton, 1974, 2194

74J   A.P. Joshi, J. Indian Chem. Soc., 1974, 51, 643

74K   M. Kodama, Bull. Chem. Soc. Japan, 1974, 47, 1547

74KG   A.I. Kapustnikov and I.P. Gorelov, Russ. J. Inorg. Chem., 1974, 19, 1742 (3183)

74KH   J. Kollmann and E. Hoyer, J. Prakt. Chem., 1974, 316, 119

74KI   H. Kakihana and S. Ishiguro, Bull. Chem. Soc. Japan, 1974, 47, 1665

74KJ   Y.D. Kane and D.M. Joshi, Curr. Sci. (India), 1974, 43, 332

74KM   N.N. Krot and M.P. Mefodeva, Bull. Acad. Sci. USSR, Chem., 1974, 23, 2052 (2133)

74KMS   G.N. Kupriyanova, L.I. Martynenko, and V.I. Spitsyn, Dokl. Phys. Chem., 1974
          215, 368 (912)

74KNP   E.I. Klabunovskii, V.I. Neupokoev, and V.A. Pavlov, Russ. J. Phys. Chem., 1974,
          48, 1135 (1917)

74KNT   N.A. Kostromina, L.B. Novikova, and T.V. Ternovaya, Russ. J. Inorg. Chem., 1974,
          19, 1450 (2654)

74KPM   V.I. Kornev, N.I. Pechurova, and L.I. Martynenko, Russ. J. Inorg. Chem., 1974,
          19, 146 (265)

74KPS   N.I. Kurkina, N.Yu. Petrova, and N.A. Skorik, Russ. J. Inorg. Chem., 1974, 19,
          358 (661)

74KR   A.Yu. Kireeva, M.V. Rudomino, and N.M. Dyatlova, J. Gen. Chem. USSR, 1974, 44,
          2594 (2637)

74KS   J. Klepetar and K. Stulik, J. Electroanal. Chem., 1974, 55, 255

74KSa   F. Kai and Y. Sadakane, J. Inorg. Nucl. Chem., 1974, 36, 1404

74KT   N.A. Kostromina and R.V. Tikhonova, Russ J. Inorg. Chem., 1974, 19, 1096 (2000)

74KU   F.Ya. Kulba, V.G. Ushakova, and Yu.B. Yakovlev, Russ. J. Inorg. Chem., 1974, 19,
          972 (1785)

74KZ   N.P. Komar and G.S. Zaslavskaya, Russ. J. Phys. Chem., 1974, 48, 292 (494)

74L   S. Lal, Monat. Chem., 1974, 105, 974

74LA   R. Lundqvist and J.E. Andersson, Acta Chem. Scand., 1974, A28, 700

74LK   V.I. Levin and G.E. Kodina, Russ. J. Inorg. Chem., 1974, 19, 1128 (2060)

74LKB   B. Lenarcik, J. Kulig, and B. Barszcz, Rocz. Chem., 1974, 48, 2111

74LKL   B. Lenarcik, J. Kulig, and P. Laidler, Rocz. Chem., 1974, 48, 1151

74LKS   S.J. Lau, T.P.A. Kruck, and B. Sarkar, J. Biol. Chem., 1974, 249, 5878

74LP   C.L. Liotta, E.M. Perdue, and H.P. Hopkins, Jr., J. Amer. Chem. Soc., 1974, 96,
          7308

74LPa   P. Lumme and I. Pitkanen, Acta Chem. Scand., 1974, A28, 1106

74LV   P. Lumme and P. Virtanen, Acta Chem. Scand., 1974, A28, 1055

74M   P. Mirti, Anal. Chim. Acta, 1974, 69, 69

74MA   V. Madison, M. Atreyi, C.M.Deber, and E.R. Blout, J. Amer. Chem. Soc., 1974, 96,
          6725

74MB   S.H. Mehdi and B.W. Budesinsky, J. Coord. Chem., 1974, 3, 287

74MJ   A.K. Maheswari, D.S. Jain, H.C. Saraswat, and J.N. Gaur, J. Electrochem. Soc.
          India, 1974, 23, 155

74MK   P.K. Migal and E.I. Koptenko, Russ. J. Inorg. Chem., 1974, 19, 175 (322)

74MKa   P.K. Migal and E.P. Koptenko, Russ. J. Inorg. Chem., 1974, 19, 1263 (2313)

74ML   O. Makitie, L.H.J. Lajunen, and H. Saarinen, Finn. Chem. Lett., 1974, 96

74MM   R.J. Motekaitis and A.E. Martell, Inorg. Chem., 1974, 13, 550

74MMK   M.H. Mihailov, V.Ts. Mihailova, and V.A. Khalkin, J. Inorg. Nucl. Chem., 1974,
          36, 141

74MMN   K.K. Mui, W.A.E. McBryde, and E. Nieboer, Canad. J. Chem., 1974, 52, 1821

74MMS   N. Motohashi, I. Mori, Y. Sugiura, and H. Tanaka, Chem. Pharm. Bull. (Japan),
          1974, 22, 654

74MP   N.A.A. Mumallah and W.J. Popiel, Anal. Chem., 1974, 46, 2055

74MR   M.M. Muir and P.R. Rechani, Inorg. Chim. Acta, 1974, 11, 137

74MS   A. Meretoja, R. Salvela, and E.J. Hakoila, Finn. Chem. Lett., 1974, 267

74MT   P.N. MohanDas and C.P. Trivedi, J. Indian Chem. Soc., 1974, 51, 522

74NA   R. Nakon and R.J. Angelici, J. Amer. Chem. Soc., 1974, 96, 4178

74NB   K. Nag and P. Banerjee, J. Inorg. Nucl. Chem., 1974, 36, 2145

74NBA   R. Nakon, E.M. Beadle, Jr., and R.J. Angelici, J. Amer. Chem. Soc., 1974, 96, 719

74NF   V.A. Nazarenko and G.V. Flyantikova, Soviet Progr. Chem., (Ukr. Khim. Zh.), 1974,
          40, No. 11, 62 (1188)

74NFa   V.A. Nazarenko and G.V. Flyantikova, Soviet Progr. Chem., (Ukr. Khim. Zh.), 1974,
          40, No. 12, 1 (1235)

74NG   L.V. Nikitina, A.I. Grigorev, and N.M. Dyatlova, J. Gen. Chem. USSR, 1974, 44,
          1568 (1598)

74NGa   L.V. Nikitina, A.I. Grigorev, and N.M. Dyatlova, J. Gen. Chem. USSR, 1974, 44,
          1641 (1669)

74NGF   I. Nagypal, A. Gergely, and E. Farkas, J. Inorg. Nucl. Chem., 1974, 36, 699

74NJ   M.L. Narwade and D.V. Jahagirdar, J. Indian Chem. Soc., 1974, 51, 1008

74NKD   L.V. Nikitina, L.D. Karmazina, and N.M. Dyatlova, Russ. J. Inorg. Chem., 1974,
          19, 1671 (3058)

74NKK   P. Nenova, B. Karadakov and D. Kantcheva, J. Inorg. Nucl. Chem., 1974, 36, 464

74NP   J. Novak and J. Podlaha, J. Inorg. Nucl. Chem., 1974, 36, 1061

74NS   V.B. Nikolaevskii, V.P. Shilov, and N.N. Krot, Soviet Radiochem., 1974, 16, 57
          (61)

74NT   R. Nasanen, P. Tilus, and T. Teikari, Finn. Chem. Lett., 1974, 263

74P   J. Podlahova, Coll. Czech. Chem. Comm., 1974, 39, 2724

74PB   P.C. Parikh and P.K. Bhattacharya, Indian J. Chem., 1974, 12, 402

74PC   I.G. Perkov, P.E. Chaplanov, and I.T. Polkovnichenko, Russ. J. Phys. Chem., 1974,
          48, 583 (1005)

74PH   R.M. Pytkowicz and J.E. Hawley, Limnol. Oceanogr., 1974, 19, 223

74PI   N.I. Pechurova, O.P. Ioanisiani, R.P. Tishchenko, and V.I. Spitsyn, Bull. Acad.
          Sci. USSR, 1974, 23, 1395 (1472)

74PK   A.T. Pilipenko, P.P. Kish, and G.M. Vitenko, Soviet Progr. Chem., (Ukr. Khim.
          Zh.), 1974, 40, No. 8, 61 (850)

74PKL   G.A. Prik, B.E. Kozer, and A.A. Lopatina, Tr. Ivanovskii Khim.-Tekhnol. Inst.,
          1974, No. 17, 66

74PM	S.P. Pande and K.N. Munshi, <u>J. Indian Chem. Soc.</u>, 1974, <u>51</u>, 339
74PN	V.E. Plyushchev, G.V. Nadezhdina, G.S. Loseva, V.V. Melnikova, and T.S. Parfenova, <u>J. Gen. Chem. USSR</u>, 1974, <u>44</u>, 2274 (2319)
74PP	B.N. Palmer and H.K.J. Powell, <u>J. Chem. Soc. Dalton</u>, 1974, 2086
74PPa	B.N. Palmer and H.K.J. Powell, <u>J. Chem. Soc. Dalton</u>, 1974, 2089
74PPF	J. Pinart, C. Petitfaux, and J. Faucherre, <u>Bull. Soc. Chim. France</u>, 1974, 1786
74PPS	V.M. Plakhotnik, A.B. Prokhorov, O.I. Samoilova, L.I. Tikhonova, and V.G. Yashunskii, <u>J. Gen. Chem. USSR</u>, 1974, <u>44</u>, 574 (600)
74PR	A.T. Pilipenko, O.P. Ryabushko, L.A. Krivokhizhina, and N.L. Emchenko, <u>Soviet Progr. Chem.</u>, <u>(Ukr. Khim. Zh.)</u>, 1974, <u>40</u>, No. 1, 69 (73)
74PS	O.N. Puplikova and M.A. Savich, <u>J. Gen. Chem. USSR</u>, 1974, <u>44</u>, 1136 (1179)
74R	A. Ricard, <u>Ann. Chim.</u>, 1974, <u>9</u>, 203
74RK	J.C. Ringen and R.E. Kirby, <u>J. Inorg. Nucl. Chem.</u>, 1974, <u>36</u>, 199
74RM	S. Ramamoorthy and P.G. Manning, <u>Inorg. Nucl. Chem. Letters</u>, 1974, <u>10</u>, 623
74RO	D.L. Rabenstein, R. Ozubko, S. Libich, C.A. Evans, M.T. Fairhurst, and C. Suvanprakorn, <u>J. Coord. Chem.</u>, 1974, <u>3</u>, 263
74RP	O.P. Ryabushko, A.T. Pilipenko, and N.L. Emchenko, <u>Soviet Progr. Chem. (Ukr. Khim. Zh.)</u>, 1974, <u>40</u>, No. 2, 75 (190)
74SB	F. Sprta and M. Bartusek, <u>Coll. Czech. Chem. Comm.</u>, 1974, <u>39</u>, 2023
74SC	B.S. Sekhon and S.L. Chopra, <u>Israel J. Chem.</u>, 1974, <u>12</u>, 917
74SCa	B.S. Sekhon and S.L. Chopra, <u>Indian J. Chem.</u>, 1974, <u>12</u>, 1322
74SG	A.P. Samsonov and I.P. Gorelov, <u>Russ. J. Inorg. Chem.</u>, 1974, <u>19</u>, 1159 (2115)
74SGP	A.K. Sinha, J.C. Ghosh, and B. Prasad, <u>J. Indian Chem. Soc.</u>, 1974, <u>51</u>, 586
74SJ	K.H. Schroder and B.G. Johnsen, <u>Talanta</u>, 1974, <u>21</u>, 671
74SK	G.M. Sergeev and I.A. Korshunov, <u>Soviet Radiochem.</u>, 1974, <u>16</u>, 767 (783)
74SKT	V.A. Shormanov, G.A. Krestov, and E.A. Trupikov, <u>Russ. J. Inorg. Chem.</u>, 1974, <u>19</u>, 1007 (1848)
74SM	S.K. Srivastava and H.B. Mathur, <u>Indian J. Chem.</u>, 1974, <u>12</u>, 1289
74SMN	T. Sekine, R. Murai, M. Niitsu, and N. Ihara, <u>J. Inorg. Nucl. Chem.</u>, 1974, <u>36</u>, 2569
74SRL	H. Saarinen, T. Raikas, and L. Lajunen, <u>Finn. Chem. Lett.</u>, 1974, 109
74SRR	H. Saarinen, T. Raikas, and A. Rauhala, <u>Finn. Chem. Lett.</u>, 1974, 104
74SS	R. Sundaresan and A.K. Sundaram, <u>Proc. Indian Acad. Sci.</u>, 1974, <u>79A</u>, 161
74SSa	M. Salomen and B.K. Stevenson, <u>J. Chem. Eng. Data</u>, 1974, <u>19</u>, 42
74SSb	W. Szczepaniak and J. Siepak, <u>Chem. Anal. (Warsaw)</u>, 1974, <u>19</u>, 351
74ST	O.P. Sunar, S. Tak and C.P. Trivedi, <u>J. Inorg. Nucl. Chem.</u>, 1974, <u>36</u>, 1163
74TB	S.C. Tripathi and S. Bhowmik, <u>J. Indian Chem. Soc.</u>, 1974, <u>51</u>, 605
74TN	G.S. Tereshin and E.V. Nikiforova, <u>Russ. J. Inorg. Chem.</u>, 1974, <u>19</u>, 1020 (1869)
74TS	R.C. Tewari and M.N. Srivastava, <u>Acta Chim. Acad. Sci. Hung.</u>, 1974, <u>83</u>, 259
74UP	V.V. Udovenko and G.B. Pomerants, <u>Russ. J. Inorg. Chem.</u>, 1974, <u>19</u>, 301 (556)
74VG	A. Vanni and M.C. Gennaro, <u>Ann. Chim.</u> (Rome), 1974, <u>64</u>, 397
74VK	V.P. Vasilev, L.A. Kochergina, and T.D. Yastrebova, <u>J. Gen. Chem. USSR</u>, 1974, <u>44</u>, 1346 (1371)

74VP   G.P. Vakhramova, N.I. Pechurova, and V.I. Spitsyn, Russ. J. Inorg. Chem., 1974,
        19, 1486 (2722)

74VS   V.P. Vasilev and L.D. Shekhanova, Russ. J. Inorg. Chem., 1974, 19, 1623 (2969)

74VSU   V.V. Velichko, V.I. Suprunovich, and Yu.I. Usatenko, Russ. J. Inorg. Chem., 1974,
        19, 547 (1004)

74W   O.A. Weber, J. Inorg. Nucl. Chem., 1974, 36, 1341

74Wa   R. Wojtas, Rocz. Chem., 1974, 48, 219

74Wb   Z. Warnke, Rocz. Chem., 1974, 48, 735

74Wc   Z. Warnke, Rocz. Chem., 1974, 48, 1205

74WK   J. Wojcik, F. Karczynski, and G. Kupryszewski, Rocz. Chem., 1974, 48, 1179

74WW   M.D. Walker and D.R. Williams, J. Chem. Soc. Dalton, 1974, 1186

74YA   A. Yokoyama, H. Aiba and H. Tanaka, Bull. Chem. Soc. Japan, 1974, 47, 112

74YI   T. Yoshino, H. Imada, S. Murakami and M. Kagawa, Talanta, 1974, 21, 211

74YKP   E.G. Yakovleva, L.S. Koltsova, N.I. Pechurova, V.I. Spitsyn, and L.M. Timakova,
        Russ. J. Inorg. Chem., 1974, 19, 1483 (2717)

74YKU   T. Yano, H. Kobayashi, and K. Ueno, Bull. Chem. Soc. Japan, 1974, 47, 2806

74YM   T. Yoshino, S. Murakami, M. Kagawa and T. Araragi, Talanta, 1974, 21, 79

74YMa   T. Yoshino, S. Murakami, M. Kagawa, Talanta, 1974, 21, 199

74YO   T. Yoshino, H. Okazaki, S. Murakami, and M. Kagawa, Talanta, 1974, 21, 673

74YOa   T. Yoshino, H. Okazaki, S. Murakami, and M. Kagawa, Talanta, 1974, 21, 676

74YY   T. Yano, Y. Yanagihara, H. Kobayashi, and K. Ueno, Bull. Chem. Soc. Japan,
        1974, 47, 2889

74ZG   V.K. Zolotukhin and O.M. Gnatyshin, Soviet Progr. Chem., (Ukr. Khim. Zh.), 1974,
        40, No. 8, 8 (795)

74ZK   A.D. Zuberbuhler and T.A. Kaden, Helv. Chim. Acta, 1974, 57, 1897

75A   G. Anderegg, Helv. Chim. Acta, 1975, 58, 1218

75Aa   R. Aruga, Atti Accad. Sci. Torino, 1975, 109, 361

75Ab   R. Aruga, J. Chem. Soc. Dalton, 1975, 2534

75AA   S. Ahrland and E. Avsar, Acta Chem. Scand., 1975, A29, 881

75AAa   S. Ahrland and E. Avsar, Acta Chem. Scand., 1975, A29, 890

75AB   P. Andersen, T. Berg, and J. Jacobsen, Acta Chem. Scand., 1975, A29, 381

75ABa   P. Andersen, T. Berg, and J. Jacobsen, Acta Chem. Scand., 1975, A29, 599

75ABR   A. Arevalo, A. Bazo, J.C. Rodriquez Placeres, and T. Moreno, An. Quim., 1975,
        71, 361

75AC   A. Arevalo, A. Cabrera Gonzalez, J.C. Rodriquez Placeres, and T. Moreno, An.
        Quim., 1975, 71, 367

75AM   R.S. Ambulkar and K.N. Munshi, J. Indian Chem. Soc., 1975, 52, 315

75AN   V.P. Antonovich, E.M. Nevskaya, E.I. Shelikhina, and V.A. Nazarenko, Russ. J.
        Inorg. Chem., 1975, 20, 1642 (2968)

75AP   R.P. Agarwal and D.D. Perrin, J. Chem. Soc. Dalton, 1975, 268

75APa   R.P. Agarwal and D.D. Perrin, J. Chem. Soc. Dalton, 1975, 1045

75APB   G. Anderegg, N.G. Podder, P. Blauenstein, M. Hangartner, and H. Stunzi, J. Coord.
        Chem., 1975, 4, 267

75APS   V. Almagro, M.J. Pena, and J. Sancho, An. Quim., 1975, 71, 933

75AS   F. Arnaud-Neu and M.J. Schwing-Weill, Inorg. Nucl. Chem. Letters, 1975, 11, 131

75ASa   F. Arnaud-Neu and M.J. Schwing-Weill, Inorg. Nucl. Chem. Letters, 1975, 11, 655

75AV   J.K. Agrawal and R. Vijayavargiya, J. Indian Chem. Soc., 1975, 52, 190

75AVa   J.K. Agrawal and R. Vijayavargiya, J. Indian Chem. Soc., 1975, 52, 576

75AZ   L.N. Andreeva, N.M. Zhirnova, and K.V. Astakhov, Russ. J. Phys. Chem., 1975, 49,
          652 (1108)

75B   R. Barbucci, Inorg. Chim. Acta, 1975, 12, 113

75Ba   C. Bremard, Bull. Soc. Chim. France, 1975, 953

75Bb   E.J. Billo, Inorg. Nucl. Chem. Letters, 1975, 11, 491

75BA   E.A. Belousov and A.A. Alovyainikov, Russ. J. Inorg. Chem., 1975, 20, 803 (1428)

75BC   J.T. Bulmer, T.G. Chang, P.J. Gleeson, and D.E. Irish, J. Soln. Chem., 1975, 4,
          969

75BDR   M. Barres, J.P. Dubes, R. Romanetti, H. Tachoire, and C. Zahra, Thermochim. Acta,
          1975, 11, 235

75BDT   M. Barres, J.P. Dubes, and H. Tachoire, Compt. Rend. Acad. Sci. Paris, Ser. C,
          1975, 280, 855

75BE   M.J. Blais, O. Enea, and G. Berthon, Thermochim. Acta, 1975, 12, 25

75BF   C. Bianchini, L. Fabbrizzi, and P. Paoletti, J. Chem. Soc. Dalton, 1975, 1036;
          see also Inorg. Chem., 1975, 14, 197

75BG   E. Bottari and G. Goretti, Monat. Chem., 1975, 106, 1337

75BHP   P. Bianco, J. Haladjian, and R. Pilard, J. Less-Common Metals, 1975, 42, 127

75BHS   J. Bjerrum, A.S. Halonin, and L.H. Skibsted, Acta Chem. Scand., 1975, A29, 326

75BK   E. Brucher, R. Kiraly, and I. Nagypal, J. Inorg. Nucl. Chem., 1975, 37, 1009

75BM   A. Braibanti and G. Mori, J. Chem. Soc. Dalton, 1975, 1319

75BMa   P.J. Brignac, Jr. and C. Mo, Anal. Chem., 1975, 47, 1465

75BN   G.V. Budu, L.V. Nazarova, and A.P. Tkhoryak, Russ. J. Inorg. Chem., 1975, 20,
          1608 (2904)

75BO   S. Bergstrom and G. Olofsson, J. Soln. Chem., 1975, 4, 535

75BP   G. Brookes and L.D. Pettit, J. Chem. Soc. Dalton, 1975, 2106

75BPa   G. Brookes and L.D. Pettit, J. Chem. Soc. Dalton, 1975, 2112

75BPb   G. Brookes and L.D. Pettit, J. Chem. Soc. Dalton, 1975, 2302

75BS   M. Bartusek and J. Senkyr, Scripta Fac. Sci. Nat. Univ. Purk. Brun., Chem. 1,
          1975, 5, 61

75BPV   R. Barbucci, P. Paoletti, and A. Vacca, Inorg. Chem., 1975, 14, 302

75BW   A.C. Baxter and D.R. Williams, J. Chem. Soc. Dalton, 1975, 1757

75CA   A.L. Cummings and K.P. Anderson, Anal. Chem., 1975, 47, 2310

75CB   J.L. Chabard, G. Besse, D. Pepin, J. Petit, and J.A. Berger, Bull. Soc. Chim.
          France, 1975, 1943

75CG   L. Ciavatta and M. Grimaldi, J. Inorg. Nucl. Chem., 1975, 37, 163

75CGa   M. Castillo-Martos and F.A. Gonzalez-Vilchez, An. Quim., 1975, 71, 594;
          J. Inorg. Nucl. Chem., 1975, 37, 316

75CGb   P. Cauchetier and C. Guichard, J. Inorg. Nucl. Chem., 1975, 37, 1771

75CGP   L. Ciavatta, M. Grimaldi, and R. Palombari, J. Inorg. Nucl. Chem., 1975, 37, 1685

75CK   C. Chatterjee and T.A Kaden, Helv. Chim. Acta, 1975, 58, 1881

75CM   A.M. Corrie, G.K.R. Makar, M.L.D. Touche, and D.R. Williams, J. Chem. Soc.
         Dalton, 1975, 105

75CP   Y. Couturier and C. Petitfaux, Bull. Soc. Chim. France, 1975, 141

75CPa   Y. Couturier and C. Petitfaux, Bull. Soc. Chim. France, 1975, 1043, 1545

75CPb   B.K. Choudhary and B. Prasad, J. Indian Chem. Soc., 1975, 52, 679

75CPM   E. Chiacchierini, V. Petrone, and A. Magri, Gazz. Chim. Ital., 1975, 105, 205

75CR   J.J. Cruywagen and E.F.C.H. Rohwer, Inorg. Chem., 1975, 14, 3136; J. South
         African Chem. Inst., 1976, 29, 30

75CS   P.B. Chakrawarti and H.N. Sharma, J. Indian Chem. Soc., 1975, 52, 171

75CSa   P.B. Chakraborty and H.N. Sharma, J. Indian Chem. Soc., 1975, 52, 1211

75CY   M. Chikuma, A. Yokoyama, Y. Ueda, and H. Tanaka, Chem. Pharm. Bull. (Japan),
         1975, 23, 473

75D   D.G. Dhuley, Curr. Sci. (India), 1975, 44, 491

75DB   T.F.Dorigatti and E.J. Billo, J. Inorg. Nucl. Chem., 1975, 37, 1515

75DBT   G. Duc, F. Bertin, and G. Thomas-David, Bull Soc. Chim. France, 1975, 495

75DK   D.M. Druskovich and D.L. Kepert, J. Chem. Soc. Dalton, 1975, 947

75DM   U.N. Dash and J. Mohanty, Thermochim. Acta, 1975, 12, 189

75DOa   P.G. Daniele, G. Ostacoli, and A. Vanni, Ann. Chim. (Rome), 1975, 65, 465

75DOb   P.G. Daniele, G. Ostacoli, and V. Zelano, Ann. Chem. (Rome), 1975, 65, 617

75DOc   P.G. Daniele, G. Ostacoli, and A. Vanni, Atti Accad. Sci. Torino, 1975, 109, 547

75DOd   P.G. Daniele, G. Ostacoli, and V. Zelano, Ann. Chim. (Rome), 1975, 65, 455

75DW   J.R. DeMember and F.A. Wallace, J. Amer. Chem. Soc. 1975, 97, 6240

75EA   M.S. El-Ezaby and I.E. Abdel-Aziz, J. Inorg. Nucl. Chem., 1975, 37, 2013

75EB   O. Enea, M.J. Blais, and G. Berthon, Thermochim. Acta, 1975, 12, 29

75EG   M.S. El-Ezaby and N. Gayed, J. Inorg. Nucl. Chem., 1975, 37, 1065

75EM   M. Edrissi and A. Massoumi, Talanta, 1975, 22, 693

75EW   B. Elgquist and M. Wedborg, Marine Chem., 1975, 3, 215

75F   F.H. Fisher, J. Soln. Chem., 1975, 4, 237

75Fa   V.A. Fedorov, Soviet J. Coord. Chem., 1975, 1, 751 (890)

75FB   V.A. Fedorov, I.M. Bolshakova, and T.G. Moskalenko, Russ. J. Inorg. Chem., 1975,
         20, 859 (1536)

75FBa   D.P. Feldman, V.M. Berdnikov, and B.P. Matseevskii, Russ. J. Phys. Chem., 1975,
         49, 1858 (3148)

75FBb   D.P. Feldman, V.M. Berdnikov, and B.P. Matseevskii, Russ. J. Phys. Chem., 1975,
         49, 1860 (3152)

75FC   T.B. Field, J. Coburn, J.L. McCourt, and W.A.E. McBryde, Anal. Chim. Acta, 1975,
         74, 101

75FCK   V.A. Fedorov, G.E. Chernikova, M.A. Kuznechikhina, and T.I. Kuznetsova, Russ. J.
         Inorg. Chem., 1975, 20, 1613 (2912)

75FF   V.A. Fedorov, A.V. Fedorova, and G.G. Nifanteva, Russ. J. Inorg. Chem., 1975,
         20, 1453 (2625)

75FFa   V.A. Fedorov, A.V. Fedorova, G.G. Nifanteva, and Z.P. Abdugalimov, Soviet J. Coord. Chem., 1975, 1, 1154 (1378)

75FFb   F.H. Fisher and A. P. Fox, J. Soln. Chem., 1975, 4, 225

75FK   V.A. Fedorov and T.N. Koneva, Soviet J. Coord. Chem., 1975, 1, 707 (836)

75FL   Ya.D. Fridman, M.G. Levina, and E.P. Tsoi, Soviet J. Coord. Chem., 1975, 1, 1373 (1667)

75FN   G. Folcher, C. Neveu, and P. Rigny, J. Inorg. Nucl. Chem., 1975, 37, 1537

75FR   V.A. Fedorov, A.M. Robov, and I.D. Isaev, Russ. J. Phys. Chem., 1975, 49, 1841 (3115)

75FS   W. Forsling and S. Sjoberg, Acta Chem. Scand., 1975, A29, 569

75FSa   B.E. Fischer and H. Sigel, J. Inorg. Nucl. Chem., 1975, 37, 2127

75FST   S.M. Feltch, J.E. Stuehr, and G.W. Tin, Inorg. Chem., 1975, 14, 2175

75GB   A.M. Golub, S.S. Butsko, and L.P. Dobryanskaya, Russ. J. Inorg. Chem., 1975, 20, 1510 (2728)

75GD   C. Gerard and C. Ducauze, Bull. Soc. Chim. France, 1975, 1955

75GF   E.A. Gyunner and A.M. Fedorenko, Russ. J. Inorg. Chem., 1975, 20, 841 (1502)

75GG   V.V. Grigoreva and I.V. Golubeva, Russ. J. Inorg. Chem., 1975, 20, 705 (1254)

75GK   M.B. Goldhaber and I.R. Kaplan, Marine Chem., 1975, 3, 83

75GN   I.P. Gorelov and V.M. Nikolskii, Russ. J. Inorg. Chem., 1975, 20, 966 (1722)

75GNF   A. Gergely, I. Nagypal, and E. Farkas, J. Inorg. Nucl. Chem., 1975, 37, 551

75GO   J. Gonzalez Velasco, J. Ortega, and J. Sancho, An. Quim., 1975, 71, 706

75H   L. Harju, Talanta, 1975, 22, 1029

75HA   I. Haq and A. Aziz Khan, J. Indian Chem. Soc., 1975, 52, 1096

75HK   M.B. Hafez and M.A. Khalifa, Ann. Chim., 1975, 10, 135

75HKa   I. Haq and A.A. Khan, Indian J. Chem., 1975, 13, 298

75HL   D.K. Hazra and S.C. Lahiri, Anal. Chim. Acta, 1975, 79, 335

75HM   W.R. Harris, R.J. Motekaitis, and A.E. Martell, Inorg. Chem., 1975, 14, 974

75HMa   M. Hariharan, R.J. Motekaitis, and A.E. Martell, J. Org. Chem., 1975, 40, 470

75HS   H. Hennig, K. Schulze, M. Muhlstadt, E. Hoyer, R. Kirmse, and L.S. Emeljanowa, Z. Anorg. Allg. Chem., 1975, 413, 10

75HV   G.J.M. Heijne and W.E. van der Linden, Talanta, 1975, 22, 923

75IH   A. Ivaska and L. Harju, Talanta, 1975, 22, 1051

75IKV   L.N. Intskirveli, I.V. Kolosov, and G.M. Varshal, Russ. J. Inorg. Chem., 1975, 20, 1323 (2388)

75IP   M. Israeli and L.D. Pettit, J. Inorg. Nucl. Chem., 1975, 37, 999

75IT   S.P. Ivashchenko, V.Ya. Temkina, A.Ya. Fridman, and O.Yu. Lavrova, Soviet J. Coord. Chem., 1975, 1, 420 (520)

75IV   A.A. Ivakin, E.M. Voronova, and L.D. Kurbatova, Russ. J. Inorg. Chem., 1975, 20, 700 (1246)

75IY   J.I. Itoh, T. Yotsuyanagi, and K. Aomura, Anal. Chim. Acta, 1975, 76, 471

75IYa   J.I. Itoh, T. Yotsuyanagi, and K. Aomura, Anal. Chim. Acta, 1975, 77, 229

75J   J.B. Jensen, Acta Chem. Scand., 1975, A29, 250

75Ja   L. Johansson, Acta Chem. Scand., 1975, A29, 365

75Jb   A.P. Joshi, Curr. Sci. (India), 1975, 44, 504

75JB   J.D. Joshi and P.K. Bhattacharya, Indian J. Chem., 1975, 13, 88

75JBa   P.C. Jain and S.P. Banerjee, J. Indian Chem. Soc., 1975, 52, 458

75JS   P.T. Joseph, K.J. Scariah, O.F. Thomas, T. Radhakrishnan, and P.N.M. Das, Indian
          J. Chem., 1975, 13, 742

75JT   R.M. Jellish and L.C. Thompson, J. Coord. Chem., 1975, 4, 199

75K   E.J. King, J. Chem. Soc. Faraday I, 1975, 71, 88

75Ka   R. Karlicek, Coll. Czech. Chem. Comm., 1975, 40, 3825

75Kb   M. Kunaszewska, Rocz. Chem., 1975, 49, 821

75Kc   I.V. Kolosov, Soviet J. Coord. Chem., 1975, 1, 282 (357)

75Kd   J.G. Kloosterboer, Inorg. Chem., 1974, 14, 536

75Ke   J. Kragten, Talanta, 1975, 22, 505

75Kf   R. Karlicek, Coll. Czech. Chem. Comm., 1975, 40, 78

75Kg   V.I. Kornev, Russ. J. Inorg. Chem., 1975, 20, 1534 (2772)

75KA   P.V. Khadikar and R.L. Ameria, Acta Chim. Acad. Sci. Hung., 1975, 85, 131

75KAL   V.I. Kornev, O.A. Artemeva, and N.Ya. Laptev, Russ. J. Phys. Chem., 1975, 49, 61
          (111)

75KAM   V.I. Kornev, L.G. Alekseeva, and I.P. Mukanov, Russ. J. Phys. Chem., 1975, 49,
          43 (81)

75KB   P. Kondziela and J. Biernat, J. Electroanal. Chem., 1975, 61, 281

75KBR   N.A. Kostromina, N.V. Beloshitskii, and V.F. Romanov, Soviet J. Coord. Chem.,
          1975, 1, 1144 (1367)

75KG   A.I. Kapustnikov and I.P. Gorelov, Russ. J. Inorg. Chem., 1975, 20, 506 (904)

75KH   R.D. Keen, D.A House, and H.K.J. Powell, J. Chem. Soc. Dalton, 1975, 688

75KI   I.V. Kolosov, L.N. Intskirveli, and G.M. Varshal, Russ. J. Inorg. Chem., 1975,
          20, 1179 (2121)

75KIa   L.D. Kurbatova, A.A. Ivakin, and E.M. Voronova, Soviet J. Coord. Chem., 1975,
          1, 1230 (1481)

75KK   R.C. Kapoor and J. Kishan, Indian J. Chem., 1975, 13, 1078

75KKa   P.V. Khadikar and M.G. Kanungo, J. Inorg. Nucl. Chem., 1975, 52, 473

75KKI   V.I. Kornev, S.L. Kharitonova, and L.B. Ionov, Russ. J. Phys. Chem., 1975, 49,
          1812 (3058)

75KKK   N.P. Komar, Yu.M. Khoroshevskii, and V.P. Kolesnik, Russ. J. Phys. Chem., 1975,
          49, 949 (1598)

75KKP   A.I. Kirillov, G.N. Koroleva, and N.S. Poluektov, Russ. J. Inorg. Chem., 1975,
          20, 1784 (3228)

75KL   M.Ya. Kutuzova, L.E. Leshchishina, and V.A. Fedorov, Russ. J. Inorg. Chem., 1975,
          20, 459 (817)

75KM   A. Kaneda and A.E. Martell, J. Coord. Chem., 1975, 4, 137

75KMa   A. Kaneda and A.E. Martell, J. Amer. Chem. Soc., 1975, 99, 1586

75KMA   H.C. Kaehler, S. Mateo, J. Ascanio, and F. Brito, An. Quim., 1975, 71, 763

75KMB   H.C. Kaehler, S. Mateo, and F. Brito, An. Quim., 1975, 71, 689

75KP   V.P. Khramov and I.G. Panchenko, Russ. J. Inorg. Chem., 1975, 20, 832 (1484)

75KS   R.J. Knight and R.N. Sylva, J. Inorg. Nucl. Chem., 1975, 37, 779

75KST   B. Kopecka, V. Springer, and J. Turan, Chem. Zvesti, 1975, 29, 727

75KU   F.Ya. Kulba, V.G. Ushakova, and Yu.B. Yakovlev, Russ. J. Inorg. Chem., 1975,
           20, 43 (79)

75KY   F.Ya. Kulba, Yu.B. Yakovlev, and D.A. Zenchenko, Russ. J. Inorg. Chem., 1975,
           20, 994 (1781)

75KZ   F.Ya. Kulba, D.A. Zenchenko, and Yu.B. Yakolev, Russ. J. Inorg. Chem., 1975,
           20, 1314 (2372)

75L   T. Lengyel, Acta Chim. Acad. Sci. Hung., 1975, 85, 125

75LB   W.E. van der Linden and C. Beers, Talanta, 1975, 22, 89

75LBa   N.M. Lukovskaya and T.A. Bogoslovskaya, Soviet Progr. Chem. (Ukr. Khim. Zh.),
           1975, 41, No. 5, 79 (529)

75LBH   J.C. Landry, J. Buffle, W. Haerdi, M. Levental, and G. Nembrini, Chimia (Switz.),
           1975, 29, 253

75LG   K. Lal and S.P. Gupta, Indian J. Chem., 1975, 13, 973

75LK   M. Langova, V. Kuban, and D. Nonova, Coll. Czech. Chem. Comm., 1975, 40, 1694

75LL   J. Lagrange and P. Lagrange, Bull. Soc. Chim. France, 1975, 1455

75LM   D.J. Leggett and W.A.E. McBryde, Anal. Chem., 1975, 47, 1065

75LMa   W.H. Leung and F.J. Millero, J. Soln. Chem., 1975, 4, 145

75LMb   G.C. Lalor and H. Miller, J. Inorg. Nucl. Chem., 1975, 37, 1832

75LN   K.I. Litovchenko, V.N. Nikitenko, and V.S. Kublanovskii, Soviet J. Coord. Chem.,
           1975, 1, 1152 (1376)

75LP   B. Lenarcik, J. Pioch, and Z. Warnke, Rocz. Chem., 1975, 49, 1511

75LS   J.M. Lehn and J.P. Sauvage, J. Amer. Chem. Soc., 1975, 97, 6700

75LT   Z. Libus and H.Tialowska, J. Soln. Chem., 1975, 4, 1011

75LW   P. Letkeman and J.B. Westmore, J. Chem. Soc. Dalton, 1975, 480

75LWa   B. Lenarcik and M. Wisniewski, Rocz. Chem., 1975, 49, 497

75M   L. Meites, Talanta, 1975, 22, 733

75MA   V.M. Masalovich, A.E. Aleshechkina, and B.P. Sereda, Russ. J. Inorg. Chem.,
           1975, 20, 1652 (2987)

75MB   A.A. Menkov and R.I. Bocharova, Russ. J. Inorg. Chem., 1975, 20, 186 (336)

75MG   Yu.I. Mikhailyuk and V.I. Gordienko, Russ. J. Inorg. Chem., 1975, 20, 1617
           (2921)

75MH   K.A. McGee and P.B. Hostetler, Amer. J. Sci., 1975, 275, 304

75MHB   N. Moulin, M. Hussonnois, L. Brillard, and R. Guillaumont, J. Inorg. Nucl.
           Chem., 1974, 37, 2521

75MJ   A.K. Maheshwari, D.S. Jain, and J.N. Gaur, Monat. Chem., 1975, 106, 1033

75MM   J.P. McCann and A. McAuley, J. Chem. Soc. Dalton, 1975, 783

75MMa   D.T. MacMillan, I. Murase, and A.E. Martell, Inorg. Chem., 1975, 14, 468

75MMb   G. McLendon, R.J. Motekaitis, and A.E. Martell, Inorg. Chem., 1975, 14, 1993

75MMc   G. McLendon and A.E. Martell, J. Coord. Chem., 1975, 4, 235

75MMH   G. McLendon, D.T. MacMillan, M. Hariharan, and A.E. Martell, Inorg. Chem., 1975,
           14, 2322

75MN   S.S. Morozova, L.V. Nikitina, N.M. Dyatlova, and G.V. Serebryakova, Russ.
          J. Inorg. Chem., 1975, 20, 228 (413)

75MS   M.S. Masoud and T.M. Salem, J. Electroanal. Chem., 1975, 66, 117

75MSa   Yu.A. Maletin and I.A. Sheka, Soviet J. Coord. Chem., 1975, 1, 696 (824)

75MSb   S.D. Makhijani and S.P. Sangal, J. Indian Chem. Soc., 1975, 52, 788

75MSB   Yu.A. Makashev, M.I. Shalaevskaya, V.V. Blokhin, and V.E. Mironov, Russ. J.
          Phys. Chem., 1975, 49, 493 (837)

75MV   O.I. Martynova, L.G. Vasina, and I.S. Krotova, Doklady Phys. Chem., 1975, 225,
          1289 (862)

75N   A. Napoli, Gazz. Chim. Ital., 1975, 105, 1073

75NA   K.G. Nelson and K.N. Amin, J. Pharm. Sci., 1975, 64, 350

75NF   A.M. Neduv, A.Ya. Fridman, and O.Yu. Lavrova, Soviet J. Coord. Chem., 1975,
          1, 406 (505)

75NG   V.M. Nikolskii and I.P. Gorelov, Russ. J. Inorg. Chem., 1975, 21, 1764 (3191)

75NL   R. Nasanen and E. Lindell, Finn. Chem. Lett., 1975, 38

75NM   N. Nakasuka, R.P. Martin, and J.P. Scharff, Bull. Soc. Chim. France, 1975, 1973

75NT   N.M. Nikolaeva and L.D. Tsvelodub, Russ. J. Inorg. Chem., 1975, 20, 1677 (3033)

75NW   G. Nakagawa, H. Wada, and T. Hayakawa, Bull. Chem. Soc. Japan, 1975, 48, 424

75O   A. Olin, Acta Chem. Scand., 1975, A29, 907

75Oa   G. Olofsson, J. Chem. Thermodyn., 1975, 7, 507

75OD   G. Ostacoli, P.G. Daniele, and A. Vanni, Ann. Chim. (Rome), 1975, 65, 197

75OH   G. Olofsson and L.G. Hepler, J. Soln. Chem., 1975, 4, 127

75OS   R. Osterberg and B. Sjoberg, J. Inorg. Nucl. Chem., 1975, 37, 815

75OSH   S. O'Cinneide, J.P. Scanlan, and M.J. Hynes, J. Inorg. Nucl. Chem., 1975, 37,
          1013

75OT   M. Orama, P. Tilus, E. Lindell, L. Hakola, and L. Adler, Finn. Chem. Lett., 1975,
          35

75OW   H. Ogino, T. Watanabe, and N. Tanaka, Inorg. Chem., 1975, 14, 2093

75P   L. Pettersson, Acta Chem. Scand., 1975, A29, 677

75PB   P.C. Parikh and P.K. Bhattacharya, Indian J. Chem., 1975, 13, 190

75PBa   T. Paal and L. Barcza, Acta Chim. Acad. Sci. Hung., 1975, 87, 33

75PBb   T. Paal and L. Barcza, Acta Chim. Acad. Sci. Hung., 1975, 85, 139

75PJ   V. Parthasarathy and C.K. Jorgensen, Chimia (Switz.), 1975, 29, 210

75PJa   N.G. Palaskar, D.V. Jahagirdar, and D.D. Khanolkar, J. Indian Chem. Soc., 1975,
          52, 134

75PM   S.P. Pande and K.N. Munshi, J. Indian Chem. Soc., 1975, 52, 498

75PN   V.E. Plyushchev, G.V. Nadezhdina, G.S. Loseva, V.V. Melnikova, and
          T.S. Parfenova, Russ. J. Inorg. Chem., 1975, 20, 33 (60)

75PP   B.T. Pham and J. Podlahova, Coll. Czech. Chem. Comm., 1975, 40, 347

75PPa   J. Pouradier and A. Pailliotet, Comp. Rend. Acad. Sci. Paris, Ser. C, 1975, 280,
          1049

75PPF   J. Pinart, C. Petitfaux and J. Faucherre, Compt. Rend. Acad. Sci. Paris, Ser. C,
          1975, 281, 519

75PR   S.K. Patil and V.V. Pamakrishna, Inorg. Nucl. Chem. Letters, 1975, 11, 421

75PSb   A.A. Popel and V.A. Shchukin, Russ. J. Inorg. Chem., 1975, 20, 1068 (1917)

75PT   W.J. Popiel and E.H. Tamimi, J. Chem. Eng. Data, 1975, 20, 246

75PTa   I.G. Prisyagina and G.S. Tereshin, Russ. J. Inorg. Chem., 1975, 20, 36 (66)

75Q   I. Qvarfort-Dahlman, Chem. Scripta, 1975, 8, 112

75R   J.H. Ritsma, Rec. Trav. Chim. Pays-Bas, 1975, 94, 174

75Ra   J.H. Ritsma, Rec. Trav. Chim. Pays-Bas, 1975, 94, 210

75Rb   E.J. Reardon, J. Phys. Chem., 1975, 79, 422

75Rc   A.J. Read, J. Soln. Chem., 1975, 4, 53

75RB   E.F.C.H. Rohwer, J.A. Brink, and J.J. Cruywagen, J. South African Chem. Inst.,
          1975, 28, 1

75RM   S. Ramamoorthy and P.G. Manning, J. Inorg. Nucl. Chem., 1975, 37, 363

75RMa   R.A. Reichle, K.G. McCurdy, and L.G. Hepler, Canad. J. Chem., 1975, 53, 3841

75RR   R. Raghavan, V.V. Ramakrishna, and S.K. Patil, J. Inorg. Nucl. Chem., 1975, 37,
          1540

75RS   C.B. Riolo, T.F. Soldi, G. Gallotti, M. Pesavento, and G. Spini, Gazz. Chim.
          Ital., 1975, 105, 221

75S   H. Sigel, Inorg. Chem., 1975, 14, 1535

75Sa   A.V. Stepanov, Russ. J. Phys. Chem., 1975, 49, 1025 (1741)

75Sb   R. Smied, J. Inorg. Nucl. Chem., 1975, 37, 318

75Sc   H. Sigel, J. Amer. Chem. Soc., 1975, 97, 3209

75SA   N.P. Samsonova, T.T. Adler, E.A. Sinyakova, V.G. Sharypova, and V.A. Fedorov,
          Soviet J. Coord. Chem., 1975, 1, 1393 (1691)

75SC   B.S. Sekhon and S.L. Chopra, Ann. Chim., 1975, 10, 21

75SCC   M.M. Santos, J.R.F.G. deCarvalho, and R.A.G. deCarvalho, J. Soln. Chem., 1975, 4,
          25

75SCH   L. Sucha, J. Cadek, K. Hrabek, and J. Vesely, Coll. Czech. Chem. Comm., 1975,
          40, 2020

75SG   J.P. Scharff and R. Genin, Anal. Chim. Acta, 1975, 78, 201

75SGa   J.M. Suarez Cardeso and S. Gonzalez Garcia, An. Quim., 1975, 71, 625

75SGP   I. Sovago, A. Gergely, and J. Posta, Acta Chim. Acad. Sci. Hung., 1975, 85, 153

75SH   Y. Sugiura, Y. Hirayama, H. Tanaka, and H. Sakurai, J. Inorg. Nucl. Chem., 1975,
          37, 2367

75SK   W. Steinmann and T.A. Kaden, Helv. Chim. Acta, 1975, 58, 1358

75SKa   O.P. Sharma and R.B. Kharat, J. Indian Chem. Soc., 1975, 52, 174

75SL   M. Saiki and F.W. Lima, Radiochem. Radioanal. Letters, 1975, 21, 371

75SLB   A.V. Serdyukova, A.N. Lazarev, G.M. Baranov, and V.V. Perekalin, Russ. J. Inorg.
          Chem., 1975, 20, 299 (536)

75SM   G.C. Shivahare, S. Mathur, and M.K. Mathur, J. Indian Chem. Soc., 1975, 52, 170

75SP   H. Sigel and B. Prijs, Chimia (Switz.), 1975, 29, 134

75SPF   M.M. Santos, J.D.L. Pinto, J.B. Ferreira, and R.A. Guedes deCarvalho, J. Soln.
          Chem., 1975, 4, 31

75SS   I. Sostaric and V. Simeon, Monat. Chem., 1975, 106, 169

75SSa   R. Sundaresan and A.K. Sundaram, Indian J. Chem., 1975, 13, 299

75SSD   D. Satyanarayana, G. Sahu, and R.C. Das, J. Chem. Soc. Dalton, 1975, 2236

75ST   S.P. Singh and J.P. Tandon, Monat. Chem., 1975, 106, 271

75STa   S.P. Singh and J.P. Tandon, Monat. Chem., 1975, 106, 871

75STb   S.P. Singh and J.P. Tandon, Acta Chim. Acad. Sci. Hung., 1975, 84, 419

75SW   H. Silber and P. Wehner, J. Inorg. Nucl. Chem., 1975, 37, 1025

75T   R.C. Turner, Canad. J. Chem., 1975, 53, 2811

75TK   T.V. Ternovaya and N.A. Kostromina, Soviet Progr. Chem. (Ukr. Khim. Zh.), 1975,
           41, No. 7, 5 (679)

75TKM   H. Tsukuda, T. Kawai, M. Maeda, and H. Ohtaki, Bull. Chem. Soc. Japan, 1975, 48,
           691

75TP   R.P. Tishchenko, N.I. Pechurova, and V.I. Spitsyn, Bull. Acad. Sci. USSR, 1975,
           24, 201 (261)

75TR   M.M. Taqui Khan and P.R Reddy, J. Inorg. Nucl. Chem., 1975, 37, 771

75TRY   V.Ya. Temkina, M.N. Rusina, G.F. Yaroshenko, M.Z. Branzburg, L.M. Timakova,
           and N.M. Dyatlova, J. Gen. Chem. USSR, 1975, 45, 1530 (1564)

75UK   Yu.I. Usatenko, E.A. Klimkovich, and I.S. Panchenko, Russ. J. Inorg. Chem., 1975,
           20, 221 (401)

75VA   V.P. Vasilev, S.A. Aleksandrova, and E.G. Zaburdaeva, Russ. J. Inorg. Chem.,
           1975, 20, 488 (871)

75VB   V.P. Vasilev and A.K. Belonogova, Russ. J. Inorg. Chem., 1975, 20, 16 (30)

75VH   J. Votava, J. Havel, and M. Bartusek, Chem. Zvesti, 1975, 29, 734

75VHa   J. Votava, J. Havel, and M. Bartusek, Scripta Fac. Sci. Nat. Univ. Brno, Chem. 1,
           1975, 5, 71

75VK   V.P. Vasilev and E.V. Kozlovskii, Russ. J. Inorg. Chem., 1975, 20, 672 (1196)

75VKa   V.P. Vasilev and B.T. Kunin, Russ. J. Inorg. Chem., 1975, 20, 1050 (1881)

75VKb   A. Vesala and V. Koskinen, Finn. Chem. Lett., 1975, 145

75VKF   S.N. Vinogradov, L.N. Khurtova, and L.M. Firyulina, Russ. J. Inorg. Chem., 1975
           20, 370 (664)

75WO   N. Watanabe, S. Ohe, and S. Takamoto, Nippon Kagaku Kaishi, 1975, 298

75WT   N. Watanabe and S. Takamoto, Bull. Chem. Soc. Japan, 1975, 48, 2211

75YF   S. Yamada, S. Funahashi, and M. Tanaka, J. Inorg. Nucl. Chem., 1975, 37, 835

75YY   H. Yokoyama and H. Yamatera, Bull. Chem. Soc. Japan, 1975, 48, 2708, 2719

75ZG   N.I. Zinenvich and L.A. Garmash, Russ. J. Inorg. Chem., 1975, 20, 1571 (2838)

75ZK   G.D. Zegzhda, V.N. Kabanova, and F.M. Tulyupa, Russ. J. Inorg. Chem., 1975, 20,
           1289 (2325)

75ZL   S. Zielinski, L. Lomozik, and J. Andrzejewska, Russ. J. Inorg. Chem., 1975, 20,
           182 (329)

75ZN   G.D. Zegzhda, S.I. Neikovskii, and F.M. Tulyupa, Soviet J. Coord. Chem., 1975, 1,
           1240 (1494)

76A   G. Anderegg, Z. Naturforsch., 1976, 31b, 786

76AA   S. Ahrland and E. Avsar, Acta Chem. Scand., 1976, A30, 15

76AAA   M.K. Akhmedli, A.M. Ayubova, and S.R. Azimova, Russ. J. Inorg. Chem., 1976, 21,
           101 (188)

76ABa   M.D. Agrawal, C.S. Bhandari, M.K. Dixit, and N.C. Sogani, Monat. Chem., 1976,
         107, 75

76ABb   Ya.V. Ashak, Yu.A. Bankovskii, and A.M. Deme, J. Anal. Chem. USSR, 1976, 31, 706
         (865)

76ABc   S.A. Abbasi, B.G. Bhat, and R.S. Singh, Inorg. Nucl. Chem. Letters, 1976, 12,
         391

76ABd   S.A. Abbasi, B.G. Bhat, and R.S. Singh, Indian J. Chem., 1976, 14A, 718

76AC   G. Arena, R. Cali, E. Rizzarelli, and S. Sammartano, Thermochim. Acta, 1976, 17,
         155

76ACa   G. Arena, R. Cali, E. Rizzarelli, and S. Sammartano, Thermochim. Acta, 1976, 16,
         315

76ACP   E. Atlas, C. Culberson, and R.M. Pytkowicz, Marine Chem., 1976, 4, 243

76AD   T. Almgren, D. Dyrssen, B. Elgquist, and O. Johansson, Marine Chem., 1976, 4,
         289

76AE   L. Abello, A. Ensuque, and G. Lapluye, J. Chim. Phys., 1976, 73, 268

76AH   M. Aplincourt and R. Hugel, Bull Soc. Chim. France, 1976, 1793

76AK   R.C. Agarwal, R.P. Khandelwal, and P.C. Singhal, J. Indian Chem. Soc., 1976, 53,
         977

76AM   G. Anderegg and S.C. Malik, Helv. Chim. Acta, 1976, 59, 1498

76AMa   R.S. Ambulkar and K.N. Munshi, Indian J. Chem., 1976, 14A, 424

76AMA   A.E. Aleshechkina, V.M. Masalovich, P.K. Agasyan, and B.P. Sereda, Russ. J.
         Inorg. Chem., 1976, 21, 973 (1775)

76AMP   L. Antolini, L. Menabue, and G.C. Pellacani, Anal. Chim. Acta, 1976, 83, 337

76AN   I.I. Alekseeva, I.I. Nemzer, L.I. Yuranova, and V.V. Borisova, Russ. J. Inorg.
         Chem., 1976, 21, 1817 (3298)

76AP   R.P. Agarwal and D.D. Perrin, J. Chem. Soc. Dalton, 1976, 89

76AS   M. Kh. Akhmetov, L.A. Sapozhnikova, E.A. Savelev, and G.N. Altshuler, Russ. J.
         Phys. Chem., 1976, 50, 128 (227)

76B   J.A. Bishop, Anal. Chim. Acta, 1976, 87, 255

76Ba   L. Barcza, J. Phys. Chem., 1976, 80, 821

76BB   T. Berg and G. Berthon, Bull.Soc. Chim. France, 1976, 1065

76BBa   G. Berthon and T. Berg, J. Chem. Thermodyn., 1976, 8, 1145

76BBb   R. Barbucci and M. Budini, J. Chem. Soc. Dalton, 1976, 1321

76BBc   A.K. Basak and D. Banerjea, Indian J. Chem., 1976, 14A, 184

76BBd   N.L. Babenko, A.I. Busev, and N.V. Sukhorukova, Russ. J. Inorg. Chem., 1976, 21,
         540 (990)

76BBe   N.L. Babenko, A.I. Busev, N.V. Sukhorukova, and O.S. Frolova, Russ. J. Inorg.
         Chem., 1976, 21, 377 (701)

76BBC   R. Bedetti, U. Biader Ceipidor, V. Carunchio, and M. Tomassetti, J. Inorg. Nucl.
         Chem., 1976, 38, 1391

76BBP   P. DiBernardo, A. Bismondo, and R. Portanova, Inorg. Chim. Acta, 1976, 18, 47

76BC   S.C. Baghel, K.K. Choudhary, and J.N. Gaur, J. Electrochem. Soc. India, 1976,
         25, 167

76BCa   P. Bentler, K. Christen, and H. Gamsjager, Chimia (Switz.), 1976, 30, 104

76BD   R.G. Bidkar, D.G. Dhuley, and R.A. Bhobe, Curr. Sci. (India), 1976, 45, 168

76BEF   G. Berthon, O. Enea, and Y. Fuseau, Thermochim. Acta, 1976, 16, 323

76BEK   A. Bonsen, F. Eggers, and W. Knoche, Inorg. Chem., 1976, 15, 1212

76BF   I.M. Batyaev and R.S. Fogileva, Russ. J. Inorg. Chem., 1976, 21, 653 (1199)

76BG   T.V. Bettger, M.I. Gromova, M.N. Rusina, and O.V. Shornikova, Soviet J. Coord.
          Chem., 1976, 2, 396 (530)

76BH   P. Bianco, J. Haladjian, and R. Pilard, J. Chim. Phys., 1976, 73, 280

76BHa   P. Bianco, J. Haladjian, and R. Pilard, J. Electroanal. Chem., 1976, 72, 341

76BHb   A.M. Bond and G. Hefter, J. Electroanal. Chem., 1976, 68, 203

76BHS   H. Bilinski, R. Huston, and W. Stumm, Anal. Chim. Acta, 1976, 84, 157

76BK   R.H. Byrne and D.R. Kester, Marine Chem., 1976, 4, 255

76BKa   R.H. Byrne and D.R. Kester, Marine Chem., 1976, 4, 275

76BM   R.H. Busey and R.E. Mesmer, J. Soln. Chem., 1976, 5, 147

76BMa   R.P. Bonomo, S. Musumeci, E. Rizzarelli, and S. Sammartano, J. Inorg. Nucl.
          Chem., 1976, 38, 1851

76BMb   R.P. Bonomo, S. Musumeci, E. Rizzarelli, and S. Sammartano, Talanta, 1976, 23,
          253

76BMc   L.N. Balyatinskaya and Yu.F. Milyaev, Soviet J. Coord. Chem., 1976, 2, 1225
          (1594)

76BMD   A. Braibanti, G. Mori, and F. Dallavalle, J. Chem. Soc. Dalton, 1976, 826

76BP   G. Brookes and L.D. Pettit, J. Chem. Soc. Dalton, 1976, 42

76BPa   G. Brookes and L.D.Pettit, J. Chem. Soc. Dalton, 1976, 1224

76BR   S.V. Bagawde, V.V. Ramkrishna, and S.K. Patil, J. Inorg. Nucl. Chem., 1976, 38,
          1339

76BRa   S.V. Bagawde, V.V. Ramakrishna, and S.K. Patil, J. Inorg. Nucl. Chem., 1976, 38,
          1669

76BRb   S.V. Bagawde, V.V. Ramakrishna, and S.K. Patil, J. Inorg. Nucl. Chem., 1976, 38,
          2085

76BS   L.G. Bogachuk and I.A. Sheka, Soviet Progr. Chem., (Ukr. Khim. Zh.), 1976, 42,
          No. 9, 1 (899)

76BSB   N.L. Babenko, N.V. Sukhorukova, A.I. Busev, and L.K. Simakova, Russ. J. Inorg.
          Chem., 1976, 21, 1668 (3022)

76BT   A.L. Bogacheva, T.V. Ternovaya, V.P. Khramov, and N.A. Kostromina, Soviet J.
          Coord. Chem., 1976, 2, 237 (321)

76BTa   A.L. Bogacheva, T.V. Ternovaya, V.P. Khramov, and N.A. Kostromina, Soviet J.
          Coord. Chem., 1976, 2, 793 (1036)

76BV   A.G. Brits, R. van Eldik, and J.A. van den Berg, Z. Phys. Chem. (Frankfurt),
          1976, 99, 107

76C   M.C. Chattopadhyaya, J. Inorg. Nucl. Chem., 1976, 38, 569

76CC   E.E. Chao and K.L. Cheng, Anal. Chem., 1976, 48, 267

76CD   E. Chiacchierini, G. D'Ascenzo, G. De Angelis, and A. Marino, Gazz. Chim. Ital.,
          1976, 106, 19

76CF   Y. Couturier, R. Fournaise, and C. Petitfaux, Bull. Soc. Chim. France, 1976, 697

76CG   E. Campi and M.C. Gennaro, Atti Accad. Sci. Torino, 1976, 110, 343

76CGP   L. Ciavatta, M. Grimaldi, and R. Polombari, J. Inorg. Nucl. Chem., 1976, 38, 823

76CH   J.J. Cruywagen, J.B.B. Heyns, and E.F.C.H. Rohwer, J. Inorg. Nucl. Chem., 1976,
         38, 2033

76CJ   R.L. Coates and M.M. Jones, J. Inorg. Nucl. Chem., 1976, 38, 1549

76CJG   K.K. Choudhary, D.S. Jain, and J.N. Gaur, Indian J. Chem., 1976, 14A, 773

76CJK   K.K. Chaturvedi, P. Jain, and R. Kaushal, J. Indian Chem. Soc., 1976, 53, 335

76CK   P.K. Chattopadhyay and B. Kratochvil, Inorg. Chem., 1976, 15, 3104

76CKR   W.C. Copenhafer, M.W. Kendig, T.P. Russell, and P.H. Rieger, Inorg. Chim. Acta,
         1976, 17, 167

76CL   J.P. Collin and P. Lagrange, Bull. Soc. Chim. France, 1976, 1304

76CM   A.K. Covington and R.A. Matheson, J. Soln. Chem., 1976, 5, 781

76CP   Y. Couturier and C. Pettitfaux, Bull. Soc. Chim. France, 1976, 704

76CPF   J.L. Colin, J. Pinart, and J. Faucherre, Bull. Soc. Chim. France, 1976, 399

76CS   M.C. Chattopadhyaya and R.S. Singh, Indian J. Chem., 1976, 14A, 118

76CU   G. Choppin and P.J. Unrein, in "Transplutonium Elements 1975", W. Muller and
         R. Lindner, Eds., North-Holland, Amsterdam, 1976, p.97

76CW   B. Carlsson and G. Wettermark, J. Inorg. Nucl. Chem., 1976, 38, 1525

76CWa   A.M. Corrie and D.R. Williams, J. Chem. Soc. Dalton, 1976, 1068

76CWW   A.M. Corrie, M.D. Walker, and D.R. Williams, J. Chem. Soc. Dalton, 1976, 1012

76D   I. Dellien, Acta Chem. Scand., 1976, A30, 576

76DBT   G. Duc, F. Bertin, and G. Thomas-David, Bull. Soc. Chim. France, 1976, 414

76DCD   K. Dwivedi, M. Chandra, and A.K. Dey, Indian J. Chem., 1976, 14A, 900

76DCG   H.F. de Brabander, H.S. Creyf, A.M. Goeminne, and L.C. Van Poucke, Talanta, 1976,
         23, 405

76DG   V.M. Drozdova and I.P. Gorelov, Russ. J. Inorg. Chem., 1976, 21, 204 (377)

76DGa   V.M. Drozdova and I.P. Gorelov, Russ. J. Inorg. Chem., 1976, 21, 1295 (2355)

76DGN   N.K. Dutt, S. Gupta, and K. Nag, Indian J. Chem., 1976, 14A, 1000

76DH   I. Dellien and L.G Hepler, Canad. J. Chem., 1976, 54, 1383

76DK   A. D'Aprano, J. Komiyama, and R.M. Fuoss, J. Soln. Chem., 1976, 5, 279

76DMH   I. Dellien, K.G. McCurdy, and L.G Hepler, J. Chem. Thermodyn., 1976, 8, 203

76DMN   R.C. Das, M.K. Misra, and B.K. Nanda, Indian J. Chem., 1976, 14A, 624

76DMP   U.N. Dash, J. Mohanty, and K.N. Panda, Thermochim, Acta, 1976, 16, 55

76DO   P.G. Daniele and G. Ostacoli, Ann. Chim. (Rome), 1976, 66, 387

76DOb   P.G. Daniele and G. Ostacoli, Ann. Chim. (Rome), 1976, 66, 537

76DOC   P.G. Daniele, G. Ostacoli, and P.A. Caldoro, Ann. Chim. (Rome), 1976, 66, 127

76DR   U.N. Dash and P.C. Rath, Thermochim. Acta, 1976, 16, 407

76E   L.I. Elding, Inorg. Chim. Acta, 1976, 20, 65

76EE   A.E. El-Hilaly and M.S. El-Ezaby, J. Inorg. Nucl. Chem., 1976, 38, 1533

76EEa   M.S. El-Ezaby and F.R. Eziri, J. Inorg. Nucl. Chem., 1976, 38, 1901

76EN   N.G. Elenkova and T.K. Nedelcheva, J. Electroanal. Chem., 1976, 69, 395

76ES   V.E. Ermakova and M.S. Shapnik, Russ. J. Inorg. Chem., 1976, 21, 1172 (2130)

76EW   W.J. Eilbeck and M.S. West, J. Chem. Soc. Dalton, 1976, 274

76F   G. Ferroni, Electrochim. Acta, 1976, 21, 283

76FA   V.V. Fomin, L.I. Averbakh, R.A. Leman, and S.A. Konovalova, Russ. J. Inorg.
              Chem., 1976, 21, 551 (1008)

76FB   J.J. Fardy and J.M. Buchanan, J. Inorg. Nucl. Chem., 1976, 38, 579

76FE   Y. Fuseau, O. Enea, and G. Berthon, Thermochim. Acta, 1976, 16, 39

76FF   V.A. Fedorov, A.V. Fedorova, and G.G. Nifanteva, J. Gen. Chem. USSR, 1976, 46,
              896 (900)

76FJ   G.V. Fazakerley, G.E. Jackson, and P.W. Linder, J. Inorg. Nucl. Chem., 1976, 38,
              1397

76FK   V.A. Fedorov, A.I. Khokhlova, and G.E. Chernikova, J. Inorg. Nucl. Chem., 1976,
              21, 184 (344)

76FKa   V.A. Fedorov, A.I. Khokhlova, and G.E. Chernikova, Soviet J. Coord. Chem., 1976,
              2, 785 (1027)

76FKM   V.A. Fedorov, L.I. Kiprin, and V.E. Mironov, Russ. J. Phys. Chem., 1976, 50,
              1658 (2777)

76FKS   V.A. Fedorov, T.N. Kalosh, and L.I. Shmydko, Russ. J. Inorg. Chem., 1976, 21,
              465 (853)

76FL   J.G. Frost, M.B. Lawson, and W.G. McPherson, Inorg. Chem., 1976, 15, 940

76FM   L. Fabbrizzi, M. Micheloni, and P. Paoletti, Inorg. Chem., 1976, 15, 1451

76FRI   V.A. Fedorov, A.M. Robov, and I.D. Isaev, Russ. J. Phys. Chem., 1976, 50, 58
              (104)

76FRS   V.A. Fedorov, A.M. Robov, I.I. Shmydko, T.N. Koneva, L.S. Simaeva, and
              V.A. Kukhtina, Russ. J. Phys. Chem., 1976, 50, 1330 (2213)

76FS   V.A. Fedorov and L.P. Shishin, Russ. J. Phys. Chem., 1976, 50, 210 (356)

76FSH   M. Frydman, L.G. Sillen, E. Hogfeldt, and D. Ferri, Chem. Scripta, 1976, 10,
              152

76G   S. Gobom, Acta Chem. Scand., 1976, A30, 745

76Ga   S. Gobom, Acta Chem. Scand., 1976, A30, 771

76GA   A.K. Garg, S.V. Arya, and W.U. Malik, Indian J. Chem., 1976, 14A, 994

76GC   M.F. Grigoreva, G.V. Chernyavskaya, and I.A. Tserkovnitskaya, Russ. J. Inorg.
              Chem., 1976, 21, 230 (431)

76GD   I.P. Gorelov, V.M. Drozdova, and M.Kh. Kolosova, J. Anal. Chem. USSR, 1976, 31,
              1271 (1697)

76GJ   J.N. Gaur, D.S. Jain, and K.K. Choudhary, J. Electrochm. Soc. India, 1976, 25,
              169

76GK   A. Gergely and T. Kiss, Inorg. Chim. Acta, 1976, 16, 51

76GKa   I.P. Gorelov and A.I. Kapustnikov, Russ. J. Inorg. Chem., 1976, 21, 182 (339)

76GKb   I.P. Gorelov and A.I. Kapustnikov, Russ. J. Inorg. Chem., 1976, 21, 1405 (2554)

76GKc   H.S. Gowda and B. Keshavan, Indian J. Chem., 1976, 14A, 293

76GMa   J.M. Gatez, E. Merciny, and G. Duychaerts, Anal. Chim. Acta, 1976, 84, 383

76GMb   A.K. Guru, and A.V. Mahajan, Curr. Sci. (India), 1976, 45, 492

76GMc   C.P. Gupta and K.N. Munshi, Indian J. Chem., 1976, 14A, 510

76GMd   N.N. Ghosh and G.N. Mukhopadhyay, Indian J. Chem., 1976, 14A, 264

76GMe   N.N. Ghosh and S.K. Mukhopadhyay, Indian J. Chem., 1976, 14A, 413

76GMf   N.N. Ghosh and S.K. Mukhopadhyay, J. Indian Chem. Soc., 1976, 53, 233

76GN   N.N. Ghosh and M.M. Nandi, J. Indian Chem. Soc., 1976, 53, 317

76GO   J. Gonzalez Velasco, J. Ortega, and J. Sancho, J. Inorg. Nucl. Chem., 1976, 38,
         889

76GP   M. Gold and H.K.J. Powell, J. Chem. Soc. Dalton, 1976, 230

76GPP   H.L. Girdhar, S. Parveen, and M.K. Puri, Indian J. Chem., 1976, 14A, 1021

76GS   A. Gergely and I. Sovago, Inorg. Chim. Acta, 1976, 20, 19

76GSa   R. Good and D.T. Sawyer, Inorg. Chem., 1976, 15, 1427

76GSb   R.I. Gelb and L.M. Schwartz, J. Chem. Soc. Perkin II, 1976, 930

76GSc   A.A. Gundorina and A.N. Sergeeva, Russ. J. Inorg. Chem., 1976, 21, 500 (918)

76GT   A.M. Golub and B. Tanirbergenov, Soviet Progr. Chem., (Ukr. Khim. Zh.), 1976,
         42, No. 2, 10 (124)

76GTa   H.S. Gowda and K.N. Thimmaiah, Indian J. Chem., 1976, 14A, 821

76H   J.L. Hall, J. Electroanal. Chem., 1976, 68, 217

76HB   B. Hurnik and E. Banaszak, Rocz. Chem., 1976, 50, 2035

76HF   R.D. Hancock, N.P. Finkelstein, and A. Evers, J. Inorg. Nucl. Chem., 1976, 38,
         343

76HH   S. Hietanen and E. Hogfeldt, Chem. Scripta, 1976, 9, 24

76HHa   S. Hietanen and E. Hogfeldt, Chem. Scripta, 1976, 10, 37

76HHb   S. Hietanen and E. Hogfeldt, Chem. Scripta, 1976, 10, 39

76HHc   S. Hietanen and E. Hogfeldt, Chem. Scripta, 1976, 10, 41

76HM   W.R. Harris and A.E. Martell, Inorg. Chem., 1976, 15, 713

76HMa   B.F. Hitch and R.E. Mesmer, J. Soln. Chem., 1976, 5, 667

76HS   Y. Hojo, Y. Sugiura, and H. Tanaka, J. Inorg. Nucl. Chem., 1976, 38, 641

76IB   L. Ilcheva and J. Bjerrum, Acta Chem. Scand., 1976, A30, 343

76IBG   A.A. Ivakin, S.V. Borobeva, A.M. Gorelov, and E.M. Gertman, Russ. J. Inorg.
         Chem., 1976, 21, 1641 (2973)

76IC   A.A. Ivakin, I.G. Chufarova, N.I. Petunina, and V.A. Gurevich, Russ. J. Inorg.
         Chem., 1976, 21, 971 (1770)

76IM   B.N. Ivanov-Emin, S.G. Malyugina, L.D. Borzova, V.P. Dolganev, B.G. Zaitsev,
         and A.M. Egorov, Russ. J. Inorg. Chem., 1976, 21, 657 (1208)

76IS   B.N. Ivanov-Emin, T.N. Susanin, A.M. Egorov, V.P. Pyzhikov, B.E. Zaitsev,
         and A.I. Ezhov, Soviet J. Coord. Chem., 1976, 2, 687 (898)

76ITH   R.M. Izatt, R.E. Terry, B.L. Haymore, L.D. Hansen, N.K. Dalley, A.G. Avondet,
         and J. J. Christensen, J. Amer. Chem. Soc., 1976, 98, 7620

76ITN   R.M. Izatt, R.E. Terry, D.P. Nelson, Y. Chan, D.J. Eatough, J.S. Bradshaw,
         L.D. Hansen, and J.J. Christensen, J. Amer. Chem. Soc., 1976, 98, 7626

76IV   A.A. Ivakin, S.V. Vorobeva, E.M. Gertman, and E.M. Voronova, Russ. J. Inorg.
         Chem., 1976, 21, 237 (442)

76JB   P.D. Jadhav, R.G. Bidkar, D.G. Dhuley, and R.A. Bhobe, J. Indian Chem. Soc.,
         1976, 53, 451

76JGC   M. deJesus Tavares, M.A. Gouveia, and R.G. de Carvalho, J. Inorg. Nucl. Chem.,
         1976, 38, 1363

76JP   M.B. Jones and L. Pratt, J. Chem. Soc. Dalton, 1976, 1207

76JPB   D.S. Jain, O. Prakash, and S.K. Bhasin, J. Electrochem. Soc. India, 1976, 25, 139

76K   S. Katayama, J. Soln. Chem., 1976, 5, 241

76KB   O.B. Khachaturyan and A.A. Belyakova, Russ. J. Phys. Chem., 1976, 50, 1424 (2382)

76KBa   O.B. Khachaturyan and A.A. Belyakova, Russ. J. Phys. Chem., 1976, 50, 1574 (2641)

76KBM   L.A. Kulvinova, V.V. Blokhin, and V.E. Mironov, Russ. J. Phys. Chem., 1976, 50, 773 (1287)

76KD   Ts.B. Konunova, T.B. Denisova, G.M. Prizhilevskaya, and L.A. Arnaut, Soviet J. Coord. Chem., 1976, 2, 584 (772)

76KF   A.I. Khokhlova and V.A. Fedorov, Russ. J. Inorg. Chem., 1976, 21, 1652 (2995)

76KFA   B. Khan, O. Farooq, and N. Ahmad, J. Electroanal. Chem., 1976, 74, 239

76KG   A.I. Kapustnikov and I.P. Gorelov, Russ. J. Inorg. Chem., 1976, 21, 72 (136)

76KGa   A.I. Kapustnikov and I.P. Gorelov, Russ. J. Inorg. Chem., 1976, 21, 814 (1489)

76KI   B.P. Karadakov and Khr. R. Ivanova, Russ. J. Inorg. Chem., 1976, 21, 56 (106)

76KIa   A.S. Kereichuk and L.M. Ilicheva, Russ. J. Inorg. Chem., 1976, 21, 205 (380)

76KK   R. Karlsson and L. Kullberg, Chem. Scripta, 1976, 9, 54

76KKa   E.E. Kriss and G.T. Kurbatova, Russ. J. Inorg. Chem., 1976, 21, 1302 (2368)

76KKb   M. Kodama and E. Kimura, J. Chem. Soc. Dalton, 1976, 116; J. Chem. Soc. Chem. Comm., 1975, 326

76KKc   M. Kodama and E. Kimura, J. Chem. Soc. Dalton, 1976, 1720; J. Chem. Soc. Chem. Comm., 1975, 891

76KKd   M. Kodama and E. Kimura, J. Chem. Soc. Dalton, 1976, 2335

76KKe   M. Kodama and E. Kimura, J. Chem. Soc. Dalton, 1976, 2341

76KKf   M. Kodama and E. Kimura, Bull. Chem. Soc. Japan, 1976, 49, 2465

76KL   T.R. Khan and C.H. Lanford, Canad. J. Chem., 1976, 54, 3192

76KLa   E. Kauffmann, J.M. Lehn, and J.P. Sauvage, Helv. Chim. Acta, 1976, 59, 1099

76KLb   T.P.A. Kruck, S.J. Lau, and B. Sarkar, Canad. J. Chem., 1976, 54, 1300

76KM   A.S. Kereichuk and N.V. Mokhnatova, Russ. J. Inorg. Chem., 1976, 21, 651 (1195)

76KMM   P.K. Kanungo, M.R. Mali, and R.K. Mehta, Curr. Sci. (India), 1976, 45, 757

76KMY   N.F. Kosenko, T. V. Malkova, and K.B. Yatsimirskii, Russ. J. Inorg. Chem., 1976, 21, 1654 (2999)

76KN   N.A. Kostromina and L.B. Novikova, Soviet J. Coord. Chem., 1976, 2, 691 (903)

76KNa   T. Kitagawa and K. Nomura, Bull. Chem. Soc. Japan, 1976, 49, 3518

76KNI   B.P. Karadakov, P.P. Nenova, and Khr. R. Ivanova, J. Inorg. Nucl. Chem., 1976, 38, 1033

76KP   T.S. Kee and H.K.J. Powell, Aust. J. Chem., 1976, 29, 921

76KPa   B.E. Kozer and G.A. Prik, Russ. J. Phys. Chem., 1976, 50, 367 (626)

76KS   N. Kojima, Y. Sugiura, and H. Tanaka, Bull. Chem. Soc. Japan, 1976, 49, 1294

76KSa   N. Kojima, Y. Sugiura, and H. Tanaka, Bull. Chem. Soc. Japan, 1976, 49, 3023

76KSb   M.A. Khan and M.J. Schwing-Weill, Inorg. Chem., 1976, 15, 2202

76KSc   J.L. Kurz and M.A. Stein, J. Phys. Chem., 1976, 80, 154

76KSA   G.A. Krestov, V.A. Shormanov, and V.N. Afanasev, Russ. J. Inorg. Chem., 1976, 21, 397 (738)

76KT   M. Kopacz, J. Terpilowski, and R. Manczyk, Rocz. Chem., 1976, 50, 3

76KU   E.A. Klimkovich, Yu.I. Usatenko, and N.I. Levchenko, J. Anal. Chem. USSR, 1976
         31, 973 (1195)

76KVP   Ts.B. Konunova, A.S. Venichenko, and M.S. Popov, Russ. J. Inorg. Chem., 1976, 21,
         53 (100)

76KVS   N.P. Komar, S.I. Vovk, and T.I. Sinyuta, Russ. J. Phys. Chem., 1976, 50, 1439
         (2403)

76KY   F.Ya. Kulba, Yu.B. Yakovlev, and D.A. Zenchenko, Russ. J. Phys. Chem., 1976, 50,
         1332 (2216)

76L   M. Lamache-Duhameaux, J. Inorg. Nucl. Chem., 1976, 38, 1979

76La   Y.H. Lee, Acta Chem. Scand., 1976, A30, 586

76Lb   Y.H. Lee, Acta Chem. Scand., 1976, A30, 586, 593

76Lc   L.H.J. Lajunen, Finn. Chem. Letters, 1976, 31

76Ld   L.H.J. Lajunen, Finn. Chem. Letters, 1976, 36

76Le   L.H.J. Lajunen, Finn. Chem. Letters, 1976, 53

76Lf   L.H.J. Lajunen, Finn. Chem. Letters, 1976, 58

76Lg   L.H.J. Lajunen, Finn. Chem. Letters, 1976, 63

76LH   E.A. Lewis, L.D. Hansen, E.J. Baca, and D.J. Temer, J. Chem. Soc. Perkin II,
         1976, 125

76LK   B. Lenarcik, J. Kulig, and B. Barszcz, Rocz. Chem., 1976, 50, 183

76LL   R.J. Lemire and M.W. Lister, J. Soln. Chem., 1976, 5, 171

76LM   Z. Libus and W. Maciejewski, Rocz. Chem., 1976, 50, 1661

76LMG   F.I. Lobanov, A. Mirzada, and I.M. Gibalo, Moscow Univ. Chem. Bull., 1976, 31,
         No. 4, 41 (436)

76LMP   J.E. Land, W.R. Mountcastle, H.T. Peters, and C. Sinkule, Auburn Univ. WRRI
         Bull. No. 24, Jan. 1976

76LMR   S.T. Lobo, T.S.S.R. Murty, and R.E. Robertson, Canad. J. Chem., 1976, 54, 3607

76LP   S.C. Lal and B. Prasad, J. Indian Chem. Soc., 1976, 53, 136

76LS   P.W. Linder, M.J. Stanford, and D.R. Williams, J. Inorg. Nucl. Chem., 1976, 38,
         1847

76LU   D. Linke and E. Uhlig, Z. Anorg. Allg. Chem., 1976, 422, 243

76LW   B. Lenarcik, M. Wisniewski, and M. Gabryszewski, Rocz. Chem., 1976, 50, 407

76M   L. Monsted, Acta Chem. Scand., 1976, A30, 599

76Ma   M.R. Masson, J. Inorg. Nucl. Chem., 1976, 38, 545

76Mb   B. Michaylova, Acta Chim. Acad. Sci. Hung., 1976, 90, 111

76MD   T.Ya. Medved, N.M. Dyatlova, V.P. Markhaeva, M.V. Rudomino, N.V. Churilina,
         Yu.M. Polikarpov, and M.I. Kabachnik, Bull. Acad. Sci. USSR, Chem. Sci.,
         1976, 25, 992 (1018)

76MFG   M.R. Melardi, G. Ferroni, and J. Galea, Bull. Soc. Chim. France, 1976, 1004

76MFT   T.V. Malkova, L.A. Fomina, and I.A. Tsvetkova, Russ. J. Inorg. Chem., 1976, 21,
         1477 (2686)

76MG   A.C. Mishra and C.M. Gupta, Indian J. Chem., 1976, 14A, 450

76MGD   E. Merciny, J.M. Gatez, and G. Duyckaerts, Anal. Chim. Acta, 1976, 86, 247

76MH   F. Marsicano and R.D. Hancock, J. Coord. Chem., 1976, 6, 21

76MI   H. Matsushita and N. Ishikawa, Nippon Kagaku Kaishi, 1976, 1322

76MK   G.R. Miller and D.R. Kester, Marine Chem., 1976, 4, 67

76MKD   W.J. McDowell, O.L. Keller, Jr., P.E. Dittner, J.R. Tarrant, and G.N. Case,
          J. Inorg. Nucl. Chem., 1976, 38, 1207

76MKS   R. Murai, K. Kurakane, and T. Sekine, Bull. Chem. Soc. Japan, 1976, 49, 335

76MM   L. Monsted and O. Monsted, Acta Chem. Scand., 1976, A30, 203

76MMa   T.H. Mhaske and K.N. Munshi, Indian J. Chem., 1976, 14A, 421

76MMM   R.J. Motekaitis, I. Murase, and A.E. Martell, Inorg. Chem., 1976, 15, 2303

76MMR   R. Maggiore, S. Musumeci, E. Rizzarelli, and S. Sammartano, Inorg. Chim. Acta,
          1976, 18, 155

76MP   N.K. Mohanty and R.K. Patnaik, Indian J. Chem., 1976, 14A, 448

76MPa   N.K. Mohanty and R.K. Patnaik, J. Indian Chem. Soc., 1976, 53, 337

76MPb   M. Meloun and J. Pancl. Coll. Czech. Chem. Comm., 1976, 41, 2365

76MPc   A.V. Mahajani and K.S Pitre, J. Indian Chem. Soc., 1976, 53, 97

76MPR   I.N. Maksimova, N.N. Pravidin, and V.E. Razuvaev, Soviet Progr. Chem., (Ukr.
          Khim. Zh.), 1976, 42, No. 10,9 (1019)

76MPS   E. Mentasti, E. Pelizzetti, and G. Saini, J. Inorg. Nucl. Chem., 1976, 38, 785

76MR   V.P. Markhaeva, M.V. Rudomino, Yu.M. Polikarpov, T.Ya. Medved, N.M. Dyatlova,
          and M.I. Kabachnik, Bull. Acad. Sci. USSR, Chem. Sci., 1976, 25, 998 (1024)

76MRa   S. Mukerjee and N.S. Rawat, Curr. Sci. (India), 1976, 45, 330

76MRb   S. Mukerjee and N.S. Rawat, J. Indian Chem. Soc., 1976, 53, 523

76MS   W.L. Marshall and R. Slusher, J. Inorg. Nucl. Chem., 1976, 38, 279

76MSa   N. Mahadevan and R.M. Sathe, J. Indian Chem. Soc., 1976, 53, 97

76MSb   S.D. Makhijani and S.P. Sangal, J. Indian Chem. Soc., 1976, 53, 448

76MT   G.K.R. Makar, M.L.D Touche, and D.R. Williams, J. Chem. Soc. Dalton, 1976, 1016

76MTa   H.C. Malhotra and S.K. Thareja, Indian J. Chem., 1976, 14A, 223

76NC   A. Napoli and P.L. Cignini, J. Inorg. Nucl. Chem., 1976, 38, 2013

76NCW   G. Nowogrocki, J. Canonne, and M. Wozniak, Bull. Soc. Chim. France, 1976, 1369

76ND   R. Nayan and A.K. Dey, J. Coord. Chem., 1976, 6, 13

76NF   C. Neveu, G. Folcher, and A.M. Laurent, J. Inorg. Nucl. Chem., 1976, 38, 1223

76NG   H.C. Nelson and D.E. Goldberg, Inorg. Chim. Acta, 1976, 19, L23

76NGa   V.M. Nikolskii and I.P. Gorelov, Russ. J. Inorg. Chem., 1976, 21, 461 (846)

76NGb   V.M. Nikolskii and I.P. Gorelov, Russ. J. Inorg. Chem., 1976, 21, 889 (1628)

76NGT   T.V. Nepostaeva, M.I. Gromova, and L.M. Timakova, Soviet J. Coord. Chem., 1976,
          2, 392 (524)

76NM   N.A. Nepomnyashchaya, A.A. Menkov, and A.S. Lenskii, J. Anal. Chem. USSR, 1976,
          31, 925 (1138)

76NV   V.A. Nazarenko, L.I. Vinarova, and N.V. Lebedeva, Russ. J. Inorg. Chem., 1976,
          21, 1298 (2360)

76OC   E. Orebaugh and G.R. Choppin, J. Coord. Chem., 1976, 5, 123

76OD   G. Ostacoli, P.G. Daniele, and A. Vanni, Ann. Chim. (Rome), 1976, 66, 305

76OSM   N. Oyama, T. Shirato, H. Matsuda, and H. Ohtaki, Bull. Chem. Soc. Japan, 1976, 49, 3047

76OST   J.F. Ojo, Y. Sasaki, R.S. Taylor, and A.G. Sykes, Inorg. Chem., 1976, 15, 1006

76OT   K. Ogura, K. Takatu, and T. Yosino, Talanta, 1976, 23, 872

76P   J. Podlahova, Coll. Czech. Chem. Comm., 1976, 41, 1485

76Pa   T. Paal, Acta Chim. Acad. Sci. Hung., 1976, 91, 393

76PA   S. Parthasarathy, and S. Ambujavalli, Inorg. Chim. Acta, 1976, 17, 213

76PB   E. Cs. Porzsolt, M.T. Beck, and A. Bitto, Inorg. Chim. Acta, 1976, 19, 173

76PBP   V.M. Peshkova, Yu.A. Barbalat, and R.B. Polenova, Russ. J. Inorg. Chem., 1976, 21, 1182 (2149)

76PC   J.M. Pislot and S. Combet, Anal. Chim. Acta, 1976, 85, 149

76PCB   M.M. Petit-Ramel, G. Chottard, and J. Bolard, J. Chim. Phys., 1976, 73, 181

76PD   D.P. Poe and H. Diehl, Talanta, 1976, 23, 141

76PG   A.I. Postoronko, F.P. Gorbenko, and L.S. Gvozdeva, Russ. J. Inorg. Chem., 1976, 21, 1831 (3324)

76PJB   A.C.M. Paiva, L. Juliano, and P. Boschcov, J. Amer. Chem. Soc., 1976, 98, 7645

76PJK   N.G. Palaskar, D.V. Jahagirdar, and D.D. Khanolkar, J. Inorg. Nucl. Chem., 1976, 38, 1673

76PK   J.E. Powell and S. Kulprathipanja, Inorg. Chem., 1976, 15, 493

76PM   I.V. Pyatnitskii, T.L. Makarchuk, and E.F. Gavrilova, Soviet Progr. Chem., (Ukr. Khim. Zh.), 1976, 42, No. 11,71 (1191)

76PN   N.V. Petrov, V.S. Nabokov, B.V. Zhadanov, I.A. Savich, and V.I. Spitsyn, Russ. J. Phys. Chem., 1976, 50, 1328 (2208)

76PP   C. Panda and R.K. Patnaik, Indian J. Chem., 1976, 14A, 446

76PPa   J. Pouradier and A. Pailliotet, Comp. Rend. Acad. Sci. Paris, Ser. C, 1976, 283, 53

76PPb   C. Panda and R.K. Patnaik, J. Indian Chem. Soc., 1976, 53, 718

76PPF   J. Pinart, C. Petitfaux, and J. Faucherre, Bull. Soc. Chim. France, 1976, 691

76PPR   J. Pinart, C. Petitfaux, and A. Roy, Bull. Soc. Chim. France, 1976, 683

76PR   S.K. Patil and V.V. Ramakrishna, J. Inorg. Nucl. Chem., 1976, 38, 1075

76PS   L.D. Pettit and J.L.M. Swash, J. Chem. Soc. Dalton, 1976, 588

76PSa   L.D. Pettit and J.L.M. Swash, J. Chem. Soc. Dalton, 1976, 2416

76PSb   B. Perlmutter-Hayman and R. Shinar, Inorg. Chem., 1976, 15, 2932

76PSc   K.S. Pitzer and L.F. Silvester, J. Soln. Chem., 1976, 5, 269

76PSd   V. Pandu Ranga Rao and V.V. Sharma, J. Inorg. Nucl. Chem., 1976, 38, 1179

76PT   B. Perlmutter-Hyman and E. Tapuhi, J. Coord. Chem., 1976, 6, 31

76PTS   N.I. Pechurova, R.P. Tishchenko, and V.I. Spitsyn, Bull. Acad. Sci. USSR, Chem. Sci., 1976, 25, 1 (7)

76R   J.H. Ritsma, J. Inorg. Nucl. Chem., 1976, 38, 907

76Ra   S. Raman, J. Inorg. Nucl. Chem., 1976, 38, 1741

76RC   R.L. Reeves, H.L. Cohen, S.A. Harkaway, and C.J. Kaiser, J. Soln. Chem., 1976, 5, 709

76RD   K.S. Rajan and J.M. Davis, J. Inorg. Nucl. Chem., 1976, 38, 897

76RDa   S.W. Rajbhoj and D.G. Dhuley, Curr. Sci. (India), 1976, 45, 205

76RG   P.C. Rawat and C.M. Gupta, Indian J. Chem., 1976, 14A, 716

76RL   N.V. Raghavan and D.L. Leussing, J. Amer. Chem. Soc., 1976, 98, 723

76RM   K.S. Rajan, A.A. Manian, J.M. Davis, and H. Dekirmenjian, Brain Res., 1976, 107, 317

76RN   P.R. Rechani, R. Nakon, and R.J. Angelici, Bioinorg. Chem., 1976, 5, 329

76RP   P.W. Roller and W.F. Pickering, Aust. J. Chem., 1976, 29, 2395

76RR   S.C. Rustagi and G.N. Rao, Indian J. Chem., 1976, 14A, 112

76RS   D.L. Rabenstein and T.L. Sayer, Anal. Chem., 1976, 48, 1141

76RSR   D. Reddy, B. Sethuram, and T.N. Rao, Indian J. Chem., 1976, 14A, 67

76RT   D.L. Rabenstein, M.C. Tourangeau, and C.A. Evans, Canad. J. Chem., 1976, 54, 2517

76RV   P. Rabindra Reddy, K. Venugopal Reddy, and M.M. Taqui Khan, J. Inorg. Nucl. Chem., 1976, 38, 1923

76S   R.S. Sandhu, Thermochim. Acta, 1976, 16, 398

76Sa   R.S. Sandhu, Thermochim. Acta, 1976, 17, 270

76Sb   R.S. Sandhu, Indian J. Chem., 1976, 14A, 1020

76Sc   V.D. Salikhov, Russ. J. Inorg. Chem., 1976, 21, 39 (72)

76Sd   K.N. Sahu, J. Indian Chem. Soc., 1976, 53, 657

76SA   H. Stunzi and G. Anderegg, Helv. Chim. Acta, 1976, 59, 1621

76SB   R.I. Stearns and A.F. Berndt, J. Phys. Chem., 1976, 80, 1060

76SC   K.K. Sen Gupta and A.K. Chatterjee, J. Inorg. Nucl. Chem., 1976, 38, 875

76SF   K. Sugasaka and A. Fujii, Bull. Chem. Soc. Japan, 1976, 49, 82

76SG   I. Sovago and A. Gergely, Inorg. Chim. Acta, 1976, 20, 27

76SH   Y. Sugiura and Y. Hirayama, Inorg. Chem., 1976, 15, 679

76SJ   D.N. Shelke and D.V. Jahagirdar, Bull. Chem. Soc. Japan, 1976, 49, 2142

76SJa   D.N. Shelke and D.V. Jahagirdar, Curr. Sci. (India), 1976, 45, 163

76SK   R.S. Saxena and G.L. Khandelwal, J. Indian Chem. Soc., 1976, 53, 870

76SKS   S.S. Sindhu, J.N. Kumaria, and R.S. Sandhu, Indian J. Chem., 1976, 14A, 817

76SKa   T.P. Shpak, I.V. Kolosov, and M.M. Senyavin, Russ. J. Inorg. Chem., 1976, 21, 1823 (3309)

76SL   V.M. Samoilenko, V.I. Lyashenko, and T.V. Poltoratskaya, Russ. J. Inorg. Chem., 1976, 21, 1804 (3274)

76SM   T. Sato and S. Murakami, Anal. Chim. Acta, 1976, 82, 217

76SN   H. Sigel and C.F. Naumann, J. Amer. Chem. Soc., 1976, 98, 730

76SO   H. Sakurai, H. Okumura, and S. Takeshima, Yakugaku Zasshi, 1976, 96, 242

76SP   J.L.M. Swash and L.D. Pettit, Inorg. Chim. Acta, 1976, 19, 19

76SPa   Y. Suzuki and J.E. Powell, Bull. Chem. Soc. Japan, 1976, 49, 2327

76SPb   R.S. Sindhu, K.B. Pandeya, and R.P. Singh, Indian J. Chem., 1976, 14A, 913

76SR   V.A. Shenderovich and V.I. Ryaboi, Russ. J. Inorg. Chem., 1976, 21, 982 (1791)

76SS   P.V. Selvaraj and M. Santappa, J. Inorg. Nucl. Chem., 1976, 38, 837

76SSa   S.S. Sandhu, R.S. Sandhu, and J.N. Kumaria, Indian J. Chem., 1976, 14A, 366

76SSb   S.S. Sandhu, R.S. Sandhu, J.N. Kumaria, J. Singh, and N.S. Sekhon, J. Indian
         Chem. Soc., 1976, 53, 114

76SSc   S.S. Sharma and M. Singh, J. Indian Chem. Soc., 1976, 53, 693

76SSd   C.L. Sharma and S. Sharma, Indian J. Chem., 1976, 14A, 534

76SSe   J.P.N. Srivastava and M.N. Srivastava, Indian J. Chem., 1976, 14A, 818

76SSf   R.S. Sandhu and S. Singh, J. Indian Chem. Soc., 1976, 53, 1242

76ST   J. Springborg and H. Toftlund, Acta Chem. Scand., 1976, A30, 171

76T   P.B. Trinderup, Acta Chem. Scand., 1976, A30, 47

76Ta   R.C. Thompson, Inorg. Chem., 1976, 15, 1080

76TB   V.V. Tsibanov, I.O. Bogatyrev, and K.B. Zaborenko, Soviet J. Coord. Chem., 1976,
         2, 178 (234)

76TC   G. Terzian, A. Cormons, M. Asso, and D. Benlian, J. Chim. Phys., 1976, 73, 146

76TCI   C. Tsonopoulos, D.M. Coulson, and L.B. Inman, J. Chem. Eng. Data, 1976, 21, 190

76TD   R.S. Taylor and H. Diebler, Bioinorg. Chem., 1976, 6, 247

76TG   V.I. Tikhomirov and N.K. Gornovskaya, Russ. J. Inorg. Chem., 1976, 21, 1082
         (1970)

76TI   V.Ya. Temkina, S.P. Ivashchenko, N.V. Tsirulnikova, and R.P. Lastovskii,
         J. Gen. Chem. USSR., 1976, 46, 500 (501)

76TM   Ya.I. Turyan and L.M. Makarova, Russ. J. Inorg. Chem., 1976, 21, 361 (670)

76TN   B. Thumerel and J. Nicole, Compt. Rend. Acad. Sci. Paris, Ser. C, 1976, 282, 327

76TR   M.M. Taqui Khan and P. Rabindra Reddy, J. Inorg. Nucl. Chem., 1976, 38, 1234

76TS   S. Tribalat and L. Schriver, J. Inorg. Nucl. Chem., 1976, 38, 145

76TT   L.I. Tikhonova and G.I. Tkacheva, Russ. J. Inorg. Chem., 1976, 21, 1799 (3264)

76TW   M.L.D. Touche and D.R. Williams, J. Chem. Soc. Dalton, 1976, 1355

76TZ   L.P. Tikhonova and V.Ya. Zayats, Russ. J. Inorg. Chem., 1976, 21, 1184 (2154)

76V   L.C. van Pouche, Talanta, 1976, 23, 161

76VB   V.P. Vasilev and A.K. Belonogova, Russ. J. Inorg. Chem., 1976, 21, 31 (55)

76VBa   V.P. Vasilev and A.K. Belonogova, Russ. J. Inorg. Chem., 1976, 21, 350 (649)

76VBb   V.P. Vasilev and A.K. Belonogova, Russ. J. Inorg. Chem., 1976, 21, 1646 (2982)

76VBc   C.A. Vega and R.G. Bates, Anal. Chem., 1976, 48, 1293

76VG   V.P. Vasilev, N.K. Grechina, and A.A. Ikonnikov, Russ. J. Inorg. Chem., 1976, 21,
         667 (1225)

76VK   P. Vanura and L. Kuca, Coll. Czech. Chem. Comm., 1976, 41, 2857

76VKK   V.P. Vasilev, E.V. Kozlovskii, and G.L. Kokurina, Russ. J. Inorg. Chem., 1976,
         21, 1826 (3314)

76VKM   V.P. Vasilev, V.D. Korableva, and P.S. Mukhina, J. Gen. Chem. USSR, 1976, 46,
         1191 (1207)

76VKO   V.P. Vasilev, L.A. Kochergina, and T.D. Orlova, J. Gen. Chem. USSR, 1976, 46,
         2109 (2192)

76VKV   V.P. Vasilev, N.I. Kokurin, and V.N. Vasileva, Russ. J. Inorg. Chem., 1976, 21,
         218 (407)

76VN   Y. Vandewalle and J. Nicole, Comp. Rend. Acad. Sci. Paris, Series C, 1976, 282,
         1073

76VP   V.V. Vekshin, N.I. Pechurova, and V.I. Spitsyn, Russ. J. Inorg. Chem., 1976, 21, 530 (972)

76VPa   G.P. Vakhramova, N.I. Pechurova, and V.I. Spitsyn, Bull. Acad. Sci. USSR, Chem. Sci., 1976, 25, 1385 (1448)

76VKS   V.P. Vasilev, L.D. Shekhanova, and L.A. Kochergina, J. Gen. Chem. USSR, 1976, 46, 729 (730)

76W   R. Wojtas, Rocz. Chem., 1976, 50, 619

76Wa   Z. Warnke, Rocz. Chem., 1976, 50, 1801

76WC   L.F. Wong, J.C. Cooper, and D.W. Margerum, J. Amer. Chem. Soc., 1976, 98, 7268

76WL   M.H. West and J.I. Legg, J. Amer. Chem. Soc., 1976, 98, 6945

76Y   M. Yamada, Bull. Chem. Soc. Japan, 1976, 49, 1023

76YC   S.S. Yun and G.R. Choppin, J. Inorg. Nucl. Chem., 1976, 38, 332

76YCB   S.S. Yun, G.R. Choppin, and D. Blakeway, J. Inorg. Nucl. Chem., 1976, 38, 587

76YK   E.N. Yurchenko, G.R. Kolonin, G.P. Shironosova, and T.P. Aksenova, Russ. J. Inorg. Chem., 1976, 21, 1682 (3050)

76YN   S. Yamada, J. Nagase, S. Funahashi, and M. Tanaka, J. Inorg. Nucl. Chem., 1796, 38, 617

76YZ   R. Yang and L.J. Zompa, Inorg. Chem., 1976, 15, 1499

76ZB   O.S. Zhuravleva and V.M. Berdnikov, Soviet J. Coord. Chem., 1976, 2, 6 (9)

76ZK   A.P. Zharkov, F.Ya. Kulba, and N.A. Babkina, Russ. J. Inorg. Chem., 1976, 21, 522 (957)

76ZL   S. Zelinski and L. Lomozik, Russ. J. Inorg. Chem., 1976, 21, 372 (692)

76ZN   G.D. Zegzhda, S.I. Neikovskii, and F.M. Tulyupa, Soviet J. Coord. Chem., 1976, 2, 124 (162)

76ZNa   G.D. Zegzhda, S.I. Neikovskii, and F.M. Tulyupa, Russ. J. Inorg. Chem., 1976, 21, 264 (490)

76ZP   K. Zutshi and S.K. Pathak, J. Electrochem. Soc. India, 1976, 25, 83

77A   R. Aruga, J. Inorg. Nucl. Chem., 1977, 39, 2159

77Aa   G. Anderegg, "Critical Survey of Stability Constants of EDTA Complexes," Pergamon Press, Oxford, 1977

77Ab   R. Aruga, Atti Accad. Sci. Torino, 1977, 111, 555

77Ac   R. Aruga, Atti Accad. Sci. Torino, 1977, 111, 193

77Ad   R. Aruga, Ann. Chim. (Rome), 1977, 67, 21

77AA   M. Aguilar, S. Alegret, and E. Casassas, J. Inorg. Nucl. Chem., 1977, 39, 733

77ABJ   P. Andersen, T. Berg, and J. Jacobsen, Acta Chem. Scand., 1977, A31, 219

77ABP   L.V. Alekseeva, N.L. Burde, I.V. Podgornaya, and A.A. Ivakin, J. Gen. Chem. USSR, 1977, 47, 632 (695)

77AC   G. Arena, R. Cali, E. Rizzarelli, S. Sammartano, R. Burbucci, and M.J.M Campbell, J. Chem. Soc. Dalton, 1977, 581

77ACM   Z. Amjad, J.G. Chambers, and A. McAuley, Canad. J. Chem., 1977, 55, 3575

77AF   A. Anichini, L. Fabbrizzi, P. Paoletti, and R.M. Clay, Inorg. Chim. Acta, 1977, 24, L21

77AG   J.A. Adejumobi and D.R. Goddard, J. Inorg. Nucl. Chem., 1977, 39, 912

77AGK   R.Ya. Aliev, M.N. Guseinov, and N.G. Klyuchnikov, Russ. J. Inorg. Chem., 1977, 22, 1403 (2587)

77AH   K.G. Ashurst and R.D. Hancock, J. Chem. Soc. Dalton, 1977, 1701

77AHa   A.J. Aarts and M.A. Herman, Bull. Soc. Chim. Belg., 1977, 86, 757

77AM   Z. Amjad and A. McAuley, J. Chem. Soc. Dalton, 1977, 304

77AMa   E.H. Abbott and J.M. Mayer, J. Coord. Chem., 1977, 6, 135

77AMb   Y.K. Agrawal and A. Mudaliar, J. Indian Chem. Soc., 1977, 54, 757

77AN   V.P. Antonovich, E.M. Nevskaya, and E.I. Shelikhina, Russ. J. Inorg. Chem., 1977,
          22, 197 (363)

77ANa   V.P. Antonovich, E.M. Nevskaya, and E.N. Suvorova, Russ. J. Inorg. Chem., 1977,
          22, 696 (1278)

77ANY   I.I. Alekseeva, I.I. Nemzer, L.I. Yuranova, V.V. Borisova, and
          Z.N. Prozorovskaya, Russ. J. Inorg. Chem., 1977, 22, 58 (111)

77AP   R.P. Agarwal and D.D. Perrin, J. Chem. Soc. Dalton, 1977, 53

77AR   S. Aditya, A.K. Roy, and S.C. Lahiri, Z. Phys. Chem. (Leipzig), 1977, 258,
          1033

77ARM   A. Arevalo, J.C. Rodriquez Placeres, T. Moreno, and J. Segura, An. Real. Soc.
          Espan. Fis. Quim., 1977, 73, 784

77AS   F. Arnaud-Neu and M.J. Schwing-Weill, Inorg. Nucl. Chem. Letters, 1977, 13, 17

77ASa   D.W. Appleton and B. Sarkar, Bioinorg. Chem., 1977, 7, 211

77ASb   J. Ananthaswamy, B. Sethuram, and T.N. Rao, Indian J. Chem., 1977, 15A, 177

77ASS   F. Arnaud-Neu, B. Spiess, and M.J. Schwing-Weill, Helv. Chim. Acta, 1977, 60,
          2633

77AT   S. Ahrland and B. Tagesson, Acta Chem. Scand., 1977, A31, 615

77ATT   S. Ahrland, B. Tagesson, and D. Tuhtar, Acta Chem. Scand., 1977, A31, 625

77BA   G. Biswas, S. Aditya, and S.C. Lahiri, Thermochim. Acta, 1977, 19, 55

77BB   M.J. Blais and G. Berthon, Canad. J. Chem., 1977, 55, 199

77BBA   V.A. Bodnya, O.N. Bubelo, and I.P. Alimarin, Moscow Univ. Chem. Bull., 1977, 32,
          No. 6, 63 (714)

77BC   S.C. Baghel, K.K. Choudhary, and J.N. Gaur, Bull. Chem. Soc. Japan, 1977, 50,
          1486

77BD   R.G. Bidkar, D.G. Dhuley, and R.A. Bhobe, Indian J. Chem.,, 1977, 15A, 63

77BE   M.J. Blais, O. Enea, and G. Berthon, Thermochim. Acta, 1977, 20, 335

77BH   P. Bianco, J. Haladjian, and R. Pilard, Anal. Chim. Acta, 1977, 93, 255

77BK   N.V. Beloshitskii, N.A. Kostromina, and I.A. Sheka, Soviet J. Coord. Chem.,
          1977, 3, 1408 (1797)

77BKa   S.P. Bag and A.K. Khastagir, Indian J. Chem., 1977, 15A, 233

77BL   C. Birraux, J.C. Landry, and W. Haerdi, Anal. Chim. Acta, 1977, 90, 51

77BLa   C. Birraux, J.C. Landry, and W. Haerdi, Anal. Chim. Acta, 1977, 93, 281

77BM   J. Buffle and A.E. Martell, Inorg. Chem., 1977, 16, 2221

77BMa   E. Bottari and R. Montali, Monat. Chem., 1977, 108, 1033

77BMb   R.H. Busey and R.E. Mesmer, Inorg. Chem., 1977, 16, 2444

77BMK   V.A. Bocharov, V.V. Melnik, V.M. Korotchenko, and T.A. Blank, Russ. J. Phys.
          Chem., 1977, 51, 1742 (2978)

77BMO   K. Bukietyuska, A. Mondry, and E. Osmeda, J. Inorg. Nucl. Chem., 1977, 39, 483

77BN   G.V. Budu, L.V. Nazarova, and A.P. Tkhoryak, Russ. J. Inorg. Chem., 1977, 22, 618 (1128)

77BNC   P. DiBernardo, V. Di Napoli, A. Cassol, and L. Magon, J. Inorg. Nucl. Chem., 1977, 39, 1659

77BO   S. Bergstrom and G. Olofsson, J. Chem. Thermodyn., 1977, 9, 143

77BOS   M.L. Banon, J. Ortega, and J. Sancho, J. Electroanal. Chem., 1977, 78, 173

77BP   G. Brookes and L.D. Pettit, J. Chem. Soc. Dalton, 1977, 1918

77BR   S.A. Bedell, P.R. Rechani, R.J. Angelici, and R. Nakon, Inorg. Chem., 1977, 16, 972

77BRP   S.V. Bagawde, V.V. Ramkrishna, and S.K. Patil, Radiochem. Radioanal. Letters, 1977, 31, 65

77BS   A.P. Borisova and I.A. Savich, Russ. J. Phys. Chem., 1977, 51, 377 (641)

77BSa   J. Bjerrum and L.H. Skibsted, Acta Chem. Scand., 1977, A31, 673

77BSb   S.J. Broderius and L.L. Smith, Jr., Anal. Chem., 1977, 49, 424

77CA   E. Casassas, J.J. Arias-Leon, and F. Garcia-Montelongo, J. Chim. Phys., 1977, 74, 324

77CAa   E. Casassas, J.J. Arias-Leon, and F. Garcia-Montelongo, J. Chim. Phys., 1977, 74, 424

77CB   H.H. Christensen and O.K. Borggaard, Acta Chem. Scand., 1977, A31, 793

77CC   G. Calvaruso, F.P. Cavasino, E. Di Dio, and R. Triolo, Inorg. Chim. Acta, 1977, 22, 61

77CCa   E.E. Chao and K.L. Cheng, Talanta, 1977, 24, 247

77CD   P. Chaudhuri and H. Diebler, J. Chem. Soc. Dalton, 1977, 596

77CE   G.R. Choppin and D.D. Ensor, J. Inorg. Nucl. Chem., 1977, 39, 1226

77CF   A.K. Covington, M.I.A. Ferra, and R.A. Robinson, J. Chem. Soc. Faraday I, 1977, 73, 1721

77CGG   G.R. Choppin, M.R. Goedken, and T.F. Gritmon, J. Inorg. Nucl. Chem., 1977, 39, 2025

77CGR   K.C. Chang, E. Grunwald, and L.R. Robinson, J. Amer. Chem. Soc., 1977, 99, 3794

77CH   P.J. Cerutti and L.G. Hepler, Thermochim Acta, 1977, 20 309

77CJ   R.L. Coates and M.M. Jones, J. Inorg. Nucl. Chem., 1977, 39, 677

77CJa   R.L. Coates and M.M. Jones, J. Inorg. Nucl. Chem., 1977, 39, 943

77CM   P. Carpenter, C.B. Monk, and R.J. Whewell, J. Chem. Soc. Faraday I, 1977, 73, 553

77CP   J.C. Cases, V.B. Parker, and M.V. Kilday, J. Res. Nat. Bur. Stand., 1977, 82, 19

77CS   P. Chaudhuri and H. Sigel, J. Amer. Chem. Soc., 1977, 99, 3142

77CT   T.V. Chuiko, F.M. Tulyupa, and A.M. Arishkevich, Russ. J. Inorg. Chem., 1977, 22, 879 (1602)

77D   I. Dellien, Acta Chem. Scand., 1977, A31, 473

77DA   J.J.R.F. da Silva and M.C.T. Abreu Vaz, J. Inorg. Nucl. Chem., 1977, 39, 613

77DB   G. Duc, F. Bertin, and G. Thomas-David, Bull. Soc. Chim. France., 1977, 196

77DBa   G. Duc, F. Bertin, and G. Thomas-David, Bull. Soc. Chim. France., 1977, 645

77DBb   L.P. Dobryanskaya and S.S. Butsko, Soviet Progr. Chem. (Ukr. Khim. Zh.), 1977, 43, No. 4, 11 (349)

77DF   P. De Maria and A. Fini, Tetrahedron, 1977, 33, 553

77DFa   P. De Maria, A. Fini, and F.M. Hall, Gazz. Chim. Ital., 1977, 107, 521

77DG   V.M. Demina, T.I. Grigor, and V.A. Fedorov, Soviet J. Coord. Chem., 1977, 3, 1052
             (1349)

77DK   B.K. Deshmukh and R.B. Kharat, J. Inorg. Nucl. Chem., 1977, 39, 165

77DO   P.G. Daniele and G. Ostacoli, Ann. Chim. (Rome), 1977, 67, 37

77DOa   P.G. Daniele and G. Ostacoli, Ann. Chim. (Rome), 1977, 67, 311

77DP   D.G. Dalmais and M.M Petit-Ramel, Bull. Soc. Chim. France, 1977, 54

77DS   R.G. Deshpande, D.N. Shelke, and D.V. Jahagirdar, Indian J. Chem., 1977, 15A, 320

77E   H. Einaga, J. Chem. Soc. Dalton, 1977, 912

77EA   M.S. El-Ezaby, A.I. Abu-Shady, N. Gayed, and F.R. El-Eziri, J. Inorg. Nucl.
             Chem., 1977, 39, 169

77EB   G.I. Efremova, R.T. Buchkova, A.V. Lapitskaya, and S.B. Pirkes, Russ. J. Inorg.
             Chem., 1977, 22, 527 (954)

77EE   M.S. El-Ezaby, M.A. El-Dessouky, and N.M. Shuaib, Canad. J. Chem., 1977, 55,
             2613

77EF   W.J. Evans, V.L. Frampton, and A.D. French, J. Phys. Chem., 1977, 81, 1810

77EK   V.N. Egorov and I.E. Kuzinets, Russ. J. Inorg. Chem., 1977, 22, 681 (1249)

77EL   K.J. Ellis, A.G. Lappin, and A. McAuley, J. Soln. Chem., 1977, 6, 183

77EO   L.I. Elding and L.F. Olsson, Inorg. Chem., 1977, 16, 2789

77EP   Y. Eini, B. Perlmutter-Hayman, and M.A. Wolff, J. Coord. Chem., 1977, 7, 27

77ERG   C.A. Evans, D.L. Rabenstein, G. Geier, and I.W. Erni, J. Amer. Chem. Soc., 1977,
             99, 8106

77ERM   M.S. El-Ezaby, M. Rashad, and N.M. Moussa, J. Inorg. Nucl. Chem., 1977, 39, 175

77F   W. Forsling, Acta Chem. Scand., 1977, A31, 759, 783

77FF   F.H. Fisher and A.P. Fox, J. Soln. Chem., 1977, 6, 641

77FK   V.A. Fedorov and A.I. Khokhlova, Russ. J. Inorg. Chem., 1977, 22, 662 (1215)

77FKa   V.A. Fedorov and A.I. Khokhlova, Soviet J. Coord. Chem., 1977, 3, 751 (970)

77FM   L. Fabbrizzi, M. Micheloni, P. Paoletti, and G. Schwarzenbach, J. Amer. Chem.
             Soc., 1977, 99, 5574

77FN   A.V. Fedorova, G.G. Nifanteva, L.I. Kiprin, and V.A. Fedorov, Soviet J. Coord.
             Chem., 1977, 3, 262 (353)

77FS   S. Fan, A.C. Storer, and G.G. Hammes, J. Amer. Chem. Soc., 1977, 99, 8293

77FZ   L. Fabbrizzi and L.J. Zompa, Inorg. Nucl. Chem. Letters, 1977, 13, 287

77GA   O.E.S. Godnho and L.M. Aleixo, J. Coord. Chem., 1977, 6, 245

77GB   K.D. Gupta, S.C. Baghel, and J.N. Gaur, J. Electrochem. Soc. India, 1977, 26,
             No. 3, 35

77GD   V.G. Gontar, N.A. Dobrynina, A.M. Evseev, and N.V. Sterlikova, Moscow Univ. Chem.
             Bull., 1977, 32, No. 2, 88 (224)

77GF   J. Granot and D. Fiat, J. Amer. Chem. Soc., 1977, 99, 70

77GGC   T.F. Gritmon, M.P. Goedken, and G.R. Choppin, J. Inorg. Nucl. Chem., 1977, 39,
             2021

77GGR   I.V. Gavrilova, M.I. Gelfman, and V.V. Tazumovskii, Soviet J. Coord. Chem., 1977,
             3, 963 (1237)

77GGS   J.P. Gupta, B.S. Garg, and R.P. Singh, Indian J. Chem., 1977, 15A, 1107

77GK   A. Gergely and T. Kiss, J. Inorg. Nucl. Chem., 1977, 39, 109

77GMD   J.M. Gatez, E. Merciny, and G. Duyckaerts, Anal. Chim. Acta, 1977, 94, 91

77GMT   S.V. Gorbachev, T.G. Marchenkova, and E.G. Timofeeva, Soviet J. Coord. Chem.,
          1977, 3, 1159 (1490)

77GN   A. Gergely and I. Nagypal, J. Chem. Soc. Dalton, 1977, 1104

77GNa   I.P. Gorelov and V.M. Nikolskii, J. Gen. Chem. USSR, 1977, 47, 1473 (1606)

77GS   I.P. Gorelov, Yu.E. Svetogorov, and R.I. Gorelova, Russ. J. Inorg. Chem., 1977,
          22, 868 (1597)

77GSL   R.I. Gelb, L.M. Schwartz, D.A. Laufer, and J.O. Yardley, J. Phys. Chem., 1977,
          81, 1268

77H   P. Hambright, Inorg. Chem., 1977, 16, 2987

77HA   M.B. Hafez and A.M. Atwa, Ann. Chim., 1977, 15th Ser., 2, 61

77HC   Y. Hasegawa and G.R. Choppin, Inorg. Chem., 1977, 16, 2931

77HCa   R.W. Hay and C.R. Clark, J. Chem. Soc. Dalton, 1977, 1866

77HE   A. Huss. Jr. and C.A. Eckert, J. Phys. Chem., 1977, 81, 2268

77HF   R.D. Hancock, N.P. Finkelstein, and A. Evers, J. Inorg. Nucl. Chem., 1977, 39,
          1031

77HG   G.G. Herman and A.M. Goeminne, J. Coord. Chem., 1977, 7, 75

77HGE   G.G. Herman, A.M. Goeminne, and Z. Eechhaut, J. Coord. Chem., 1977, 7, 53

77HH   H. Holvik and H. Hoiland, J. Chem. Thermodyn., 1977, 9, 345

77HM   W.R. Harris and A.E. Martell, J. Amer. Chem. Soc., 1977, 99, 6746

77HMa   R.D. Hancock and G.J. McDougall, J. Coord. Chem., 1977, 6, 163

77HMB   D.E. Horner, J.C. Mailen, and H.R. Bigelow, J. Inorg. Nucl. Chem., 1977, 39, 1645

77HO   H. Hedstrom, A. Olin, P. Svanstrom, and E. Aslin, J. Inorg. Nucl. Chem., 1977,
          39, 1191

77HOa   A. Hammam, A. Olin, and P. Svanstrom, Acta Chem. Scand., 1977, A31, 384

77HS   Y. Hojo, Y. Sugiura, and H. Tanaka, J. Inorg. Nucl. Chem., 1977, 39, 715

77SHa   Y. Hojo, Y. Sugiura, and H. Tanaka, J. Inorg. Nucl. Chem., 1977, 39, 1859

77HSA   F.R. Hartley, G.W. Searle, R.M. Alcock, and D.E. Rogers, J. Chem. Soc. Dalton,
          1977, 469

77HZ   D.M. Higgins and L.J. Zompa, J. Coord. Chem., 1977, 7, 105

77IC   A.A. Ivakin, I.G Chufarova, and N.I. Petunina, Russ. J. Inorg. Chem., 1977, 22,
          800 (1470)

77IH   T. Ishimitsu, S. Hirose, and H. Sakurai, Talanta, 1977, 24, 555

77IK   N.D. Ivanova, K.B. Kladnitskaya, N.A. Kostromina, and N.I. Taranenko, Russ. J.
          Inorg. Chem., 1977, 22, 216 (399)

77IM   N.A. Ibrahim and L.I. Martynenko, Russ. J. Inorg. Chem., 1977, 22, 517 (935)

77JB   P.D. Jadhav and R.A. Bhobe, J. Inorg. Nucl. Chem., 1977, 39, 2290

77JKK   J.K. Jailwal, J. Kishan, and R.C. Kapoor, J. Indian Chem. Soc., 1977, 54, 1161

77JKS   U. Jain, V. Kumari, R.C. Sharma, and G.K. Chaturvedi, J. Chim. Phys., 1977, 74,
          1038

77K   K.M. Kanth, J. Indian Chem. Soc., 1977, 54, 935

77KBa   V.I. Kornev and P.N. Buev, Soviet J. Coord. Chem., 1977, 3, 343 (456)

77KC   I.V. Kotlyarova, NI.I. Chukhrova, and N.A. Skorik, <u>Russ. J. Inorg. Chem.</u>, 1977, <u>22</u>, 807 (1482)

77KD   J.I. Kim and H. Duschner, <u>J. Inorg. Nucl. Chem.</u>, 1977, <u>39</u>, 471

77KDK   A.A. Kurganov, V.A. Davankov, Yu.D. Koreshkov, and S.V. Pogozhin, <u>Soviet J. Coord. Chem.</u>, 1977, <u>3</u>, 514 (667)

77KG   F.Ya. Kulba, F.G. Gavryuchenkov, S.A. Nikolaeva, and Z.V. Reshetnikova, <u>Russ. J. Inorg. Chem.</u>, 1977, <u>22</u>, 660 (1210)

77KH   A. Komura, M. Hayashi, and H. Imanaga, <u>Bull. Chem. Soc. Japan</u>, 1977, <u>50</u>, 2927

77KJ   R. Karlicek and V. Jokl, <u>Coll. Czech. Chem. Comm.</u>, 1977, <u>42</u>, 637

77KK   M. Kodama and E. Kimura, <u>J. Chem. Soc. Dalton</u>, 1977, 1473

77KKa   M. Kodama and E. Kimura, <u>J. Chem. Soc. Dalton</u>, 1977, 2269

77KKb   V.I. Kornev and L.V. Kardapolova, <u>Russ. J. Inorg. Chem.</u>, 1977, <u>22</u>, 764 (1405)

77KKc   R.C. Kapoor and J. Kishan, <u>J. Indian Chem. Soc.</u>, 1977, <u>54</u>, 350

77KKM   V.I. Kornev, L.V. Kardapolova, and I.P. Mukanov, <u>Russ. J. Inorg. Chem.</u>, 1977, <u>22</u>, 80 (146)

77KKV   I.E. Kalinichenko, E.V. Kobylyanskii, I.M. Vasilkevich, and G.F. Dvorko, <u>J. Gen. Chem. USSR</u>, 1977, <u>47</u>, 1484 (1618)

77KL   V.S. Kublanovskii, K.J. Litovchenko, and V.N. Nikitenko, <u>Russ. J. Inorg. Chem.</u>, 1977, <u>22</u>, 973 (1795)

77KM   V.I. Kornev and O.V. Mazeina, <u>Soviet J. Coord. Chem.</u>, 1977, <u>3</u>, 633 (820)

77KMK   V.I. Kornev, I.P. Mukanov, and M.N. Konyukhov, <u>Russ. J. Phys. Chem.</u>, 1977, <u>51</u>, 812 (1380)

77KN   K.Y. Kim and G.H. Nancollas, <u>J. Phys. Chem.</u>, 1977, <u>81</u>, 948

77KP   I. Khalil and M.M. Petit-Ramel, <u>Bull. Soc. Chim. France</u>, 1977, 1127

77KPN   K.M. Kanth, K.B. Pandeya, and H.L. Nigam, <u>Indian J. Chem.</u>, 1977, <u>15A</u>, 62

77KS   V.I. Kornev and V.P. Semakin, <u>Soviet J. Coord. Chem.</u>, 1977, <u>3</u>, 1155 (1486)

77KSa   M.A. Khan and M.J. Schwing-Weill, <u>Bull. Soc. Chim. France</u>, 1977, 399

77SKS   I.V. Kotlyarova, N.A. Skorik, and V.N. Kumok, <u>Russ. J. Inorg. Chem.</u>, 1977, <u>22</u>, 1400 (2582)

77KST   N.A. Kostromina, V.P. Shelest, T.V. Ternovaya, and Ts.B. Konunova, <u>Russ. J. Inorg. Chem.</u>, 1977, <u>39</u>, 1659 (3050)

77KT   A.J. Kresge and Y.C. Tang, <u>J. Org. Chem.</u>, 1977, <u>42</u>, 757

77KTS   N.A. Kostromina, T.V. Ternovaya, M.T. Shestakova, and S.B. Pirkes, <u>Soviet J. Coord. Chem.</u>, 1977, <u>3</u>, 779 (1008)

77KV   N.P. Komar and S.I. Vovk, <u>Russ. J. Phys. Chem.</u>, 1977, <u>51</u>, 1189 (2037)

77KVa   V.I. Kornev and V.A. Valyaeva, <u>Russ. J. Inorg. Chem.</u>, 1977, <u>22</u>, 510 (920)

77KVb   N.P. Komar, S.I. Vovk, and V.N. Podnos, <u>Russ. J. Phys. Chem.</u>, 1977, <u>51</u>, 889 (1512)

77KVc   R.L. Kushwaha, C.K. Verma, O. Prakash, and S.P. Mushran, <u>J. Indian Chem. Soc.</u>, 1977, <u>54</u>, 285

77KVF   L.N. Khurtova, S.N. Vinogradov, and L.M. Firyulina, <u>Russ. J. Inorg. Chem.</u>, 1977, <u>22</u>, 1330 (2459)

77KY   V.E. Kalinina, K.B. Yatsimirskii, V.M. Lyakushina, and L.I. Tikhonova, <u>Russ. J. Inorg. Chem.</u>, 1977, <u>22</u>, 1344 (2488)

77L   M.C. Lim, <u>J. Chem. Soc. Dalton</u>, 1977, 1398

77La   M. Lamache-Duhameaux, J. Inorg. Nucl. Chem., 1977, 39, 2081

77Lb   S. Lasztity, Radiochem. Radioanal. Letters, 1977, 29, 215

77LA   R. Louis, F. Arnaud-Neu, R. Weiss, and M.J. Schwing-Weill, Inorg. Nucl. Chem.
         Letters, 1977, 13, 31

77LB   V.A. Litvinenko and N.I. Bogatyr, Soviet J. Coord. Chem., 1977, 3, 1184 (1520)

77LBK   B. Lenarcik, B. Barszcz, and J. Kulig, Rocz. Chem., 1977, 51, 1315

77LG   G. Lenarcik and M. Gabryszewski, Rocz. Chem., 1977, 51, 855

77LGa   B. Lenarcik and M. Gabryszewski, Rocz. Chem., 1977, 51, 2001

77LGL   L.I. Lukashova, I.M. Gibalo, and F.I. Lovanov, Moscow Univ. Chem. Bull., 1977,
         32, No. 4, 50 (442)

77LK   V.I. Levin, G.E. Kodina, and V.S. Novoselov, Soviet J. Coord. Chem., 1977, 3,
         1170 (1503)

77LKa   B. Lenarcik and J. Kulig, Rocz. Chem., 1977, 51, 637

77LM   L.H.J. Lajunen, O. Makitie, and A. Laakkonen, Finn. Chem. Letters, 1977, 1

77LN   B. Lenarcik and K. Nabialek, Rocz. Chem., 1977, 51, 417

77LPK   V.V. Lukachina, A.T. Pilipenko, and O.I. Kapova, Russ. J. Inorg. Chem., 1977, 22,
         694 (1275)

77LPW   V.M. Loyal, R. Pizer, and R.G. Wilkins, J. Amer. Chem. Soc., 1977, 99, 7185

77LS   S.H. Laurie and B. Sarkar, J. Chem. Soc. Dalton, 1977, 1822

77LSa   J.M. Lehn and J. Simon, Helv. Chim. Acta, 1977, 60, 141

77LW   B. Lenarcik and M. Wisniewski, Rocz. Chem., 1977, 51, 1625

77M   W. Mark, Acta Chem. Scand., 1977, A31, 157

77MB   J.B. Macaskill and R.G. Bates, J. Phys. Chem., 1977, 81, 496

77MC   C. Makridou, M. Cromer-Morin, and J.P. Scharff, Bull. Soc. Chim. France, 1977, 59

77MF   N.D. Mitrofanova, E.D. Filippova, B.M. Fedorov, and L.I. Martynenko, Russ. J.
         Inorg. Chem., 1977, 22, 673 (1235)

77MG   P. Mirti and M.C. Gennaro, J. Inorg. Nucl. Chem., 1977, 39, 1259

77MK   P. Michaille and T. Kikindai, J. Inorg. Nucl. Chem., 1977, 39, 493

77MKa   M. Meloun and S. Kotryl, Coll. Czech. Chem. Comm., 1977, 42, 2115

77ML   O. Makitie, L.H.J. Lajunen, and A. Laakkonen, Finn. Chem. Letters, 1977, 31

77MN   A.A. Menkov and N.A. Nepomnyashchaya, Russ. J. Inorg. Chem., 1977, 22, 1155
         (2135)

77MO   M.A. Makhyoun, M.M. Osman, and T.M. Salem, Z. Anal. Chem., 1977, 285, 47

77MP   L. Musani-Marazovic and Z. Pucar, Marine Chem., 1977, 5, 229

77MPa   N.K. Mohanty and R.K. Patnaik, J. Indian Chem. Soc., 1977, 54, 867

77MR   S. Mukerjee and N.S. Rawat, J. Indian Chem. Soc., 1977, 54, 439

77MRR   I.N. Marov, L.V. Ruzaikina, V.A. Ryabukhin, P.A. Korovaikov, and N.M. Dyatlova,
         Soviet J. Coord. Chem., 1977, 3, 1038 (1333)

77MS   K. Maroszynska and M. Strawiak, Rocz. Chem., 1977, 51, 3

77MSa   S.D. Makhijani and S.P. Sangal, Indian J. Chem., 1977, 15A, 841

77MSb   S.D. Makhijani and S.P. Sangal, Indian J. Chem., 1977, 15A, 565

77MSc   S.D. Makhijani and S.P. Sangal, J. Indian Chem. Soc., 1977, 54, 670

77MSV   M. Micheloni, A. Sabatini, and A. Vacca, Inorg. Chim. Acta, 1977, 25, 41

77MSW   M. Makles-Grotowska, J. Starosta, and W. Wojciechowski, Russ. J. Inorg. Chem.,
         1977, 22, 396 (719)

77MT   G.O. Morpurgo and A.A.G. Tomlinson, J. Chem. Soc. Dalton, 1977, 744

77MTV   P.K. Migal, V.A. Tsiplyakova, N.B. Vu, T.N. Nguen, and T.F.T. Nguen, Russ. J.
         Inorg. Chem., 1977, 22, 1449 (2669)

77N   A. Napoli, J. Inorg. Nucl. Chem., 1977, 39, 463

77Na   N.M. Nikolaeva, Russ. J. Inorg. Chem., 1977, 22, 1323 (2447)

77Nb   O. Navratil, Coll. Czech. Chem. Comm., 1977, 42, 2140

77Nc   O. Navratil, Coll. Czech. Chem. Comm., 1977, 42, 2778

77ND   R. Nayan and A.K. Dey, J. Indian Chem. Soc., 1977, 54, 759

77NF   A.M. Neduv, A.Ya. Fridman, and N.M. Dyatlova, Soviet J. Coord. Chem., 1977, 3,
         125 (171)

77NFK   V.A. Nazarenko, G.V. Flyantikova, L.I. Korolenko, and N.G. Sharaya, Russ. J.
         Inorg. Chem., 1977, 22, 982 (1811)

77NP   V.A. Nazarenko, E.N. Poluektova, and G.G. Shitareva, Russ. J. Inorg. Chem., 1977,
         22, 551 (998)

77NS   V.A. Nazarenko, G.G. Shitareva, and E.N. Poluektov, Russ. J. Inorg. Chem., 1977,
         22, 541 (980)

77NT   N.M. Nikolaeva and L.D. Tsvelodub, Russ. J. Inorg. Chem., 1977, 22, 205 (380)

77NZ   S.I. Neikovskii, G.D. Zegzhda, O.I. Avershina, and F.M. Tulyupa, Russ. J. Inorg.
         Chem., 1977, 22, 1306 (2413)

77OB   Omprakash, S.K. Bhasin, D.S. Jain, and J.N. Gaur, J. Electrochem. Soc. India,
         1977, 26, No. 1, 21

77OC   H. Ohtaki and K. Cho, Bull. Chem. Soc. Japan, 1977, 50, 2674

77OH   N. Oyama, M. Horie, H. Matsuda, and H. Ohtaki, Bull. Chem. Soc. Japan, 1977, 50,
         1945

77OK   C.W. Owens, E. ten-Krooden, and A.K. Grzybowski, J. Chem. Eng. Data, 1977, 22,
         244

77OM   N. Oyama, H. Matsuda, and H. Ohtaki, Bull. Chem. Soc. Japan, 1977, 50, 406

77OO   G. Olofsson and I. Olofsson, J. Chem. Thermodyn., 1977, 9, 65

77P   E.A. Polyak, J. Gen. Chem. USSR, 1977, 47, 2210 (2415)

77Pa   N.G. Palaskar, J. Indian Chem. Soc., 1977, 54, 1030

77Pb   T. Paal, Acta Chim. Acad. Sci. Hung., 1977, 95, 31

77PB   B.N. Patel and E.J. Billo, Inorg. Nucl. Chem. Letters, 1977, 13, 335

77PC   E. Pais and R.A. Guedes de Carvalho, J. Inorg. Nucl. Chem., 1977, 39, 1725

77PD   A. Peguy and H. Diebler, J. Phys. Chem., 1977, 81, 1355

77PM   J. Padmos and K.J. Metman, Rec. Trav. Chim., 1977, 96, 50

77PMG   I.V. Pyatnitskii, T.L. Makarchuk, and E.F. Gavrilova, Russ. J. Inorg. Chem.,
         1977, 22, 959 (1767)

77PP   C. Panda and R.K. Patnaik, J. Indian Chem. Soc., 1977, 54, 843

77PR   H.K.J. Powell and J.M. Russell, Aust. J. Chem., 1977, 30, 1467

77PRa   H.K.J. Powell and J.M. Russell, Aust. J. Chem., 1977, 30, 2433

77PRS   K.S. Pitzer, R.N. Roy, and L.F. Silvester, J. Amer. Chem. Soc., 1977, 99, 4930

77PS   L.D. Pettit and J.L.M. Swash, J. Chem. Soc. Dalton, 1977, 697

77PST   B. Perlmutter-Hayman, F. Secco, E. Tapuhi, and M. Venturini, J. Chem. Soc.
           Dalton, 1977, 2220

77PU   V.P. Poddymov and A.A. Ustinova, Russ. J. Inorg. Chem., 1977, 22, 877 (1617)

77RB   M. Rebstockova and M. Bartusek, Coll. Czech. Chem. Comm., 1977, 42, 627

77RC   R.W. Ramette, C.H. Culberson, R.G. Bates, Anal. Chem., 1977, 48, 867

77RG   D.L. Rabenstein, M.S. Greenberg, and R. Saetre, Inorg. Chem., 1977, 16, 1241

77RGa   W. Riesen, H. Gamsjager, and P.W. Schindler, Geochim. Cosmochim. Acta, 1977, 41,
           1193

77RK   T.J. Riedo and T.A. Kaden, Chimia (Switz.), 1977, 31, 220

77RL   N.V. Raghavan and D.L. Leussing, J. Indian Chem. Soc., 1977, 54, 68

77RLE   S.J. Rehfeld, H.F. Loken, and D.J. Eatough, Thermochim. Acta, 1977, 18, 265

77RLW   L.J. Rodriguez, G.W. Liesegang, R.D. White, M.M. Farrow, N. Purdie, and
           E.M. Eyring, J. Phys. Chem., 1977, 81, 2118

77RN   J. Rangarajan, B.I. Nemade, and R. Sundaresan, Proc. Indian Acad. Sci., Sec. A,
           1977, 85, 454

77RR   A.M.S. Raju and V.P.R. Rao, Indian J. Chem., 1977, 15A, 1005

77RRa   K. Rangaraj and V.V. Ramanujam, J. Inorg. Nucl. Chem., 1977, 39, 489

77RS   D. Reddy, B. Sethuram, and T.N. Rao, Indian J. Chem., 1977, 15A, 899

77RSa   D. Reddy, B. Sethuram, and T.N. Rao, Indian J. Chem., 1977, 15A, 333

77RSB   M. Ragot, J.C. Sari, and J.P. Belaich, Biochim. Biophys. Acta, 1977, 499, 411,
           421

77RW   F.J.C. Rossotti and R.J. Whewell, J. Chem. Soc. Dalton, 1977, 1223

77RWa   F.J.C. Rossotti and R.J. Whewell, J. Chem. Soc. Dalton, 1977, 1229

77RWb   J.T.H. Roos and D.R. Williams, J. Inorg. Nucl. Chem., 1977, 39, 367

77RWc   C.H. Rochester and D.N. Wilson, J. Chem. Soc. Faraday I, 1977, 73, 569

77S   S. Sjoberg, Acta Chem. Scand., 1977, A31, 705

77Sa   S. Sjoberg, Acta Chem. Scand., 1977, A31, 718

77Sb   S. Sjoberg, Acta Chem. Scand., 1977, A31, 729

77Sc   V.S. Kublanovskii, Russ. J. Inorg. Chem., 1977, 22, 405 (735)

77Sd   N.A. Skorik, Russ. J. Inorg. Chem., 1977, 22, 776 (1425)

77Se   J.P. Scanlan, J. Inorg. Nucl. Chem., 1977, 39, 635

77Sf   N.P. Slabbert, J. Inorg. Nucl. Chem., 1977, 39, 883

77Sg   H. Sigel, J. Inorg. Nucl. Chem., 1977, 39, 1903

77Sh   R.S. Sandhu, Monat. Chem., 1977, 108, 51

77SB   R.L. Sayer, S. Backs, C.A. Evans, E.K. Miller, and D.L. Rabenstein, Canad. J.
           Chem., 1977, 55, 3255

77SF   H. Sigel, B.E. Fisher, and B. Prijs, J. Amer. Chem. Soc., 1977, 99, 4489

77SFa   G.M. Sycheva, A.Ya. Fridman, and Yu.A. Afanasev, Soviet J. Coord. Chem., 1977, 3,
           423 (549)

77SFb   G.M. Sycheva, A.Ya. Fridman, and Yu.A. Afanasev, Soviet J. Coord. Chem., 1977, 3,
           896 (1161)

77SG   N. Sabiani, J. Galea, and G. Ferroni, Ann. Chim., 1977, 15th Ser., 2, 249

77SH   Y. Sugiura and Y. Hirayama, J. Amer. Chem. Soc., 1977, 99, 1581

77SHa   R.M. Siebert and P.B. Hostetler, Amer. J. Sci., 1977, 277, 697, 716

77SI   T. Sekine, S. Iwahori, S. Johnsson, and R. Murai, J. Inorg. Nucl. Chem., 1977, 39, 1092

77SJ   D.N. Shelke and D.V. Jahagirdar, J. Inorg. Nucl. Chem., 1977, 39, 2223

77SJH   E. Skou, T. Jacobsen, W. van der Hoeven, and S. Atlung, Electrochim. Acta, 1977, 22, 169

77SJS   J.P.N. Srivastava, R. Johri, and M.N. Srivastava, Indian J. Chem., 1977, 15A, 1109

77SK   T. Sato and T. Kato, J. Inorg. Nucl. Chem., 1977, 39, 1205

77SKa   T. Sugano and K. Kubo, Nippon Kagaku Kaishi, 1977, 500

77SL   H. Saarinen, L. Lajunen, and P. Isoluoma, Finn. Chem. Letters, 1977, 66

77SM   R. Sarin and K.N. Munshi, Indian J. Chem., 1977, 15A, 327

77MSa   R. Sarin and K.N. Munshi, J. Indian Chem Soc., 1977, 54, 659

77SMT   T. Sekine, R. Murai, K. Kakahashi, and S. Iwahori, Bull. Chem. Soc. Japan, 1977, 50, 3415

77SN   H. Sigel, C.F. Naumann, B. Prijs, D.B. McCormick, and M.C. Falk, Inorg. Chem., 1977, 16, 790

77SP   R.S. Sindhu, D.B. Pandeya, and R.P. Singh, Monat. Chem., 1977, 108, 361

77SPa   J.P. Shukla, S.A. Pai, and M.S. Subramanian, Radiochem. Radioanal. Letters, 1977, 29, 241

77SR   T.F. Soldi, C.B. Riolo, G. Gallotti, and M. Pesavento, Gazz. Chim. Ital., 1977 107, 347

77SS   O.G. Sakovich and N.A. Skorik, Russ. J. Inorg. Chem., 1977, 22, 51 (98)

77SSa   S.Y. Shetty and R.M. Sathe, J. Inorg. Nucl. Chem., 1977, 39, 1837

77SSb   J.A. Swamy, B. Sethuram, and T.N. Rao, Indian J. Chem., 1977, 15A, 9

77SSc   J.A. Swamy, B. Sethuram, and T.N. Rao, Indian J. Chem., 1977, 15A, 449

77SSd   J.P.N. Srivastava and M.N. Srivastava, Curr. Sci. (India), 1977, 46, 443

77SSM   M.P. Singh, Y.P. Singh, and W.U. Malik, J. Indian Chem. Soc., 1977, 54, 568

77SSS   W. Szczepaniak, E. Siepak, and J. Siepak, Chem. Anal. (Warsaw), 1977, 22, 283

77ST   H. Sakurai and S. Takeshima, Talanta, 1977, 24, 531

77TGE   J.J. Tombeux, A.M. Goeminne, and Z. Eeckhaut, J. Inorg. Nucl. Chem., 1977, 39, 1655

77TGS   J.J. Tombeux, A.M. Goeminne, and J. Schaubroeck, Thermochim. Acta, 1977, 19, 327

77TJ   M.M. Taqui Khan and M.S. Jyoti, Indian J. Chem., 1977, 15A, 1002

77TK   N.N. Tananaeva and N.A. Kostromina, Soviet J. Coord. Chem., 1977, 3, 1280 (1639)

77TMW   B. Tummler, G. Maass, E. Weber, W. Wehner, and F. Vogtle, J. Amer. Chem. Soc., 1977, 99, 4683

77TR   L.M. Timakova, M.N. Rusina, G.F. Yaroshenko, and V.Ya. Temkina, J. Gen. Chem. USSR, 1977, 47, 628 (691)

77TS   Ya.I. Turyan, A.B. Sukhomlinov, and P.M. Zaitsev, J. Gen. Chem. USSR, 1977, 47, 2156 (2362)

77UB   L. Urbancik and M. Bartusek, Coll. Czech. Chem. Comm., 1977, 42, 446

77V   G.M.H. Van de Velde, J. Inorg. Nucl. Chem., 1977, 39, 1357

77VB   V.P. Vasilev and A.K. Belonogova, Russ. J. Inorg. Chem., 1977, 22, 1303 (2407)

77VBa   J. Votava and M. Bartusek, Coll. Czech. Chem. Comm., 1977, 42, 620

77VJ   A.E. Van Til and D.C. Johnson, Thermochim. Acta, 1977, 20, 177

77VK   V.P. Vasilev and E.V. Kozlovskii, Russ. J. Inorg. Chem., 1977, 22, 472 (853)

77VKa   T.V. Vasileva and N.P. Komar, Russ. J. Phy. Chem., 1977, 51, 584 (985)

77VKT   R.M.H. Verbeeck, R.A. Khan, and H.P. Thun, Bull. Soc. Chim. Belg., 1977, 86, 503

77VL   V.P. Vasilev, V.P. Lymar, and A.I. Lytkin, Russ. J. Inorg. Chem., 1977, 22, 1357
           (2511)

77VLa   V.P. Vasilev, V.P. Lymar, and A.I. Lytkin, Russ. J. Inorg. Chem., 1977, 22, 1440
           (2652)

77VN   Y. Vandewalle, J. Nicole, and J.P. Thumerel, Bull. Soc. Chim. France, 1977, 593

77VNa   Y. Vandewalle, J. Nicole, and J.P. Thumerel, Bull. Soc. Chim. France, 1977, 829

77VPK   Z.A. Vladimirova, Z.N. Prozorovskaya, and L.N. Komissarova, Russ. J. Inorg.
           Chem., 1977, 22, 691 (1269)

77VPS   G.P. Vakhramova, N.I. Pechurova, and V.I. Spitsyn, Soviet J. Coord. Chem., 1977,
           3, 273 (365)

77VV   V.P. Vasilev and V.N. Vasileva, Russ. J. Inorg. Chem., 1977, 22, 633 (1160)

77VZ   G.M. Voldman, A.N. Zelikman, V.S. Kagermanyan, V.P. Pakulin, S.N. Smirnov, and
           and E.E. Kharlamova, Soviet J. Coord. Chem., 1977, 3, 1290 (1651)

77W   D.R. Williams, J. Inorg. Nucl. Chem., 1977, 39, 711

77WB   M. Wilgocki and J. Biernat, Rocz. Chem., 1977, 51, 1297

77WH   H. Waki, Y. Hisazumi, and S. Ohashi, J. Inorg. Nucl. Chem., 1977, 39, 349

77YKK   K.B. Yatsimirskii, M.I. Kabachnik, M.A. Konstantirovskaya, E.I. Sinyavskaya,
           T.Ya. Medved, and N.P. Nesterova, Russ. J. Inorg. Chem., 1977, 22, 236
           (435)

77YKP   Yu.B. Yakovlev, F.Ya. Kulba, A.G. Pusko, and M.N. Gerchikova, Russ. J. Inorg.
           Chem., 1977, 22, 27 (53)

77YKU   Yu.B. Yakovlev, F.Ya. Kulba, V.G. Ushakova, and I.A Vitkauskaite, Russ. J. Inorg.
           Chem., 1977, 22, 45 (87)

77YO   Kh.M. Yakubov, E.Ya. Offengenden, and Z.N. Yusupov, Soviet J. Coord. Chem., 1977,
           3, 1094 (1400)

77ZF   V.M. Zyathovskii, A.P. Filippov, K.B. Yatsimirskii, V.M. Belousov, and
           T.A. Palchevskaya, Russ. J. Inorg. Chem., 1977, 22, 89 (163)

77ZG   K. Zutshi and K.C. Gupta, J. Electrochem. Soc. India, 1977, 26, No. 4, 33

77ZL   S. Zelinski and L. Lomozik, Russ. J. Inorg. Chem., 1977, 22, 514 (928)

77ZLa   S. Zielinski and L. Lomozik, Russ. J. Inorg. Chem., 1977, 22, 812 (1491)

77ZT   T.V. Zakharova, T.V. Ternovaya, S.B. Pirkes, and N.A. Kostromina, Russ. J. Inorg.
           Chem., 1977, 22, 962 (1775)

77ZV   K. Zutshi, P.S. Verma, and K.C. Gupta, J. Electrochem. Soc. India, 1977, 26,
           No. 2, 37

78A   R. Aruga, J. Inorg. Nucl. Chem., 1978, 40, 1077

78Aa   R.C. Agarwal, J. Indian Chem. Soc., 1978, 55, 220

78Ab   R. Aruga, Inorg. Chem., 1978, 17, 2503

78Ac   R. Aruga, _Atti Accad. Sci. Torino_, 1978, _112_, 345

78AA   L. Asso, M. Asso, J. Mossoyan, and D. Benlian, _J. Chim. Phys._, 1978, _75_, 561

78ABD   R.D. Alexander, D.H. Buisson, A.W.L. Dudeney, and R.J. Irving, _J. Chem. Soc._
          Faraday I, 1978, _74_, 1081

78ABS   S.A. Abbasi, B.G. Bhat, and R.S. Singh, _Indian J. Chem._, 1978 _16A_, 790

78AC   G. Arena, R. Cali, E. Rizzarelli, S. Sammartano, R. Barbucci, and
          M.J.M. Campbell, _J. Chem. Soc. Dalton_, 1978, 1090

78AE   B.A. Abd-El-Nabey and M.S. El-Ezaby, _J. Inorg. Nucl. Chem._, 1978, _40_, 739

78AF   A. Anichini, L. Fabbrizzi, P. Paoletti, and R.M. Clay, _J. Chem. Soc. Dalton_,
          1978, 577; _Inorg. Chim. Acta_, 1977, _22_, L25

78AG   I.I. Alekseeva, A.D. Gromova, I.V. Dermeleva, and N.A. Khvorostukhina, _Russ. J._
          Inorg. Chem., 1978, _23_, 54 (98)

78AK   G. Arena, G. Kavu, and D.R. Williams, _J. Inorg. Nucl. Chem._, 1978, _40_, 1221

78AKD   S. Arora, H.L. Kalra, S.N. Dubey, and D.M. Puri, _J. Indian Chem. Soc._, 1978, _55_,
          445

78AM   G. Arena, S. Musumeci, E. Rizzarelli, and S. Sammartano, _Inorg. Chim. Acta_, 1978,
          _27_, 31

78AMK   R.Ya. Aliev, D.B. Musaev, and N.G. Klyuchnikov, _Russ. J. Phys. Chem._, 1976, _52_,
          1086 (1871)

78AMW   M. Alei, Jr., L.O. Morgan, and W.E. Wageman, _Inorg. Chem._, 1978, _17_, 2288

78AR   A. Arevalo, J.C. Rodriquez-Placeres, T. Moreno, and J. Segura, _J. Electroanal._
          Chem., 1978, _92_, 55

78AS   A. Avdeef, S.R. Sofen, T.L. Bregante, and K.N. Raymond, _J. Amer. Chem. Soc._,
          1978, _100_, 5362

78ASJ   F. Arnaud-Neu, M.J. Schwing-Weill, J. Juillard, R. Lewis, and R. Weiss, _Inorg._
          Nucl. Chem. Letters, 1978, _14_, 367

78B   E.J. Billo, _J. Inorg. Nucl. Chem._, 1978, _40_, 1971

78BB   J.W. Bixler and A.M. Bond, _Inorg. Chem._, 1978, _17_, 3684

78BBa   A.K. Basak and D. Banerjea, _J. Indian Chem. Soc._, 1978, _55_, 853

78BBG   K.A. Burkov, E.A. Busko, L.A. Garmash, and G.V. Khonin, _Russ. J. Inorg. Chem._,
          1978, _23_, 971 (1767)

78BBI   A.P. Borisova, V.N. Belyaev, N.A. Ibragim, and A.M. Evseev, _Russ. J. Phys. Chem._,
          1978, _52_, 212 (385)

78BG   K.A. Burkov, L.A. Garmash, and L.S. Lilich, _Russ. J. Inorg. Chem._, 1978, _23_,
          1770 (3193)

78BH   S. Bouhlassa, S. Hubert, and R. Guillaumont, _Radiochem. Radioanal. Letters_, 1978,
          _32_, 247

78BK   R.H. Byrne and D.R. Kester, _J. Soln. Chem._, 1978, _7_, 373

78BKG   K.S. Balaji, S.D. Kumar, and P. Gupta-Bhaya, _Anal. Chem._, 1978, _50_, 1972

78BKP   V.I. Belevantsev, T.I. Koroleva, and B.I. Peshchevitskii, _Soviet J. Coord._
          Chem., 1978, _4_, 47 (60)

78BKS   G.A. Bagiyan, I.K. Koroleva, and N.V. Soroka, _Russ. J. Inorg. Chem._, 1978, _23_,
          1337 (2422)

78BL   S. Bandopadhyaya and S.C. Lahiri, _J. Indian Chem. Soc._, 1978, _55_, 1286

78BM   R.H. Busey and R.E. Mesmer, _J. Chem. Eng. Data_, 1978, _23_, 175

78BMW   G. Berthon, P.M. May, and D.R. Williams, J. Chem. Soc. Dalton, 1978, 1433

78BO   S. Bergstrom and G. Olofsson, J. Soln. Chem., 1978, 7, 497

78BP   G. Biedermann and R. Palombari, Acta Chem. Scand., 1978, A32, 381

78BPG   S.K. Bhasin, O. Parkash, and J.N. Gaur, J. Electrochem. Soc. India, 1978, 27, 159

78BR   P. DiBernardo, E. Roncari, U. Mazzi, and F. Bettella, Thermochim. Acta, 1978, 23,
         293

78BS   E. Bottari and C. Severini, J. Coord. Chem., 1978, 8, 69

78BSa   A.C.M. Bourg and P.W. Schindler, Chimia (Switz.), 1978, 32, 166

78BT   N.N. Bukov, N.N. Tananaeva, V.T. Panyushkin, N.A. Kostromina, Yu.A. Afanasev,
         and V.D. Buikliskii, Soviet J. Coord. Chem., 1978, 4, 1163 (1532)

78BW   B. Banas and M. Wronska, Polish J. Chem., 1978, 52, 239

78BZ   E.A. Belousov and L.Yu. Zakharova, Russ. J. Inorg. Chem., 1978, 23, 1583 (2855)

78CI   C.M. Corilla and B.Z. Iofa, Russ. J. Inorg. Chem., 1978, 23, 1296 (2350)

78CK   N.N. Chernova and I.G. Kurskii, Russ. J. Inorg. Chem., 1978, 23, 561 (1014)

78CKM   P.J. Cerutti, H.C. Ko, K.G. McCurdy, and L.G. Hepler, Canad. J. Chem., 1978, 56,
         3084

78CKP   F. Christensson, H.C.S. Koefoed, A.C. Petersen, and K. Rasmussen, Acta Chem.
         Scand., 1978, A32, 15

78CO   G.R. Choppin and E. Orebaugh, Inorg. Chem., 1978, 17, 2300

78CP   Y. Couturier and C. Petitfaux, Bull. Soc. Chim. France, 1978, I-121

78CPa   Y. Couturier and C. Petitfaux, Bull. Soc. Chim. France, 1978, I-435

78CU   D.E. Caddy and J.H.P. Utley, J. Inorg. Nucl. Chem., 1978, 40, 1103

78CV   F. Coccioli and M. Vicedomini, J. Inorg. Nucl. Chem., 1978, 40, 2103, 2106

78CW   J.C. Cooper, L.F. Wong, and D.W. Margerum, Inorg. Chem., 1978, 17, 261

78D   K. Doi, J. Inorg. Nucl. Chem., 1978, 40, 1639

78Da   K. Doi, Talanta, 1978, 25, 97

78DB   J. Dumonceau, S. Bigot, M. Treuil, J. Faucherre, and F. Fromage, Compt. Rend.
         Acad. Sci. Paris, Ser. C, 1978, 287, 325

78DD   N. Delannoy, A. Delannoy, J. Hennion, and J. Nicole, Compt. Rend. Acad. Sci.
         Paris, Ser. C, 1978, 287, 527

78DJ   D. Dyrssen, O. Johansson, and M. Wedborg, Marine Chem., 1978, 6, 275

78DK   D.G. Dhuley and R.K. Kale, Indian J. Chem., 1978, 16A, 451

78DMP   E.A. Daniele, F.C. March, H.K.J. Powell, W.T. Robinson, and J.M. Russell, Aust.
         J. Chem., 1978, 31, 723

78DMY   N.K. Davidenko, P.A. Manorik, and K.B. Yatsimirskii, Russ. J. Inorg. Chem.,
         1978, 23, 1794 (3233)

78DP   K.P. Dubey and S. Parveen, Curr. Sci. (India), 1978, 47, 415

78DZ   P. Di Bernardo, P. Zanello, D. Curto, and R. Portanova, Inorg. Chim. Acta, 1978,
         29, L185

78E   L.I. Elding, Inorg. Chim. Acta, 1978, 28, 255

78EF   M.M. Emara, N.A. Farid, and G. Atkinson, Anal. Lett., 1978, A11, 797

78EO   L.G. Ekstrom and A. Olin, Chem. Scripta, 1978, 13, 10

78EW   B. Elgquist and M. Wedborg, Marine Chem., 1978, 6, 243

78F   W. Forsling, Acta Chem. Scand., 1978, A32, 471

78Fa   W. Forsling, Acta Chem. Scand., 1978, A32, 857

78FB   I. Filipovic, B. Bach-Dragutinovic, N. Ivicic, and Vl. Simeon, Thermochim. Acta,
         1978, 27, 151

78FBG   V.A. Fedorov, V.I. Belevantsev, and N.N. Golovneu, Soviet J. Coord. Chem., 1978,
         4, 1276 (1673)

78FD   Ya.D. Fridman, N.V. Dolgashova, and V.F. Nazorov, Soviet J. Coord. Chem., 1978,
         4, 1440 (1873)

78FF   F.H. Fisher and A.P. Fox, J. Soln. Chem., 1978, 7, 561

78FG   O. Forsberg, B. Gelland, P. Ulmgren, and O. Wahlberg, Acta Chem. Scand., 1978,
         A32, 345

78FK   A.P. Filippov and I.V Khyarsing, Russ. J. Inorg. Chem., 1978, 23, 839 (1523)

78FKK   V.A. Fedorov, M.A. Kuznechikhina, I.V. Kanarsh, G.M. Kirnyuk, and
         G.E. Chernikova, Soviet J. Coord. Chem., 1978, 4, 33 (42)

78FM   T.B. Field and W.A.E. McBryde, Canad. J. Chem., 1978, 56, 1202

78FMP   L. Fabbrizzi, M. Micheloni, and P. Paoletti, Inorg. Chem., 1978, 17, 494

78FMS   Y. Fukuda, P.R. Mitchell, and H. Sigel, Helv. Chem. Acta., 1978, 61, 638

78FN   D.W. Franco, E.A. Neves, and M.A.C. Dellatorre, Cienc. Cult. (Sao Paulo), 1978,
         30, 1450; Chem. Abstr., 1979, 90, 93239f

78FP   L. Fabbrizzi, P. Paoletti, R.M. Clay, Inorg. Chem., 1978, 17, 1042

78FR   N. Fatouros, F. Rouelle, and M. Chemla, J. Chim. Phys., 1978, 75, 476

78FS   C.M. Frey and J.E. Stuehr, J. Amer. Chem. Soc., 1978, 100, 134

78FSa   C.M. Frey and J.E. Stuehr, J. Amer. Chem. Soc., 1978, 100, 139

78FU   S. Funahashi, F. Uchida, and M. Tanaka, Inorg. Chem., 1978, 17, 2784

78GB   J. Galea, R. Beccaria, and G. Ferroni, Electrochim. Acta, 1978, 23, 647

78GC   K.D. Gupta, K.K. Choudhary, and J.N. Gaur, Indian J. Chem., 1978, 16A, 73

78GCa   K.D. Gupta, K.K. Choudhary, and J.N. Gaur, J. Electrochem. Soc. India, 1978, 27,
         257

78GD   J.F. Giuliani and T. Donohue, Inorg. Chem., 1978, 17, 1090

78GF   N.N. Golovnev and V.A. Fedorov, Russ. J. Inorg. Chem., 1978, 23, 771 (1401)

78GFB   J. Galea, G. Ferroni, and J.P. Belaich, Electrochim. Acta, 1978, 23, 103

78GFN   A. Gergely, E. Farkas, I. Nagypal, and E. Kas, J. Inorg. Nucl. Chem., 1978, 40,
         1709

78GG   O.D. Gupta, K.D. Gupta, S.C. Bhagel, and J.N. Gaur, J. Electrochem. Soc. India,
         1978, 27, 265

78GMC   H. Gross, T.Ya. Medved, B. Costisella, F.I. Belskii, and M.I. Kabachnik, J. Gen.
         Chem. USSR, 1978, 48, 1746 (1914)

78GMM   I.M. Gnutova, R.A. Maier, Yu.S. Maslennikov, and N.A. Skorik, Russ. J. Inorg.
         Chem., 1978, 23, 827 (1501)

78GN   I.P. Gorelov, V.M. Nikolskii, and A.I. Kapustnikov, J. Gen. Chem. USSR, 1978,
         48, 2357 (2596)

78GS   O.E.S. Godinho and E. Stein, Z. Phys. Chem. (Leipzig), 1978, 259, 161

78GSa   A.A. Gundorina and A.N. Sergeeva, Soviet J. Coord. Chem., 1978, 4, 389 (522)

78GSK   S.B. Gholse, O.P. Sharma, and R.B. Kharat, J. Indian Chem. Soc., 1978, 55, 778

78GSM   Ya.G. Goroshchenko, E.K. Sikorskaya, and V.P. Malitskaya, Soviet Progr. Chem. (Ukr. Khim. Zh.), 1978, 44, No. 1, 22 (24)

78GT   G. Giasson and P.H. Tewari, Canad. J. Chem., 1978, 56, 435

78HH   R.P. Hanzlik and A. Hamburg, J. Amer. Chem. Soc., 1978, 100, 1745

78HJ   O.W. Howarth and M. Jarrold, J. Chem. Soc. Dalton, 1978, 503

78HM   W.R. Harris, I. Murase, J.H. Timmons, and A.E. Martell, Inorg. Chem., 1978, 17, 889

78IH   T. Ishimitsu, S. Hirose, and H. Sakurai, Chem. Pharm. Bull. (Japan), 1978, 26, 74

78IS   N. Ivicic and Vl. Simeon, Thermochim. Acta, 1978, 25, 299

78IT   R.M. Izatt, R.E. Terry, L.D. Hansen, A.G. Avondet, J.S. Bradshaw, N.K. Dalley, T.E. Jensen, and J.J. Christensen, Inorg. Chim. Acta, 1978, 30, 1

78J   M. Jawaid, Talanta, 1978, 25, 215

78JB   P.D. Jadhav, R.G. Bidkar, D.G. Dhuley, and R.A. Bhobe, J. Inorg. Nucl. Chem., 1978, 40, 1437

78JBa   G.K. Johnson and J.E. Bauman, Jr., Inorg. Chem., 1978, 17, 2774

78JI   M. Jawaid, F. Ingman, D.H. Liem, and T. Wallin, Acta Chem. Scand., 1978, A32, 7

78JIa   M. Jawaid, F. Ingman, D.H. Liem, Acta Chem. Scand., 1978, A32, 333

78JP   K.S. Johnson and R.M. Pytkowicz, Amer. J. Sci., 1978, 278, 1428

78JPa   A.C. Jha and B. Prasad, J. Indian Chem. Soc., 1978, 55, 301

78JS   P.D. Jadhav, D.N. Shelke, and R.A. Bhobe, J. Inorg. Nucl. Chem., 1978, 40, 572

78K   V.I. Kornev, Russ. J. Phys. Chem., 1978, 52, 1054 (1813)

78Ka   I.V. Kolosov, Soviet J. Coord. Chem., 1978, 4, 397 (531)

78Kb   J. Kragten, Talanta, 1978, 25, 239

78Kc   V.N. Kumok, Russ. J. Inorg. Chem., 1978, 23, 1436 (2591)

78Kd   V.N. Kumok, Russ. J. Inorg. Chem., 1978, 23, 985 (1792)

78Ke   V.N. Kumok, Radiokhim., 1978, 20, 691; Chem. Abstr., 1979, 90, 29886n

78KA   N.P. Komar, L.P. Adamovich, V.V. Melnik, and N.O. Mchedlov-Petrosyan, J. Anal. Chem. USSR, 1978, 33, 640 (822)

78KC   A.S. Kereichuk and I.M. Churikova, Russ. J. Inorg. Chem., 1978, 23, 928 (1686)

78KCS   Y. Khayat, M. Cromer-Morin, and J.P. Scharff, Compt. Rend. Acad. Sci. Paris, Ser. C, 1978, 287, 265

78KCT   A.S. Kereichuk, I.M. Churikova, and V.I. Tikhomirov, Russ. J. Inorg. Chem., 1978, 23, 1345 (2436)

78KD   J. Kragten and L.G. Decnop-Weever, Talanta, 1978, 25, 147

78KJK   R.C. Kapoor, J.K. Jailwal, and J. Kishan, J. Inorg. Nucl. Chem., 1978, 40, 155

78KJN   Y.D. Kane, D.M. Joshi, and G.S. Natarajan, Proc. Indian Acad. Sci., 1978, 87A, 359

78KK   M. Kodama and E. Kimura, J. Chem. Soc. Dalton, 1978, 104

78KKa   M. Kodama and E. Kimura, J. Chem. Soc. Dalton, 1978, 1081

78KKb   M. Kodama and E. Kimura, Inorg. Chem., 1978, 17, 2446

78KKc   N. Kallay and I. Krznaric, Indian J. Chem., 1978, 16A, 713

78KKd   Kabir-ud-Din and I.A. Khan, Monat. Chem., 1978, 109, 1343

78KKA   Ts.B. Konunova, L.S. Kachkar, and L.A. Arnaut, Soviet J. Coord. Chem., 1978, 4, 780 (1027)

78KKH   J.J. Klingenberg, D.S. Knecht, A.E. Harrington, and R.L. Meyer, J. Chem. Eng. Data, 1978, 23, 327

78KKK   N.A. Kosrtomina, L.N. Krashnevskaya, and A.I. Kirillov, Soviet Progr. Chem. (Ukr. Khim. Zh.), 1978, 44, No. 3, 5 (231)

78KKS   V. Koblizkova, V. Kuban, and L. Sommer, Coll. Czech. Chem. Comm., 1978, 43, 2711

78KL   J. Kulig and B. Lenarcik, Polish J. Chem., 1978, 52, 477

78KLH   Z. Kleckova, M. Langova, and J. Havel, Coll. Czech. Chem. Comm., 1978, 43, 3163

78KM   V.I. Kornev and G.I. Manasheva, Soviet J. Coord. Chem., 1978, 4, 1169 (1539)

78KMD   M.I. Kabachnik, T.Ya. Medved, N.M. Dyaglova, Yu.M. Polikarpov, B.K. Shcherbakov, and F.I. Belskii, Bull. Acad. Sci. USSR, 1978, 27, 374 (433)

78KN   S.C. Khurana and V.J. Nigam, J. Inorg. Nucl. Chem., 1978, 40, 159

78KPI   V.P. Krasnov, I.V. Podgornaya, A.A. Ivakin, and L.V. Alekseeva, J. Gen. Chem. USSR, 1978, 48, 2354 (2593)

78KPJ   R. Karlicek, M. Polasek, and V. Jokl, Coll. Czech. Chem. Comm., 1978, 43, 2897

78KPS   L.V. Kurochkina, N.I. Pechurova, N.I. Snezhko, and V.I. Spitsyn, Russ. J. Inorg. Chem., 1978, 23, 1481 (2676)

78KPU   E.A. Klimkovich, S.M. Pirogov, and Yu.I. Usatenko, J. Anal. Chem. USSR, 1978, 33, 859 (1110)

78KS   M. Kimura and J. Shirai, J. Inorg. Nucl. Chem., 1978, 40, 1085

78KSa   N. Kojima, Y. Sugiura, and H. Tanaka, Chem. Pharm. Bull. (Japan), 1978, 26, 579

78KST   N. Kojima, Y. Sugiura, and H. Tanaka, Chem. Pharm. Bull. (Japan), 1978, 26, 440

78KV   V.I. Kornev, V.A. Valyaeva, and I.P. Mukanov, Russ. J. Phys. Chem., 1978, 52, 645 (1132)

78KVa   V.I. Kornev and V.A. Valyaeva, Russ. J. Phys. Chem., 1978, 52, 1053 (1815)

78KVb   V.I. Kornev, V.A. Valyaeva, and S.I. Zobnin, Russ. J. Phys. Chem., 1978, 52, 1057 (1818)

78KVc   V.I. Kornev, V.A. Valyaeva, and S.I. Zobnin, Russ. J. Phys. Chem., 1978, 52, 1561 (2707)

78KZ   A.A. Kurganov, L.Ya. Zhuchkova, and V.A. Davankov, J. Inorg. Nucl. Chem., 1978, 40, 1081

78KZa   A.A. Kurganov, L.Ya. Zuchkova, and V.A. Davankov, Soviet J. Coord. Chem., 1978, 4, 1138 (1503)

78L   M.C. Lim, J. Chem. Soc. Dalton, 1978, 726

78La   D.J. Leggett, Anal. Chem., 1978, 50, 718

78LB   D. Lesht and J.E. Bauman, Jr., Inorg. Chem., 1978, 17, 3332

78LH   A.P. Leugger, L. Hertli, and T.A. Kaden, Helv. Chim. Acta, 1978, 61, 2296

78LK   Z. Libus and G. Kowalewska, Polish J. Chem., 1978, 52, 709

78LKa   B. Lenarcik and J. Kulig, Polish J. Chem., 1978, 52, 2089

78LKb   L.H.J. Lajunen and M. Karvo, Anal. Chim. Acta, 1978, 97, 423

78LKc   L.H.J. Lajunen and M. Karvo, Acta Chem. Scand., 1978, A32, 370

78LL   J. Lagrange, P. Lagrange, and K. Zare, Bull. Soc. Chim. France, 1978, I-7

78LM   J.M. Lehn and F. Montavon, Helv. Chim. Acta, 1978, 61, 67

78LN   B. Lenarcik, K. Nabialek, and M. Gabryszewski, Polish J. Chem., 1978, 52, 401

78LP   Lutfullah and R. Paterson, J. Chem. Soc. Faraday I, 1978, 74, 484

78LR   B. Lenarcik and M. Rzepka, Polish J. Chem., 1978, 52, 447

78LRa   B. Lenarcik and M. Rzepka, Polish J. Chem., 1978, 52, 1629

78LW   B. Lenarcik and M. Wisniewski, Polish J. Chem., 1978, 52, 193

78M   W.A.E. McBryde, "A Critical Review of Equilibrium Data for Proton and Metal
          Complexes of 1,10-Phenanthroline, 2,2'-Bipyridyl and Related Compounds,"
          Pergamon Press, 1978

78MA   M.S. Mohan and E.H. Abbott, Inorg. Chem., 1978, 17, 2203

78MAa   M.S. Mohan and E.H. Abbott, Inorg. Chem., 1978, 17, 3083

78MAb   M.S. Mohan and E.H. Abbott, J. Coord. Chem., 1978, 8, 175

78MAc   C.B. Monk and M.F. Amira, J. Chem. Soc. Faraday I, 1978, 74, 1170

78MAS   E.A. Mambetkaziev, M.U. Abilova, A.M. Shaldybaeva, S.I. Zhdanov, and
          G.A. Myrzabaeva, Soviet Electrochem., 1978, 14, 1510 (1734)

78MB   M. Mikesova and M. Bartusek, Coll. Czech. Chem. Comm., 1978, 43, 1867

78MBB   W.U. Malic, R. Bembi, P.P. Bhargava, and R. Singh, J. Indian Chem. Soc., 1978,
          55, 222

78MC   M. Meloun, J. Chylkova, and J. Pancl, Coll. Czech. Chem. Comm., 1978, 43, 1027

78MCG   H.E.L. Madsen, H.H. Christensen, and C. Gottlieb-Petersen, Acta Chem. Scand.,
          1978, A32, 79

78ME   E.K. Millar, C.A. Evans, and D.L. Rabenstein, Canad. J. Chem., 1978, 56, 3104

78MG   A.K. Maheshwari and J.N. Gaur, J. Electrochem. Soc. India, 1978, 27, 175

78MGD   E. Merciny, J.M. Gatez, and G. Duyckaerts, Anal. Chim. Acta, 1978, 100, 329

78MGK   P.K. Migal, A.P. Gerbeleu, and P.G. Kalitina, Russ. J. Inorg. Chem., 1978, 23,
          882 (1602)

78MH   F. Marsicano and R.D. Hancock, J. Chem. Soc. Dalton, 1978, 228

78MJ   W.G. Mitchell and M.M Jones, J. Inorg. Nucl. Chem., 1978, 40, 1957

78MK   J.M. Malin and R.C. Koch, Inorg. Chem., 1978, 17, 752

78MM   T.H. Mhaske and K.N. Munshi, Indian J. Chem., 1978, 16A, 546

78MMa   T.H. Mhaske and K.N. Munshi, J. Indian Chem. Soc., 1978, 55, 885

78MMb   C. Musikas and M. Marteau, Radiochem. Radioanal. Letters, 1978, 33, 41

78MMc   L.I. Mitkina, N.V. Melchakova, and V.M. Peshkova, Russ. J. Inorg. Chem., 1978,
          23, 693 (1258)

78MMd   L.I. Mitkina, N.V. Melchakova, and V.M. Peshkova, Soviet J. Coord. Chem., 1978,
          4, 1286 (1684)

78MMG   B. Mayer, R. Medancic, B. Grabaric, and I. Filipovic, Croat. Chem. Acta, 1978,
          51, 151

78MMN   D.F.C. Morris, J.D. MacCarthy, and R.J. Newton, Electrochim. Acta, 1978, 23,
          1383

78MMW   M. Micheloni, P.M. May, and D.R. Williams, J. Inorg. Nucl. Chem., 1978, 40, 1209

78MN   Y. Masuda, T. Nakamori, and E. Sekido, Nippon Kagaku Kaishi, 1978, 199

78MNa   Y. Masuda, T. Nakamori, and E. Sekido, Nippon Kagaku Kaishi, 1978, 204

78MNG   E.D. Malakhaev, V.M. Nikolskii, and I.P. Gorelov, J. Gen. Chem. USSR, 1978, 48,
          2361 (2601)

78MP   J.H. Milller and J.E. Powell, Inorg. Chem., 1978, 17, 774

78MPa   A.I. Moskvin and A.N. Poznyakov, Soviet J. Coord. Chem., 1978, 4, 811 (1065)

78MPK   S.P. Mushran, O. Prakash, R.L. Kushwaha, and C.K. Verma, J. Indian Chem. Soc.,
           1978, 55, 548

78MPV   M. Micheloni, P. Paoletti, and A. Vacca, J. Chem. Soc. Perkin II, 1978, 945

78MS   B. Magyar and G. Schwarzenbach, Acta Chem. Scand., 1978, A32, 943

78MSa   P.R. Mitchell and H. Sigel, J. Amer. Chem. Soc., 1978, 100, 1564

78MSb   S.D. Makhijani and S.P. Sangal, J. Indian Chem. Soc., 1978, 55, 987

78MSc   Yu.A. Maletin and I.A. Sheka, Soviet J. Coord. Chem., 1978, 4, 1247 (1639)

78MSP   M. Micheloni, A. Sabatini, and P. Paoletti, J. Chem. Soc. Perkin II, 1978, 828

78MST   G.S. Malik, S.P. Singh, and J.P. Tandon, J. Prakt. Chem., 1978, 320, 324

78MY   M. Munakata and K. Yamada, Bull. Chem. Soc. Japan, 1978, 51, 3500

78MZ   B.D. Mulkina, S.I. Zhdanov, and E.A. Mambetkaziev, J. Gen. Chem. USSR, 1978, 48,
           1278 (1394)

78MZa   B.D. Mulkina, S.I. Zhdanov, and E.A. Mambetkaziev, J. Gen. Chem. USSR, 1978, 48,
           22 (29)

78NB   A.I. Nabil, A.P. Borisova, L.I. Martynenko, and A.M. Evseev, Russ. J. Inorg.
           Chem., 1978, 23, 203 (364)

78NF   G.G. Nifanteva, A.V. Fedorova, A.M. Robov, and V.A. Fedorov, Soviet J. Coord.
           Chem., 1978, 4, 279 (372)

78NM   B.I. Nabivanets, E.A. Mazurenko, N.V. Chernaya, and V.G. Matyashev, Russ. J.
           Inorg. Chem., 1978, 23, 619 (1119)

78NMa   O. Navratil and A. Malach, Coll. Czech. Chem. Comm., 1978, 43, 2890

78NP   A.Yu. Nazarenko and I.V. Pyatnitskii, Russ. J. Inorg. Chem., 1978, 23, 1470
           (2655)

78NPM   I. Nematov, O.M. Petrukhin, A.E. Martirosov, and Sh.T. Talipov, Russ. J. Inorg.
           Chem., 1978, 23, 715 (1299)

78NS   T. Nozaki, M. Sakamoto, K. Goto, N. Higaki, K. Ueda, T. Oi, and J. Kanazawa,
           Nippon Kagaku Kaishi, 1978, 976

78OS   A. Olin and P. Svanstrom, Acta Chem. Scand., 1978, A32, 435

78OKS   Kh.K. Ospanov, U.I. Sholtyrova, and Yu.Ya. Kharitonov, Russ. J. Inorg. Chem.,
           1978, 23, 1508 (2724), 1510 (2728)

78P   J. Podlahova, Coll. Czech. Chem. Comm., 1978, 43, 57, 64

78Pa   J. Podlahova, Coll. Czech. Chem. Comm., 1978, 43, 3007

78Pb   E.M. Perdue, Geochim. Cosmochim. Acta, 1978, 42, 1351

78PB   O. Parkash, S.K. Bhasin, and D.S. Jain, J. Less-Common Metals, 1978, 60, 179

78PBa   O. Parkash, S.K. Bhasin, and D.S. Jain, J. Electrochem. Soc. India, 1978, 27, 251

78PK   V.F. Pestrikov and Yu.P. Khranilov, Soviet J. Coord. Chem., 1978, 4, 275 (368)

78PM   L.D. Pethe and B.D. Mali, Indian J. Chem., 1978, 16A, 364

78PP   R. Petrola, K. Poppius, L. Hakkarainen, and O. Makitie, Anal. Chim. Acta, 1978,
           99, 393

78PR   H.K.J. Powell and J.M. Russell, Aust. J. Chem., 1978, 31, 2409

78PS   L.D. Pettit and J.L.M. Swash, J. Chem. Soc. Dalton, 1978, 286

78PSa   Pushparaja and M. Sudersanan, Indian J. Chem., 1978, 16A, 504

78PSb   D.R. Prasad and K. Saraswathi, Indian J. Chem., 1978, 16A, 1110

78PSc   D.R. Prasad and K. Saraswathi, Indian J. Chem., 1978, 16A, 631

78PSS   T.V. Petrova, I.I. Seifullina, and L.D. Skrylev, Russ. J. Phys. Chem., 1978, 52, 1218 (2108)

78PT   B. Perlmutter-Hayman and E. Tapuhi, J. Coord. Chem., 1978, 8, 75

78R   F. Rodante, Thermochim. Acta, 1978, 23, 311

78RB   M.P. Ryan and J.E. Bauman, Jr., Inorg. Chem., 1978, 17, 3329

78RM   K.S. Rajan, S. Mainer, and J.M. Davis, J. Inorg. Nucl. Chem., 1978, 40, 2089

78RMa   K.S. Rajan, S. Mainer, and J.M. Davis, Bioinorg. Chem., 1978, 9, 187

78RMR   L.V. Ruzaikina, I.N. Marov, V.A. Ryabukhin, A.N. Ermakov, and V.N. Filimonova, J. Anal. Chem. USSR, 1978, 33, 837 (1082)

78RP   A.M. Reznik, L.I. Pokrovskaya, and N.D. Doroshenko, Russ. J. Inorg. Chem., 1978, 23, 22 (40)

78RR   P.R. Reddy, K.V. Reddy, and M.M. Taqui Khan, J. Inorg. Nucl. Chem., 1978, 40, 1265

78RS   P.R. Reddy, J. Shamanthakamani, and M.M. Taqui Khan, J. Inorg. Nucl. Chem., 1978, 40, 1673

78RSa   A.L.J. Rao and M. Singh, Curr. Sci. (India), 1978, 47, 448

78S   Y. Sugiura, Inorg. Chem., 1978, 17, 2176

78SA   H. Stunzi and G. Anderegg, J. Coord. Chem., 1978, 7, 239

78SB   L.H. Skibsted and J. Bjerrum, Acta Chem. Scand., 1978, A32, 429

78SD   R.C. Sharma, S.S. Dhindsa, and D.N. Bhargava, Monat. Chem., 1978, 109, 179

78SG   Yu.E. Svetogorov and I.P. Gorelov, Russ. J. Inorg. Chem., 1978, 23, 668 (1211)

78SGa   Yu.E. Svetogorov and I.P. Gorelov, Russ. J. Inorg. Chem., 1978, 23, 698 (1267)

78SGb   T.I. Smirnova and I.P. Gorelov, Russ. J. Inorg. Chem., 1978, 23, 1506 (2719)

78SGc   S.K. Sarpal and A.R. Gupta, Indian J. Chem., 1978, 16A, 55

78SGd   R.S. Sharma and J.N. Gaur, Indian J. Chem., 1978, 16A, 507

78SGe   R.S. Sharma and J.N. Gaur, J. Electrochem. Soc. India, 1978, 27, 261

78SH   M. Sugawara, M. Hirota, and T. Kambara, Fresenius' Z. Anal. Chem., 1978, 293, 302

78SJ   D.N. Shelke and D.V. Jahagirdar, Indian J. Chem., 1978, 16A, 60

78SK   P. Schultz-Grunow and T.A. Kaden, Helv. Chim. Acta, 1978, 61, 2291

78SKa   G.M. Sergeev and I.M. Korenmann, Russ. J. Inorg. Chem., 1978, 23, 66 (121)

78SKG   I. Sovago, T. Kiss, and A. Gergely, J. Chem. Soc. Dalton, 1978, 964

78SKM   S.A. Shcherbakova, N.A. Krasnyanskaya, N.V. Melchakova, and V.M. Peshkova, Russ. J. Inorg. Chem., 1978, 23, 424 (770)

78SKZ   F.D. Shevchenko, L.A. Kuzina, V.N. Zinchenko, and L.N. Kuznetsova, Soviet Progr. Chem. (Ukr. Khim. Zh.), 1978, 44, No. 8,94 (877)

78SL   N.I. Shutaya, B.I. Lobov, and D.P. Dobychin, Russ. J. Inorg. Chem., 1978, 23, 1490 (2692)

78SM   R. Sarin and K.N. Munshi, Indian J. Chem., 1978, 16A, 455

78SMa   R. Sarin and K.N. Munshi, J. Indian Chem. Soc., 1978, 55, 512

78SP   N.A. Skorik and N.I. Pechurova, Russ. J. Inorg. Chem., 1978, 23, 348 (628)

78SPK   V.A. Shormanov, N.I. Pimenova, G.A. Krestov, and G.S. Bykova, Russ. J. Inorg.
          Chem., 1978, 23, 243 (438)

78SPS   M.S. Stuklova, N.I. Pechurova, and V.I. Spitsyn, Russ. J. Inorg. Chem., 1978, 23,
          1677 (3021)

78RS   K. Saraswathi and V.S Ramachandran, Indian J. Chem., 1978, 16A, 1112

78SS   J.P.N. Srivastava and M.N. Srivastava, J. Inorg. Nucl. Chem., 1978, 40, 2076

78SSa   T. Shibahara and A.G. Sykes, J. Chem. Soc. Dalton, 1978, 95, 100

78SSb   R.C. Srivastava and M.N. Srivastava, J. Inorg. Nucl. Chem., 1978, 40, 1439

78SSc   W. Szczepaniak and J. Siepak, Polish J. Chem., 1978, 52, 721

78SSK   V.M. Savostina, O.A. Shpigun, N.V. Klimova, I.Ya. Kolotyrikina, and
          V.M. Peshkova, Moscow Univ. Chem. Bull., 1978, 33, No. 6, 46 (692)

78SSR   V.A. Shenderovich, E.F. Strizhev, and V.I. Ryaboi, Russ. J. Inorg. Chem., 1978,
          23, 1484 (2681)

78SSS   I.P. Saraswat, C.L. Sharma, and A. Sharma, J. Indian Chem. Soc., 1978, 55, 757

78SYM   H. Sato, Y. Yokoyama, and K. Momoki, Anal. Chim. Acta, 1978, 99, 167

78SYN   T. Sakurai, O. Yamauchi, and A. Nakahara, Bull. Chem. Soc. Japan, 1978, 51, 3203

78TA   M.M. Taqui Khan and M. Amara Babu, J. Inorg. Nucl. Chem., 1978, 40, 2110

78TE   G.M. Toptygina, V.I. Evdokimov, N.A. Eliseeva, and V.S. Badanin, Russ. J. Inorg.
          Chem., 1978, 23, 810 (1471)

78TGK   F.M. Tulyupa, L.I. Gerasyutina, L.G. Karyaka, and V.A. Mokienko, Russ. J. Inorg.
          Chem., 1978, 23, 685 (1243)

78TGY   Ya.I. Turyan, I.N. Gnusin, and V.I. Yatsenko, Russ. J. Inorg. Chem., 1978, 23,
          1145 (2083)

78THH   K. Tamura, S. Harada, M. Hiraissh, and T. Yasunaga, Bull. Chem. Soc. Japan, 1978,
          51, 2928

78THM   J.H. Timmons, W.R. Harris, I. Murase, and A.E. Martell, Inorg. Chem., 1987, 17,
          2192

78TJ   M.M. Taqui Khan and M.S. Jyoti, J. Inorg. Nucl. Chem., 1978, 40, 1731

78TL   Ya.I. Turyan, I.N. Logvinov, and N.K. Strizhov, Russ. J. Inorg. Chem., 1978, 23,
          1082 (1970)

78TLa   R.E. Thompson, E.L.F. Li, H.O. Spivey, J.P. Chandler, A.J. Katz, and
          J.R. Appleman, Bioinorg. Chem., 1978, 9, 35

78TS   K. Timmers and R. Sternglanz, Bioinorg. Chem., 1978, 9, 145

78TSG   Ya.I. Turyan, N.K. Strizhov, V.N. Gnusin, K.M. Kardailova, and L.M. Maluka,
          Russ. J. Inorg. Chem., 1978, 23, 1132 (2061)

78TSK   T.V. Ternovaya, V.P. Shelest, N.A. Kostromina, and Ts.B. Konunova, Russ. J.
          Inorg. Chem., 1978, 23, 670 (1215)

78TT   M. Tabata and M. Tanaka, Inorg. Chem., 1978, 17, 2779

78TZ   L.M. Timakova, B.V. Zhadanov, G.F. Yaroshenko, I.A. Polyakova, V.Ya. Temkina,
          and R.P. Lastovskii, J. Gen. Chem. USSR, 1978, 48, 1684 (1846)

78UV   V.B. Ukraintsev, R.Q. Vasquez, Yu.N. Kukushkin, G.B. Avetikyan, and Yu.V. Fadeer,
          Russ. J. Inorg. Chem., 1978, 23, 1010 (1836)

78V   J.F. Verchere, J. Chem. Research (S), 1978, 178

78VB   P.R. Vasudeva Rao, S.V. Bagawde, V.V. Ramakrishna, and S.K. Patil, J. Inorg.
          Nucl. Chem., 1978, 40, 123

78VBa   P.R. Vasudeva Rao, S.V. Bagawde, V.V. Ramakrishna, and S.K. Patil, J. Inorg. Nucl. Chem., 1978, 40, 339

78VK   P. Vanura and L. Kuca, Coll. Czech. Chem. Comm., 1978, 43, 1460

78VKO   V.P. Vasilev, L.A. Kochergina, and T.D. Orlova, J. Gen. Chem. USSR, 1978, 48, 2511 (2770)

78VKP   P.R. Vasudeva Rao, M. Kusumakumari, and S.K. Patil, Radiochem. Radioanal. Letters, 1978, 33, 305

78VKS   V.P. Vasilev, L.A. Kochergina, and T.B. Sokolova, J. Gen. Chem. USSR, 1978, 48, 593 (650)

78VL   V.P. Vasilev, V.P. Lymar, and A.I. Lytkin, Russ. J. Inorg. Chem., 1978, 23, 29 (55), 525 (950)

78VLa   V.P. Vasilev, V.P. Lymar, and A.I. Lytkin, Russ. J. Inorg. Chem., 1978, 23, 683 (1238)

78VP   V.V. Vekshin, N.I. Pechurova, and V.I. Spitsyn, Soviet J. Coord. Chem., 1978, 4, 139 (187)

78VS   A.G. Vitenberg and T.P. Strukova, Soviet J. Coord. Chem., 1978, 4, 270 (361)

78VSG   A.P. Vasilev, V.I. Shorokhova, N.K. Grechina, and L.V. Katrovtseva, Russ. J. Inorg. Chem., 1978, 23, 1274 (2313)

78VV   J. Vliegen and L.C. Van Poucke, Bull. Soc. Chim. Belges, 1978, 87, 837

78VZ   V.P. Vasilev, G.A. Zaitseva, and N.V. Provorova, J. Gen. Chem. USSR, 1978, 48, 1934 (2128)

78WH   E.M. Wooley, J.O. Hill, W.K. Hannan, and L.G. Hepler, J. Soln. Chem., 1978, 7, 385

78WN   M. Wozniak and G. Nowogrocki, Talanta, 1978, 25, 633

78WNa   M. Wozniak and G. Nowogrocki, Talanta, 1978, 25, 643

78WNb   M. Wozniak and G. Nowogrocki, Bull. Soc. Chim. France, 1978, I-153

78WNc   J.K. Walker and R. Nakon, Inorg. Chem., 1978, 17, 1151

78WNA   R.D. Wood, R. Nakon, R.J. Angelici, Inorg. Chem., 1978, 17, 1088

78Y   H. Yamaoka, Z. Phys. Chem. (Leipzig), 1978, 259, 301

78YK   Yu.B. Yakovlev, F.Ya. Kulba, A.G. Pusko, and N.A. Titova, Russ. J. Inorg. Chem., 1978, 23, 229 (411)

78YO   S. Yamada, K. Ohsumi, and M. Tanaka, Inorg. Chem., 1978, 17, 2790

78YS   T.F. Young, C.R. Singleterry, and I.M. Klotz, J. Phys. Chem., 1978, 82, 671

78Z   L.J. Zompa, Inorg. Chem., 1978, 17, 2531

78ZA   V.I. Zelenov, Yu.A. Afanasev, L.A. Aslanyan, and V.T. Panyushkin, Russ. J. Inorg. Chem., 1978, 23, 1404 (2538)

78ZG   D. Zutshi and K.C. Gupta, Indian J. Chem., 1978, 16A, 453

79A   R. Aruga, J. Inorg. Nucl. Chem., 1979, 41, 845

79Aa   R. Aruga, J. Inorg. Nucl. Chem., 1979, 41, 849

79Ab   R. Aruga, Atti Accad. Sci. Torino, 1979, 113, 135

79AB   I.I. Alekseeva, V.V. Borisova, I.I. Nemzer and L.I. Yuranova, Russ. J. Inorg. Chem., 1979, 24, 1467 (2642)

79AD   P. Amico, P.G. Daniele, V. Cucinotta, E. Rizzarelli, and S. Sammartano, Inorg. Chim. Acta, 1979, 36, 1

79AE   L. Abello, A. Ensuque, R. Touiti, and G. Lapluye, J. Chim. Phys., 1979, 76, 602

79AK   D.W Appleton, T.P.A. Kruck, and B. Sarkar, J. Inorg. Biochem., 1979, 10,1

79AM   F.J. Andres Ordax and J.M. Merino de la Fuente, Bull. Soc. Chim. France, 1979,
          I-430

79AO   S.O. Ajayi, A. Olin, and P. Svanstrom, Acta Chem. Scand., 1979, A33, 93

79AOa   S.O. Ajayi, A. Olin, and P. Svanstrom, Acta Chem. Scand., 1979, A33, 97

79AOA   R. Abu-Eittah, A. Osman, and G. Arafa, J. Inorg. Nucl. Chem., 1979, 41, 555

79AP   E.A. Aksenova, I.V. Pimenova, E.G. Timofeeva, O.P. Nesterova, and
          T.G. Marchenkova, Soviet J. Coord. Chem., 1979, 5, 1266 (1629)

79AR   G. Arena, E. Rizzarelli, and B. Sarkar, Inorg. Chim. Acta, 1979, 37, L555

79AS   F. Arnaud-Neu, M.J. Schwing-Weill, R. Louis, and R. Weiss, Inorg. Chem., 1979,
          18, 2956

79ASJ   B.R. Arbad, D.N. Shelke, and D.V. Jahagirdar, J. Indian Chem. Soc., 1979, 56,
          947

79ASN   J. Ananthaswamy, B. Sethuram, and T. Navaneeth Rao, Indian J. Chem., 1979, 18A,
          123

79ASR   A.I. Andreev, N.P. Samsonova, A.M. Robov, and V.A. Fedorov, Soviet J. Coord.
          Chem., 1979, 5, 1035 (1325)

79AV   L.K. Agarwal, P.S. Verma, and D.S. Jain, J. Electrochem. Soc. India, 1979, 28,
          229

79B   E.J. Billo, Inorg. Chim. Acta, 1979, 37, L533

79BB   M.J. Blais and G. Berthon, J. Inorg. Nucl. Chem., 1979, 41, 933

79BBa   J. Bjerrum and E. Bang, Acta Chem. Scand., 1979, A33, 297

79BBG   G. Battistuzzi Gavioli, L. Benedetti, G. Grandi, G. Marcotrigiano,
          G.C. Pellacani, and M. Tonelli, Inorg. Chim. Acta, 1979, 37, 5

79BBN   P.N. Buev, S.S. Butsko, S.I. Nikitenko, N.I. Pechurova, and P.I. Shmanko,
          Russ. J. Inorg. Chem., 1979, 24, 1849 (3320)

79BC   R.P. Bonomo, R. Cali, F. Riggi, E. Rizzarelli, S. Sammartano, and G. Siracusa,
          Inorg. Chem., 1979, 18, 3417

79BE   M.J. Blais, O. Enea, and G. Berthon, Thermochim. Acta, 1979, 30, 37

79BEa   M.J. Blais, O. Enea, and G. Berthon, Thermochim. Acta, 1979, 30, 45

79BEM   A.P. Borisova, A.M. Evseev, N.M. Muratova, and L.I. Martynenko, Russ. J. Inorg.
          Chem., 1979, 24, 840 (1515)

79BG   N.N. Basargin, V.A. Golosnitskaya, I.A. Zanina, and Yu.G. Rozovskii, Russ. J.
          Inorg. Chem., 1979, 24, 201 (363)

79BH   B.A. Bilal, F. Herrmann, and W. Fleischer, J. Inorg. Nucl. Chem., 1979, 41, 347

79BK   C.M.G. van der Berg and J.R. Kramer, Anal. Chim. Acta, 1979, 106, 113

79BKa   S.D. Brown and B.R. Kowalski, Anal. Chem., 1979, 51, 2133

79BL   N.I. Bogatyr and V.A. Litvinenko, Soviet J. Coord. Chem., 1979, 5, 523 (666)

79BLa   N.I. Bogatyr and V.A. Litvinenko, Soviet J. Coord. Chem., 1979, 5, 1259 (1620)

79BM   A. Braibanti, G. Mori, and F. Dallavalle, J. Chem. Soc. Dalton, 1979, 1050

79BP   O.D. Bonner and P.R. Prichard, J. Soln. Chem., 1979, 8, 113

79BPG   S.K. Basin, O. Parkash, J.N. Gaur, and D.S. Jain, J. Electrochem. Soc. India,
          1979, 28, 103

79BPU   V.I. Belevantsev, B.I. Pechchevitskii, and K.A. Udachin, Soviet J. Coord. Chem.,
          1979, 5, 20 (27)

79BR   M. Beltowska-Brzezinska and W. Reksc, Polish J. Chem., 1979, 53, 2175

79BS   G.A. Boos, T.F. Solveva, and A.V. Zakharov, Russ. J. Inorg. Chem., 1979, 24, 1060
          (1914)

79CB   A. Cervilla, A. Beltran, and J. Beltran, Canad. J. Chem., 1979, 57, 773

79CC   C.J. Carrano, S.R. Cooper, and K.N. Raymond, J. Amer. Chem. Soc., 1979, 101,
          599

79CF   L. Ciavatta, D. Ferri, M. Grimaldi, R. Palombari, and F. Salvatore, J. Inorg.
          Nucl. Chem., 1979, 41, 1175

79CK   H. Cohen, L.J. Kirschenbaum, E. Zeigerson, M. Jaaoobi, E. Fuchs, G. Ginzburg,
          and D. Meyerstein, Inorg. Chem., 1979, 18, 2763

79CM   R. Clay, J. Murray-Rust, and P. Murray-Rust, J. Chem. Soc. Dalton, 1979, 1135

79CP   F. Chouaib and C. Poitrenaud, Anal. Chim. Acta, 1979, 108, 333

79CS   M. Cromer-Morin, J.P. Scharff, M. Claude, and M.R. Paris, Anal. Chim. Acta, 1979,
          104, 299

79CSV   R. Corigli, F. Secco, and M. Venturini, Inorg. Chem., 1979, 18, 3184

79DB   E.S. Domnina, L.V. Baikalova, L.E. Protasova, N.M. Deriglazov, N.N. Chipanina,
          D.D. Taryashinova, V.I. Skorobogatova, and G.G. Skvortsova, Soviet J.
          Coord. Chem., 1979, 5, 10 (14)

79DD   R.C. Das, U.N. Dash, and K.N. Panda, Thermochim. Acta, 1979, 32, 301

79DDH   J.M. Degorre, A. Delannoy, J. Hennion, and J. Nicole, Bull. Soc. Chim. France,
          1979, I-471, I-477

79DP   H.C. Dasgupta and D.K. Pathak, Indian J. Chem., 1979, 17A, 427

79DR   A.G. Dickson and J.P. Riley, Marine Chem., 1979, 7, 89

79DRa   A.G. Dickson and J.P. Riley, Marine Chem., 1979, 7, 101

79DS   S.N. Dubey, A. Singh, H.L. Kalra, and D.M. Puri, J. Indian Chem. Soc., 1979,
          56, 451

79DZ   N.K. Davidenko and N.N. Zinich, Soviet J. Coord. Chem., 1979, 5, 1 (3)

79DZa   N.K. Davidenko and N.N. Zinich, Russ. J. Inorg. Chem., 1979, 24, 891 (1608)

79DZb   N.K. Davidenko and N.N. Zinich, Russ. J. Inorg. Chem., 1979, 24, 1352 (2439)

79DZc   N.K. Dzyuba and G.D. Zegzhda, Russ. J. Inorg. Chem., 1979, 24, 542 (978)

79E   H. Einaga, J. Chem. Soc. Dalton, 1979, 1917

79EB   O. Enea, G. Berthon, M. Cromer-Morin, and J.P. Scharff, Thermochim. Acta, 1979,
          33, 311

79EE   M.S. El-Ezaby, M.A. El-Dessouky, and N.M. Shuaib, J. Inorg. Nucl. Chem., 1979,
          41, 1765

79EF   S.D. Ershova, A.Ya. Fridman, and N.M. Dyatlova, Russ. J. Inorg. Chem., 1979, 24,
          301 (541)

79EFa   S.D. Ershova, A.Ya. Fridman, and N.M. Dyatlova, Russ. J. Inorg. Chem., 1979, 24,
          685 (1231)

79EFb   M.M. Emara and N.A. Farid, Egypt. J. Chem., 1979, 22, 77

79EFc   M.M. Emara and N.A. Farid, Egypt. J. Chem., 1979, 22, 89

79EM   M.S. El-Ezaby, H.M. Marafie, and S. Fareed, J. Inorg. Biochem., 1979, 11, 317

79ES   M.I. Ermakova, I.A. Shikhova, T.A. Sinitsyna, and N.I. Latosh, J. Gen. Chem.
          USSR, 1979, 49, 1216 (1387)

79EW   B. Elgquist and M. Wedborg, Marine Chem., 1979, 7, 273

79F   W. Forsling, Acta Chem. Scand., 1979, A33, 641

79Fa   L. Fabbrizzi, J. Chem. Soc. Dalton, 1979, 1857

79FD   Ya.D. Fridman, N.V. Dolgashova, and T.Zh. Zhusupbekov, Russ. J. Inorg. Chem.,
         1979, 24, 1852 (3325)

79FF   F.H. Fisher and A.P. Fox, J. Soln. Chem., 1979, 8, 309

79FFG   V.A. Fedorov, A.V. Fedorova, N.N. Golovnev, G.G. Nifanteva, and N.V. Glazunova,
         Russ. J. Inorg. Chem., 1979, 24, 80 (146)

79FK   V.I. Fadeeva and S.K. Kochetkova, Russ. J. Inorg. Chem., 1979, 24, 1175 (2122)

79FKD   V.A. Fedorov, T.N. Kalosh, and N.R. Deryagina, Russ. J. Inorg. Chem., 1979, 24,
         1285 (2317)

79FKR   M. Fischer, W. Knoche, B.H. Robinson, and J.H.M. Wedderburn, J. Chem. Soc.
         Faraday I, 1979, 75, 119

79FL   Ya.D. Fridman, M.G. Levina, and Z.M. Pulatova, Soviet J. Coord. Chem., 1979, 5,
         632 (807)

79FM   L. Fabbrizzi, M. Micheloni, and P. Paoletti, J. Chem. Soc. Dalton, 1979, 1581

79FS   A.Ya. Fridman, G.M. Sycheva, and Yu.A. Afanasev, Soviet J. Coord. Chem., 1979,
         5, 888 (1132)

79FW   H. Frye and G.H. Williams, J. Inorg. Nucl. Chem., 1979, 41, 591

79GA   J. Gonzales Velasco, S. Ayllon, and J. Sancho, J. Inorg. Nucl. Chem., 1979, 41,
         1075

79GB   H. Gamsjager and P. Beutler, J. Chem. Soc. Dalton, 1979, 1415

79GBa   F. Guay and A.L. Beauchamp, J. Amer. Chem. Soc., 1979, 101, 6260

79GBb   P.K. Govil and S.K. Banerji, Indian J. Chem., 1979, 17A, 624

79GBG   K.D. Gupta, S.C. Baghel, and J.N. Gaur, Monat. Chem., 1979, 110, 657

79GBS   J.N. Gaur, S.C. Baghel, and R.S. Sharma, J. Indian Chem. Soc., 1979, 56, 255

79GC   A. Goswami, K.K. Choudhary, and J.N. Gaur, Indian J. Chem., 1979, 17A, 202

79GG   K.D. Gupta, O.D. Gupta, K.K. Choudhary, and J.N. Gaur, J. Electrochem. Soc.
         India, 1979, 28, 107

79GGP   A.K. Ghosh, J.C. Ghosh, and B. Prasad, J. Indian Chem. Soc., 1979, 56, 489

79GGS   J.P. Gupta, B.S. Garg, and R.P. Singh, J. Indian Chem. Soc., 1979, 56, 145

79GKD   A. Gergely, T. Kiss, and G. Deak, Inorg. Chim. Acta, 1979, 36, 113

79GKN   I.P. Gorelov, A.I. Kapustnikov, and V.M. Nikolskii, J. Gen. Chem. USSR, 1979,
         49, 576 (663)

79GM   R.J. Gualtieri, W.A.E. McBryde, and H.K.J. Powell, Canad. J. Chem., 1979, 57,
         113

79GO   M. Granberg, A. Olin, and P. Svanstrom, Acta. Chem. Scand., 1979, A33, 561

79GP   R. Griffith, Jr., L. Pillai, and M.S. Greenberg, J. Soln. Chem., 1979, 8, 601

79GR   R. Guevremont and D.L. Rabenstein, Canad. J. Chem., 1979, 57, 466

79GS   I. Granberg and S. Sjoberg, Acta Chem. Scand., 1979, A33, 531

79GSK   V.D. Gusev, V.A. Shormanov, and G.A. Krestov, Soviet J. Coord. Chem., 1979, 5,
         555 (706)

79GSN   I.P. Gorelov, A.P. Samsonov, V.M. Nikolskii, V.A. Babich, Yu.E. Svetogorov,
         T.I. Smirnova, E.D. Malakhaev, Yu.M. Kozlov, and A.I. Kapustnikov,
         J. Gen. Chem. USSR, 1979, 49, 573 (659)

79H   I. Haq, Monat. Chem., 1979, 110, 1205

79Ha   R.D. Hancock, S. Afr. J. Chem., 1979, 32, 49

79HCC   W.R. Harris, C.J. Carrano, S.R. Cooper, S.R. Sofen, A.E. Avdeef, J.V. McArdle,
         and K.N. Raymond, J. Amer. Chem. Soc., 1979, 101, 6097

79HCR   W.R. Harris, C.J. Carrano, and K.N. Raymond, J. Amer. Chem. Soc., 1979, 101, 2722

79HG   G.G. Herman and A.M. Goeminne, J. Coord. Chem., 1979, 8, 231

79HGD   G.G. Herman, A.M. Goeminne, and H.F. DeBrabander, Thermochim. Acta, 1979, 32, 27

79HGE   G.G. Herman, A.M. Goeminne, and Z. Eeckhaut, J. Coord. Chem., 1979, 9, 1

79HJ   R.D. Hancock, G. Jackson, and A. Evers, J. Chem. Soc. Dalton, 1979, 1384

79HM   R.D. Hancock, G.J. McDougall, and F. Marsicano, Inorg. Chem., 1979, 18, 2847

79HMa   D.L. Hoo and B. McConnell, J. Amer. Chem. Soc., 1979, 101, 7470

79HN   E. Hogfeldt and S. Nilsson, Acta Chem. Scand., 1979, A33, 559

79HP   P.H.C. Huebel and A.I. Popov, J. Soln. Chem., 1979, 8, 615

79HR   W.R. Harris and K.N. Raymond, J. Amer. Chem. Soc., 1979, 101, 6534

79HRB   H. Hoiland, J.A. Ringseth, and T.S. Brun, J. Soln. Chem., 1979, 8, 779

79HW   R.K. Hessley, S. Waykole, and R.L. Sublett, Canad. J. Chem., 1979, 57, 2292

79IC   A.A. Ivakin, I.G. Chufarova, and N.I. Petunina, Russ. J. Inorg. Chem., 1979, 24,
         389 (695)

79IO   S. Ishiguro and H. Ohtaki, Bull. Chem. Soc. Japan, 1979, 52, 3198

79IR   A.K. Ilyasova and Kh.M. Rakhimbekova, Soviet J. Coord. Chem., 1979, 5, 300 (395)

79IV   A.A. Ivakin, S.V. Vorobeva, and E.M. Gertman, Russ. J. Inorg. Chem., 1979, 24,
         19 (36)

79IVa   A.A. Ivakin, S.V. Vorobeva, A.M. Gorelov, and E.M. Gertman, Russ. J. Inorg.
         Chem., 1979, 24, 1089 (1965)

79JA   D.S. Jain and L.K. Agarwal, Indian J. Chem., 1979, 18A, 83

79JAa   D.S. Jain and L.K. Agarwal, J. Electrochem. Soc. India, 1979, 28, 227

79JB   P.D. Jadhav and R.A. Bhobe, J. Inorg. Nucl. Chem., 1979, 41, 853

79JBa   P.D. Jadhav and R.A. Bhobe, Indian J. Chem., 1979, 17A, 311

79JK   S.L. Jain, J. Kishan, and R.C. Kapoor, Indian J. Chem., 1979, 18A, 133

79JPC   T.J. Janjic, L.B. Pfendt, and M.B. Celap, J. Inorg. Nucl. Chem., 1979, 41, 1019

79JPP   T.J. Janjic, L.B. Pfendt, and V. Popov, J. Inorg. Nucl. Chem., 1979, 41, 63

79JW   O. Johansson and M. Wedborg, Marine Chem., 1979, 8, 57

79KA   U.G. Krishnam Raju, J. Ananthaswamy, B. Sethuram, and T. Navaneeth Rao, Indian
         J. Chem., 1979, 18A, 221

79KB   Yu.M. Kozlov and V.A. Babich, Russ. J. Inorg. Chem., 1979, 24, 769 (1386)

79KBa   Yu.M. Kozlov and V.A. Babich, Russ. J. Inorg. Chem., 1979, 24, 1493 (2690)

79KBb   M.I. Kabachnik, F.I. Belskii, M.P. Komarova, B.K. Shcherbabov, E.I. Matrosov,
         Yu.M. Polikarpov, N.M. Dyatlova, and T.Ya. Medved, Bull. Acad. Sci. USSR,
         1979, 28, 1591 (1726)

79KBc   I. Krznaric, J. Bozic, and N. Kallay, Croat. Chem. Acta, 1979, 52, 183

79KC   Y. Khayat, M. Cromer-Morin, and J.P. Scharff, J. Inorg. Nucl. Chem., 1979, 41,
         1498

79KCa   M.S. Mayadeo and A.M. Chaubal, J. Indian Chem. Soc., 1979, 56, 854

79KD   J. Kragten and L.G. Decnop-Weever, Talanta, 1979, 26, 1105

79KF   I. Kruhak and I. Filipovic, Croat. Chem. Acta, 1979, 52, 207

79KG   T. Kiss and A. Gergely, Inorg. Chim. Acta, 1979, 36, 31

79KGT   L.G. Karyaka, L.I. Gerasyutina, and F.M. Tulyupa, Russ. J. Inorg. Chem., 1979,
        24, 1036 (1869)

79KH   J. Komarek, J. Havel, and L. Sommer, Coll. Czech. Chem. Comm., 1979, 44, 3241

79KK   M. Kodama and E. Kimura, J. Chem. Soc. Dalton, 1979, 325

79KKa   Ts.B. Konunova and S.A. Kudritskaya, Soviet J. Coord. Chem., 1979, 5, 516 (657)

79KKb   C.S. Kallianou and T.A. Kaden, Helv. Chim. Acta, 1979, 62, 2562

79KKK   A.I. Kublanovskaya, G.E. Kuzminskaya, and V.S. Kublanovskii, Soviet Progr. Chem.
        (Ukr. Khim. Zh.), 1979, 45, No. 1,29 (26)

79KKT   N.A. Kostromina, G.S. Kholodnaya, N.N. Tananaeva, N.V. Beloshitskii, and
        A.I. Kirillov, Soviet J. Coord. Chem., 1979, 5, 1403 (1802)

79KM   V.I. Kazbanov, G.D. Malchikov, A.K. Starkov, and G.V. Khrustaleva, Russ. J.
        Inorg. Chem., 1979, 24, 73 (134)

79KN   S.S. Kelkar and B.I. Nemade, Indian J. Chem., 1979, 18A, 534

79KNa   M. Kopach and D. Novak, Russ. J. Inorg. Chem., 1979, 24, 869 (1566)

79KV   V.I. Kornev and V.A. Valyaeva, Soviet J. Coord. Chem., 1979, 5, 80 (103)

79KW   M. Koganemaru, H. Waki, S. Ohashi, and G. Kura, J. Inorg. Nucl. Chem., 1979, 41,
        1457

79KZ   S.P. Kounaves and A. Zirino, Anal. Chim. Acta, 1979, 109, 327

79LB   B. Lenarcik and B. Barszcz, Polish J. Chem., 1979, 53, 963

79LG   B. Lenarcik, M. Gabryszewski, and M. Wisniewski, Polish J. Chem., 1979, 53, 2429

79LJ   M. Leban, D. Jeffries, and J. Fresco, Canad. J. Chem., 1979, 57, 3190

79LK   L.H.J. Lajunen, A. Kostama, and M. Karvo, Acta Chem. Scand., 1979, A33, 681

79LM   P. Letkeman and A.E. Martell, Inorg. Chem., 1979, 18, 1284

79LN   L.H.J. Lajunen, H. Nurmeshiemi, and K. Raisanen, Finn. Chem. Lett., 1979, 112

79LP   S.H. Laurie, D.H. Prime, and B. Sarkar, Canad. J. Chem., 1979, 57, 1411

79LPa   L.H.J. Lajunen and S. Parhi, Inorg. Nucl. Chem. Letters, 1979, 15, 311

79LR   B. Lenarcik, M. Rzepka, and J. Glowacki, Polish J. Chem., 1979, 53, 2199

79LS   H. Lakusta and B. Sarkar, J. Inorg. Biochem., 1979, 11, 303

79M   S. Murakami, J. Inorg. Nucl. Chem., 1979, 41, 209

79Ma   P. Mirti, J. Inorg. Nucl. Chem., 1979, 41, 323

79Mb   E. Mentasti, Inorg. Chem., 1979, 18, 1512

79MA   M. Maeda, R. Arnek, and G. Biedermann, J. Inorg. Nucl. Chem., 1979, 41, 343

79MB   M.S. Mohan, D. Bancroft, and E.H. Abbott, Inorg. Chem., 1979, 18, 1527

79MBa   M.S. Mohan, D. Bancroft, and E.H. Abbott, Inorg. Chem., 1979, 18, 2468

79MBb   M.S. Mohan, D. Bancroft, and E.H. Abbott, Inorg. Chem., 1979, 18, 344

79MBc   V.V. Melnik and V.A. Bocharov, Russ. J. Phys. Chem., 1979, 53, 122 (223)

79MBd   M. Mikesova and M. Bartusek, Coll. Czech. Chem. Comm., 1979, 44, 3256

79MBe   G. McLendon and M. Bailey, Inorg. Chem., 1979, 18, 2120

79MC   B. Monzyk and A.L. Crumbliss, J. Amer. Chem. Soc., 1979, 101, 6203

79MD   J. Maslowska and A. Dorabialski, Polish J. Chem., 1979, 53, 917

79MG   M.R. Melardi, J. Galea, G. Ferroni, A. Belaich, and M. Ragot, Bull. Soc. Chim. Belg., 1979, 88, 1015

79MM   N.M. Muratova and L.I. Martynenko, Russ. J. Inorg. Chem., 1979, 24, 192 (347)

79MMa   D.H. Macartney and A. McAuley, Inorg. Chem., 1979, 18, 2891

79MMb   N.M. Muratova and L.I. Martynenko, Russ. J. Inorg. Chem., 1979, 24, 855 (1543)

79MMK   G.Kh. Makhmeeva, L.I. Martynenko, and G.N. Kupriyanova, Russ. J. Inorg. Chem., 1979, 24, 138 (248)

79MN   M. Miyazaki, S. Nishimura, A. Yoshida, and N. Okubo, Chem. Pharm. Bull. (Japan), 1979, 27, 532

79MP   A.I. Moskvin and A.N. Poznyakov, Russ. J. Inorg. Chem., 1979, 24, 1357 (2449)

79MPa   M. Meloun and J. Pancl, Coll. Czech. Chem. Soc., 1979, 44, 2032

79MPS   E. Mentasti, E. Pelizzetti, F. Secco, and M. Venturini, Inorg. Chem., 1979, 18 2007

79MR   S. Mukherjee and N.S. Rawat, J. Indian Chem. Soc., 1979, 56, 413

79MS   H.C. Malhotra and L.K. Sharma, Gazz. Chim. Ital., 1979, 109, 113

79MSI   H. Manohar, D. Schwarzenbach, W. Iff, and G. Schwarzenbach, J. Coord. Chem., 1979, 8, 213

79MST   M.Machtinger, M. Sloim-Bombard, and B. Tremillon, Anal. Chim. Acta, 1979, 107, 349

79MT   M.S. Mohan and M.M Taqui Khan, J. Coord. Chem., 1979, 8, 207

79MTN   M. Maeda, Y. Tanaka, and G. Nakagawa, J. Inorg. Nucl. Chem., 1979, 41, 705

79N   E. Neher-Neumann, Acta Chem. Scand., 1979, A33, 421

79ND   S. Nagpal, S.N. Dubey, H.L. Kalra, and D.M. Puri, Indian J. Chem., 1979, 18A, 270

79NKC   T.A. Neubecker, S.T. Kirksey, Jr., K.L. Chellappa, and D.W. Margerum, Inorg. Chem., 1979, 18, 444

79NKK   P.P. Nenova, D.St. Kancheva, and B.P. Karadakov, Soviet J. Coord. Chem., 1979, 5, 1280

79NL   V.N. Nikitenko, K.I. Litovchenko, and V.S. Kublanovskii, Russ. J. Inorg. Chem., 1979, 24, 369 (662)

79NM   C.Y. Ng, R.J. Motekaitis, and A.E. Martell, Inorg. Chem., 1979, 11, 2982

79NMa   B.I. Nabivanets, E.A. Mazurenko, V.G. Matyashev, and N.V. Chernaya, Russ. J. Inorg. Chem., 1979, 24, 380 (681)

79NS   T. Nozaki, M. Sakamoto, J. Nakagi, and M. Miyake, Nippon Kagaku Kaishi, 1979, 891

79NSa   P.K.R. Nair and K. Srinivasulu, J. Inorg. Nucl. Chem., 1979, 41, 251

79NT   S. Nushi, M.T. Kalcec, I. Filipovic, and I. Piljac, Croat. Chem. Acta, 1979, 52, 17

79O   W.A. de Oliveira, J. Coord. Chem., 1979, 9, 7

79OL   R. Osterberg, R. Ligaarden, and D. Persson, J. Inorg. Biochem., 1979, 10, 341

79P   D.D. Perrin, "Stability Constants of Metal-ion Complexes Part B Organic Ligands," IUPAC Chemical Data Series No. 22, Pergamon Press, Oxford, 1979

79PB   J.E. Poldoski and T.J. Bydalek, J. Inorg. Nucl. Chem., 1979, 41, 205

79PBC   L. Pillai, R.D Boss, and M.S Greenberg, J. Soln. Chem., 1979, 8, 635

79PBJ   O. Parkash, S.K. Bhasin, and D.S. Jain, J. Electrochem. Soc. India, 1979, 28, 173

79PG   M.M. Palrecha and J.N. Gaur, J. Electrochem. Soc. India, 1979, 28, 233

79PGa   M.M. Palrecha and J.N. Gaur, J. Electrochem. Soc. India, 1979, 28, 231

79PGR   R.M. Pogranichnaya, U.F. Goloborodko, and B.E. Reznik, J. Anal. Chem. USSR, 1979, 34, 708 (917)

79PGS   S.N. Poddar, S. Ghosh, B. Sur, S. Roy Chaudhuri, S.M. Bhattacharyya, and A.K. Das, Indian J. Chem., 1979, 18A, 252

79PK   M. Pollak and V. Kuban, Coll. Czech. Chem. Comm., 1979, 44, 725

79PKT   G.A. Prik, B.E. Kozer, and T.A. Tselyapina, Russ. J. Phys. Chem., 1979, 53, 493 (872)

79PM   L.D. Pethe and B.D. Mali, Indian J. Chem., 1979, 18A, 170

79PP   J. Podlahova and J. Podlaha, Coll. Czech. Chem. Comm., 1979, 44, 321

79PPa   J. Podlahova and J. Podlaha, Coll. Czech. Chem. Comm., 1979, 44, 1343

79PPb   C. Panda and R.K. Patnaik, J. Indian Chem. Soc., 1979, 56, 133

79PPc   C. Panda and R.K. Patnaik, J. Indian Chem. Soc., 1979, 56, 951

79PPB   J.E. Powell, M.W. Potter, and H.R. Burkholder, J. Inorg. Nucl. Chem., 1979, 41, 1771

79PS   N.S. Poonia, S.K. Sarad, A. Jayakumar, and G. Chandra Kumar, J. Inorg. Nucl. Chem., 1979, 41, 1759

79PT   B. Perlmutter-Hayman and E. Tapuhi, Inorg. Chem., 1979, 18, 875

79PTa   B. Perlmutter-Hayman and E. Tapuhi, J. Coord. Chem., 1979, 9, 177

79PV   J.M. Poirier and J.F. Verchere, Talanta, 1979, 26, 341, 349

79PZ   S. Pathak and D. Zutshi, Indian J. Chem., 1979, 18A, 84

79PZa   S. Pathak and K. Zutshi, J. Electrochem. Soc. India, 1979, 28, 163

79R   F. Rodante, Thermochim. Acta, 1979, 32, 293

79RJ   R.W. Renfrew, R.S. Jamison, and D.C. Weatherburn, Inorg. Chem., 1979, 18, 1584

79RK   R.J. Riedo and T.A. Kaden, Helv. Chim. Acta, 1979, 62, 1089

79RKS   E. Russeva, V. Kuban, and L. Sommer, Coll. Czech. Chem. Comm., 1979, 44, 374

79RM   J.L. Roberts, R.E. McClintock, Y. El-Omrani, and J.W. Larson, J. Chem. Eng. Data, 1979, 24, 79

79RP   S. Randhawa, B.S. Pannu, and S.L. Chopra, Thermochim. Acta, 1979, 32, 111

79RPa   S. Randhawa, B.S. Pannu, and S.L. Chopra, Thermochim. Acta, 1979, 33, 335

79RRS   V.V. Ramanujam, K. Rengaraj, and B. Sivasankar, Bull. Chem. Soc. Japan, 1979, 52, 2713

79RRT   P.R. Reddy, K.V. Reddy, and M.M. Taqui Khan, J. Inorg. Nucl. Chem., 1979, 41, 423

79RS   A.L.J. Rao and M. Singh, Indian J. Chem., 1979, 18A, 86

79RW   F.J.C. Rossotti and R.J. Whewell, J. Chem. Soc. Dalton, 1979, 257

79RZ   E.N. Rizkalla and M.T.M. Zaki, Talanta, 1979, 26, 507

79S   D.N. Shelke, Inorg. Chim. Acta, 1979, 32, L45

79Sa   G.S. Sinyakova, Russ. J. Inorg. Chem., 1979, 24, 1342 (2419)

79Sb   E. Still, Anal. Chim. Acta, 1979, 107, 105

79Sc   G.S. Sinyakova, Russ. J. Inorg. Chem., 1979, 24, 1487 (2677)

79Sd   A.K. Sinha, J. Indian Chem. Soc., 1979, 56, 677, 927

79SA   H. Stunzi and G. Anderegg, Helv. Chim. Acta, 1979, 62, 223

79SB   A. Sarpotdar and J.G. Burr, J. Inorg. Nucl. Chem., 1979, 41, 549

79SBa   R.S. Saxena and S.P. Bansal, J. Electrochem. Soc. India, 1979, 28, 185

79SBb   P.C. Srivastava and B.K. Banerjee, Indian J. Chem., 1979, 17A, 583

79SBc   P.C. Srivastava and B.K. Banerjee, J. Indian Chem. Soc., 1979, 56, 779

79SBd   R.S. Saxena and S.P. Bansal, J. Indian Chem. Soc., 1979, 56, 831

79SD   R.N. Sylva and M.R. Davidson, J. Chem., Soc. Dalton, 1979, 232

79SDa   R.N. Sylva and M.R. Davidson, J. Chem., Soc. Dalton, 1979, 465

79SDD   Yu.I. Salnikov, F.V. Devyatov, and I.M. Davletbaeva, Russ. J. Inorg. Chem., 1979,
         24, 1018 (1838)

79SDK   A. Singh, S.N. Dubey, H.L. Kalra, and D.M. Puri, Indian J. Chem., 1979, 17A, 623

79SF   L. Sabatini and L. Fabbrizzi, Inorg. Chem., 1979, 18, 438

79SFa   V.N. Sosnitskii and G.M. Fofanov, Russ. J. Inorg. Chem., 1979, 24, 947 (1708)

79SFD   E.V. Shemyakina, A.Ya. Fridman, and N.M. Dyatlova, Soviet J. Coord. Chem., 1979,
         5, 715 (905)

79SFK   E.I. Stepanovskikh, G.M. Fofanov, and G.A. Kitnev, Russ. J. Inorg. Chem., 1979,
         24, 522 (941)

79SG   I. Sovago and A. Gergely, Inorg. Chim. Acta, 1979, 37, 233

79SGa   N.P. Sachan and C.M. Gupta, Indian J. Chem., 1979, 18A, 82

79SGH   I. Sovago, A. Gergely, B. Harman, and T. Kiss, J. Inorg. Nucl. Chem., 1979, 41,
         1629

79SJ   D.N. Shelke and D.V. Jahagirdar, J. Inorg. Nucl. Chem., 1979, 41, 925

79SJa   D.N. Shelke and D.V. Jahagirdar, J. Inorg. Nucl. Chem., 1979, 41, 929

79SJb   D.N. Shelke and D.V. Jahagirdar, J. Inorg. Nucl. Chem., 1979, 41, 1635

79SJc   D.N. Shelke and D.V. Jahagirdar, J. Indian Chem. Soc., 1979, 56, 344

79SK   R.S. Sandhu and R.K. Kalia, Thermochim. Acta, 1979, 30, 351

79SKa   R.S. Sandhu, R. Kumar, and R.K. Kalia, Thermochim. Acta, 1979, 30, 355

79SKb   V.A. Shormanov, Yu.S. Koryagin, and G.A. Krestov, Soviet J. Coord. Chem., 1979,
         5, 194 (251)

79SKF   L.E. Strong, T. Kinney, and P. Fischer, J. Soln, Chem., 1979, 8, 329

79SN   T. Sakurai and A. Nakahara, Inorg. Chim. Acta, 1979, 34, L245

79SO   L.N. Savoskina, E.R. Oskotskaya, and M.Z. Yampolskii, J. Anal. Chem. USSR, 1979,
         34, 1133 (1465)

79SP   H. Stunzi and D.D. Perrin, J. Inorg. Biochem., 1979, 10, 309

79SPa   M.I. Shtokalo and E.E. Perepechenko. Russ. J. Inorg. Chem., 1979, 24, 1665 (2996)

79SPM   U. Strahm, R.C. Patel, and E. Matijevic, J. Phys. Chem., 1979, 83, 1689

79SPT   H. Stunzi, D.D. Perrin,   T. Teitei, and R.L.N. Harris, Aust. J. Chem., 1979,
         32, 21

79SR   H. Sigel, V.M. Rheinberger, and B.E. Fisher, Inorg. Chem., 1979, 18, 3334

79SS   K.M. Suyan, N.P. Sachan, S.K. Shah, and C.M. Gupta, Indian J. Chem., 1979, 18A,
         81

79SSa   N.P. Sachan, S.K. Shah, and C.M. Gupta, Indian J. Chem., 1979, 17A, 622

79SSb   I.P. Saraswat, C.L. Sharma, and A. Sharma, J. Indian Chem. Soc., 1979, 56, 257

79SSc   K.M. Suyan, S.K. Shah, and C.M. Gupta, J. Indian Chem. Soc., 1979, 56, 923

79SSd   J.R. Siefker and R.D. Shah, Talanta, 1979, 26, 505

79SSe   W. Swczepaniak and J. Siepak, Polish J. Chem., 1979, 53, 1715

79SSf   I.P. Saraswat, C.L. Sharma, and A. Sharma, J. Indian Chem. Soc., 1979, 56, 928

79T   P.B. Trinderup, Acta Chem. Scand., 1979, A33, 7

79TB   F.M. Tulyupa, E.Ya. Bairova, V.V. Movchan, and M.V. Povstyanoi, Soviet J. Coord.
           Chem., 1979, 5, 120 (159)

79TG   Ya.I. Turyan and V.N. Gnusin, J. Gen. Chem. USSR, 1979, 49, 217 (246)

79TK   G.C. Tereshin and O.B. Kuznetsova, Soviet J. Coord. Chem., 1979, 5, 1275 (1639)

79TKK   I.N. Tananaeva, G.S. Kholodnaya, N.A. Kostromina, and A.I. Kirillov, Russ. J.
           Inorg. Chem., 1979, 24, 1014 (1832)

79TM   E.G. Timofeeva, T.G. Marchenkova, O.P. Nesterova, and L.M. Chizhikova, Soviet
           J. Coord. Chem., 1979, 5, 345 (450)

79TMa   E.G. Timofeeva, T.G. Marchenkova, and O.P. Nesterova, Soviet J. Coord. Chem.,
           1979, 5, 1308 (1679)

79TP   H.H. Trimm and R.C. Patel, Inorg. Chim. Acta, 1979, 35, 15

79TS   L.I. Tikhonov, O.I. Samoilova, and V.G. Yashunskii, Russ. J. Inorg. Chem., 1979,
           24, 688 (1237)

79TZ   L.P. Tikhonova, V.Ya. Zayats, and I.P. Svarkovskaya, Russ. J. Inorg. Chem., 1979,
           24, 1502 (2706)

79VG   P.R. Vasudeva Rao, N.M. Gudi, S.V. Bagawde, and S.K. Patil, J. Inorg. Nucl.
           Chem., 1979, 41, 235

79VH   R.E. Viola, C.R. Hartzell, and J.J. Villafranca, J. Inorg. Biochem., 1979, 10,
           281, 293

79VK   V.P. Vasilev and L.A. Kochergina, J. Gen. Chem. USSR, 1979, 49, 1795 (2042)

79VKO   V.P. Vasilev, L.A. Kochergina, and T.D. Orlova, J. Gen. Chem. USSR, 1979, 49,
           1440 (1649)

79VKP   V.P. Vasilev, L.A. Kochergina, and L.V. Pokalyaeva, J. Gen. Chem. USSR, 1979,
           49, 1799 (2047)

79VKR   V.P. Vasilev, L.A. Kochergina, M.V. Rudomino, and N.V. Nichugina, J. Gen. Chem.
           USSR, 1979, 49, 1803 (2052)

79VN   V.V. Vekshin, S.I. Nikitenko, and N.I. Pechurova, Moscow Univ. Chem. Bull., 1979,
           34, No. 6, 43 (555)

79VS   V.P. Vasilev, V.I. Shorokhova, and A.V. Katrovtseva, Russ. J. Inorg. Chem., 1979,
           24, 1473 (2652)

79WN   M. Wozniak and G. Nowogrocki, Talanta, 1979, 26, 381

79WNa   M. Wozniak and G. Nowogrocki, Talanta, 1979, 26, 1135

79WNb   D. Waysbort and G. Navon, Inorg. Chem., 1979, 18, 9

79WNc   H. Waki and T. Nakashima, J. Inorg. Nucl. Chem., 1979, 41, 113

79YM   T. Yoshino, S. Murakami, and K. Ogura, J. Inorg. Nucl. Chem., 1979, 41, 1011

79YR   Yu.B. Yakovlev and L.I. Ravlenko, Russ. J. Inorg. Chem., 1979, 24, 1167 (2107)

79YY   A. Yuchi, S. Yamada, and M. Tanaka, Bull. Chem. Soc. Japan, 1979, 52, 1643

79ZK   A.D. Zuberbuhler and T.A. Kaden, Talanta, 1979, 26, 1111

79ZKT   V.F. Zolin, L.G. Koreneva, and L.I. Tikhonova, Soviet J. Coord. Chem., 1979, 5,
          1116 (1440)

79ZKV   A.P. Zharkov, F.Ya. Kulva, and V.N. Volkov, Soviet J. Coord. Chem., 1979, 5, 34
          (45)

79ZL   K. Zare, P. Lagrange, and J. Lagrange, J. Chem. Soc. Dalton, 1979, 1372

79ZN   G.D. Zegzhda, S.I. Neikovskii, F.M. Tulyupo, and A.P. Tulya, Soviet J. Coord.
          Chem., 1979, 5, 494 (632)

79ZP   B.V. Zhadanov, I.A. Polyakova, N.V. Tsirulnikova, T.M. Sushitskaya, and
          V.Ya. Temkina, Soviet J. Coord. Chem., 1979, 5, 1254 (1614)

79ZT   T.V. Zakharova, T.V. Ternovaya, S.B. Pirkes, and N.A. Kostromina, Russ. J.
          Inorg. Chem., 1979, 24, 1012 (1827)

80BH   A.M. Bond and G.T. Hefter, "Critical Survey of Stability Constants and Related
          Thermodynamic Data of Fluoride Complexes in Aqueous Solution," IUPAC
          Chemical Data Series, No. 27, Pergamon Press, Oxford, 1980

81IS   N. Ivicic and Vl. Simeon, J. Inorg. Nucl. Chem., 1981, 43, 2581

LIGAND FORMULA INDEX

Order of elements: C,H,O,N, others in alphabetical order.

$C_4H_{11}NS^+$	462	$C_5H_8O_4$	310,314,469	$C_5H_{13}NS$	139,170
$C_4H_{11}NSe$	432,434,436	$C_5H_8O_7$	469	$C_5H_{13}N_3$	166
$C_4H_{11}N_3S$	391	$C_5H_8N_2$	205,209	$C_5H_{13}N_3S$	391
$C_4H_{12}ON_2$	148,160	$C_5H_8N_4$	463	$C_5H_{14}O_3NP$	265,273
$C_4H_{12}O_3NP$	266,442	C_5H_9O	376	$C_5H_{14}O_4NP$	467
$C_4H_{12}O_4NP$	467	$C_5H_9ONS_2$	475	$C_5H_{14}N_2$	437,461,461
$C_4H_{12}N_2$	144,145,147, 172,435	$C_5H_9O_2N$	8,37	$C_5H_{14}N_2S$	150,161
		$C_5H_9O_3N$	427,456	$C_5H_{15}O_7NP_2$	271,275
$C_4H_{12}N_2S$	150	$C_5H_9O_3NS$	304,443,443	$C_5H_{15}N_3$	152,163,461
$C_4H_{12}N_2Se$	433	$C_5H_9O_3NS_2$	304		
$C_4H_{13}O_3NP^+$	283	$C_5H_9O_4N$	13,61	$C_6H_2O_4Br_2$	473
$C_4H_{13}O_7NP_2$	268	$C_5H_9O_4NS$	25	$C_6H_2O_4Cl_2$	473
$C_4H_{13}N_3$	163	$C_5H_9O_5N$	455,455	$C_6H_2O_6$	473
$C_4H_{14}O_6N_2P_2$	278,467	$C_5H_9N_3$	213	$C_6H_3O_7N_3$	447
		$C_5H_9N_3S$	439	$C_6H_3O_8N_3$	449
$C_5H_2O_2F_3$	358	$C_5H_{10}ON_2S$	388	$C_6H_4OBr_2$	447
$C_5H_2O_5$	364	$C_5H_{10}O_2$	286,467,467	$C_6H_4OI_2$	447
$C_5H_4ON_4$	259,259	$C_5H_{10}O_2N_2$	476	$C_6H_4O_5N_2$	447,447,447, 447
$C_5H_4O_2S$	468	$C_5H_{10}O_3N_2$	19,92,93,95, 100,101	$C_6H_4O_8Br_2S_2$	471
$C_5H_4O_3$	302,364			$C_6H_4N_2$	230,440
$C_5H_4O_3N_2S$	468	$C_5H_{10}O_4N_2$	40,97,102	$C_6H_4N_3Cl$	463
$C_5H_4O_4O_4N_2$	468	$C_5H_{10}O_5N_2$	455	$C_6H_4N_6Fe$	477
C_5H_4NBr	224,440	$C_5H_{10}N_2S$	387	C_6H_5ON	236
C_5H_4NCl	224,440	$C_5H_{11}ON$	436	C_6H_5OBr	446,446
C_5H_4NI	225	$C_5H_{11}O_2N$	4,5,11,456	C_6H_5OCl	446,446,446, 446
C_5H_5ON	231,464	$C_5H_{11}O_2NS$	22,24		
C_5H_5ONS	464	$C_5H_{11}O_8P$	335	C_6H_5OI	447,447
$C_5H_5OF_3S$	475	$C_5H_{11}N$	155,434	$C_6H_5O_2N$	129,447,459,459
$C_5H_5O_2N$	464	$C_5H_{11}N_3S$	388	$C_6H_5O_2N_2$	464
$C_5H_5O_2F_3$	358	$C_5H_{12}ON_2S_2$	462	$C_6H_5O_3N$	447,447,447
$C_5H_5O_4N_3$	455	$C_5H_{12}O_2N_2$	28	$C_6H_5O_4N$	341,342
C_5H_5N	219	$C_5H_{12}O_2N_2S$	26	$C_6H_5N_3$	439
$C_5H_5N_5$	466	$C_5H_{12}O_3S_4$	474	C_6H_6	475
$C_5H_6ON_2$	464	$C_5H_{12}O_4S_3$	474	C_6H_6O	445
$C_5H_6O_2N_2$	456,466	$C_5H_{12}O_5S_4$	474	$C_6H_6ON_2$	233,237,237
$C_5H_6O_4$	311,444,469	$C_5H_{12}N_2$	437,461	$C_6H_6OS_3$	474
$C_5H_6N_2$	239,239,240, 256	$C_5H_{12}N_2S$	386,386	$C_6H_6O_2$	340
		$C_5H_{12}N_3S$	390	$C_6H_6O_2N_2$	460,460
$C_5H_7O_4N$	468	$C_5H_{13}O_3As$	445	$C_6H_6O_3$	349,362,472
C_5H_8OS	475	$C_5H_{13}O_7NP_2$	275	$C_6H_6O_3N_2$	466
$C_5H_8O_2$	357	$C_5H_{13}O_{14}P_3$	470	$C_6H_6O_4$	362
$C_5H_8O_3$	468				

LIGAND NAME INDEX

SUPPLEMENTAL INDEX FOR VOL. I
for protonation values for other ligands and ligands considered but not included
(For pp. 395-410)

ERRATTA FOR VOLUMES 1,2,3,4

Volume 1

p. xii   Last paragraph, for $\log K_2 = \log K_1 + \Delta H(T_2 - T_1)/0.00256$,
          read $\log K_2 = \log K_1 + \Delta H(T_2 - T_1)(0.00246)$

p. xiii   Under Entropy values, for $\Delta S = 3.36(1.364 \log K + \Delta H)$,
          read $\Delta S = 3.36(1.363 \log K + \Delta H)$.

p. 20   Name of ligand, for 3-Aminorpropanoic, read 3-Aminopropanoic.

p. 47   Under H^+, for H_3L/H_2LH, read $H_3L/H_2L.H$.

p. 47   Under Bibliography, Ni^{2+}, for 73SF, read 73SR.

p. 93   Under Bibliography, for Zn^{2+},Cd^{2+},Pb^{2+}, read $Zn^{2+},Cd^{2+},Fe^{3+},Pb^{2+}$

p. 96   Under Bibliography, Fe^{2+}, for 64S, read 64Sa.
          Under Fe^{3+}, for 64AL,64Sa, read 64AL,64S.

p. 134   In structural formula, for N=C, read N≡C.

p. 281   Name of ligand, for Diethylenetrinitrilotetraacetic,
          read Diethylenetrinitrilopentaacetic.

p. 285   Under Bibliography, Fe^{3+}, delete 68CL.

p. 287   Under Fe^{2+}, for ML.M.L, read ML/M.L.

p. 288   Under Co^{2+}, for HL/H.L, read ML/M.L.

p. 304   In structure, for CH_2SCH_3, read $CH_2CH_2SCH_3$.
          In formula, for $C_6H_{12}O_3N_2S$, read $C_7H_{14}O_3N_2S$.

p. 320   In name of ligand, for glumatic, read glutamic.

p. 367   Under Y^{3+}, for -0.8^d, read -0.8^a.

p. 395   For DL-3-(4-Sulfamylphenyl)alanine, correct lines to match the following:

DL-3-(4-Sulfamylphenyl)alanine, HL	10.26			49NS
	8.64			
	1.99			
L-3-(4-Methoxyphenyl)alanine, HL	9.15^g			58MEW
DL-2-Aminohexanedioic acid, H_2L	9.73^a			68HL
DL-2-Amino-3-hydroxypentanoic acid, HL	9.096^b	-10.4	7	42SG
	2.108^b	-0.8	7	
DL-3-Methoxyalanine, HL	9.176^b	-10.2	8	60K
	2.037^b	-0.8	7	
DL-2-Amino-3-mercapto-3-methylpentanoic acid, H_2L	10.5^c			65FCW
	7.96^c			
DL-2-Amino-4-mercaptobutanoic acid, H_2L	10.32^c			65FCW
	8.56^c			
L-3-(2-Cyanoethylthio)alanine, HL	8.46^g			65FCW

L-3-(3-Ethoxy-3-oxopropylthio)alanine, HL	8.51^c	65FCW
L-Aspartic acid 4-methyl ester, HL	8.50^g	65CM
DL-Glutamic acid 5-ethyl ester, HL	9.19^b 2.15^b	63N
L-Glutamic acid 5-benzyl ester, HL	8.89 2.06	52E

p. 399   For N-[2-(Phenylsulfonylamino)ethyl]iminodaicetic,
         read N-[2-(Phenylsulfonylamino)ethyl]iminodiacetic.

         For N-(2-Acetamidoethyl)iminodaicetic,
         read N-(2-Acetamidoethyl)iminodiacetic.

p. 401   Delete Glycyl-DL-Phosphoserylglycine, H_3L 8.07^b 59FO
                                                            2.96^b

p. 401   Delete DL-Phosphoserylglycine, H_3L 6.95^b 59FO
                                                            3.18^b

p. 403   For DL-Leucyglycylglycine, read DL-Leucylglycylglycine.

p. 415   Under 55TKS, for L. M. Kolthoff, read I. M. Kolthoff.

p. 447   Under 71P, for 7383, read 1383.

p. 454   Delete $C_6H_{12}O_3N_2S$ 304. To $C_7H_{14}O_3N_2S$ 51, add 304.

Volume 2

p. xii   Last paragraph, for log K = log K_1 + $\Delta H(T_2 - T_1)/0.00256$.
         read log K_2 = log K_1 + $\Delta H(T_2 + T_1)(0.00246)$.

p. xii   Under Entropy Values, for ΔS = 3.36 (1.364 log K + ΔH),
         read ΔS = 3.36 (1.363 log K + ΔH).

p. 63 and 64   Under Bibliography, for 68GF, read 68GFa.

p. 77   Under Bibliography, for 72KMa, read 72KM.

p. 88   In structural formula, for $CH_3NHCH_2CH_2NHCH_2$, read $CH_3NHCH_2CH_2NHCH_3$.

p. 127   Under Bibliography, for 72KM, raed 72KMa.

p. 159   Under Bibliography, for Ng^{2+}, read Hg^{2+}.

p. 169   In structural formula, for CH_2, read CH_3.

p. 209   In the first column, delete the symbol Cu^+.

p. 235   Under H^+, for $M_2L/$, read $H_2L/$.

p. 304   In structural formula, for $H_2NCH_2CH_2NHCH_2PO_3H$, read $H_2NCH_2CH_2NHCH_2PO_3H_2$.

p. 374   In tenth entry, for 54ZL, read 65ZL.

p. 389   Under 73BB, for 2669, read 2969.

p. 403   Under Benzo[b]pyridine, for 341, read 222.

p. 406   For 2,6-Dimethylpiperazine, 76, read 2,6-Dimethylpiperidine, 76.

p. 413   For 2-(2-Pryidyl)-1,3-thiazole, read 2-(2-Pyridyl)-1,3-thiazole.

Volume 3

p. 12   In first table, under Bibliography, under other references, for 68REa, read 68RSa.

p. 27   In heading of third column, for 25°, 1.0, read 25°, 0.1.

p. 45   In name of second ligand, for -4-methyl, read -3-methyl.

p. 91   In structural formula, for =0, read =S.

p. 117   In second table, under H^+, for 5.15^d, read 5.15^e.

p. 125   Under Y^{3+} through Yb^{3+}, delete superscript h.

p. 125   Under Gd^{3+}, for 7.00, read 8.00.

p. 126   Under Lu^{3+}, delete superscript h.

p. 163   Under Pb^{2+}, for $M(OH)_2L.OH/M(OH)_3L$, read $M(OH)_2L.OH/M(OH)_3.L$.

p. 164   Under Bibliography, for Gd^{2+}, read Ga^{2+}.

p. 188   In second table name, for Haligeno, read Halogeno.

p. 191   Under Bibliography, under UO_2^{2+}, add 60BS, 64RM.

p. 201   Under Mo(VI), for ML_2L_2, read MO_2L_2.

p. 213   In name of second ligand, for 2,4,6-, read 2,3,4-.

p. 225   In first table, under Zn^{2+}, for 6.88^d, read 6.88^c.

p. 250   In structural formula of first ligand, for $\overset{O}{\overset{\|}{C}}\overset{O}{\overset{\|}{C}}H_2CF_3$, read $\overset{O}{\overset{\|}{C}}\overset{O}{\overset{\|}{C}}H_2CCF_3$.

p. 314   Under Bibliography, under Cu_2^{2+}, reference 73EB should be under Ag^+.

p. 388   Under 60SV, for V. G. Spidsyn, read V.G. Spitsyn.

p. 467   Under N-Acetylcysteine, for 359, read 337.

p. 484   Under 2-Mercaptoethanol, for 380, read 280.

Volume 4

p. 9   Under Zn^{2+}, under $M_2L/M^2.L$, for 5.5, read 5.5^e.

p. 13   Under Sn^{2+}, for 59T, read 58T.

p. 33   Under footnotes, add $^i20°$, 0.6; $^o23°$, 0.

p. 39   In structural formula, for =OH, read -OH.

p. 63   Under H^+, for 9.54, read 6.54.

p. 102   Under Bibliography, under Pb^{2+}, for 64MH, read 63MH.

p. 108   Under Zn^{2+}, $ML_2/M.L^2$ should be moved from the fourth line to the third line.

p. 112   Under Bi^{3+}, for 63KMa, read 63MKa.

p. 211   In fourth line, for 65Wb, read 65HWb.